# THE COMPLETE OFFICIAL MGB

## Model Years 1962 - 1974

# Selected Titles From Bentley Publishers

## *Engineering*

**Robert Bentley's Repair Manual for British Cars** *John Organ*
ISBN 978-0-8376-0041-3

**Vintage Racing British Sports Cars** *Terry Jackson*
ISBN 978-0-8376-0153-3

**Scientific Design of Exhaust & Intake Systems** *Philip H. Smith and John C. Morrison*
ISBN 978-0-8376-0309-4

**The Design and Tuning of Competition Engines** *Philip H. Smith* ISBN 978-0-8376-0140-3

**New Directions in Suspension Design: Making the Car Faster** *Colin Campbell*
ISBN 978-0-8376-0150-2

**Racing Car Design and Development** *Len Terry with Alan Baker* ISBN 978-0-8376-0080-2

## *Austin-Healey*

**Austin-Healey 100/6 NS 3000 Workshop Manual: 1956-1968**
*British Leyland Motors*
ISBN 978-0-8376-1743-5
Hardcover, 416 pages

**The Complete Official 1275 cc Sprite/Midget: 1967-1974**
British Leyland Motors
ISBN 978-0-8376-1748-0
Hardcover, 400 pages

**The Complete Official 948 cc & 1098 cc Sprite/Midget**
British Leyland Motors
ISBN 978-0-8376-1747-3
Hardcover, 256 pages

## *MG*

**The Complete Official MGB: 1962-1974** *British Leyland Motors*
ISBN 978-0-8376-1761-9
Hardcover, 480 pages
**Includes:** Driver's Handbook, Workshop Manual, Special Tuning Manual

**The Complete Official MGB: 1975-1980** *British Leyland Motors*
ISBN 978-0-8376-1760-2
Hardcover, 304 pages

**The Complete Official MG Midget 1500: 1975-1979**
*British Leyland Motors*
ISBN 978-0-8376-1744-2
Hardcover, 384 pages

**The MG Workshop Manual: 1929-1955** *W.E. Blower*
ISBN 978-0-8376-1746-6
Hardcover, 602 pages

## *Triumph*

**The Complete Official Triumph GT6, GT6+ & GT6 MK III: 1967-1973** *British Leyland Motors*
ISBN 978-0-8376-1749-7
Hardcover, 480 pages
**Includes:** Driver's Handbook, Workshop Manual

**The Complete Official Triumph TR6 & TR250: 1967-1976**
*British Leyland Motors*
ISBN 978-0-8376-1762-6
Hardcover, 608 pages
**Includes:** Driver's Handbook, Workshop Manual

**The Complete Official Triumph Spitfire MK III, MK IV & 1500: 1968-1974** *British Leyland Motors*
ISBN 978-0-8376-1763-3
Hardcover, 480 pages
**Includes:** Driver's Handbook, Workshop Manual

**The Complete Official Triumph Spitfire 1500: 1975-1980**
*British Leyland Motors*
ISBN 978-0-8376-1745-9
Hardcover, 536 pages
**Includes:** Driver's Handbook, Workshop Manual

**The Complete Official Triumph TR7: 1975-1981**
*British Leyland Motors*
ISBN 978-0-8376-1742-8
Hardcover, 544 pages
**Includes:** Driving Manual, Workshop Manual

**The Complete Official Triumph TR4 & TR4A: 1961-1968**
*British Leyland Motors*
ISBN 978-0-8376-1741-1
Hardcover, 404 pages
**Includes:** Driving Manual, Workshop Manual, Competition Preparation Manual by R.W. "Kas" Kastner

**The Complete Official Triumph TR2 & TR3: 1953-1961**
*British Leyland Motors*
ISBN 978-0-8376-1750-3
Hardcover, 456 pages
**Includes:** Driver's Instruction Book, Service Instruction Manual

## *Jaguar*

**The Complete Official Jaguar "E" Series 1 & Series 2: 1961-1971**
*British Leyland Motors*
ISBN 978-0-8376-1764-0
Hardcover, 790 pages
**Includes:** Driver's Handbook, Workshop Manual, Special Tuning Manual

**B BentleyPublishers®**
**Automotive Reference**

Bentley Publishers has published service manuals and automobile books since 1950. For more information, please contact Bentley Publishers at 1734 Massachusetts Avenue, Cambridge, MA 02138 USA, or visit our web site at **BentleyPublishers.com**

# THE COMPLETE OFFICIAL MGB

## Model Years 1962–1974

*Comprising the official*

Driver's Handbook

Workshop Manual

BentleyPublishers.com

BENTLEY PUBLISHERS® | Automotive Reference™

Bentley Publishers, a division of Robert Bentley, Inc.
1734 Massachusetts Avenue
Cambridge, MA 02138 USA
800-423-4595 / 617-547-4170

Information that makes the difference®

BentleyPublishers®.com

**Technical contact information**

We welcome your feedback. Please submit corrections and additions to our technical discussion forum at:

**http://www.BentleyPublishers.com**

**Updates and corrections**

We will evaluate submissions and post appropriate editorial changes online as updates or tech discussion. Appropriate updates and corrections will be added to the book in future printings. Check for updates and corrections for this book before beginning work on your vehicle. See the following web address for additional information:

**http://www.BentleyPublishers.com/updates/**

**WARNING—Important Safety Notice**

Do not use this manual for repairs unless you are familiar with basic automotive repair procedures and safe workshop practices. This manual illustrates the workshop procedures for some maintenance and service work. It is not a substitute for full and up-to-date information from the vehicle manufacturer or for proper training as an automotive technician. Note that it is not possible for us to anticipate all of the ways or conditions under which vehicles may be serviced or to provide cautions as to all of the possible hazards that may result.

We have endeavored to ensure the accuracy of the information in this manual. Please note, however, that considering the vast quantity and the complexity of the service information involved, we cannot warrant the accuracy or completeness of the information contained in this manual.

**FOR THESE REASONS, NEITHER THE PUBLISHER NOR THE AUTHOR MAKES ANY WARRANTIES, EXPRESS OR IMPLIED, THAT THE INFORMATION IN THIS MANUAL IS FREE OF ERRORS OR OMISSIONS, AND WE EXPRESSLY DISCLAIM THE IMPLIED WARRANTIES OF MERCHANTABILITY AND OF FITNESS FOR A PARTICULAR PURPOSE, EVEN IF THE PUBLISHER OR AUTHOR HAVE BEEN ADVISED OF A PARTICULAR PURPOSE, AND EVEN IF A PARTICULAR PURPOSE IS INDICATED IN THE MANUAL. THE PUBLISHER AND AUTHOR ALSO DISCLAIM ALL LIABILITY FOR DIRECT, INDIRECT, INCIDENTAL OR CONSEQUENTIAL DAMAGES THAT RESULT FROM ANY USE OF THE EXAMPLES, INSTRUCTIONS OR OTHER INFORMATION IN THIS MANUAL. IN NO EVENT SHALL OUR LIABILITY WHETHER IN TORT, CONTRACT OR OTHERWISE EXCEED THE COST OF THIS MANUAL.**

Before attempting any work on your MGB, read the Warnings and Cautions inside the front cover and any warning or caution that accompanies a procedure in the manual. Review the warnings and cautions each time you prepare to work on your MGB.

Your common sense and good judgment are crucial to safe and successful service work. Read procedures through before starting them. Think about whether the condition of your car, your level of mechanical skill, or your level of reading comprehension might result in or contribute in some way to an occurrence which might cause you injury, damage your car, or result in an unsafe repair. If you have doubts for these or other reasons about your ability to perform safe repair work on your car, have the work done at an authorized MGB dealer or other qualified shop.

Part numbers listed in this manual are for identification purposes only, not for ordering. Always check with your authorized MGB dealer to verify part numbers and availability before beginning service work that may require new parts.

Special tools required to perform certain service operations are identified in the manual and are recommended for use. Use of improper tools may be detrimental to the car's safe operation as well as the safety of the person servicing the car.

The vehicle manufacturer will continue to issue service information updates and parts retrofits after the editorial closing of this manual. Some of these updates and retrofits will apply to procedures and specifications in this manual. We regret that we cannot supply updates to purchasers of this manual.

**ACKNOWLEDGEMENT**

*Special thanks are due M.J.R. Barratt, British Leyland Motors, Inc.*

This manual is prepared, published and distributed by Bentley Publishers, 1734 Massachusetts Avenue, Cambridge, Massachusetts 02138 USA. All information contained in this manual is based on the information available to the publisher at the time of editorial closing. MGB has not reviewed and does not vouch for the accuracy or completeness of the technical specifications and work procedures described and given in this manual.



**ISBN 978-0-8376-1761-9**

**Job Code: X115-19**

**Library of Congress Catalog Card No. 75-7766**

The paper used in this publication is acid free and meets the requirements of the National Standard for Information Sciences-Permanence of Paper for Printed Library Materials. ∞

Manufactured in the United States of America.

# Preface

This set of publications, namely the *Driver's Handbook* (owner's manual), *Workshop Manual,* and *Special Tuning Handbook,* was assembled with the sports car enthusiast in mind. However, the appeal of this manual is not confined to such a limited group; anyone who owns an MGB will derive great benefit from the material presented here. The *Driver's Handbook* is the same manual provided with every new MGB. It includes lubrication instructions as well as minor tune up and maintenance procedures. The *Workshop Manual* is the official factory service manual and was intended primarily for use by dealer service departments. This explains the liberal reference to special service tools, although the instructions start at quite a basic level and presuppose only limited knowledge of the subject. Included here are descriptions of how each part of the car functions. The *Special Tuning Handbook* was written for the serious enthusiast who wishes to race his car or improve its performance for hot street use. Included here are listings of competition parts as available from authorized MGB dealers.

There are several reasons why a person will work on his own car. To begin with, a great deal of money can be saved, although often at the expense of an equal amount of time and effort. The automobile enthusiast takes pride in and derives pleasure from driving a well-tuned machine and will go to great lengths to keep his car in such condition. He is quite often extremely knowledgeable about the workings of his car and firmly believes that the care and attention he can provide far outweigh the experience of a professional mechanic who is merely doing his job. Furthermore, he realizes that an intimate knowledge of his car's behavior helps to make him a better driver.

What follows is intended to provide some useful information and hints for the less experienced individual who would like to work on his own car. The most important thing to keep in mind is that there is nothing mysterious about the functioning of a car. Even though some cars have been known to exhibit a definite personality, a fault becomes quite logical once its cause has been found.

**Maintenance and General Care**

An automobile, like any piece of machinery, responds well to regular maintenance and careful use. There is much to be gained by conscientiously following the manufacturer's recommendations concerning lubrication and adjustments, as described in the *Driver's Handbook*. (The letters in parentheses in the *Driver's Handbook* refer to columns in the Table of Recommended Lubricants.) This is one area where an ounce of prevention is really worth a pound of cure. Whether or not this maintenance is carried out by the individual owner, it is essential that attention be given at periodic intervals. It is worthwhile to keep an accurate record of all maintenance and repairs performed.

Although simple maintenance procedures can be carried out with a minimum of tools, the car owner will sooner or later find it desirable to obtain a fairly complete set. The basic tools include an assortment of open and box end wrenches and a selection of screwdrivers, pliers, one or two adjustable wrenches and other assorted items like feeler gauges, spark plug socket, etc. If any extensive work is contemplated, a ⅜″ or ½″ drive socket set will be most useful. Also, a torque wrench is a very worthwhile purchase. A good set of tools can be a good investment; they will last a lifetime if properly used and cared for. Avoid buying cheaply made tools as they will break or wear out quickly and will often damage bolts, fittings, etc.

Whenever possible, use the wrench correctly suited for the job; excessive use of pliers and adjustable wrenches will roughen and deform the parts they are removing, making later use of the correct tool impossible. It is important that all bolts, nuts and spark plugs be tightened correctly. A fastener that is too tight can be worse than one which is too loose. Broken bolts and distorted parts are the trademarks of the musclebound mechanic. An experienced mechanic knows by feel how to tighten things and he develops this knowledge by long use of a torque wrench. If you have one, by all means use it. It is a good idea to refrain from purchasing specialized tools, such as gear pullers, bushing and seal drivers and odd wrenches, until the need actually arises. In this way the initial expense is reduced and one is always certain of getting the correct sized tool for his needs.

It is appropriate to mention here that the bolts and nuts used on the MGB are for the most part fitted by standard American wrenches, although in certain instances the older Whitworth pattern wrenches are called for. However, these instances are rare, usually involving some accessory component such as the fuel pump, and careful use of adjustable wrenches will almost always get the job done.

In any work done on a car, it is of the utmost importance to keep everything as clean as possible. Dirt will cause endless trouble if not kept out.

In many cases, there are several ways of doing a job in addition to that suggested by the factory. For example, a commercially available timing light, triggered by the spark to no. 1 cylinder, can be used to set the timing with equal or better accuracy. Also, any of the presently available "ignition analyzers" as used in garages and service stations can be used to test the electrical system.

A final note in this section concerns the oil and gas to be used. Although certain brands of oil are recommended by the factory, any premium grade oil of the recommended weight will prove satisfactory. Do not attempt to use a racing oil on the street as it is not designed for prolonged use. As regards the grade of gasoline to be used, it should be realized that this car, and all imported cars for that matter, was designed to use locally available fuels. Since American gasolines are somewhat superior in octane rating (knock resistance), there is no reason why the car cannot use a Regular grade of fuel. If this is to be done successfully, however, the engine will have to be kept in careful tune.

The presence of knocking or detonation, commonly called "pinging" and sounding like loose marbles in the crankcase, indicates the need for a spark adjustment or the presence of too much carbon in the combustion chambers. Carbon is a normal by-product of combustion and, under prolonged city driving conditions, can build up to the point where the compression ratio is significantly raised. This increase in compression will raise the octane required of the fuel, making the engine more likely to ping. This condition is remedied by removing the cylinder head and scraping out the carbon, described in the *Workshop Manual* under the heading of "decarbonizing". An alternative to the above procedure is simply to use a fuel of higher octane rating. This is the approach used by American car manufacturers and decarbonizing is practically unheard of in this country. In addition, the spark can be advanced slightly with Premium fuel resulting in more power and often better fuel economy. Occasional fast driving or highway use will keep the carbon buildup below a troublesome level; and for a car that is principally driven in such a manner, there is no reason why a Regular grade of gasoline cannot be used.

**Major Repairs**

When the need to make a major repair arises, there are several factors to consider before deciding whether or not to undertake the job. The most important point is to understand exactly what the repair entails. This can best be done by reading thoroughly the appropriate section or sections in the *Workshop Manual* and by talking with experts to resolve any unanswered questions. When such an understanding has been reached, the problem can be expressed in terms of the amount of money to be saved vs. the labor and equipment involved.

Most repairs involve the use of a series of special tools to make the job easier, more efficient or even possible. Without a doubt, to achieve the best results the procedures should be carried out exactly as described, using all the special tools. However, in many instances similar or alternate methods can be used which will remove the need to purchase the factory tools. For example, seals can often be installed by the careful use of an arbor press or a hammer and drift punch; most of the commercially available gear and bearing pullers will work quite satisfactorily.

There are, however, certain jobs which should only be attempted by experienced people using the correct equipment. There is no other way to adjust properly a differential; and engine boring, bushing reaming, etc., require the services of a competent machine shop. One, therefore, needs to distinguish between what can and cannot be done. It is usually not worthwhile to purchase expensive equipment to do a job only once.

One will find that a majority of the major repairs involve only one or two operations which require skilled personnel or special equipment. It is well within the abilities of the average person to perform all operations up to this point, go to an expert for the required steps, and to continue from there. For example, one can save a great deal of money in an engine overhaul by performing the disassembly, assembly and cleaning operations himself.

In repairs of this sort, cleanliness is very important. Any time an engine block or gear case is disassembled, it and its contents should be cleaned thoroughly inside and out, including all oil passages. It is common practice to have an engine block "boiled" in a weak acid solution. This process results in a very clean block but also destroys the pressed-in camshaft bearings. As the fitting of new camshaft bearings involves the fairly expensive process of align boring, it is often advisable to use a cold solvent instead.

The importance of clearly understanding every step of the job cannot be overemphasized. Always read and understand the entire procedure before starting. It is also generally advisable to use only genuine replacement parts when making repairs of a major nature.

**Competition Preparation**

The *Special Tuning Handbook* and the competition parts lists were developed by the factory for use in sports car racing. They are proven modifications and should yield the results indicated. The parts and procedures described are usually more than enough to make the car competitive in its class. It is generally inadvisable to proceed too far if the car is to be driven only on the street, as additional performance comes only at the expense of reliability, flexibility, increased noise and reduced fuel economy.

If the decision has been made to prepare the car for competition work, there are several factors to be considered. The most obvious of these is that any such modifications will immediately void the warranty. Secondly, all such work is expensive. Thirdly, all performance modifications work as a team and the various components must be compatible for best results. The teaming of modifications is fully described in the various "stages" of the tuning handbook. Careful and meticulous work has a great value here, as the car must be running at the end of a race in order to win. The little things separate a winner from the also-rans.

It is important to note that the presence of parts in the parts listings does not necessarily mean that they are legal for all types of racing. It's up to the individual to determine just what is allowed when planning the car, not after being protested. In non-sanctioned racing, such as gymkhanas and autocrosses, almost anything goes, but club racing and events regulated by national organizations generally have very strict limits concerning which equipment is permissible.

An intelligent thing to do when planning any competition modifications is to talk to others running your brand in the type of racing you wish to enter. They will be able to tell you not only what is legal, but also which modifications produce the best results. In such matters, there is no substitute for experience.

*Massachusetts Institute of Technology*
*Cambridge, Massachusetts*

*Richard Roberts, S. M.*
*Mechanical Engineering*

## Special Note

Editor's Notes are provided when additional information is helpful to the reader's understanding of the material. These notes are found at the end of Part 1, at the end of individual sections in *The Workshop Manual,* and at the end of Part 3.

## Note on the Fourth Revised Edition

The fourth revised and updated edition of *The Complete Official MGB* has been expanded to include the most recent additions to the *Driver's Handbook*, the *Workshop Manual* supplements issued for the 1974 model year, and the engine emission control data applicable to 1972 through 1974 cars with the SU HIF 4 carburetors.

The second edition included a section on automatic transmission repairs. This has been omitted from both the third and fourth editions, since cars with an automatic gearbox have never been imported into the U.S. or Canada. In its place we have continued to substitute new material that is applicable to MGB cars sold in North America. Among other topics new with this edition is an appendix covering the sequential seat belt system that was legally mandatory on cars sold in the U.S. during 1974.

*Special thanks are due David G. Head*

*New England Parts Representative*

*Leyland Motor Sales*

*Leonia, N.J.*

# English-American Equivalents

| *English* | *American* |
|---|---|
| **GENERAL** | |
| L.H.S. | left hand side (viewed from drivers seat) |
| R.H.S. | right hand side (viewed from drivers seat) |
| L.H.D. | left hand drive (viewed from drivers seat) |
| R.H.D. | right hand drive (viewed from drivers seat) |
| bush | bushing (bronze, rubber, etc.) |
| circlip | snap ring |
| distance piece | spacer |
| end float | end clearance |
| engine revolution counter | tachometer |
| extractor | gear or bearing puller |
| fraze | burr from cutting, drilling, etc. |
| grub screw | dog screw, locating screw |
| joint washer | gasket |
| jointing compound | gasket cement, sealing compound |
| laden | loaded |
| methylated spirits | denatured alcohol |
| paraffin | kerosene |
| perished | rotted (from oil, etc.) |
| petrol | gasoline |
| renew | replace |
| set screw | bolt |
| spanner | end wrench |
| spigot | pilot |
| split pin | cotter pin |
| spring washer | lock washer |
| swarf | chips from cutting, drilling, etc. |
| **ENGINE** | |
| choke tube | venturi |
| cotters | split valve locks |
| float chamber | carburetor bowl |
| gudgeon pin | piston pin, wrist pin |
| oil sump | oil pan |
| silencer | muffler |
| valve crash speed | valve float rpm, redline |
| welch plug | water jacket plug, core plug, freeze plug |
| **CLUTCH** | |
| clutch housing | bell housing |
| clutch release bearing | throwout bearing |
| clutch withdrawal fork | throwout arm |
| spigot bearing | pilot bearing |
| **GEARBOX** | |
| bulk ring | synchronizing ring, synchro cone |
| first motion shaft | input shaft |
| laygear | counter gear, cluster gear |
| layshaft | countershaft, cluster gear shaft |
| propeller shaft | driveshaft |
| third motion shaft | output shaft |
| **REAR AXLE** | |
| crown wheel | ring gear |

| English | American |
|---|---|
| **ELECTRICAL** | |
| control box | voltage regulator |
| distributor suction advance | vacuum advance |
| dynamo | generator |
| earth | ground (positive earth = positive ground) |
| H.T. | high tension |
| Megger | ohm meter |
| micro adjuster | octane selector (changes vacuum advance) |
| **SUSPENSION AND STEERING** | |
| hydraulic damper | shock absorber |
| swivel axle | spindle, stub axle |
| swivel pin | pivot pin, king pin |
| **BODY** | |
| bonnet | hood, engine compartment cover |
| boot | trunk compartment |
| bulkhead | firewall |
| fascia | dashboard |
| hood | top, roof |
| mono construction body | unit construction body (no frame) |
| over rider | bumperette |
| seat squab | seat back, upright portion of seat |
| wing | fender |

# Contents

# 3 OFFICIAL MGB WORKSHOP MANUAL

# 4 OFFICIAL SPECIAL TUNING MANUAL FOR THE MGB TOURER AND GT



# APPENDIX

## *Please read these warnings and cautions before proceeding with maintenance and repair work.*

### *WARNING—*

• Do not re-use any fasteners that are worn or deformed in normal use. Many fasteners are designed to be used only once and become unreliable and may fail when used a second time. This includes, but is not limited to, nuts, bolts, washers, self-locking nuts or bolts, circlips, cotter pins. For replacements always use new parts.

• Never work under a car unless it is solidly supported on stands designed for the purpose. Do not support a car on cinder blocks, hollow tiles, or other props that may crumble under continuous load. Do not work under a car that is supported solely by a jack.

• If you are going to work under a car on the ground, make sure that the ground is level. Block the wheels to keep the car from rolling. Never work under the car while the engine is running. Disconnect the battery ground strap to prevent others from starting the car while you are under it.

• Never run the engine unless the work area is well ventilated. Carbon monoxide kills.

• Friction materials such as brake or clutch discs may contain asbestos fibers. Do not create dust by grinding, sanding, or by cleaning with compressed air. Avoid breathing asbestos fibers and asbestos dust. Breathing asbestos can cause serious diseases such as asbestosis or cancer, and may result in death.

• Tie long hair behind your head. Do not wear a necktie, a scarf, loose clothing, or a necklace when you work near machine tools or running engines. If your hair, clothing or jewelry were to get caught in the machinery, severe injury could result.

• Disconnect the battery ground strap whenever you work on the fuel system or the electrical system. When you work around fuel, do not smoke or work near heaters or other fire hazards. Keep an approved fire extinguisher handy.

• Illuminate your work area adequately but safely. Use a portable safety light for working inside or under the car. Make sure the bulb is enclosed by a wire cage. The hot filament of an accidentally broken bulb can ignite spilled fuel or oil.

• Catch draining fuel, oil or brake fluid in suitable containers. Do not use food or beverage containers that might mislead someone into drinking from them. Store flammable fluids away from fire hazards. Wipe up spills at once, but do not store the oily rags, which can ignite and burn spontaneously.

• Finger rings should be removed so that they cannot cause electrical shorts, get caught in running machinery or be crushed by heavy parts.

• Keep sparks, lighted matches and open flame away from the top of the battery. If hydrogen gas escaping from the cap vents is ignited, it will ignite gas trapped in the cells and cause the battery to explode.

• Always observe good workshop practices. Wear goggles when you operate machine tools or work with battery acid. Gloves or other protective clothing should be worn whenever the job requires it.

• Do not attempt to work on your car if you do not feel well. You increase the danger of injury to yourself and others if you are tired, upset or have taken medicine or any other substance that may impair you from being fully alert.

• Some aerosol tire inflators are highly flammable. Use extreme caution when repairing a tire that may have been inflated using an aerosol tire inflator. Keep sparks, open flame or other sources of ignition away from the tire repair area. Inflate and deflate the tire at least four times before breaking the bead from the rim. Completely remove the tire from the rim before attempting any repair.

• Greases, lubricants and other automotive chemicals contain toxic substances, many of which are absorbed directly through the skin. Read manufacturer's instructions and warnings carefully. Use hand and eye protection. Avoid direct skin contact.

### *CAUTION—*

• If you lack the skills, tools and equipment, or a suitable workshop for any procedure described in this Manual, we suggest you leave such repairs to an authorized dealer or other qualified shop. We especially urge you to consult an Authorized Dealer before beginning repairs on any car that may still be covered wholly or in part by the new-car warranty.

• British Leyland Motors was constantly improving its cars and sometimes these changes, both in parts and specifications, were made applicable to earlier models. Therefore, part numbers listed in this manual are for reference only. Always check with your Authorized Parts Department for the latest information.

• Before starting a job, make certain that you have all the necessary tools and parts on hand. Read all the instructions thoroughly, do not attempt shortcuts. Use tools appropriate to the work and use only replacement parts meeting specifications. Makeshift tools, parts and procedures will not make good repairs.

• Use pneumatic and electric tools only to loosen threaded parts and fasteners. Never use these tools to tighten fasteners, especially on light alloy parts.

• Be mindful of the environment and ecology. Before you drain the crankcase, find out the proper way to dispose of the oil. Do not pour oil onto the ground, down a drain, or into a stream, pond or lake. Consult local ordinances that govern the disposal of wastes.

# MGB
# TOURER AND GT
# DRIVER'S HANDBOOK
# 1962-1970

## PART 1

# FOREWORD

YOU will find in these pages an introduction to your vehicle and information essential to satisfactory use and maintenance.

A Passport to Service containing service vouchers is provided, and the regular use of the vouchers in sequence is the best safeguard against the possibility of abnornal repair bills at a later date. Replacement Passport to Service books are obtainable free of charge from Distributors or Dealers. Prevent rather than cure.

Completed voucher counterfoils are proof of regular servicing and could well enhance the value of your vehicle in the eyes of a prospective buyer.

By keeping the Passport to Service, signed by the Distributor, Dealer or vendor, in the vehicle you can quickly establish the date of purchase and provide the necessary details if adjustments are required to be carried out under warranty.

Claims for the replacement of parts under warranty must be submitted to the supplying Distributor or Dealer or, when this is not possible, to the nearest Distributor or Dealer, informing them of the vender's name and address. Except in cases of emergency, warranty work should always be carried out by a BMC appointed Distributor or Dealer.

**When Service Parts are required insist on BMC GENUINE PARTS as these are designed and tested for your vehicle and in addition have the full backing of the BMC Factory Warranty. ONLY WHEN GENUINE PARTS ARE USED CAN BMC ACCEPT RESPONSIBILITY.**

# A. GENERAL DATA

**Engine**

| | |
|---|---|
| Bore | 3·16 in. (80·26 mm.) |
| Stroke | 3·5 in. (89 mm.) |
| No. of cylinders | 4 |
| Capacity | 1798 c.c. (109·8 cu. in.) |
| Compression ratio: High compression | 8·8 : 1 |
| Low compression | 8·0 : 1 |
| Firing order | 1, 3, 4, 2 |
| Valve clearance (cold) | ·015 in. (·38 mm.) |
| Oil pressure: Idling | 10 to 25 lb./sq. in. (·7 to 1·7 kg./cm.$^2$) |
| Normal | 50 to 80 lb./sq. in. (3·5 to 5·6 kg./cm.$^2$) |
| Idling setting | 500 r.p.m. |

**Ignition**

| | |
|---|---|
| Static ignition timing: High compression | 10° B.T.D.C. |
| Low compression | 8° B.T.D.C. |
| Stroboscopic ignition timing | 14° B.T.D.C. at 600 r.p.m. |
| Contact breaker gap | ·014 to ·016 in. (·36 to ·41 mm.) |
| Sparking plugs | Champion N-9Y |
| Plug gap | ·025 in. (·64 mm.) |

**Fuel system**

| | |
|---|---|
| Carburetters | Two S.U. type HS4 |
| Needle | FX. |
| Fuel pump | S.U. Type HP or AUF 300 electric |

**Gearbox and overdrive**

| | | |
|---|---|---|
| Overdrive ratio | ·802 : 1 | |
| Overall ratios: First | 14·214 : 1 | |
| Second | 8·656 : 1 | *Overdrive* |
| Third | 5·369 : 1 | 4·306 : 1 |
| Fourth | 3·909 : 1 | 3·135 : 1 |
| Reverse | 18·588 : 1 | |
| Top gear speed per 1,000 r.p.m.: | | |
| Standard | 17·9 m.p.h. (27·3 km.p.h.) | |
| Overdrive | 22·3 m.p.h. (35·5 km.p.h.) | |

**Wheels**

| | | |
|---|---|---|
| Type | *Tourer* | *GT* |
| Ventilated disc | 4J × 14 | 5J × 14 |
| Wire (optional) | 4½ × 14 (60-spoke) | |

**Tyres**

| | | |
|---|---|---|
| Standard: | | |
| Size | 5·60—14 (tubed C41) | |
| Optional: | *Tourer* | *GT* |
| Size | 155—14 (SP) | 165—14 (SP) |
| Standard tyres: | | |
| Pressures (set cold). | *Tourer* | *GT* |
| Front | 18 lb./sq. in. (1·3 kg./cm.$^2$) | 20 lb./sq. in. (1·4 kg./cm.$^2$) |
| Rear | 18 lb./sq. in. (1·3 kg./cm.$^2$) | 24 lb./sq. in. (1·7 kg./cm.$^2$) |
| Sustained speeds in excess of 90 m.p.h. (145 km.p.h.): | | |
| Front | 24 lb./sq. in. (1·7 kg./cm.$^2$) | 26 lb./sq. in. (1·8 kg./cm.$^2$) |
| Rear | 24 lb./sq. in. (1·7 kg./cm.$^2$) | 30 lb./sq. in. (2·1 kg./cm.$^2$) |
| Optional tyres (SP): | | |
| Pressures (set cold): | *Tourer* | *GT* |
| Front | 21 lb./sq. in. (1·5 kg./cm.$^2$) | 21 lb./sq. in. (1·5 kg./cm.$^2$) |
| Rear | 24 lb./sq. in. (1·7 kg./cm.$^2$) | 24 lb./sq. in. (1·7 kg./cm.$^2$) |
| Sustained speeds in excess of 90 m.p.h. (145 km.p.h.): | | |
| Front | 27 lb./sq. in. (1·9 kg./cm.$^2$) | 28 lb./sq. in. (2·0 kg./cm.$^2$) |
| Rear | 31 lb./sq. in. (2·2 kg./cm.$^2$) | 31 lb./sq. in. (2·2 kg./cm.$^2$) |

**NOTE.—Rear tyre pressures should be increased by 2 lb./sq. in. (·14 kg./cm.$^2$) when touring with a full luggage compartment.**

**Capacities**

| | |
|---|---|
| Fuel tank: | |
| Early Tourer cars | 10 gallons (45·4 litres, 12 U.S. gal.) |
| Later Tourer and GT cars | 12 gallons (54 litres, 14 U.S. gal.) |
| Cooling system | 9½ pints (5·4 litres, 11·4 U.S. pints) |
| Cooling system with heater | 10 pints (5·6 litres, 12 U.S. pints) |
| Sump | 7½ pints (4·26 litres, 9 U.S. pints) |
| Oil cooler (when fitted) | ¾ pint (·42 litre, ·9 U.S. pint) |
| Gearbox | 4½ pints (2·56 litres, 5·6 U.S. pints) |
| Gearbox and overdrive | 5¼ pints (3·0 litres, 6 U.S. pints) |
| Rear axle: | |
| Tourer | 2¼ pints (1·3 litres, 2·7 U.S. pints) |
| GT | 1½ pints (·85 litres, 2 U.S. pints) |

**Dimensions**

| | |
|---|---|
| Length (overall) | 12 ft. 8½ in. (3·8 m.) |
| Length (overall) with over-riders | 12 ft. 9 $\frac{7}{16}$ in. (3·9 m.) |
| Width (overall) | 4 ft. 11 $\frac{15}{16}$ in. (152·3 cm.) |
| Height (overall), hood erected | 4 ft. 1⅜ in. (125·4 cm.) |
| Ground clearance (minimum) | 5 in. (12·7 cm.) |
| Track: Front: Disc wheels | 4 ft. 1 in. (124·4 cm.) |
| Wire wheels | 4 ft. 1¼ in. (125·0 cm.) |
| Rear: Disc wheels | 4 ft. 1¼ in. (125·0 cm.) |
| Wire wheels | 4 ft. 1¼ in. (125·0 cm.) |
| Wheelbase | 7 ft. 7 in. (231·1 cm.) |
| Turning circle | 32 ft. (9·75 m.) |
| Toe-in | $\frac{1}{16}$ to $\frac{3}{32}$ in. (1·5 to 2·3 mm.) |

**Weights**

| | *Tourer* | *GT* |
|---|---|---|
| Unladen | 1,920 lb. (871 kg.) | 2,190 lb. (993 kg.) |
| Kerbside | 2,030 lb. (920 kg.) | 2,310 lb. (1048 kg.) |
| Max. permissible gross | 2,430 lb. (1102 kg.) | 2,660 lb. (1206 kg.) |

# B. IDENTIFICATION

**When communicating with your Distributor or Dealer always quote the car and engine numbers. When the communication concerns the transmission units or body details it is necessary to quote also the transmission casing and body numbers.**

**Car number.** Stamped on a plate secured to the top left-hand side of the front bulkhead.

**Engine number.** On a metal plate fixed to the right-hand side of the cylinder block.

**Gearbox number.** Stamped on the top of the gearbox to the left of the dipstick and filler plug.

**Rear axle number.** Stamped on the front of the axle tube on the left-hand side adjacent to the spring seat.

**Ignition key number. To reduce the possibility of theft ignition switches are not marked with a number. Owners are advised to make a note of the number appearing on the ignition key in case of loss.**

**NOTE.—References to right- or left-hand in this Handbook are made when viewing the car from the rear.**

# C. CONTROLS AND INSTRUMENTS

**Hand brake**

The hand brake is of the pull-up lever type, operating mechanically on the rear wheels only. To release the hand brake, pull it upwards to take the load, press the ratchet release button located in the end of the lever with the thumb and push the lever down into the 'off' position. The hand brake is automatically adjusted at the same time as the foot brake and requires no separate adjustment.

**Pedals**

The left-hand pedal operates the clutch, the centre pedal the brakes, and the right-hand pedal the accelerator. Keep the foot clear of the clutch pedal except when engagement or disengagement of any gear is intended, or when in heavy traffic. Driving with the foot resting on the pedal will lead to rapid clutch wear.

**Gear lever**

The four forward gears and the reverse gear are engaged by moving the lever to the positions illustrated.

To engage reverse gear move the lever to the left of the neutral position until resistance is felt, apply side pressure to the lever to overcome resistance and then pull it backwards to engage the gear.

Synchromesh engagement is provided on second, third and fourth gears.

Always ensure that the gear lever is in the neutral position before attempting to start the engine.

**Horn switch**

The horns are sounded by a horn-push located in the steering-wheel hub.

# CONTROLS AND INSTRUMENTS

**Ignition and starter switch**

The ignition and starter are both controlled by a single switch operated by a removable key. To switch on the ignition insert the key and turn it in a clockwise direction until a slight resistance is felt. Further movement in the same direction will operate the starter motor. Release the key immediately the engine starts. If the engine fails to start first time wait until it has come to rest before using the starter again.

**Headlight beam dipping switch**

The headlight main-beam dipping switch is located on the toeboard to the left of the clutch pedal. It is of the single-acting repeating type, lowering the beams on one application and raising them on the next. A warning light on the face of the speedometer glows when the beams are in the raised position.

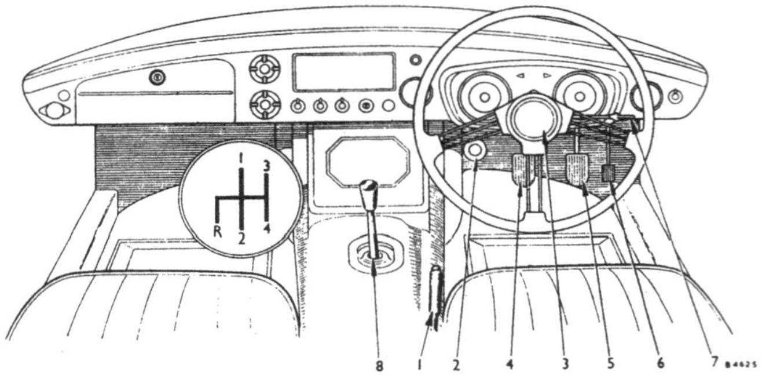

*Driving controls—right-hand drive*

1. Hand brake.
2. Headlight dip switch.
3. Horn switch.
4. Clutch pedal.
5. Brake pedal.
6. Accelerator pedal.
7. Direction indicator.
8. Gear lever.

**Choke or mixture control**

To enrich the mixture and assist starting when the engine is cold pull out the knob marked 'C'. The control, when turned half a turn clockwise, will hold in any position, giving a progressively richer mixture as it is pulled out.

**On no account should the engine run for any length of time with the knob pulled fully out.** It should be returned to the 'off' position (pushed in) as soon as possible as the engine warms up.

The first ¼ in. (6 mm.) approx. of movement operates only the throttle control. This initial movement can be used to give a fast engine idling speed and prevent stalling when driving at low speeds before the engine has fully warmed up.

**Heater and ventilating controls**

These controls provide a means of regulating the heating and ventilating system. Full operating instructions are given on page 5. .

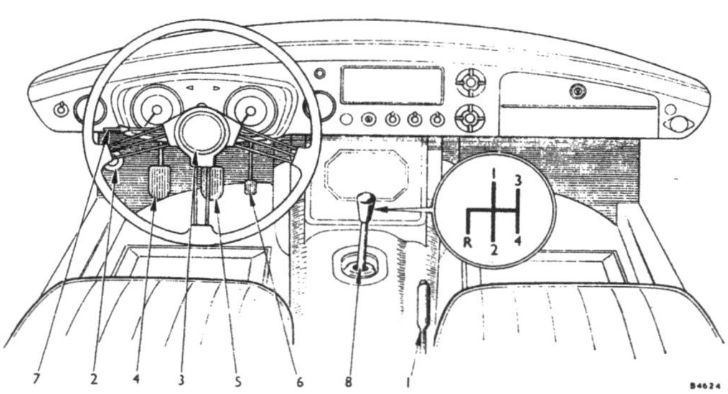

*Driving controls—left-hand drive*

1. Hand brake.
2. Headlight dip switch.
3. Horn switch.
4. Clutch pedal.
5. Brake pedal.
6. Accelerator pedal.
7. Direction indicator.
8. Gear lever.

**Ignition warning light**

The ignition warning light serves the dual purpose of reminding the driver to switch off the ignition, and of acting as a no-charge indicator. With the ignition switched on the warning light should only be illuminated when the engine is not running, or is running at a very low speed. As the engine speed increases the light should dim and then go out at a fairly low engine speed.

If the light fails to go out until higher engine speeds are reached or remains alight at all times, inspect the dynamo driving belt for correct tension or breakage. If the belt is in order the charging system must be overhauled by a Distributor or Dealer.

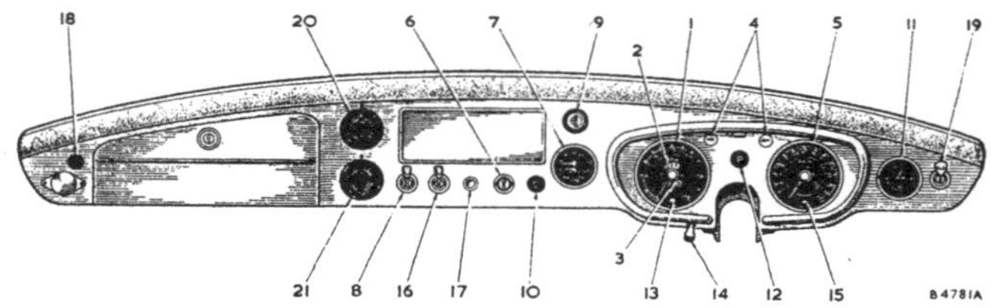

*Instruments and switches—right-hand drive*

1. Speedometer.
2. Trip mileage indicator.
3. Total mileage indicator.
4. Direction indicator warning lights.
5. Tachometer.
6. Ignition and starter switch.
7. Oil pressure and water temperature gauge.
8. Blower switch.
9. Lighting switch.
10. Choke (mixture control).
11. Fuel gauge.
12. Panel light switch.
13. Headlight main-beam warning light.
14. Trip mileage resetting knob.
15. Ignition warning light.
16. Windscreen wiper switch.
17. Windscreen washer.
18. Map light switch.
19. Overdrive switch.
20. Heater control.
21. Air control.

**Lighting switch**

*Earlier cars*

A lever-type switch. Move the lever downward to the half-way position for the side and tail lights, and into the fully down position for the headlights.

*Later cars*

A push/pull switch. Pull out to the first position for the side and tail lights, and fully out for headlights.

**Map-reading light switch**

The map-reading light is controlled by a switch adjacent to the lamp. The lamp will only operate when the side lights are switched on.

**Fuel gauge**

A few seconds after the ignition is switched on the fuel gauge indicates the quantity of fuel in the tank. An important note on filling up with fuel is given on page 9.

**Panel light switch**

The panel light switch is located between the speedometer and the engine revolution indicator. To illuminate the instruments turn the knob clockwise. The first movement of the knob switches on the lights and further turning to the right dims them.

The panel lights will only function when the side lights are switched on.

**Overdrive switch**

The overdrive switch is mounted on the outer edge of the fascia adjacent to the instrument panel. The switch bezel is marked 'NORMAL' and 'OVERDRIVE'.

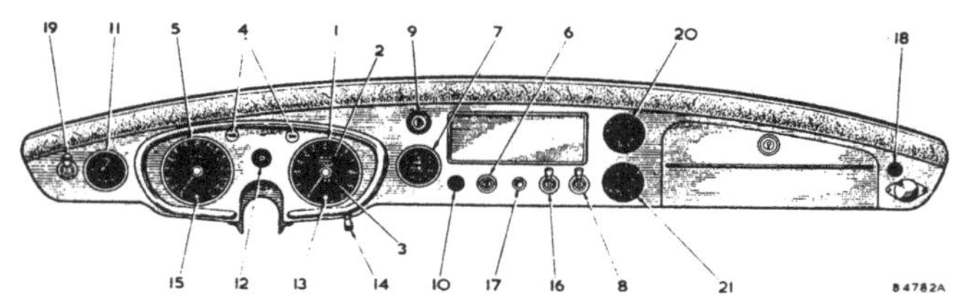

*Instruments and switches—left-hand drive*

1. Speedometer.
2. Trip mileage indicator.
3. Total mileage indicator.
4. Direction indicator warning lights.
5. Tachometer.
6. Ignition and starter switch.
7. Oil pressure and water temperature gauge.
8. Blower switch.
9. Lighting switch.
10. Choke (mixture control).
11. Fuel gauge.
12. Panel light switch.
13. Headlight main-beam warning light
14. Trip mileage resetting knob.
15. Ignition warning light.
16. Windscreen wiper switch.
17. Windscreen washer.
18. Map light switch.
19. Overdrive switch.
20. Heater control.
21. Air control.

**Oil pressure gauge**

The engine oil pressure should be between 50 and 80 lb./sq. in. (3·5 and 5·6 kg./cm.²) under normal running conditions. Approximately 10 to 25 lb./sq. in. (·7 to 1·7 kg./cm.²) should be shown when the engine is idling. Should the gauge fail to register any pressure at all, stop the engine immediately and investigate the cause.

**Temperature gauge**

When the engine is running the gauge indicates the temperature of the coolant leaving the cylinder head.

As overheating may cause serious damage, the readings should be noted and after the initial rise in temperature during the warming-up period any sudden upward change in the reading calls for immediate investigation.

When the ignition is switched off the needle returns to the 'cold' position.

**Direction indicator and headlight flasher switch**

The direction indicator switch is mounted on an arm on the steering-column below the steering-wheel. The indicators operate only when the ignition is switched on, and the indicator warning lights in the instrument panel flash when they are in use.

Lifting the lever towards the steering-wheel flashes the headlight beams on later Tourers and all GT cars.

**Speedometer**

In addition to recording the road speed this instrument also records the trip and total distances. The trip recorder enables the length of a particular journey to be recorded, and can be set to zero by pushing upwards and turning the knob located beneath the instrument.

**Tachometer**

This instrument indicates the revolutions per minute of the engine, and this assists the driver to use the most effective engine speed range for maximum performance in any gear (see 'RUNNING INSTRUCTIONS').

**Windscreen wiper switch**

The two windscreen wiper blades are operated by a lever-type switch. Flick down the switch to start the wiper motor. The blades park automatically when the motor is switched off.

**Windscreen washer**

To wash the windscreen press the control knob. When following other vehicles, particularly under dirty road conditions, the washer should be operated before the wiper blades are set in motion.

In cold weather the reservoir should be filled with a mixture of water and recommended washer solvent to prevent the water freezing in the reservoir and on the windscreen.

**Do not use radiator anti-freeze solution in the windscreen-washing equipment.**

(See supplementary notes at end of Part 1.)

# D. HEATING AND VENTILATING

**Fresh-air system (standard)**

Provision is made for admitting fresh air to the car interior for ventilation purposes.

The fresh air is ducted from the bonnet grille to an air box and passes to the car interior through an air outlet vent on the dash panel.

The fresh-air control is located on the dash panel to the left of the speaker panel. Pull the control to open the vent to one of the two positions.

**Heating system (optional)**

When a heating and demist unit is fitted, fresh air is ducted from the bonnet grille, through the air box to the heater matrix and then to the demist ducts or to the car interior. Two manually operated spring-loaded air outlet doors are located on the inner foot-well sides to admit air to the interior of the car.

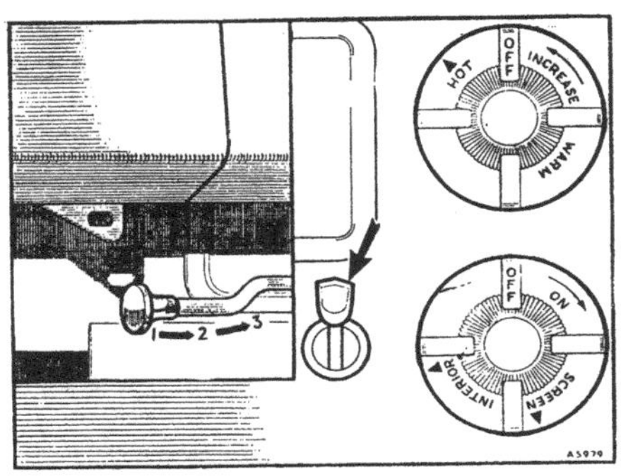


*The heater controls and blower switch with the fresh-air control shown inset*

The ram effect caused by the forward motion of the car will provide an adequate supply of air above speeds of approximately 25 m.p.h. (40 km.p.h.). An electric blower in the heater unit controlled by a switch on the fascia panel augments the supply when a greater quantity of air is required or when the car is travelling at low speeds or is stationary.

**Controls**

The upper control operates a water valve connected to the engine cooling system; the valve controls the rate of flow of the hot water to the heater matrix.

Move the control in the direction of the arrow to increase the heat supply to 'WARM' or 'HOT'.

The lower control operates an air flap in the heater outlet box which deflects the air to the screen or to the air outlet doors and car interior.

Turn the control in the direction of the arrow to 'INTERIOR' or 'SCREEN' to deflect air to the car interior or to the windscreen.

Both controls may be set to intermediate positions and thus provide a wide variety of settings to meet varying conditions.

**Use of controls**

NOTE.—**The full heat output will not be available until the engine has reached its operating temperature.**

To obtain maximum heat supply inside the car, turn the air control to 'INTERIOR', the heat control to 'HOT', open the air outlet doors and switch on the blower.

To defrost the windscreen turn the air control to 'SCREEN', the heat control to 'HOT', close the air outlet doors and switch on the blower.

To obtain a supply of unheated fresh air turn the air control to 'INTERIOR', the heat control to 'OFF', open the air outlet doors and switch on the blower.

# E. BODY DETAILS

**Bonnet lock**

To open the bonnet pull the release inside the car and press up the safety catch. Disengage the bonnet stay from its clip and place the free end in the bracket attached to the front wheel arch.

To close, secure the stay in its clip and lower the bonnet. Apply light pressure with the palms of the hands at the front corners of the bonnet and press down quickly. As the bonnet is aluminium, handle it with care and do not use unnecessary force. The safety catch and lock will be heard to engage.

*Bonnet lock control and safety catch*

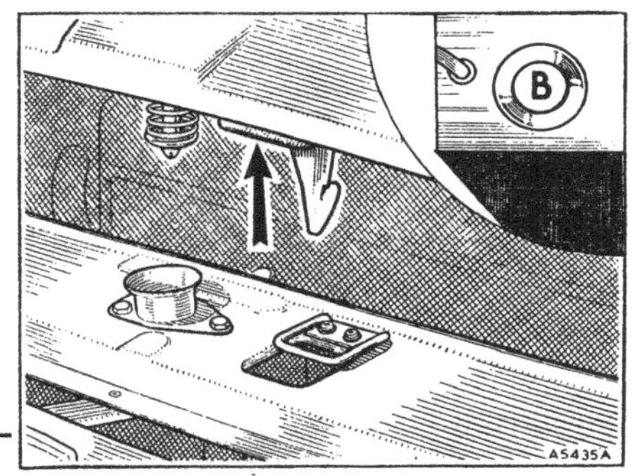


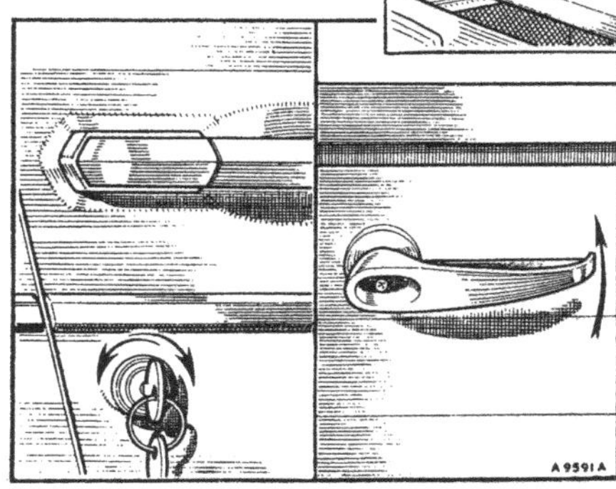


*The external and internal door locks*

**Door locks**

Both doors may be locked from the outside with the key or from the inside with the internal door handle.

To lock the doors from the outside turn the key slightly towards the front of the car. To unlock the doors turn the key slightly towards the rear of the car. After locking or unlocking the doors return the key to the vertical position and withdraw it.

To lock the doors from inside move the door handle upwards with the minimum amount of pressure to the handle.

Early cars: Only the passenger door may be locked from the inside by turning the locking knob up. Turn the knob down to unlock the door.

**Doors**

Both doors are provided with door-pulls.

A window regulator handle is fitted to the inside of each door.

A draughtproof ventilator panel adjacent to each window may be opened by releasing the catch.

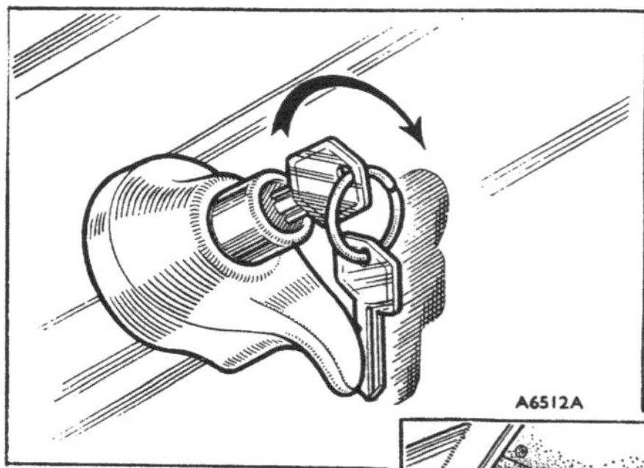


*The luggage compartment lock*

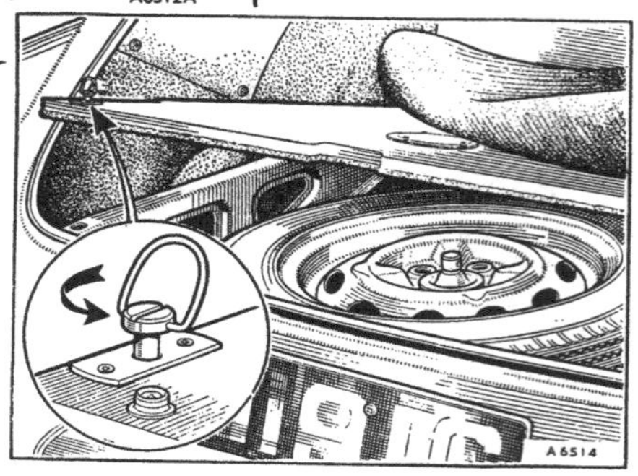


*The spare wheel*

**Luggage compartment**

The luggage compartment is locked with the key provided.

*Tourer*

To open, depress the button catch and raise the lid. Unclip the stay from underneath the lid and secure the free end in the bracket attached to the side of the luggage compartment.

*GT*

To open, depress the button catch and raise the counterpoised lid.

**Spare wheel**

The spare wheel is stowed in the luggage compartment.

*GT*

Turn back the luggage compartment mat and unscrew the quick-release screws. Lift the luggage compartment floor.

*Tourer and GT*

Unscrew the clamp plate bolt, and lift out the spare wheel.

**Seat adjustment**

Both front seats are adjustable and may be moved forwards or backwards. To adjust a seat, move the seat catch outwards, slide the seat to the required position and ensure that the catch engages the nearest slot in the seat runner.

The seat squabs are also adjustable for rake by loosening the squab to frame bracket securing screw, loosening the two adjusting screw locknuts and then adjusting the screws.

After adjustment lock the screws with their locknuts and tighten the bracket screw.

The bracket may be reversed to enable the seat squab to tip for the purpose of loading or unloading luggage into the rear cockpit.

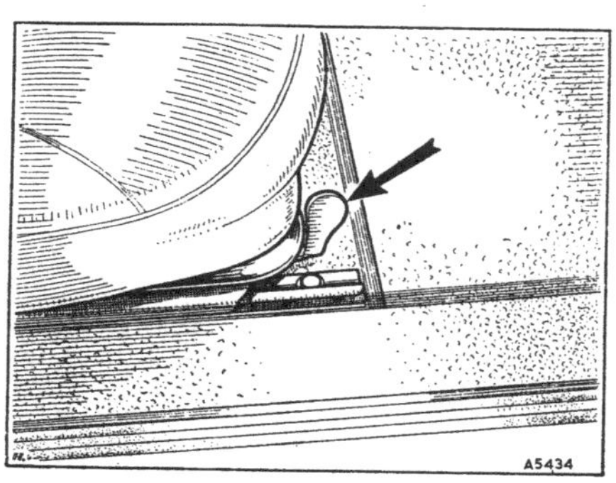


*A seat catch and slide*

**Glovebox**

The glovebox is located on the passenger's side of the fascia panel and can be locked with a separate key.

**Fuel filler**

The fuel filler pipe is fitted with a bayonet-type filler cap which may be removed by turning it in an anti-clockwise direction.

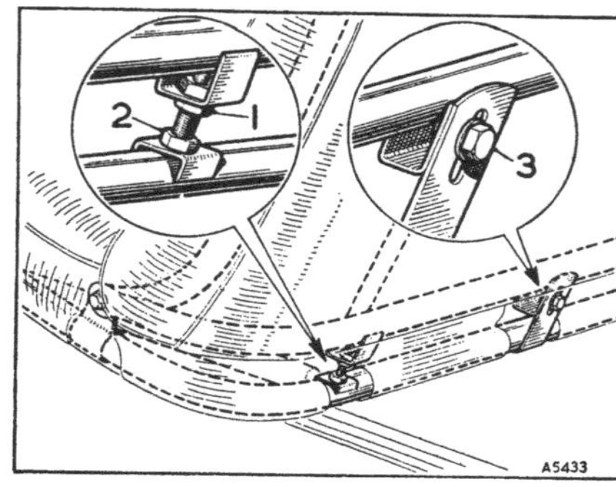


*A seat squab, showing:*

1. **Adjusting screw locknut.**
2. **Adjusting screw.**
3. **Bracket securing screw.**

**Hood (standard)**

*Erecting the hood*

Remove the hood from its stowed position (see page 7).

Erect the collapsible frame and pull the front stick forward; leave the rear stick in the collapsed position until the canopy is fitted.

Place the ends of the frame in the support sockets that are fitted one to each rear quarter panel. The long stick faces forward.

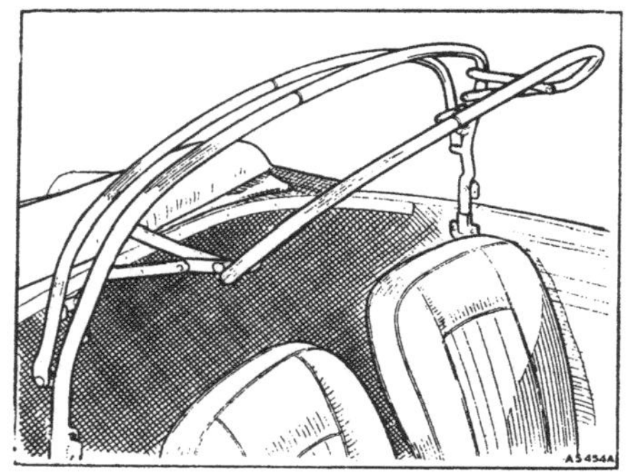


*The hood sticks fitted and erected*

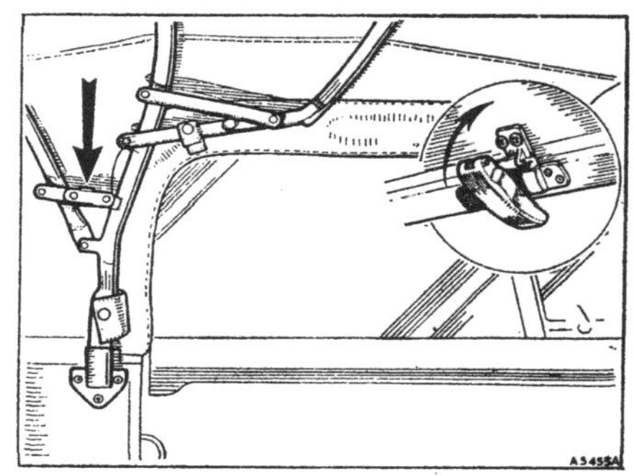

*Extend the rear stick and ensure that the links (one shown arrowed) pass over their centres, inset is the windscreen toggle catches*

Unfold the hood over the frame and engage the two locating sockets on the lower rear edge of the canopy with the retaining plates on the rear tonneau panel. Engage the hood quarter panel retainers in the locating sockets to the rear of each door opening. Engage the fasteners around the rear edge of the canopy. Secure the hood header rail to the windscreen frame with the toggle lever catches and the two fastener assemblies.

Extend the rear stick to tension the canopy and ensure that the connecting links pass over their centres. Then secure the hood to the foot of each socket and the rear quarter of the frame linkages.

*Removing the hood*

Unlock the over-centre links on the rear stick to slacken the canopy. Undo the fasteners and release the toggle lever catches on the windscreen. Release all fasteners around the rear of the hood and the retainers to the rear of each door opening. Slide the rear hood rail backwards clear of the two slotted fasteners on the tonneau panel, and lift the hood from the frame.

**NOTE.—It is most important that the instructions given should be followed when folding and stowing the hood in order to obviate damage to the quarter- and back-lights. Never fold the hood when it is wet or damp.**

*Folding the hood*

Folding the hood correctly is of utmost importance; lay the hood on a flat surface with the lining upwards. Fold the hood in the way shown in the illustration and then roll it up carefully, avoiding kinking.

*Stowing the hood and hood sticks*

Withdraw the frame from the support sockets, fold the sticks and separate the two halves of the frame. Place the canopy and the frame in their respective stowage bags and stow in the luggage compartment.

The hood may suffer damage if stowed without the protection of stowage bag.

# BODY DETAILS

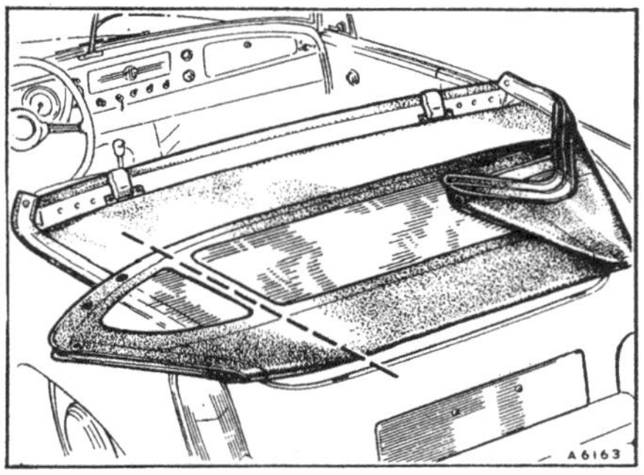

*Fold the quarter-lights inwards, folding on a line between the quarter-light and back-light*

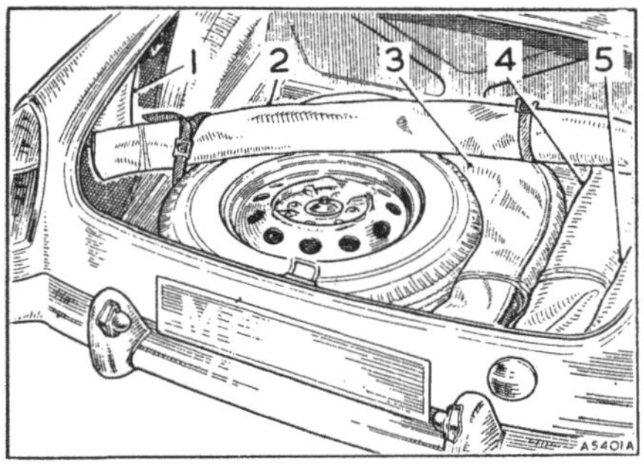


*The luggage compartment, showing the stowage positions for:*

1. Tonneau rail.
2. Standard hood canopy.
3. Hood frame.
4. Tool bag.
5. Tonneau cover and hood cover.

**Hood (optional)**

*Removing the hood cover*

To remove the hood cover undo the two lift-dot fasteners on one of the inner side panels of the rear cockpit, release the four fasteners from the rear quarter-panel, and disengage the hood clip from its socket. Release the other side in a similar manner.

Disengage the back of the cover from the anchor plates on the rear tonneau panel, remove and fold the cover, and place it in its stowage bag.

*Lift the hood from the rear cockpit and lay it on the tonneau panel. Engage the swivel links (inset) in their respective slots and unroll the back-light (early models only)*

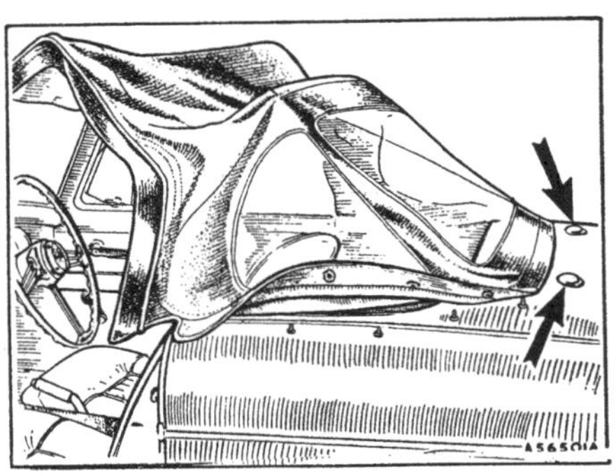

*Lift the hood and engage the bottom of the back-light in the anchor plates on the tonneau panel. Pull the hood forward and engage the hood clips and then do up the four fasteners on the rear quarter-panel*

Remove and dismantle the tonneau rail and place it in its stowage bag. Stow the cover and rail in the luggage compartment.

*Raising the hood*

To raise the hood, release the hood-securing straps, lift the hood from the rear cockpit, and pull it back onto the tonneau panel. Alternatively, lower the seat squabs (this is only possible when the squab-retaining brackets have been reversed) and swing the hood forward until it rests on the rear of the seat squabs.

Pivot each swivel link in turn to its upper position and drop them into their respective slots in the base of the hood frame fitting (early models).

Lift the hood from the seat squabs or tonneau panel and allow the back-light to unroll. Engage the bottom of the back-light in the anchor plates on the tonneau panel and pull the header rail forward to prevent the hood falling backwards.

Unfold a quarter-light, engage the hood clip in its socket, and then do up the four fasteners on the quarter-panel. Secure the other quarter-light in a similar manner.

*Extend the hinge links and (inset) do up the toggle fasteners*

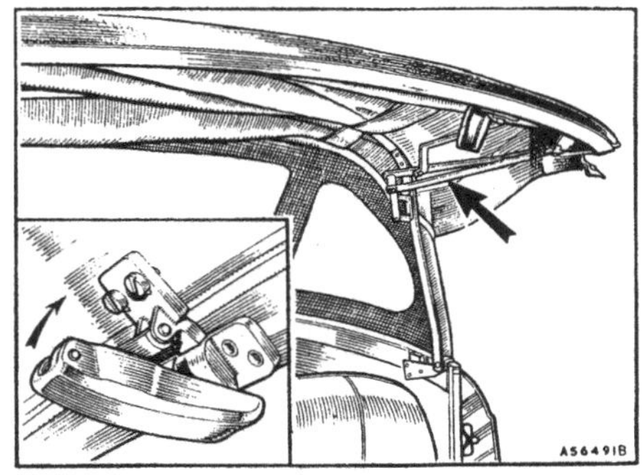

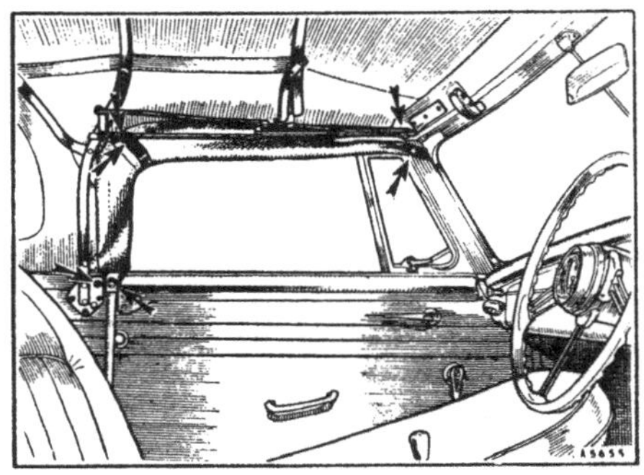

*Do up the six lift-dot fasteners (three shown arrowed)*

From inside the car fully extend the hinge links and engage the toggle fasteners on the header rail with their sockets on the top of the windscreen frame.

Secure the six lift-dot fasteners situated one at the top of each windscreen pillar, one on the rear of each hinge link, and one on each swivel link.

Link each of the two hood-retaining straps together.

*Lowering the hood*

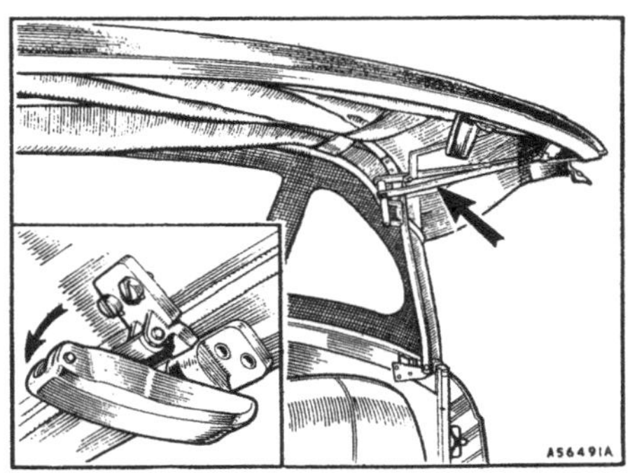

*Undo the six lift-dot fasteners, release the toggle fasteners, and collapse the hinge links*

*Fold the quarter-lights and side overhang inwards, release the back-light from the anchor plates, and lay the hood on the tonneau panel or rear of the seat squabs. Roll up the back-light and cover it with the surplus canopy. Then disengage the swivel links (later cars)*

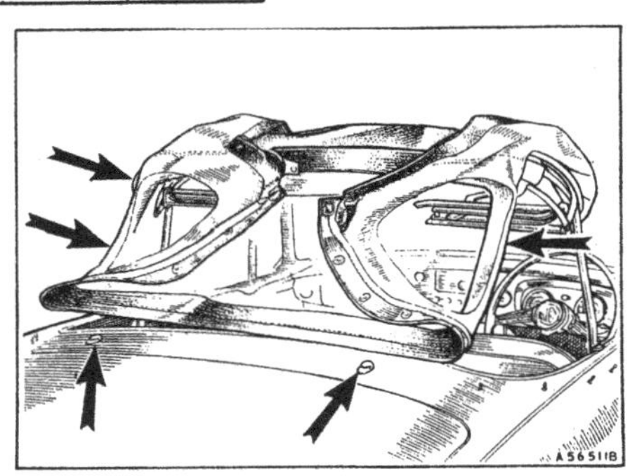

## BODY DETAILS

Never fold the hood when it is wet or damp.

From inside the car release the six lift-dot fasteners and the toggle fasteners. Lift the header rail clear of the windscreen frame and then collapse the hinge links.

On one side of the car release the four fasteners on the rear quarter-panel, disengage the hood clip, and fold the quarter-light onto the back-light; ensure that the fold is made along the hood material, continue this fold along the side edge of the hood to the front.

From the other side of the car release and fold the remaining quarter-light and side overhang in a similar manner.

Bring the header rail towards the rear frame and pull out the surplus hood material clear of the frame. Support the hood with one hand and with the other release the bottom of the hood from the anchor plates. Allow the hood to rest on the tonneau panel or the rear of the seat squabs.

Disengage each swivel link in turn (early models) and then roll up the back-light and quarter-lights. Fold the surplus canopy over the back-light to protect it.

Lift the hood from the rear of the seat squabs or the tonneau panel and pivot it down into the rear cockpit. Pull the hood back as far as it will go (early models) and secure it with the two straps.

*Fitting the hood cover*

Assemble the tonneau rail and fit it to the body sockets with the cross-rod of the rail towards the rear. Spread the cover over the rail and engage the back of the cover with the anchor plates on the tonneau panel. Engage the hood cover clip with its socket and do up the four fasteners on the rear quarter panel. Do up the two lift-dot fasteners inside the rear cockpit.

Secure the other side of the hood cover in a similar manner.

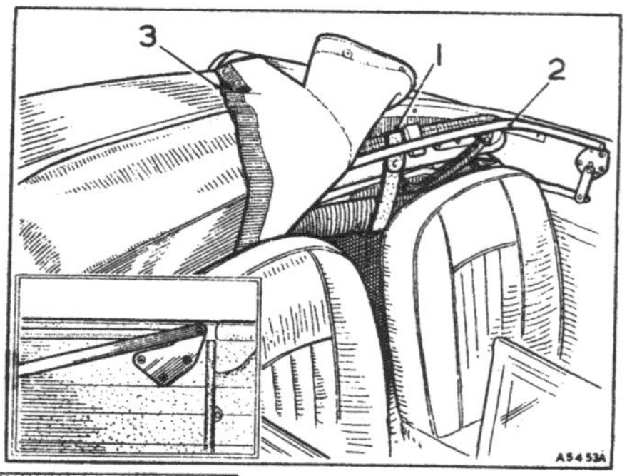


*The optional hood in the stowed position, showing:*

1. A hood securing strap.
2. The tonneau rail.
3. Hood cover.

*Inset: later type head stick*

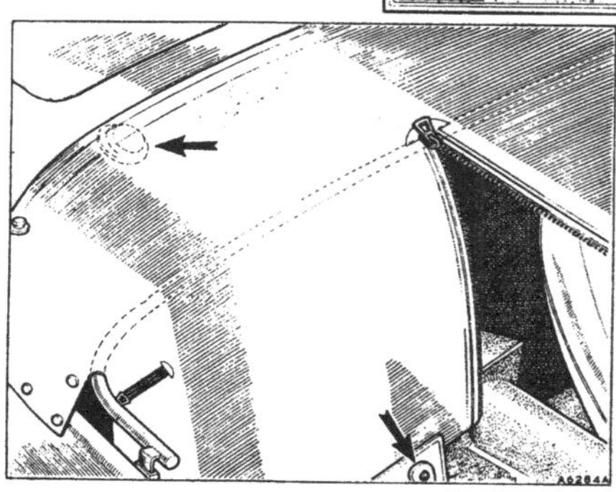

*The tonneau cover fitted, with the cover from the driving position folded down behind the seat and secured to the heelboard*

**Tonneau cover**

*Fitting*

Remove the tonneau cover and rail from its stowed position in the luggage compartment. Assemble the rail into the existing hood stick sockets. Engage the rear edge of the cover with the retaining plates and fasteners on the tonneau and quarter panels.

Extend the cover forward to cover the seats and attach the front of the cover to the fasteners on the fascia panel top.

The front of the cover may be unzipped and folded back to give access to the driver's seat or folded back to give access to both front seats. When folded back ensure that the cover is folded under and that the flaps are securely attached to the fasteners on the heelboard.

Zip fasteners in the cover behind the tonneau rail permit seat belts to be used when the cover is in the folded-back position. Ensure that the flaps adjacent to these fasteners are secured.

*Removing*

When removing the tonneau cover reverse the fitting procedure. Dismantle the tonneau cover rail, fold the tonneau cover and stow it in the luggage compartment.

**Hard top**

*Fitting*

Assemble the quarter panel side brackets to the side sockets on the car, using a spring washer under the head of the bolt. Tighten the single bolt securing each bracket and remove the hard top securing bolts complete with their four washers; keep these ready to hand.

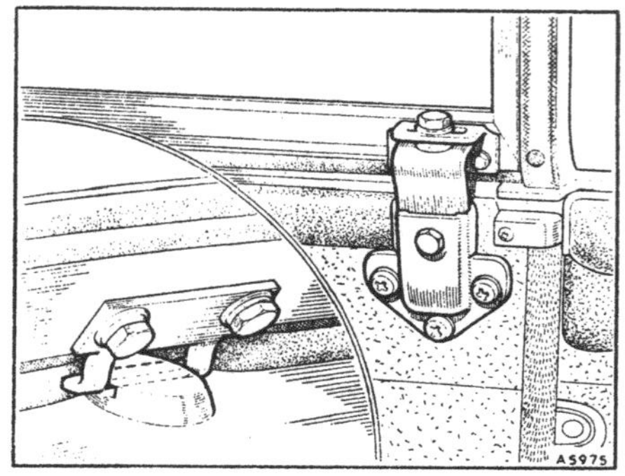

*One of the two hard top fixing brackets and the rear anchor bracket shown inset*

*The windscreen toggle catches in the open and closed position*

With one person at each side, lift the hard top into position over the rear of the car.

Lower the rear end of the hard top and engage the anchor plate brackets with the car anchor plates. Difficulty will be met if the rear sealing rubber is not clear of the anchor plate brackets.

Centralize the hard top to the car by lining up the hard top drip moulding with the rear wing top beading.

Apply hand pressure evenly to the rear of the hard top and check that the front toggle fastener tongues are in their sockets on the windscreen with the front sealing rubber forward of the windscreen top. In this position, with pressure still applied from the rear, insert the quarter panel side bracket bolts, one each side with one washer under the head of the bolt and the other three between the bracket on the hard top and the bracket on the side socket of the car. **DO NOT TIGHTEN THE BOLTS DOWN FULLY AT THIS STAGE.**

Centralize the hard top onto the windscreen and ensure that the front sealing rubber fits down snugly and evenly. With the hard top pushed forward as far as possible, adjust the front toggle links to give an adequate tension to the over-centre action; this is achieved by trial. Tighten the bolts securing the toggles to the hard top. Fasten the toggle links and apply the safety catches.

Loosen the hard top side securing bolts and check that the quarter panel sealing rubber is correctly positioned to pull down onto the car body.

Whilst pressing down on the rear of the hard top, examine the gap between the two side brackets. Determine the quantity of washers necessary to fill this gap and fit one less than this. Repeat this on the other side of the car.

Keeping the rest of the hard top centralized tighten the hard top side securing bolts fully. With pressure applied downwards on the rear of the hard top, tighten the anchor plate securing bolts.

Final check of the correct hard top position is obtained by winding up the door windows. A gap of approximately $\frac{5}{16}$ in. (8 mm.) should exist between the rear vertical edge of the door windows and the quarter light channel, adjustment may be achieved by loosening the side securing bolts and moving the hard top forward or to the rear as required. Check also that an even and adequate seal is made between the window and the hard top rubber. Retighten the side securing bolts and carefully check the door opening and closing actions.

*Removing*

Release the toggle fastener safety catches by unscrewing the Phillips-headed screws. Swing the catches clear of the fasteners and release the toggles.

Remove the bolts, complete with washers, securing the hard top to the quarter panel side brackets.

Loosen the bolts securing the rear anchor plate brackets to the hard top.

Raise the front of the hard top to clear the toggle fastener tongues, move the hard top to the rear to disengage the anchor plates, and lift it clear of the car.

Remove the quarter panel side brackets from their sockets on the car and stow them in the inverted position on the hard top complete with fixing bolts and washers.

**Vent windows**

To open, pull the catch forward and push outwards. Close the window by pulling the centre of the catch inwards and then pushing backwards until the catch is felt to snap over into the locked position.

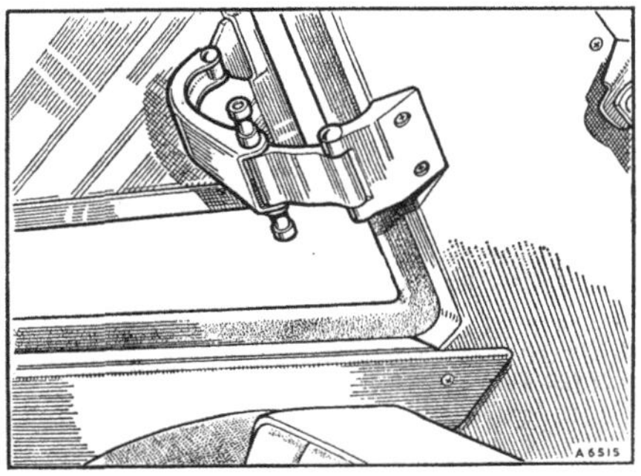

*The rear ventilator window toggle catch*

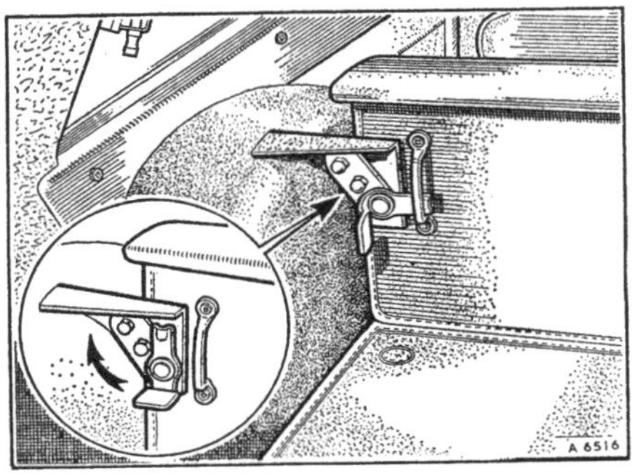

*The rear seat lock*

**Rear seat (increased luggage capacity)**

Release the lock, and pull the seat forward at the top, and lower it until the floor is level. A ledge is provided at the top of the seat to stop the luggage moving forward during heavy braking.

# F. RUNNING INSTRUCTIONS

**Starting**

Before starting the engine ensure that the gear lever is in neutral and the hand brake is applied. If the engine is cold pull out the choke control.

Switch on the ignition, ensure that the ignition warning light glows and that the fuel gauge registers, then operate the starter. Release it if the engine fails to start within five or six seconds, wait for the crankshaft to stop rotating, and then operate the starter again.

Should the engine fail to start after a reasonable number of attempts, check for possible causes.

The continued use of the starter when the engine fails to start will not only discharge the battery but also damage the starter.

If the pinion fails to disengage when the engine starts the starter will emit a high-pitched whine and the engine must be stopped immediately.

As soon as the engine starts, release the starter and warm up the engine at a fairly fast idling speed. Should the oil pressure gauge fail to register any pressure or if the pressure is very low, stop the engine immediately and investigate the cause. Failure to do so may result in serious damage to the engine. Also check that the ignition warning light goes out when the engine is running above idling speed; failure of the light to behave in this fashion indicates a broken fan belt or other fault in the system.

Push in the mixture control knob completely as soon as the engine will run evenly without its use.

**Warming up**

Research has proved that the practice of warming up an engine by allowing it to idle slowly is definitely harmful. The correct procedure is to let the engine run fairly fast, at approximately 1,000 r.p.m., corresponding to a speed of about 18 m.p.h. (27 km.p.h.) in top gear, so that it attains its correct working temperature **as quickly as possible.** Allowing the engine to work slowly in a cold state leads to excessive cylinder wear, and far less damage is done by driving the car straight on the road from cold than by letting the engine idle slowly in the garage.

**Wet brakes**

When the vehicle is being washed or driven through water the brake linings may become wet. To dry them, apply the brakes several times with the vehicle moving slowly. Driving with wet brakes can be dangerous.

**Filling up with fuel**

When filling up with fuel avoid filling the tank until fuel is visible in the filler intake tube. Should this be done and the car left in the sun, there will be a considerable risk of fuel leakage due to expansion, and consequent danger from exposed fuel. If inadvertently overfilled, take care to park the car in the shade with the filler intake as high as possible.

**Running in**

The treatment given to a new car will have an important bearing on its subsequent life, and engine speeds during this early period must be limited. The following instructions should be strictly adhered to.

*During the first* 500 *miles* (800 *km.*)

DO NOT exceed 45 m.p.h. (72 km.p.h.).
DO NOT operate at full throttle in any gear.
DO NOT allow the engine to labour in any gear.

**Tachometer**

For normal road work, and to obtain the most satisfactory service from your engine, select the appropriate gear to maintain engine speeds of between 2,000 and 4,500 r.p.m.

When maximum acceleration is required, upward gear selections should be made when the needle reaches the yellow sector (5,500–6,000 r.p.m.). Prolonged or excessive use of the highest engine speeds will tend to shorten the life of the engine. Allowing the engine to pull hard at low engine speeds must be avoided as this also has a detrimental effect on the engine.

The beginning of the red sector (6,000 r.p.m.) indicates the maximum safe speed for the engine. **Never allow the needle to enter the red sector.**

**Overdrive (optional)**

The overdrive can be engaged at any speed within the third or fourth gear range at any throttle opening between light and full throttle. **DO NOT** depress the clutch pedal to select 'overdrive' or 'normal'.

To prevent overspeeding the engine **DO NOT** select 'normal' top from 'overdrive' top at speeds in excess of the direct drive maximum or change from 'overdrive' third to 'normal' third at speeds in excess of 'normal' third speed maximum.

If increased acceleration is required when in 'overdrive' top the overdrive can be switched out to give normal top gear.

**Towing eye**

If it is necessary to tow the car, use the towing eye provided on the front cross-member.

**High-compression engine (8·8 : 1)**

This engine is a highly developed unit and it is essential that you should know something about the specialized maintenance it requires if you are to maintain it at the peak of its mechanical efficiency. Special recommendations on the sparking plugs, ignition settings, and fuel to be used are given by the manufacturers, and it is stressed that failures are bound to occur if these are not strictly adhered to. Particular care is needed with this engine owing to its high compression ratio, which makes it extremely sensitive to variations in fuel, ignition timing, and the heat range of the sparking plugs.

In lower-compression engines a much wider range of fuels can be tolerated without causing serious damage to the engine, and ignition settings will stand variations of a reasonable amount. Also, even if the incorrect sparking plugs are used, no more damage may be incurred than burnt-out plugs or leaky valves. But with an engine having a very high compression ratio the range of fuels, sparking plugs, and ignition settings is much narrower and it is essential that the mixture should always be correct, and particularly never overweak at maximum load or power.

High-compression engines are very sensitive to variations in spark advance (over-advance) and to fuel/air ratio (mixture). Variations in these settings will increase the combustion temperature, and if the variation is excessive pre-ignition will cause high shock waves, resulting in damage to the engine.

The engine should be decarbonized at regular intervals as excessive deposits of ash from the combustion of lubricating oil and fuel can cause pre-ignition difficulties.

**Choice of fuel**

*When fitted with H.C. engine* (*compression ratio* 8·8 : 1)

The octane number of a motor fuel is an indication given by the fuel technicians of its knock resistance. High-octane fuels have been produced to improve the efficiency of engines by allowing them to operate on high compression ratios, resulting in better fuel economy and greater power. Owing to the high compression ratio of the engine, fuels with an octane rating **below 98** are **not** suitable; should it be necessary to use a fuel with a lower octane number, the car must be used very carefully until the correct fuel can be obtained.

It is necessary to use Super grade fuels in the 100-octane range unless premium fuels of minimum 98-octane (Research method) are available.

*When fitted with L.C. engine* (*compression ratio* 8·0 : 1)

Premium fuels of minimum 93-octane (Research method) up to 97-octane are required, with preference to 95/97.

Super grade fuels in the 100-octane range can be used if preferred.

# COOLING SYSTEM

A pressurized cooling system is used on this vehicle and the pressure must be released gradually when removing the radiator filler cap while the system is hot. It is advisable to protect the hands against escaping steam and then turn the cap slowly anti-clockwise until the resistance of the safety stop is felt. Leave the cap in this position until all pressure is released. Press the cap downwards against the spring to clear the safety stops and continue turning until it can be lifted off.

*The radiator drain tap on the left-hand side of the radiator bottom tank*

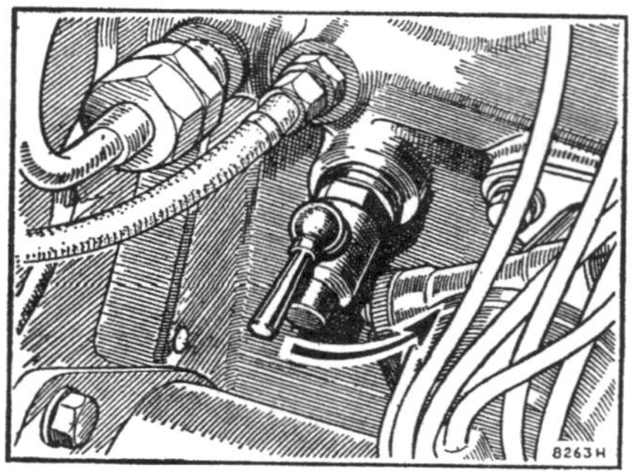


*The cylinder block drain tap is on the right-hand side of the block at the rear. Turn in the direction of the arrow to open*

**Frost precautions**

Water, when it freezes, expands, and if precautions are not taken there is considerable risk of bursting the radiator, cylinder block, or heater (where fitted). Such damage may be avoided by draining the cooling system when the vehicle is left for any length of time in frosty weather, or by adding anti-freeze to the water.

**Warning**

**When a heater unit is fitted an anti-freeze solution must be added in the cooling system since no provision is made for draining the heater.**

**Do not use radiator anti-freeze solution in the windscreen-washing equipment (where fitted).**

**(See supplementary notes at end of Part 1.)**

**Anti-freeze solutions**

Anti-freeze can remain in the cooling system for two years provided that the specific gravity of the coolant is checked periodically and anti-freeze added as necessary. This operation should be carried out by an authorized Distributor or Dealer.

After the second winter the system should be drained and refilled with fresh water, and the appropriate amount of anti-freeze added when required.

Only anti-freeze of the ethylene glycol or glycerine type is suitable for use in the cooling system. We recommend owners to use Bluecol Anti-freeze (non-corrosive) in order to protect the cooling system during frosty weather and reduce corrosion to a minimum. We also approve the use of any anti-freeze which conforms to Specification B.S.3151 or B.S.3152.

The correct quantities of anti-freeze for different degrees of frost protection are:

| *Anti-freeze* | *Commences to freeze* | | *Frozen solid* | | *Amount of anti-freeze* | | |
|---|---|---|---|---|---|---|---|
| % | ° C. | ° F. | ° C. | ° F. | Pts. | U.S. Pts. | Litres |
| 25 | —13 | 9 | —26 | —15 | 2½ | 3 | 1·4 |
| 33⅓ | —19 | —2 | —36 | —33 | 3½ | 4 | 2·0 |
| 50 | —36 | —33 | —48 | —53 | 5 | 6 | 2·8 |

Before adding anti-freeze mixture to the radiator it is advisable to clean out the cooling system thoroughly by flushing out the passages with a hose inserted in the filler cap while keeping the drain taps open.

Only top up when the cooling system is at its normal running temperature, in order to avoid losing anti-freeze due to expansion.

Make sure that the cooling system is water-tight, examine all joints, and replace any defective rubber hose with new.

(See Editor's note at end of Part 1.)

**Draining the cooling system**

There are two drain taps provided, one on the radiator bottom tank and the other on the right-hand side of the cylinder block. To drain the coolant stand the car on level ground and open both taps.

When draining in freezing weather, do so when the engine is hot. Run the engine slowly for one minute when the water has ceased flowing to clear any water from the pump and other places where it might collect. Finally, leave a reminder on the car to the effect that the cooling system has been drained.

If the system contains anti-freeze remember to collect it in a clean container for future use.

**Filling the cooling system**

To avoid wastage by overflow add just sufficient coolant to cover the bottom of the header tank. Run the engine until it is hot and add sufficient coolant to bring the surface to the level of the indicator positioned inside the header tank below the filler neck.

NOTE.—**If a heater is fitted, ensure that the heater temperature control is set to 'HOT' when draining or filling the system.**

# H. IGNITION

**Static ignition timing**

The point where ignition should start is given in **'GENERAL DATA'**. With the crankshaft stationary at this position, the contact breaker points should be just beginning to open. When the engine is running timing is varied by a centrifugal advance mechanism and a vacuum control.

**Checking static ignition timing**

The information given below describes a method of checking the ignition timing; it does not detail the resetting of the timing when the distributor has been removed from the engine.

Check that the contact points are set to the correct gap when on the peak of the distributor cam (see page 11).

The rim of the crankshaft pulley has a small groove which will correspond with the long pointer on the timing cover when Nos. 1 and 4 pistons are at T.D.C.; the other two pointers indicate 5° and 10° B.T.D.C. To turn the pulley to the required position remove the sparking plugs, engage top gear, and push the car forward until the groove in the pulley is in the correct position (see **'GENERAL DATA'**).

*(Right) The distributor adjusting nut (arrowed). (Left) The timing cover pointers and crankshaft pulley groove*

With the crankshaft in this position the contact points should be just about to open. If the points are open, turn the knurled nut towards 'R' until they are closed; if they are closed, turn the nut towards 'A'. In both cases turn the nut until the points are just parting.

A simple electrical method may be used to ensure an accurate check. Connect a 12-volt bulb between the low-tension terminal on the side of the distributor and a good earth point on the engine. Switch on the ignition. If the bulb lights, turn the knurled nut towards 'R' until the light goes out and then back towards 'A' until it just lights. This will give the correct static timing.

If pinking should occur due to the use of a fuel of a lower range than our recommendations, retarding the ignition 2 to 3° can be tolerated. In no circumstances should the ignition be advanced beyond the correct setting.

If this adjustment cannot be made with the knurled nut, consult your Distributor/Dealer.

(See Editor's note at end of Part 1.)

**Distributor (lubrication)**

*Cam bearing*

Remove the rotor arm from the top of the distributor drive spindle and add a few drops of oil Ref. E (page 21) to the cam bearing. Do not remove the screw that is exposed to view; there is a clearance between the screw and the inner face of the cam spindle for the oil to pass. Replace the rotor arm with its drive lug correctly engaging the spindle slot and push it onto the spindle as far as it will go.

*Cam spindle and contact breaker pivot*

Smear a very small amount of grease to Ref. C (page 21) on the cam spindle and also on the contact breaker pivot. Do not allow the grease to get on the contacts; lubricate sparingly.

*Automatic advance control*

Add a few drops of oil to Ref. E (page 21) through the hole in the contact breaker base plate to lubricate the automatic advance mechanism. Do not over-oil or allow any oil to get on or near the contacts. Carefully wipe away any surplus oil and see that the contact breaker points are clean.

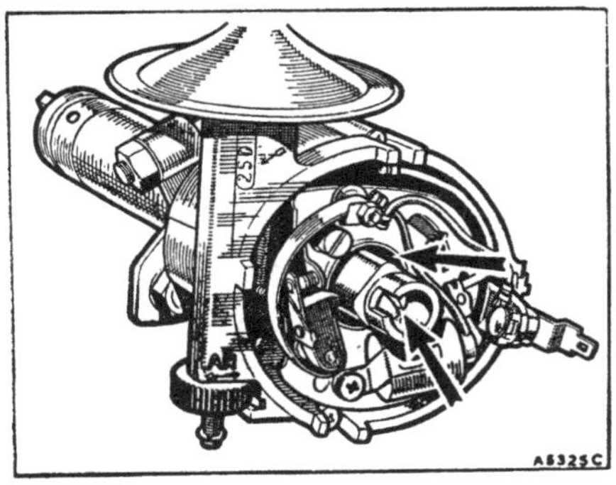


*The distributor cam bearing and automatic timing control lubricating points*

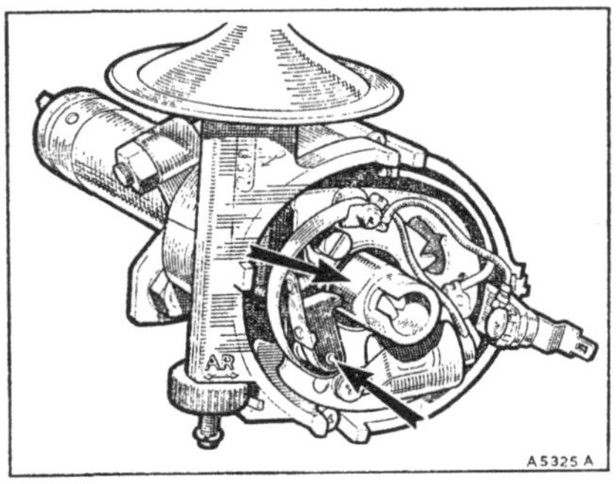


*The distributor cam spindle and contact breaker lubricating points*

**Distributor (mechanical check)**

Check the functioning of the automatic advance and retard mechanism as follows.

*Centrifugal advance mechanism*

Remove the distributor cap and grasp the rotor firmly. Turn the rotor arm in the direction of rotation and release it. The rotor arm should return to its original position without showing any tendency to stick.

*Vacuum advance*

Use a screwdriver to check the movement of the moving plate. Where a modified cap having a window cut in the side is available fit the cap and start the engine. Open the throttle and observe the movement of the contact breaker plate.

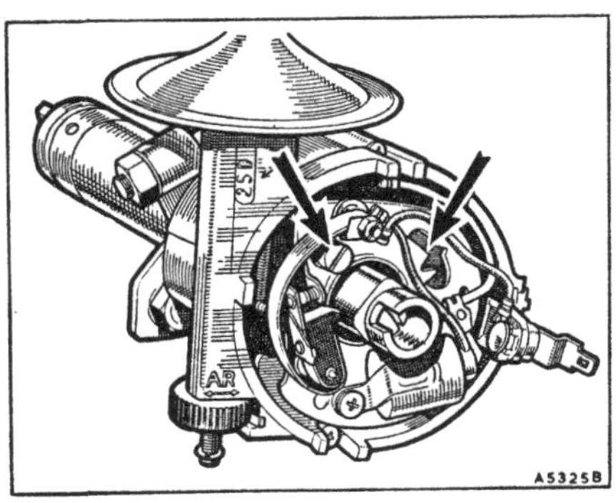


*The distributor points, contact plate securing screws, and the (screwdriver) adjusting slot, indicated by arrows*

*Contact breaker*

Remove the distributor cap and turn the crankshaft until the contacts are fully open. Check the gap with a feeler gauge (see **'GENERAL DATA'**); the gauge should be a sliding fit in the gap. If the gap varies appreciably from the gauge thickness, slacken the contact plate securing screw (see illustration) and adjust the contact gap by inserting a screwdriver in the notched hole at the end of the plate and turning clockwise to decrease and anti-clockwise to increase the gap. Retighten the securing screw.

If the contact breaker points are burned or blackened, clean them with a fine carborundum stone or with fine emery-cloth.

Cleaning the contacts is made easier if the contact breaker lever carrying the moving contact is removed. To do this unscrew the nut securing the end of the spring, remove the spring washer, flat washer, and both lead terminals, and lift off the lever complete with spring. After cleaning refit the contact breaker and check the gap.

The high-tension cables connecting the distributor to the sparking plugs may, after long use, also show signs of perishing. They must then be replaced by the correct type of ignition cable. Cut the cables to length, fill the holes in the cap with silicone grease, push the cables well home in the cap, and secure with the pointed screws.

**Sparking plugs**

Remove the plugs and clean off all carbon deposit from the electrodes and plug threads with a stiff brush dipped in paraffin (kerosene).

Check the plug gaps, and reset if necessary to the recommended clearance (see **'GENERAL DATA'**). When resetting, bend the side electrode only—never bend the centre electrode as this may split the insulator tip.

When refitting the plugs, make sure that the copper washers are not defective in any way. If they have become worn and flattened, fit new ones to ensure a gastight joint. Screw the plug down by hand as far as possible, then use a spanner for tightening only. Always use a tubular box spanner to avoid possible damage to the insulator, and do not under any circumstances use a movable wrench. Never overtighten a plug, but ensure that a good joint is made between the plug body, washer, and cylinder head.

*Use a Champion sparking plug gauge and setting tool. Move the side and never the centre electrode*

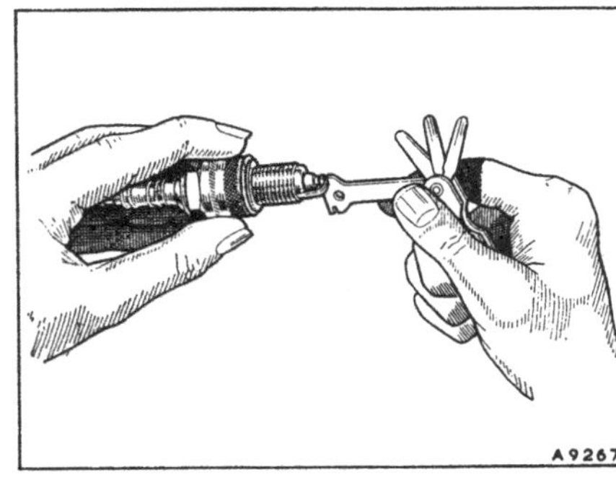


The correct grade of sparking plug for use under normal driving conditions is the Champion N–9Y. Plugs of a lower heat range (hotter running) should not be used, otherwise pre-ignition will occur, with consequent rise in combustion temperature and resulting engine damage. For competition work or hard driving where high output is consistently sustained the Champion N3 sparking plug should be used. This is a cooler-running plug and will ensure lower combustion temperatures and an increased margin of safety. Accumulated deposits of carbon, leaking or cracked insulators, and thin electrodes are all causes of pre-ignition. The plugs should therefore be examined, cleaned, and adjusted at the specified intervals and defective ones renewed.

**Coil**

The coil requires no attention beyond keeping its exterior clean, particularly between the terminals, and occasionally checking that the terminal connections are quite tight.

If the high-tension cable needs renewal it must be replaced by the correct type. Bare the end of the cable for ¼ in. (6 mm.), pass it through its moulded terminal and washer, and spread out the strands to ensure good contact.

# I. ELECTRICAL

(See Editor's note at end of Part 1.)

**Batteries**

The batteries are located in trays beneath the rear floor panel. To gain access to them remove the moulded carpet, release the five screws in the floor panel, and remove the panel.

The battery electrolyte must be maintained at the correct level.

Remove the filler plugs weekly and examine the level of the electrolyte in each cell. If necessary, add sufficient distilled water until the perforated separator guard in each cell is just covered. Do not overfill. More frequent topping up may be necessary in hot climates or if long daily runs are made.

Do not use tap-water and do not use a naked light when examining the condition of the cells. Wipe away all dirt and moisture from the top of the battery.

**Checking the specific gravity**

Check the condition of the battery by taking hydrometer readings of the specific gravity of the electrolyte in each of the cells. Readings should not be taken immediately after topping up the cells. The hydrometer must be held vertically and the readings taken at eye-level. Check that the float is free and take care not to draw in too much electrolyte. The specific gravity readings and their indications are as follows:

| | *For climates below 27° C. (80° F.)* | *For climates above 27° C. (80° F.)* |
|---|---|---|
| Battery fully charged .. .. | 1·270 to 1·290 | 1·210 to 1·230 |
| Battery about half-charged .. | 1·190 to 1·210 | 1·130 to 1·150 |
| Battery fully discharged .. .. | 1·110 to 1·130 | 1·050 to 1·070 |

These figures are given assuming that the temperature of the solution is about 16° C. (60° F.). If the temperature of the electrolyte exceeds 16° C. (60° F.) ·002 must be added to the hydrometer reading for each 3° C. (5° F.) rise to give the true specific gravity. Similarly, ·002 must be subtracted from the hydrometer reading for every 3° C. (5° F.) below 16° C. (60° F.). The readings for all cells should be approximately the same. If one cell gives a reading very different from the rest the battery should be examined by an authorized Distributor or Dealer.

Never leave the battery in a discharged condition for any length of time. Have it fully charged, and every fortnight give it a short refreshing charge to prevent any tendency for the plates to become permanently sulphated.

(See Editor's note at end of Part 1.)

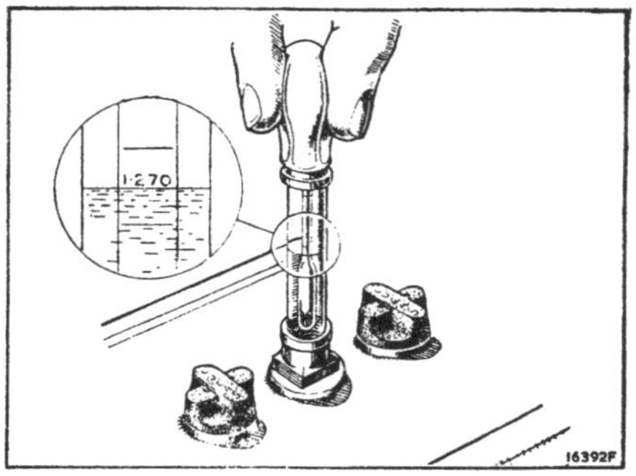


*Taking hydrometer readings*

**Fuel pump**

The fuel pump is mounted on a bracket secured to the heelboard adjacent to the front of the right-hand rear spring.

**Windscreen wiper**

To reposition a wiper arm on the spindle, the arm can be withdrawn when the small spring clip is held clear of the retaining groove. Replace the arm in the required position and push it hard down onto the spindle until it is secured in position by the retaining clip.

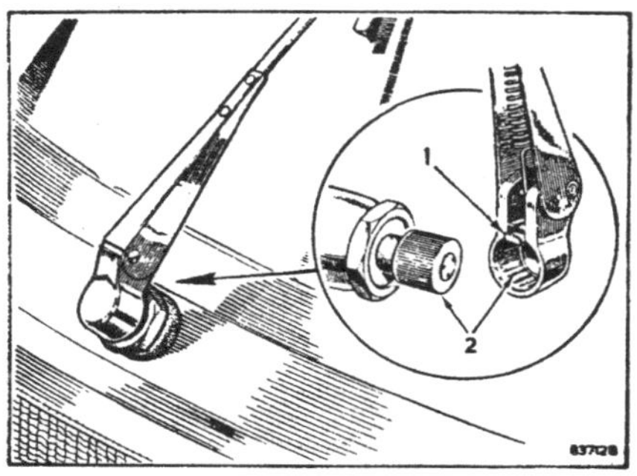


*Windscreen wiper arm*

1. Retaining clip.
2. Splined shaft.

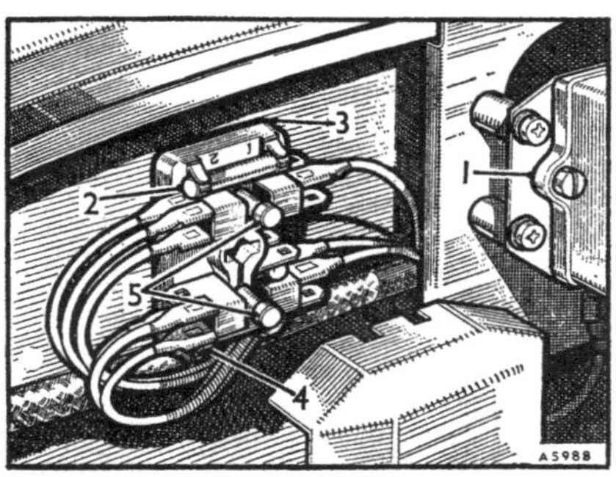


*Regulator and fuse block*

1. Regulator.
2. AUX. IGN. fuse (35-amp.).
3. Fuse block.
4. AUX. fuse (35-amp.).
5. Spare fuses.

**Fuses**

The fuses are housed in a separate fuse block on the engine bulkhead.

The fuse connecting terminals 'A1' and 'A2' protects the accessories that operate irrespective of whether the ignition is on or off.

The fuse connecting terminals 'A3' and 'A4' protects the accessories that operate only when the ignition is switched on (stop lights, direction indicators, etc.).

**Blown fuses**

The units which are protected by the fuses can readily be identified on the wiring diagram. A blown fuse is indicated by the failure of all the units protected by it, and is confirmed by examination of the fuse when withdrawn.

Before renewing a blown fuse inspect the wiring of the units that have failed or evidence of a short circuit or other fault.

Spare fuses are provided and it is important to use only the correct replacement fuse. The fusing value is marked on a coloured paper slip inside the glass tube of the fuse.

**Voltage regulator**

This is a sealed unit located on the right-hand front wheel arch which controls the charging rate of the dynamo in accordance with the needs of the batteries. It requires no attention and should not be disturbed.

**Jammed starter pinion**

If the starter pinion becomes jammed in mesh with the flywheel, it can usually be freed by turning the starter armature with a spanner applied to the shaft extension at the commutator end.

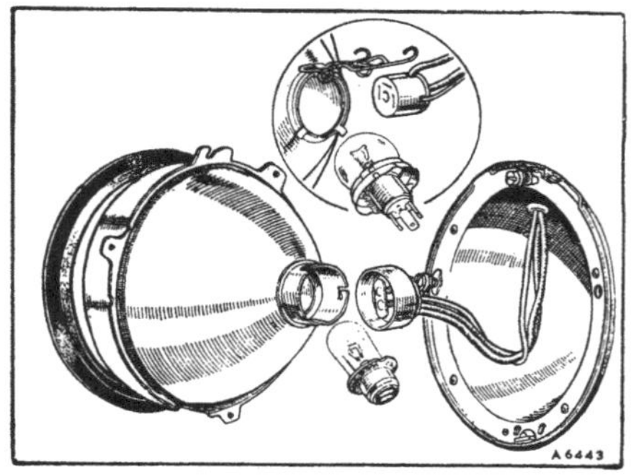


*The headlamp light unit (L.H.D. except Europe and North America) removed showing the bulb holder and back-shell, etc., with the European-type lamp inset*

**Headlamps (European type)**

To change a bulb ease the rim off the bottom of the lamp, pull the bottom of the rim forward and lift off the retaining lugs at the top of the lamp. Remove the three inner rim retaining screws, remove the rim and pull the light unit forward from the back-shell.

*Early Tourer cars:* Press the light unit and turn anti-clockwise to release it from the back-shell.

The bulb is released by withdrawing the three-pin socket and pinching the two ends of the wire retaining clip to clear the bulb flange. When replacing the bulb care must be taken to see that the rectangular pip on the bulb flange engages the slot in the reflector seating. Replace the spring clip with its coils resting in the base of the bulb flange and engaging the two retaining lugs on the reflector seating for the bulb.

**Headlamps (L.H.D. except Europe and North America)**

Access to the bulb is obtained in the same manner as that described for European-type headlamps. Twist the back-shell anti-clockwise and pull it off. The bulb can then be withdrawn from its holder.

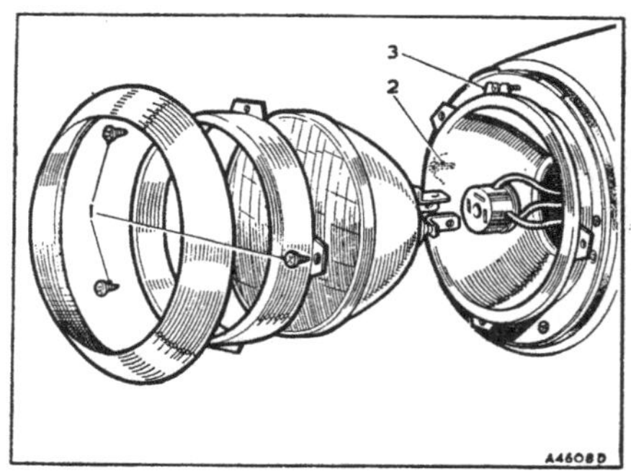


*The U.K. sealed-beam headlamp, showing:*

1. Retaining plate screws.
2. Horizontal adjustment screw.
3. Vertical adjustment screw.

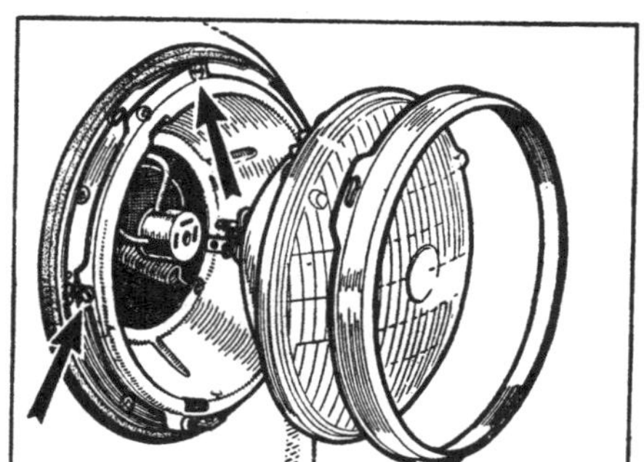


*The North American sealed-beam headlamp with the beam adjusting screws indicated by the arrows*

**Headlamps (U.K. sealed-beam type)**

To change a sealed-beam light unit ease the outer rim off the bottom of the lamp, pull the bottom of the rim forward and lift off the retaining lugs at the top of the lamp.

# J. WIRING DIAGRAM

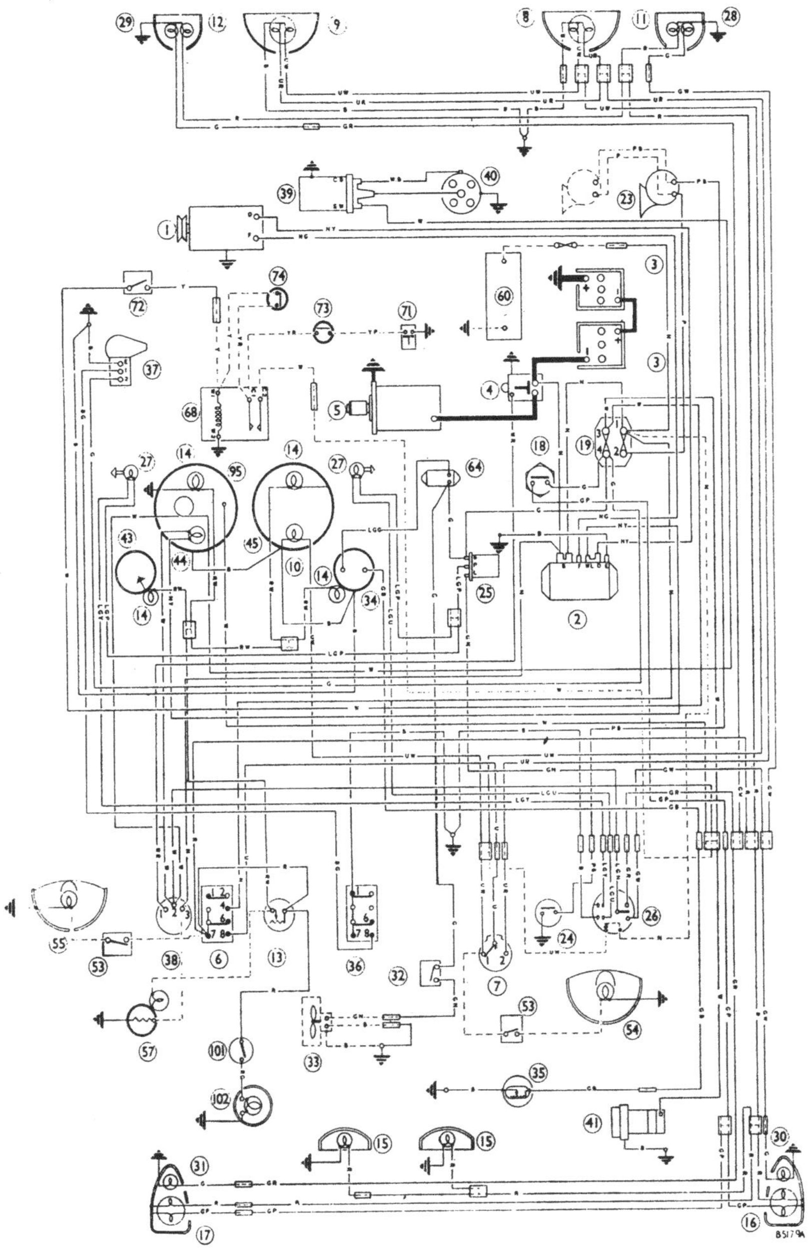

**CABLE COLOUR CODE**

| | | |
|---|---|---|
| N. Brown. | P. Purple. | W. White. |
| U. Blue. | G. Green. | Y. Yellow. |
| R. Red. | LG. Light green. | B. Black. |

When a cable has two colour code letters the first denotes the main colour and the second denotes the tracer colour

## KEY TO WIRING DIAGRAM

1. Dynamo.
2. Control box.
3. Battery—6-volt.
4. Starter solenoid.
5. Starter motor.
6. Lighting switch.
7. Headlight dip switch.
8. R.H. headlamp.
9. L.H. headlamp.
10. Main-beam warning light.
11. R.H. side lamp.
12. L.H. side lamp.
13. Rheostat—panel lights.
14. Panel lights.
15. Number-plate lamps.
16. R.H. stop/tail lamp.
17. L.H. stop/tail lamp.
18. Stop light switch.
19. Fuse unit.
23. Horn (twin when fitted*).
24. Horn-push.
25. Flasher unit.
26. Direction indicator (and headlight flasher*) switch.
27. Direction indicator warning lights.
28. R.H. front flasher lamp.
29. L.H. front flasher lamp.
30. R.H. rear flasher lamp.
31. L.H. rear flasher lamp.
32. Heater or fresh-air motor switch.
33. Heater or fresh-air motor.*
34. Fuel gauge.
35. Fuel gauge tank unit.
36. Windscreen wiper switch.
37. Windscreen wiper motor.
38. Ignition/starter switch.
39. Ignition coil.
40. Distributor.
41. Fuel pump.
43. Oil pressure gauge.
44. Ignition warning lamp.
45. Speedometer.
53. Fog or driving light switch.*
54. Driving lamp.*
55. Fog lamp.*
57. Cigar-lighter—illuminated.*
59. Map light switch.
60. Radio.*
63. Flasher relay.
64. Bi-metal instrument voltage stabilizer.
68. Overdrive relay unit.*
71. Overdrive solenoid.*
72. Overdrive manual control switch.*
73. Overdrive gear switch.*
74. Overdrive throttle switch.*
95. Tachometer.
101. Map light switch.
102. Map light.

Items marked thus * may be fitted as an optional extra.

Remove the three inner rim retaining screws and remove the rim. Pull the light unit from the back-shell and disconnect the three-pin plug to release it from the back-shell.

**Headlamps (North American sealed-beam type)**

To change a sealed-beam light unit, lift the bottom of the outer rim forwards and upwards, and detach the rim. Slacken the three Phillips screws securing the light unit retaining rim and turn the rim anti-clockwise to remove, supporting the lens of the light unit at the same time. Pull off the three-pin plug from the rear of the light unit.

**Setting the headlight beams**

The headlight beams must be set so that the main driving beams are straight ahead and parallel with the road surface and with each other, or in accordance with the local regulations. To adjust, remove the lamp rim and set each lamp to the correct position in the vertical plane by turning the adjusting screw at the top of the light unit in a clockwise direction to raise and anti-clockwise to lower the beam. Horizontal adjustment is made by turning the adjustment screw on the right-hand side of the light unit. On early European and L.H.D. except Europe and North American type units two horizontal adjusting screws are fitted.

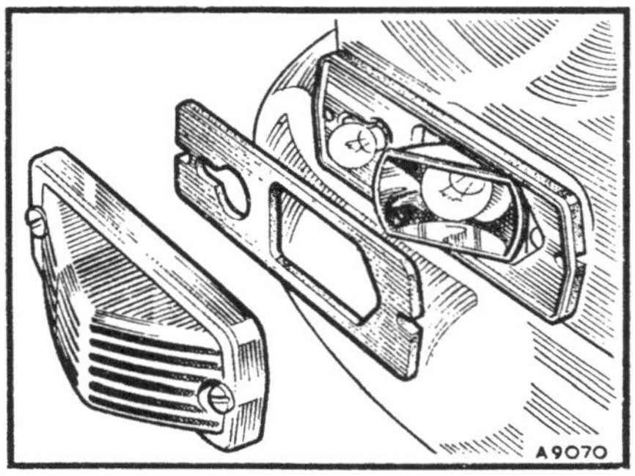


*The side and direction indicator lamp*

**Checking and resetting should be carried out at the beginning of each winter. This work is best entrusted to a Distributor Dealer, who will have specialist equipment available for this purpose.**

**Side and direction indicator lamps**

Extract the two retaining screws to release the plated rim and the lens to gain access to both the light and direction indicator bulbs.

**Stop and tail and direction indicator lamps**

Extract the one retaining screw from the bottom of the lamp and slide the lens upwards to release it from the retaining tongue at the top of the lamp.

**Panel and warning lights**

The panel and warning light bulb holders are a push fit in their housings and are accessible from below the fascia panel.

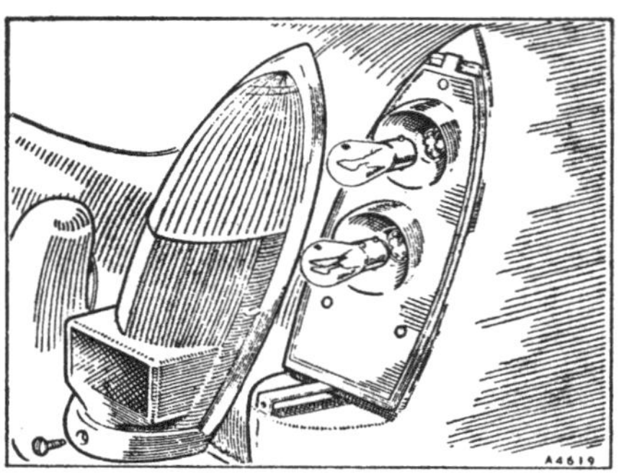


*The stop/tail and flashing indicator lights*

**Number-plate lamps**

The rear number-plate lamps are fitted one to each inside face of the rear bumper over-riders.

To gain access to a lamp bulb remove the two securing screws and retain the nuts and distance pieces from the inside of the over-rider. Take off the lamp hood and remove the bulb.

When refitting the screws and nuts ensure that the thick edge of the wedge-shaped distance piece is fitted so that it faces the front of the car.

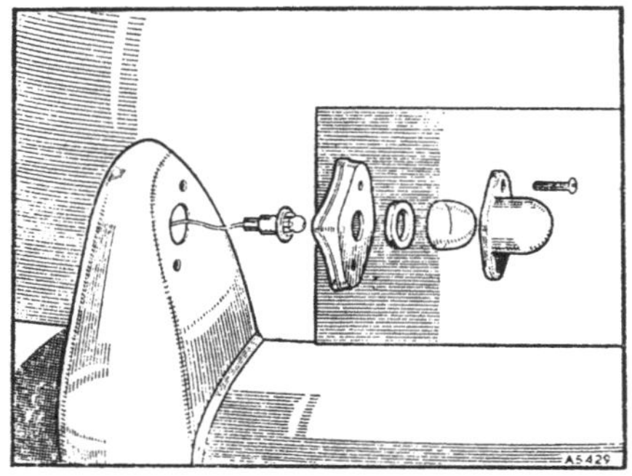


*The number-plate lamps*

**Dynamo bearing**

Add two or three drops of engine oil to Ref. E (page 22) to the dynamo bearing through the central hole in the rear end bearing plate. Do not over-oil.

| **Replacement bulbs** | *Volts* | *Watts* | *BMC Part No.* |
|---|---|---|---|
| Headlamp, L.H.D. (except North America and Europe) | 12 | 50/40 | BFS 415 |
| Headlamps (Europe except France) | 12 | 45/40 | BFS 410 |
| Headlamp (France only) | 12 | 45/40 | BFS 411 |
| Sidelamp | 12 | 6 | BFS 989 |
| Direction indicator, front | 12 | 21 | BFS 382 |
| Direction indicator, rear | 12 | 21 | BFS 382 |
| Stop and tail lamps | 12 | 6/21 | BFS 380 |
| Number-plate illumination lamp | 12 | 6 | BFS 207 |
| Panel and warning lights | 12 | 2·2 | BFS 987 |

**(See supplementary notes at end of Part 1.)**

# K. ENGINE/TRANSMISSION

**Checking engine oil level (A)** (See Editor's note at end of Part 1.)

The level of the oil in the engine sump is indicated by the dipstick on the right-hand side of the engine. Maintain the level at the 'MAX' mark on the dipstick and never allow it to fall below the 'MIN' mark.

The filler is on the forward end of the rocker cover and is provided with a quick-action cap. The filler cap also incorporates a filter for the closed-circuit crankcase breathing intake (later cars).

The oil level should always be checked before a long run.

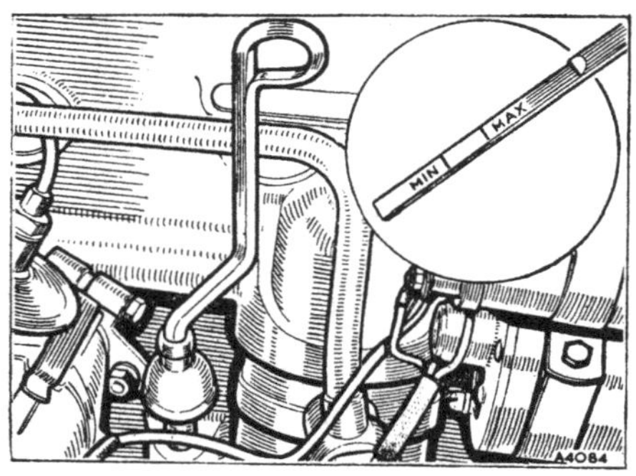


***The engine oil dipstick with the markings shown in the inset***

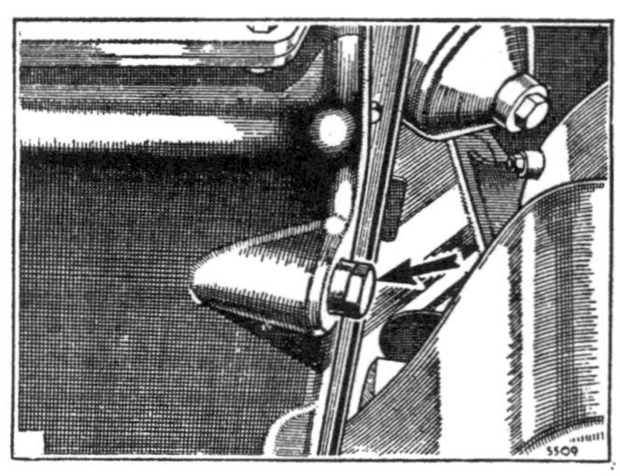


***The engine sump drain plug is located on the right-hand side of the sump***

# ENGINE/TRANSMISSION

**Draining the sump**

**Drain the oil from the engine sump and refill with new oil. The drain plug is on the right-hand side of the sump and should be removed after a journey while the oil is still warm and will drain easily.**

The sump capacity is shown in **'GENERAL DATA'**.

**Oil filter**

The external oil filter is of the renewable-element type and is located on the right-hand side of the cylinder block. The filter is released by undoing the central bolt securing the filter body to the filter head. Wash out the casing with petrol (gasoline) and dry it before fitting a new felt-type element. Ensure that the filter head to casing sealing ring is correctly positioned, clean, and serviceable. Hold the casing against spring pressure before finally tightening the central fixing bolt and ascertain that the casing is seating centrally on the seal.

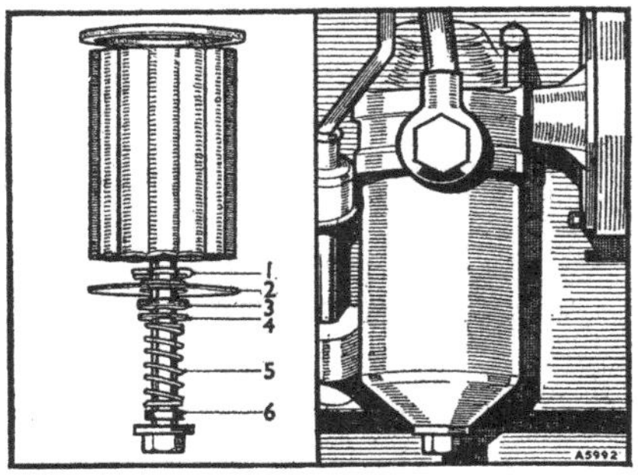


*The engine oil filter, showing the positions in which the washers are fitted below the element*

1. Retaining clip.
2. Element pressure plate.
3. Felt washer.
4. Steel washer.
5. Pressure spring.
6. Rubber washer.

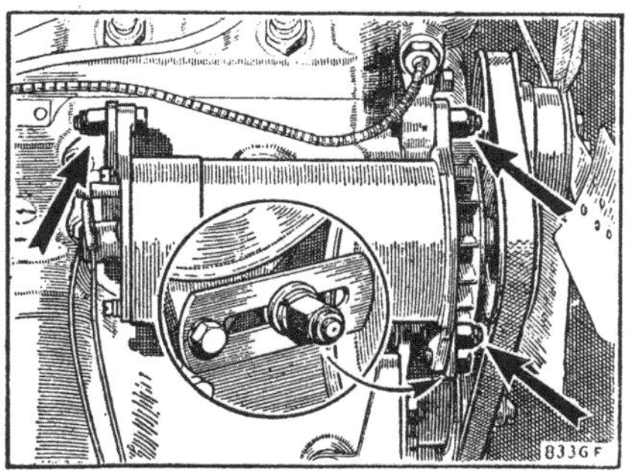


*The four dynamo securing bolts, all of which must be slackened for belt adjustment*

**Dynamo driving belt**

To adjust the belt tension slacken the four dynamo-securing bolts and move the dynamo the required amount by hand. Care should be taken to avoid overtightening the belt. Tighten up the bolts thoroughly, particularly the one passing through the slotted adjusting link.

**Propeller shaft**

Lubrication nipples are provided on the front and rear universal joints (early Tourer cars) and on the sliding yoke at the front end of the propeller shaft. To lubricate, give three or four strokes of the grease gun filled to Ref. C (page 21).

**Water pump**

A lubrication plug is provided on the water pump casing. Remove the plug and lubricate sparingly with grease to Ref. C (page 21).

**Valve rocker clearance**

To check the valve rocker clearance remove the rocker cover, and insert a ·015 in. (·38 mm.) feeler gauge between the valve rocker arms and valve stems. The gauge should be a sliding fit when the engine is cold. Check the clearances in the following order:

*The water pump lubrication plug*

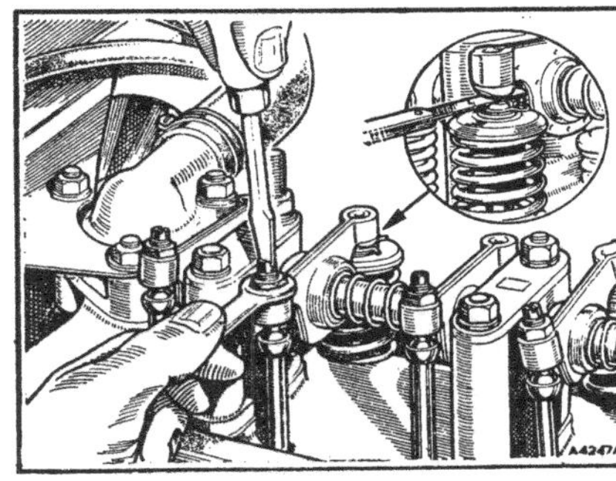


*The method of setting the valve clearance, and (inset) using a feeler gauge to check the clearance*

| | |
|---|---|
| Check No. 1 valve with No. 8 fully open. | Check No. 8 valve with No. 1 fully open. |
| ,, ,, 3 ,, ,, ,, 6 ,, ,, | ,, ,, 6 ,, ,, ,, 3 ,, ,, |
| ,, ,, 5 ,, ,, ,, 4 ,, ,, | ,, ,, 4 ,, ,, ,, 5 ,, ,, |
| ,, ,, 2 ,, ,, ,, 7 ,, ,, | ,, ,, 7 ,, ,, ,, 2 ,, ,, |

To adjust the clearance slacken the adjusting screw locknut on the opposite end of the rocker arm and rotate the screw clockwise to reduce the clearance or anti-clockwise to increase it. Retighten the locknut when the clearance is correct, holding the screw against rotation with a screwdriver.

**Closed-circuit breathing**

An air filter is incorporated in the oil filler cap. The cap is serviced only as a complete assembly.

To test the control valve, run the engine at idling speed while at normal running temperature. Remove the oil filler cap. If the valve is functioning correctly the engine speed will rise by approximately 200 r.p.m. as the cap is removed, the change being audibly noticeable. If no change in speed occurs, service the valve as follows:

Remove the spring clip and dismantle the valve. Clean all metal parts with a solvent (trichlorethylene, fuel, etc.). **Do not use an abrasive.** If deposits are difficult to remove immerse in boiling water before applying the solvent. Clean the diaphragm with detergent or methylated spirits.

Reassemble the valve, making sure that the metering needle is in the cruciform guides and the diaphragm is seated correctly.

NOTE—**The first-type valve assembly (without the cruciform guides) is serviced only as an assembly.**

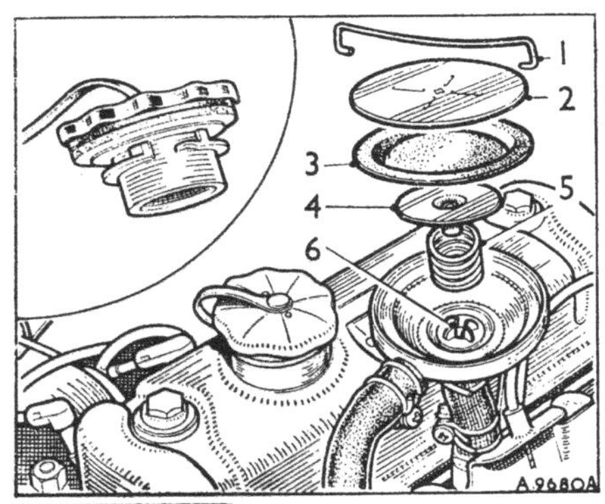


*The breather control valve with the combined air filter/oil filler cap shown inset (second type)*

1. Spring clip.
2. Cover.
3. Diaphragm.
4. Metering needle.
5. Spring.
6. Cruciform guides.

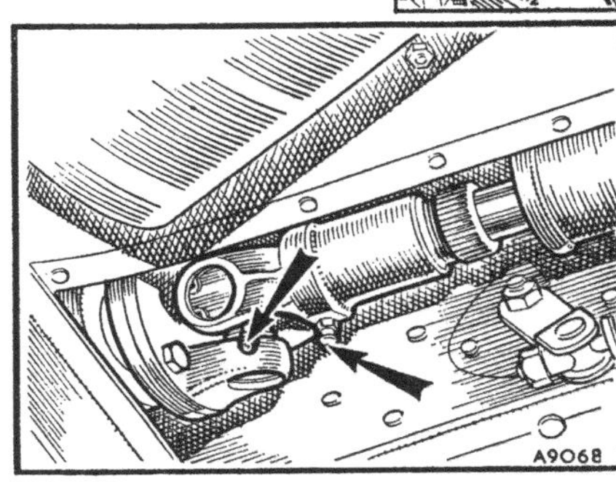


*The lubrication nipples for the front universal joint (early Tourer cars) and the sliding yoke*

**Gearbox and overdrive**

When topping up or refilling ensure that the level does not rise above the 'HIGH' mark on the dipstick.

The combined filler plug and dipstick for the gearbox is located beneath the rubber plug on the tunnel behind the speaker panel.

When the overdrive is fitted, remove the overdrive drain plug and the gearbox drain plug and drain off the oil. Remove the overdrive filter cover plate and gasket from the left-hand side of the unit, and withdraw the filter gauze seal and magnetic rings. Clean, replace, and refill the gearbox with oil to Ref. A (page 21).

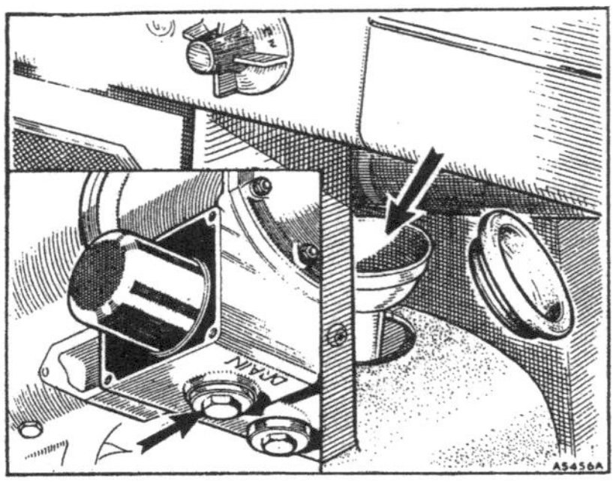

*The gearbox oil filler plug and (arrowed) a funnel passing through the tunnel to the gearbox. The overdrive filter is shown inset*

*The rear axle filler and level plug*

**Rear axle**

A combined oil filler and level plug is located on the rear of the axle. The oil level should be maintained at the bottom of the plug aperture. Ensure that the car is standing level when checking.

After topping up the oil level allow sufficient time for any surplus oil, which may have been added accidentally, to run out of the aperture before replacing the plug. On no account must the axle be overfilled.

**NOTE.—It is essential that only Hypoid oil be used in the rear axle.**

# L. CARBURETTERS

**Slow-running adjustment and synchronization**

When the engine is fully run in the slow running may require adjustment. This must only be carried out when the engine has reached its normal running temperature.

As the needle size is determined during engine development, tuning of the carburetters is confined to correct idling setting. Slacken the actuating arms on the throttle spindle interconnection. Close both throttles fully by unscrewing the throttle adjusting screws, then open each throttle by screwing down each screw one turn.

Remove the suction chamber and piston assemblies, marking each to ensure replacement in their original positions, remove air cleaners and disconnect the mixture control cable. Screw the jet adjusting nuts until each jet is flush with the bridge of its carburetter, or as near to this as possible (both jets being in the same relative position to the bridge of their respective carburetters). Replace the pistons and suction chamber assemblies, and check that the pistons fall freely onto the bridge of the carburetters (by means of the piston lifting pins). Turn down each jet adjusting nut two complete turns (12 flats).

*The carburetter showing:*

1. Jet adjusting nut.
2. Jet link.
3. Jet head.
4. Float-chamber securing nut.
5. Fast idling adjusting screw.
6. Throttle adjusting screw.
7. Vacuum ignition take-off.

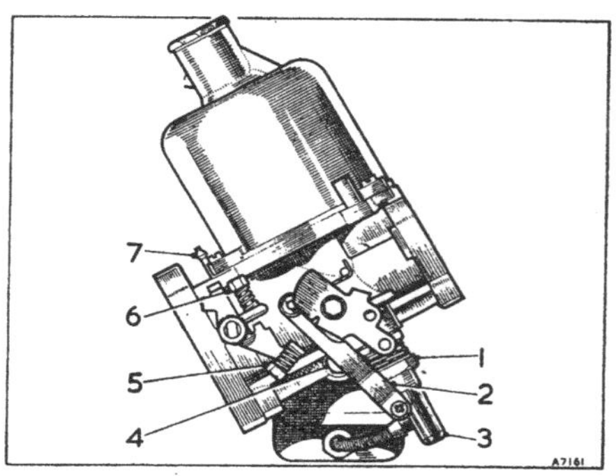


Restart the engine, and turn the throttle adjusting screws to give the desired idling speed by moving each screw an equal amount. By listening to the hiss in the intakes, adjust the throttle adjusting screws until the intensity of the hiss is similar on both intakes. This will synchronize the throttles.

When this is satisfactory, the mixture should be adjusted by screwing each jet adjusting nut, up to weaken, or down to enrich, by the same amount until the fastest idling speed consistent with even firing is obtained. During this adjustment it is necessary to press the jets upwards and ensure that they are in contact with the adjusting nuts.

As the mixture is adjusted the engine will probably run faster and it may therefore be necessary to unscrew the throttle adjusting screws a little, each by the same amount, to reduce the speed.

Now check the mixture strength by lifting the piston of the front carburetter by approximately $\frac{1}{32}$ in. (·75 mm.) when:

(1) If the engine speed increases, the mixture strength of the front carburetter is too rich.
(2) If the engine speed immediately decreases, the mixture strength of the front carburetter is too weak.
(3) If the engine speed momentarily increases very slightly, the mixture strength of the front carburetter is correct.

Repeat the operation at the rear carburetter, and after adjustment re-check the front carburetter, since both carburetters are interdependent.

When the mixture is correct the exhaust note should be regular and even. If it is irregular, with a splashy type of misfire and colourless exhaust, the mixture is too weak. If there is a regular or rhythmical type of misfire in the exhaust beat, together with a blackish exhaust, then the mixture is too rich.

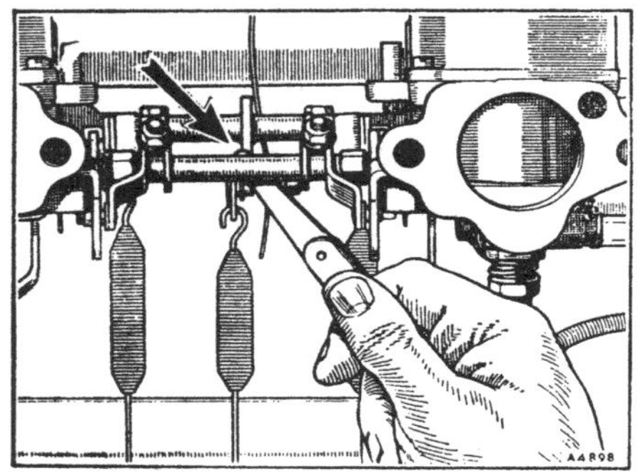

*The feeler between the throttle shaft stop and the choke interconnecting rod*

**Throttle linkage**

Each throttle is operated by a lever and pin, with the pin working in a forked lever attached to the throttle spindle. A clearance exists between the pin and fork which must be maintained when the throttle is closed and the engine idling to prevent any load from the accelerator linkage being transferred to the throttle butterfly and spindle.

To set this clearance, with the throttle shaft levers free on the throttle shaft, put a ·012 in. (·305 mm.) feeler between the throttle shaft stop and the mixture (choke) control interconnecting rod. Move the throttle shaft lever downwards until the lever pin rests lightly on the lower arm of the fork in the carburetter throttle lever. Tighten the clamp bolt of the throttle shaft lever at this position. When both carburetters have been dealt with, remove the feeler. The pins on the throttle shaft should then have clearance in the forks.

Reconnect the mixture control cable, ensuring that the jet heads return against the lower face of the jet adjusting nuts when the mixture control is pushed fully in. Pull out the mixture control knob on the dash panel until the linkage is about to move the carburetter jets a minimum of $\frac{1}{4}$ in. (6 mm.) and adjust the fast-idle adjusting screws to give an engine speed of about 1,000 r.p.m. when hot.

**Carburetter lubrication**

The damper reservoir on each carburetter must be topped up periodically with thin engine oil to Ref. E (page 21). Under no circumstances should a heavy-bodied lubricant be used. Unscrew the damper cap, withdraw the damper, and top up the reservoir until the oil level is $\frac{1}{2}$ in. (13 mm.) above the top of the hollow piston rod. Push the damper assembly back into position and screw the cap firmly into the reservoir.

*Lubricating the carburetter piston dampers*

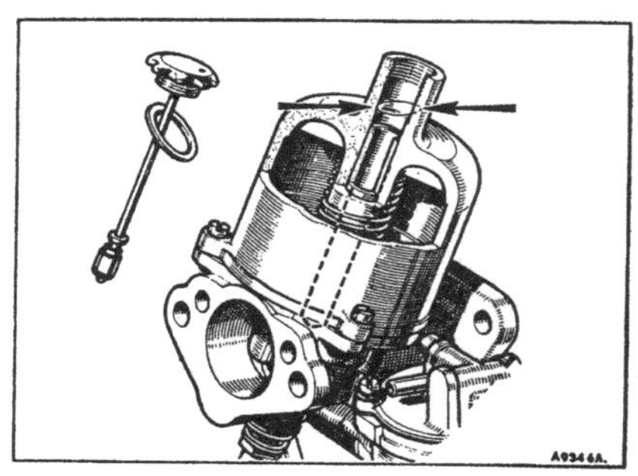

**Air cleaner (dry type)**

To renew the air filter element unscrew the bolts securing each air cleaner to the choke bracket and fixing plate respectively and remove from the car.

Remove the base plate and withdraw the paper element. Clean the inside of the casing thoroughly and reassemble, using a new element.

# BRAKES

**Inspect rear brake linings**

Block the front wheels and release the hand brake.

Remove the road wheel, slacken off the brake-shoe adjuster fully and withdraw the brake drum.

Inspect the linings for wear and blow out the backplate assembly and drum.

Replace the drum and road wheel and adjust the brake-shoes.

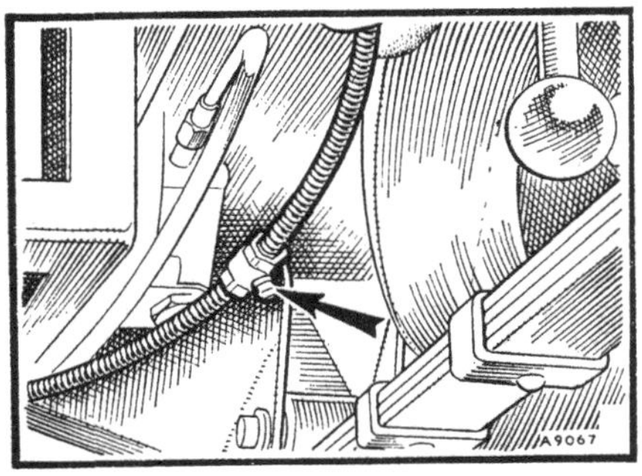


*A grease nipple is provided on the hand brake cable and will be found near the front end of the right-hand rear spring*

*The level of the fluid in the hydraulic brake and clutch master cylinders should be maintained at $\frac{1}{2}$ in. (13 mm.) below the bottom of the filler cap*

**Hand brake cable**

A lubrication nipple is provided on the hand brake cable. To lubricate charge the nipple with grease to Ref. C (page 21).

**Brake and clutch master cylinder**

To check the fluid level in the hydraulic brake and clutch master cylinders remove the filler caps (see illustration above). When topping up use only Lockheed Disc Brake Fluid (Series II).

(See Editor's note at end of Part 1.)

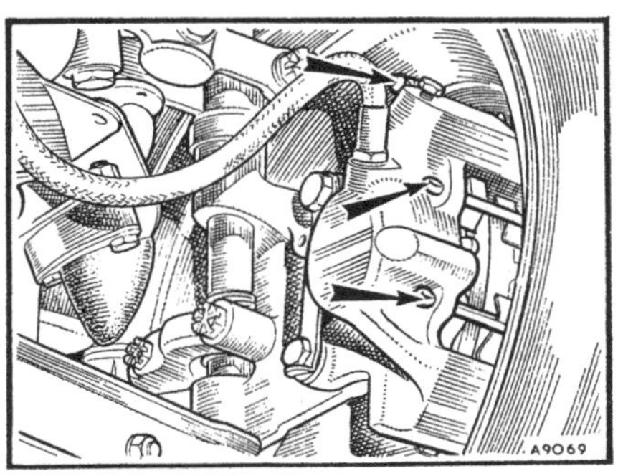


*The upper arrow indicates the bleed nipple and the lower arrows the spring clip retaining pins*

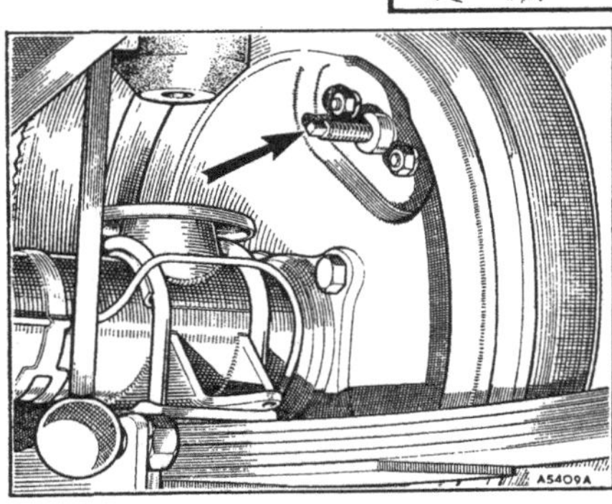


*The rear brake-shoes are adjusted by a square-headed adjuster behind the brake backplate*

**Disc brakes**

Wear on the disc brake friction pads is automatically compensated for during braking operations and manual adjustment is not therefore required. If the wear on one pad is greater than on the other their operating positions should be changed over.

When the pads have worn to a minimum thickness of $\frac{1}{16}$ in. (1·6 mm.) they must be renewed.

Entrust this work to an authorized Distributor or Dealer.

**Rear brake adjustment**

Excessive brake pedal travel is an indication that the rear brake-shoes require adjusting. The brakes on both rear wheels must be adjusted to regain even and efficient braking. Block the front wheels and jack up each rear wheel in turn. Fully release the hand brake. Turn the adjuster in a clockwise direction until the wheel is locked, then turn back until the wheel is free to rotate without the shoe rubbing. Adjust the other rear brake in a similar way. The hand brake is automatically adjusted at the same time. If, however, hand brake lever movement is excessive consult your Distributor/Dealer.

**(See supplementary notes at end of Part 1.)**

# N. STEERING AND SUSPENSION

**Steering rack (early Tourer cars)**

A lubricating nipple is provided on the steering pinion housing. The nipple is accessible from above on R.H.D. cars and from below the radiator on L.H.D. cars. To lubricate give not more than 10 strokes of an oil gun filled with oil to Ref. B (page 21).

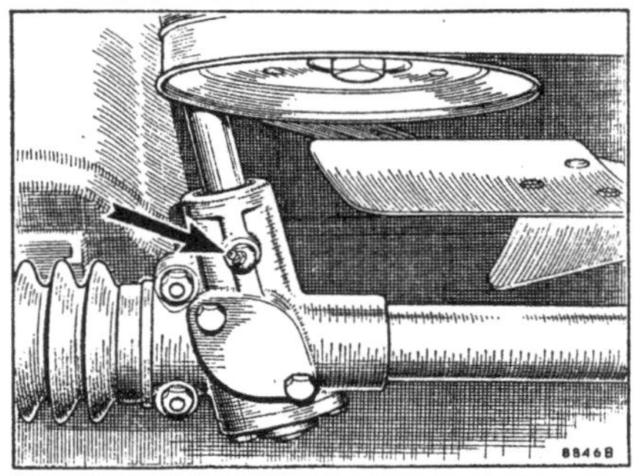


*The arrow indicates the pinion nipple (early Tourer cars)*

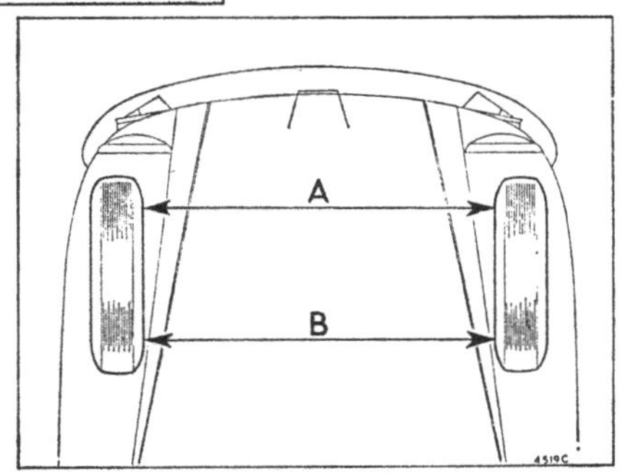


*The front wheels must be set so that dimension* (A) *is between $\frac{1}{16}$ and $\frac{3}{32}$ in. (1·5 and 2·3 mm.) smaller than dimension* (B) *when the wheels are in the straight-ahead position*

**Front wheel alignment**

Excessive and uneven tyre wear is usually caused by faulty wheel alignment The front wheels should be set with between $\frac{1}{16}$ and $\frac{3}{32}$ in. (1·5 and 2·3 mm.) toe-in, and care should be taken that the measurements are taken at axle level and that the rims run true.

Since correct alignment is so important and entails the use of a special gauge, this work should be entrusted to an authorized Distributor or Dealer.

**Front suspension (C)**

Lubrication nipples are provided on the top and bottom swivel pin bushes and in the base of each swivel pin; each nipple should be charged with grease.

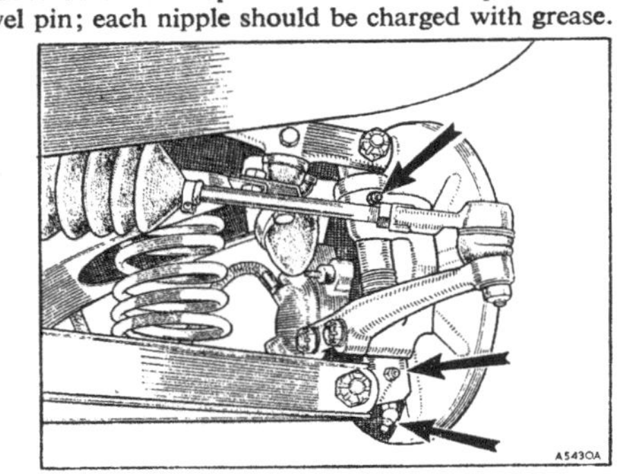

*The front suspension lubrication nipples, which require regular attention. This illustration shows the left-hand side only*

# 0. WHEELS AND TYRES

**Jacking up the car**

The jack is designed to lift one side of the car at a time. Apply the hand brake, and if the car is on an incline it is advisable to scotch one of the wheels on the opposite side of the car to the one being jacked.

Insert the lifting arm of the jack into the socket. **Make certain that the jack lifting arm is pushed fully into the socket and that the base of the jack is on firm ground.** The jack should lean slightly outwards at the top to allow for the radial movement of the car as it is raised.

*Jacking position*

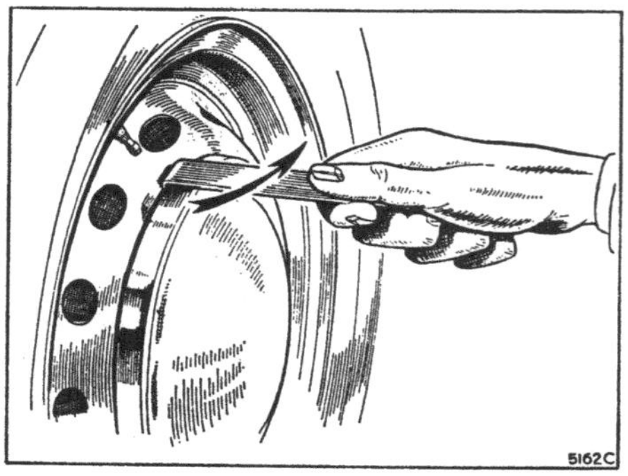

*Removing a wheel disc*

**Jack maintenance**

If the jack is neglected it may be difficult to use in a roadside emergency. Examine it occasionally, clean off accumulated dust, and lightly oil the thread to prevent the formation of rust.

**Removing the wheel discs (pressed-steel wheels)**

Insert the wheel disc lever in the recess provided in the road wheel and lever off the disc, using a sideways motion.

To refit the hub disc, place the rim over two of the buttons on the wheel centre and give the outer face a sharp blow with the hand over the third button.

**Removing and replacing the wheels (pressed type)**

Slacken the four nuts securing the road wheel to the hub; turn anti-clockwise to loosen and clockwise to tighten. Raise the car with the jack to lift the wheel clear of the ground and remove the nuts. Withdraw the road wheel from the hub. When refitting the road wheel locate the wheel on the hub, lightly tighten the nuts with the wheel nut spanner (securing nuts must be fitted with the **taper side towards the wheel**), and lower the jack. Fully tighten the wheel nuts, tightening them diagonally and progressively, at the same time avoid overtightening.

Replace the hub cover.

**Removing and replacing the wheels (wire type)**

Use the mallet to slacken the winged hub nut or the spanner to slacken the octagonal hub nut used.

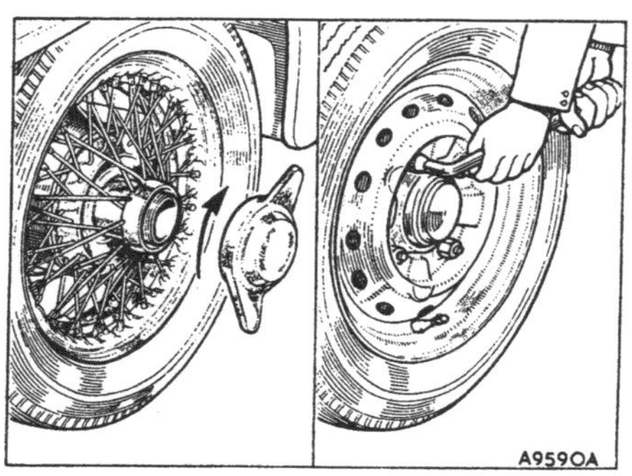

*(Left) Turn the winged hub nuts clockwise to unscrew on the right-hand side of the car and anti-clockwise on the left-hand side*

*(Right) Removing the road wheel securing nuts (pressed type)*

**Care of wire wheels**

(1) When the car is new, after the first long run or after 50 miles (80 km.) of short runs, jack up the wheels and hammer the nuts to make sure that they are tight.

(2) Always jack up a wheel before using the hammer, and always hammer the nuts tight.

(3) Locknuts are marked 'LEFT' or 'RIGHT' to show to which side of the car they must be fitted, and also with the word 'UNDO' and an arrow.

(4) Before replacing a wheel wipe all serrations, threads, and cones of the wheel and hub and then lightly coat them with grease. If a forced change is made on the road, remove, clean, and grease as soon as convenient.

(5) Once a year remove the wheels for examination and regreasing.

**Tyre maintenance**

To obtain the best tyre mileage and to suppress the development of irregular wear on the tyres the wheels can be interchanged diagonally bringing the spare wheel into use.

Excessive local distortion as a result of striking a kerb, a loose brick, a deep pot-hole, etc., may cause the casing cords to fracture.

Tyres, including the spare, must be maintained at the pressures recommended on page 3; check with an accurate tyre gauge at least once a week, and regulate as necessary. Pressures should be checked when the tyres are cold; do not reduce the pressure in warm tyres where the increase above the normal pressure is due to temperature.

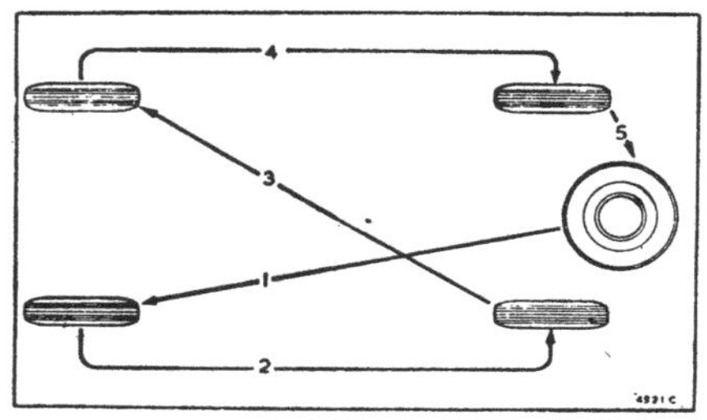

*Two jacking operations only are required to interchange the road wheels in this order*

See that the valve caps are screwed down firmly by hand. The cap prevents the entry of dirt into the valve mechanism and forms an additional seal on the valve, preventing any leakage if the valve core is damaged.

Flints and other sharp objects should be removed with a penknife or similar tool. If neglected, they may work through the cover.

Any oil or grease which may get onto the tyres should be cleaned off by using fuel sparingly. Do not use paraffin (kerosene), which has a detrimental effect on rubber.

When repairing tubes have punctures or injuries vulcanized. Ordinary patches should only be used for emergencies.

Vulcanizing is absolutely essential in the case of tubes manufactured from synthetic rubber.

**Tyre replacement**

Radial-ply (SP) tyres are optional equipment and if used they must be fitted in sets of four.

**Tyre removal and refitting**

Inextensible wires are incorporated in the edges of tyres. Do not attempt to stretch the edges of the tyre cover over the rim. Force is entirely unnecessary and dangerous, as it merely tends to damage the cover edges. Fitting or removing will be quite easy if the wire edges are carefully adjusted into the rim base. If the cover edge fits tightly on the rim seating it should be freed by using the tyre levers.

Remove all valve parts to deflate the tyre completely and push both cover edges into the base of the rim at the point diametrically opposite to the valve, then lever the cover edge near the valve over the rim edge (see illustration above).

This permits the tyre valve to be pushed through the hole in the rim and the inner tube to be withdrawn for attention when required.

A similar technique has to be employed when replacing the tyre, first fitting the tyre into the rim at a point opposite to the valve and finishing the fitting in the region of the valve, keeping the beaded edge in the well-base of the rim.

## WHEELS AND TYRES

*The cover beads should be pushed into the well-base of the rim opposite to the valve*

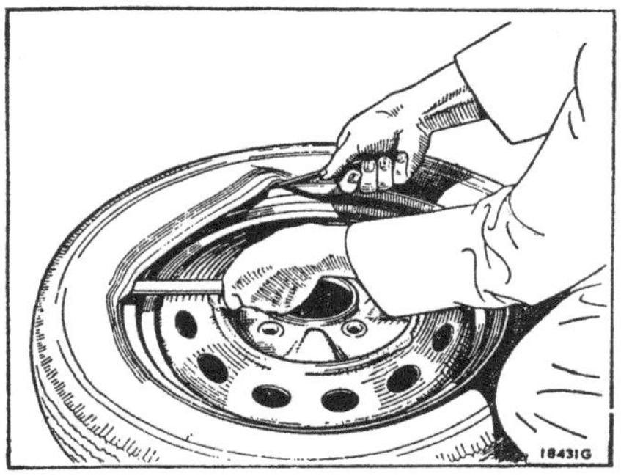

*The cover edge can then be levered over the rim close to the valve position to remove it, or replace it, as required*

# P. MAINTENANCE SUMMARY

**Weekly**

Inspect engine oil level, and top up as necessary.
Check water level in radiator, and top up if necessary.
Check battery and top up to correct levels.
Test tyre pressures.

**3,000 miles (5000 km.) or 3 months service**

1. *Engine*
   Top up carburetter piston dampers.
   Check the coolant level in radiator.
   Top up windscreen washer bottle.
2. *Clutch*
   Check level of fluid in the hydraulic clutch master cylinder.
3. *Brakes*
   Check brakes, and adjust if necessary.
   Make visual inspection of brake lines and pipes.
   Check level of fluid in the hydraulic brake master cylinder.
4. *Electrical*
   Check batteries, and top up to correct level.
   Check headlamp alignment.
5. *Lubrication*
   Check and top up engine oil level.
   Change engine oil (if using monograde only).
   Lubricate all grease nipples (except steering rack and pinion).
6. *Wheels and tyres*
   Check tyre pressures.

**6,000 miles (10000 km.) or 6 months service**

1. *Engine*
   Top up carburetter piston dampers.
   Check fan belt tension.
   Check valve rocker clearances, and adjust if necessary.
   Check the coolant level in radiator.
   Top up windscreen washer bottle.
2. *Ignition*
   Check automatic retard and advance mechanism.
   Check distributor contact points, and adjust if necessary.
   Lubricate all distributor parts as necessary.
   Clean and adjust sparking plugs.
3. *Clutch*
   Check level of fluid in the hydraulic clutch master cylinder.
4. *Steering*
   Check wheel alignment, and adjust if necessary.
5. *Brakes*
   Check brakes, and adjust if necessary.
   Make visual inspection of brake lines and pipes.
   Check level of fluid in the hydraulic brake master cylinder.
   Inspect the disc brake friction pads and report if attention is required.
6. *Electrical*
   Check battery cell specific gravity readings, and top up to correct level.
   Check all lamps for correct functioning.
   Check headlamp alignment.
7. *Lubrication*
   Change oil in engine.
   Lubricate dynamo end bearing.
   Top up gearbox, overdrive (if applicable), and rear axle oil levels.
   Fit new oil filter element.
   Lubricate all grease nipples (except steering rack and pinion).
   Lubricate door locks and hinges.
   Clean overdrive filter element.
8. *Wheels and tyres*
   Check tyre pressures.

**9,000 miles (15000 km.) or 9 months service**

Carry out the 3,000 miles (5000 km.) or 3 months service.

**12,000 miles (20000 km.) or 12 months service**

1. *Engine*
   Top up carburetter piston dampers.
   Renew air filter elements.
   Check valve rocker clearances, and adjust if necessary.
   Check fan belt tension.
   Lubricate water pump sparingly.
   Check coolant level in radiator, and top up if necessary.
   Top up windscreen washer bottle.
   Crankcase closed-circuit breathing system; change engine oil filler cap and clean crankcase breather valve.
2. *Ignition*
   Check automatic retard and advance mechanism.
   Clean, and adjust distributor contact points.
   Lubricate all distributor parts as necessary.
   Fit new sparking plugs.
3. *Clutch*
   Check level of fluid in the hydraulic clutch master cylinder.
4. *Steering*
   Check steering and suspension moving parts for wear.
   Check wheel alignment, and adjust if necessary.
5. *Brakes*
   Check brakes, and adjust if necessary.
   Make visual inspection of brake lines and pipes.
   Check level of fluid in the hydraulic brake master cylinder.
   Inspect disc brake friction pads and report if attention is required.
   Inspect and blow out brake linings and drums.
6. *General*
   Tighten rear road spring seat bolts.
7. *Electrical*
   Check battery cell specific gravity readings and top up to correct level.
   Check all lamps for correct functioning.
   Check headlamp alignment.
8. *Lubrication*
   Change oil in engine.
   Lubricate dynamo bearing.
   Top up gearbox, overdrive (if applicable), and rear axle oil levels.
   Fit new oil filter element.
   Lubricate steering rack and pinion.
   Lubricate steering rack and pinion (early Tourer cars).
   Lubricate all grease nipples.
   Clean overdrive filter element.
9. *Wheels and tyres*
   Check tyre pressures.

NOTE. Take the advice of your Distributor/Dealer on:
1. The need for more frequent engine oil changes.
2. When to change round road wheels.
3. When to check and adjust headlight beams.

# Q. HYDRAULIC BRAKE SYSTEMS PREVENTIVE MAINTENANCE

In addition to the recommended periodical inspection of brake components it is advisable as the car ages and as a precaution against the effects of wear and deterioration, to make a more searching inspection and renew parts as necessary. It is recommended that:

(1) Disc brake pads, drum brake linings, hoses, and pipes should be examined at intervals no greater than those laid down in the Passport to Service.
(2) Brake fluid should be changed completely every 18 months or 24,000 miles (40000 km.) whichever is the sooner.

(3) All fluid seals in the hydraulic system and all flexible hoses should be examined and renewed if necessary every 3 years or 40,000 miles (65000 km.) whichever is the sooner. At the same time the working surface of the pistons and the bores of the master cylinder, wheel cylinders, and other slave cylinders should be examined and new parts fitted where necessary.

Care must be taken always to observe the following points:

(*a*) At all times use the recommended brake fluid.

(*b*) Never leave fluid in unsealed containers. It absorbs moisture quickly and this can be dangerous.

(*c*) Fluid drained from the system or used for bleeding is best discarded.

(*d*) The necessity for absolute cleanliness throughout cannot be over-emphasized.

## R. BODY ATTENTION

**Coachwork**

Regular care of the body finish is necessary if the new appearance of the car exterior is to be maintained against the effects of air pollution, rain, and mud.

Wash the bodywork frequently, using a soft sponge and plenty of water containing a mild detergent. Large deposits of mud must be softened with water before using the sponge. Smears should be removed by a second wash in clean water, and with the sponge if necessary. When dry, clean the surface of the car with a damp chamois-leather. In addition to the regular maintenance, special attention is required if the car is driven in extreme conditions such as sea spray or on salted roads. In these conditions and with other forms of severe contamination an additional washing operation is necessary, which should include underbody hosing. Any damaged areas should be immediately covered with paint and a complete repair effected as soon as possible. Before touching-in light scratches and abrasions with paint thoroughly clean the surface. Use petrol/white spirit (gasoline/hydrocarbon solvent) to remove spots of grease or tar.

The application of BMC Car Polish is all that is required to remove traffic film and to ensure the retention of the new appearance.

**Bright trim**

Never use an abrasive on stainless, chromium, aluminium, or bright plastic parts, and on no account clean them with metal polish. Remove spots of grease or tar with petrol/white spirit (gasoline/hydrocarbon solvent) and wash frequently with water containing a mild detergent. When the dirt has been removed polish with a clean dry cloth or chamois-leather until bright. Any slight tarnish found on stainless or plated components which have not received regular attention may be removed with BMC Chrome Cleaner. An occasional application of mineral light oil or grease will help to preserve the finish, particularly during winter, when salt may be used on the roads, but these protectives must not be applied to plastic finishes.

**Windscreen**

If smearing has occurred it can be removed with BMC Screen Cleaner.

**Interior**

Clean the carpets with a stiff brush or vacuum cleaner, preferably before washing the outside of the car. The most satisfactory way to give carpets a thorough cleaning is to apply BMC 2-way Cleaner with a semi-stiff brush, brush vigorously and remove the surplus with a damp cloth or sponge. Carpets should not be cleaned by the 'Dry-Clean' process. The upholstery and roof lining may be treated with BMC 2-way Cleaner applied with a damp cloth and a light rubbing action.

A razor blade will remove transfers from the window glass.

**Cleaning the hood**

To clean the hood it is only necessary to use soap and water, with a soft brush to remove any ingrained dirt. Frequent washing with soap and water considerably improves the appearance and wearing qualities of the hood, and it should be washed at least as often as the rest of the car.

**Do not use caustic soaps, detergents, or spirit cleaners to clean the hood or the hood back-light.**

**The BMC approved products mentioned above are obtainable from your Distributor or Dealer.**

## S. THE BMC SERVICE FACTORY EXCHANGE UNIT SCHEME

**The BMC Service Factory Exchange Unit Scheme**

The BMC Exchange Scheme—the most comprehensive in Europe—has been designed specifically to **save you money.**

Briefly, the scheme covers practically every major assembly on any BMC car marketed in the last 10 years, and includes components such as heaters and servo units for brakes as well as a wide range of instruments.

If, for example, you want another engine, the Distributor returns the old one to us, and we issue one which has been fully reconditioned in one of our own specialist factories.

By using this technique the cost is considerably reduced but **not the quality,** and each replacement unit carries the same factory warranty as a brand-new one.

Your BMC Distributor or Dealer will be pleased to give you full details and comparative examples of the money which you can save by taking advantage of this scheme.

**Units available**

- Engines and Ancillaries
- Clutches
- Gearboxes
- Rear Axles and Differential Assemblies
- Braking System Units
- Steering Gears
- Instruments
- Electrical Units
- Bumper Bars
- Fuel Pumps
- Shock Absorbers
- Heaters

## T. BODY DETAILS

### SEAT BELTS

Seat belts are obtainable from authorized Distributors and Dealers, and should only be fitted by them to the attachment points incorporated in the body structure.

The approved 'Kangol Magnet' belts consist of a long belt attached at one end to the rear wheel arch and the other end to the sill, and a short belt attached to the side of the drive shaft tunnel farthest from the wearer.

*A driver's seat belt*

**To fasten**—lift up the magnetic buckle tongue and engage the hook into the hinged part of the tongue.

**To release**—lift the magnetic buckle tongue.

**To adjust**—tighten the short belt with the adjuster at the buckle until the buckle rests on the side of the hip (see illustration). With the adjuster at the sill tighten the belt until lap belt fits comfortably and a hand clearance between the diagonal belt and the chest is just possible.

**To stow**—hook the buckle on the long belt into the slot in the stowing bracket mounted under the hood centre locating socket. Attach the magnetic buckle on the short belt to the seat frame.

## U. SUPPLEMENTARY TOOL KIT

To supplement the tool kit a waterproof canvas roll containing the following is obtainable from all Distributors. Part number AKF 1596 should be quoted.

6 spanners:
- $\frac{5}{16}$ in. × $\frac{3}{8}$ in. A.F.
- $\frac{7}{16}$ in. × $\frac{1}{2}$ in. A.F.
- $\frac{1}{2}$ in. × $\frac{9}{16}$ in. A.F.
- $\frac{9}{16}$ in. × $\frac{5}{8}$ in. A.F.
- $\frac{11}{16}$ in. × $\frac{13}{16}$ in. A.F.
- $\frac{3}{4}$ in. × $\frac{7}{8}$ in. A.F.

1 pair 6 in. pliers.

1 7 in. × $\frac{3}{8}$ in. diameter tommy-bar.

1 $\frac{1}{2}$ in. × $\frac{9}{16}$ in. A.F. tubular spanner.

2 screwdrivers.

## KEY TO RECOMMENDED LUBRICANTS

| | A | | | B | | C | D | E |
|---|---|---|---|---|---|---|---|---|
| Component | Engine and Gearbox | | | Rear Axle and Steering Gear | | All Grease Points | Upper Cylinder Lubricant | Oilcan and Carburetter |
| Climatic conditions predominating | Tropical and temperate down to 5° C. (41° F.) | Extreme cold temperatures between 5° C. (41° F.) and −12° C. (10° F.) | Arctic conditions temperatures consistently below −12° C. (10° F.) | All conditions down to −12° C. (10° F.) | Arctic consistently below −12° C. (10° F.) | All conditions | All conditions | All conditions |
| MOBIL | Mobil Special 20W/40 Mobil A.F. | Mobil Arctic or Mobiloil Special 10W/30 or 10W/40 | Mobiloil Special 10W/30 Mobiloil 10W Mobiloil Super 10W/40 | Mobilube G.X. 90 | Mobilube G.X. 80 | Mobilgrease M.P. | Mobil Upperlube | Mobiloil Special 10W/30 Mobiloil Super 10W/40 |
| ESSO | Esso Motor Oil 40/50 Esso Motor Oil 40 Esso Extra Motor Oil 20/40 | Esso Motor Oil 20 or 20W/30 Esso Extra Motor Oil 10W/30 | Esso Motor Oil 10W Esso Extra Motor Oil | Esso Gear Oil G.P. 90/40 or G.P. 90 | Esso Gear Oil G.P.80 | Esso Multi-purpose Grease H | Esso Upper Cylinder Lubricant | Esso Extra Motor Oil 10W/30 |
| FILTRATE | Filtrate Heavy Filtrate 20W/50 | Filtrate Zero or Filtrate 10W/30 | Filtrate Sub-Zero 10W or Filtrate 10W/30 | Filtrate E.P. Grease 90 | Filtrate E.P. Gear 80 | Filtrate Super Lithium Grease | Filtrate Petroyle | Filtrate 10W/30 Multigrade |
| BP | Energol S.A.E. 40 or Super Visco-Static 20W/50 | Energol S.A.E. 20W Super Visco-Static 10W/40 or Visco-Static | Energol S.A.E. 10W Super Visco-Static 10W/40 or Visco-Static | BP Gear Oil S.A.E. 90 E.P. | BP Gear Oil S.A.E. 80 E.P. | Energrease L. 2 | BP Upper Cylinder Lubricant | Visco-Static or Super Visco-Static 10W/40 |
| SHELL | Shell Super Motor Oil Shell X—100 40 Shell X—100 Multigrade 20W/40 or 20W/50 | Shell Super Motor Oil Shell X—100 20W Shell X—100 Multigrade 10W/30 or 20W/40, or 20W/50 | Shell Super Motor Oil or Shell X—100 10W Shell X—100 Multigrade 10W/30 | Spirax 90 E.P. | Spirax 80 E.P. | Shell Retinax A | Shell Upper Cylinder Lubricant | Shell Super Motor Oil |
| CASTROL | Castrol X.L. | Castrolite or Castrol X.L. | Castrol Z or Castrolite | Castrol Hypoy | Castrol Hypoy Light | Castrolease L.H. | Castrollo | Castrolite |
| DUCKHAM'S | Q. 20/50 | Q. 20/50 or Q. 5500 | Q. 5500 | Duckham's Hypoid 90 | Duckham's Hypoid 90 | Duckham's L.B.10 Grease | Duckham's Adcoid Liquid | Q. 5500 |
| STERNOL | Sternol W.W. 40 or W.W. Multigrade 20W/50 | Sternol W.W. 20 or W.W. Multigrade 10W/40 or 20W/50 | Sternol W.W. 10 or W.W. Multigrade 10W/40 | Ambroleum E.P. 90 | Ambroleum E.P. 80 | Ambroline L.H.T. | Sternol Magikoyl | Sternol W.W. Multi-grade 10W/40 |

# LUBRICATION CHART

## KEY TO LUBRICATION CHART

**Weekly**

(1) ENGINE. Check oil level with the dipstick, and replenish if necessary with new oil to Ref. A.

**Every 3,000 miles (5000 km.) or 3 months**

(2) CARBURETTERS. Remove the cap from each suction chamber and insert a small quantity of oil to Ref. E. Lubricate the carburetter controls.

(3) FRONT SUSPENSION. Charge each upper and lower swivel pin bush and the lower swivel pin nipple with grease to Ref. C.

(4) PROPELLER SHAFT. Charge each universal joint nipple (early Tourer cars) and sliding yoke nipple with grease to Ref. C.

(5) HAND BRAKE CABLE Charge the nipple on the hand brake cable with grease to Ref. C.

**Every 6,000 miles (10000 km.) or 6 months**

(6) ENGINE. Drain the used oil from the sump and refill to the 'MAX' mark on the dipstick with new oil to Ref. A.

(7) GEARBOX AND OVERDRIVE. Check the oil level with the dipstick, and replenish if necessary with new oil to Ref. A.

(8) REAR AXLE. Replenish if necessary to the level of the filler plug with new oil to Ref. B.

(9) DYNAMO. Add two drops of oil to Ref. E to the hole in the rear end bearing plate.

(10) OIL FILTER. Renew the element and wash the bowl in fuel.

(11) DISTRIBUTOR. Withdraw rotating arm and add a few drops of oil to Ref. E to opening and also to advance mechanism through gap around cam spindle. Smear cam and pivot pin with grease or oil.

**Every 12,000 miles (20000 km.) or 12 months**

(12) STEERING. (Early Tourer cars.) Apply an oilgun containing oil to Ref. B to the nipple and give several strokes.

(13) WATER PUMP. Remove the plug on the body and add a small quantity of grease to Ref. C.

**The engine oil change periods are those recommended when a multigrade oil is used. Monograde or single viscosity oil should be changed at 3,000 mile (5000 km.) or 3 month intervals.**

## LUBRICATION CHART

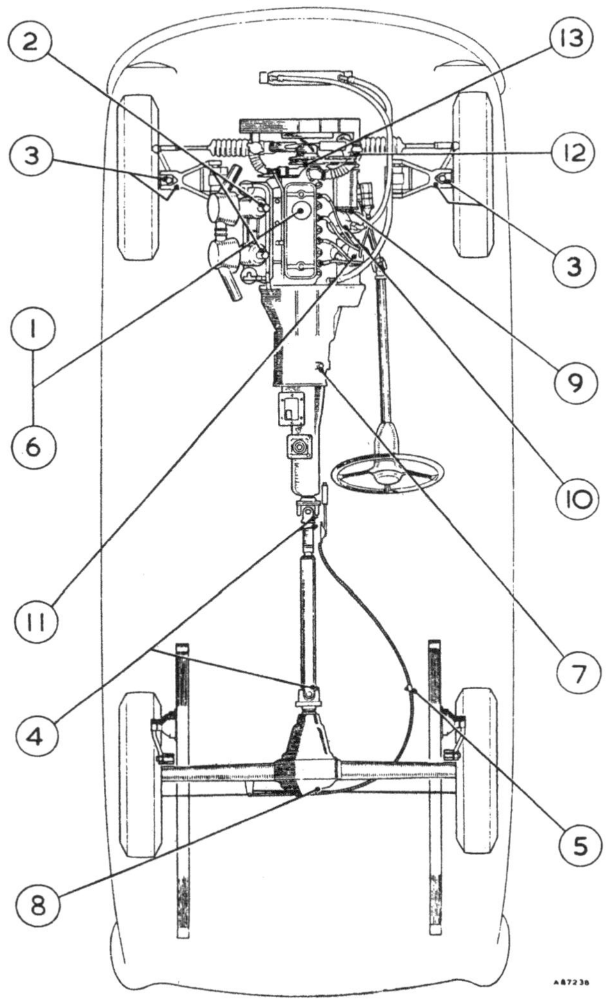

## EDITOR'S NOTES

**G. Cooling System**

*Anti-freeze solutions*

You can check the specific gravity of the coolant solution by using an anti-freeze hydrometer. With this device, a small quantity of the coolant solution is sucked into a cylinder and the position of a glass float indicates the specific gravity, usually in terms of the temperature at which ice will begin to form.

Before adding the anti-freeze mixture the cooling system should be flushed. A more thorough job can be done if the thermostat is removed from its housing and the housing replaced before flushing.

**H. Ignition**

*Checking static ignition timing*

The ignition timing can also be set quite accurately with any of the commercially available timing lights. The light flashes every time the number one cylinder fires, illuminating the timing pointers and pulley. The engine is run at 600 rpm and the knurled nut is adjusted until the pointer is in the correct position. The correct timing setting is given in the "General Data" under "stroboscopic timing setting."

**I. Electrical**

*Battery*

Do not use a "naked" light", or open flame, when examining the condition of the cells. It is dangerous to bring an open flame near a battery because the normal charging process produces hydrogen and oxygen gases which are explosive when mixed in the proper proportion.

*Checking the specific gravity*

When checking the specific gravity of a battery be sure to use a battery hydrometer and not an anti-freeze hydrometer, which looks quite similar.

**K. Engine/Transmission**

*Checking the oil level and all lubrication headings*

The letters in parentheses following a lubrication heading refer to the "Key to Recommended Lubricants" found on page 21.

**M. Brakes**

*Brakes*

Standard brake fluids must not be used as they will boil under hard use in a disc brake system, resulting in a "spongy" pedal and much reduced stopping ability. Any heavy duty fluid specified for disc brake systems will give satisfactory results.

## SUPPLEMENTARY NOTES

### THERMOSTAT

For maximum efficiency the engine operating temperature is maintained within certain limits by a bellows-type thermostat fitted in the water outlet at the front of the cylinder head.

When the engine is cold this valve is closed, and when the engine is started the flow of water to the radiator is temporarily restricted. Due to this, the temperature of the water in the cylinder head and cylinder jackets will quickly rise, thus ensuring rapid warming up. The heat so generated will gradually expand the bellows, so opening the valve and ultimately permitting a full flow of water to the radiator.

The thermostat is detachable.

A small hole drilled in the head of the valve provides a by-pass to cope with any expansion of the cooling water. When the system has been completely drained it is essential, when refilling, to allow sufficient time for any trapped air to escape through the by-pass hole in the valve before finally topping up.

The thermostat opening is set by the manufacturer and cannot be altered. It opens at the temperature marked on the body of the thermostat. During decarbonizing it is policy to test this opening by immersing the thermostat in water raised to the requisite temperature. The valve should open under these conditions, but if it fails to open a new unit must be fitted.

# MGB
# TOURER AND GT
# DRIVER'S HANDBOOK
# 1971-1974

## GHN5, GHN4, GHN 5UC, and GHN 5UD
## GHD5, GHD4, GHD 5UC, and GHD 5UD

## PART 2

# FOREWORD

This Handbook provides an introduction to your car, together with information on the care and periodic maintenance required to combine trouble-free motoring with minimal running costs.

Claims for the replacement of parts under warranty must be submitted to the supplying Distributor or Dealer, or when this is not possible, to the nearest Distributor or Dealer, informing them of the vendor's name and address. Except in emergency, warranty work should always be carried out by an appointed Distributor or Dealer.

By keeping the Passport to Service, signed by the Distributor, Dealer, or vendor in the vehicle, you can quickly establish the date of purchase and provide the necessary details if adjustments are required to be carried out under warranty.

Regular use of the Passport to Service Maintenance Scheme is the best safeguard against the possibility of abnormal repair bills at a later date. Failure to have your car correctly maintained could invalidate the terms of the warranty and may result in unsatisfactory operation of the emission control systems.

Safety features embodied in the car may be impaired if other than genuine parts are fitted. In certain territories, legislation prohibits the fitting of parts not to the vehicle manufacturer's specification. Owners purchasing accessories while travelling abroad should ensure that the accessory and its fitted location on the car conform to mandatory requirements existing in their country of origin.

Your Distributor or Dealer is provided with the latest information concerning special service tools and workshop techniques. This enables him to undertake your service and repairs in the most efficient and economic manner. The operations carried out by your Distributor or Dealer will be in accordance with current recommendations and may be subject to revision from time to time.

Further details on service parts will be found under 'SERVICE' on page 68. **Please note that references to right- or left-hand in this Handbook are made when viewing the car from the rear.**

Specification details set out in this Handbook apply to a range of vehicles and not to any particular vehicle. For the specification of any particular vehicle owners should consult their Distributor or Dealer.

The Manufacturers reserve the right to vary their specifications with or without notice, and at such times and in such manner as they think fit. Major as well as minor changes may be involved in accordance with the Manufacturer's policy of constant product improvement.

Whilst every effort is made to ensure the accuracy of the particulars contained in this Handbook, neither the Manufacturer nor the Distributor or Dealer, by whom this Handbook is supplied, shall in any circumstances be held liable for any inaccuracy or the consequences thereof.

**Emission Controls**

Your car is fitted with emission controls and devices required by the United States Clean Air Act and the Canadian Federal Motor Vehicle Safety Act.

Please read carefully the **'EMISSION CONTROL SYSTEMS'** section of the Handbook which contains information on the emission control systems fitted to your car and recognition of symptoms of malfunctions which could affect emissions.

It is imperative that you familiarize yourself with the contents of this section and ensure that the car you have purchased will remain in compliance with the intentions of the above Acts.

All maintenance checks and adjustments showing this sign should be entrusted to your Austin MG Dealer.

## CONTROLS

*Fig. 1* **Synchromesh gearbox**

**Pedals** (1) (2) (3) The pedals are arranged in the conventional positions.

The brake pedal operates the dual hydraulic braking system applying the brakes on all four wheels, also when the ignition is switched on, bringing the stop warning lights into operation.

**Gear lever** (4) The gear positions are indicated on the lever knob. To engage reverse gear move the lever to the left in the neutral position until resistance is felt, apply further side pressure to overcome the resistance and then pull the lever back to engage the gear. The reverse lights operate automatically when reverse is selected with the ignition switched on.

Synchromesh engagement is provided on all forward gears.

**Hand brake** (5) The hand brake is of the pull-up lever type, operating mechanically on the rear wheels only. To release the hand brake pull the lever up slightly, depress the button on the end of the lever and push the lever down.

**Mixture control (choke)** (6) Pull out the knob marked 'choke' to enrich the fuel mixture when starting the engine from cold. Lock the control in the desired position by turning the control knob a quarter turn clockwise.*

* Also see 'RUNNING INSTRUCTIONS'.

*Fig. 1*

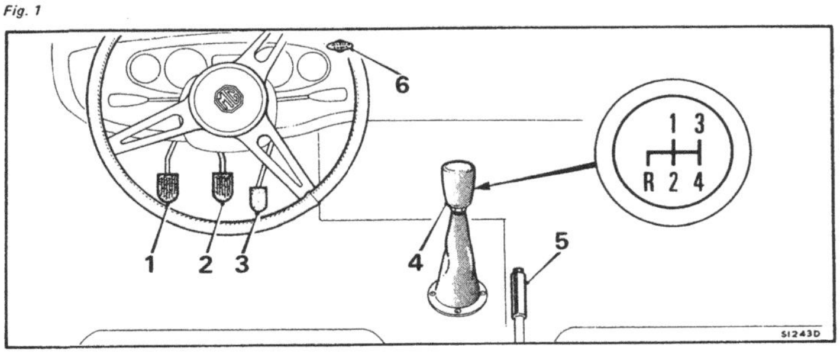


## WARNING SYSTEMS

**Anti-theft warning buzzer** A combined ignition and steering lock with warning buzzer is fitted to the car. The warning buzzer will sound if the driver's door is opened while the key is in the lock. The buzzer will not operate if the key is removed from the lock.

When leaving the car unattended always:

Set the hand brake.

Lock the steering by removing the key from the ignition steering lock.

Lock the car doors and remove the key.

**Brakes** *Fig. 1* **Pressure failure warning.** The lamp (1) in the switch will glow, when the brake pedal is pressed, if any part of the hydraulic system is inoperative or considerable adjustment of the rear brakes is required. If this occurs and in your judgement you can drive safely with braking on two wheels only, proceed at reduced speed to the nearest service facility for immediate repairs. The vehicle should not be driven in this condition except in cases of real emergency, and when in your judgement you can proceed safely at reduced speed. Extreme care must be taken and heavy braking avoided.

To test the warning lamp, press the switch (2). If the bulb is functioning the lamp will glow and will go out as the switch is released. To test the hydraulic system, apply normal foot pressure to the brake pedal. The lamp will remain off if the hydraulic system is functioning satisfactorily. Check the bulb and the system frequently.

If the warning lamp glows at any time except when the bulb is being tested, **the cause must be investigated IMMEDIATELY.**

**Seat belt warning** *Fig. 2* The seat belt warning system fitted to the car consists of a warning lamp (1) on the centre console illuminating the words **'FASTEN BELTS'**, and a warning buzzer.

The warning system will operate when the ignition is switched on (position **'II'** on the ignition switch), when a gear is selected and either the driver's or passenger's seat belt is not fastened by the wearer.

**PRECAUTION:** A heavy parcel placed on the passenger's seat **may** operate the warning system. To prevent this happening it would be necessary to fasten the passenger's seat belt.

*Fig. 1*

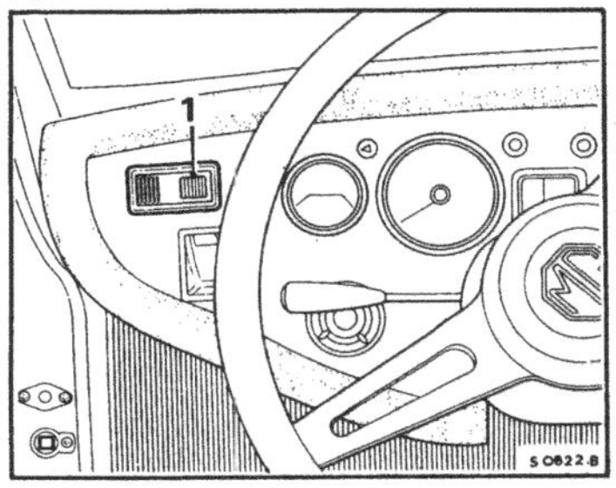


*Fig. 2*

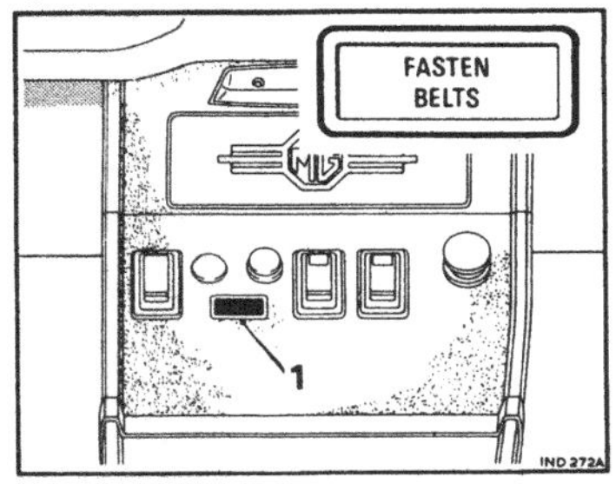


## LOCKS

**Keys** To reduce the possibility of theft, locks and ignition switches are not marked with a number, Owners are advised to make a note of the numbers stamped on the keys, or on the numbered tag supplied with the keys, or on a label attached to the windscreen. It is most important that **OWNERS MAKE A NOTE OF THE KEY NUMBERS IMMEDIATELY** on taking delivery of the car. Consult your Distributor or Dealer regarding steering lock key replacements.

**Steering lock** *Fig. 1* The lock face (4) is marked **'O'** (off), **'I'** (auxiliary), **'II'** (ignition), **'III'** (start).

**To lock the steering,** turn the key to position **'I'**, press the key in, and while maintaining pressure, turn anti-clockwise to position **'O'** and withdraw the key. The steering lock is set during withdrawal of the key and rotation of the steering-wheel engages the lock. When unlocking, turn the steering to assist disengagement of the locking plunger.

Under no circumstances must the key be moved from the **'I'** position towards the **'O'** position **WHEN THE CAR IS IN MOTION**. The car may be towed for recovery with the key in the lock at position **'I'**.

**WARNING.**—The lock fitted to the steering-column works in conjunction and is integral with the ignition starter switch. The designed operating sequence prevents the engine being started with the steering LOCKED. **Serious consequences may result from alterations or substitution of the ignition start switch which would permit the engine to be started with the LOCK ENGAGED. Under no circumstances must the ignition switch or the ignition engine start function be separated from the steering lock.**

**Ignition and starter.** The combined ignition and steering lock face is marked **'O'** (off), **'I'** (auxiliary), **'II'** (ignition), **'III'** (start). Insert the key in the lock, and turn to position **'I'**. In this position the ignition is off but electrical items not wired through the ignition switch may be operated, viz, radio, windscreen wipers and screen washer. Turn the key to position **'II'** to switch on the ignition; further movement to **'III'** operates the starter.

*Fig. 1*

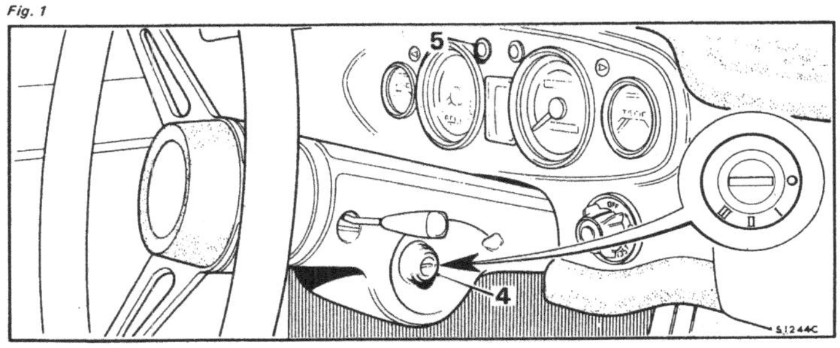


The fuel gauge and direction indicators will not operate unless the ignition switch is at position **'II'**.

To remove the key from the switch, turn the key to position **'I'**, press the key in, and while maintaining pressure turn anti-clockwise to position **'O'** and withdraw the key.

**(5) Ignition warning light (red). The ignition warning light serves the dual purpose of reminding the driver to switch off the ignition and of acting as a no-charge indicator. The light should glow when the ignition is switched on, and go out and stay out at all times while the engine is running above normal idling speed.**

**Door locks** *Fig. 2* Both doors may be locked from the outside with the key or from the inside with the internal door handle.

To lock the doors from the outside, turn the key slightly towards the front of the car. To unlock the doors, turn the key slightly towards the rear of the car. After locking or unlocking the doors return the key to the vertical position and withdraw it.

To lock the doors from inside the car, move the locking plate towards the rear of the car (1, inset Fig. 2).

**Luggage compartment** *Fig. 3* To open, depress the lock plunger and raise the lid. When fully raised the support stay will automatically spring into engagement and the lid will be held in the open position.

To close, raise the lid slightly, push the catch (1) on the bonnet stay forward to release the locking mechanism, and lower the lid. Raising the luggage compartment lid automatically switches on the light.

Springs retain the GT lid in the open position.

*Fig. 2*

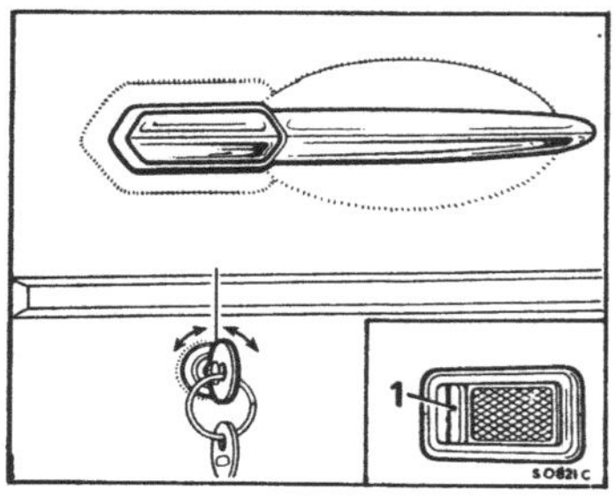


*Fig. 3*

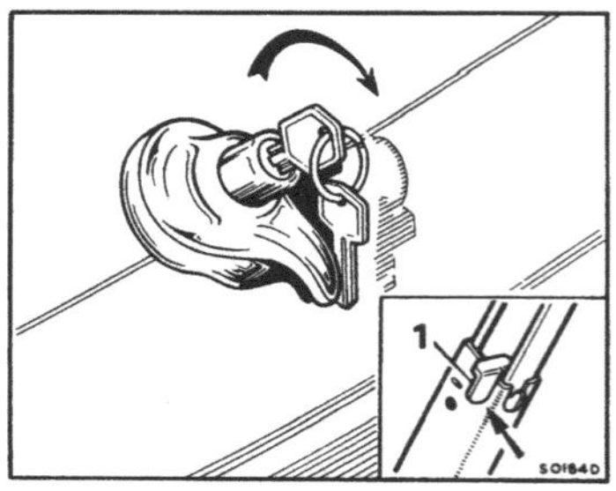


## INSTRUMENTS AND SWITCHES

**Instruments** *Fig. 1* (1) **Speedometer.** In addition to recording the road speed this instrument also records the total distance (3), and the distance travelled for any particular trip (2). To reset the trip recorder, push the knob (4) upwards and turn it clockwise; ensure that all the counters are returned to zero.

(5) **Tachometer.** The instrument indicates the revolutions per minute of the engine and assists the driver to use the most effective engine speed range for maximum performance in any gear.*

(6) **Oil.** The gauge indicates the pressure of the oil in the engine lubrication system.*

(7) **Water.** The gauge is marked 'C' (cold), 'N' (normal), and 'H' (hot), indicating the temperature of the coolant as it leaves the cylinder head.*

(8) **Fuel.** When the ignition is switched on the gauge indicates approximately the amount of fuel in the tank.*

* Also see 'RUNNING INSTRUCTIONS'.

*Fig. 1*

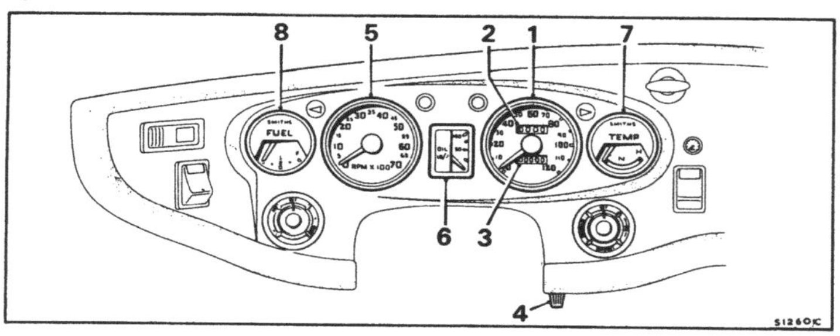


**Switches** *Fig. 2* (1) **Lighting switch.** Press the lower end of the switch rocker to the first position to operate the parking and tail lamps and to the second position to operate the headlamps. The marking on the switch is illuminated when the panel lamps are switched on.

(2) **Headlamp low beam—(4) Flasher.** With the headlamps switched on at the lighting switch, move the lever down away from the steering-wheel to operate the high beam (3). Lifting the lever towards the steering-wheel from the low-beam position will flash (4) the headlamp high-beams irrespective of whether the lighting switch is on or off.

(5) **Headlamp high-beam warning lamp (blue).** The warning lamp glows when the headlamps are switched on and the beam is in the raised position. The lamp goes out when the beam is lowered.

(6) **Panel lamp switch.** With the parking lamps switched on, illumination of the instruments and switches may be varied by rotating the panel lamp switch knob. Turning the switch knob clockwise from the off position immediately illuminates the panel lamps and further clockwise movement will increase the light brilliance.

**Reverse lamps.** The reverse lamps operate automatically if reverse gear is selected and the ignition switch is at position 'II' (ignition).

*Fig. 2*

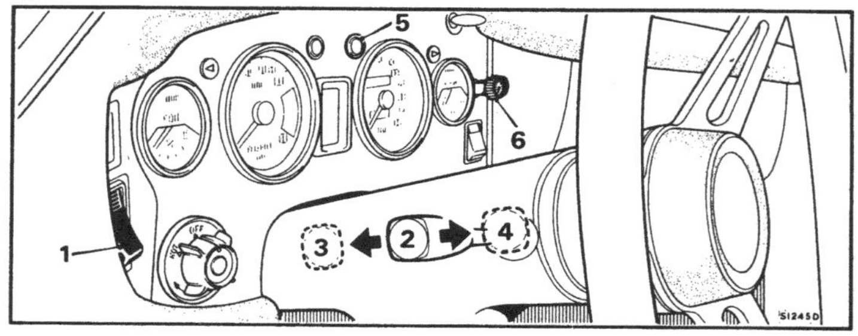


**Switches** *Fig. 3* (1) **Direction indicators.** The switch is self-cancelling and operates the indicators only when the ignition is switched on. A visual warning of a front or rear bulb failure is given when, after switching on an indicator, the warning lamp and the serviceable bulb on the affected side give a continuous light.

(2) **Direction indicator warning lamps (green).** The arrow-shaped lamps show the direction selected and each operates with the appropriate flashing direction indicators.

(3) **Horn.** The horn is sounded by pressing the centre motif of the steering-wheel.

(4) **Overdrive.** Move the lever towards the steering-wheel to engage overdrive; move the lever away from the steering-wheel to return to direct drive.

(5) **Windscreen wiper.** Move the switch lever down to operate the windscreen wipers at slow speed; further movement in the same direction will operate the wipers at fast speed. The wiper blades park automatically when the switch lever is returned to the off position.

(6) **Windscreen washer.** Press the knob on the end of the switch lever to operate the windscreen washer. When the windscreen is dirty, operate the washer before setting the wipers in motion.

In cold weather the washer reservoir should be filled with a mixture of water and a recommended washer solvent to prevent the water freezing. On no account should radiator anti-freeze or methylated spirits (denatured alcohol) be used in the windscreen washer.

*Fig. 3*

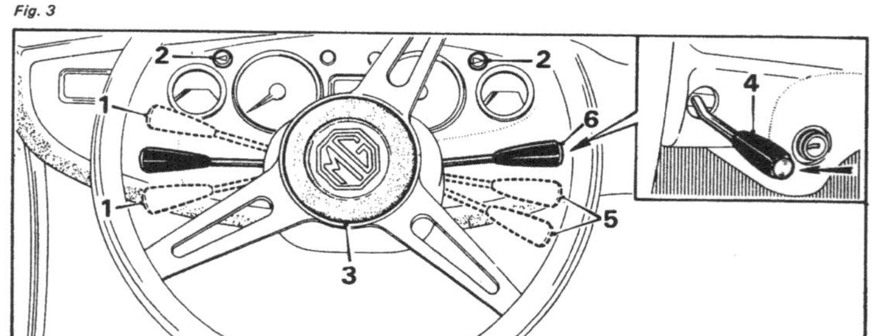

**Switches**
*Fig. 4*

(1) **Courtesy lamp.** The courtesy lamp is controlled by a switch (2) on the lamp and also by switches operated by the doors. With both doors closed the lamp may be switched on or off using the switch on the lamp. Opening either door switches on the lamp and closing the door extinguishes it.

(3) **Fog and auxiliary long-range lamp (spot lamp) switch.** The two lamps use one common switch (3) which is provided with two positions as follows:

To operate the fog lamp, press the lower end of the switch rocker to the first position. The fog lamp will only illuminate when the headlamps are switched to low beam, and will automatically extinguish when the headlamps are switched to main beam.

To operate the long-range lamp, press the switch rocker to the second position. The lamp will only illuminate when the main headlamps are on high beam, and will extinguish automatically when the main headlamps are switched to low beam.

(5) **Hazard warning.** To use the direction indicators as hazard warning lights, press the lower end of the switch rocker; all direction indicators and the warning lamp (4) will operate together, irrespective of whether the ignition is switched on or off. The marking on the switch is illuminated when the panel lamps are switched on.

(6) **Heated back-light (GT).** Press the lower end of the switch rocker to operate the heated back-light when the ignition is switched on. The warning lamp in the switch glows when the heated back-light is switched on.

(7) **Cigar-lighter (if fitted).** To operate, press the knob inwards. When ready for use the lighter will partially eject itself and may then be withdrawn. The rim of the cigar-lighter is illuminated when the panel lamps are switched on.

**Radio (if fitted).** Full operating instructions are supplied with the radio.

*Fig. 4*

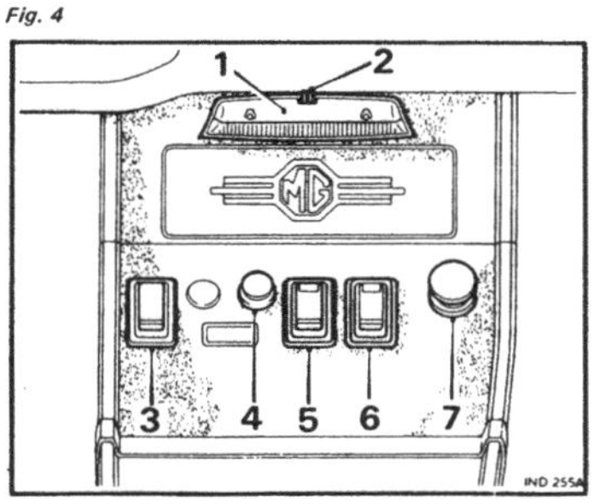


## BODY FITTINGS

**Windows and ventilators**

Turn the handle on each door to open or close the windows. The ventilation panels adjacent to each window may be opened after releasing the catch.

**Rear ventilators (GT)**

To open, pull the catch forward and then push outwards.

To close, pull the centre of the catch inwards and then push it backwards until the catch snaps over into the locked position.

**Luggage compartment (GT)**
*Fig. 1*

To increase the luggage capacity the back of the rear seat may be lowered. Release the locks and pull the seat forward.

*Fig. 1*

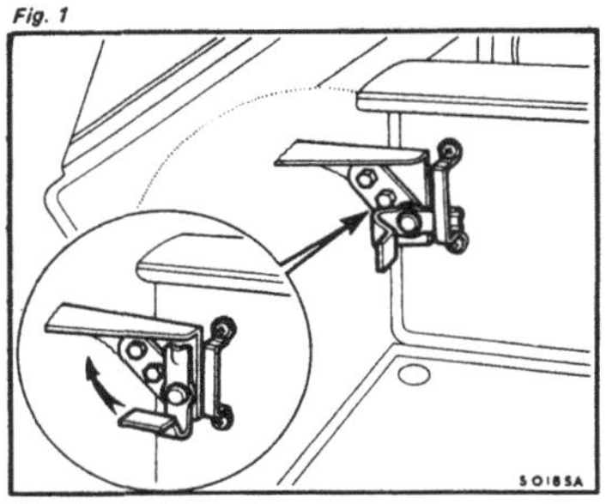


*Fig. 2*

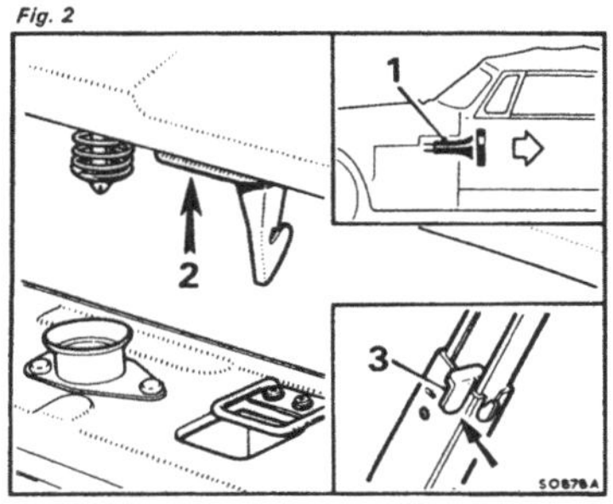


**Bonnet (Hood)**
*Fig. 2*

To raise the bonnet, pull the knob (1) located inside the car on the left-hand side below the fascia panel.

Press up the safety catch (2) under the front of the bonnet. Raise the bonnet and when fully raised the support stay will automatically spring into engagement and the bonnet will be held in the open position.

To close, raise the bonnet slightly, push the catch (3) on the bonnet stay rearwards to release the locking mechanism, and lower the bonnet. Apply light pressure with the palms of the hands at the front corners of the bonnet and press down quickly; undue force is not necessary and may cause damage. The safety catch and lock will be heard to engage.

**Mirrors**

**External** (Fig. 3). The mirror head may be adjusted from the seat position when the window is open. To adjust, slacken the clamp nut and move the mirror head to the position illustrated. Tighten the clamp nut after adjustment.

**Interior** (Fig. 4). The mirror stem with anti-dazzle head is designed to break away from the mounting bracket on impact. The stem may be refitted in the mounting bracket as follows. Align the stem ball (1) with the bracket cup (2), ensuring that the small protrusion (3) on the stem aligns with the indent of the mounting bracket. Give them a smart tap with a soft instrument to join the two components.

**Anti-dazzle.** To reduce mirror dazzle, press the lever (4) towards the windscreen.

*Fig. 3*

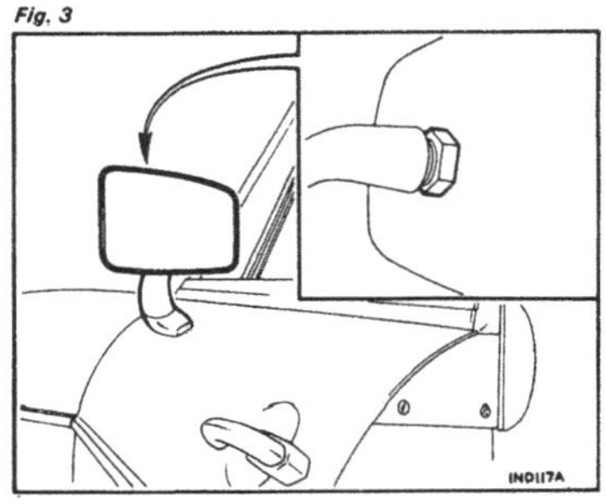


*Fig. 4*

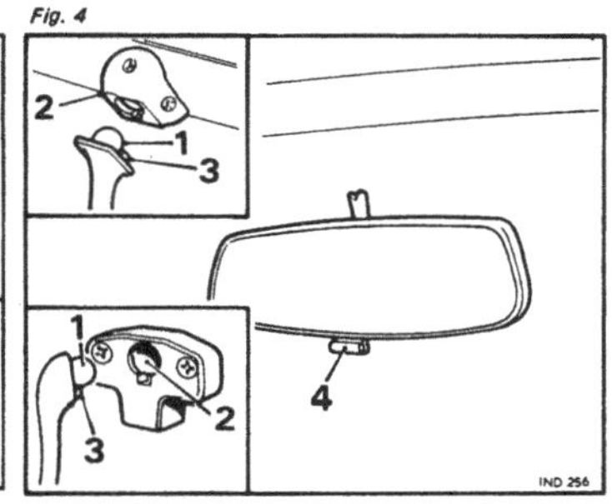


**Roof rack GT only**

The weight carried on the GT roof rack must not exceed 50 lb. Weight carried on the roof of a vehicle has an adverse effect on its handling; therefore drive with due discretion. Cornering and the behaviour of the car in a cross-wind will be particularly affected.

**Body and door drainage points**
*Fig. 5*

Periodic examination of the drain holes should be made to ensure that they are clear of obstruction; use a piece of stiff wire to probe the apertures.

Careless application of underseal can result in restricted drainage. Masking tape or plugs used when underseal is being applied must be removed immediately the operation is completed.

Jacking up beneath the underfloor may deform the drain apertures; always use the jacking points provided.

*Fig. 5*

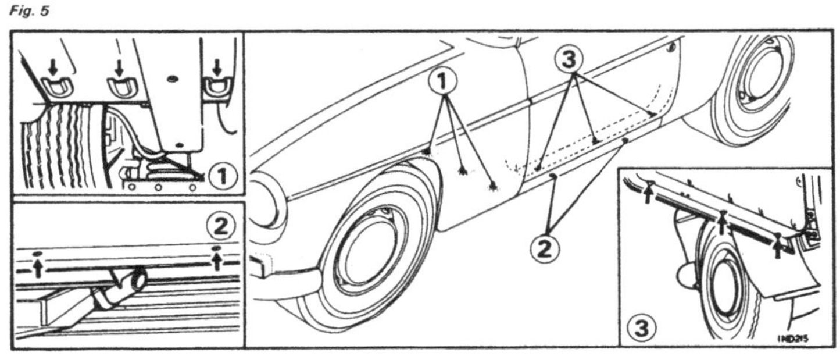

**It is most important that the instructions for raising, lowering and folding the hood are followed. Do not fold when the hood is wet or damp.**

**Lowering the hood**
*Fig. 6*

Unclip the sun visors (1) and move them to one side.

Release both windscreen frame toggle catches (2).

Release the two fasteners (3) on the windscreen rails, the two fasteners (4) on the cant rails and the two fasteners (5) on the hood mounting brackets.

*Fig. 7* Release the four fasteners from each rear quarter panel (6) and pull the hood slightly forward to disengage the hook (7) from the socket (8) on the body side panel.

Move the seat tilt catch forward and incline the seat backs towards the front of the car.

*Fig. 6*

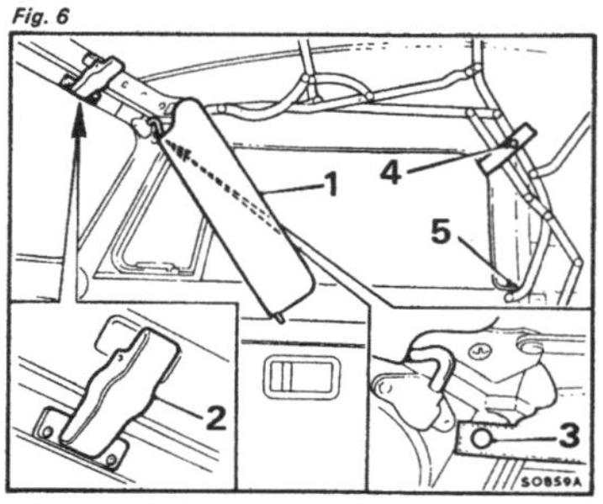


*Fig. 7*

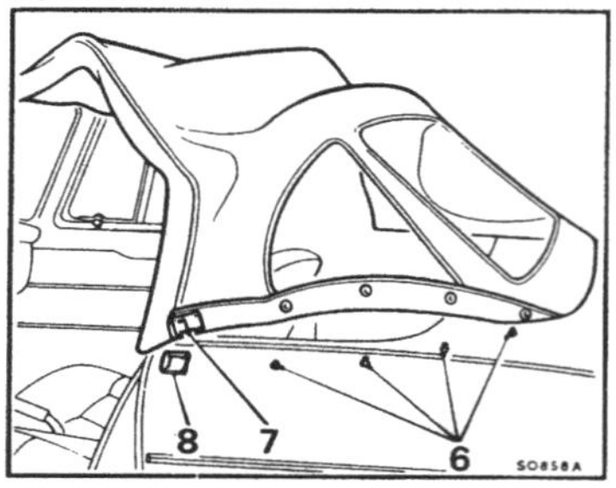


*Fig. 8* Raise the hood header rail (11) until it is poised approximately midway over the door aperture.

Disengage the hood rear rail from the anchor plates (9) on the tonneau panel.

Fold each quarter-light (10) onto the back-light and continue the fold in the material forward to the header rail (11). **ENSURE THAT THE FOLD IS MADE IN THE HOOD MATERIAL BETWEEN THE QUARTER-LIGHT AND THE BACK-LIGHT. FAILURE TO DO THIS MAY CAUSE PERMANENT DAMAGE TO THE BACK-LIGHT MATERIAL.**

Push the header rail (11) rearwards, and at the same time draw the back-light and hood material (12) out over the luggage compartment lid ensuring that the hood material does not become trapped between the hood sticks.

*Fig. 9* Fully lower the hood. Fold the two windscreen frame toggle catches (13) rearwards to prevent them damaging the back-light.

Roll the back-light and material forward over the folded hood. Position and secure the two retaining straps (14).

Replace the sun visors and return the seat back-rests to their original positions.

**Raising the hood**

Remove the hood cover. Move the seat tilt catch forwards and incline the seat backs towards the front of the car.

Unclip the sun visors (1) and move to one side.

Raise the header rail (11) and unfold the hood. Engage the rear rail in the anchor plates (9). Pull the hood slightly forwards and engage each hook (7) in its socket (8) on the body side panel. Position the header rail on the windscreen ensuring the rail seal is forward of the seal flange. Secure the windscreen frame toggle catches and fasteners (3), (4), and (5) inside the car.

Secure the fasteners (6) at each rear quarter.

Reposition the seats and sun visors.

*Fig. 8*

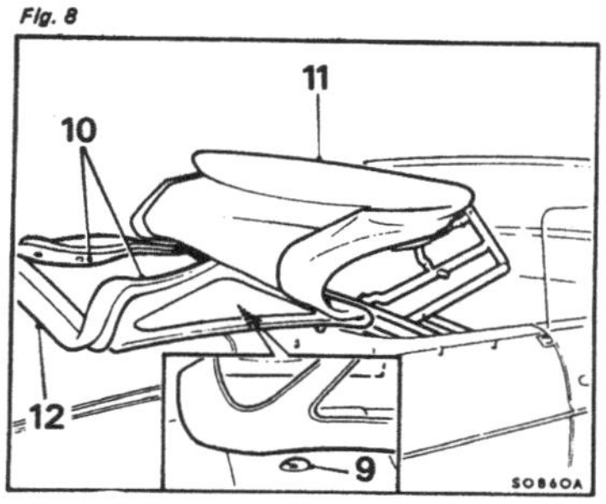


*Fig. 9*

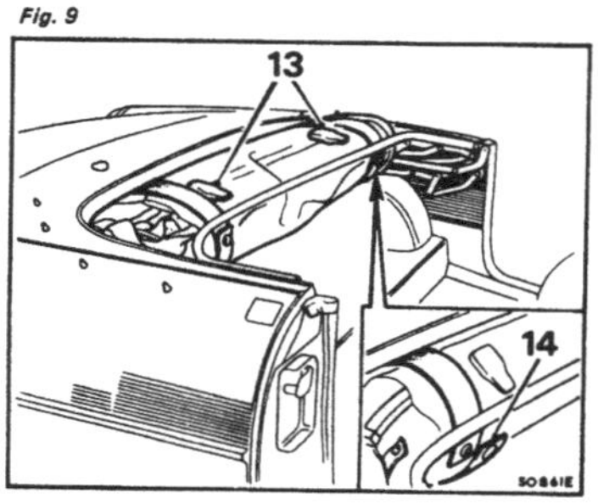


**Fitting the hood cover**
*Figs. 10 and 11*

Assemble the hood cover rail (15) and fit it into the hood support sockets with the cross-rod towards the rear.

Lay the hood over the rail.

Engage the cover rear rail in the two anchor plates (16) on the tonneau panel.

Pull the cover slightly forwards and engage each side hook in its socket (17) on the body panel.

Secure the fasteners (18) at each quarter side panel.

Secure the four fasteners (19) inside the car.

**Removing the hood cover**

Reverse the fitting procedure.

**Stowage**

Stowage bags are provided to protect the hood cover and hood cover rail. The stowage bags together with the tool bag are stowed in the luggage compartment and secured with the straps provided.

*Fig. 10*

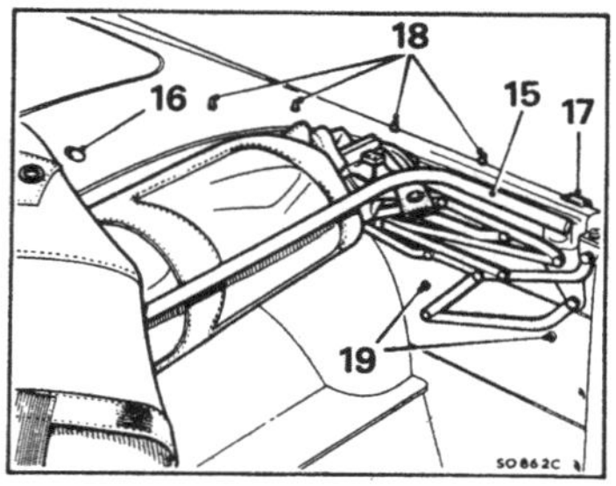


*Fig. 11*

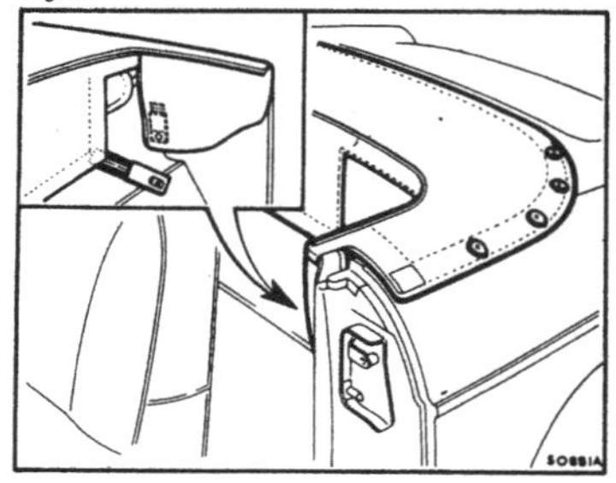

**Hard top**

**Fitting.** Remove the hood. Fit the hard top side brackets into the hood support sockets and secure with the bolts and spring washers.

Position the hard top on the car, engaging the rear securing plates with the slotted anchor plates on the tonneau panel (inset, Fig. 12) ensuring that the sealing rubber does not foul the slots.

Line up the hard top drip moulding with the rear wing top beading. Push the hard top forwards and engage the toggle catch tongues in the sockets on the windscreen frame.

Fit the bolts into the side fixing brackets; screw in but do not tighten. Ensure that the front sealing rubber is correctly positioned forward of the windscreen frame. Adjust the toggle catches to give adequate tension when fastened the securing bolt slots allow movement), tighten the securing bolts, fasten the catches and lock them with the securing clips (inset, Fig. 13).

Check that the sealing rubbers are correctly positioned, then slowly and evenly tighten the side fixing bolts until the hard top seals evenly to the body. **AVOID OVERTIGHTENING.** Measure the gap between the hard top and body side fixing brackets (arrowed, Fig. 12). Remove the bolts and fit washers between the brackets to the thickness of the gap. Refit and tighten the bolts.

Wind up both windows and check that a gap of approximately $\frac{1}{16}$ in. exists between the rear edge of the window and the hard top quarter channel. Adjust if necessary by loosening the side fixing bolts and repositioning the hard top. Ensure that there is an adequate seal between the window and hard top rubber and that the doors, when opened with the windows up, do not foul the opening surround.

**Removing.** Unlock and release the windscreen toggle fasteners. Remove the side fixing bolts. Raise the front of the hard top to disengage the toggle fastener tongues from the windscreen sockets, move the hard top to the rear to disengage the anchor plates, then lift it clear of the car. Remove the side fixing brackets from the hood support sockets. Assemble the fittings loosely to the hard top to prevent loss.

*Fig. 12*

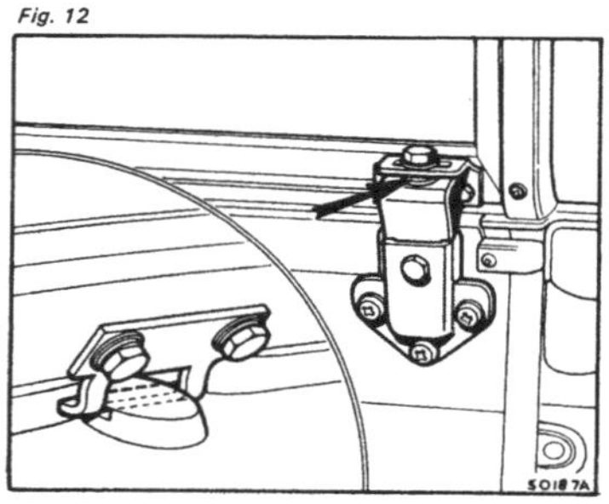


*Fig. 13*

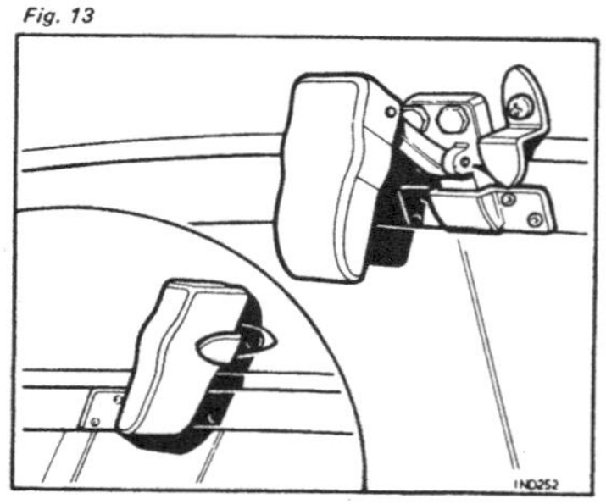

**Tonneau cover**
*Fig. 14*

**Fitting.** Lay the tonneau cover over the cockpit and place the pockets in the cover over the head restraints on the seats.

Engage the rear edge retainer with the slotted plates on the tonneau panel, and secure to each quarter panel with the fasteners.

Extend the tonneau cover forward and attach the front of the cover to the fasteners on the fascia panel top and windscreen pillars.

*Fig. 15* **Usage.** The centre zip allows the cover to be folded down to give access to the driving seat or to both seats. Fold the cover down behind the seat and secure it with the fasteners on the flap to the heelboard.

The short side zips allow the seat belt to be used.

**Removing.** Reverse the fitting procedure.

*Fig. 14*

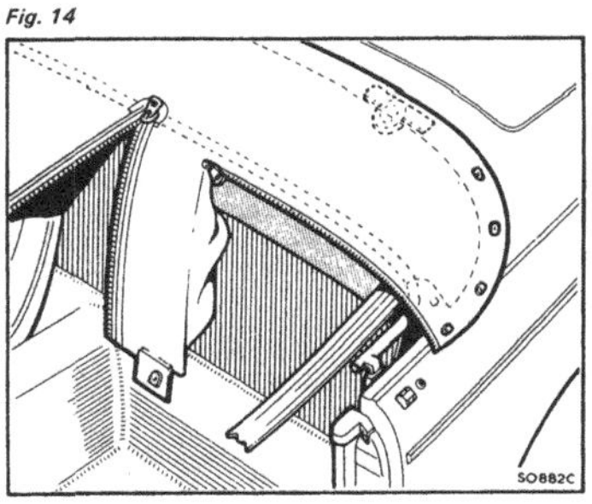


*Fig. 15*

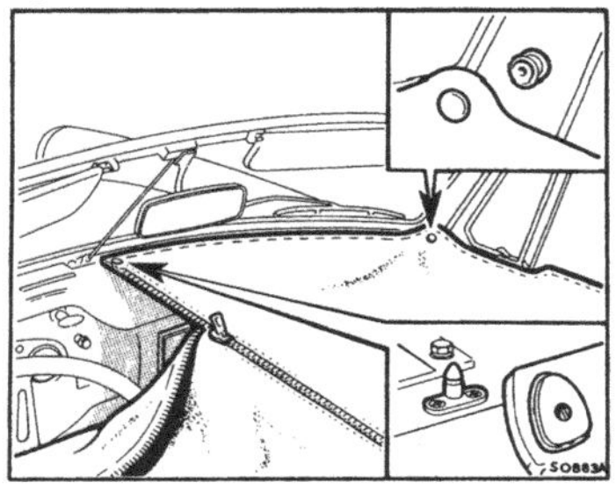


## SEATS AND SEAT BELTS

**SEATS** *Fig. 1*

**Adjustment** Both seats are adjustable and can be moved easily into the most comfortable position. Move the lever (1) located beneath the front of the seat outwards; hold the lever in this position while the seat position is adjusted. The locking pin is spring-loaded and will automatically lock the seat in the required position when the lever is released.

The rake of the back or squab of the seat may also be adjusted. Ease the body weight from the seat back, move the lever (2) in the direction of the arrow. Release the lever and ensure the seat back is fully locked in position; check by applying back pressure on the seat.

To gain access to the rear compartment, move the seat tilt catch (3) forward, and fold the back of the seat against the steering-wheel. The tilt catch will automatically re-engage when the rear of the seat is moved back to the correct driving position.

**SEAT BELTS** *Fig. 2*

**To fasten** *Later cars:* Lift the engagement tongue (1) and draw the belt from the automatic reel retractor over the shoulder and across the chest and push it into the locking clip (2) of the short belt nearest the wearer.

*Early cars:* Lift the magnetic buckle tongue (8) and engage the hook (9) into the hinged part of the tongue.

**To release** *Later cars:* Press the release button (3) on the short belt.

*Early cars:* Lift the magnetic buckle tongue (8).

**Warning system** *Later cars:* See page 25 for details of the seat belt warning system which provides an audible and visual warning reminder to the driver and passenger to fasten their belts.

*Fig. 1*

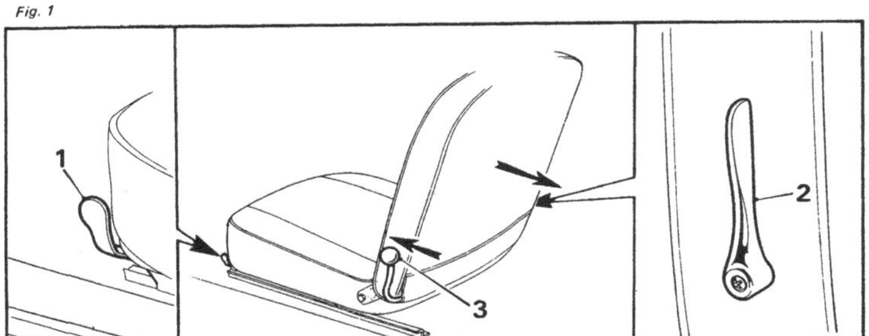


**To adjust** *Early cars:* Tighten the short belt with the adjuster at the buckle until the buckle rests on the side of the hip. With the adjuster at the sill tighten the belt until the lap belt fits comfortably and a hand clearance between the diagonal belt and the chest is just possible. Slight re-adjustment may be necessary during use.

**To stow** *Later cars:* Place the engagement tongue into the stowing pocket (5) mounted under the hood centre locating socket.

*Early cars:* Hook the buckle on the long belt into the slot (10) in the stowing bracket mounted under the hood centre locating socket. Attach the magnetic buckle on the short belt to the seat frame.

**Rear fixing (Early cars only)** The long belt is attached at the rear end by a quick-release fastener to the tonneau panel.

**To release the belt,** press down the black plastic locking plate (6) and slide the belt bracket (7) out of engagement with the securing stud slot.

**To secure the belt,** position the large hole in the belt bracket over the securing stud, slide the bracket forward until it engages in the securing stud slot and locks in position with the locking plate. Check that the belt webbing is not twisted.

**Wearing** Never attempt to wear the belt other than as a complete diagonal and lap assembly. Do not try to use the belt for more than one person at any one time, even children.

**Care of the belts** No unauthorized alterations or additions to the belts should be made. Do not bleach or re-dye the belt webbing. Inspect the webbing periodically for signs of abrasion, cuts, fraying, and general wear; pay particular attention to the fixing points and adjusters. Replace belts that are defective or have been subjected to severe strain.

*Later cars:* After releasing the belt, allow the webbing to retract into the automatic reel. Ensure that while the belt is retracted the engagement tongue has not moved on the belt to a point near the sill mounting; this can be remedied by moving the tongue and belt clip (4) towards the reel.

*Early cars:* If at any time the webbing is unthreaded from the brackets and adjuster, ensure that it is rethreaded correctly as shown (11).

*Fig. 2*

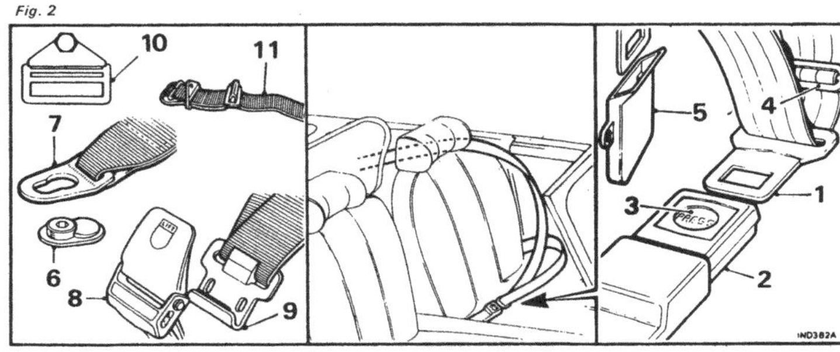


## HEATING AND VENTILATING

**FRESH AIR** *Fig. 1* Fresh air is admitted to the car for cooling and ventilation through a vent behind the centre console and through the face-level vents mounted on the fascia. The flow of air through the lower vent may be adjusted by moving the knob (1) backwards to one of the three open positions or forwards to close the vent. The air flow from the face-level vents may be adjusted by turning the serrated wheels (2) outwards to open the vents, and the flow may be directed by moving the shutter control (3).

**Air distribution** Air enters the car through the two doors located one each side of the gearbox tunnel in the foot wells. Air distribution for heating is independent of the fresh-air system; the control knob (1) should be in the closed position when heated air is being distributed.

**HEATER** *Fig. 2* The heating and ventilating system is designed to provide fresh air either heated by the engine cooling system or at outside temperature to the car at floor level and for demisting and defrosting to the windscreen.

**Controls** *Fig. 2* **Heater.** Turn the knob (1) in the direction of the arrow to raise the air temperature.

**Air flow.** Turn the knob (2) in the direction of the arrow to direct the air distribution.

**Booster.** Press the lower end of the switch rocker (3) to boost the air flow.

**Usage** The heater and air flow controls may be set at the position marked on the control knobs or to any other intermediate positions. By varying the control settings, and utilizing the booster blower, a wide range of settings can be obtained to suit prevailing conditions.

*Fig. 1*

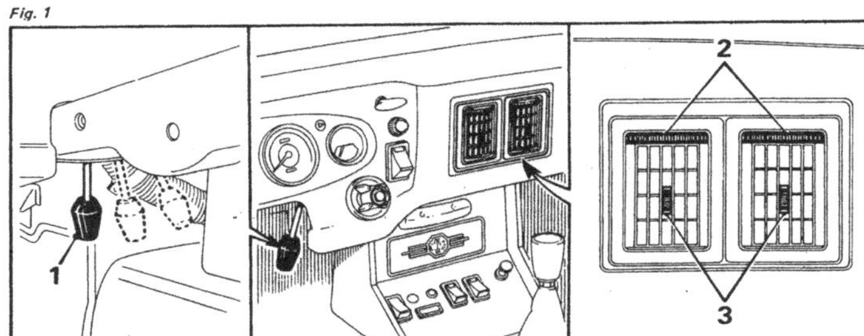

**Illumination** The markings on the booster switch and the control dials and the position indicators on the rotary control knobs are illuminated when the panel lamps are switched on.

**Heated back-light (GT)** (*when fitted*) *Fig. 2* The heated back-light can be switched on only when the ignition is switched on. A warning lamp in the switch glows when the switch rocker (4) is pushed to the 'on' position.

The window has the heating element on the surface of the glass and with reasonable care will last indefinitely. The following practices will damage the circuit and must be avoided:

1. Scratching off labels and advertising stickers.
2. Wiping the glass with the back of a ringed hand.
3. Stowing hard and metal objects so that they abrade the glass.
4. Cleaning with harsh abrasives.

*Fig. 2*

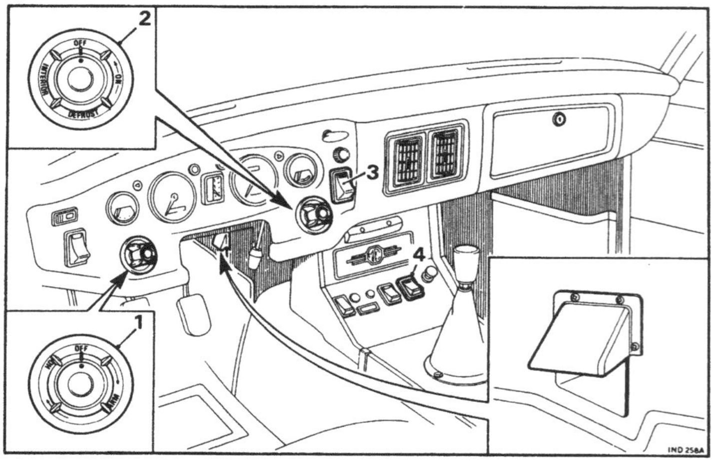


## RUNNING INSTRUCTIONS

The following instructions are a guide for starting, running and loading the car, and include notes on the use of the controls and the indications of the instruments.

**Choice of fuel** Our MG engines have been designed to operate on fuels of 91 octane rating or above and have not been developed for the regular use of unleaded or low lead gasolines. The use of such fuels cannot be recommended as they could have a detrimental effect on engine components, resulting in loss of performance, excess exhaust emissions and, possibly, complete engine failure.

**Starting** Check that the gear lever is in the neutral position.

Pull out the **mixture control (choke).**

Switch on the ignition (page 25) and check:
That the **ignition warning lamp** glows.
That the fuel gauge registers.

Operate the **starter.**

After the engine has started, check:
That the **oil pressure gauge** registers
That the **ignition warning light** has gone out.

Push the mixture control (choke) in to the minimum setting.

Check the **temperature gauge reading.**

**Mixture control (Choke)** The function of this control is to enrich the air/fuel mixture for cold engine starting and to provide a faster idle speed without enrichment during the warm-up period.

The amount which the control knob must be pulled out to achieve easy starting will be dependent on engine temperature and prevailing conditions.

To lock the control in the required position, turn the control knob a quarter of a turn clockwise.

After the engine has been started with the aid of the choke, unlock the control and push it in gradually as the engine warms, until only about ¼ in. of travel remains. With the control in this position the engine will run at a faster idle speed and attain its correct working temperature as quickly as possible.

Do not warm up the engine by allowing it to idle slowly or by leaving it to idle with the control pulled out. Driving the car onto the road while the engine is cold with the control partly pulled out is preferable to allowing the engine to idle, or run with the control pulled out, in the garage or on the driveway prior to moving off.

**Ignition warning lamp** The lamp should glow when the engine is switched on, and go out and stay out at all times when the engine is running above normal idling speed. Failure to do so indicates a fault in the battery charging system. Check that the fan belt is correctly tensioned before consulting your Distributor or Dealer.

**Starter** Do not operate the starter for longer than five to six seconds.

To prevent damage the starter cannot be operated while the engine is running. If the engine fails to start, the ignition key must be returned to the off position before the starter can be operated again.

If after a reasonable number of attempts the engine should fail to start, switch off the ignition and investigate the cause. Continued use of the starter when the engine will not start not only discharges the battery but may also damage the starter.

**Oil pressure gauge** The gauge should register a pressure as soon as the engine is started up. The pressure may rise above 80 lb./sq. in. when the engine is started from cold, and as the oil is circulated and warmed the pressure should then drop to between 50 and 80 lb./sq. in. at normal running speeds and to between 10 and 25 lb./sq. in. at idling speed.

Should the gauge fail to register any pressure, stop the engine immediately and investigate the cause. Start by checking the oil level.

**Temperature gauge** Normal operating temperature is reached when the pointer is in the 'N' sector.

Overheating may cause serious damage. Investigate any upward change in the temperature gauge reading immediately. Check coolant level and fan belt tension.

When the ignition is switched off the needle returns to the 'cold' position.

**Tachometer** For normal road work, and to obtain the most satisfactory service from your engine, select the appropriate gear to maintain engine speeds of between 2,000 and 4,500 r.p.m.

When maximum acceleration is required upward gear selections should be made when the needle reaches the yellow sector (5,500–6,000 r.p.m.). Prolonged or excessive use of the highest engine speeds will tend to shorten the life of the engine. Allowing the engine to pull hard at low engine speeds must be avoided as this also has a detrimental effect on the engine.

The beginning of the red sector (6,000 r.p.m.) indicates the maximum safe speed for the engine.

**Never allow the needle to enter the red sector.**

**Running in** The treatment given to a new car will have an important bearing on its subsequent life, and engine speeds during this early period must be limited. The following instructions should be strictly adhered to.

During the first 500 miles:
**DO NOT** exceed 45 m.p.h.
**DO NOT** operate at full throttle in any gear.
**DO NOT** allow the engine to labour in any gear.

**Wet brakes** If the car has been washed, driven through water, or over wet roads for prolonged periods full braking power may not be available. Dry the brakes by applying the foot brake lightly several times while the car is in motion. Keep the hand brake applied while using high-pressure washing equipment.

**Vehicle loading** Due consideration must be given to the overall weight carried when fully loading the car. Any loads carried on a roof rack or downward load from a towing hitch must also be included in the maximum loading.

*Towing* **The towing weight of 1,680 lb. is the maximum that is permissible.** When using bottom gear a gradient of up to 1 in 8 can be ascended while towing a weight not exceeding this figure. It may be necessary to adjust the maximum towing weight to comply with local conditions and regulations. The recommended downward load of a trailer or caravan on the towing hitch is 75 to 100 lb., but this may be reduced or exceeded at the discretion of the driver. Any load carried on the roof or downward load from a towing hitch must also be included in the maximum loading of the vehicle.

**Towing for recovery** Should it become necessary to tow the car, use the towing-eyes provided.

For recovery the car should be towed with the key in the ignition/steering lock at position **'I'**. For tow starting the key should be at position **'II'**.

**Overdrive** (*when fitted*) The overdrive unit, controlled by a switch on the steering-column, provides a higher driving ratio, for use with third or fourth gear. To engage overdrive move the switch lever towards the steering-wheel; to disengage move the lever away from the steering-wheel. Accelerator pedal pressure should be maintained and it is not necessary to depress the clutch pedal during engagement or disengagement.

Overdrive can be engaged at any throttle opening when in third or top gear. If increased acceleration is required the overdrive can be 'switched out' without alteration to the throttle setting. Do not 'switch out' the overdrive when travelling at speeds exceeding normal third or top gear road speeds.

In certain driving conditions while travelling in third gear, the overdrive can be switched in to provide a top gear ratio or out to provide third gear acceleration without the necessity of manually moving the gear-change lever.

**If for any reason the overdrive does not disengage, do not reverse the car otherwise extensive damage may result.**

## CLEANING

**Interior** Clean the carpets with a semi-stiff brush or a vacuum cleaner, preferably before washing the outside of the car. The most satisfactory way to give carpets and nylon faced upholstery a thorough cleaning is with **UNIPART Upholstery Cleaner,** diluted one part with eight parts warm water. Apply vigorously with semi-stiff brush, and remove the surplus with a damp cloth or sponge. Carpets should not be cleaned by the 'dry-clean' process. The plastic faced upholstery and roof lining may be treated with undiluted **UNIPART Upholstery Cleaner** spread thinly over the surface to be cleaned with a brush or cloth. Leave for five minutes, then wipe off with a moist sponge or cloth.

**UNIPART Upholstery Cleaner** can be used for cleaning and renovating all the usual upholstery materials, and rubber, but it should not be used on painted surfaces.

**Body** Regular care of the body finish is necessary if the new appearance of the car exterior is to be maintained against the effects of air pollution, rain, and mud.

Wash the bodywork frequently, using a soft sponge and plenty of water containing **UNIPART Car Shampoo.** Large deposits of mud must be softened with water before using the sponge. Smears should be removed by a second wash in clean water, and with the sponge if necessary. When dry, clean the surface of the car with a damp chamois-leather. In addition to the regular maintenance, special attention is required if the car is driven in extreme conditions such as sea spray or on salted roads. In these conditions and with other forms of severe contamination an additional washing operation is necessary which should include underbody hosing. Any damaged areas should be immediately covered with paint and a complete repair effected as soon as possible. Before touching-in light scratches and abrasions with paint, thoroughly clean the surface. Use petrol/white spirit (gasoline/hydrocarbon solvent) to remove spots of grease or tar.

The application of **UNIPART Car Polish** is all that is required to remove traffic film and to ensure the retention of the new appearance.

**Bright trim** Never use an abrasive on stainless, chromium, aluminium, or plastic bright parts and on no account clean them with metal polish. Remove spots of grease or tar with petrol/white spirit (gasoline/hydrocarbon solvent) and wash frequently with water containing **UNIPART Car Shampoo.** When the dirt has been removed polish with a clean dry cloth or chamois-leather until bright. Any slight tarnish found on stainless or plated components which have not received regular attention may be removed with **UNIPART Chrome Cleaner.** An occasional application of light mineral oil or grease will help to preserve the finish, particularly during winter when salt may be used on the roads, but these protectives must not be applied to plastic finishes.

**Windscreen** If windscreen smearing has occurred it can be removed with **UNIPART Screen Cleaner.**

**UNIPART** products mentioned above are obtainable from your Distributor or Dealer.

## COOLING SYSTEM

**Radiator filler cap** *Fig. 1* (1) The system is pressurized to 10 lb./sq. in. when hot, and the pressure must be released gradually when the filler cap is removed. It is advisable to protect the hands against escaping steam and turn the cap slowly anti-clockwise until the resistance of the safety stops is felt. Leave the cap in this position until all pressure is released. Press the cap downwards against the spring to clear the safety stops, and continue turning until it can be lifted off.

**Draining the cooling system** To drain the cooling system, slacken the hose clip and remove the bottom hose at its connection to the radiator. Remove the drain plug (2) on the engine cylinder block.

When draining in freezing weather, do so when the engine is hot. Run the engine slowly for one minute when the water has ceased flowing to clear any water from the pump and other places where it might collect. Finally, leave a reminder on the vehicle to the effect that the cooling system has been drained.

If the system contains anti-freeze, remember to collect it in a clean container for future use.

**Filling the cooling system** To avoid wastage by overflow add just sufficient coolant to cover the bottom of the header tank. Run the engine until it is hot and add sufficient coolant to bring the surface to the level of the indicator positioned inside the header tank below the filler neck.

**NOTE.—The heater control must be set to 'HOT' when draining or filling the cooling system.**

**Frost precautions** Water expands when it freezes, and if precautions are not taken there is considerable risk of bursting the radiator, cylinder block, or heater. Such damage may be avoided by adding anti-freeze to the water.

**Do not use radiator anti-freeze solution in the windscreen washer.**

**Anti-freeze solutions** Anti-freeze can remain in the cooling system for two years provided that the specific gravity of the coolant is checked periodically and anti-freeze added as necessary. The specific gravity check should be carried out by an authorized Distributor or Dealer.

Top up only when the cooling system is at its normal running temperature in order to avoid losing anti-freeze due to expansion.

After the second winter the system should be drained and flushed. Refer to the instructions given for draining the cooling system, then clean out the system thoroughly by flushing water through the radiator passages using a hose inserted in the radiator filler orifice.

Before adding the recommended anti-freeze make sure that the cooling system is watertight; examine all joints and replace any defective hose with new.

We recommend owners to use **UNIPART Frostbeat or Bluecol Anti-freeze** to protect the cooling system during frosty weather and reduce corrosion to the minimum. We also approve the use of anti-freeze which conforms to specification B.S. 3151 or B.S. 3152.

The correct quantities of anti-freeze for different degrees of frost protection are:

**(See Editor's notes at the end of Part 2.)**

| *Anti-freeze* | *Commences to freeze* | | *Frozen solid* | | *Amount of anti-freeze* |
|---|---|---|---|---|---|
| % | °C. | °F. | °C. | °F. | *U.S. Pts.* |
| 25 | −13 | 9 | −26 | −15 | 3 |
| 33⅓ | −19 | − 2 | −36 | −33 | 4 |
| 50 | −36 | −33 | −48 | −53 | 6 |

## WHEELS AND TYRES

**Jacking up** *Fig. 1* The jack is designed to lift one side of the car at a time. Apply the hand brake, and block the wheels on the opposite side to that being jacked; use a wood block jammed tight against the tyre tread.

Insert the lifting arm of the jack into the socket located in the door sill panel. **Make certain that the jack lifting arm is pushed fully into the socket and that the base of the jack is on firm ground.** The jack should lean slightly outwards at the top to allow for the radial movement of the car as it is raised.

**WARNING.—Do not work beneath the vehicle with the lifting jack as the sole means of support.** Place suitable supports under the front side-members or rear axle to give adequate support and safety while working.

**Jack maintenance** If the jack is neglected it may be difficult to use in a roadside emergency. Examine it occasionally, clean off accumulated dust, and lightly oil the thread to prevent the formation of rust.

**WHEELS**
**Preventive maintenance** Owners are recommended to check wheel nuts for tightness each week in addition to checking the other items listed. Take care not to overtighten (torque wrench setting 60 to 65 lb. ft.).

**Pressed type** *Removing and refitting* *Fig. 2* Slacken the four nuts securing the road wheel to the hub; turn anti-clockwise to loosen and clockwise to tighten. Raise the car with the jack to lift the wheel clear of the ground and remove the nuts. Withdraw the road wheel from the hub.

When refitting the road wheel locate the wheel on the hub, lightly tighten the nuts (1) with the wheel nut spanner (securing nuts must be fitted with the **taper side towards the wheel**), and lower the jack. Fully tighten the wheel nuts, tightening them diagonally and progressively, at the same time avoid over-tightening.

The wheel centre trim (2) must be removed and fitted to the wheel in use.

Replace the wheel disc and jack socket plug.

*Fig. 1*

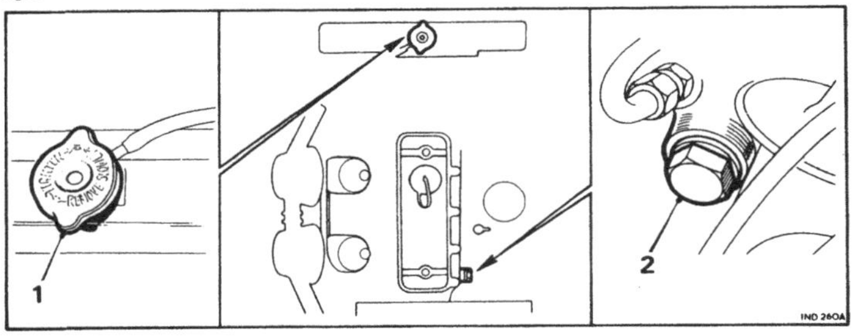


*Fig. 1*

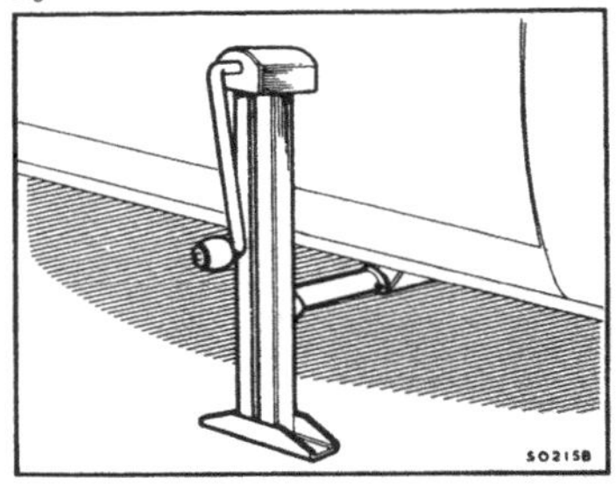

*Fig. 2*

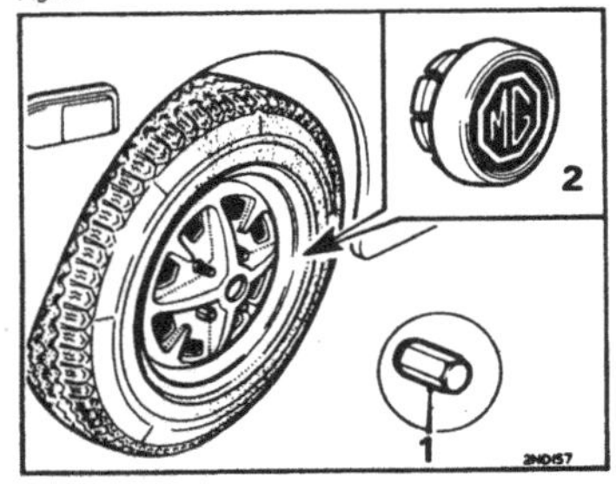

**Wire type**
*Removing and refitting*
*Fig. 3*

Use the spanner to slacken the octagonal hub nuts.

Always jack up a wheel before using the hammer, and always hammer the nuts tight.

Locknuts are marked 'LEFT' or 'RIGHT' to show to which side of the car they must be fitted, and also with the word 'UNDO' and an arrow.

Before replacing a wheel wipe all serrations, threads, and cones of the wheel and hub and then lightly coat them with grease. If a forced change is made on the road, remove, clean, and grease as soon as convenient.

*Maintenance* When the car is new, after the first long run or after 50 miles of short runs, jack up the wheels and hammer the nuts to make sure that they are tight.

Once a year remove the wheels for examination and regreasing.

**Spare wheel** The spare wheel is secured to the floor of the luggage compartment on tourer cars and below the luggage compartment floorboard on GT cars. To gain access to the spare wheel on GT cars (Fig. 4) turn back the luggage compartment floor covering, unscrew the two quick-release screws and lift the floorboard.

The spare wheel supplied with new cars is inflated above the recommended running pressure. The pressure must be checked and adjusted before use.

*Fig. 3*

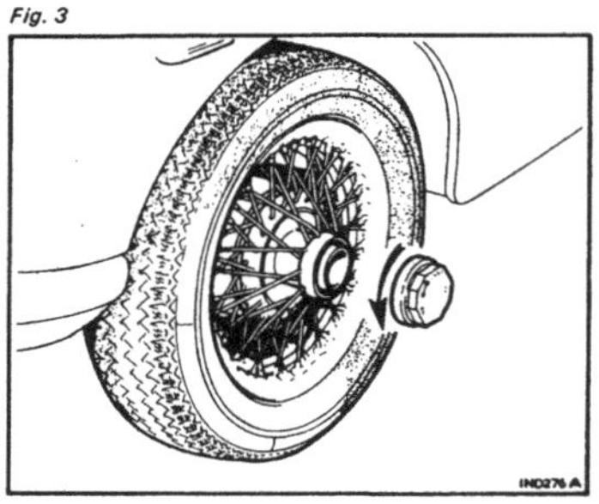

*Fig. 4*

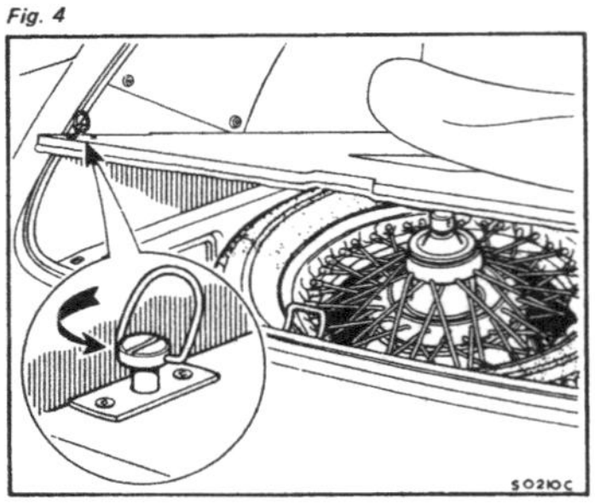

## TYRES

**Markings** Tyres are marked with the maximum load and inflation pressure figures. When fitting replacement tyres ensure that they are to the same specification and marking. **The permissible load and tyre pressures are shown on page 40 of this handbook.**

**Wear indicator** Tyres fitted as original equipment have wear indicators incorporated in their tread pattern. When the tyre tread has worn down until 0·06 in. of the tread is remaining the wear indicator bar will appear across the full width of the tread pattern.

**Maintenance** Tyres, including the spare, must be maintained at the pressures recommended (see **'GENERAL DATA'**); check with an accurate tyre gauge at least once a week, and regulate as necessary. Pressures should be checked when the tyres are cold; do not reduce the pressure in warm tyres where the increase above the normal pressure is due to temperature. See that the valve caps are screwed down firmly by hand. The cap prevents the entry of dirt into the valve mechanism and forms an additional seal on the valve, preventing any leakage if the valve core is damaged. The spare wheel supplied with new cars is inflated above the recommended running pressure. The pressure must be checked and adjusted before use.

Excessive local distortion can cause the casing of a tyre to fracture and may lead to premature tyre failure. Tyres should be examined, especially for cracked walls, exposed cords, etc. Flints and other sharp objects should be removed from the tyre tread; if neglected, they may work through the cover. Any oil or grease which may get onto the tyres should be cleaned off by using fuel sparingly. Do not use paraffin (kerosene), which has a detrimental effect on rubber.

**Repairs** When repairing tubes, have punctures or injuries vulcanized. Ordinary patches should only be used for emergencies. Vulcanizing is absolutely essential for tubes manufactured from synthetic rubber. **(See Editor's notes at the end of Part 2.)**

**Replacement** Radial-ply tyres are standard equipment and **replacements must be of the radial-ply type.**

**Wheel and tyre balancing** Unbalanced wheel and tyre assemblies may be responsible for abnormal wear of the tyres and vibration in the steering. Consult your Distributor/Dealer.

## BRAKES

**Brake and clutch master cylinder**
*Fig. 1*

The level of the fluid in the brake master cylinder reservoir is visible through the plastic reservoir (1); the level must be maintained up to the position marked (2) on the side of the reservoir. To check the level of the clutch master cylinder reservoir (3), remove the plastic filler cap. The fluid level must be maintained up to the bottom of the filler neck.

Use only **Lockheed Disc Brake Fluid (Series 329S)** or **Castrol Girling Brake Fluid;** alternatively, use a brake fluid conforming to **F.M.V. S.S. D.O.T.3 specification with a minimum boiling-point of 260° C. (500° F.).** Before refitting the filler caps check that the breather holes (indicated by the arrows) in the caps are clear. The centre disc (4) of the brake reservoir cap may be removed for cleaning. **(See Editor's notes at the end of Part 2.)**

**Brake pedal**
*Fig. 2*

A free movement of $\frac{1}{8}$ in. (A), measured at the pedal pad must be maintained on the pedal. To adjust the free movement, slacken the stop light switch locknut (1) and turn the switch (2) clockwise to decrease or anti-clockwise to increase the clearance. Tighten the stop light switch locknut.

**Front brakes** Wear of the disc brake friction pads is automatically compensated for and manual adjustment is therefore not required. Before the lining material has worn down to the minimum permissible thickness of $\frac{1}{16}$ in. the brake pads must be renewed. Special equipment is required, and new pads should be fitted by an authorized Distributor or Dealer.

**Rear brakes** Excessive brake pedal travel is an indication that the rear brake-shoes require adjusting. The brakes on both rear wheels must be adjusted to regain even and efficient braking.

*Adjusting*
*Fig. 3*

Block the front wheels, fully release the hand brake and jack up each rear wheel in turn. Turn the adjuster (arrowed) in a clockwise direction (viewed from the centre of the car) until the wheel is locked, then turn the adjuster back until the wheel is free to rotate without the shoes rubbing. Repeat the adjustment on the other rear brake.

**Inspecting rear brake linings** Block the front wheels, release the hand brake, and jack up each rear wheel in turn. Remove the road wheel and slacken off the brake-shoe adjuster fully. Remove the two countersunk screws (pressed wheels) or the four nuts (wire wheels) and withdraw the brake-drum.

*Fig. 1*

*Fig. 2*

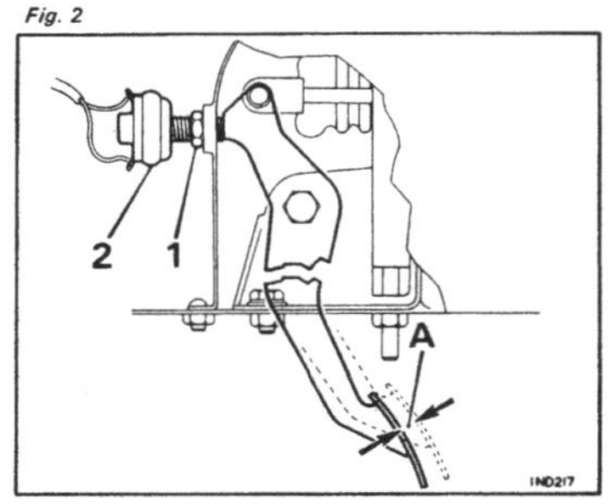


Inspect the linings for wear, and clean out the dust from the backplate assembly and drum. Refit the drum and road wheel and adjust the brake-shoes.

**Replacing brake-shoes or pads** When it becomes necessary to renew brake-shoes or pads it is essential that only genuine shoes or pads, with the correct grade of lining, are used. Always fit new shoes or pads as complete axle sets, never individually or as a single wheel set. Serious consequences could result from out-of-balance braking due to mixing of linings.

Replacement brake-shoes or pads are obtainable from your Distributor or Dealer.

**Hand brake**
*Fig. 4*

The hand brake is automatically adjusted with the rear brakes. If there is excessive movement of the hand brake lever, consult your Distributor or Dealer. To lubricate charge the nipple on the hand brake cable with one of the recommended greases.

**Preventive maintenance** In addition to the recommended periodical inspection of brake components it is advisable as the car ages, and as a precaution against the effects of wear and deterioration, to make a more searching inspection and renew parts as necessary.

It is recommended that:

(1) Disc brake pads, drum brake linings, hoses, and pipes should be examined at intervals no greater than those laid down in the Passport to Service.
(2) Brake fluid should be changed completely every 18 months or 18,000 miles whichever is the sooner.
(3) All fluid seals in the hydraulic system should be renewed, and all flexible hoses should be examined and renewed if necessary every 3 years or 36,000 miles whichever is the sooner. At the same time the working surface of the piston and of the bores of the master cylinder, wheel cylinders, and other slave cylinders should be examined and new parts fitted where necessary.

Care must be taken always to observe the following points:

(*a*) At all times use the recommended brake fluid.
(*b*) Never leave fluid in unsealed containers. It absorbs moisture quickly and this can be dangerous.
(*c*) Fluid drained from the system or used for bleeding is best discarded.
(*d*) The necessity for absolute cleanliness throughout cannot be over-emphasized.

*Fig. 3*

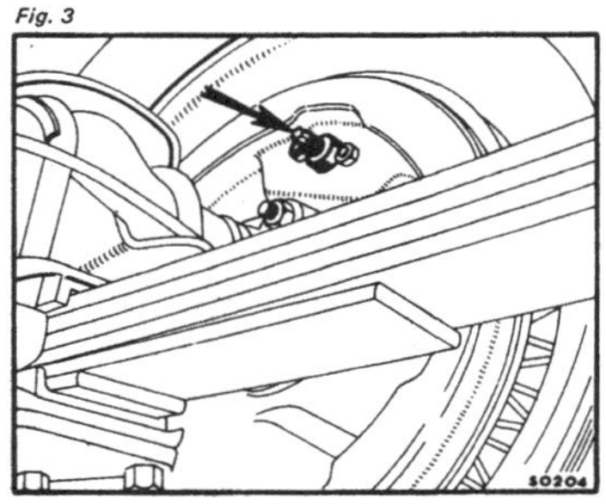

*Fig. 4*

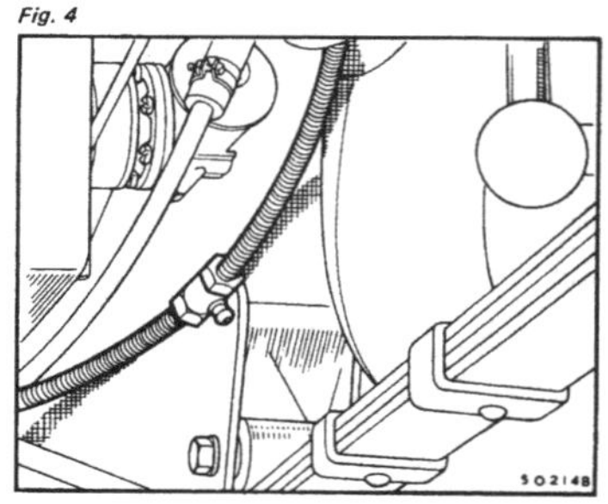

# ELECTRICAL

**POLARITY** The electrical installation on this car is **NEGATIVE** (−) earth return and the correct polarity must be maintained at all times. Reversed polarity will permanently damage semi-conductor devices in the alternator and tachometer, and the radio transistors (when fitted).

Before fitting a radio or any other electrical equipment, make certain that it has the correct polarity for installation in this vehicle.

**BATTERIES**
**Access**
*Fig. 1*
Remove the carpet covering the rear compartment floor. Turn the five quick-release fasteners (1) anti-clockwise one half turn and remove the battery compartment cover panel (2).

**Checking, topping-up**
*Fig. 1*
The vehicle must be on level ground when the electrolyte is being checked.

**DO NOT USE A NAKED LIGHT WHEN CHECKING THE LEVELS.**
**(See Editor's notes at the end of Part 2.)**

Remove the battery vent cover; use the grip at the centre of the cover (3), this will ensure that the filling valves are operated correctly. If no electrolyte is visible inside the battery pour distilled water into the filling trough until the three tubes (4), and the connecting trough (5), are filled. Refit the vent cover. Check the second battery.

**General** The batteries must be kept dry and clean; cable and battery terminals should be smeared with petroleum jelly.

**Precautions Type CA** Except when topping-up or taking hydrometer readings the vent cover must be fitted. The electrolyte will flood if the cover is removed for long periods during or within thirty minutes of the battery being normal (5 amp.), or fast charged (30–40 amps.). Single-cell discharge testers cannot be used on these batteries. Operation of the filling device will be destroyed if the battery case is drilled or punctured.

*Fig. 1*

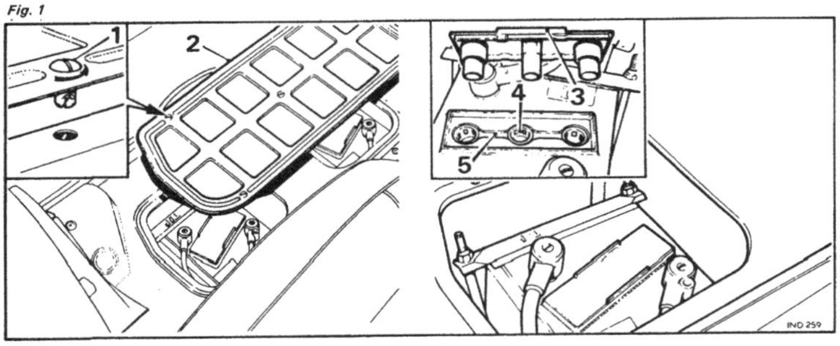


**FUSES**
*Fig. 2*
The fuses are housed in a fuse block (1) mounted in the engine compartment body on the oil filter side of the engine.

**Fuse connecting 1–2.** The fuse (2) protects one parking lamp, one tail lamp, one number-plate lamp, and one front and rear side marker lamp.

**Fuse connecting 3–4.** The fuse (3) protects one parking lamp, one tail lamp, one number-plate lamp, and one front and rear side-marker lamp.

**Fuse connecting 5–6.** The fuse (4) protects the circuits which operate only when the ignition is switched on. These circuits are for the direction indicators, brake stop lamps, reverse lamps and seat belt warning. The heated back-light (when fitted) on GT models is also protected by this fuse.

**Fuse connecting 7–8.** The fuse (5) protects the equipment which operates independently of the ignition switch, namely horns, interior and luggage compartment lamps, headlamp flasher, brake failure warning lamp, door and seat belt audible warning, and the cigar-lighter (if fitted).

Two spare fuses (6) are provided and it is important to use the correct replacement fuse. The fusing value, current rated 17 amp. (35 amp. blow rated), is marked on a coloured slip of paper inside the glass tube of the fuse.

**Line fuses**
*Fig. 2*
**Auxiliary equipment—white/green and green/pink wiring.** The 35 amp. line fuse (7) protects the windscreen wiper, windscreen washer, heater blower motor and radio circuits when the ignition switch is in position 'I'.

**Running-on control valve—slate and slate/purple wiring.** The 35 amp. line fuse (8) protects the running-on control valve circuit which operates when the ignition is switched off.

**Hazard warning.** The 35 amp. line fuse (9) protects the hazard warning lamps and is located behind the hazard warning switch. It is accessible only when the centre console is withdrawn (see page 34).

**Radio.** A separate additional line fuse protects the radio (if fitted). See the instructions supplied with the radio for the correct fuse ratings.

*Fig. 2*

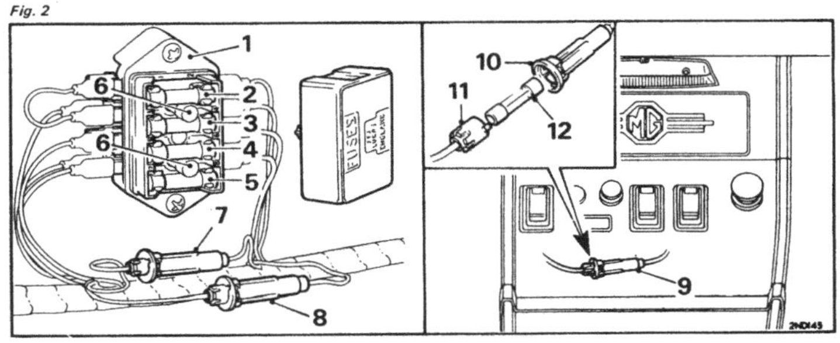


**To change a line fuse,** hold one end of the cylindrical fuse holder (10), push in and twist the other end (11). Remove the fuse (12) from the cylindrical holder.

**Blown fuses** A blown fuse is indicated by the failure of all the units protected by it, and is confirmed by examination of the fuse when withdrawn. Before renewing a blown fuse inspect the wiring of the units that have failed for evidence of a short-circuit or other fault.

**Accessories**
*Fig. 2*
If an electrical accessory is being fitted and it is required to operate independently of the ignition circuit it should be connected to terminal 'B' on the fuse block; if it is required to operate only when the ignition is switched on, connect it to terminal '6'. The terminal numbers are marked on the fuse block.

**HEADLAMPS**
**Light unit**
*Fig. 3*
To remove a light unit, ease the bottom of the outer rim (1) forwards away from the lamp. Unscrew the three inner rim retaining screws (2), remove the inner rim (3), withdraw the light unit (4), and disconnect the three-pin plug (5).

To fit a light unit, connect the three-pin plug, position the light unit in the headlamp body ensuring that the three lugs formed on the outer edge of the light unit engage in the slots formed in the body, and fit the inner retaining rim. Position the outer rim on the retaining lugs with the cut-away portion of the rim at the bottom of the lamp, press the rim downwards and inwards.

**Beam setting** Two adjusting screws are provided on each headlamp for setting the main beams. The screw (6) is for adjusting the beam in the vertical plane, and the screw (7) is for horizontal adjustment. The beams must be set in accordance with local regulations; resetting and checking should be entrusted to your Distributor or Dealer, who will have special equipment available for this purpose.

*Fig. 3*

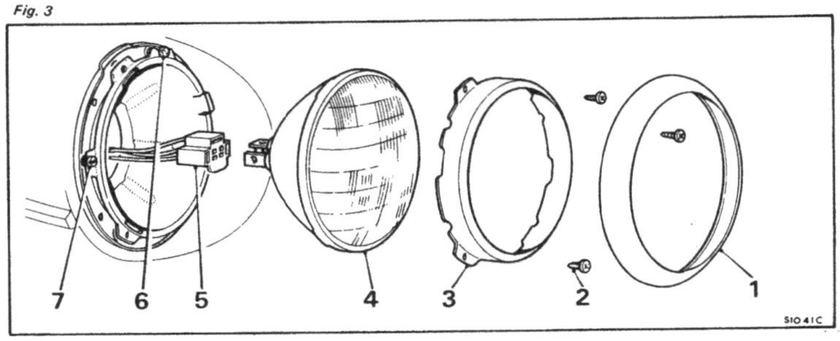


**LAMPS**
**Parking and direction indicator**
*Fig. 4*
To gain access to the parking and direction indicator bulb (1), unscrew the two retaining screws (2) and withdraw the rim and lens.

**Stop, tail and direction indicator**
*Fig. 5*
Remove the lens retaining screw (1) and remove the lens to gain access to the direction indicator and stop/tail bulbs.

*Fig. 4*

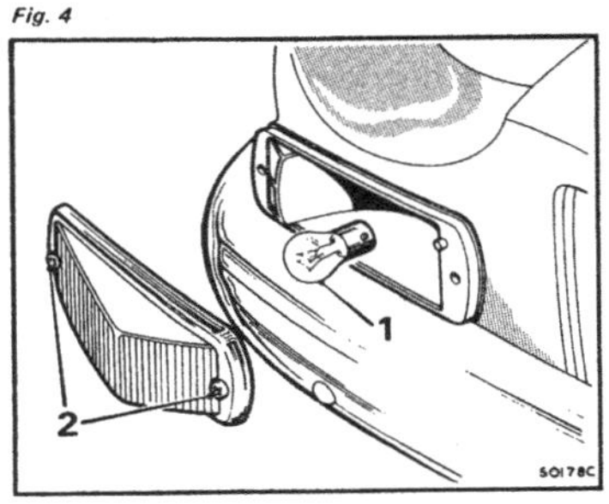


*Fig. 5*

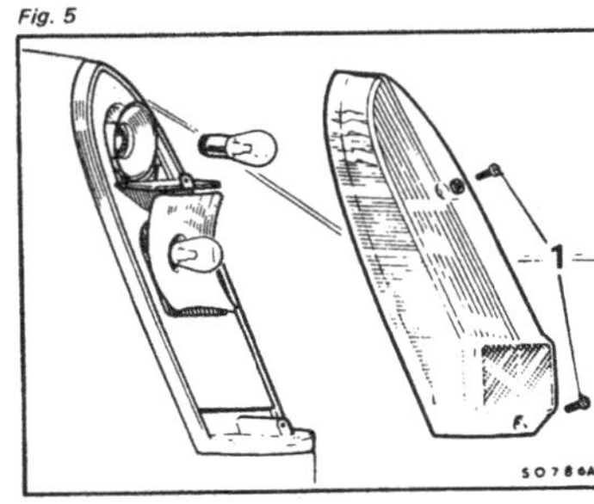

**Number-plate** *Fig. 6* To change a bulb, remove the two securing screws and nuts, lift off the lamp hood (1) and lens (2). When refitting, ensure that the wedge-shaped distance piece (3) is fitted with its thickest edge towards the rear.

**Side marker** *Fig. 7* To renew a bulb, remove the securing screw (1) and lift off the lamp lens, noting that one end is secured by a locating tab (2). When refitting, ensure that the sealing rubber is positioned correctly and that the lens tab (2) is located beneath the lamp body rim before refitting the securing screw.

*Fig. 6*

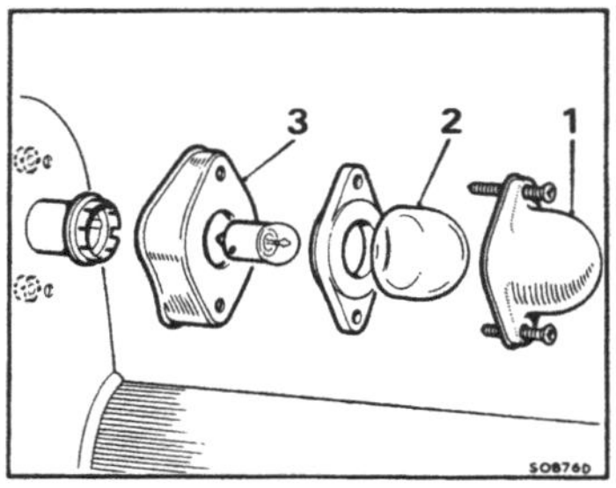


*Fig. 7*

**Reverse** *Fig. 8* To renew a bulb, remove the two securing screws and withdraw the lens. Press the bulb down towards the lower contact and withdraw it from the lamp. Fit one end of the new bulb into the hole in the lower contact, then press the top of the bulb into the lamp until the point of the cap engages in the hole in the upper contact.

**Courtesy** *Fig. 9* To renew a bulb, remove the two screws (1) retaining the lamp bezel and remove the bezel and lens. The bulb may then be withdrawn from its contacts.

**Luggage Compartment (Tourer)** *Fig. 10* The lens is held in the lamp by four locating lugs. To gain access to the bulb, gently squeeze the sides of the lens together and withdraw it from the lamp. Remove the bulb from its contacts.

**Trunk lid or interior (GT)** *Fig. 11* The lens in the lamp is held in position by four locating lugs. To gain access to the bulb, gently squeeze the lens sides and pull outwards. Remove the bulb from the retaining clips.

*Fig. 8*

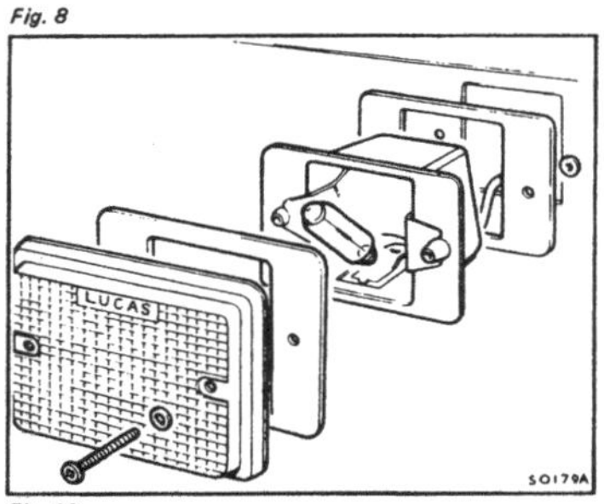


*Fig. 9*

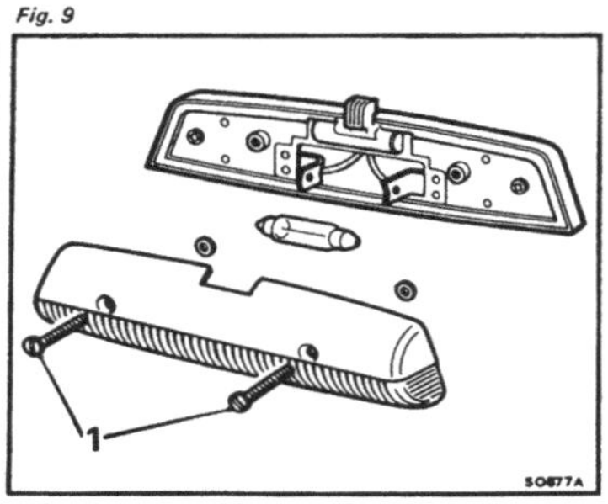


*Fig. 10*

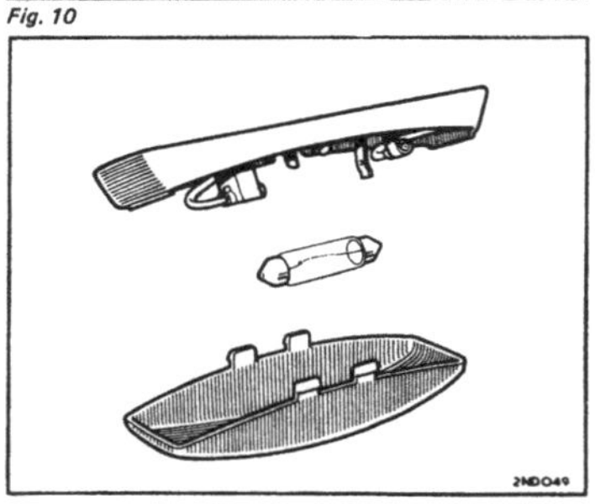


*Fig. 11*

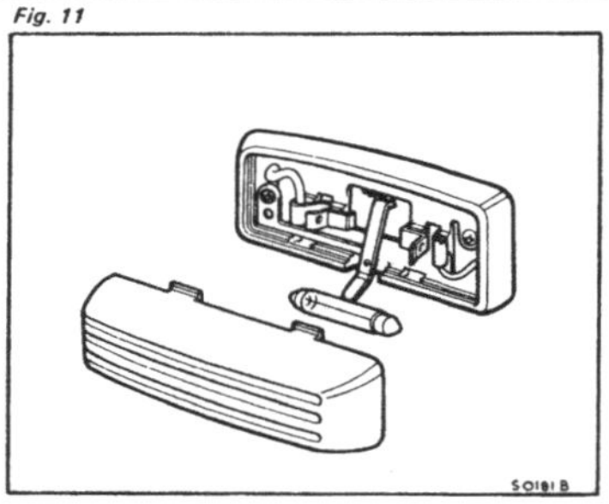


**Warning, panel and illumination** *Figs. 12 and 13* To gain access to the bulbs the fascia bottom panel must be removed and/or the centre console withdrawn.

**Fascia bottom panel.** Unscrew the three screws (1) securing the bottom panel (2) and pull the panel forward from its retaining clips (3) at the rear.

**Centre console.** Remove the gear-lever knob and locknut (15). Unscrew the four screws (16) securing the gaiter retaining ring. Raise the hinged arm-rest, unscrew the retaining screw (17) and remove the arm-rest complete with the gaiter. Unscrew the four screws (18) retaining the console and withdraw the console rearwards.

**Heater control lamp bulbs.** Remove the fascia bottom panel. Remove the push-fit bulb holders (4) from the controls and remove the bayonet fixing type bulb (5). To remove the air flow control illumination bulb the centre console must also be withdrawn.

**Instrument panel lamp bulbs.** Remove the fascia bottom panel. Remove the push-fit bulb holders (6) from the instruments and unscrew the bulbs (7).

**Lights and heater booster switch bulbs.** Remove the fascia bottom panel. Remove the push-fit bulb holders (8) from the switches and remove the bayonet fixing type bulbs (9). To remove the heater booster switch bulb the centre console must also be withdrawn.

**Warning lamp bulbs.** Remove the fascia bottom panel. Remove the push-fit bulb holders (10) from the lamps and remove the bayonet fixing type bulbs (11).

*Fig. 12*

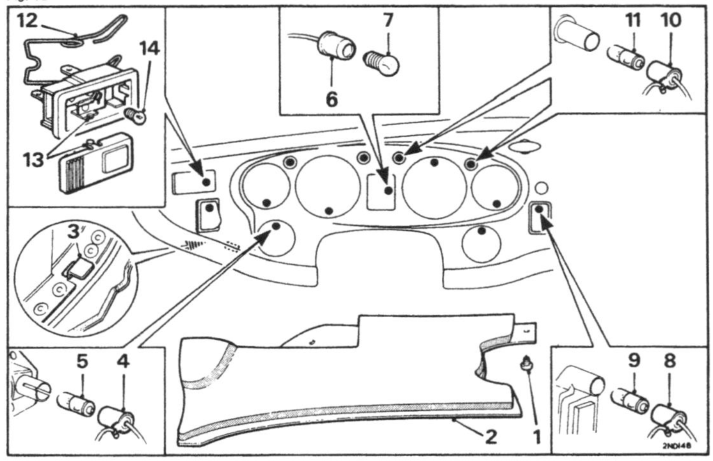


**Brake failure warning lamp bulb.** Remove the fascia bottom panel. Remove the retaining spring clip (12) and withdraw the holder/test-push assembly from the fascia. Through the two pivot holes in the holder depress the pivot legs (13) and remove the test-push rocker from its holder. Unscrew and remove the bulb (14).

**Hazard switch bulb.** Withdraw the centre console. Remove the push-fit bulb holder (19) from the switch, and remove the bayonet fixing type bulb (20).

**Heated back-light warning lamp bulb.** Withdraw the centre console. Remove the push-fit bulb holder (19) from the switch and remove the bayonet fixing type bulb (20)

**Hazard and seat belt warning lamp bulbs.** Withdraw the centre console. Remove the push-fit bulb holders (21) from the lamps and remove the bayonet fixing type bulbs (22).

**Cigar-lighter (if fitted) illumination bulb.** Withdraw the centre console. Squeeze the sides of the bulb hood (23) and remove the hood. Remove the bulb holder (24) from the hood clip and remove the bayonet fixing type bulb (25).

*Fig. 13*

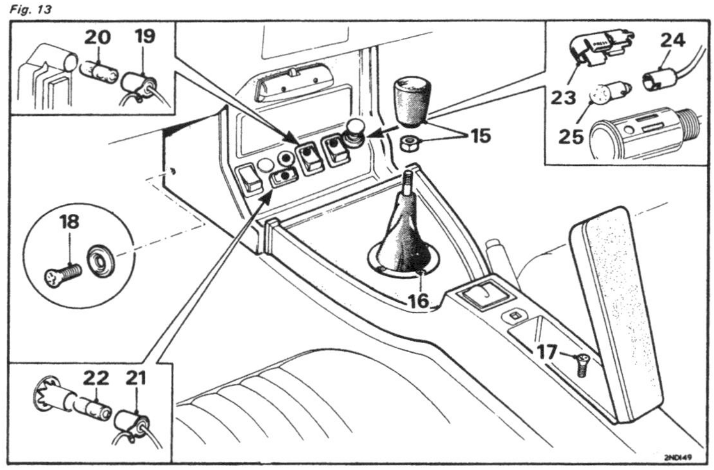


**Replacement bulbs**

| | *Volts* | *Watts* | *Part No.* |
|---|---|---|---|
| Headlamp (sealed beam) | 12 | 50/40 | — |
| Sidelamp (with flasher) | 12 | 5/21 | GLB 380 |
| Stop/tail | 12 | 5/21 | GLB 380 |
| Reverse | 12 | 18 | BFS 273 |
| Number-plate lamp | 12 | 6 | GLB 989 |
| Direction indicator | 12 | 21 | GLB 382 |
| Side-marker lamp, front and rear | 12 | 5 | BFS 222 |
| Ignition warning | 12 | 2 | GLB 281 |
| Main beam | 12 | 2 | GLB 281 |
| Direction indicator warning lamp | 12 | 2 | GLB 281 |
| Brake warning lamp | 12 | 1·5 | GLB 280 |
| Panel illumination lamp | 12 | 2·2 | GLB 987 |
| Cigar-lighter illumination | 12 | 2·2 | BFS 643 |
| Luggage compartment lamp | 12 | 6 | GLB 254 |
| Courtesy lamp | 12 | 6 | GLB 254 |
| Hazard warning lamp | 12 | 2 | GLB 281 |
| Heated back-light warning lamp | 12 | 2·2 | GLB 987 |
| Seat belt warning lamp | 12 | 2 | GLB 281 |
| Switch illumination | 12 | 2 | GLB 281 |
| Heater rotary control illumination | 12 | 2 | GLB 281 |

**ALTERNATOR** The following precautions must be observed to prevent inadvertent damage to the alternator and its control equipment.

**Polarity.** Ensure that the correct battery polarity is maintained at all times; reversed battery or charger connections will damage the alternator rectifiers.

**Battery connections.** The battery must never be disconnected while the engine is running.

**WINDSCREEN WIPER**

**Arms** *Fig. 14* To reposition a wiper arm, hold the spring clip (1) clear of the retaining groove in the spindle (2) and withdraw the arm. Place the arm in the required position and press it on to the spindle until it is secured by the clip.

**Blades** *Fig. 14* To renew a wiper blade, pull the wiper arm (3) away from the windscreen, press the arm against the spring fastener (4) and pull the blade from the arm. Insert the end of the wiper arm into the spring fastener of the new blade, and push the blade into engagement with the arm.

**To ensure efficient wiping it is recommended that wiper blades are renewed annually.**

**Washer** *Fig. 15* The pump for the windscreen washer is mounted on the right-hand wing panel in the engine compartment in the position as illustrated. The fluid flow is indicated on the pump just above the fluid connections.

*Fig. 14*

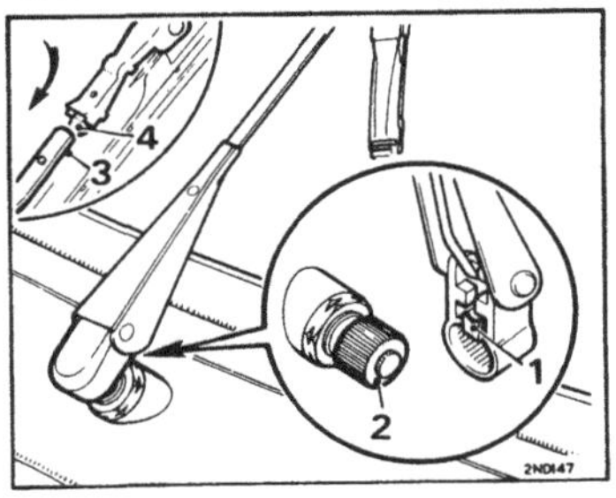


*Fig. 15*

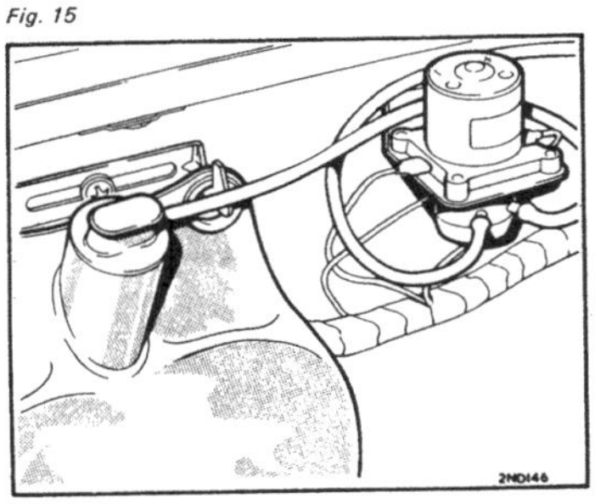


## IGNITION

**Ignition timing** The ignition timing is set dynamically to give optimum engine performance with efficient engine emission control. Electronic test equipment must be used to check the ignition timing setting and the automatic advance (see **'GENERAL DATA'**). Checking and adjustment to the ignition timing setting should be carried out by your Austin MG Dealer control service station.

The dynamic ignition timing must be checked after cleaning, re-setting, or renewing of the distributor contacts. (**See Editor's notes at the end of Part 2.**)

**Distributor** *Fig. 1* **Cleaning contacts.** Inspect the contact points (1) and, if burned, clean with fine emery cloth or fine carborundum stone. Wipe the contacts clean with a fuel-moistened cloth. Renew pitted or worn points.

**Lubrication.** Very lightly smear the pivot post (2) and around the cam (3) with grease. Add a few drops of oil through the hole in the contact breaker plate to lubricate the centrifugal weights and around the screw (5) in the centre of the cam spindle (do not remove this screw as clearance is provided for oil to pass). **Avoid over-lubricating.** Carefully wipe away all surplus lubricant and see that the contact breaker points are perfectly clean and dry.

**Contact breaker gap.** Turn the crankshaft until the points are fully open. Check the contact gap (1) with a feeler gauge (see **'GENERAL DATA'**); the gauge should be a sliding fit in the gap.

If the gap varies appreciably from the gauge thickness, slacken the contact plate securing screw (6) and adjust the gap by inserting a screwdriver in the notched hole (7) at the end of the plate and turning clockwise to decrease and anti-clockwise to increase the gap. Tighten the securing screw. Turn the crankshaft and recheck the gap. Refit the rotor, wipe the inside of the distributor cover clean and refit.

**Contact set renewing.** Remove the nut (8) and lift the top insulating bush and both leads from the stud. Remove the securing screw (6) with its spring and plain washers and lift off the one-piece contact set. If removal of the moving contact only is required, leave the securing screw (6) in position.

*Fig. 1*

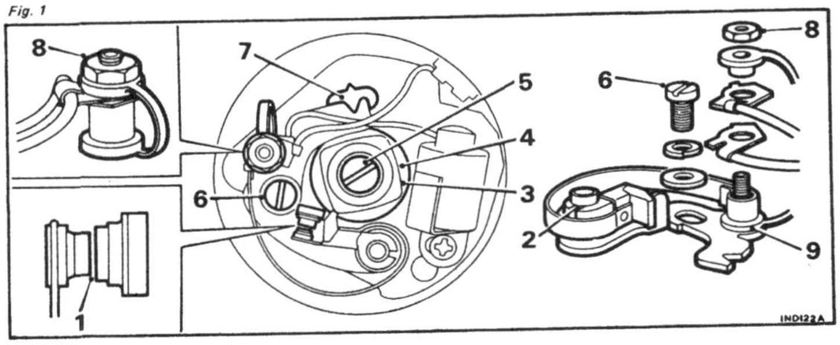


**Fitting.** Before fitting a new contact set, wipe the points clean with fuel or spirit. Lubricate the pivot post (2) and check that the insulating bush (9) is correctly positioned below the spring loop. Position the contact set on the distributor plate and lightly tighten the securing screw (6). Locate the lead terminals around the insulating bush so that they make contact with the spring and tighten the nut (8). Set the contact gap.

Whenever a new contact set has been fitted, recheck the gap after the first 500 miles. During this period, the heel of the contact will have bedded in and reduced the gap.

**Sparking plugs** *Fig. 2* The sparking plugs should be cleaned, preferably with an air-blast service unit.

Check the plug gaps, and reset if necessary to the recommended gap (see **'GENERAL DATA'**). To reset, use a special Champion sparking plug gauge and setting tool; move the side electrode, never the centre one.

When refitting the plugs make sure that the washers are not defective in any way.

Screw the plug down by hand as far as possible, then use a spanner for tightening only. Always use a tubular box spanner to avoid possible damage to the insulator, and do not under any circumstances use a movable wrench. Never overtighten a plug, but ensure that a good joint is made between the plug body, washer, and cylinder head. Wipe clean the outside of the plugs before reconnecting the H.T. leads.

When fitting new sparking plugs ensure that only the recommended type and grade are used (see **'GENERAL DATA'**).

*Fig. 2*

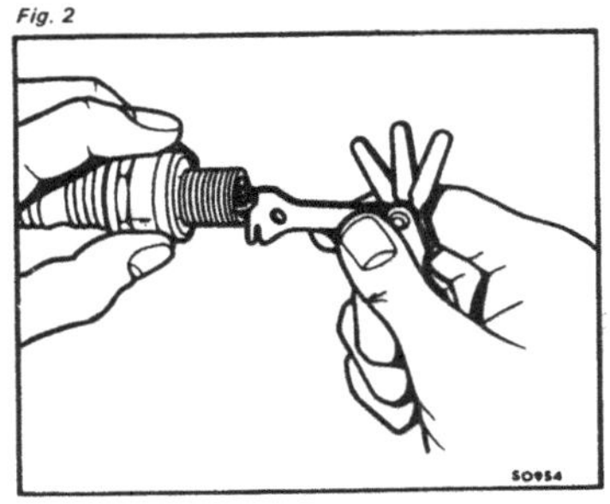


## ENGINE

**LUBRICATION Checking** *Fig. 1* The level of the oil in the engine sump is indicated by the dipstick (1) on the right-hand side of the engine. Maintain the level at the 'MAX' mark on the dip-stick and never allow it to fall below the 'MIN' mark.

The filler (2) is on the forward end of the rocker cover and is provided with a quick-action cap.

The oil level should always be checked before a long journey.

**Draining** To drain the engine oil, remove the drain plug (3) located on the right-hand side at the rear of the sump. This operation should be carried out while the engine is warm.

Clean the drain plug; check that its copper sealing washer is in a satisfactory condition, and refit.

**Filling** Fill the engine with the correct quantity (see **'GENERAL DATA'**) of a recommended oil. Run the engine for a short while then allow it to stand for a few minutes before checking the level with the dipstick.

*Fig. 1*

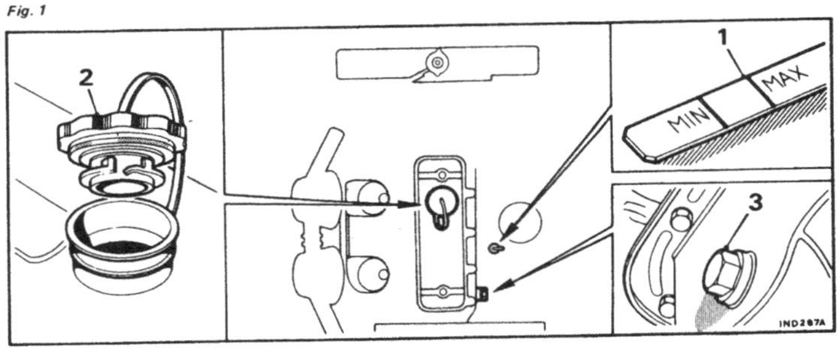

**Oil filter** The external oil filter is of the disposable cartridge type and is located on the right-hand side of the engine. Early cars are fitted with the renewable element type of filter.

*Fig. 4* **To change the filter cartridge.** Unscrew the old filter cartridge (1) from the filter head (2) and discard. Smear the seal (3) of the new filter cartridge with oil, locate on the filter head (2) and tighten with the hands only. **DO NOT USE A SPANNER TO TIGHTEN. DO NOT OVERTIGHTEN.**

*Fig. 5* **To renew the early-type filter element,** drain the filter by removing the plug (1); unscrew the filter body securing bolt (2) and lift off the body. Wash out the casing with gasoline (fuel) and dry it before fitting a new element. Check that the sealing rings (3) and (5) and the rubber washer (4) are in a satisfactory condition.

Reassemble the filter, ensuring that the components are correctly positioned.

*Fig. 4*

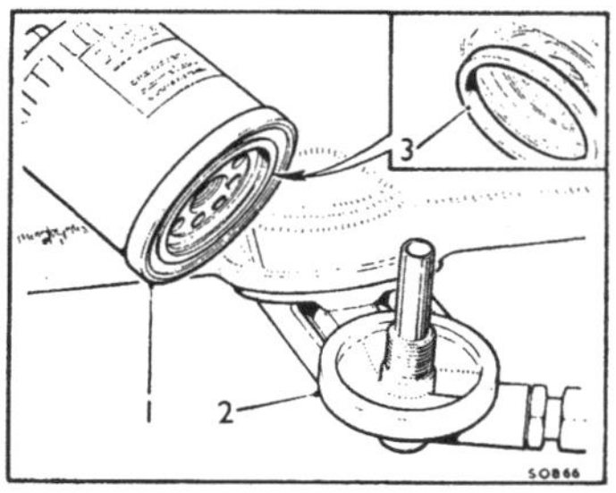


*Fig. 5*

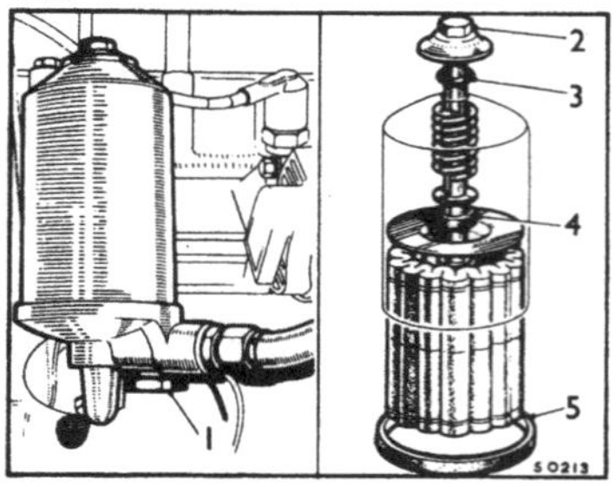


## DRIVE BELT

**Alternator** *Fig. 3* **Tension.** When correctly tensioned, a total deflection of ½ in. under moderate hand pressure, should be possible at the midway point of the longest belt run between the pulleys.

**Adjusting.** To adjust the belt tension, slacken the securing bolts (1) and adjusting link nuts (2), and move the alternator to the required position. Apply any leverage necessary to the alternator end bracket (3) only and not to any other part; to avoid damaging the drive-end bracket the lever should preferably be of wood or soft metal. Tighten the bolts and re-check the belt tension. **DO NOT OVERTIGHTEN** as this will impose an excess loading on the drive bearings.

## VALVE ROCKER CLEARANCE

**Checking**

Remove the rocker cover and insert a ·015 in. feeler gauge between the valve rocker arms and valve stems (inset). The gauge should be a sliding fit when the engine is cold. Check each clearance in the following order:

Check No. 1 valve with No. 8 fully open. Check No. 8 valve with No. 1 fully open.
„ „ 3 „ „ „ 5 „ „ „ „ 6 „ „ „ 3 „ „
„ „ 5 „ „ „ 4 „ „ „ „ 4 „ „ „ 5 „ „
„ „ 2 „ „ „ 7 „ „ „ „ 7 „ „ „ 2 „ „

**Adjusting** *Fig. 6* Slacken the adjusting screw locknut on the opposite end of the rocker arm and rotate the screw clockwise to reduce the clearance or anti-clockwise to increase it. Retighten the locknut when the clearance is correct, holding the screw against rotation with a screwdriver.

*Fig. 3*

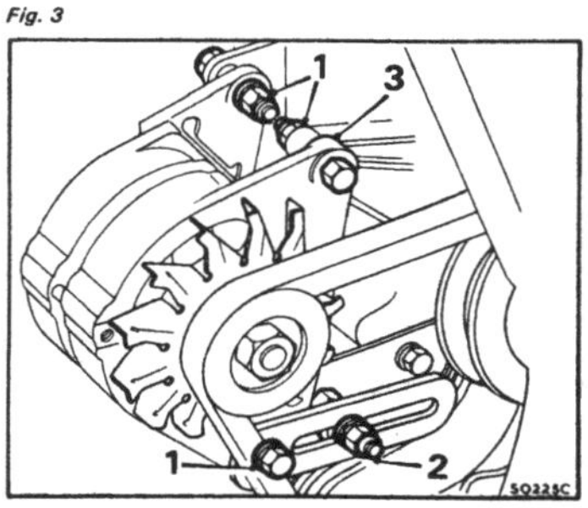


*Fig. 6*

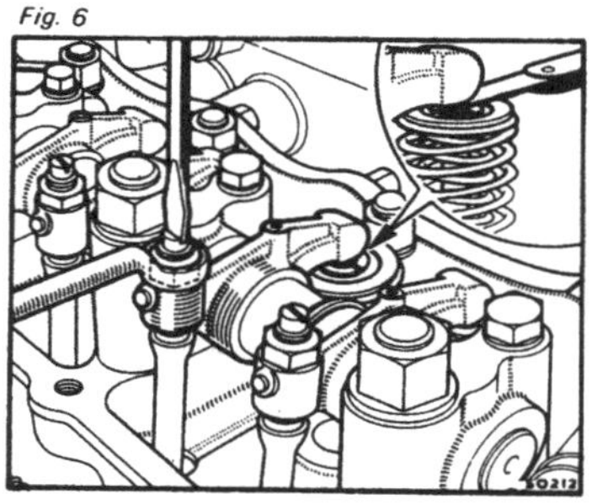


## EMISSION CONTROL SYSTEMS

You and each subsequent owner of the car are urged to make sure that the recommended maintenance procedures are carried out at the intervals specified. For the emission controls to continue to function effectively, it is strongly recommended that you arrange for regular maintenance inspections to be carried out by your Austin MG Dealer or by any other qualified service outlet which regularly performs such service on British Leyland cars.

You have been provided with a Passport to Service which contains an inspection and maintenance schedule up to 50,000 miles or 5 years. You should have the maintenance record completed by your Austin MG Dealer (or by other dealer or station equipped to render such service) at the regular mileage intervals indicated in the Schedule. The Handbook and Passport to Service should be handed to subsequent purchasers of the vehicle at the time of sale so that the maintenance instructions are available and that the record of maintenance can be continued.

You are also urged to study with care the section covering **'MALFUNCTION IDENTIFICATION'.** Study of this section will be of aid to you in detecting possible malfunctions of the emission controls so that necessary service measures can immediately be taken.

**IMPORTANT**

Your attention is particularly drawn to the following:

1. Maintenance and service charges applicable to the emission control system are not covered by the warranty and are not reimbursable, unless shown to have been caused by defects in materials and workmanship covered by the warranty.
2. Our MG engines have not been designed for the regular use of unleaded or low lead gasoline and the use of such fuels cannot be recommended as they could have a detrimental effect on engine components, resulting in loss of performance, excess exhaust emissions and possibly complete engine failure.

**General description** This section gives a general description of the crankcase, exhaust and fuel evaporative emission control systems fitted to this vehicle and the function of their individual components. It must be emphasized that correct carburetter adjustment and ignition timing which have been pre-set at the factory are

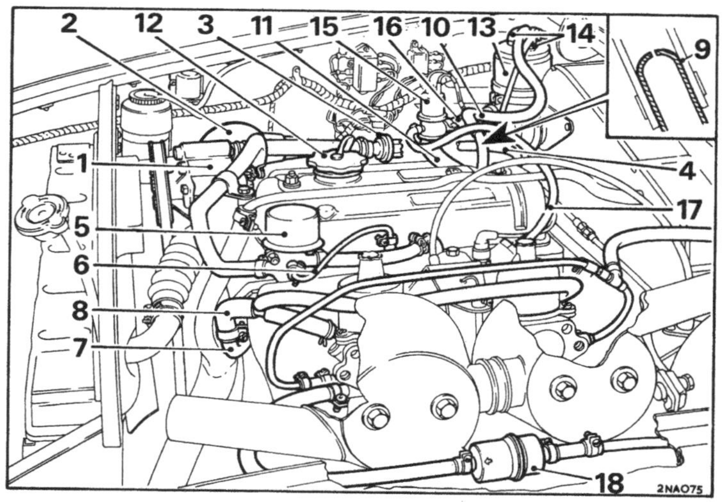


*Fig. 1 The emission control components*

1. Air pump
2. Air pump air cleaner
3. Check valve
4. Air manifold
5. Gulp valve
6. Sensing pipe
7. Oil separator/flame trap
8. Breather pipe
9. Restricted connection
10. Purge line
11. Air vent pipe
12. Sealed oil filler cap
13. Charcoal adsorption canister
14. Vapour lines
15. Running-on control valve
16. Running-on control hose
17. Running-on control pipe
18. Fuel line filter

essential for the efficient functioning of the exhaust emission controls. Should it become necessary to check these settings this work should be carried out by an Austin MG or British Leyland Dealer who has the specialist equipment and training to undertake these adjustments.

The basic engine tuning data will be found on the emission control information label located in the front of the engine compartment.

**Crankcase emission control** The engine crankcase breather outlet incorporates an oil separator flame-trap which is connected by hoses to the controlled depression chamber between the piston and the throttle disc valve of the carburetter(s). Piston blowby fumes are drawn into the chamber where they combine with the engine inlet charge for combustion in the engine cylinders in the normal way. Fresh filtered air is supplied to the engine crankcase through a hose connected between the engine valve rocker cover and the charcoal canister of the fuel evaporative emission control system.

**Exhaust emission control** The exhaust emission control system is designed to give the required degree of control of the carbon monoxide, unburnt hydrocarbons and oxides of nitrogen content of exhaust gases.

The main feature of the exhaust emission control system is a combination of the engine modification and air injection techniques and consists of modified carburetters, modified ignition timing, and air injection into the exhaust ports.

The quantity of air-polluting elements in the gases leaving the exhaust pipe is reduced by adding air to the hot gases immediately they leave the combustion chambers of the engine. The injection of air into the exhaust gases promotes a continued conversion of the undesirable hydrocarbon and carbon monoxide components of the exhaust gases to relatively harmless carbon dioxide and water.

An air pump mounted on the front of the engine, and belt-driven from the water pump pulley, supplies air under pressure through hoses and a check valve and distribution manifold to injectors in each exhaust port in the engine cylinder head. The check valve prevents high pressure exhaust gases from blowing back into the air pump due to, for example, pump drive failure.

Air from the pump is also supplied to a gulp valve, the outlet of which is connected to the engine inlet manifold. A small-bore sensing pipe connected between the inlet manifold and the diaphragm chamber of the gulp valve relays changes in manifold depression to the valve which will open under certain conditions such as those created by deceleration or engine overrun.

When the gulp valve opens, a small quantity of air is admitted directly into the inlet manifold to lean off the rich air/fuel mixture which is present in the manifold under conditions immediately following throttle closure. This mixture, having been reduced to a burnable condition, combines with engine inlet charge for combustion in the engine cylinders in the normal way.

The carburetters are manufactured to a special exhaust emission specification and are tuned to give the maximum emission control consistent with retaining vehicle performance and drivability. The metering needle is arranged in such a manner that it is always lightly spring loaded against the side of the jet to ensure consistency of fuel metering. A spring loaded valve incorporated in the throttle disc limits the inlet manifold depression and ensures that during conditions of engine overrun the air/fuel mixture enters the engine cylinders in a burnable condition consistent with low emission levels.

The ignition distributor is tuned with slightly retarded ignition at low engine r.p.m. compared with non-emission equipped vehicles, and the timing is balanced between mechanical and vacuum characteristics to give optimum timing consistent with low emissions.

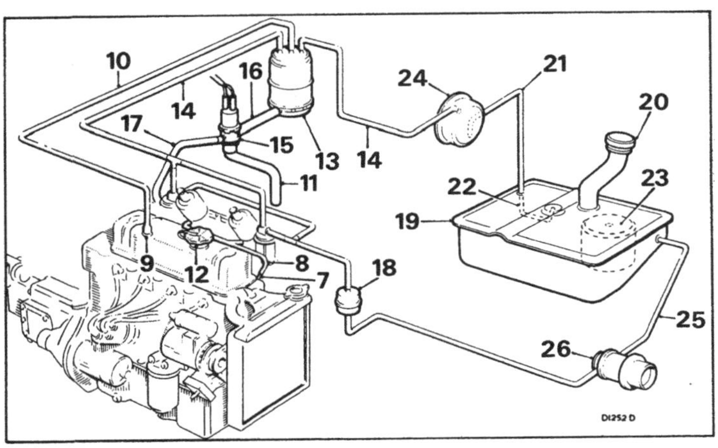


*Fig. 2 The layout of the fuel evaporative loss control system*

7. Oil separator/flame trap
8. Breather pipe
9. Restrictor connection
10. Purge line
11. Air vent pipe
12. Sealed oil filler cap
13. Charcoal adsorption canister
14. Vapour lines
15. Running-on control valve
16. Running-on control hose
17. Running-on control pipe
18. Fuel line filter
19. Fuel tank
20. Sealed fuel filler cap
21. Vapour line
22. Vapour tube
23. Capacity limiting tank
24. Separation tank
25. Fuel pipe
26. Fuel pump

**Fuel evaporative loss control** To prevent air pollution by vapours from the fuel tank and carburetter vents, the control equipment stores the vapour in a charcoal-filled canister while the engine is stopped and disposes of it via the engine crankcase emission control system when the engine is running.

The fuel tank venting is designed to ensure that vapours are vented through the control system even when the car is parked on an inclined surface.

A capacity-limiting device in the fuel tank ensures sufficient free capacity is available after filling to accommodate fuel which would otherwise be displaced as a result of a high temperature rise. The inclusion of a small separation tank in the vapour line prevents liquid fuel from being carried with the vapour to the storage canister.

## MALFUNCTION IDENTIFICATION

Check the following items regularly for visual signs of a malfunction and also if any of the Driving Symptoms listed should persistently occur. **If you are unable to locate and/or correct the malfunction you are advised to contact your Austin MG Dealer immediately.**

**Visual checks**

1. Condition and adjustment of drive belts.
2. Baked or overheated hose between air pump and check valve.
3. All hoses for security, damage and deterioration.
4. Fuel leakage.
5. Oil filler cap for sealing.
6. Fuel filler cap for sealing.

**Driving symptoms**

1. Violent backfire in exhaust system.
2. Hesitation to accelerate on re-opening the throttle after sudden throttle closure.
3. Engine surges (erratic operation at varying throttle openings).
4. Engine idles erratically or stalls.
5. Noisy air pump.
6. Ignition warning light on above idle speed (slack or broken fan belt).
7. Smell of fuel vapours.
8. Engine stops after short running periods (fuel starvation).
9. Lack of power.
10. High fuel consumption.
11. Engine misfires (engine jerks on cruise and acceleration).
12. High temperature indicated (overheating of coolant).

## MAINTENANCE OPERATIONS

All items marked † in the **MAINTENANCE SUMMARY** given on pages 65 to 67 are emission control related.

**Adsorption canister** *Fig. 3*

The charcoal adsorption canister must be renewed every 24,000 miles or 24 months.

To renew the canister, disconnect the air vent pipe (1) vapour pipes (2) and purge pipe (3) from their connections on the canister. Remove the securing bracket nut and bolt (4), collecting the spacer, and remove the canister.

When refitting, ensure that all connections to the canister are tightened securely.

To prevent the engine running on after the ignition has been switched off a control valve is fitted to the air vent pipe of the adsorption canister. The valve is a self-contained unit and requires no regular maintenance. Care should be taken when renewing the adsorption canister not to disturb the valve or its connections.

**Gulp valve** *Fig. 4*

To renew, disconnect the hoses (1) and sensing pipe (1) from the gulp valve. Unscrew the mounting screws and nuts (2) and remove the gulp valve (3). Fit the new valve, reconnect the hoses and sensing pipe ensuring that all joints are made secure and airtight.

**Fuel line filter**

The filter must be renewed every 12,000 miles or 12 months.

**Purge line restrictor**

To check, disconnect the purge line from the rocker cover elbow. Examine the orifice of the restriction formed in the elbow for obstruction. Clear any dirt or deposits from the restrictor orifice, using a length of wire.

*Fig. 3*

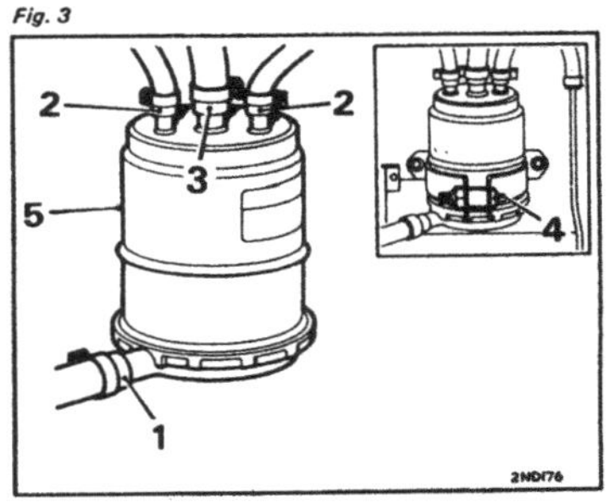


*Fig. 4*

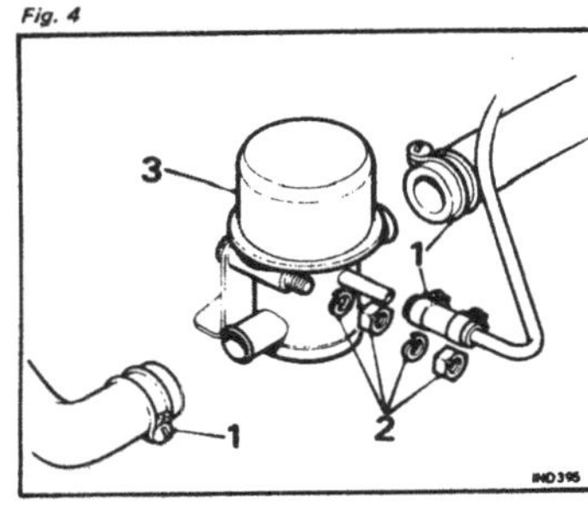

**AIR PUMP** The air pump should be checked for correct functioning every 24,000 miles or 24 months.

**Drive belt** *Fig. 5* **Tension.** When correctly tensioned, a total deflection of ½ in. under moderate hand pressure, should be possible at the midway point of the longest belt run between the pulleys.

**Adjusting.** To adjust the belt tension, slacken the securing bolts (1) and adjusting link nuts (2), and move the air pump to the required position. Tighten the bolts and re-check the belt tension, **DO NOT OVERTIGHTEN** as this will impose an excess loading on the drive bearings.

**Air cleaner** *Fig. 6* The element of the air pump air cleaner must be renewed every 12,000 miles or 12 months; more frequent changes may be necessary in dusty operating conditions.

Unscrew the self-locking nut (1), withdraw the cover (2) and discard the element (3). Clean the inside of the cover thoroughly and re-assemble using a new element.

**Filler caps** Both the engine oil filler cap and the fuel tank filler cap are non-venting and form a seal on the filling apertures.

**IT IS ESSENTIAL TO THE SATISFACTORY OPERATION OF THE EVAPORATIVE LOSS SYSTEM THAT BOTH CAPS ARE ALWAYS REFITTED CORRECTLY AND TIGHTENED FULLY. A DEFECTIVE CAP OR CAP SEAL MUST BE REPLACED.**

*Fig. 5*

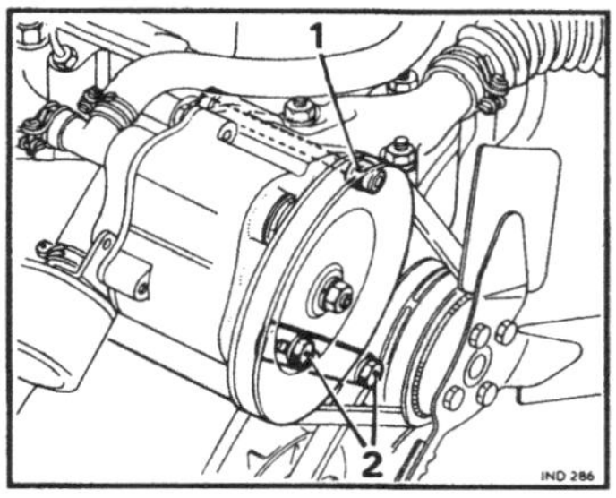


*Fig. 6*

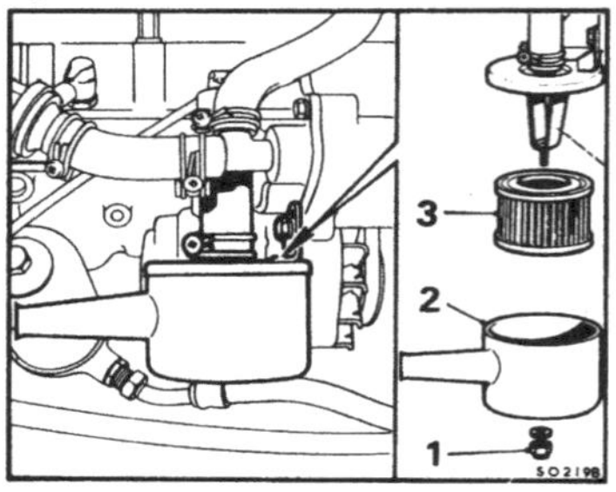


## FUEL SYSTEM

**AIR CLEANERS** The elements of both the carburetter air cleaners must be renewed every 12,000 miles or 12 months; more frequent changes may be necessary in dusty operating conditions.

**Carburetter** *Fig. 1* Unscrew the bolts (1) securing the air cleaner assembly to the carburetter and lift it from the car. Remove the base plate (2) and withdraw the element (3). Clean the inside of the casing (4) thoroughly and reassemble using a new element.

**CARBURETTERS**

**Air pollution control** The carburetter incorporates features which assist in reducing exhaust emissions. Maladjustment or the fitting of parts not to the required specification may render these features ineffective.

**Lubrication** *Fig. 2* **Carburetter.** Unscrew the damper cap at the top of the carburetter and withdraw the damper. Top up with clean engine oil to bring the level 1 inch below the top of the carburetter damper tube. Push the damper assembly back into position and screw in the cap. Under no circumstances should heavy bodied lubricant be used. Failure to lubricate the piston may cause the piston to flutter and reduce acceleration and have an adverse effect on exhaust emission.

**Throttle.** Lubricate the carburetter throttle and choke control linkages and cables, and the accelerator pedal fulcrum.

**Tuning** *Figs. 3 and 4* The efficient operation of the engine and exhaust emission control equipment depends not only on correct carburetter settings but also on correct ignition timing, contact breaker and spark plug gaps and valve rocker clearances. It is essential that these items are checked before adjusting the carburetters.

Tuning of the carburetters is confined to setting the idle and fast idle speeds and the mixture setting at idle speed. Adjustments should only be undertaken on cars required to conform with exhaust emission control regulations if the use of a reliable tachometer, carburetter balance meter and an exhaust gas analyser (CO meter) is available.

1. Remove the air cleaners.
2. Top up the carburetter piston dampers with recommended engine oil to the correct level if necessary.
3. Check the throttle control for correct functioning.
4. Ensure that the mixture control (choke) will return fully, that the cable has $\frac{1}{16}$ in. free play (1) before it starts to pull on the lever and a small clearance (2) exists between the fast idle screws and their cams when the cams are against their stops (3).
5. Raise each carburetter piston with the lifting pin (4), release the pin and check that the piston falls freely onto the bridge of the carburetter, indicated by a distinct metallic click. If the piston fails to fall freely, consult your Distributor or Dealer.
6. Connect a reliable tachometer.
7. Start the engine and run it at a fast idle speed until it attains normal running temperature then run it for a further five minutes.
8. Increase the engine speed to 2,500 r.p.m. for thirty seconds.

**NOTE.**—Tuning can now be commenced. If delay prevents the adjustment being completed within three minutes, increase the engine speed to 2,500 r.p.m. for thirty seconds and then continue tuning. Repeat this clearing procedure at three-minute intervals until tuning is completed.

9. Check the idle speed (tachometer) see **'GENERAL DATA'** and check the carburetters for balanced air intake using a balance meter.
10. If the balance is not correct, adjust as follows: release a throttle connector (9) between the carburetters and turn the adjusting screw (5) on one of the carburetters until the balance is correct. Then adjust the idle speed by turning the throttle adjusting screw (5) on each carburetter by the same amount. Re-check the carburetter balance.

    Check the throttle shaft pin clearance and adjust if necessary—see paragraph (18).

*Fig. 3*

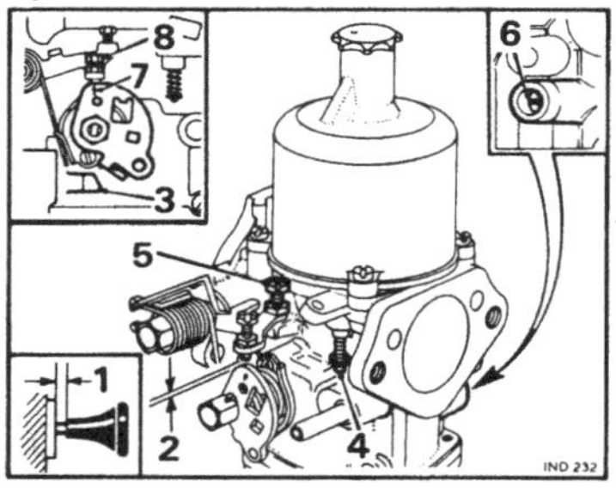


*Fig. 4*

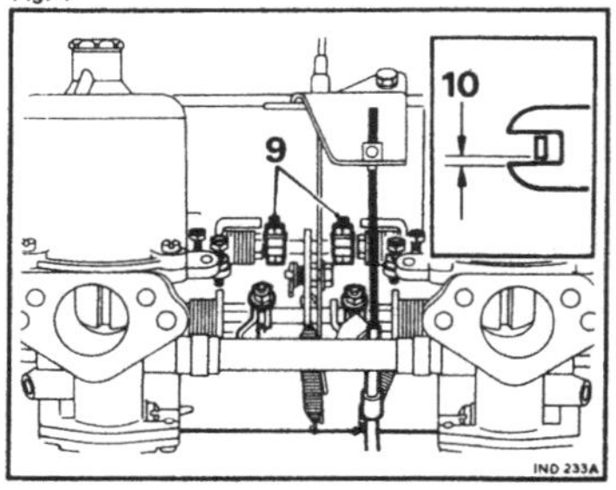


If a smooth idle at the correct speed and balance is not obtainable, adjust the idle speed mixture setting as follows.

11. Stop the engine. Remove each suction chamber and piston and screw the jet adjusting screw (6) in until the jets are flush with the bridge of the carburetter, or in as far as possible. Turn the jet adjusting screw (6) on each carburetter out two complete turns. Refit the piston and suction chambers and top up the piston damper oil levels.

    **NOTE.**—This operation need not be carried out if it is known that the jets are in the same relative position.
12. Restart the engine and run it at idle speed.
13. Turn the jet adjusting screw (6) on both carburetters in the same direction, out to weaken or in to enrich, by the same amount until the fastest speed is registered on the tachometer, turn both screws slowly out (weaken) until the speed just commences to fall, then turn each screw to the weakest position for maximum speed.
14. Using the exhaust gas analyser, check that the percentage CO reading is within the prescribed limits. If the reading falls outside the limits, reset both jet adjusting screws by the minimum amount necessary to bring the reading just within the limits.
15. Re-check the idle speed and carburetter balance and adjust as necessary with the throttle adjusting screws.
16. Set the throttle interconnection clamping levers so that the link pin is ·012 in. away from the lower edge of the forks (see inset 10) as follows:
17. Stop the engine and slacken both clamping bolts (9) on the throttle spindle interconnection.
18. Insert a ·060 in. feeler gauge between the throttle shaft lever and the cold start interconnection rod. Move each lost motion pick-up lever in turn until the tag of the lever rests on the lower arm of the carburetter throttle fork. Tighten the clamping bolts (9, Fig. 4) on each fork, ensuring that there is approximately $\frac{1}{32}$ in. end-float on the interconnection rod. Remove the feeler gauge. The pins on the throttle spindle lever should then have clearance (10) in the throttle fork.
19. With the fast idle cams of both carburetters against their respective stops, set the cold start interconnection so that both cams commence to move simultaneously.
20. With the fast idle cams against their stops check that a $\frac{1}{16}$ in. free movement of the mixture control (choke) cable exists before the cable moves the cams.
21. Run the engine at 1,500 r.p.m. and check the carburetters for balance.

*Fig. 1*

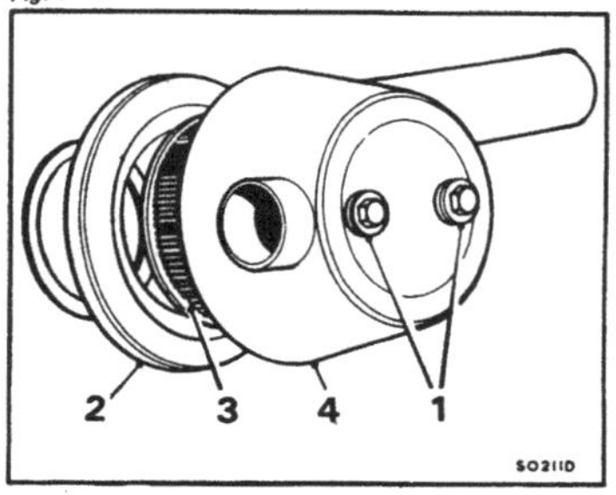


*Fig. 2*

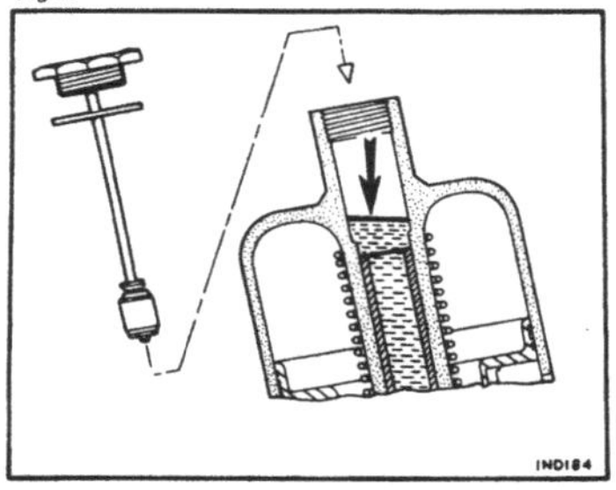

22. Pull out the mixture control (choke) until the arrow (7) marked on the cam is positioned under the fast idle adjusting screw (8) of each carburetter.
23. Using the balancing meter to ensure equal adjustment, turn the fast idle adjusting screws to give the correct fast idle speed (see **'GENERAL DATA'**). Stop the engine.
24. Refit the air cleaners.

## TRANSMISSION

**GEARBOX Checking** *Fig. 1*
To gain access to the gearbox combined oil filler and dipstick (1), lift the carpet on the right-hand side of the gearbox cover at the rear of the control console and remove the rubber plug (2). Clean around the dipstick before removing it.

The oil level must be maintained at the 'HIGH' mark on the dipstick.

**OVERDRIVE Draining** *Fig. 1*
Remove the plug (3) to drain the oil from the gearbox and overdrive unit.

**Sump filter** Drain the gearbox and overdrive unit.

Clean the sump cover and its surroundings. Remove the cover securing screws, withdraw the cover (4) and the filter (5). Clean all metallic particles from the two magnets fitted to the inside of the cover, wash the cover and filter in gasoline. Refit the filter and cover.

**Relief valve filter** Remove the plug and the seal (6); withdraw the relief valve approximately ½ in. and remove the filter (7). Wash the filter, plug and seal in gasoline.

Fit the filter to the relief valve, push the valve fully home and refit the plug and seal.

**Filling** Fill the gearbox and overdrive unit through the combined oil filler (1) with the correct quantity (see **'GENERAL DATA'**) of one of the recommended oils. Run the car for a short distance, allow it to stand for a few minutes, then re-check the level with the dipstick. **Anti-friction additives must not be used in the gearbox or overdrive.**

*Fig. 1*

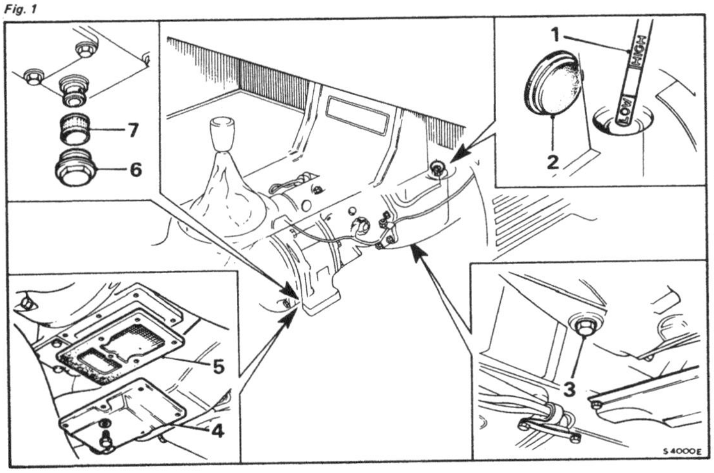


**REAR AXLE Checking** *Fig. 2*
A combined oil filler and level plug (1) is located on the rear of the axle. The oil level must be maintained at the bottom of the plug aperture; ensure that the car is standing level when checking. After topping up the oil level, allow sufficient time for any surplus oil which may have been added to run out of the aperture before replacing the plug.

**Do not drain the rear axle when the After-sales Service is carried out.**

**PROPELLER SHAFT Lubrication** *Fig. 3*
A nipple (1) is provided at the front end of the propeller shaft for lubricating the sliding yoke. To lubricate, give three or four strokes of a gun filled with a recommended grease.

*Fig. 2*

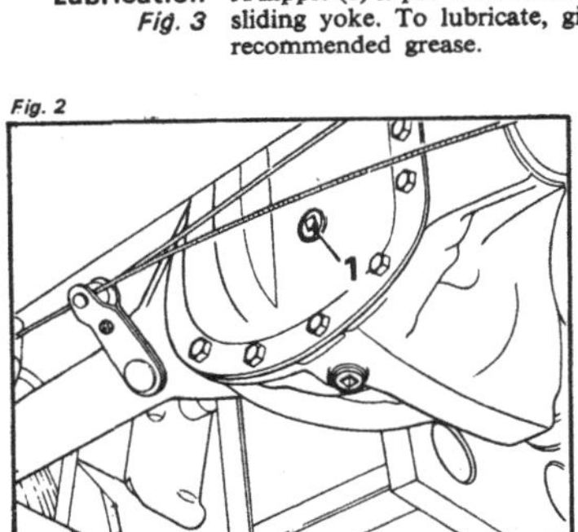


*Fig. 3*

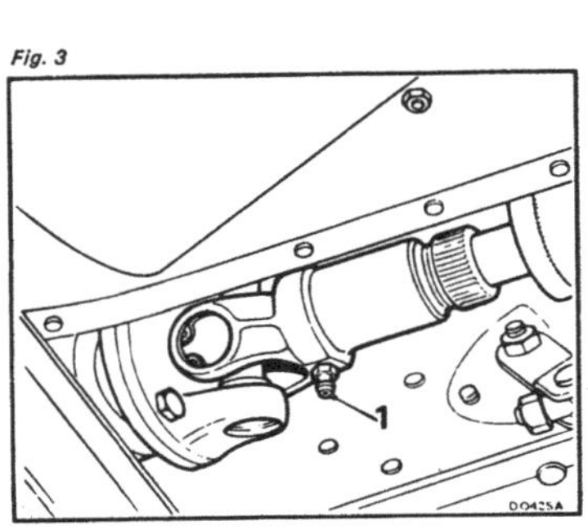


## STEERING/SUSPENSION

**Wheel alignment** *Fig. 1*
Incorrect wheel alignment can cause excessive and uneven tyre wear. The front wheels must be set so that the distance 'A' is $\frac{1}{16}$ in. to $\frac{3}{32}$ in. (toe in) less than the distance 'B'.

Wheel alignment requires the use of a special gauge and this work should be entrusted to your Distributor or Dealer.

**Lubrication** *Fig. 2*
The three lubricating nipples (arrowed) on each of the swivel pins should be charged periodically with one of the recommended greases.

*Fig. 1*

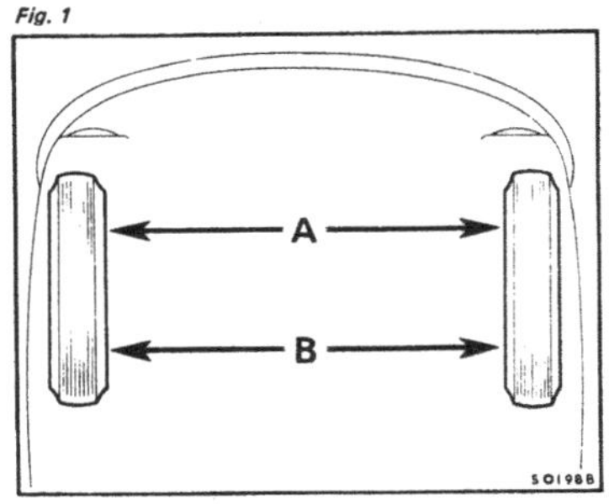


*Fig. 2*

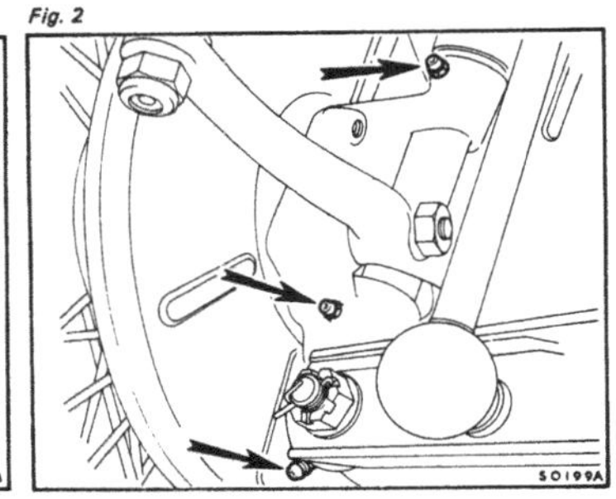

## GENERAL DATA

| | | | |
|---|---|---|---|
| **Engine** | Type: | 18V | |
| | Early cars—with evaporative loss | 18GK and 18GJ | |
| | Early cars—non evaporative loss | 18GH and 18GF | |
| | Bore | 3·16 in. (80·26 mm.) | |
| | Stroke | 3·5 in. (89 mm.) | |
| | No. of cylinders | 4 | |
| | Capacity | 1798 c.c. (109·8 cu. in.) | |
| | Compression ratio | 8 : 1 | |
| | Firing order | 1, 3, 4, 2 | |
| | Valve clearance (cold) | ·015 in. (·38 mm.) | |
| | Oil pressure: Idling | 10 to 25 lb./sq. in. (·7 to 1·7 kg./cm.²) | |
| | Normal | 50 to 80 lb./sq. in. (3·5 to 5·6 kg./cm.²) | |
| | Idle speed: | | |
| | 18V | 850 r.p.m. | |
| | 18GF, 18GH, 18GJ, 18GK | 900 r.p.m | |
| | Fast idle speed | 1,300 to 1,400 r.p.m. | |
| **Ignition** (1971-1972) | Stroboscopic ignition timing: | | |
| | 18V | 16° B.T.D.C. at 1,500 r.p.m. | |
| | 18GK | 15° B.T.D.C. at 1,500 r.p.m. | |
| | 18GJ, 18GH, 18GF | 20° B.T.D.C. at 1,000 r.p.m. | |
| **Ignition** (1973-1974) | Stroboscopic ignition timing (vacuum disconnected) | 11° B.T.D.C. at 1,500 r.p.m. | |
| | Timing marks | Notch on crankshaft pulley, pointers on timing chain cover | |
| | Contact breaker gap | ·014 to ·016 in. (·36 to ·41 mm.) | |
| | Sparking plugs | Champion N-9Y | |
| | Plug gap | ·025 in. (·64 mm.) | |
| **Fuel system** (1971-1972) | Recommended octane rating | Premium 98 or above | |
| | Carburetters: 18V | Two S.U. type HIF4 | |
| | 18GK, 18GH, 18GF | Two S.U. type HS4 | |
| | Needle: 18V | AAU (spring-loaded) | |
| | 18GK | AAL (spring-loaded) | |
| | 18GJ | AAE (spring-loaded) | |
| | 18GH, 18FG | FX | |
| **Fuel system** (1973-1974) | Recommended octane rating | Regular 91 or above (see page 24) | |
| | Carburetters | Two S.U. type HIF 4 | |
| | Needle | ABD | |
| | Fuel pump | S.U. type AUF 300 electric | |
| **Gearbox and overdrive** | Overdrive ratio | ·82 : 1 | |
| | Overall ratios: First | 13·446 : 1 | *Overdrive* |
| | Second | 8·47 : 1 | |
| | Third | 5·40 : 1 | 4·43 : 1 |
| | Fourth | 3·909 : 1 | 3·20 : 1 |
| | Reverse | 12·098 : 1 | |
| | Top gear speed per 1,000 r.p.m.: | | |
| | Standard | 18 m.p.h. (29 km.p.h.) | |
| | Overdrive | 22 m.p.h. (35 km.p.h.) | |

**Capacities**

| | |
|---|---|
| Fuel tank | 10 gallons (45·4 litres; 12 U.S. gal.) |
| Fuel tank (non evaporative loss) | 12 gallons (54 litres; 14 U.S. gal.) |
| Cooling system | 9½ pints (5·4 litres, 11·4 U.S. pints) |
| Cooling system with heater | 10 pints (5·6 litres, 12 U.S. pints) |
| Sump | 5¼ pints (3 litres; 6 U.S. pts.) |
| Sump with oil cooler | 6 pints (3·4 litres; 7¼ U.S. pts.) |
| Sump (early cars) | 7½ pints (4·26 litres, 9 U.S. pints) |
| Sump with oil cooler (early cars) | 8½ pints (4·5 litres, 9·6 U.S. pints) |
| Gearbox | 5¼ pints (3 litres, 6 U.S. pints) |
| Gearbox and overdrive | 6 pints (3·4 litres, 7 U.S. pints) |
| Rear axle | 1½ pints (·85 litre, 2 U.S. pints) |

**Dimensions**

| | |
|---|---|
| Length | 12 ft. 8½ in. (3·8 m.) |
| Length with over-riders | 12 ft. 9 3/16 in. (3·89 m.) |
| Width | 4 ft. 11 15/16 in. (152·3 cm.) |
| Height, hood erected | 4 ft. 1⅜ in. (125·4 cm.) |
| Ground clearance (minimum) | 5 in. (12·7 cm.) |
| Track: Front | 4 ft. 1 in. (124·4 cm.) |
| Track: Rear | 4 ft. 1¼ in. (125·0 cm.) |
| Wheelbase | 7 ft. 7 in. (231·1 cm.) |
| Turning circle | 32 ft. (9·75 m.) |
| Toe-in | 1/16 to 3/32 in. (1·5 to 2·3 mm.) |

**Wheels** Type:

| | |
|---|---|
| Press spoked | 5J FH × 14 |
| Pressed disc | 4J × 14 (GT) 5J × 14 |
| Wire | 4½J × 14 (60-spoke) |

**Tyre pressures**

| Tourer | 5·60—14S Cross-ply tyres | | 155SR—14 Radial-ply tyres | |
|---|---|---|---|---|
| | *Front* | *Rear* | *Front* | *Rear* |
| Normal car weight | 19 lb./sq. in. (1·34 kg./cm.²) | 22 lb./sq. in. (1·55 kg./cm.²) | 21 lb./sq. in. (1·48 kg./cm.²) | 24 lb./sq. in. (1·69 kg./cm.²) |
| Maximum weight | 19 lb./sq. in. (1·34 kg./cm.²) | 24 lb./sq. in. (1·69 kg./cm.²) | 21 lb./sq. in. (1·48 kg./cm.²) | 26 lb./sq. in. (1·83 kg./cm.²) |
| **GT** | | | | |
| Normal car weight | 24 lb./sq. in. (1·4 kg./cm.²) | 24 lb./sq. in. (1·69 kg./cm.²) | **Tyre size 165SR—14** **Pressures as Tourer** | |
| Maximum weight | 20 lb./sq. in. (1·4 kg./cm.²) | 26 lb./sq. in. (1·83 kg./cm.²) | | |

**It is recommended that for sustained speeds at or near the maximum the above tyre pressures are increased by 6 lb./sq. in. (·42 kg./cm.²).**

**Weights**

| | | Total weight | | Distribution | | | |
|---|---|---|---|---|---|---|---|
| | | | | *Front* | | *Rear* | |
| | | *Tourer* | *GT* | *Tourer* | *GT* | *Tourer* | *GT* |
| **Kerbside** | Including full fuel tank, all optional extras and accessories. | 2,303 lb. (1044 kg.) | 2,401 lb. (1088 kg.) | 1,127 lb. (512 kg.) | 1,162 lb. (527 kg.) | 1,176 lb. (534 kg.) | 1,239 lb. (562 kg.) |
| **Normal** | Kerbside weight including driver, passenger and 50 lb. (22 kg.) luggage | 2,653 lb. (1203 kg.) | 2,751 lb. (1248 kg.) | 1,235 lb. (560 kg.) | 1,269 lb. (576 kg.) | 1,418 lb. (643 kg.) | 1,482 lb. (672 kg.) |
| **Maximum** | Normal weight including tow-bar hitch load | 2,753 lb. (1249 kg.) | 2,851 lb. (1293 kg.) | 1,193 lb. (541 kg.) | 1,231 lb. (558 kg.) | 1,560 lb. (708 kg.) | 1,620 lb. (735 kg.) |
| **Maximum permissible towing weight** | | 1,680 lb. (762 kg.) | 1,680 lb. (762 kg.) | | | | |
| **Maximum towbar hitch load** | | 100 lb. (45 kg.) | 100 lb. (45 kg.) | | | | |
| **Maximum roof rack load** | | — | 50 lb. (23 kg.) | | | | |

## MAINTENANCE SUMMARY

**Basic engine tuning data will be found on the Vehicle Emission Control Information label located in the engine compartment.**

**Weekly or before a long journey**

Check oil level in engine, and top up if necessary.
Check battery, and top up to correct level if necessary.
Check coolant level, and top up if necessary.
Check windscreen washer, and top up if necessary.
Check tyre pressures, including spare, and adjust if necessary.

**MAINTENANCE INTERVALS**

† **These items are emission related**

Carry out the services indicated by **X** in column.

The Lubrication Service at **3,000-mile or 3-month** intervals.
**A at 6,000-mile or 6-month** intervals
**B at 12,000-mile or 12-month** intervals

| | Lubrication Service | A | B |
|---|---|---|---|
| **LUBRICATION** | | | |
| Change engine oil | X | X | X |
| Fit new oil filter | | X | X |
| Check level of all fluid reservoirs, brake, clutch, rear axle, transmission, battery, windshield washer | X | X | X |
| Check all tyre pressures, including spare | X | X | X |
| Lubricate all grease fittings | X | X | X |
| Lubricate all locks and hinges | X | X | X |
| †Lubricate all throttle and choke controls and cables | X | X | X |
| **ENGINE** | | | |
| †Check all drive belts; adjust if required (Renew all belts at 24,000 miles or 24 months) | | X | X |
| †Check all hoses, vacuum, air and water for conditions and tightness | | X | X |
| †Renew all air filter cleaner elements (air pump and carburetter) | | | X |
| †Adjust valve rocker clearances | | X | X |
| †Tighten all manifold nuts | | | X |
| †Visually check exhaust and intake systems for leaks | | X | X |
| Check heating and cooling system for leaks | | X | X |
| †Check restrictor in rocker cover purge line for obstruction | | | X |
| †Check air injection manifold for damage, leaks and security | | | X |
| †Inspect the check and gulp valves, and renew if necessary | | | X |
| †Inspect air pump, correct or replace if necessary | | At 24,000 mile or 24-month intervals | |
| **IGNITION** | | | |
| †Renew spark plugs | | | X |
| †Renew breaker points | | | X |
| †Lubricate distributor cam, breaker pivots and advance mechanism | | X | X |
| †Wipe clean and examine distributor cap for cracks and tracking, carbon brush for sticking and wear | | | X |
| †Renew distributor cap every 24,000 miles or 24 months) | | | |
| †Wipe clean and examine high tension leads for damage, deterioration and security | | | X |
| †(Renew high tension leads every 24,000 miles or 24 months) | | | |
| **FUEL SYSTEM** | | | |
| †Top up carburetter piston damper | | X | X |
| †Renew fuel line filter | | | X |
| ‡Renew charcoal canister | | At 24,000-miles or 24-month intervals | |
| †Check condition of fuel filler cap seal, and renew if necessary | | | X |
| **OSCILLOSCOPE CHECK** | | | |
| †Check/adjust breaker points, resistance and dwell | | X | X |
| †Check/adjust ignition timing and distributor advance characteristics (vacuum disconnected) | | X | X |
| †Check distributor vacuum advance at idle | | X | X |
| †Check/adjust idle speed and mixture setting | | X | X |
| †Check/adjust choke and carburetter fast idle speed setting | | X | X |
| †Power check, engine cylinder comparison | | | X |
| †Check spark plugs (cruise and unload condition); clean, and adjust if necessary | | X | |
| Check charging system output | | X | X |
| **TRANSMISSION** | | | |
| Clean overdrive filter at 24,000-mile or 24-month intervals only | | | X |
| **STEERING AND SUSPENSION** | | | |
| Check for oil leaks | | X | X |
| Check steering joints for security, backlash and gaiter condition | | X | X |
| Check/adjust front wheel alignment | | X | X |
| **BRAKES** | | | |
| Check visually hydraulic pipes and unions for chafing, leaks and corrosion | | X | X |
| Check/adjust brake pedal travel; check hand brake operation | | X | X |
| Inspect brake pads for wear and discs for condition | | X | |
| Inspect brake linings and pads for wear, drums and discs for condition | | | X |
| **ELECTRICAL** | | | |
| Check function of lamps, horns, indicators and windscreen wipers | | X | X |
| Grease battery connections | | X | X |
| Check headlamp beam alignment, and adjust if necessary | | | X |
| Check windscreen wiper blades; replace if necessary | | X | X |
| **EXHAUST, FUEL AND CLUTCH PIPES** | | | |
| †Check visually fuel and clutch pipes and unions for chafing, leaks and corrosion | | X | X |
| †Check exhaust for leakage and security | | X | X |

| | A | B |
|---|---|---|
| **WHEELS AND TYRES** | | |
| Check that tyres comply with manufacturer's specification (63) | | X |
| Check visually and report depth of tread, cuts in tyre fabric, exposure of ply or cord structure, lumps or bulges | X | X |
| Check tightness of road wheel nuts | X | X |
| **BODY** | | |
| Check condition and security of seats and seat belts | X | X |
| Check rear view mirror for looseness, cracks and crazing | X | X |

The schedule on the preceding pages is the minimum service required to maintain your vehicle under normal driving conditions. For other than normal driving conditions, and those caused by seasonal changes, we recommend that you consult your Distributor or Dealer.

## SERVICE

**Service parts and accessories** **Genuine BRITISH LEYLAND and UNIPART parts and accessories are designed and tested for your vehicle and have the full backing of the British Leyland Factory Warranty.**

For more information on **UNIPART** see your British Leyland Distributor or Dealer.

**Genuine British Leyland and UNIPART parts and accessories are supplied in cartons and packs bearing either or both of these symbols.**

**Identification** **When communicating with your Distributor or Dealer always quote the commission and engine numbers. When the communication concerns the transmission units or body details it is necessary to quote also the transmission casing and body numbers.**

**Commission number.** Stamped on a plate secured to the left-hand side of the bonnet lock platform.

**Engine number.** Stamped on a plate secured to the right-hand side of the cylinder block.

**Gearbox number.** Stamped on the left-hand side of the gearbox casing.

**Rear axle number.** Stamped on the front of the left-hand rear axle tube near the spring seating.

**Supplementary tool kit** To supplement the tool kit a waterproof canvas roll containing the following is obtainable from all Distributors. Part No. AKF 1596 should be quoted.

6 spanners:
- $\frac{5}{16}$ in. × $\frac{3}{8}$ in. A.F.
- $\frac{7}{16}$ in. × $\frac{1}{2}$ in. A.F.
- $\frac{1}{2}$ in. × $\frac{9}{16}$ in. A.F.
- $\frac{9}{16}$ in. × $\frac{5}{8}$ in. A.F.
- $\frac{11}{16}$ in. × $\frac{13}{16}$ in. A.F.
- $\frac{3}{4}$ in. × $\frac{7}{8}$ in. A.F.

1 pair 6 in. pliers.
1 7 in. × $\frac{3}{8}$ in. diameter tommy-bar.
1 $\frac{1}{2}$ in. × $\frac{9}{16}$ in. A.F. tubular spanner.
2 screwdrivers.

**British Leyland Motors Inc.**
**600 Willowtree Road, Leonia**
**New Jersey 07605**
*Telephone:* (201) 461/7300 *Telex.* 135491

**British Leyland Motors**
**Canada Limited**
**4445 Fairview Street**
**P.O. Box 5033**
**Burlington · Ontario · Canada**
*Telephone:* (416) 632/3040 *Telex:* 021678

## TUNING MODIFICATIONS

**Full details are given in Tuning Booklet C-AKD 4034 G of varying stages of tune. The booklet also explains the fitting of the following range of parts:**

> **Warning:** The car is delivered from the factory in its standard form tuned to give maximum performance with complete reliability and meeting the U.S. and Canadian emission control and safety legal requirements.
>
> Super-tuned cars should be utilized only in off-road competition driving. Super-tuning, modifications, alterations or additions to the car from its factory-delivered standard form may prohibit its use on public roads.
>
> **REMINDER:** The warranty provides that it is invalidated if the car is used for racing.

**Engine** Half-race and full-race camshafts. Steel timing gears. Special hardened crankshaft. Special racing clearance bearings. Competition cylinder head gasket. High compression flat top pistons, competition inlet and exhaust valves. Bronze valve guides. Competition twin and triple valve spring sets and collars. Competition rocker shaft brackets and distance pieces. Large capacity, deep sump. Competition oil cooler.

**Carburetter** Twin $1\frac{3}{4}$ in. or 2 in. S.U. carburetters and full installation kit. Weber twin-choke carburetters and full installation kit. Long range fuel tank.

**Ignition** Competition distributor with a special advance curve. Racing sparking plugs.

**Transmission** Competition clutch assembly and driven plate. Wide range of axle ratios. Front and rear road springs. Limited slip differential.

**Suspension** Front anti-roll bar and installation kit. Competition setting shock absorbers.

**Wheels and brakes** Wide-rim wire wheels. Competition brake pads and shoes. Fly-off hand brake parts. Brake servo kit.

**Miscellaneous** Passenger foot-rest, lightweight bucket seats. Leather bonnet straps. Plastic headlamp cowl kit. Competition lightweight exhaust system.

*For further details see your Distributor or Dealer or write to:*

**Special Tuning Department**
**Austin Morris Group** BRITISH LEYLAND UK LIMITED
**Body and Assembly Division** MG Plant
ABINGDON-ON-THAMES BERKSHIRE ENGLAND
*Telephone:* Abingdon 251 *Telex:* 83123 *Telegrams:* Emgee, Abingdon

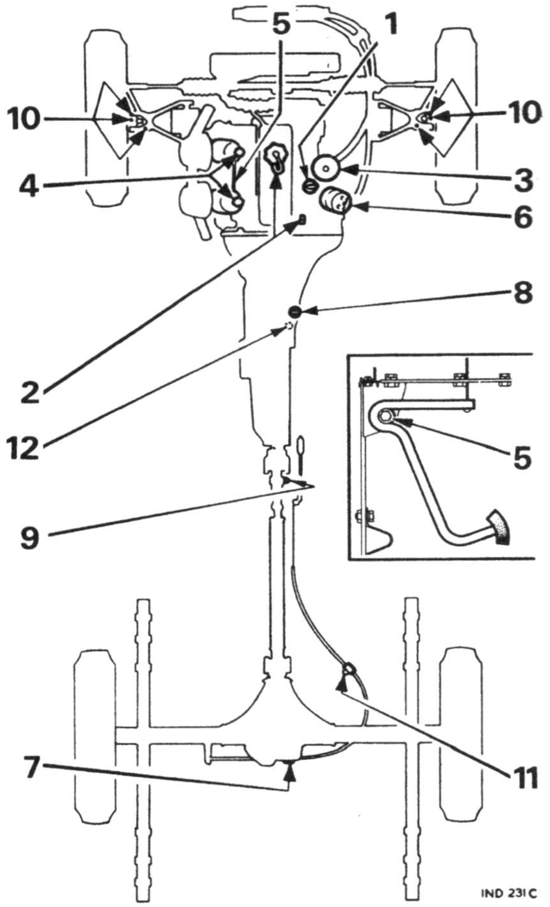


NOTE.—**Ensure that the vehicle is standing on a level surface when checking the oil levels.**

**Weekly**

(1) ENGINE. Check oil level, and top up if necessary.

**Every 6,000 miles or 6 months**

(2) ENGINE. Drain and refill with new oil.
(3) ENGINE OIL FILTER. Remove disposable cartridge, fit new.
(4) CARBURETTERS. Top up carburetter piston dampers.
(5) THROTTLE AND CHOKE. Lubricate throttle and choke control linkages, cables, and accelerator pedal fulcrum.
(6) DISTRIBUTOR. Lubricate all parts as necessary.
(7) REAR AXLE. Check oil level, and top up if necessary.
(8) GEARBOX AND OVERDRIVE. Check oil level and top up if necessary.
(9) PROPELLER SHAFT (1 nipple)
(10) FRONT SUSPENSION (6 nipples) } Give three or four strokes with a grease gun.
(11) HAND BRAKE CABLE (1 nipple)

Lubricate all door, bonnet, boot locks and hinges and hand brake mechanical linkage.

**Every 24,000 miles or 24 months**

(12) GEARBOX WITH OVERDRIVE. Drain, clean overdrive filters—refer to page 39.

**Lubrication service at 3,000 miles or 3 months**

(2) ENGINE. Drain and refill with new oil.
(5) THROTTLE AND CHOKE. Lubricate throttle and choke control linkages, cables, and accelerator pedal fulcrum.
(7) REAR AXLE. Check oil level, and top up if necessary.
(8) GEARBOX AND OVERDRIVE. Check oil level, and top up if necessary.
(9) PROPELLER SHAFT (1 nipple)
(10) FRONT SUSPENSION (6 nipples) } Give three or four strokes with a grease gun.
(11) HAND BRAKE CABLE (1 nipple)

Lubricate all door, bonnet, boot locks and hinges and hand brake mechanical linkage.

**Recommended oils and greases are given in the WORKSHOP MANUAL (Part 3), page 392.**

## EDITOR'S NOTES

**J. Cooling System**

*Anti-freeze solutions*

You can check the specific gravity of the coolant solution by using an anti-freeze hydrometer. With this instrument, a quantity of coolant is sucked in and the position of a glass float indicates the specific gravity of the solution, usually in terms of the temperature at which ice begins to form.

A more thorough job of flushing can be done if the thermostat is removed from its housing and the housing reinstalled before flushing.

**K. Wheels and Tires**

*Tire maintenance*

"Vulcanizing" is a process of tire repair in which a patch is applied under heat and pressure.

**L. Brakes**

*Brake and clutch master cylinder*

Only the brake fluids specified should be used. Other brake fluids may have a lower boiling point and will boil in disc brake systems under hard usage. This can cause temporary malfunction until the fluid has cooled (pedal will feel spongy).

**M. Electrical**

*Battery*

Do not use a naked light, or open flame, when examining the condition of the cells. The reason for this is that hydrogen and oxygen gases are produced under normal charging conditions and the mixture of these two gases in the proper proportions is explosive.

**N. Ignition**

*Checking ignition timing*

The spark timing may be set with a commercially available timing light. A flash occurs every time the No. 1 cylinder fires, illuminating the timing marks at the crankshaft pulley. The engine should be run at the rpm specified in "General Data" and the knurled nut adjusted until the setting corresponds to that listed in "General Data" for the engine type you are working with.

# MGB
# WORKSHOP MANUAL

## PART 3

### SPECIFICATION

Purchasers are advised that the specification details set out in this Manual apply to a range of vehicles and not to any particular vehicle. For the specification of any particular vehicle Purchasers should consult their Distributor or Dealer.

The Manufacturers reserve the right to vary their specifications with or without notice, and at such times and in such manner as they think fit. Major as well as minor changes may be involved in accordance with the Manufacturer's policy of constant product improvement.

Whilst every effort is made to ensure the accuracy of the particulars contained in this Manual, neither the Manufacturer nor the Distributor or Dealer, by whom this Manual is supplied, shall in any circumstances be held liable for any inaccuracy or the consequences thereof.

# INTRODUCTION

## DESCRIPTION

This manual is intended to assist the skilled mechanic in carrying out repairs and replacements in a minimum of time.

References to left- or right-hand side in this manual are made when viewing the car from the rear.

## MANUAL ARRANGEMENT

The first part of the Manual includes the General Data, and Engine Tuning Data.

A Service Tools section and the Recommended Lubricants Chart are featured at the end of the Manual.

The remainder of the manual is divided into sections and each section carries a reference letter that identifies the section with an assembly or a major component. Each section is preceded by a contents page and is sub-divided numerically. The pages and illustrations are numbered consecutively within each section and the section title and letter are shown at the top of each page.

Maintenance items within the Sections should be carried out at the intervals specified in the Passport to Service or Driver's Handbook.

## IMPORTANT **Canada and U.S.A. only**

**On cars fitted with vehicle emission control equipment service operations and adjustments showing this symbol must be followed by an exhaust emission check.**

**Servicing and adjusting vehicle emission control equipment must be carried out in accordance with the instructions given in Workshop Manual Supplement AKD 4957.**

## REPAIRS AND REPLACEMENTS

When replacements are required it is essential that only genuine **British Leyland** and **Unipart** parts are used.

Attention is particularly drawn to the following points concerning repairs and the fitting of replacement parts and accessories:

Safety features embodied in the car may be impaired if other than genuine parts are fitted. In certain territories, legislation prohibits the fitting of parts not to the vehicle manufacturers specification.

Torque wrench setting figures given in the Manual must be strictly adhered to.

Locking devices, where specified, must be fitted. If the efficiency of a locking device is impaired during removal it must be renewed.

Owners purchasing accessories while travelling abroad should ensure that the accessory and its fitted location on the car conform to mandatory requirements existing in their country of origin. **The terms of the Owner's Service Statement may be invalidated by the fitting of other than genuine British Leyland and Unipart parts.**

All **British Leyland** parts and **Unipart** replacements have the full backing of the Owner's Service Statement.

Genuine parts are supplied in cartons bearing one or both of these symbols.

# CONTENTS

# GENERAL DATA

## (18G/18GA)

### ENGINE

| | |
|---|---|
| Type | 18G, 18GA. |
| Number of cylinders | 4. |
| Bore | 3·16 in. (80·26 mm.). |
| Stroke | 3·5 in. (89 mm.). |
| Capacity | 1798 c.c. (109·8 cu. in.). |
| Combustion chamber volume (valves fitted) | 42·5 to 43·5 c.c. (2·59 to 2·65 cu. in.). |
| Firing order | 1, 3, 4, 2. |
| Valve operation | Overhead by push-rod. |
| Compression ratio: H.C. | 8·8 : 1. |
| L.C. | 8·0 : 1. |
| Compression pressure: H.C. | 160 lb./sq. in. (11·25 kg./cm.²). |
| L.C. | 130 lb./sq. in. (9·15 kg./cm.²). |
| Torque: H.C. | 110 lb. ft. (15·2 kg. m.) at 3,000 r.p.m. |
| L.C. | 105 lb. ft. (14·5 kg. m.) at 3,000 r.p.m. |
| Engine idle speed (approx.) | 500 r.p.m. |
| Oversize bore: First | ·010 in. (·254 mm.). |
| Max. | ·040 in. (1·016 mm.). |

**Crankshaft**

| | |
|---|---|
| Main journal diameter | 2·1262 to 2·127 in. (54·01 to 54·02 mm.). |
| Crankpin journal diameter | 1·8759 to 1·8764 in. (47·648 to 47·661 mm.). |
| Crankshaft end-thrust | Taken on thrust washers at centre main bearing. |
| Crankshaft end-float | ·004 to ·005 in. (·10 to ·13 mm.). |

**Main bearings**

| | |
|---|---|
| Number and type | Three thinwall. |
| Material | Steel-backed copper-lead. |
| Length | 1⅛ in. (28·5 mm.). |
| Diametrical clearance | ·001 to ·0027 in. (·0254 to ·068 mm.). |
| Undersizes | —·010, —·020, —·030 and — 040 in. (—·254, —·508, —·762 and —1·016 mm.). |

**Connecting rods**

| | |
|---|---|
| Type | Angular-split big-end, split clamp small-end. |
| Length between centres | 6·5 in. (165·1 mm.). |

**Big-end bearings**

| | |
|---|---|
| Type | Shell. |
| Material | Steel-backed copper-lead. |
| Length | ·995 to 1·005 in. (25·2 to 25·52 mm.). |
| Diametrical clearance | ·001 to ·0027 in. (·0254 to ·068 mm.). |
| Undersizes | —·010, —·020, —·030 and —·040 in. (—·254, —·508, —·762 and —1·016 mm.). |
| End-float on crankpin (nominal) | ·008 to ·012 in. (·20 to ·30 mm.). |

**Pistons**

| | |
|---|---|
| Type | Aluminium solid skirt. |
| Clearance in cylinder: Top | ·0036 to ·0045 in. (·091 to ·121 mm.). |
| Bottom | ·0018 to ·0024 in. (·045 to ·060 mm.). |
| Number of rings | 4 (3 compression, 1 oil control). |
| Width of ring grooves: Top | } ·064 to ·065 in. (1·625 to 1·651 mm.). |
| Second | |
| Third | |
| Oil control | ·1578 to ·1588 in. (4·008 to 4·033 mm.). |
| Gudgeon pin bore | ·7501 to ·7503 in. (19·052 to 19·057 mm.). |

**Piston rings**

*Compression*

| | |
|---|---|
| Type: Top | Parallel } cast iron—molybdenum filled. |
| Second and third | Tapered } cast iron—molybdenum filled. |
| Width: Top / Second and third | ·0615 to ·0625 in. (1·562 to 1·587 mm.). |
| Fitted gap: Top / Second and third | ·012 to ·017 in. (·304 to ·431 mm.). |
| Ring to groove clearance: Top / Second and third | ·0015 to ·0035 in. (·038 to ·088 mm.). |

*Oil control*

| | |
|---|---|
| Type | Slotted scraper. |
| Width | ·1552 to ·1562 in. (3·94 to 3·96 mm.). |
| Fitted gap | ·012 to ·017 in. (·304 to ·431 mm.). |
| Ring to groove clearance | ·0016 to ·0036 in. (·04 to ·09 mm.). |

**Gudgeon pin**

| | |
|---|---|
| Type | Semi-floating. |
| Fit in piston | Free fit at 20° C. (68° F.). |
| Diameter (outer) | ·7499 to ·7501 in. (19·04 to 19·05 mm.). |

**Camshaft**

| | |
|---|---|
| Journal diameters: Front | 1·78875 to 1·78925 in. (45·424 to 45·437 mm.). |
| Centre | 1·72875 to 1·72925 in. (43·910 to 43·923 mm.). |
| Rear | 1·62275 to 1·62325 in. (41·218 to 41·230 mm.). |
| Bearing liner inside diameter (reamed after fitting): Front | 1·79025 to 1·79075 in. (45·472 to 45·485 mm.). |
| Centre | 1·73025 to 1·73075 in. (43·948 to 43·961 mm.). |
| Rear | 1·62425 to 1·62475 in. (41·256 to 41·269 mm.). |
| Diametrical clearance | ·001 to ·002 in. (·0254 to ·0508 mm.). |
| End-thrust | Taken on locating plate. |
| End-float | ·003 to ·007 in. (·076 to ·178 mm.). |
| Cam lift | ·250 in. (6·35 mm.). |
| Drive | Chain and sprocket from crankshaft. |
| Timing chain | $\frac{3}{8}$ in. (9·52 mm.) pitch × 52 pitches. |

**Tappets**

| | |
|---|---|
| Type | Barrel with flat base. |
| Outside diameter | $\frac{13}{16}$ in. (20·64 mm.). |
| Length | 2·293 to 2·303 in. (58·25 to 58·5 mm.). |

**Rocker gear**

*Rocker shaft*

| | |
|---|---|
| Length | 14$\frac{1}{32}$ in. (356 mm.). |
| Diameter | ·624 to ·625 in. (15·85 to 15·87 mm.). |

*Rocker arm*

| | |
|---|---|
| Bore | ·7485 to ·7495 in. (19·01 to 19·26 mm.). |
| Rocker arm bush inside diameter | ·6255 to ·626 in. (15·8 to 15·9 mm.). |
| Ratio | 1·4 : 1. |

**Valves**

| | |
|---|---|
| Seat angle: Inlet and exhaust | 45$\frac{1}{2}$°. |
| Head diameter: Inlet | 1·562 to 1·567 in. (38·67 to 38·80 mm.). |
| Exhaust | 1·343 to 1·348 in. (34·11 to 34·23 mm.). |
| Stem diameter: Inlet | ·3422 to ·3427 in. (8·68 to 8·69mm.). |
| Exhaust | ·3417 to ·3422 in. (8·660 to 8·661 mm.). |
| Stem to guide clearance: Inlet | ·0015 to ·0025 in. (·0381 to ·0778 mm.). |
| Exhaust | ·002 to ·003 in. (·0508 to ·0762 mm.). |
| Valve lift: Inlet and exhaust | ·3645 in. (9·25 mm.). |

**Valve guides**

| | |
|---|---|
| Length: Inlet | $1\frac{5}{8}$ in. (41·275 mm.). |
| Exhaust | $2\frac{13}{64}$ in. (55·95 mm.). |
| Outside diameter: Inlet and exhaust | ·5635 to ·5640 in. (14·30 to 14·32 mm.). |
| Inside diameter: Inlet and exhaust | ·3442 to ·3447 in. (8·73 to 8·74 mm.). |
| Fitted height above head: Inlet and exhaust | $\frac{5}{8}$ in. (15·875 mm.). |
| Interference fit in head: Inlet and exhaust | ·0005 to ·00175 in. (·012 to ·044 mm.). |
| (Later cars) | |
| Length : Inlet | $1\frac{7}{8}$ in. (47·63 mm.). |
| Fitting height above head: Inlet | $\frac{3}{4}$ in. (19 mm.). |

**Valve springs**

| | |
|---|---|
| Free length: Inner | $1\frac{31}{32}$ in. (50·0 mm.). |
| Outer | $2\frac{9}{64}$ in. (54·4 mm.). |
| Fitted length: Inner | $1\frac{7}{16}$ in. (36·5 mm.). |
| Outer | $1\frac{9}{16}$ in. (39·7 mm.). |
| Load at fitted length: Inner | 28 to 32 lb. (12·7 to 14·5 kg.). |
| Outer | 72 lb. (32·7 kg.). |
| Load at top of lift: Inner | 48 to 52 lb. (21·7 to 23·6 kg.). |
| Outer | 117 lb. (53 kg.). |
| Valve crash speed | 6,200 r.p.m. |

**Valve timing**

| | |
|---|---|
| Timing marks | Dimples on camshaft and crankshaft wheels. |
| Rocker clearance: Running | ·015 in. (·38 mm.) cold. |
| Timing | ·055 in. (1·4 mm.). |
| Inlet valve: Opens | 16° B.T.D.C. |
| Closes | 56° A.B.D.C. |
| Exhaust valve: Opens | 51° B.B.D.C. |
| Closes | 21° A.T.D.C. |

**ENGINE LUBRICATION**

| | |
|---|---|
| System | Wet sump, pressure-fed. |
| System pressure: Running | Between 50 and 80 lb./sq. in. (3·51 and 5·6 kg./cm.$^2$). |
| Idling | Between 10 and 25 lb./sq. in. (·7 and 1·7 kg./cm.$^2$). |
| Oil pump | Hobourn-Eaton or eccentric rotor. |
| Capacity | $3\frac{1}{4}$ gal./min. at 2,000 r.p.m. |
| Oil filter | Tecalemit full-flow felt element. |
| By-pass valve opens | 13 to 17 lb./sq. in. (·9 to 1·1 kg./cm.$^2$). |
| Oil pressure relief valve | 70 lb./sq. in. (4·9 kg./cm.$^2$). |
| Relief valve spring: Free length | 3 in. (76·2 mm.). |
| Fitted length | $2\frac{5}{32}$ in. (54·7 mm.). |
| Load at fitted length | 15·5 to 16·5 lb. (7·0 to 7·4 kg.). |

**FUEL SYSTEM**

| | |
|---|---|
| **Carburetters** | Twin S.U. Type HS4. |
| Choke diameter | $1\frac{1}{2}$ in. (38·1 mm.). |
| Jet size | ·090 in. (2·2 mm.). |
| Needles | No. 5 (Standard), No. 6 (Rich), No. 21 (Weak). |
| Piston spring | Red. |
| **Air cleaners** | Cooper paper element. |

**Fuel pump**

| | |
|---|---|
| Type (Early cars) | S.U. electric HP. |
| Minimum flow | 7 gal./hr. (31·8 litres/hr., 8·4 U.S. gal./hr.). |

| | |
|---|---|
| Suction head | 2 ft. 6 in. (76·2 cm.). |
| Delivery head | 4 ft. (122 cm.). |
| Minimum starting voltage | 9·5 volts. |
| Type (Later cars) | S.U. electric AUF 300. |
| Minimum flow | 15 gal./hr. (68·2 litre/hr. 18 U.S. gal./hr.). |
| Suction head | 18 in. (457 mm.). |

## COOLING SYSTEM

| | |
|---|---|
| Type | Pressurized. Pump-impeller- and fan-assisted. |
| Thermostat setting | |
| Standard | 74° C. (165° F.). |
| Hot climate | 74° C. (165° F.). |
| Cold climate | 82° C. (180° F.). |
| Pressure cap | 7 lb. (3·175 kg.). |
| Fan blades | 3 at 24°. |
| Fan belt: Width | $\frac{3}{8}$ in. (9·5 mm.). |
| Outside length | 35$\frac{1}{2}$ in. (90·2 cm.). |
| Thickness | $\frac{5}{16}$ in. (7·9 mm.). |
| Tension | $\frac{1}{2}$ in. (12·8 mm.) movement. |
| Type of pump | Centrifugal. |
| Pump drive | Belt from crankshaft pulley. |

## IGNITION SYSTEM

| | | |
|---|---|---|
| **Coil** | HA.12 (oil-filled). | |
| Resistance at 20° C. (68° F.): Primary winding | 3·1 to 3·5 ohms (cold). | |
| Consumption: Ignition switch on | 3·9 amps. | |
| At 2,000 r.p.m. | 1·4 amps. | |
| **Distributor** | 25D4. | |
| Rotation of rotor | Anti-clockwise. | |
| Cam form | 4-cylinder high-lift. | |
| Cam closed period | 60°±3°. | |
| Cam open period | 30°±3°. | |
| Automatic advance | Centrifugal and vacuum. | |
| | *Serial Number* 40897 (*H.C.*) | *Serial Number* 40916 (*L.C.*) |
| Automatic advance commences | 400 r.p.m. | 400 r.p.m. |
| Maximum advance (crankshaft degrees) | 20° at 2,200 r.p.m. | 24° at 4,400 r.p.m. |
| Vacuum advance (crankshaft degrees) | 20° at 13 in. (33·3 cm.) Hg. | 16° at 12 in. (30 5 cm.) Hg. |
| Decelerating check (crankshaft degrees, engine r.p.m.) | 20° at 2,200 r.p.m. | 24° at 4,400 r.p.m. |
| | 15° at 1,600 r.p.m. | 18° at 3,000 r.p.m. |
| | 9° at 900 r.p.m. | 9° at 1,000 r.p.m. |
| | 6° at 700 r.p.m. | 8° at 800 r.p.m. |
| | 6° at 600 r.p.m. | 6° at 600 r.p.m. |
| Contact point gap setting | ·014 to ·016 in. (·35 to ·40 mm.). | |
| Breaker spring tension | 18 to 24 oz. (510 to 680 gm.). | |
| Condenser capacity | ·18 to ·24 mF. | |
| Timing marks | Pointer on timing chain case and notch on crankshaft pulley. | |
| Static ignition timing: H.C. | 10° B.T.D.C. (98/100-octane fuel). | |
| L.C. | 8° B.T.D.C. (95/97-octane fuel). | |
| Stroboscopic ignition timing: H.C. | 14° B.T.D.C. at 600 r.p.m. | |
| L.C. | 12° B.T.D.C. at 600 r.p.m. | |
| Suppressors | Lucas W55 Type L.2. Fitted in plug leads. | |
| Sparking plugs | Champion N-9Y. | |
| Size | 14 mm. $\frac{3}{4}$ in. (19·0 mm.) reach. | |
| Gap | ·024 to ·026 in. (·625 to ·660 mm.). | |

## CLUTCH

| | |
|---|---|
| Make and type | Borg & Beck 8 in. DS.G diaphragm spring. |
| Diaphragm spring colour | Dark blue. |
| Clutch plate diameter | 8 in. (20·32 cm.). |
| Facing material | Wound yarn. |
| Number of damper springs | 6. |
| Damper spring load | 110 to 120 lb. (49·8 to 54·3 kg.). |
| Damper spring colour | Black/light green. |
| Clutch release bearing | Graphite (MY3D). |
| Clutch fluid | Lockheed Disc Brake Fluid (Series II). |

## GEARBOX AND OVERDRIVE

| | | *Overdrive* |
|---|---|---|
| Number of forward gears | 4. | |
| Gearbox ratios: Reverse | 4·76 : 1. | |
| First | 3·64 : 1. | |
| Second | 2·21 : 1. | |
| Third | 1·37 : 1. | |
| Fourth | 1·00 : 1. | |
| Overdrive ratio | ·802 : 1. | |
| Overall gear ratios: Reverse | 18·588 : 1. | |
| First | 14·214 : 1. | |
| Second | 8·656 : 1. | |
| Third | 5·369 : 1. | 4·306 : 1. |
| Fourth | 3·909 : 1. | 3·135 : 1. |
| Top gear speed per 1,000 r.p.m.: Standard | 17·9 m.p.h. (27·3 km.p.h.). | |
| Overdrive | 22·3 m.p.h. (35·5 km.p.h.). | |
| Speedometer gear ratio: Standard | 9/28. | |
| Overdrive | 5/16. | |
| Synchromesh hub springs: Free length | ½ in. (12·7 mm.). | |
| Fitted length | 5/16 in. (7·9 mm.). | |
| Load at fitted length | 4 to 5 lb. (1·8 to 2·2 kg.). | |
| Shaft and reverse plunger detent springs: Free length | 1 3/16 in. (30·16 mm.). | |
| Fitted length | ¾ in. (19·0 mm.). | |
| Load at fitted length | 18 to 20 lb. (8·16 to 9·07 kg.). | |
| Reverse plunger spring: Free length | 1 in. (25·4 mm.). | |
| Fitted length | 13/16 in. (20·63 mm.). | |
| Load at fitted length | 91½ to 92½ lb. (41·4 to 41·9 kg.). | |
| Mainshaft second and third gear end-float | ·004 to ·006 in. (·10 to ·15 mm.). | |
| Laygear end-float | ·002 to ·003 in. (·05 to ·08 mm.). | |
| Overdrive | | |
| Pump spring: Free length | 2·000 in. (50·8 mm.). | |
| Rate | 11 lb. in. (12·7 kg. cm.). | |
| Clutch spring: Free length | 1·510 in. (38·3 mm.). | |
| Rate | 154 lb. in. (178 kg. cm.). | |

## PROPELLER SHAFT

| | |
|---|---|
| Type | Open tubular, telescopic. |
| Universal joints | Hardy Spicer needle roller. |
| Angular movement | 18° to 20°. |
| Overall length: Fully extended: Standard | 30¾ in. (78·1 cm.). |
| Overdrive | 31⅞ in. (81 cm.). |
| Fully compressed: Standard | 29 1/16 in. (74 cm.). |
| Overdrive | 30 3/16 in. (76·5 cm.). |
| Length of shaft assembly: Standard | 25 11/32 in. (64·3 cm.). |
| Overdrive | 26 15/32 in. (67 cm.). |
| Tube diameter | 2 in. (50·8 mm.). |

## REAR AXLE

| | |
|---|---|
| Type | Hypoid, three-quarter-floating. |
| Ratio | 3·909 : 1 (11/43). |
| Differential bearing preload | ·002 in. (·05 mm.) 'nip' per bearing. |
| Pinion bearing preload | 7 to 9 lb. in. (·8 to 1·0 kg. m.). |
| Backlash adjustment: Crown wheel | Shims. |
| Pinion | Head washer. |

## STEERING

| | |
|---|---|
| Type | Rack and pinion. |
| Steering-wheel diameter | $16\frac{1}{2}$ in. (419·10 mm.). |
| Turns—lock to lock | 2·93. |
| Turning circle | 32 ft. (9·75 m.). |
| Universal joint | Hardy Spicer KO518, GB166. |
| Pinion end-float | ·002 to ·005 in. (·05 to ·12 mm.). |
| Damper end-float | ·0005 to ·003 in. (·012 to ·076 mm.) (unladen). |
| Toe-in | $\frac{1}{16}$ to $\frac{3}{32}$ in. (1·5 to 2·3 mm.) (unladen). |
| Angle of outer wheel with inner wheel at 20° | $19° \pm 1°$. |

## FRONT SUSPENSION

| | |
|---|---|
| Type | Independent. Coil spring and wishbone. |
| Spring: Coil diameter (mean) | 3·238 in. (82·2 mm.). |
| Free height | $9{\cdot}9 \pm \frac{1}{16}$ in. ($251{\cdot}4 \pm 1{\cdot}5$ mm.). |
| Static length at 1,030 lb. (467·2 kg.) load | $7 \pm \frac{1}{32}$ in. ($178 \pm {\cdot}8$ mm.). |
| Number of free coils | 7·5. |
| Camber angle | Nominal 1° positive ($+\frac{1}{4}°$, $-1\frac{1}{4}°$) $= 1\frac{1}{4}°$ positive, $\frac{1}{4}°$ negative (unladen). |
| Castor angle | Nominal 7° ($+\frac{1}{4}°$, $-2°$) $= 5°$ to $7\frac{1}{4}°$ (unladen). |
| King pin inclination | Nominal 8° ($+1°$, $-\frac{3}{4}°$) $= 7\frac{1}{4}°$ to 9° (unladen). |
| Dampers | Armstrong piston type. |
| Arm centres | 8 in. (203·2 mm.). |
| Wheel bearing end-float | ·002 to ·004 in. (·05 to ·10 mm.). |

## REAR SUSPENSION

| | |
|---|---|
| Type | Semi-elliptic leaf spring. |
| Number of spring leaves | 5+bottom plate. Interleaving 1/2, 2/3, 3/4. |
| Width of spring leaves | $1\frac{3}{4}$ in. (44·4 mm.). |
| Gauge of leaves | |
| Early cars | $\frac{7}{32}$ in. (5·56 mm.). |
| Later cars | 3 at $\frac{7}{32}$ in. (5·6 mm.), 3 at $\frac{3}{16}$ in. (4·8 mm.). |
| Working load ($\pm$15 lb. [7 kg.]) | |
| Early cars | 400 lb. (181·5 kg.). |
| Later cars | 450 lb. (204·1 kg.). |
| Dampers | Armstrong piston type. |
| Arm centres | $5\frac{1}{4}$ in. (133 mm.). |

## ELECTRICAL EQUIPMENT

| | |
|---|---|
| System | 12-volt, positive earth. |
| Charging system | Current/voltage control. |
| Batteries—two 6-volt | Lucas SG9E or STGZ9E. Lucas BT9E or BTZ9E—later cars. |
| Capacity: 10-hour rate | 51 amp.-hr. |
| 20-hour rate | 58 amp.-hr. |
| Plates per cell | 9. |
| Electrolyte to fill one cell | 1 pint (570 c.c.: 1·2 U.S. pints). |

| | |
|---|---|
| **Regulator** | RB.340. |
| Voltage setting at 3,000 r.p.m.: 10° C. (50° F.) | 14·9 to 15·5 volts. |
| 20° C. (68° F.) | 14·7 to 15·3 volts. |
| 30° C. (86° F.) | 14·5 to 15·1 volts. |
| 40° C. (104° F.) | 14·3 to 14·9 volts. |
| Current setting at 4,000 r.p.m. | 22 amps. |
| Cut-out relay: Cut-in voltage | 12·7 to 13·3 volts. |
| Drop-off voltage | 9·5 to 11·0 volts. |
| Reverse current | 3·0 to 5·0 amps. |
| Current regulator | 22±1 amp. |
| **Dynamo** | C40/1. 12-volt two-brush. |
| Maximum output | 22 amps. |
| Cut-in speed | 1,585 r.p.m. at 13·0 volts. |
| Field resistance | 6·0 ohms. |
| Brush spring tension | 22 to 25 oz. (623 to 708 gm.). |
| Drive | Belt from crankshaft. |
| Drive adjustment | Swinging link on dynamo. |
| **Starter motor** | M418G four-brush inertia type. |
| Lock torque | 15 lb. ft. (2·1 kg. m.) at 425 amps. |
| Torque at 1,000 r.p.m. | 8 lb. ft. (1·11 kg. m.) at 250 to 270 amps. |
| Brush spring tension | 32 to 40 oz. (907 to 1133 gm.). |
| Starter gear ratio | 13·3 : 1. |
| **Wiper motor** | DR.3A single speed. |
| Drive to wheelboxes | Rack and cable. |
| Armature end-float | ·008 to ·012 in. (·20 to ·30 mm.). |
| Running current | 2·7 to 3·4 amps. |
| Wiping speed | 45 to 50 cycles per minute. |
| **Horns** | |
| Type | 9H 12-volt. |
| Maximum current consumption | 3½ amps. |

## BRAKES

| | |
|---|---|
| Type | Lockheed hydraulic. Disc front, drum rear. |
| Brake fluid | Lockheed Disc Brake Fluid (Series II). |
| **Front** | |
| Disc diameter | 10¾ in. (27·3 cm.). |
| Pad material | Don 55—FF. |
| Swept area | 203·2 sq. in. (1311 cm.$^2$). |
| **Rear** | |
| Drum diameter | 10 in. (25·4 cm.). |
| Lining material | Don 24—FE. |
| Swept area | 106·8 sq. in. (683·9 cm.$^2$). |
| Lining dimensions | $9\frac{7}{16} \times 1\frac{3}{4} \times \frac{3}{16}$ in. (240×44·4×4·76 mm.). |

## WHEELS

| | |
|---|---|
| **Type** | Ventilated disc, 4-stud fixing. Wire (optional). |
| **Size: Disc** | 4J×14. |
| **Wire** | 4½J×14. |

**TYRES**

Standard: Size .. .. .. .. .. .. .. 5·60—14 (tubed).
Rolling radius .. .. .. .. .. 11·65 in. (29·5 cm.) at 30 m.p.h. (48 km.p.h.).
Optional: Size .. .. .. .. .. .. .. 155—14.

**TYRES PRESSURES**

| Condition | Standard tyres 5·60—14 | | Optional tyres 155—14 | |
|---|---|---|---|---|
| | *Front* | *Rear* | *Front* | *Rear* |
| Normal car weight | 18 lb./sq. in. (1·3 kg./cm.²) | 18 lb./sq. in. (1·3 kg./cm.²) | 21 lb./sq. in. (1·5 kg./cm.²) | 24 lb./sq. in. (1·7 kg./cm.²) |
| Sustained speeds in excess of 90 m.p.h. (145 km.p.h.) | 24 lb./sq. in. (1·7 kg./cm.²) | 24 lb./sq. in. 1·7 kg./cm.²) | 27 lb./sq. in. (1·9 kg./cm.²) | 31 lb./sq. in. (2·2 kg./cm.²) |

**NOTE: Rear tyre pressures may be increased by 2 lb./sq. in. (·14 kg./cm.²) with advantage when towing with a laden boot.**

**CAPACITIES**

Fuel tank: Early cars .. .. .. .. .. .. 10 gallons (45·4 litres, 12 U.S. gallons).
Later cars .. .. .. .. .. .. 12 gallons (54 litres, 14 U.S. gallons).
Cooling system .. .. .. .. .. .. .. 9½ pints (5·4 litres, 11·4 U.S. pints).
Heater .. .. .. .. .. .. .. .. ½ pint (·28 litre, ·6 U.S. pint).
Engine sump .. .. .. .. .. .. .. 7½ pints (4·26 litres, 9 U.S. pints).
Oil cooler .. .. .. .. .. .. .. .. ¾ pint (·42 litre, ·9 U.S. pint).
Gearbox .. .. .. .. .. .. .. .. 4½ pints (2·56 litres, 5·6 U.S. pints).
Gearbox and overdrive .. .. .. .. .. .. 5¾ pints (3·36 litres, 6 U.S. pints).
Rear axle .. .. .. .. .. .. .. .. 2¼ pints (1·28 litres, 2·75 U.S. pints).
Steering rack .. .. .. .. .. .. .. ⅓ pint (·19 litre, ·39 U.S. pint).

**DIMENSIONS**

Overall length .. .. .. .. .. .. .. 12 ft. 8½ in. (3·874 m.).
Overall length (with over-riders) .. .. .. .. 12 ft. 9$\frac{3}{16}$ in. (3·897 m.).
Overall width .. .. .. .. .. .. .. 4 ft. 11$\frac{15}{16}$ in. (152·3 cm.).
Overall height (hood erected) .. .. .. .. .. 4 ft. 1⅜ in. (125·4 cm.).
Ground clearance (minimum) .. .. .. .. .. 5 in. (12·7 cm.).
Wheelbase .. .. .. .. .. .. .. 7 ft. 7 in. (231·1 cm.).
Track: Front (disc wheels) .. .. .. .. .. 4 ft. 1 in. (124·4 cm.).
Rear (disc wheels) .. .. .. .. .. 4 ft. 1¼ in. (125 cm.).
Front (wire wheels) .. .. .. .. .. 4 ft. 1¼ in. (125 cm.).
Rear (wire wheels) .. .. .. .. .. 4 ft. 1¼ in. (125 cm.).

**WEIGHTS**

Unladen weight .. .. .. .. .. .. .. 1,920 lb. (871 kg.).
Engine (dry, with clutch) .. .. .. .. .. 358 lb. (163·3 kg.) approx.
Gearbox .. .. .. .. .. .. .. .. 78 lb. (35·5 kg.) approx.
Rear axle: Disc wheels .. .. .. .. .. .. 117½ lb. (53·26 kg.) approx.
Wire wheels .. .. .. .. .. .. 123 lb. (55·79 kg.) approx.

## TORQUE WRENCH SETTINGS

### Engine

| | |
|---|---|
| Main bearing nuts | 70 lb. ft. (9·7 kg. m.). |
| Flywheel set screws | 40 lb. ft. (5·5 kg. m.). |
| Gudgeon pin clamp bolt | 25 lb. ft. (3·4 kg. m.). |
| Big-end bolts | 35 to 40 lb. ft. (4·8 to 5·5 kg. m.). |
| Cylinder head nuts | 45 to 50 lb. ft. (6·2 to 6·9 kg. m.). |
| Rocker bracket nuts | 25 lb. ft. (3·4 kg. m.). |
| Oil pump to crankcase | 14 lb. ft. (1·9 kg. m.). |
| Sump to crankcase | 6 lb. ft. (·8 kg. m.). |
| Cylinder side cover screws | 2 lb. ft. (·28 kg. m.). |
| Second type—deep pressed cover | 5 lb. ft. (·7 kg. m.). |
| Timing cover—$\frac{1}{4}$ in. screws | 6 lb. ft. (·8 kg. m.). |
| Timing cover—$\frac{5}{16}$ in. screws | 14 lb. ft. (1·9 kg. m.). |
| Rear plate—$\frac{5}{16}$ in. screws | 20 lb. ft. (2·8 kg. m.). |
| Rear plate—$\frac{3}{8}$ in. screws | 30 lb. ft. (4·1 kg. m.). |
| Water pump to crankcase | 17 lb. ft. (2·4 kg. m.). |
| Water outlet elbow nuts | 8 lb. ft. (1·1 kg. m.). |
| Rocker cover nuts | 4 lb. ft. (·56 kg. m.). |
| Manifold nuts | 15 lb. ft. (2·1 kg. m.). |
| Oil filter centre-bolt | 15 lb. ft. (2·1 kg. m.). |
| Clutch to flywheel | 25 to 30 lb. ft. (3·4 to 4·1 kg. m.). |
| Carburetter stud nuts | 2 lb. ft. (·28 kg. m.). |
| Distributor clamp bolt (nut trapped) | 4·16 lb. ft. (·57 kg. m.). |
| Distributor clamp nut (bolt trapped) | 2·5 lb. ft. (·35 kg. m.). |
| Fan blade fixing screws | 7·3 to 9·3 lb. ft. (1 to 1·3 kg. m.). |

### Rear Axle

| | |
|---|---|
| Crown wheel to differential carrier | 55 to 60 lb. ft. (7·6 to 8·3 kg. m.). |
| Differential bearing cap | 60 to 65 lb. ft. (8·3 to 8·9 kg. m.). |
| Pinion bearing nut | 135 to 140 lb. ft. (18·6 to 19·3 kg. m.). |
| Rear brake adjuster securing nuts | 5 to 7 lb. ft. (·69 to ·97 kg. m.). |
| Bearing retaining nut | 180 lb. ft. (24·8 kg. m.). |

### Rear suspension

| | |
|---|---|
| Rear shock absorber bolts | 55 to 60 lb. ft. (7·6 to 8·3 kg. m.). |

### Front suspension

| | |
|---|---|
| Front shock absorber bolts | 43 to 45 lb. ft. (5·9 to 6·2 kg. m.). |
| Brake disc to hub | 40 to 45 lb. ft. (5·5 to 6·2 kg. m.). |
| Brake calliper mounting | 40 to 45 lb. ft. (5·5 to 6·2 kg. m.). |
| Bearing retaining nut | 40 to 70 lb. ft. (5·5 to 9·7 kg. m.). |
| Cross-member to body | 54 to 56 lb. ft. (7·5 to 7·7 kg. m.). |

### Steering

| | |
|---|---|
| Steering arm bolts | 60 to 65 lb. ft. (8·3 to 8·9 kg. m.). |
| Steering-wheel nut | 41 to 43 lb. ft. (5·7 to 5·9 kg. m.). |
| Steering tie-rod locknut | 33·3 to 37·5 lb. ft. (4·6 to 5·2 kg. m.). |
| Steering lever ball joint nut | 34 to 35 lb. ft. (4·7 to 4·8 kg. m.). |
| Steering universal joint bolt | 20 to 22 lb. ft. (2·8 to 3·0 kg. m.). |

### Road wheels

| | |
|---|---|
| Road wheel nuts | 60 to 65 lb. ft. (8·3 to 9 kg. m.). |

# GENERAL DATA

## (18GB)

**The following information is applicable to the 18GB-engined cars and should be used in conjunction with the preceding specification for the 18G/18GA-engined car.**

### ENGINE

| | |
|---|---|
| Type | 18GB. |
| **Main bearings** | |
| Number and type | 5 thin-wall. |
| Length: Front, centre and rear | $1\frac{1}{8}$ in. (28·5 mm.). |
| Intermediate | $\frac{7}{8}$ in. (22·23 mm.). |
| **Connecting rods** | |
| Type | Angular-split big-end, bushed small-end. |
| **Big-end bearings** | |
| Length | ·775 to ·785 in. (19·68 to 19·94 mm.). |
| **Pistons** | |
| Gudgeon pin bore | ·8126 to ·8129 (20·610 to 20·167 mm.). |
| Clearance in cylinder skirt: Top | ·0012 to ·0006 in. (·030 to ·015 mm.). |
| Bottom | ·0033 to ·0021 in. (·080 to ·050 mm.). |
| **Gudgeon pin** | |
| Type | Fully floating. |
| Fit in piston | ·0001 to ·00035 in. (·0025 to ·007 mm.). |
| Diameter (outer) | ·8124 to ·8127 (20·608 to 20·615 mm.). |

### FUEL SYSTEM

| | |
|---|---|
| Carburetter needles | FX (Standard); No. 6 (Rich); No. 21 (Weak). |

### REAR AXLE (GT)

| | |
|---|---|
| Type | Hypoid, semi-floating. |
| Ratio | 3·909 : 1 (11·43). |
| Differential bearing preload | ·002 in. (·05 mm.) nip per bearing. |
| Pinion bearing preload. New collapsible spacer only | 14 to 18 lb. in. (·161 to ·207 kg. m.). |
| Pinion bearing preload. Oil seal change spacer collapsed | Adjust to preload, refer Section Ha.4. |
| Backlash adjustment: | |
| Crown wheel | Shims. |
| Pinion | Head washer. |

### FRONT SUSPENSION (GT)

| | |
|---|---|
| Spring coil diameter (mean) | 3·28 in. (83·3 mm.). |
| Free height | $9{\cdot}1 \pm \frac{1}{16}$ in. ($231 \pm 1{\cdot}6$ mm.). |
| Number of free coils | 7·2. |
| Static length at load of 1,193 lb. (541·5 kg.)±20 lb. (9·1 kg.) | 6·6 in. (168 mm.). |

### REAR SUSPENSION (GT)

| | |
|---|---|
| Spring detail | |
| Working load (±15 lb. 7 kg.) | 510 lb. (321·6 kg.). |

### WHEELS (GT)

| | |
|---|---|
| Size: Disc | 5J×14 in. |

**TYRES (GT)**

Optional:
Size and type—Early cars .. .. .. .. .. 165—14 SP41.
Later cars .. .. .. .. .. 165—14 SP68.
Standard tyres:
Pressures (set cold):
Front .. .. .. .. .. .. .. 20 lb./sq. in. (1·4 kg./cm.$^2$).
Rear .. .. .. .. .. .. .. 24 lb./sq. in. (1·7 kg./cm.$^2$).
Sustained speeds in excess of 90 m.p.h. (145 km.p.h.):
Front .. .. .. .. .. .. .. 26 lb./sq. in. (1·8 kg./cm.$^2$).
Rear .. .. .. .. .. .. .. 30 lb./sq. in. (2·1 kg./cm.$^2$).
Optional tyres:
Pressures (set cold):
Front .. .. .. .. .. .. .. 21 lb./sq. in. (1·5 kg./cm.$^2$).
Rear .. .. .. .. .. .. .. 24 lb./sq. in. (1·7 kg./cm.$^2$).
Sustained speeds in excess of 90 m.p.h. (145 km.p.h.):
Front .. .. .. .. .. .. .. 28 lb./sq. in. (2·0 kg./cm.$^2$).
Rear .. .. .. .. .. .. .. 31 lb./sq. in. (2·2 kg./cm.$^2$).

**CAPACITIES**

Rear axle (semi-floating type) .. .. .. .. .. 1½ pints (·85 litre, 2 U.S. pints).

**DIMENSIONS**

Overall height (GT) .. .. .. .. .. .. 4 ft. 1¾ in. (126·3 cm.).

**WEIGHTS**

Unladen weight (GT) .. .. .. .. .. .. 2,190 lb. (993 kg.).

**TORQUE WRENCH SETTINGS**

**Rear Axle**

Half-shaft nut (semi-floating axle) .. .. .. .. 150 lb. ft. (20·75 kg. m.).
Differential bearing cap bolts.. .. .. .. .. 50 to 55 lb. ft. (6·9 to 7·6 kg. m.).
Crown wheel bolts .. .. .. .. .. .. 60 to 65 lb. ft. (8·3 to 8·9 kg. m.).
Pinion nut, new spacer only .. .. .. .. .. 180 to 220 lb. ft. (28·89 to 30·42 kg. m.).
Pinion nut, oil seal change .. .. .. .. .. Adjust to preload. See Section Ha.
Axle shaft nut .. .. .. .. .. .. .. 150 lb. ft. (20·6 kg. m.) and aligned to next split pin hole.

# GENERAL DATA

## (GHN4/GHD4 CARS, 18GD ENGINE—GHN5/GHD5 CARS, 18GG ENGINE)

**The following information is applicable to the GHN4 and GHD4 series cars fitted with 18GD engines and GHN5/GHD5 cars fitted with 18GG engines and should be used in conjunction with the preceding specifications.**

### ENGINE

| | |
|---|---|
| Type | 18GD—18GG. |

**Pistons**

| | |
|---|---|
| Type | Aluminium solid skirt, three rings. |
| Clearance of skirt in cylinder: Top | ·0021 to ·0033 in. (·053 to ·084 mm.). |
| Bottom | ·006 to ·0012 in. (·015 to ·030 mm.). |
| Number of rings | Three; 2 compression, 1 oil control. |
| Width of ring grooves: Compression | ·0064 to ·065 in. (1·625 to 1·651 mm.). |
| Oil control | ·1578 to ·1588 in. (4·01 to 4·033 mm.). |

**Piston rings**

| | |
|---|---|
| Type: Compression: Top | Plain, sintered alloy |
| Second | Tapered, sintered alloy—marked 'TOP'. |
| Oil control | Two chrome-faced rings with expander, Apex. |
| Fitted gap: Compression | ·012 to ·022 in. (·305 to ·600 mm.). |
| Oil control | ·015 to ·045 in. (·38 to 1·14 mm.). |
| Width: Compression | ·0615 to ·0625 in. (1·56 to 1·59 mm.). |
| Oil control | ·152 to ·158 in. (3·86 to 4·01 mm.). |
| Thickness: Compression: Top | ·124 to ·127 in. (3·14 to 3·22 mm.). |
| Second | ·104 to ·111 in. (2·64 to 2·81 mm.). |

**Gudgeon pin**

| | |
|---|---|
| Type | Press fit in connecting rod. |
| Diameter | ·8125 to ·8127 in. (20·63 to 20·64 mm.). |
| Fit in: Piston | Hand push fit at 16° C. (60° F.). |
| Small end | 12 lb. ft. (1·7 kg. m.) minimum using 18G 1150 and adaptor type C. |

**Connecting rods**

| | |
|---|---|
| Type | Horizontal split big-end, solid small-end. |
| Length between centres | 6·5 in. (165·1 mm.). |
| Locking method big-end | Multi-sided nut. |

### COOLING SYSTEM

| | |
|---|---|
| Fan belt: Width | $\frac{15}{32}$ in. (11·9 mm.). |
| Length | 35½ in. (90·2 cm.) at ⅜ in. width equivalent. |
| Thickness | $\frac{27}{64}$ in. (10·7 mm.). |
| Tension | ½ in. (12·8 mm.) movement at midway of longest run. |

Thermostat:

| | |
|---|---|
| Normal | 82° C. (180° F.). |
| Hot climates | 74° C. (165° F.). |
| Cold climates | 88° C. (190° F.). |

### FUEL SYSTEM

| | |
|---|---|
| Carburetter needles | FX (Standard), No. 6 (Rich), GZ (Weak). |

## GENERAL DATA—*continued*

### (GHN4/GHD4 CARS, 18GD ENGINE—GHN5/GHD5 CARS, 18GG ENGINE)—*continued*

**CLUTCH**

| | |
|---|---|
| Fluid | Lockheed Disc Brake Fluid (Series 329). |

**GEARBOX AND OVERDRIVE**

| | | |
|---|---|---|
| Number of forward gears | 4. | |
| Synchromesh | All forward gears. | |
| Gearbox ratios: Reverse | 3·095 : 1. | |
| First | 3·44 : 1. | |
| Second | 2·167 : 1. | |
| Third | 1·382 : 1. | |
| Fourth | 1·000 : 1. | |
| Overdrive: Type | Laycock L.H. | |
| Ratio | ·82 : 1. | |
| Overall gear ratios: Reverse | 12·098 : 1. | |
| First | 13·45 : 1. | |
| Second | 8·47 : 1. | *Overdrive* |
| Third | 5·4 : 1 | 4·43 : 1. |
| Fourth | 3·909 : 1. | 3·2 : 1. |
| Top gear speed per 1,000 r.p.m.: Standard | 18 m.p.h. (29 km.p.h.). | |
| Overdrive | 22 m.p.h. (35 km.p.h.). | |
| Speedometer gear ratio: Standard | 10 : 26. | |
| Overdrive | 8 : 21. | |
| Synchromesh hub springs: Free length | ·72 in. (18·3 mm.). | |
| Fitted length | ·385 in. (9·8 mm.). | |
| Load at fitted length | 5·5 to 6 lb. (2·5 to 2·7 kg.). | |
| First and third speed gear end-float | ·005 to ·008 in. (·13 to ·20 mm.). | |
| Second gear end-float | ·005 to ·008 in. (·13 to ·20 mm.). | |
| Laygear end-float | ·002 to ·003 in. (·05 to ·08 mm.). | |

**AUTOMATIC GEARBOX**

| | |
|---|---|
| Torque converter ratio | 2·2 : 1 to 1 : 1. |
| Gearbox ratios: Reverse | 2·09 : 1. |
| First | 2·39 : 1. |
| Second | 1·45 : 1. |
| Top | 1·00 : 1. |
| Overall gear ratio (converter at 1 : 1): Reverse | 8·17 : 1. |
| First | 9·34 : 1. |
| Second | 5·668 : 1. |
| Top | 3·909 : 1. |
| Speedometer gear ratio | 10 : 26. |

**REAR AXLE**

| | |
|---|---|
| Type | Hypoid; semi-floating. |
| Ratio | 3·909 : 1 (11/43). |
| Differential bearing preload | ·002 in. (·05 mm.) nip per bearing. |
| Pinion bearing preload | Refer to Section Ha.4. |
| Backlash adjustment: Crown wheel | Shims. |
| Pinion | Head washer. |

## (GHN4/GHD4 CARS, 18GD ENGINE—GHN5/GHD5 CARS, 18GG ENGINE)—*continued*

**ELECTRICAL**

| | | |
|---|---|---|
| System | 12-volt, NEGATIVE earth. | |
| **Batteries** | | *(Later cars)* |
| Type (two 6-volt) | Lucas BT9E or BTZ9E | Lucas CA9E. |
| Capacity: 10-hour rate | 51 amp.-hour | 53 amp.-hour. |
| 20-hour rate | 58 amp.-hour | 60 amp.-hour. |
| Plates per cell | 9 | 9. |
| Electrolyte to fill one cell | 1 pint (570 c.c., 1·2 U.S. pints). | |
| **Control box** | | |
| Type | 4TR Electronic. | |
| Voltage setting | 14·3 to 14·7 volts at 5,000 r.p.m. | |
| **Alternator** | **18GD engine** | **18GG engine** |
| Type | Lucas 16AC | Lucas 16ACR |
| Output at 14 volts and 6,000 r.p.m. | 34 amps. | *data as 16AC.* |
| Maximum permissible rotor speed | 12,500 r.p.m. | |
| Pulley ratio—alternator/crankshaft | 1·795 : 1. | |
| Stator phases | 3. | |
| Rotor poles | 12. | |
| Rotor winding resistance | 4·33 ohms ±5% at 20° C. (68° F.). | |
| Brush length (new) | ½ in. (12·6 mm.). | |
| Brush spring tension | 7 to 10 oz. (198-283 gm.) with brush face flush with brush box. | |
| **Starter** | | |
| Type | Lucas M418G Pre-engaged. | |
| Light running current | 70 amp. at 5,800 to 6,500 r.p.m. | |
| Lock torque | 17 lb. ft. (2·35 kg. m.) at 465 amps. | |
| Torque at 1,000 r.p.m. | 7 lb. ft. (·97 kg. m.) at 260 amps. | |
| Solenoid: Closing coil resistance | ·13 to ·15 ohm. | |
| Hold-on coil resistance | ·63 to ·73 ohm. | |
| Brush spring tension | 36 oz. (1·02 kg.). | |
| **Starter motor relay** | | |
| Type | Lucas Model 6RA-33243. | |
| Winding resistance | 76 ohms. | |
| Bobbin core to underside of armature air gap: | | |
| Contacts open | 0·030±0·005 in. (0·76±0·13 mm.). | |
| Contacts closed | 0·010±0·003 in. (0·25±0·08 mm.). | |
| Cut-in voltage | 4·0 to 7·5 volts. | |
| Drop-off voltage | 5 volts (maximum). | |
| **Wiper motor** | | |
| Type | Lucas, 14W (two-speed). | |
| Armature end-float | ·004 to ·008 in. (·1 to ·21 mm.). | |
| Light running current: Normal speed | 1·5 amp. | |
| High speed | 2·0 amp. | |
| Light running speed: Normal speed | 46 to 52 r.p.m. | |
| High speed | 60 to 70 r.p.m. | |

**BRAKES**

| | |
|---|---|
| Brake fluid | Lockheed Disc Brake Fluid (Series 329). |

**CAPACITIES**

| | |
|---|---|
| Gearbox .. .. .. .. .. .. .. .. | 5¼ pints (3 litres, 6 U.S. pints). |
| Gearbox and overdrive .. .. .. .. .. .. | 6 pints (3·4 litres, 7 U.S. pints). |
| Automatic gearbox .. .. .. .. .. .. | 10½ pints (6 litres, 12·7 U.S. pints). |
| Rear axle .. .. .. .. .. .. .. .. | 1½ pints (·85 litres, 2 U.S. pints). |

**TYRE PRESSURES**

| **Tourer** | **5·60—14S Cross-ply tyres** | | **155SR—14 Radial-ply tyres** | |
|---|---|---|---|---|
| | *Front* | *Rear* | *Front* | *Rear* |
| Normal car weight | 19 lb./sq. in. (1·34 kg./cm.$^2$) | 22 lb./sq. in. (1·55 kg./cm.$^2$) | 21 lb./sq. in. (1·48 kg./cm.$^2$) | 24 lb./sq. in. (1·69 kg./cm.$^2$) |
| Maximum weight | 19 lb./sq. in. (1·34 kg./cm.$^2$) | 24 lb./sq. in. (1·69 kg./cm.$^2$) | 21 lb./sq. in. (1·48 kg./cm.$^2$) | 26 lb./sq. in. (1·83 kg./cm.$^2$) |
| **GT** | | | | |
| Normal car weight | 20 lb./sq. in. (1·4 kg./cm.$^2$) | 24 lb./sq. in. (1·69 kg./cm.$^2$) | **Tyre size 165SR—14 Pressures as Tourer** | |
| Maximum weight | 20 lb./sq. in. (1·4 kg./cm.$^2$) | 26 lb./sq. in. (1·83 kg./cm.$^2$) | | |

**It is recommended that for sustained speeds at or near the maximum the above tyre pressures are increased by 6 lb./sq. in. (·42 kg./cm.$^2$).**

**WEIGHTS**

| | | **Total weight** | | **Distribution** | | | |
|---|---|---|---|---|---|---|---|
| | | | | *Front* | | *Rear* | |
| | | *Tourer* | *GT* | *Tourer* | *GT* | *Tourer* | *GT* |
| **Kerbside** | Including full fuel tank, all optional extras and accessories | 2,303 lb. (1044 kg.) | 2,401 lb. (1088 kg.) | 1,127 lb. (512 kg.) | 1,162 lb. (527 kg.) | 1,176 lb. (534 kg.) | 1,239 lb. (562 kg.) |
| **Normal** | Kerbside weight including driver, passenger and 50 lb. (22 kg.) luggage | 2,653 lb. (1203 kg.) | 2,751 lb. (1248 kg.) | 1,235 lb. (560 kg.) | 1,269 lb. (576 kg.) | 1,418 lb. (643 kg.) | 1,482 lb. (672 kg.) |
| **Maximum** | Normal weight including tow-bar hitch load | 2,753 lb. (1249 kg.) | 2,851 lb. (1293 kg.) | 1,193 lb. (541 kg.) | 1,231 lb. (558 kg.) | 1,560 lb. (708 kg.) | 1,620 lb. (735 kg.) |
| **Maximum permissible towing weight** | | 1,680 lb. (762 kg.) | 1,680 lb. (762 kg.) | | | | |
| **Maximum towbar hitch load** | | 100 lb. (45 kg.) | 100 lb. (45 kg.) | | | | |

## (GHN4/GHD4 CARS, 18GD ENGINE—GHN5/GHD5 CARS, 18GG ENGINE)—*continued*

### WHEELS

| | |
|---|---|
| Type: Pressed disc | 4J × 14 (Tourer) 5J × 14 (GT) |
| Pressed spoked | 5J F/H × 14 |
| Wire (optional) | 4½J × 14 (60-spoke) |

### TORQUE WRENCH SETTINGS

**Automatic gearbox**

| | |
|---|---|
| Converter to drive plate bolts | 25 to 30 lb. ft. (3·46 to 4·15 kg. m.). |
| Transmission case to converter housing bolts | 8 to 13 lb. ft. (1·1 to 1·8 kg. m.). |
| Extension housing to transmission case bolts | 8 to 13 lb. ft. (1·1 to 1·8 kg. m.). |
| Oil pan to gearbox bolts | 8 to 13 lb. ft. (1·1 to 1·8 kg. m.). |
| Front servo bolts | 8 to 13 lb. ft. (1·1 to 1·8 kg. m.). |
| Rear servo bolts | 13 to 27 lb. ft. (1·8 to 3·73 kg. m.). |
| Pump adaptor to housing screw | 2 to 3 lb. ft. (·28 to ·41 kg. m.). |
| Pump adaptor to housing bolts | 17 to 32 lb. ft. (2·35 to 4·43 kg. m.). |
| Pump adaptor to transmission case bolts | 8 to 18·5 lb. ft. (1·1 to 2·55 kg. m.). |
| Manual shaft locknut | 7 to 9 lb. ft. (·97 to 1·24 kg. m.). |
| Pressure adaptor plug | 4 to 5 lb. ft. (·55 to ·69 kg. m.). |
| Drain plug | 8 to 10 lb. ft. (1·1 to 1·38 kg. m.). |
| Upper valve body to lower valve body screws | 20 to 30 lb. in. (·23 to ·345 kg. m.). |
| Lower valve body to upper valve body screws | 20 to 30 lb. in. (·23 to ·345 kg. m.). |
| Oil tube and end plate to valve bodies screws | 20 to 30 lb. in. (·23 to ·345 kg. m.). |
| Valve bodies to transmission case bolts | 4·5 to 9 lb. ft. (·61 to 1·24 kg. m.). |
| Cam bracket screws | 20 to 40 lb. in. (·23 to ·46 kg. m.). |
| Governor to counter-weight screws | 4 to 5 lb. ft. (·55 to ·69 kg. m.). |
| Governor cover-plate screws | 20 to 48 lb. in. (·23 to ·552 kg. m.). |
| Front servo adjusting screw locknut | 15 to 20 lb. ft. (2·07 to 2·77 kg. m.). |
| Rear servo adjusting screw locknut | 25 to 30 lb. ft. (3·46 to 4·15 kg. m.). |
| Starter inhibitor switch locknut | 4 to 6 lb. ft. (·55 to ·83 kg. m.) |
| Downshift cable adaptor bolts | 8 to 9 lb. ft. (1·1 to 1·24 kg. m.). |
| Filler tube connector sleeve to transmission case | 20 to 30 lb. ft. (2·76 to 4·15 kg. m.). |
| Filler tube to connector sleeve nut | 17 to 18 lb. ft. (2·35 to 2·49 kg. m.). |
| Stone-guard screws | 17 to 19 lb. in. (·196 to ·219 kg. m.). |
| Driving flange nut | 55 to 60 lb. ft. (7·6 to 8·3 kg. m.). |
| Centre support bolts | 10 to 18 lb. ft. (1·38 to 2·49 kg. m.). |

**General**

| | |
|---|---|
| Stiffnut to cross-member mounting bolt | 44 to 46 lb. ft. (6·08 to 6·36 kg. m.). |
| Gearbox remote control cover to tunnel | 7·5 to 9·5 lb. ft. (1·04 to 1·31 kg. m.) |
| Hydraulic brake pipe connection—$\frac{3}{8}$ in. UNF. | 5 to 7 lb. ft. (·69 to ·96 kg. m.). |
| Hydraulic brake pipe connection—$\frac{7}{16}$ in. UNF. | 7 to 10 lb. ft. (·96 to 1·38 kg. m.). |
| Steering-wheel nut—$\frac{9}{16}$ in. UNF. | 27 to 29 lb. ft. (3·73 to 4·01 kg. m.). |
| Steering-wheel nut—$\frac{11}{16}$ in. UNF. | 41 to 43 lb. ft. (5·66 to 5·94 kg. m.). |

**Electrical**

| | |
|---|---|
| Alternator shaft nut | 25 to 30 lb. ft. (3·46 to 4·15 kg. m.). |

# GENERAL DATA

**The information given in this section refers specifically to service operations on, or affected by, equipment fitted to the MGB (GHN5/GHN4 and GHD5/GHD4) in conformity with local and territorial requirements, and must be used in conjunction with the preceding DATA.**

## ENGINE

| | |
|---|---|
| Type (1967–69) | 18GF (EEC). |
| Type (1969–70) | 18GH (EEC, Carburetter crankcase ventilation). |
| Type (1969–70) | 18GJ (EEC, ELC, Carburetter crankcase ventilation). |
| Type (1970–1) | 18GK (EEC, ELC). |
| Type (1971–2) | 18V 584Z (EEC, ELC—Non overdrive). |
| Type (1971–2) | 18V 585Z (EEC, ELC—Overdrive). |
| Type (1972–4) | 18V 672Z (EEC, ELC—Non overdrive). |
| Type (1972–4) | 18V 673Z (EEC, ELC—Overdrive). |

Abbreviations: EEC, Exhaust Emission Control; ELC, Evaporative Loss Control.

**Servicing, adjusting and tuning the above type engines and cars fitted with Vehicle Emission Control Equipment must be carried out in accordance with the instructions given in Workshop Manual Supplement AKD 4957.**

## FUEL SYSTEM

| | |
|---|---|
| Fuel tank capacity (Engine types 18GJ and 18GK) | 12 U.S. gallons (45·4 litres). |
| Carburetter needle: 18GF | FX. |
| 18GH–18GJ | AAE. |
| 18GK | AAL. |
| 18V 584Z and 18V 585Z | AAU. |
| 18V 672Z and 18V 673Z | ABD. |

## DISTRIBUTOR

| | |
|---|---|
| Make/type | Lucas 25D4. |
| Serial number: 18GF | 40897. |
| 18GH–18GJ | 41155. |
| 18GK | 41339. |
| 18V 584Z and 18V 585Z | 41370. |
| 18V 672Z and 18V 673Z | 41491. |

## TORQUE WRENCH SETTINGS

| | |
|---|---|
| Interior mirror special screw | 5 lb. in. (·058 kg. m.). |
| Steering-column top fixing bolts | 12 to 17 lb. ft. (1·66 to 2·35 kg. m.). |
| Steering-column universal joint bolts | 20 to 22 lb. ft. (2·8 to 3·04 kg. m.). |
| Steering-wheel nut | 36 to 38 lb. ft. (4·98 to 5·26 kg. m.). |
| Steering-wheel nut—$\frac{9}{16}$ in. UNF. | 27 to 29 lb. ft. (3·73 to 4·00 kg. m.). |
| Master cylinder port adaptors | 33 lb. ft. (4·56 kg. m.). |
| Master cylinder reservoir fixing bolts | 5 lb. ft. (·69 kg. m.). |
| Brake pressure failure switch (nylon) | 15 lb. in. (·173 kg. m.). |
| Brake pressure failure switch assembly end plug | 200 lb. in. (2·3 kg. m.). |
| Hydraulic brake pipe connection—$\frac{3}{8}$ in. UNF. | 5 to 7 lb. ft. (·69 to ·96 kg. m.). |
| Hydraulic brake pipe connection—$\frac{7}{16}$ in. UNF. | 7 to 10 lb. ft. (·96 to 1·38 kg. m.). |

# GENERAL DATA

## (GHN5/GHD5 CARS, 18V type ENGINE)

**The following information is applicable to the GHN5 and GHD5 series cars from Car No. 258001 fitted with the 18V type engine, and should be used in conjunction with the preceding specifications.**

**ENGINE**

| | |
|---|---|
| Type: Synchromesh gearbox | 18V 581F—18V 581Y. |
| Synchromesh gearbox with overdrive | 18V 582F—18V 582Y. |
| Automatic transmission | 18V 583F—18V 583Y. |
| **Valves** | |
| Inlet valve: Seat angle | $45\frac{1}{2}°$. |
| Head diameter | 1·625 to 1·630 in. (41·27 to 41·40 mm.). |
| Stem diameter | 0·3429 to 0·3434 in. (8·70 to 8·72 mm). |
| Stem to guide clearance | 0·0008 to 0·0018 in. (0·020 to 0·046 mm.). |
| **Valve spring** | |
| Free length | 1·92 in. (48·77 mm.). |
| Fitted length | 1·44 in. (36·58 mm.). |
| Load at fitted length | 82 lb. (37·2 kg.). |
| Load at top of lift | 142 lb. (64·4 kg.). |

**ELECTRICAL**

| | |
|---|---|
| **Starter motor** | |
| Type | Lucas 2M100 Pre-engaged. |
| Brush spring tension | 36 oz. (1·02 kg.). |
| Minimum brush length | $\frac{3}{8}$ in. (9·5 mm.). |
| Minimum commutator thickness | 0·140 in. (3·5 mm.). |
| Lock torque | 14·4 lb. ft. (2·02 kg. m.) with 463 amps. |
| Torque at 1,000 r.p.m. | 7·3 lb. ft. (1·02 kg. m.) with 300 amps. |
| Light running current | 40 amp. at 6,000 r.p.m. (approx.) |
| Maximum armature end-float | 0·010 in. (0·25 mm.). |
| Solenoid: Closing (series) winding resistance | 0·25 to 0·27 ohm. |
| Hold-on (shunt) winding resistance | 0·76 to 0·80 ohm. |

| | **Tourer** | **GT** |
|---|---|---|
| **FRONT SUSPENSION—From Car No.** | 293446. | 296196. |
| Spring coil diameter (mean) | 3·238 in. (82·24 mm.). | 3·28 in. (83·31 mm.). |
| Free height | $10{\cdot}20 \pm 0{\cdot}06$ in. ($259{\cdot}08 \pm 1{\cdot}5$ mm.). | $9{\cdot}32 \pm \frac{1}{16}$ in. ($236{\cdot}73 \pm 1{\cdot}6$ mm.). |
| Static length at load of $1{,}030 \pm 20$ lb. ($467 \pm 9$ kg.) | 7·24 in. (187·70 mm.). | —— |
| Static length at load of $1{,}193 \pm 20$ lb. ($541 \pm 9$ kg.) | —— | 6·84 in. (173·74 mm.). |
| No. of free coils | 7·5. | 7·2. |

# ENGINE TUNING DATA

## 18G, 18GA, AND 18GB ENGINES

**ENGINE**

| | |
|---|---|
| Type | 18G, 18GA, 18GB. |
| Displacement | 109·8 cu. in. (1798 c.c.). |
| Firing order | 1, 3, 4, 2. |
| Compression ratio: H.C. | 8·8 : 1. |
| L.C. | 8·0 : 1. |
| Compression pressure: H.C. | 160 lb./sq. in. (11·25 kg./cm.²). |
| L.C. | 130 lb./sq. in. (9·15 kg./cm.²). |
| Idle speed | 500 r.p.m. |
| Valve rocker clearance | ·015 in. (·38 mm.) set cold. |
| Static ignition timing: H.C. | 10° B.T.D.C. |
| L.C. | 8° B.T.D.C. |
| Stroboscopic ignition timing: H.C. | 14° B.T.D.C. at 600 r.p.m. |
| L.C. | 12° B.T.D.C. at 600 r.p.m. |
| Timing mark location | Pointer on timing case, notch on crankshaft pulley. |

**DISTRIBUTOR**

| | |
|---|---|
| Make/Type | Lucas/25D4. |
| Contact breaker gap | ·014 to ·016 in. (·35 to ·40 mm.). |
| Contact spring tension | 18 to 24 oz. (510 to 680 gm.). |
| Rotation of rotor | Anti-clockwise. |
| Dwell angle | 60°±3°. |

**Centrifugal advance**

| | *Serial number* 40897 (*H.C.*) | *Serial number* 40916 (*L.C.*) |
|---|---|---|
| Crankshaft degrees/speed (vacuum pipe disconnected) | 6° at 600 r.p.m. | 6° at 600 r.p.m. |
| | 6° at 700 r.p.m. | 8° at 800 r.p.m. |
| | 9° at 900 r.p.m. | 9° at 1,000 r.p.m. |
| | 15° at 1,600 r.p.m. | 18° at 3,000 r.p.m. |
| | 20° at 2,200 r.p.m. | 24° at 4,400 r.p.m. |

**Vacuum advance**

| | | |
|---|---|---|
| Starts | 5 in. Hg (127 mm. Hg). | 4 in. Hg (101·6 mm. Hg). |
| Finishes | 13 in. Hg (330·2 mm. Hg). | 12 in. Hg (304·8 mm. Hg). |
| Total crankshaft degrees | 20°±2°. | 16°±2°. |

**IGNITION COIL**

| | |
|---|---|
| Make/Type | Lucas/HA12. |
| Resistance: Primary | 3·1 to 3·5 ohms at 20° C. (68° F.). |

**SPARKING PLUGS**

| | |
|---|---|
| Make/Type | Champion/N9Y. |
| Gap | ·024 to ·026 in. (·625 to ·66 mm.). |

**CARBURETTERS**

| | |
|---|---|
| Make/Type | S.U./HS4. |
| Jet | ·090 in. (2·2 mm.). |
| Needle: 18G and 18GA engines | No. 5 (Std.), No. 6 (Rich), No. 21 (Weak). |
| 18GB engine | FX (Std.), No. 6 (Rich), No. 21 (Weak). |
| Piston spring | Red. |

**ELECTRICAL**

| | |
|---|---|
| Dynamo | Lucas C40/1. |
| Battery | Lucas BT9E or BTZ9E. |
| Volts/Polarity | 12-volt/Positive earth. |
| Starter | Lucas M418G Inertia-type. |
| Control unit | Lucas RB.340. |

# ENGINE TUNING DATA

## 18GD AND 18GG ENGINES

**ENGINE**

| | |
|---|---|
| Type | 18GD—18GG. |
| Displacement | 109·8 cu. in. (1798 c.c.). |
| Firing order | 1, 3, 4, 2. |
| Compression ratio: H.C. | 8·8 : 1. |
| L.C. | 8·0 : 1. |
| Compression pressure: H.C. | 160 lb./sq. in. (11·25 kg./cm.$^2$). |
| L.C. | 130 lb./sq. in. (9·15 kg./cm.$^2$). |
| Idle speed | 500 r.p.m. |
| Valve rocker clearance | ·015 in. (·38 mm.) set cold. |
| Static ignition timing: H.C. | 10° B.T.D.C. |
| L.C. | 8° B.T.D.C. |
| Stroboscopic ignition timing: H.C. | 14° B.T.D.C. at 600 r.p.m. |
| L.C. | 12° B.T.D.C. at 600 r.p.m. |
| Timing mark location | Pointer on timing case, notch on crankshaft pulley. |

**DISTRIBUTOR**

| | |
|---|---|
| Make/Type | Lucas/25D4. |
| Contact breaker gap | ·014 to ·016 in. (·35 to ·40 mm.). |
| Contact spring tension | 18 to 24 oz. (510 to 680 gm.). |
| Rotation of rotor | Anti-clockwise. |
| Dwell angle | 60°±3°. |

| **Centrifugal advance** | *Serial number* 40897 (*H.C.*) | *Serial number* 40916 (*L.C.*) |
|---|---|---|
| Crankshaft degrees/Speed (vacuum pipe disconnected) | 6° at 600 r.p.m. | 6° at 600 r.p.m. |
| | 6° at 700 r.p.m. | 8° at 800 r.p.m. |
| | 9° at 900 r.p.m. | 9° at 1,000 r.p.m. |
| | 15° at 1,600 r.p.m. | 18° at 3,000 r.p.m. |
| | 20° at 2,200 r.p.m. | 24° at 4,400 r.p.m. |
| **Vacuum advance** | | |
| Starts | 5 in. Hg (127 mm. Hg). | 4 in. Hg 101·6 mm. Hg). |
| Finishes | 13 in. Hg (330·2 mm. Hg). | 12 in. Hg (304·8 mm. Hg). |
| Total crankshaft degrees | 20°±2°. | 16°±2°. |

**IGNITION COIL**

| | |
|---|---|
| Make/Type | Lucas/HA12. |
| Resistance: Primary | 3·1 to 3·5 ohms at 20° C. (68° F.). |

**SPARKING PLUGS**

| | |
|---|---|
| Make/Type | Champion/N9Y. |
| Gap | ·024 to ·026 (·625 to ·66 mm.). |

**CARBURETTERS**

| | |
|---|---|
| Make/Type | S.U./HS4. |
| Jet | ·090 in. (2·2 mm.). |
| Needle | Standard, FX; Rich, No. 5; Weak, GZ. |
| Piston spring | Red. |

**ELECTRICAL**

| | |
|---|---|
| Alternator | 18GD, Lucas 16AC. 18GG, Lucas 16ACR. |
| Battery | Lucas CA9E early cars, Lucas BT9E or BTZ9E. |
| Volts/Polarity | 12-volt/NEGATIVE earth. |
| Starter | Lucas M418G Pre-engaged. |
| Regulator | Lucas 4TR (18GD only). |

# ENGINE TUNING DATA

## 18GF, 18GH, 18GK and 18V 584Z, 18V 585Z 18V 672Z and 18V 673Z ENGINES

Servicing, adjusting and tuning the above type engines and cars fitted with Vehicle Emission Control Equipment must be carried out in accordance with the instructions given in Workshop Manual Supplement (see page 393).

# ENGINE TUNING DATA

## 18V 581F, 18V 582F and 18V 583F ENGINES

| | |
|---|---|
| **ENGINE** | |
| Type | 18V. |
| Displacement | 109·8 cu. in. (1798 c.c.). |
| Firing order | 1, 3, 4, 2. |
| Compression ratio: H.C. | 9·0 : 1. |
| L.C. | 8·0 : 1. |
| Cranking pressure | Nominal 170 lbf./sq. in. (11·95 kgf./cm.$^2$) at 275 r.p.m. |
| Idle speed | 750 to 800 r.p.m. |
| Valve rocker clearance: Set cold | ·015 in. (·38 mm.). |
| Set hot | ·013 in. (·33 mm.). |
| Static ignition timing: H.C. | 10° B.T.D.C. |
| L.C. | 10° B.T.D.C. |
| Stroboscopic ignition timing: H.C. | 13° B.T.D.C. at 600 r.p.m. |
| L.C. | 13° B.T.D.C. at 600 r.p.m. |
| Timing mark location | Pointer on timing case, notch on crankshaft pulley. |
| **DISTRIBUTOR** | |
| Make/Type | Lucas/25D4. |
| Contact breaker gap | ·014 to ·016 in. (·35 to ·40 mm.). |
| Contact spring tension | 18 to 24 oz. (510 to 680 gm.). |
| Rotation of rotor | Anti-clockwise. |
| Dwell angle | 60°±3°. |

| **Centrifugal advance** | *Serial number* 41288 (*H.C.*) | *Serial number* 41290 (*L.C.*) |
|---|---|---|
| Crankshaft degrees/Speed (vacuum pipe disconnected) | 0 to 6° at 600 r.p.m. | 0 to 6° at 600 r.p.m. |
| | 4 to 9° at 700 r.p.m. | 6 to 10° at 800 r.p.m. |
| | 7 to 11° at 900 r.p.m. | 7 to 11° at 1,000 r.p.m. |
| | 13 to 17° at 1,600 r.p.m. | 16 to 20° at 3,000 r.p.m. |
| | 18 to 22° at 2,200 r.p.m. | 22 to 26° at 4,400 r.p.m. |
| **Vacuum advance** | | |
| Starts | 5 in. Hg (127 mm. Hg). | 4 in. Hg (101·6 mm. Hg). |
| Finishes | 13 in. Hg (330·2 mm. Hg) | 12 in. Hg (304·8 mm. Hg). |
| Total crankshaft degrees | 20°±2°. | 16°±2°. |

| | |
|---|---|
| **IGNITION COIL** | |
| Make/Type | Lucas/HA12. |
| Resistance: Primary | 3·1 to 3·5 ohms at 20° C. (68° F.). |
| **SPARKING PLUGS** | |
| Make/Type | Champion/N9Y. |
| Gap | ·024 to ·026 (·625 to ·66 mm.). |
| **CARBURETTERS** | |
| Make/Type | S.U./HS4. |
| Jet | ·090 in. (2·2 mm.). |
| Needle | AAU. |
| Piston spring | Red. |
| **ELECTRICAL** | |
| Alternator | Lucas 16ACR. |
| Battery | Lucas BT9E, BTZ9E, CA9E, SG9E or Exide 3XCK9L. |
| Volts/Polarity | 12 volt/NEGATIVE earth. |
| Starter | Lucas 2M100 Pre-engaged. |

# ENGINE TUNING DATA

## 18V 581Y, 18V 582Y, AND 18V 583Y ENGINES

**ENGINE**

| | |
|---|---|
| Type | 18V. |
| Displacement | 109·8 cu. in. (1798 c.c.). |
| Firing order | 1, 3, 4, 2. |
| Compression ratio: H.C. | 9·0 : 1. |
| L.C. | 8·0 : 1. |
| Cranking pressure | Nominal 170 lbf./sq. in. (11·95 kgf./cm.²) at 275 r.p.m. |
| Idle speed | 750 to 800 r.p.m. |
| Valve rocker clearance: Set cold | ·015 in. (·38 mm.). |
| Set hot | ·013 in. (·33 mm.). |
| Static ignition timing: H.C. | 10° B.T.D.C. |
| L.C. | 10° B.T.D.C. |
| Stroboscopic ignition timing: H.C. | 13° B.T.D.C. at 600 r.p.m. |
| L.C. | 13° B.T.D.C. at 600 r.p.m. |
| Timing mark location | Pointer on timing case, notch on crankshaft pulley. |

**DISTRIBUTOR**

| | |
|---|---|
| Make/Type | Lucas/25D4. |
| Contact breaker gap | ·014 to ·016 in. (·35 to ·40 mm.). |
| Contact spring tension | 18 to 24 oz. (510 to 680 gm.). |
| Rotation of rotor | Anti-clockwise. |
| Dwell angle | 60°±3°. |

| **Centrifugal advance** | *Serial number* 41288 (*H.C.*) | *Serial number* 41290 (L.C.) |
|---|---|---|
| Crankshaft degrees/Speed (vacuum pipe disconnected) | 0 to 6° at 600 r.p.m. | 0 to 6° at 600 r.p.m. |
| | 4 to 9° at 700 r.p.m. | 6 to 10° at 800 r.p.m. |
| | 7 to 11° at 900 r.p.m. | 7 to 11° at 1,000 r.p.m. |
| | 13 to 17° at 1,600 r.p.m. | 16 to 20° at 3,000 r.p.m. |
| | 18 to 22° at 2,200 r.p.m. | 22 to 26° at 4,400 r.p.m. |
| **Vacuum advance** | | |
| Starts | 5 in. Hg (127 mm. Hg). | 4 in. Hg (101·6 mm. Hg). |
| Finishes | 13 in. Hg (330·2 mm. Hg). | 12 in. Hg (304·8 mm. Hg). |
| Total crankshaft degrees | 20°±2°. | 16°±2°. |

**IGNITION COIL**

| | |
|---|---|
| Make/Type | Lucas/HA12. |
| Resistance: Primary | 3·1 to 3·5 ohms at 20° C. (68° F.). |

**SPARKING PLUGS**

| | |
|---|---|
| Make/Type | Champion/N9Y. |
| Gap | ·024 to ·026 (·625 to ·66 mm.). |

**CARBURETTERS**

| | |
|---|---|
| Make/Type | S.U./HIF4. |
| Jet | ·090 in. (2·2 mm.). |
| Needle | AAU. |
| Piston spring | Red. |

**ELECTRICAL**

| | |
|---|---|
| Alternator | Lucas 16ACR. |
| Battery | Lucas BT9E, BTZ9E, CA9E, SG9E or Exide 3XCK9L. |
| Volts/Polarity | 12 volt/NEGATIVE earth. |
| Starter | Lucas 2M100 Pre-engaged. |

# ENGINE TUNING DATA

18V ENGINES—To European emission control requirements (ECE 15)

| | |
|---|---|
| **ENGINE** | |
| Type | 18V. |
| Displacement | 109·8 cu. in. (1798 c.c.). |
| Firing order | 1, 3, 4, 2. |
| Compression ratio | 9·0 : 1. |
| Cranking pressure | Nominal 170 to 190 lbf./sq. in. (12 to 13·4 kgf./cm.$^2$). |
| Valve rocker clearance: set cold | ·015 in. (·38 mm.). |
| Idle speed | 750 r.p.m. |
| Fast idle speed | 1,100 to 1,200 r.p.m. |
| Stroboscopic ignition timing | 15° B.T.D.C. at 1,000 r.p.m. |
| Static ignition timing | 5° B.T.D.C. |
| Timing mark location | Pointer on timing case, notch on crankshaft pulley. |
| **DISTRIBUTOR** | |
| Make/Type | Lucas/25D4. |
| Contact breaker gap | ·014 to 0·16 in. (·35 to ·40 mm.). |
| Contact spring tension | 18 to 24 oz. (510 to 680 gm.). |
| Rotation of rotor | Anti-clockwise. |
| Dwell angle | 60°±3° |
| Serial number | 41032. |
| **Centrifugal advance** | |
| Crankshaft degrees/Speed (vacuum pipe disconnected) | 0 to 3° at 600 r.p.m. |
| | 0 to 4° at 700 r.p.m. |
| | 2 to 7° at 900 r.p.m. |
| | 10 to 14° at 1,600 r.p.m. |
| | 17 to 21° at 2,200 r.p.m. |
| **Vacuum advance** | |
| Starts | 3 in. Hg. (76 mm. Hg.). |
| Finishes | 8 in. Hg. (203 mm. Hg.). |
| Total crankshaft degrees | 14°±2°. |
| **IGNITION COIL** | |
| Make/Type | Lucas/HA12. |
| Resistance: Primary | 3·1 to 3·5 ohms at 20° C. (68° F.). |
| **SPARKING PLUGS** | |
| Make/Type | Champion/N9Y. |
| Gap | ·024 to ·026 (·625 to ·66 mm.). |
| **CARBURETTERS** | |
| Make/Type | S.U./HIF4. |
| Specification | AUD 434. |
| Jet | ·090 in. (2·2 mm.). |
| Needle | AAU. |
| Piston spring | Red. |
| **EXHAUST EMISSION** | |
| Exhaust gas analyser reading at engine idle speed | 3 to 4·5% CO. |

# SECTION A

## THE ENGINE



**† These operations must be followed by an exhaust emission check.** 

## THE ENGINE LUBRICATION SYSTEM

1. RESTRICTOR.
2. OIL SQUIRT TO CYLINDER WALLS.
3. ROCKER FEED HOLE.
4. OIL FILLER.
5. OIL RETURN TO SUMP.
6. EXTERNAL PIPE.
7. LOW PRESSURE GALLERY.
8. RELIEF VALVE.
9. FEED TO GEARS.
10. MAIN FEED.
11. OIL PUMP.
12. MAIN HIGH PRESSURE GALLERY
13. FEED TO CHAIN TENSIONER.
14. OIL BLEED TO CHAIN.
15. FULL FLOW FILTER (SEE DETAIL.)
16. OIL BLEED FOR VALVE STEM TIP.
17. PRESSURE FEED TO BALL TIP.
18. INLET.
19. OIL GALLERY.
20. TO MAIN BEARING.
21. RELEASE VALVE.
22. FEED TO ROCKER GEAR (INTERMITTENT.)

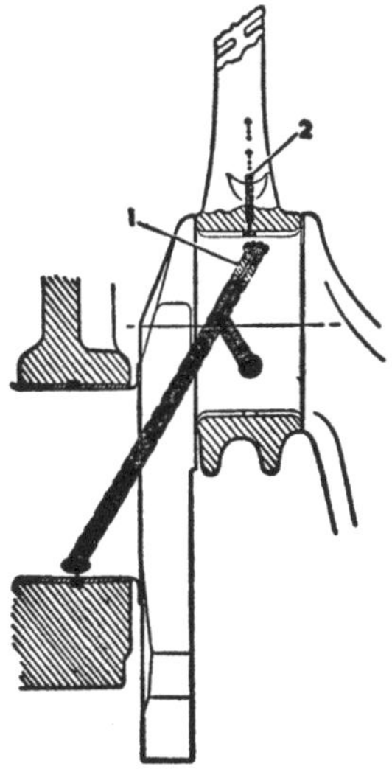


DETAIL OF BIG END LUBRICATION

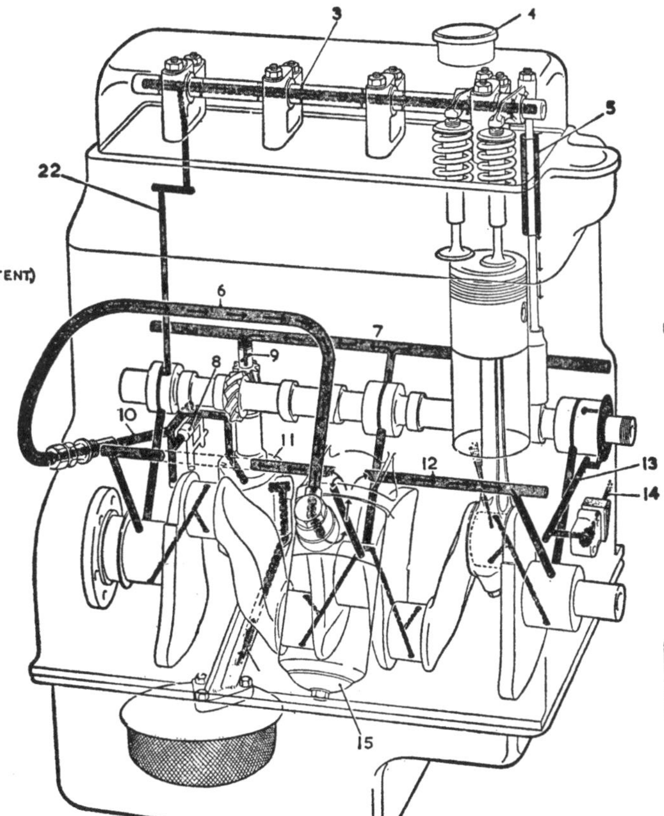


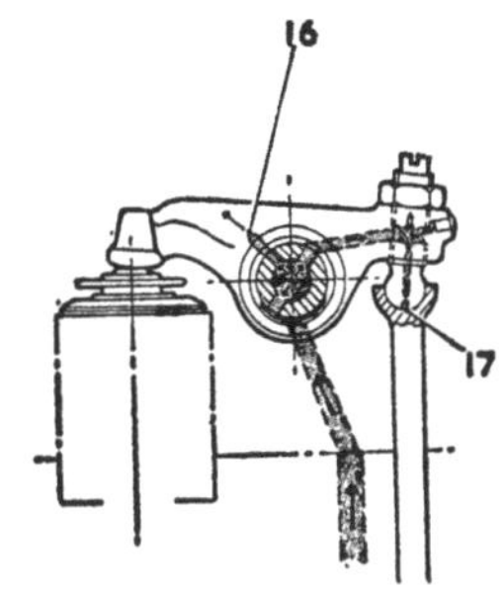


DETAIL OF ROCKER GEAR LUBRICATION

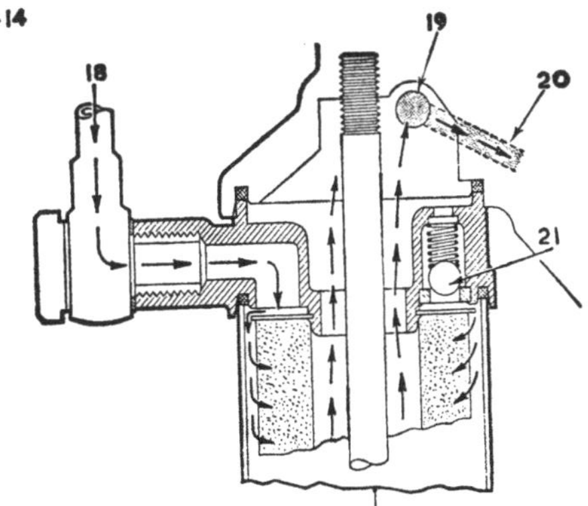


DETAIL. FULL FLOW FILTER

A0007B

## GENERAL DESCRIPTION

The overhead-valve engine is built in unit construction, with single-plate dry clutch. Cylinder bores are an integral part of the block and are water-jacketed their full length. The valves are set vertically in the detachable cylinder head and are operated through the medium of the rocker gear, push-rods, and tappets from the camshaft in the left-hand side of the cylinder block. Renewable, thinwall bearings support the counterbalanced crankshaft, while end-thrust is taken by the thrust washers fitted to the centre main bearing. The camshaft is driven by a duplex roller chain which is automatically adjusted by a spring-hydraulic-operated rubber slipper. The oil pump, distributor, and engine revolution indicator are driven from the camshaft; each component has its own drive shaft.

(See Editor's notes at end of section.)

## Section A.1

### OIL PRESSURE

Under normal running conditions the oil pressure should not drop below 50 lb./sq. in. (3·5 kg./cm.²) on the gauge at normal road speeds, whilst approximately 15 lb./sq. in. (1·05 kg./cm.²) should be shown when the engine is idling. New engines with new oil will give considerably higher readings at low speeds.

Should there be a noticeable drop in pressure, the following points should be checked:

(1) That there is a good supply of the correct grade of oil in the engine sump.

(2) That the strainer in the sump is clean and not choked with sludge.

(3) That the bearings, to which oil is fed under pressure, have the correct working clearances. Should the bearings be worn and the clearances excessive, the oil will escape more readily from the sides of the bearings, particularly when the oil is warm and becomes more fluid. This will cause a drop in pressure on the gauge as compared with that shown when the bearings are in good order.

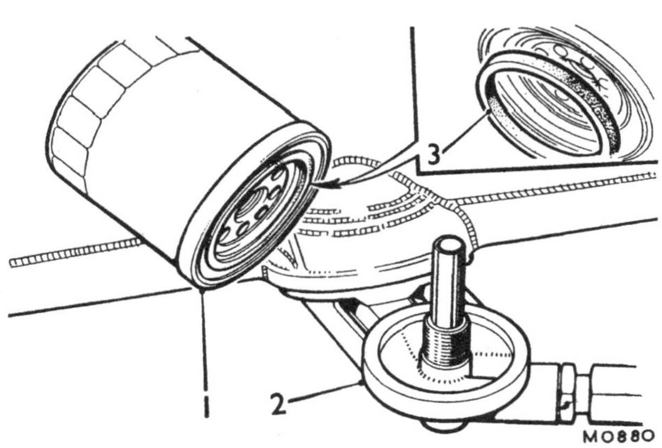


*Later type filter (disposable cartridge)*

1. Filter cartridge. 2. Filter head. 3. Seal.

## Section A.2

### LUBRICATION

**Checking the engine oil level**

Inspect the oil level in the engine, and top up if necessary to the 'MAX' mark on the dipstick. The oil filler cap is on the top of the engine valve cover and is released by turning it anti-clockwise.

**Changing the engine oil**

Drain the oil from the engine by removing the drain plug on the right-hand side of the engine sump.

The oil will flow more readily if drained when the engine is hot; allow at least 10 minutes for draining before replacing the plug.

**NOTE.—Disconnect the battery cable from its terminal on the starter before starting work on the filter.**

**Engine oil filter**

The external oil filter is of the disposable cartridge type and is located on the right-hand side of the engine. Early cars are fitted with the renewable element type of filter.

**To change the filter cartridge.** Unscrew the old filter cartridge from the filter head and discard. Smear the seal of the new filter cartridge with oil, locate on the filter head and tighten with the hands only. **DO NOT USE A SPANNER TO TIGHTEN. DO NOT OVER-TIGHTEN.**

**To renew the filter element,** drain the filter by removing the plug; unscrew the filter body securing bolt and lift off the body. Wash out the casing with gasoline (fuel) and dry it before fitting a new element. Check that the sealing rings and the rubber washer are in a satisfactory condition.

Reassemble the filter, ensuring that the components are correctly positioned.

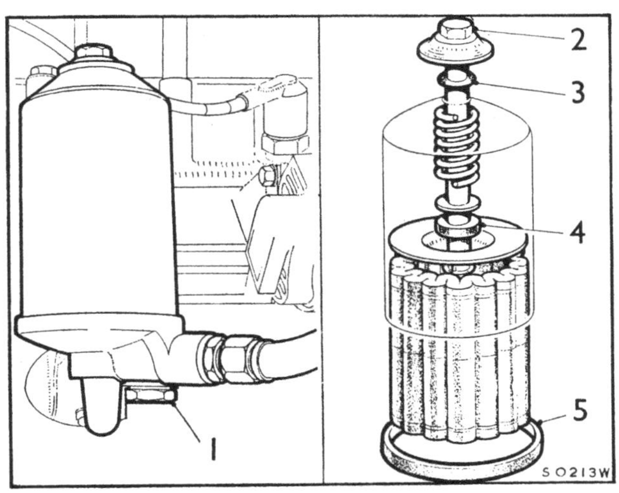


*Early type filter (renewable element)*

1. Drain plug. 2. Bolt. 3. Sealing ring. 4. Rubber washer. 5. Sealing ring.

**Oil pressure relief valve**

The automatic relief valve in the lubrication system deals with any excessive oil pressure when starting from cold. When hot, the pressure drops as the oil becomes more fluid.

Should the oil filter become blocked, two relief valves in the filter blow off to enable the oil to by-pass the filter and pass direct into the main gallery.

Continuous cold-running and unnecessary use of the mixture control are often the causes of serious oil dilution by fuel, with a consequent drop in pressure.

Particular attention is called to the recommended changes of oil.

The non-adjustable oil pressure relief valve is situated at the rear of the left-hand side of the cylinder block and is held in position by a domed hexagon nut sealed by two fibre washers or one copper washer. The relief valve spring maintains a valve cup against a seating machined in the block.

The valve should be examined to ensure that the cup is seating correctly and that the relief spring has not lost its tension. The latter can be checked by measuring the length of the spring to give the correct relief pressure (see 'GENERAL DATA'). Fit a new cup and spring if necessary.

## Section A.3

### WATER PUMP

The water pump is of the centrifugal impeller type and is mounted on a common spindle with the fan in a casting bolted to the front of the cylinder block.

The water pump and fan assembly is attached to the front of the cylinder block by four bolts and may be withdrawn and serviced as detailed in Section C.7.

If the gasket is damaged as the pump body is withdrawn from the cylinder block, ensure that all traces of it are removed before a new gasket is fitted and the pump replaced.

## Section A.4

### MANIFOLDS

Remove the air cleaners and carburetters as detailed in Section D.

Disconnect the distributor vacuum pipe and, on early cars, the heater pipe from the manifold.

Remove the six exhaust pipe to manifold clamp bolts and spring washers and release the pipe.

Six studs and nuts secure the manifolds to the cylinder head.

The four centre nuts have large washers, enabling them to secure both the inlet and exhaust manifolds. The two remaining nuts, one at each end of the manifolds, have small washers and secure the exhaust manifold only.

Replacements of the manifolds is a reversal of these instructions.

Use a new gasket.

## Section A.5

### ROCKER ASSEMBLY

Drain the cooling system—Section C.4.

Disconnect the breather pipe from the rocker cover, on 18GD and 18GF engines, unscrew the nuts securing the heater pipe to the rocker studs and move the pipe clear of the rocker cover.

Release the breather pipe from the front of the rocker cover.

Withdraw the throttle cable from the lever and outer cable abutment. Unscrew the two nuts and lift off the rocker cover, taking care not to damage the cork gasket or lose the washers and rubber seals. Notice that under the right-hand rear rocker stud nut is a special locking plate. Unscrew the eight rocker shaft bracket fixing nuts gradually, a turn at a time, in the order shown in Fig. A.2 until all load has been released.

**It is necessary to drain the radiator and slacken the seven external cylinder head securing nuts because four of the rocker shaft bracket fixing nuts also secure the cylinder head, and if the seven external cylinder head fixing nuts are not slackened distortion may result and water find its way from the cooling system into the cylinders and sump.**

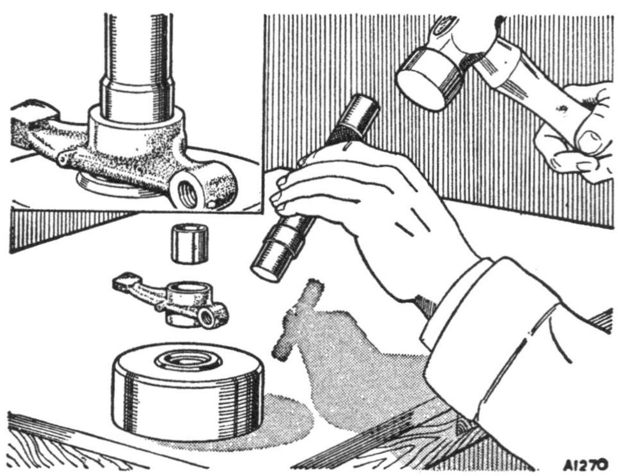


*Fig. A.1*
*Fitting a valve rocker bush, using Service tool 18G 226*

Completely unscrew the eight rocker shaft bracket nuts and remove the rocker assembly, complete with brackets and rockers. Withdraw the eight push-rods, storing them carefully so that they may be replaced in the same positions. To dismantle the rocker shaft assembly, first remove the grub screw which locates the rocker shaft in the rear rocker mounting bracket and remove the split pins, flat washers, and spring washers from each end of the shaft. Slide the rockers, brackets, and springs from the shaft.

Unscrew the plug from the front end of the shaft and clean out the oilways.

Reassembly and replacement is a reversal of the above procedure, replacing the rockers and springs in their original positions on the shaft. Remember to replace the rocker shaft locating screw lock plate. Replace the rocker cover with the vent pipe to the front. Check that the two cap nut rubber bushes and the rocker cover cork gasket are undamaged; if they are found to be faulty, fit new ones or oil leaks may result.

**Refitting rocker bushes**

To remove and replace worn rocker bushes the use of Service tool 18G 226 is recommended; the bushes and the rockers can be very easily damaged by using improvised drifts. Place the rocker on the anvil and drive out the worn bush.

Place a new bush on the driver and position the bush with the butt joint at the top of the rocker bore and the oil groove in the bush at the bottom of the rocker bore, as shown in Fig. A.3.

It will be necessary to drill the oil holes in the bush to coincide with the oilways in the rocker. The oil hole to the adjuster end can be drilled before the bush is fitted, extra care being taken to keep the holes of the bush and rocker in line during the pressing-in operation.

If the holes are drilled after fitting, the following procedure must be adopted. Remove the adjuster screw and use a ·093 in. (2·36 mm.) drill to drill out the end plug and to continue the oilway through the bush. Replug the end after this operation with a rivet (Part No. 5C 2436) and weld the plug into position. The oil hole in the top of the rocker barrel must be continued through the bush with a No. 47 drill, ·0785 in. (1·98 mm.).

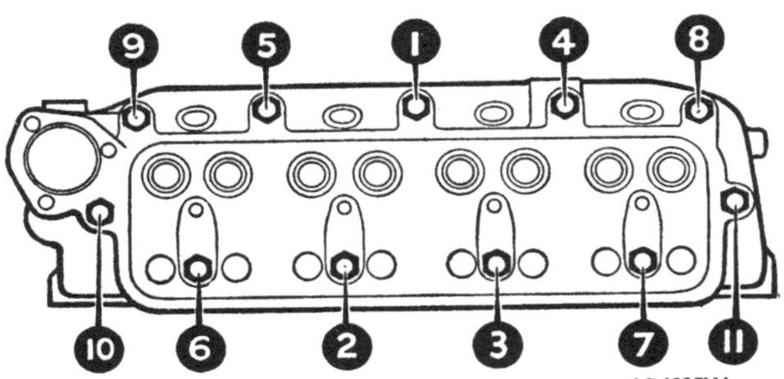


*Fig. A.2*
*Cylinder head nut slackening and tightening sequence*

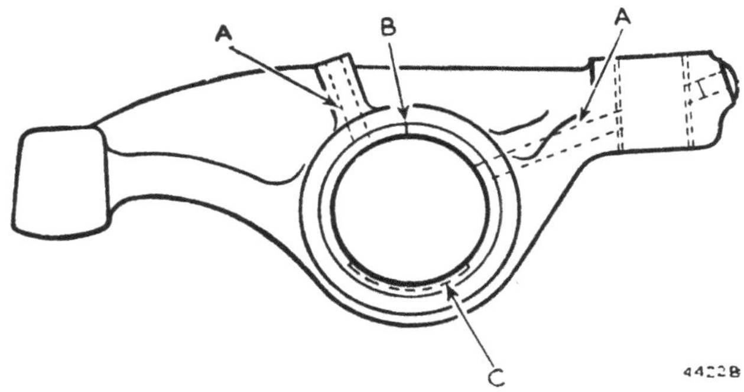


*Fig. A.3*
*Showing the correct position for a valve rocker bush*

A. Oilways. B. Joint in rocker bush. C. Oil groove.

Finally, burnish-ream the bush to the dimensions given in the '**GENERAL DATA**' section.

## Section A.6

### CYLINDER HEAD ASSEMBLY

Drain the water from the cooling system. One drain tap is at the base of the radiator on the left-hand side and at the rear of the engine on the right-hand side. On later 18GD and 18GG engines the cylinder block drain tap is replaced by a removable plug and sealing washer. If anti-freeze mixture is being used it should be drained into a suitable clean container and carefully preserved for future use.

Remove the top water hose. Remove the three thermostat housing securing nuts and washers and remove the housing and thermostat.

Remove the air cleaners and carburetters as detailed in Section D.

Remove the inlet and exhaust manifolds as detailed in Section A.4.

Remove the rocker assembly as detailed in Section A.5 and remove the seven external cylinder head nuts at the same time. Withdraw the push-rods, keeping them in the order of their removal.

Disconnect the high-tension cables and remove the sparking plugs. Disconnect the heater hose and heater control cable from the water valve.

Unscrew the thermal transmitter from the front of the cylinder head and release the conductor from its supporting clip.

Slacken the clips and disconnect the hoses from the water pipe on the left-hand side of the cylinder head and remove the pipe. Release the ignition vacuum control pipe from the rear cylinder head stud and remove the cylinder head.

Break the cylinder head joint by levering at one end and withdraw the head evenly up the studs.

## THE CYLINDER HEAD COMPONENTS

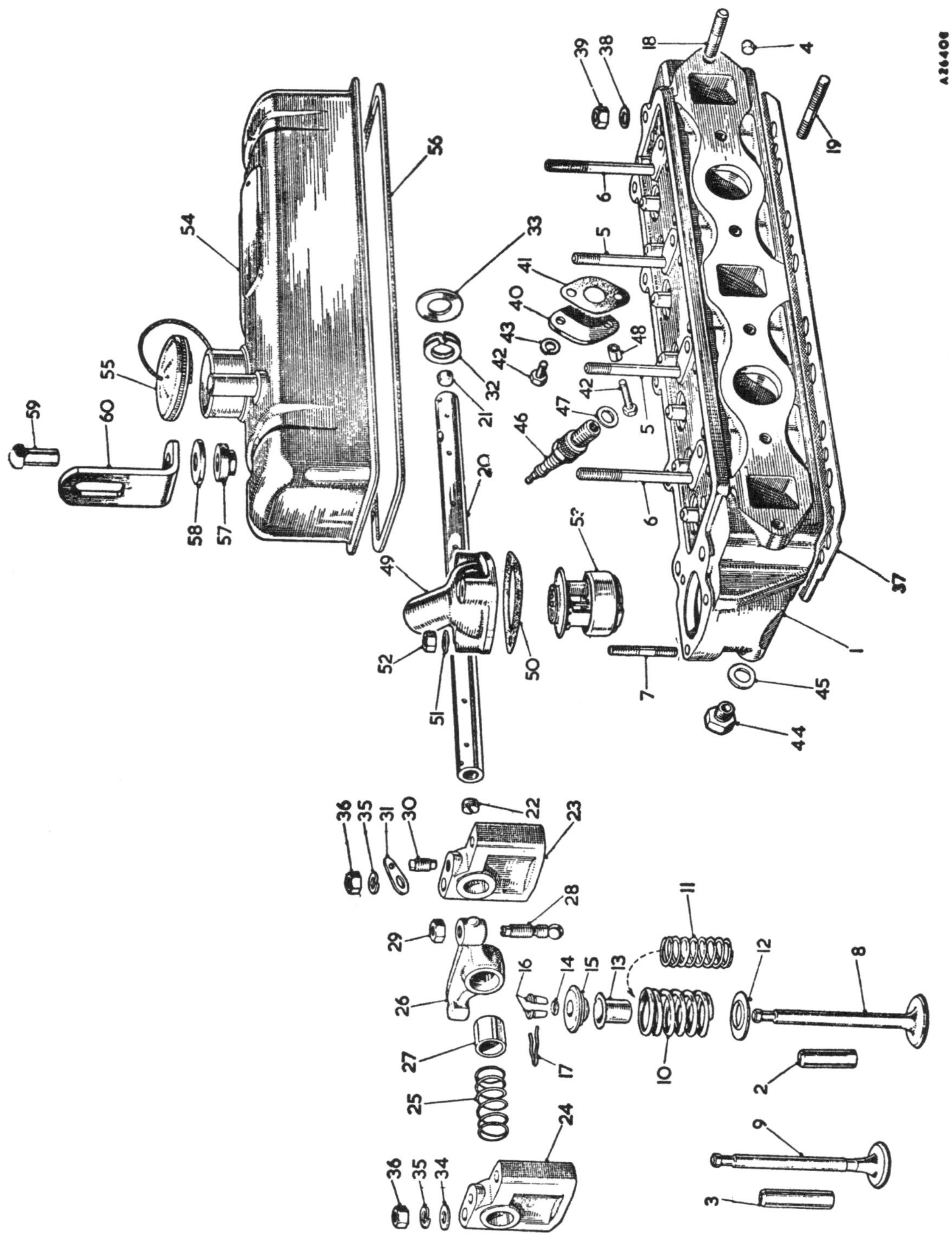

## KEY TO THE CYLINDER HEAD COMPONENTS

| No. | Description | No. | Description | No. | Description |
|---|---|---|---|---|---|
| 1. | Cylinder head assembly—with guides. | 21. | Plug—plain—valve rocker. | 41. | Joint washer blanking plate. |
| 2. | Guide—inlet valve. | 22. | Plug—screwed—valve rocker. | 42. | Screw—plate to cylinder head. |
| 3. | Guide—exhaust valve. | 23. | Bracket—tapped—rocker shaft. | 43. | Washer—spring—screw. |
| 4. | Plug—oil hole. | 24. | Bracket—plain—rocker shaft. | 44. | Plug—thermal transmitter boss. |
| 5. | Stud—short—rocker bracket. | 25. | Spring—rocker spacing. | 45. | Washer—plug. |
| 6. | Stud—long—rocker bracket. | 26. | Rocker—valve—bushed. | 46. | Plug—sparking. |
| 7. | Stud—water outlet elbow to head. | 27. | Bush—valve rocker. | 47. | Gasket—steel—plug. |
| 8. | Valve—inlet. | 28. | Screw—adjusting—tappet. | 48. | Distance piece—screw longer. |
| 9. | Valve—exhaust. | 29. | Nut—tappet adjusting screw. | 49. | Elbow—outlet—water. |
| 10. | Spring—outer—valve. | 30. | Screw—locating—rocker shaft. | 50. | Joint washer—elbow. |
| 11. | Spring—inner—valve. | 31. | Plate—lock—locating screw. | 51. | Washer—plain—stud in cylinder head. |
| 12. | Collar—valve spring. | 32. | Washer—double-coil rocker shaft. | 52. | Nut—stud. |
| 13. | Shroud—valve guide (up to Engine No. H4385 and L2815). | 33. | Washer—rocker shaft. | 53. | Thermostat. |
| 14. | Ring—valve packing. | 34. | Washer—plain—rocker shaft stud. | 54. | Cover—valve rocker. |
| 15. | Cup—valve spring. | 35. | Washer—spring—rocker shaft stud. | 55. | Cap and cable—oil filler. |
| 16. | Cotter—valve (halves). | 36. | Nut—rocker bracket stud. | 56. | Joint washer—valve rocker cover. |
| 17. | Circlip—valve cotter. | 37. | Joint washer—cylinder head. | 57. | Bush—rubber—cover. |
| 18. | Stud—exhaust manifold to head. | 38. | Washer—cylinder head nut. | 58. | Washer—cup. |
| 19. | Stud—inlet and exhaust manifold to head. | 39. | Nut—cylinder head stud. | 59. | Nut—cap—rocker cover. |
| 20. | Shaft—valve rocker—plugged. | 40. | Plate—blanking—heater outlet elbow. | 60. | Bracket—engine sling. |

**Refitting the cylinder head**

Make sure that the surfaces of both the cylinder head and the cylinder block are clean. It will be noticed that the cylinder head gasket is marked 'FRONT' and 'TOP' to assist in replacing it correctly with the copper side uppermost. Having slipped the gasket over the studs, next lower the cylinder head into position. Replace the vacuum control pipe clip and fit the seven cylinder head external nuts finger tight.

Replace the push-rods in the positions from which they were taken. Replace the rocker assembly and securing nuts finger tight. Tighten the 11 cylinder head nuts, a turn at a time, in the order given in Fig. A.2. Finally, tighten the four rocker assembly nuts.

Reassembly continues in the reverse order to the dismantling procedure.

Switch on the ignition and check the fuel system for leaks. Start the engine and run it until the normal working temperature is reached. Remove the rocker cover and check the valve clearances (see Section A.7). Replace the rocker cover and connect the breather hose.

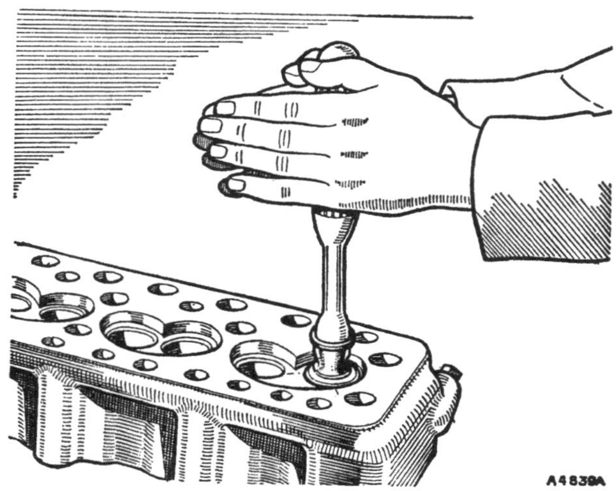

*Fig. A.5*

*Grinding in a valve, using Service tool 18G29. Occasionally lift the valve off its seat to spread the grinding paste evenly over the seat and valve face*

## Section A.7

### VALVE GEAR AND VALVES

**Removing and replacing valves**

Remove the cylinder head as in Section A.6.

Remove the valve circlip. Compress the double valve springs, using Service tool 18G45, and remove the two valve cotters. Release the valve springs and remove the compressor, valve spring cap, shroud (early engines only), inner and outer springs, and bottom collar.

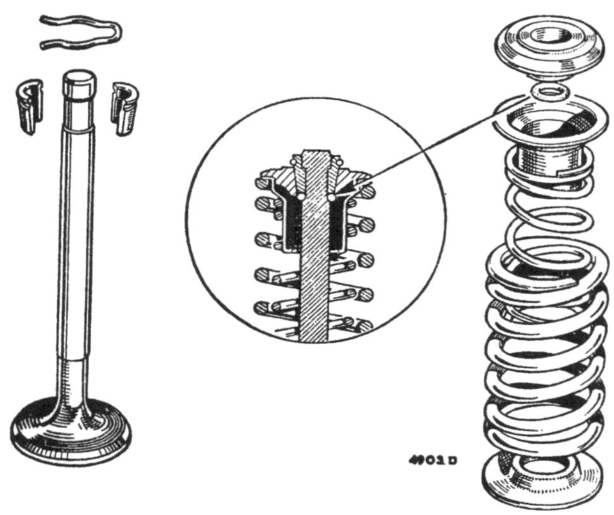

*Fig. A.4*

*Parts of the valve assembly, showing the valve, cotters, circlip, bottom collar, double valve springs, shroud (early engines only), packing ring, and spring cap. The inset shows the valve packing ring fitted correctly at the bottom of the cotter groove below the cotters*

Remove the valve packing ring from the cotter groove and withdraw the valve from the guide.

Keep the valves in their relative positions when removed from the head to ensure replacement in their original valve guides. The exhaust valve heads are concave and are smaller in diameter than the inlet valves.

To replace the valves place each valve into its guide and fit the bottom collars, inner and outer valve springs, shrouds (early engines only), and caps. Compress the valve springs and push a new synthetic rubber packing ring over the tip of the valve stem down to the bottom of the cotter groove (see Fig. A.4). Refit the two valve cotters and remove the compressor. Replace the valve circlip.

**NOTE.—Do not fit old valve packing rings, or oil sealing may suffer. The rings are fitted more easily if they have been soaked in clean engine oil for a short period before use.**

(See Editor's notes at end of section.)

**Removing and replacing valve guides**

Rest the cylinder head with its machined face downwards on a clean surface and drive the valve guide downwards into the combustion space with a suitable-sized drift. This should take the form of a hardened steel punch $\frac{9}{16}$ in. (14 mm.) in diameter and not less than 4 in. (10 cm.) in length, with a locating spigot $\frac{5}{16}$ in. (7·9 mm.) diameter machined on one end for a length of 1 in. (2·5 cm.) to engage the bore of the guide.

When fitting new valve guides these should be driven in from the top of the cylinder head. The valve guides must be inserted with the end having the largest chamfer at the top. The valve guides should be driven into the combustion spaces until they are the required height above the machined surface of the valve spring seating (see Fig. A.6 and **'GENERAL DATA'**).

**(See Editor's notes at end of section.)**

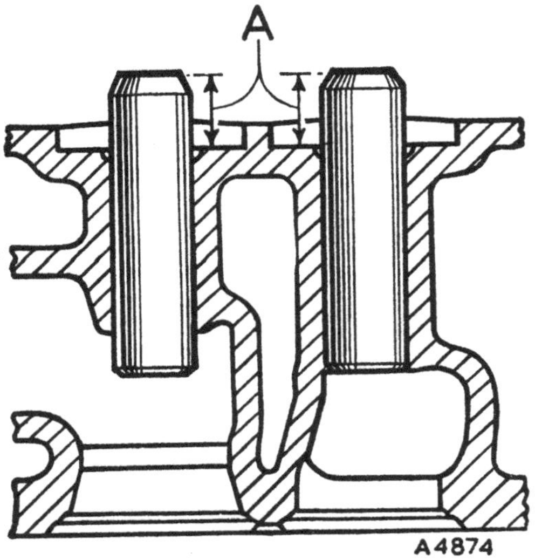


*Fig. A.6*

*When fitting valve guides they must be driven in until they are the required height above the machined surface of the valve spring seating* (A)

**Grinding and testing valves and seatings**

Each valve must be cleaned thoroughly and carefully examined for pitting. Valves in a pitted condition should be refaced with a suitable grinder or new valves should be fitted.

Pitted or uneven valve seats must be re-faced using a grinder or cutter, care being taken that only the minimum quantity of metal necessary to restore the seat is removed. If the seats are very worn or damaged refer to Section A.28.

When grinding a valve onto its seating the valve face should be smeared lightly with fine- or medium-grade carborundum paste and then lapped in with a suction (See Editor's notes at end of section.)

*Fig. A.7*

*Pitted valve seats should be re-faced, using Service tools,* 18*G* 25, 18*G* 25 A, 18*G* 174 B, 18*G* 25 *C*, 18*G* 27, 18*G* 28, 18*G* 28 *A*, 18*G* 28 *B*, 18*G* 28 *C*, and 18*G* 174 *D*

grinder (Service tool 18G 29). Avoid the use of excessive quantities of grinding paste and see that it remains in the region of the valve seating only.

A light coil spring placed under the valve head will assist considerably in the process of grinding. The valve should be ground to its seat with a semi-rotary motion and occasionally allowed to rise by the pressure of the light coil spring. This assists in spreading the paste evenly over the valve face and seat. It is necessary to carry out the grinding operation until a dull, even, mat surface, free from blemish, is produced on the valve seat and valve face.

On completion, the valve seat and ports should be cleaned thoroughly with a rag soaked in paraffin (kerosene), dried, and then thoroughly cleaned by compressed air. The valves should be washed in paraffin (kerosene) and all traces of grinding paste removed.

Fit a new valve packing ring when refitting the valves (see Fig. A.4).

**Checking valve timing**

Set No. 1 cylinder inlet valve to ·055 in. (1·4 mm.) clearance with the engine cold, and then turn the engine until the valve is about to open.

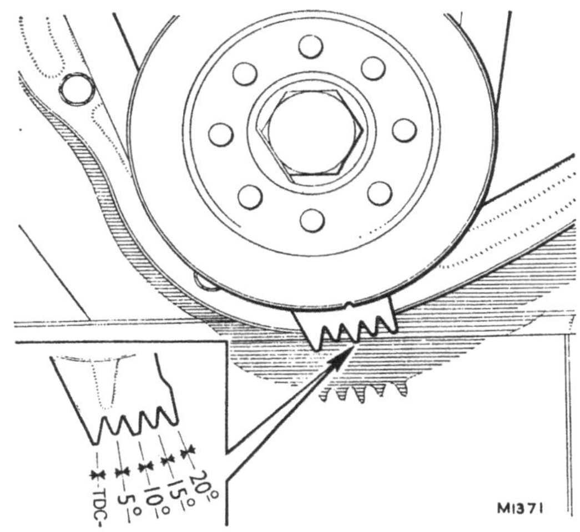


*Fig. A.8*

*The notch in the pulley approaching the T.D.C. position for pistons 1 and 4. The inset shows the timing scale 5° marks*

The indicating notch in the flange of the crankshaft pulley should then be opposite the longest of the three pointers on the timing cover, i.e. the valve should be about to open at T.D.C. and No. 4 piston will be at T.D.C. on its compression stroke.

**Do not omit to reset the inlet valve clearance to the recommended clearance (see 'GENERAL DATA') when the timing check has been completed. The clearance of ·055 in. (1·4 mm.) is necessary to bring the opening position of the valve to T.D.C. It is not possible to check the valve timing accurately with the normal running valve clearance.**

**Adjusting valve rocker clearances**

If the engine is to give its best performance and the valves are to retain their maximum useful life it is

## THE CYLINDER BLOCK COMPONENTS

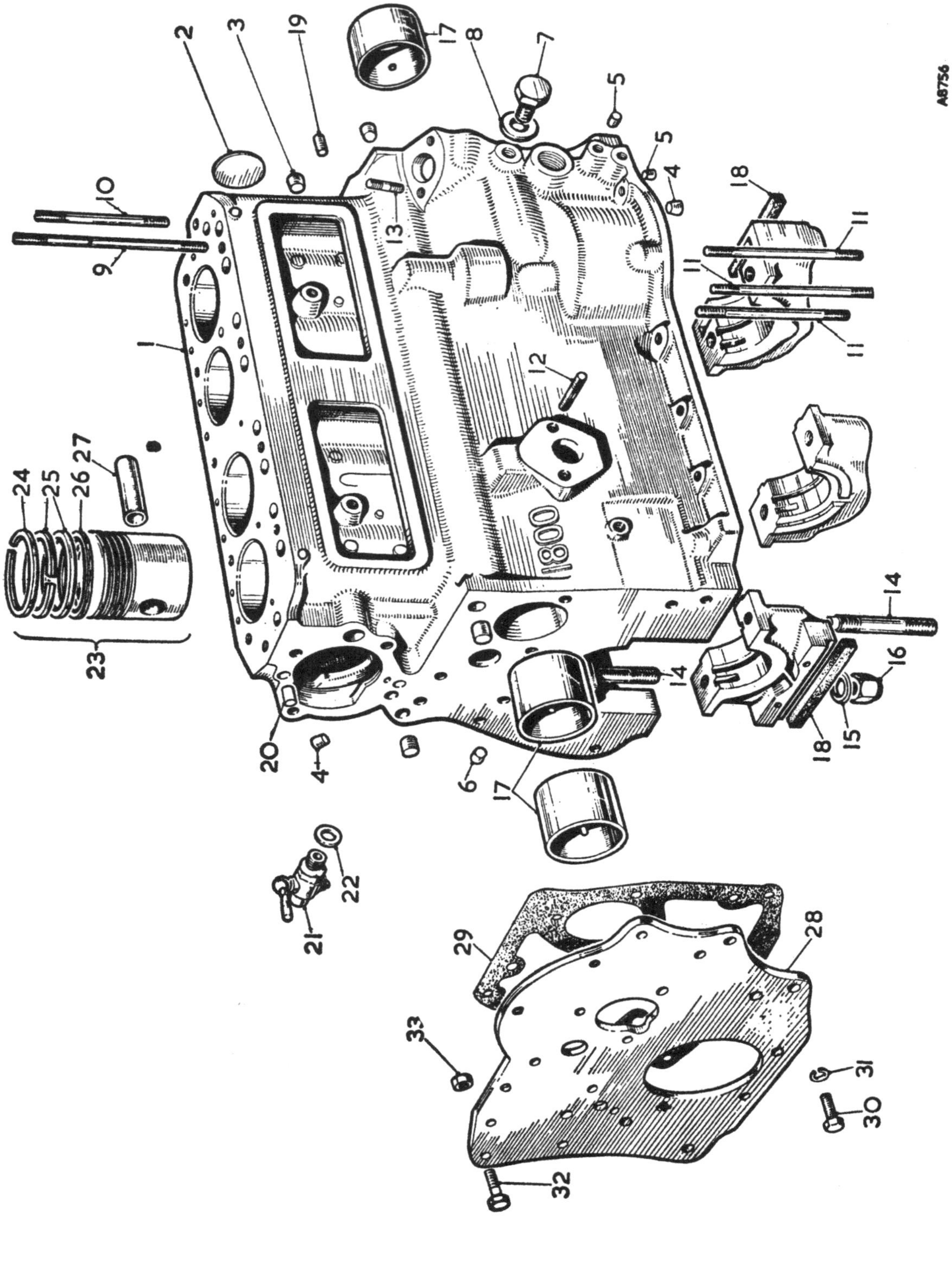

## KEY TO THE CYLINDER BLOCK COMPONENTS

| No. | Description | No. | Description | No. | Description |
|---|---|---|---|---|---|
| 1. | Cylinder block assembly. | 12. | Stud—crankcase vent pipe clip. | 23. | Piston assembly—standard H.C. |
| 2. | Plug—welch—large. | 13. | Stud—tachometer spindle housing. | 24. | Ring—compression—top—standard. |
| 3. | Plug—crankcase oil gallery. | 14. | Stud—main bearing cap. | 25. | Rings—compression—2nd and 3rd—standard. |
| 4. | Plug—taper—crankcase oil hole. | 15. | Washer—plain—main bearing stud. | 26. | Ring—scraper—standard. |
| 5. | Plug—oil relief valve vent hole. | 16. | Nut—main bearing stud. | 27. | Pin—gudgeon. |
| 6. | Plug—chain tensioner oil feed. | 17. | Liner—camshaft bearing. | 28. | Plate—engine mounting. |
| 7. | Plug—oil hole—screwed. | 18. | Joint—front and rear main bearing cap. | 29. | Joint—washer—mounting plate. |
| 8. | Washer—screwed plug. | 19. | Dowel—gearbox mounting plate. | 30. | Screw—mounting plate to crankcase. |
| 9. | Stud—cylinder head—long. | 20. | Dowel—water pump. | 31. | Washer—spring—screw. |
| 10. | Stud—cylinder head—short. | 21. | Tap—drain—cylinder block. | 32. | Bolt—engine mounting bracket to front plate. R.H. top. |
| 11. | Stud—oil pump—short. | 22. | Washer—drain tap. | 33. | Nut—bolt. |

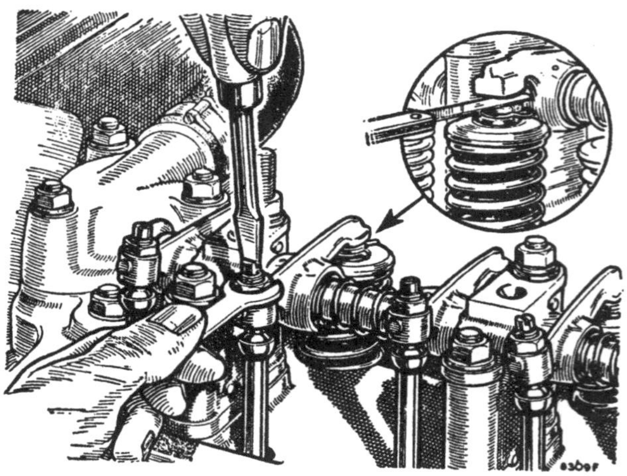

*Fig. A.9*
*The method of adjusting the valve rocker clearance and the correct position for measuring it*
**(See Editor's notes at end of section.)**

essential to maintain the correct valve clearance. Accordingly it is recommended that the clearance be checked at regular intervals and any necessary adjustments made.

For the correct valve rocker clearance refer to **'GENERAL DATA'.** The engine has been designed to operate with this clearance and no departure from it is permissible.

Provision for adjusting the valve clearance is made in the rocker arm by an adjustable screw and locknut.

The rocker adjusting screw is released by slackening the hexagon locknut with a spanner while holding the screw against rotation with a screwdriver. The valve clearance can then be set by carefully rotating the rocker screw while checking the clearance with a feeler gauge. This screw is then re-locked by tightening the hexagon locknut while again holding the screw against rotation.

It is important to note that while the clearance is being set the tappet of the valve being operated upon is on the back of its cam, i.e. opposite to the peak.

As this cannot be observed accurately, the rocker adjustment is more easily carried out in the following order, and this also avoids turning the engine over more than is necessary:

Adjust No. 1 rocker with No. 8 valve fully open.
" No. 3 " " No. 6 " " "
" No. 5 " " No. 4 " " "
" No. 2 " " No. 7 " " "
" No. 8 " " No. 1 " " "
" No. 6 " " No. 3 " " "
" No. 4 " " No. 5 " " "
" No. 7 " " No. 2 " " "

## Section A.8

### TAPPETS

Remove the carburetters (see Section D) and the rocker cover.

Remove the manifolds (see Section A.4).

Disconnect the high-tension leads from the sparking plugs.

Remove the rocker assembly as in Section A.5 and withdraw the push-rods, keeping them in their relative positions to ensure their replacement onto the same tappets. Release the breather pipe, remove the tappet covers, and lift out the tappets, also keeping them in their relative positions.

New tappets should be fitted by selective assembly so that they just fall into their guides under their own weight when lubricated.

Assembly is a reversal of the above procedure, but care should be taken to see that the tappet cover joints are oil-tight and that the rockers are adjusted to give the correct valve clearance.

## Section A.9

### DECARBONIZING

Remove the cylinder head as described in Section A.6.

Withdraw the valves as described in Section A.7

Remove the cylinder head gasket and plug the waterways with a clean rag.

If special equipment is not available for decarbonizing it will be necessary to scrape the carbon deposit from the piston crowns, cylinder block, and cylinder head, using a blunt scraper.

A ring of carbon should be left round the periphery of the piston crown, and the rim of carbon round the top of the cylinder bore should not be touched. To facilitate this an old piston ring can be sprung into the bore so that it rests on top of the piston.

The cylinder head is next given attention. The sparking plugs must be cleaned and adjusted. Clean off the carbon deposit from the valve stems, valve ports, and combustion spaces of the cylinder head. Remove all traces of carbon dust with compressed air or by the vigorous use of a tyre pump and then thoroughly clean with paraffin (kerosene) and dry off.

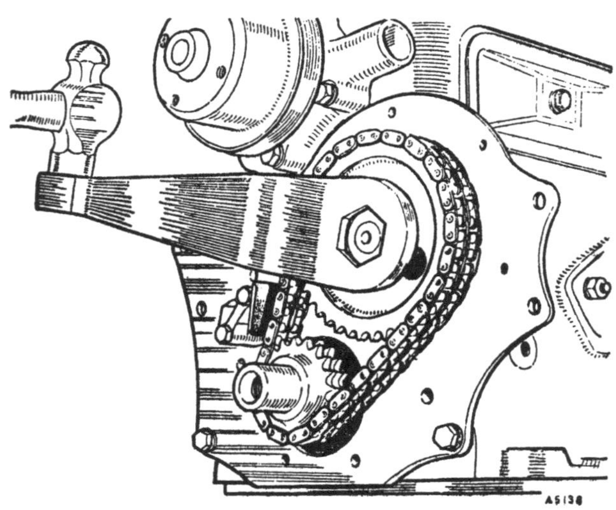

*Fig. A.10*
*Unscrew the camshaft chain wheel securing nut with spanner 18G98A*
**(See Editor's notes at end of section.)**

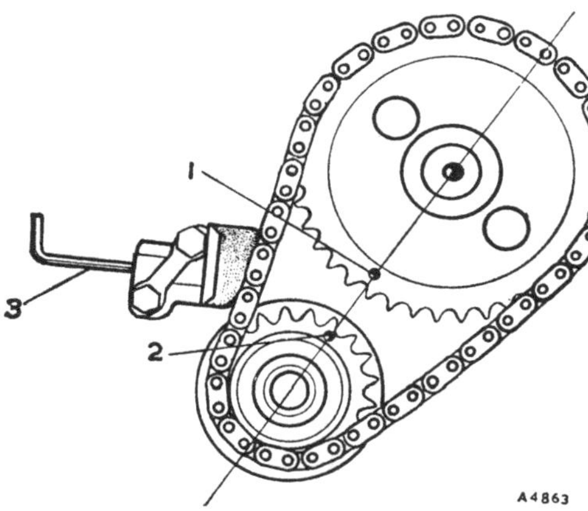


*Fig. A.11*
*Crank the engine until the timing dimples* (1) *and* (2) *are opposite each other before removing the timing chain and chain wheels. The chain tensioner is retracted into the unloaded position by turning the Allen key* (3) *in a clockwise direction*

Fit a new cylinder head gasket when replacing the head if the old has been damaged, noting that the gasket is marked to indicate the top face and the front end.

## Section A.10

### VALVE TIMING GEAR AND CHAIN TENSIONER

**Timing chain cover**

Drain the cooling system and remove the radiator—Section C.7.

Slacken the dynamo or alternator securing bolts and remove the fan belt.

On 18GB and later engines, remove the bolts securing the steering-rack to the body and ease the rack assembly forward sufficiently to allow the crankshaft pulley to be withdrawn.

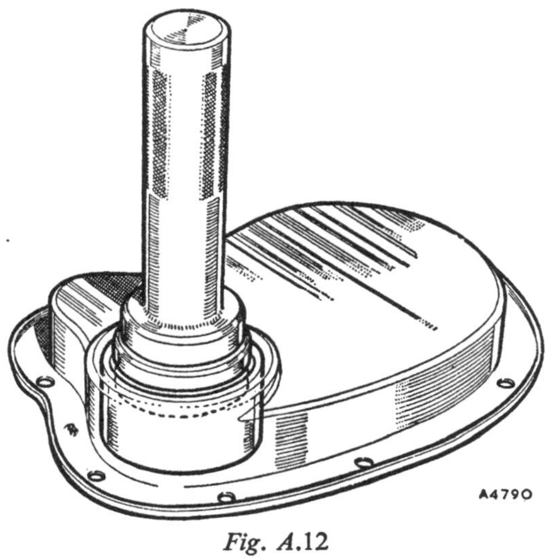


*Fig. A.*12
*Replacing the oil seal in the crankcase front cover using Service tool* 18G 134 *and adaptor* 18G 134 *BD*

Unlock the crankshaft pulley retaining bolt, unscrew the bolt using tool 18G 98 A and withdraw the pulley.

Unscrew the timing cover retaining bolts and remove the cover. Use a new gasket when refitting the timing cover.

The felt washer or oil seal (later cars) in the timing cover should also be renewed if necessary.

Ensure that the oil thrower behind the crankshaft pulley is fitted with the face marked 'F' away from the engine.

Replacement of the timing cover is a reversal of the above procedure.

When refitting the cover it is important to ensure that the seal, rubber or felt, is centralized on the crankshaft, and a Service tool (18G 1046) is available for the purpose.

**NOTE.—The early-type front cover and oil thrower must be used together. When refitting, ensure that the oil thrower is fitted with its concave side facing away from the engine. Use Service tool 18G 3 to centralize the rubber seal on the crankshaft, or use the crankshaft pulley as follows:**

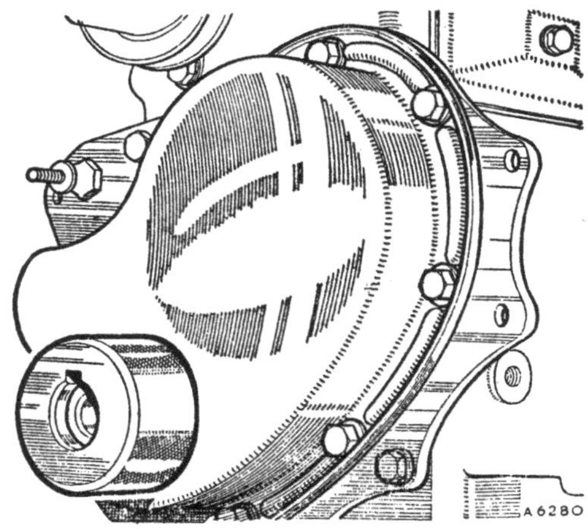


*Fig. A.*13
*Install the securing screws finger tight and centralize the crankcase front cover with the crankshaft, using Service tool* 18G 1046

If a rubber seal is fitted fill the annular groove between the lips with grease. Lubricate the hub of the pulley and push it into the seal, at the same time turning it to avoid damaging the felt or the lips of a rubber seal. Slide the pulley onto the shaft with the keyway in line with the key in the crankshaft. Turn the cover as necessary to align the set screw holes with those in the crankcase, taking care not to strain the cover against the flexibility of the seal.

Insert the set screws and tighten up.

Refit and tighten the pulley securing screw.

**Timing chain**

Remove the bottom plug from the chain tensioner, insert a ⅛ in. (3·18 mm.) Allen key in the cylinder, and turn the key clockwise until the rubber slipper head is fully retracted and locked behind the limit peg.

Unscrew and remove the two securing screws and then remove the tensioner and its backplate.

Unlock and remove the camshaft chain wheel nut and remove the nut and lock washer. Note that the locating tag on the lock washer fits into the keyway of the camshaft chain wheel.

The camshaft and crankshaft chain wheels may now be removed, together with the timing chain, by easing each wheel forward, a fraction at a time, with suitable smaller levers.

As the crankshaft gear wheel is withdrawn care must be taken not to lose the gear packing washers immediately behind it. When reassembling replace the same number of washers as was found when dismantling, unless new camshaft or crankshaft components have been fitted which will disturb the alignments of the two gear wheels. To determine the thickness of washers required place a straight-edge across the sides of the camshaft wheel teeth and measure with a feeler gauge the gap between the straight-edge and the crankshaft gear. Subtract ·005 in. (·13 mm.) from the feeler gauge reading and add the resultant thickness of crankshaft gear packing washers.

When replacing the timing chain and gears, set the crankshaft with its keyway at T.D.C., and the camshaft with its keyway approximately at the one o'clock position when seen from the front. Assemble the gears into the timing chain with the two marks on the gear wheels opposite to each other, as in Fig. A.11. Keeping the gears in this position, engage the crankshaft gear keyway with the key on the crankshaft and rotate the camshaft until the camshaft gear keyway and key are aligned. Push the gears onto the shafts as far as they will go and secure the camshaft gear with the lock washer and nut.

Replace the oil thrower, with the face marked 'F' or the concave side (early-type) away from the engine, and the remaining components.

A4866

*Fig. A.4*

*Checking the chain wheel alignment with a straight-edge and feeler gauge*

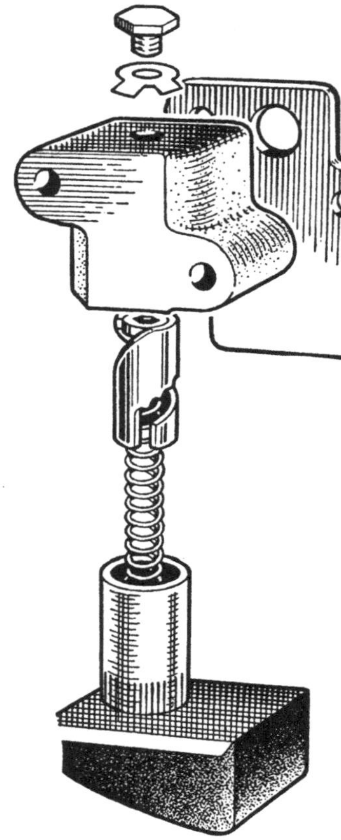

8676B

*Fig. A.15*

*The chain tensioner components*

**Timing chain tensioner**

Remove the tensioner assembly from the block as described above.

Withdraw the plunger and slipper assembly from the tensioner body and engage the lower end of the cylinder with the Allen key. Turn the key clockwise, holding the key and plunger securely until the cylinder and spring are released from inside the plunger.

Clean the components in petrol, and blow out the oil holes in the slipper and spigot.

Check the bore in the adjuster body for ovality. If the ovality is greater than ·003 in. (·076 mm.) when measured on diameters near the mouth of the bore, then a complete new adjuster must be fitted. If within the acceptable limit, then fit a new slipper head and cylinder assembly in the existing body.

It is important that dirt is not allowed to enter the adjuster, so ensure that all parts are clean before reassembly.

When the tensioner is in operation and the engine is running, oil from the lubrication system enters the spigot on the back face under pressure and lubricates the bearing surface through a hole in the slipper pad. The pad is held against the chain by the coil spring.

Should the chain stretch with use, the slipper plunger rises and the limiting peg, bearing on the top of the helical slot, rotates the cylinder until the next recess in the lower edge of the slot comes into line with the limiting peg and prevents the plunger returning to its original position and allowing the timing chain to become slack again.

When reassembling, insert the spring in the plunger and place the cylinder on the other end of the spring.

Compress the spring until the cylinder enters the plunger bore, engaging the helical slot with the peg in the plunger. Hold the assembly compressed in this position and engage the Allen key. Turn the cylinder clockwise until the end of the cylinder is below the peg and the spring is held compressed. Withdraw the key and insert the plunger assembly in the body. Replace the backplate and secure the assembly to the cylinder block.

After refitting the tensioner, check the slipper head for freedom of movement and ensure that it does not bind on the backplate when it is moved in the body.

When the timing chain is in position the tensioner is released for operation by inserting the key and turning it clockwise until the slipper head moves forward under spring pressure against the chain.

**Do not attempt to turn the key anti-clockwise or force the slipper head into the chain by external pressure.**

Secure the bolts with the locking plate, replace the bottom plug, and lock with a tab washer.

(See Editor's notes at end of section.)

## Section A.11

### SUMP AND STRAINER

Drain the radiator and disconnect the hoses, drain the sump, and then release the engine front mounting bolts. Sling the engine and lift it sufficiently to gain access to the front sump bolts.

Remove all the bolts and withdraw the sump from the crankcase.

To remove the oil strainer remove the two bolts securing it to the pump cover.

The strainer may be dismantled for cleaning by removing the centre-nut and bolt and the two delivery pipe flange bolts. Note that there is a locating tongue on the side of the cover which must be positioned correctly when replacing. Remember also to replace the distance tube.

Clean out the sump and strainer with paraffin (kerosene) and a stiff brush; never use rag.

When refitting the sump to the engine give particular attention to the sealing gaskets for the crankcase face and the two oil seal packings for the crankcase which fit into recesses in the crankcase.

If the gaskets are in good condition and have not been damaged during removal of the sump they may be used again, but it is always advisable to fit new ones. Before fitting new gaskets remove all traces of the old ones from the sump and crankcase faces. Smear the faces of the crankcase joint with grease and fit the two halves of the large gasket. Lift the sump into position on the crankcase, insert the 19 bolts, and tighten them evenly.

Refit the engine to its mountings and the hoses to the radiator. Refill the radiator and the sump with coolant and fresh oil respectively.

## Section A.12

### OIL PUMP

Two bolts secure the oil pump cover and three studs secure the pump to the crankcase. Unscrew the stud nuts and remove the pump and drive shaft.

When refitting the pump use a new joint washer.

Unscrew the two securing screws and carefully withdraw the cover, which is located on the base of the oil pump body by two dowels.

Withdraw the outer rotor, and the inner rotor complete with oil pump shaft, from the pump body.

Thoroughly clean all parts in paraffin (kerosene) and inspect them for wear. The rotor end-float and lobe clearances should be checked as follows:

(1) Install the rotors in the pump body, place a straight-edge across the joint face of the pump body, and measure the clearance between the top face of the rotors and the under side of the straight-edge. The clearance should not exceed ·005 in. (·127 mm.). In cases where the clearance is excessive this may be remedied by removing the two cover locating dowels and carefully lapping the joint face of the pump body.

(2) Check the diametrical clearance between the outer rotor and the rotor pocket in the pump body. If this exceeds ·010 in. (·254 mm.) and cannot be remedied by the renewal of either the pump body or the rotors, then the pump assembly should be renewed.

(3) With the rotors installed in the pump body measure the clearance between the rotor lobes when they are in the positions shown in Fig. A.17. If the clearance is in excess of ·006 in. (·152 mm.) the rotors must be renewed.

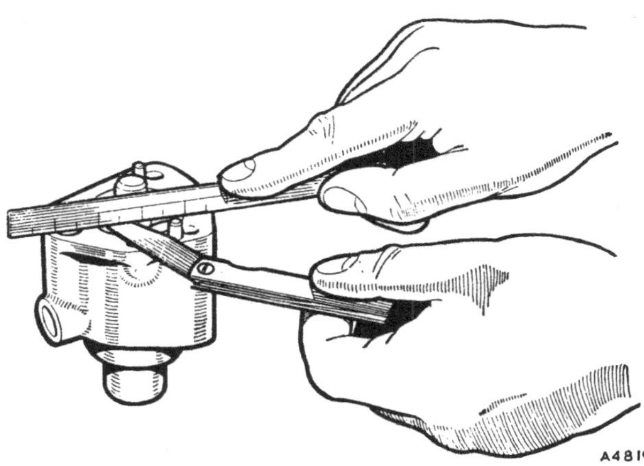


*Fig. A.16*

*Checking the oil pump rotor end-float, which should not exceed ·005 in. (·127 mm.)*

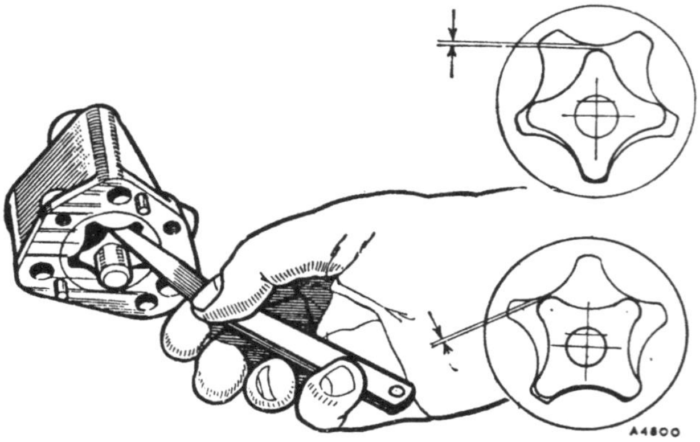


*Fig. A.*17

*The lobe clearance should not exceed* ·006 *in.* (·152 *mm.*) *when the oil pump rotors are in the positions illustrated*

Reassembly is a reversal of the dismantling procedure noting the following points:

(1) Lubricate all parts with clean engine oil.
(2) Ensure that the outer rotor is installed in the pump body with its chamfered end at the driving end of the rotor pocket in the pump body.
(3) After reassembling check the pump for freedom of action.

## Section A.13

### MAIN AND BIG-END BEARINGS

Unless the bearing journals are badly worn the big-end bearings may be renewed without removing the crankshaft. To renew the main bearings it is necessary to withdraw the crankshaft. Liners are used for both the main and the big-end bearings, which are of the shimless type and therefore non-adjustable.

**Big-end bearings**

Drain the engine oil and remove the sump as in Section A.11.

As the bearings are of the shimless type it is essential that no attempt should be made to adjust bearings which are worn. Always fit new bearings in place of worn parts. If the crankshaft journals are found to be in a worn condition it is advisable to fit a Service crankshaft, complete with main and big-end bearings, as supplied by the Factory.

Both the big-end and main bearing liners are located in the bearing housings by a small tag on one side of each half-bearing; it should be noted that the bearings are fitted so that the tags come on the same joint edge of the bearing housing, although on opposite corners.

To detach the big-end bearings, bend down the locking strips so that the bolts may be removed. Remove the connecting rod caps and extract the bearings. Care should be taken to see that the bearing journals are thoroughly cleaned before installing new bearings. No scraping is required, as the bearings are machined to give the correct diametrical clearance of ·0016 in. (·04 mm.).

**Main bearings**

Remove the engine from the car and remove the flywheel and clutch. the timing chain (Section A.10), the sump and strainer (Section A.11), and the rear engine mounting plate.

Note that a thrust washer is fitted on each side of the centre main bearing to take the crankshaft end-thrust. These thrust washers each consist of two semicircular halves, one having a lug which is located in a recess in the detachable half of the bearing and the other being plain.

Before refitting the crankshaft check the end-float (see **'GENERAL DATA'**) and select and fit new upper and lower thrust washers as required. The washers are available in standard thickness and ·003 in. (·076 mm.) oversize.

Remove the two bolts and locking plate securing the front main bearing cap to the engine front bearer plate.

Remove the main bearing cap retaining nuts and locking plates (early engines) or the self-locking nuts and plain washers (later engines).

When fitting new bearings no scraping is required as the bearings are machined to give the correct diametrical clearance of ·001 to ·0027 in. (·025 to ·067 mm.).

In the case of a 'run' bearing it is always essential to clean out thoroughly all the oilways in the crankshaft and block, wash out the engine sump with paraffin (kerosene), and clean the oil pump and sump strainer to ensure that no particles of metal are left anywhere in the lubricating system. The rear main bearing cap horizontal joint surfaces should be thoroughly cleaned and lightly covered with Hylomar Jointing Compound before the cap is fitted to the cylinder block. This will ensure a perfect oil seal when the cap is bolted down to the block. Refit each main bearing and cap, refitting the thrust washers in their correct positions at the centre main bearing with the oil grooves away from the bearing.

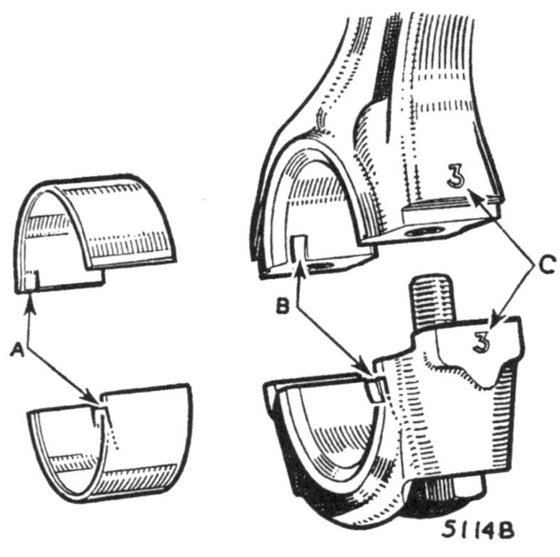


*Fig. A.*18

*The connecting rod big-end bearing locating tags* (A) *and grooves* (B). *The figures* (C) *indicate the cylinder from which the rod and cap were removed*

Lubricate the main bearing cap joint seal liberally with oil before refitting. When refitting the bearing caps on early engines use a new locking plate to lock the nuts; on later engines check that the self-locking nuts lock to the stud threads efficiently. Tighten the bearing cap nuts to the torque figure given in **'GENERAL DATA'.**

**(See Editor's notes at end of section.)**

## Section A.14

### PISTONS AND CONNECTING RODS

Remove the cylinder head as in Section A.6. Drain and remove the sump and oil strainer as in Section A.11.

The pistons and connecting rods must be withdrawn from the top of the cylinder block.

Unlock and remove the big-end bolts and remove the bearing caps. Release the connecting rod from the crankshaft.

Withdraw the piston and connecting rod from the top of the cylinder block and refit the bearing cap. The big-end bearing caps are offset. When used parts are replaced after dismantling it is essential that they should be fitted in their original positions. In order to ensure this, mark the caps and connecting rods on their sides which are fitted together with the number of the cylinder from which each was taken.

Replacement of the piston and connecting rod is a direct reversal of the above, but the piston ring gaps should be set at 90° to each other.

It is essential that the connecting rod and piston assemblies should be replaced in their own bores and fitted the same way round, i.e. with the gudgeon pin clamp screw on the camshaft side of the engine. The piston crowns are marked 'FRONT' to assist correct assembly to the connecting rods.

Refit the big-end bearings in their original positions.

**Pistons and gudgeon pins (18G/18GA)**

The gudgeon pin is rigidly held in the split little-end of the connecting rod by a clamp bolt engaging the central groove of the gudgeon pin.

Before the piston and gudgeon pin can be dismantled

**(See Editor's notes at end of section.)**

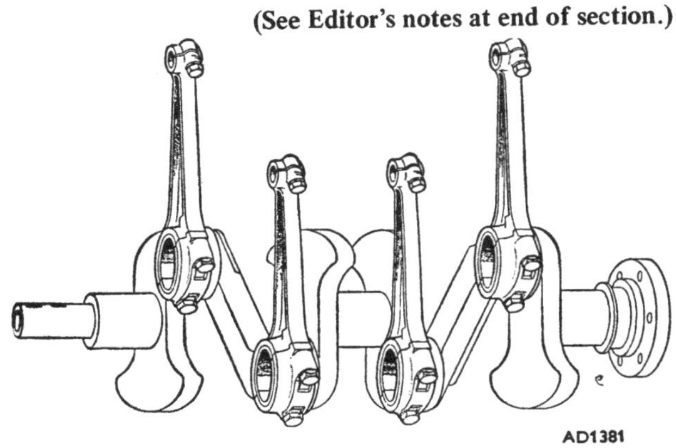


*Fig. A.19*

*Assembly positions of the connecting rods, showing the offsets*

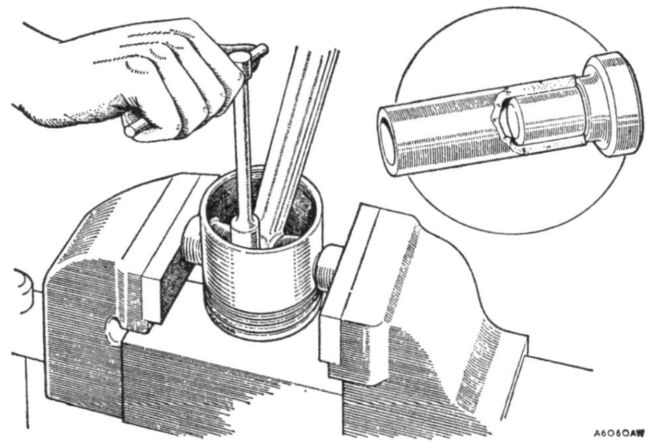


*Fig. A.20*

*Method of loosening and tightening the gudgeon pin clamp bolts* (18*G*/18*GA*)

from the connecting rod it is necessary to remove the clamp screw. To enable the assembly to be held in a vice for this operation without damage holding plugs should be inserted in each end of the gudgeon pin.

Unscrew the gudgeon pin clamp and remove it completely.

Push out the gudgeon pin.

Reassembly is a reversal of the above.

**IMPORTANT.—Attention must be given to the following points when assembling the piston to the connecting rod:**

(1) That the piston is fitted the correct way round on the connecting rod. The crown of the piston is marked 'FRONT' to assist this, and the connecting rod is fitted with the gudgeon pin clamp screw on the camshaft side.

(2) That the gudgeon pin is positioned in the connecting rod so that its groove is in line with the clamp screw hole.

(3) That the clamp screw spring washer has sufficient tension.

(4) That the clamp screw will pass readily into its hole and screw freely into the threaded portion of the little-end, and also that it will hold firmly onto the spring washer.

A certain amount of selective assembly must be used when fitting new gudgeon pins. They must be a thumb-push fit for three-quarters of their travel, to be finally tapped in with a raw-hide mallet. This operation should be carried out with the piston and gudgeon pin cold.

**Piston rings**

If no special piston ring expander is available, use a piece of thin steel such as a smoothly ground hacksaw blade or a disused ·020 in. (·50 mm.) feeler gauge.

Raise one end of the ring out of its groove. Insert the steel strip between the ring and the piston. Rotate the strip around the piston, applying slight upward pressure to the raised portion of the ring until it rests on the land

## THE ENGINE INTERNAL COMPONENTS

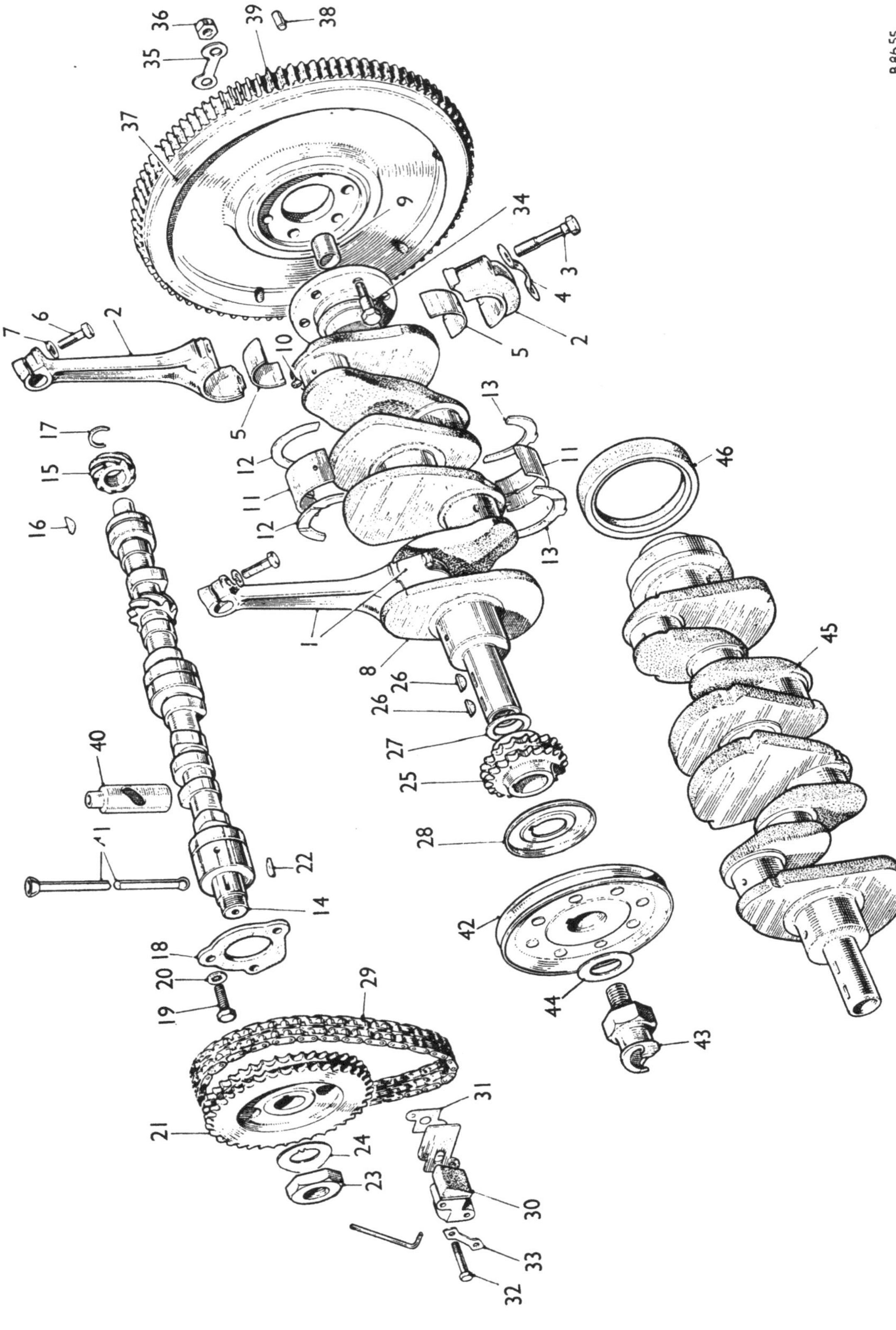

## KEY TO THE ENGINE INTERNAL COMPONENTS

| No. | Description | No. | Description | No. | Description |
|---|---|---|---|---|---|
| 1. | Connecting rod and cap—Nos. 1 and 3 cylinders. | 17. | Ring—spring—tachometer gear. | 32. | Screw—tensioner to crankcase. |
| 2. | Connecting rod and cap—Nos. 2 and 4 cylinders. | 18. | Plate—camshaft locating. | 33. | Washer—lock—screw. |
| 3. | Screw—cap. | 19. | Screw—locating plate to crankcase. | 34. | Bolt—flywheel to crankshaft. |
| 4. | Washer—lock—screw. | 20. | Washer—lock—screw. | 35. | Washer—lock—bolt. |
| 5. | Bearing—connecting rod—standard. | 21. | Gear—camshaft. | 36. | Nut—bolt. |
| 6. | Screw—connecting rod clamping } (18G/18GA). | 22. | Key—camshaft gear. | 37. | Flywheel. |
| 7. | Washer—spring—screw } (18G/18GA). | 23. | Nut—camshaft gear. | 38. | Dowel—clutch to flywheel. |
| 8. | Crankshaft } (18G/18GA). | 24. | Washer—lock-nut. | 39. | Ring gear—starting. |
| 9. | Bush—1st motion shaft. | 25. | Gear—crankshaft. | 40. | Tappet. |
| 10. | Plug. | 26. | Key—crankshaft gear and pulley. | 41. | Push-rod. |
| 11. | Bearing—main—standard. | 27. | Washer—packing—crankshaft gear. | 42. | Pulley—crankshaft. |
| 12. | Washer—thrust—upper. | 28. | Thrower—oil—front—crankshaft. | 43. | Nut—starting. |
| 13. | Washer—thrust —lower. | 29. | Chain—timing. | 44. | Washer—lock—starting nut. |
| 14. | Camshaft. | 30. | Tensioner—chain. | 45. | Crankshaft. } (18GB). |
| 15. | Gear—tachometer driving. | 31. | Joint washer—tensioner. | 46. | Oil seal. } (18GB). |
| 16. | Key—tachometer gear. | | | | |

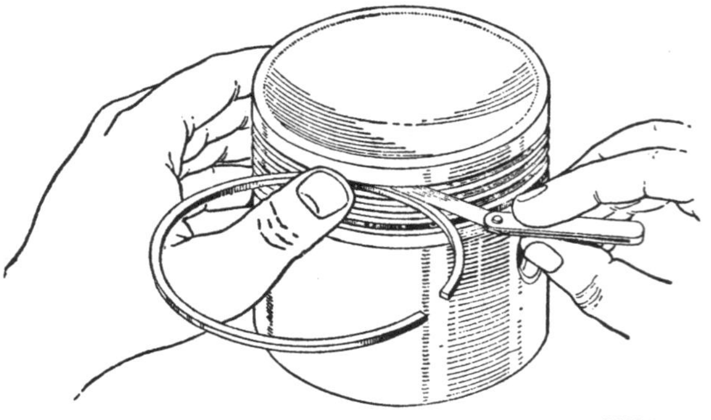


*Fig. A.21*
*Checking the piston ring to groove clearance*

above the ring grooves. It can then be eased off the piston.

Do not remove or replace the rings over the piston skirt, but always over the top of the piston.

Before fitting new rings clean the grooves in the piston to remove any carbon deposit. Care must be taken not to remove any metal or sideplay between the ring and the groove will result, with consequent excessive oil consumption and loss of gas-tightness.

When refitting the rings note that the second and third compression rings are tapered and marked with the letter 'T' (top) for correct reassembly.

**Examination**

Thoroughly clean the bearing shells and the faces of the connecting rods and the bearing caps. Examine the shells for wear, pitting, or cracking, and renew them if necessary. New shells are finished with the correct diametrical clearances and do not require bedding in.

Blow the oil passages in the connecting rods with an air blast and ensure that they are free from obstruction.

Check the connecting rods for twist and bow as shown in Fig. A.22. Do not file the caps or rods.

Check the ring to groove clearance as shown in Fig. A.21.

**(See Editor's notes at end of section.)**

*Fig. A.22*
*Checking the connecting rod alignment*

New rings must be tested in the cylinder bore to ensure that the ends do not butt together. The best way to do this is to insert the piston approximately 1 in. (2·54 cm.) into the cylinder bore, push the ring down onto the top of the piston, and hold it there in order to keep it square with the bore. The correct ring gap is ·008 to ·013 in. (·20 to ·33 mm.).

## Section A.15

### PISTON SIZES AND CYLINDER BORES

In addition to the standard pistons there is a range of oversize pistons available for service purposes.

Oversize pistons are marked with the actual oversize dimensions enclosed in an ellipse. A piston stamped ·020 is only suitable for a bore ·020 in. (·508 mm.) larger

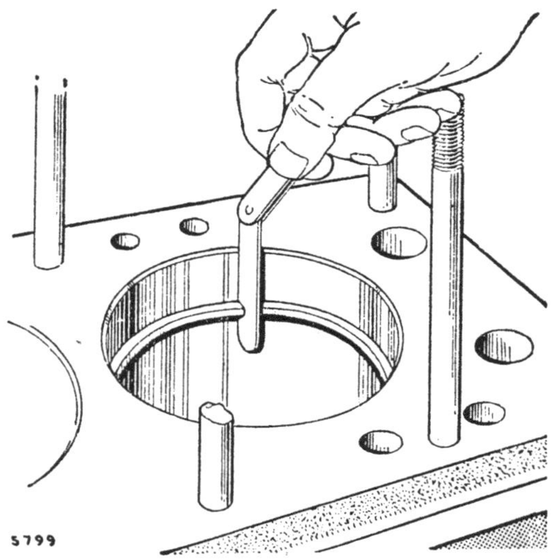


*Fig. A.23*
*Checking a piston ring gap*

than the standard bore; similarly, pistons with other markings are only suitable for the oversize bore indicated.

The piston markings indicate the actual bore size to which they must be fitted, the requisite running clearance being allowed for in the machining.

After reboring an engine, or whenever fitting pistons differing in size from those removed during dismantling, ensure that the size of the piston fitted is stamped clearly on the top of the cylinder block alongside the appropriate cylinder bore.

## Section A.16

### CAMSHAFT

Disconnect the battery.

Remove the inlet and exhaust manifold assembly (see Section A.4).

Remove the push-rods and take out the tappets (see Section A.8).

Remove the timing cover, timing chain, and gears (see Section A.7).

Disconnect the suction advance unit pipe from the distributor and take out the two bolts with flat washers securing the distributor to the housing. Do not slacken the clamping plate bolt or the ignition timing setting will be lost.

Remove the distributor assembly—Sections A.25 and A.26.

Remove the sump, oil pump, and oil pump drive shaft (see Section A.11).

If a drive-type tachometer is fitted, disconnect the drive, remove the securing nuts and washers and withdraw the drive gear.

Take out the three set screws and shakeproof washers which secure the camshaft locating plate to the cylinder block and withdraw the camshaft.

Before reassembly, which is a reversal of the dismantling procedure, assemble the camshaft retaining plate and the chain wheel to the camshaft, and check the camshaft end-float against the dimensions given in the **'GENERAL DATA'** section by measuring the clearance between the retaining plate and the thrust face of the camshaft front journal.

If the end-float is excessive the retaining plate should be renewed.

## Section A.17

### RENEWING THE CAMSHAFT BEARINGS

While the camshaft is removed it is advisable to check the bearing liners for damage and wear. If these are not in good condition they must be removed and new ones fitted.

**(See Editor's notes at end of section.)**

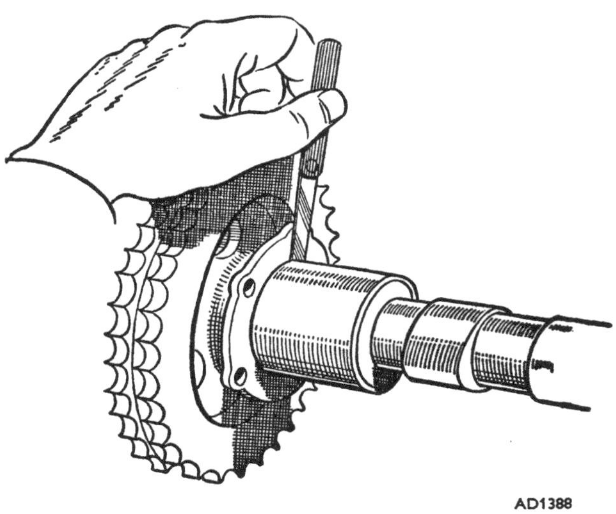


*Fig. A.24*

*Checking the camshaft end-float*

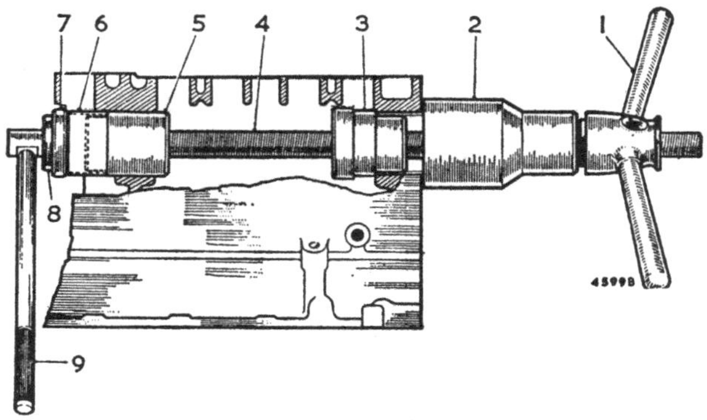


*Fig. A.25*

*Replacing the camshaft centre bearing liner, using Service tool* 18G 124 *A with adaptors* 18G 124 *H and* 18G 124 *C*

1. Wing nut.
2. Body.
3. Adaptor 18G 124 H.
4. Centre screw.
5. Adaptor 18G 124 C.
6. Bearing liner.
7. Large 'C' washer.
8. Slotted washer.
9. Tommy-bar.

The old bearings can be punched out. The new ones must be tapped into position. These bearings are easily damaged and the use of Service tool 18G 124 A is recommended.

This tool comprises a body with built-in thrust race, screw wing nut, stop plate, 'C' washer and handle, and must be used in conjunction with the following adaptors: 18G 124 B, 18G 124 C, 18G 124 F, 18G 124 H.

**Removing the front and rear liners.**

Insert the small end of the adaptor 18G 124 F into the camshaft front liner from the inside of the cylinder block, thread the body of the tool onto the centre screw, and pass the screw through the adaptor from the front of the block. Place the slotted washer on the flat at the rear of the centre screw and insert the tommy-bar into the centre screw behind the slotted washer.

Tighten up the wing nut to withdraw the worn liner.

The rear liner is withdrawn by the same method, using the adaptor 18G 124 B and withdrawing the liner from the rear of the block.

**Removing the centre liner**

Insert the stepped pilot adaptor 18G 124 H into the camshaft liner front bore from the inside of the block and the adaptor 18G 124 C into the centre liner from the rear, small end first.

With the body of the tool positioned on the centre screw, pass the screw through the pilot adaptor and the adaptor in the centre liner.

Place the slotted washer on the flat at the rear of the centre screw and insert the tommy-bar into the screw behind the slotted washer.

Tighten up the wing nut to withdraw the liner.

**Replacing the front and rear liners**

Place the new liner on the smallest diameter of the adaptor 18G 124 F and insert the adaptor into the cam-

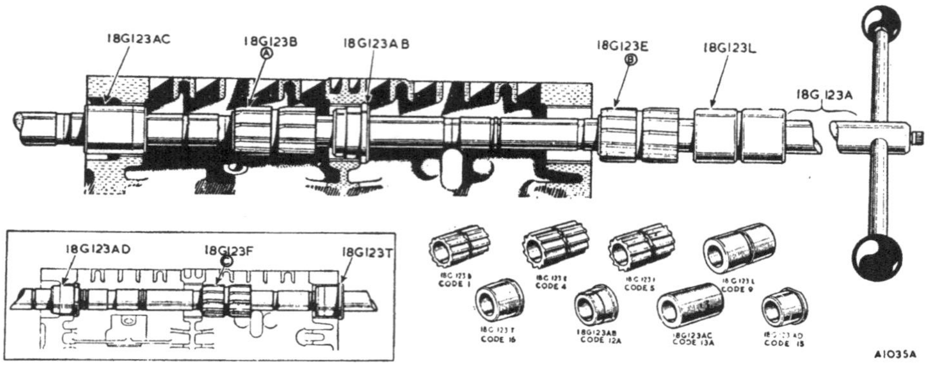


*Fig. A.26*

*Showing the cutters and pilots positioned for reaming the front and rear liners. Inset are the pilots and cutter positioned for reaming the centre liner*

A. Position No. 6. B. Position No. 10. C. Position No. 9.

shaft front liner bore from the inside of the block, largest diameter first.

Line up the oil holes in the liner and the cylinder block and make certain that they remain correctly positioned during the whole operation.

Thread the body of the tool onto the centre screw and pass the screw through the adaptor located in the front liner from the front of the block.

Position the larger of the two 'C' washers on the centre screw with the cut-away portion turned away from the butt joint of the liner; this joint **must** be covered by the washer.

Place the slotted washer on the flat at the rear of the centre screw and insert the tommy-bar into the screw behind the slotted washer.

Tighten the wing nut to pull the liner squarely into position.

The rear liner is replaced by the same method, using the adaptor 18G 124 B and pulling the liner into position from the rear of the block.

**Replacing the centre liner**

Insert the stepped pilot adaptor into the camshaft front liner from the inside of the block.

Place a new centre liner on the small end of the adaptor 18G 124 C and position the adaptor in the centre liner bore from the rear, largest diameter first. Ensure that the oil holes in the liner and the cylinder block are lined up and remain so during the whole operation.

With the body of the tool positioned on the centre screw insert the screw through the pilot adaptor and the adaptor in the centre liner bore.

Position the larger of the two 'C' washers on the centre screw with the cut-away portion turned away from the butt joints of the liner; this joint **must** be covered by the washer.

Place the slotted washer and the tommy-bar in the centre screw and tighten up the wing nut to pull the liner into position.

**Reaming the liners**

Before the camshaft can be reassembled the liners must be reamed in line in order to obtain the correct clearance between the shaft journals and their bearings For this purpose use tool 18G 123 A, which comprises an arbor with tommy-bar and Allen key, and must be used with the following adaptors: 18G 123 B, 18G 123 E, 18G 123 F, 18G 123 L, 18G 123 T, 18G 123 AB, 18 G 123 AC, 18G 123 AD.

**Reaming the front and rear liners**

Insert the taper pilots 18G 123 AB and 18G 123 AC into the centre and rear liners respectively.

Place the plain pilot 18G 123 L on the arbor, followed by the cutter 18G 123 E.

Pass the arbor through the front liner and the pilot located in the centre liner.

Place the cutter 18G 123 B on the arbor and push the arbor through the taper pilot on the rear liner.

Secure the cutters and pilots in their respective positions (see illustration), ensuring that the cutter locating pins are engaged in the correct numbered hole provided in the arbor.

The cutter for the front liner will cut first with the arbor piloting in the centre and rear liners. Clear away the swarf frequently during the operation. The cutter for the rear liner will follow with the arbor piloting in the front and centre liners. Clear away all the swarf before the plain pilot is allowed to enter the front liner.

When the cut in the rear liner is finished free the cutters and withdraw the arbor.

**Reaming the centre liner**

Set up for the second part of the operation by inserting the pilots 18G 123 T and 18G 123 AD in the front and rear liners.

Pass the arbor through the pilot in the front liner and place the cutter for the centre liner on the arbor. Push the arbor through the centre liner and the pilot located in the rear liner.

Secure the cutter and pilots in position, ensuring that the locating pin of the cutter engages the correct numbered hole in the arbor.

Ream the centre liner, release the cutter, and withdraw the arbor.

**IMPORTANT.—It is essential that the cutter flutes are kept clear of swarf at all times during the cutting operation, preferably with air-blast equipment. The cutter should be withdrawn from the liner half-way through the cut and the swarf removed from the cutter and the liner.**

**Feed the reamer very slowly, and keep the cutters dry.**

**The arbors should be lightly lubricated before assembling the cutters and pilots.**

## Section A.18

### REMOVING AND REPLACING THE CRANKSHAFT (Engine Out of Car)

Take off the clutch and the flywheel (see Section A.19), the timing cover, the timing wheels and chain (see Section A.10), the sump and the oil pump strainer (see Section A.11), and the rear engine mounting plate.

Remove the big-end bearing cap and then take off the main bearing caps (see Section A.13).

**Mark each big-end bearing cap and bearing to ensure that it is reassembled to the correct journal, taking care, in the case of the bearings, that they are not damaged or distorted when marking. Punches should not be used for this purpose.**

Lift the crankshaft out of the bearings.

Replacement of the crankshaft is a reversal of the above operations.

**Before replacing the crankshaft thoroughly clean out all oilways.**

Check the condition of the rear engine mounting plate oil seal (18GB engines) and renew if necessary.

Note that each main bearing cap is stamped with a common number which is also stamped on the centre web of the crankcase near the main bearing.

## Section A.19

### REMOVING AND REPLACING THE FLYWHEEL (Engine Out of Car)

Remove the clutch by unscrewing the six bolts and spring washers securing it to the flywheel. Release the bolts a turn at a time to avoid distortion of the cover flange. Three dowels locate the clutch cover on the flywheel.

Unlock and remove the six nuts and three lock plates which secure the flywheel to the crankshaft and remove the flywheel.

When replacing the flywheel ensure that the 1 and 4 timing mark on the periphery of the flywheel is in line with and on the same side as the first and fourth throws of the crankshaft.

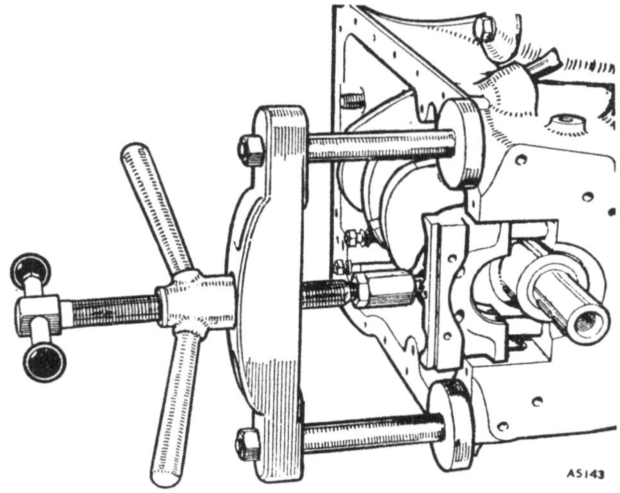


*Fig. A.27*

*Withdrawing the front main bearing cap, using Service tool 18G 42 A and adaptor 18G 42 B*

**(See Editor's notes at end of section.)**

To assist correct location of the flywheel the depression in the crankshaft flange face is stamped with a similar timing mark which should be in line with the one on the flywheel periphery.

To release the special flywheel bolts the engine sump and rear main bearing cap must also be removed.

## Section A.20

### FITTING FLYWHEEL STARTER RINGS

To remove the old starter ring from the flywheel flange split the ring gear with a cold chisel, taking care not to damage the flywheel. Make certain that the bore of the new rings and its mating surface on the flywheel are free from burrs and are perfectly clean.

To fit the new ring it must be heated to a temperature of 300 to 400° C. (572 to 752° F.), indicated by a light-blue surface colour. If this temperature is exceeded the temper of the teeth will be affected. The use of a thermostatically controlled furnace is recommended. Place the heated ring on the flywheel with the lead of the ring teeth facing the flywheel register. The expansion will allow the ring to be fitted without force by pressing or tapping it lightly until the ring is hard against its register.

This operation should be followed by natural cooling, when the 'shrink fit' will be permanently established and no further treatment required.

## Section A.21

### REMOVING AND REFITTING THE ENGINE AND POWER UNIT

It is possible to remove the engine only from the car, but it is recommended that the engine and gearbox are removed together as a complete assembly thus avoiding possible damage to the clutch when separating and refitting the gearbox to the engine within the confines of the car.

## THE ENGINE MOUNTINGS

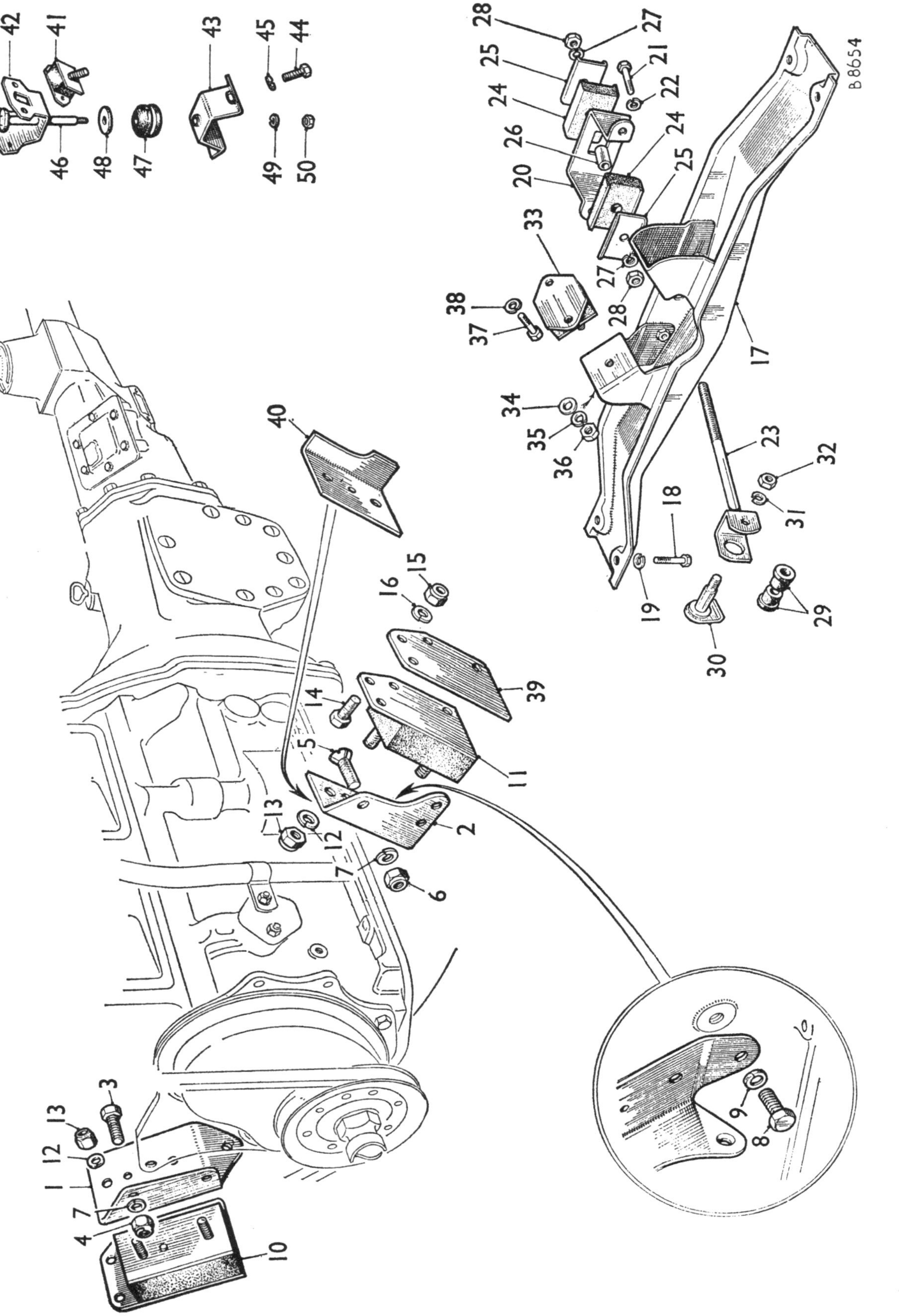

## KEY TO THE ENGINE MOUNTINGS

| No. | Description |
|---|---|
| 1. | Bracket—front engine mounting R.H. |
| 2. | Bracket—front engine mounting L.H. |
| 3. | Screw—bracket to mounting plate—R.H. |
| 4. | Nut for screw. |
| 5. | Screw—bracket to mounting plate—L.H. |
| 6. | Nut for screw. |
| 7. | Washer—spring. |
| 8. | Screw—bracket to block. |
| 9. | Washer—spring. |
| 10. | Rubber—front engine mounting—R.H. |
| 11. | Rubber—front engine mounting—L.H. |
| 12. | Washer—spring—mounting to bracket. |
| 13. | Nut—mounting to bracket. |
| 14. | Screw—mounting and coil bracket to frame bracket. |
| 15. | Nut for screw. |
| 16. | Washer—spring. |
| 17. | Cross-member assembly—rear mounting. |

| No. | Description |
|---|---|
| 18. | Screw—cross-member to frame. |
| 19. | Washer—spring. |
| 20. | Bracket—engine stay-rod. |
| 21. | Screw—bracket. |
| 22. | Washer—spring. |
| 23. | Stay-rod—engine. |
| 24. | Buffer—stay-rod. |
| 25. | Plate—buffer. |
| 26. | Distance tube for stay-rod. |
| 27. | Washer—spring—stay-rod. |
| 28. | Nut—stay-rod. |
| 29. | Bush—shouldered—gearbox rear extension. |
| 30. | Pin—stay-rod. |
| 31. | Washer—spring stay-rod pin. |
| 32. | Nut—stay-rod pin. |
| 33. | Mounting—engine—rear. |
| 34. | Washer—plain—rear mounting to cross-member. |

| No. | Description | |
|---|---|---|
| 35. | Washer—spring—rear mounting to cross-member. | |
| 36. | Nut—rear mounting to cross-member. | |
| 37. | Screw—rear mounting to gearbox. | |
| 38. | Washer—spring. | |
| 39. | Plate—packing—L.H. rubber. | |
| 40. | Control bracket—engine mounting | GT. |
| 41. | Engine mounting—rear | GT. |
| 42. | Bracket—rear engine mounting—upper | GT. |
| 43. | Bracket—rear engine mounting—lower | GT. |
| 44. | Screw—bracket to cross-member | GT. |
| 45. | Washer—spring. | GT. |
| 46. | Pin—rear engine mounting | GT. |
| 47. | Bush—pin | GT. |
| 48. | Washer—plain for pin | GT. |
| 49. | Washer—spring for pin | GT. |
| 50. | Nut for pin | GT. |

**Removing the power unit**

*All models*

(1) Disconnect the batteries.
(2) Remove the bonnet—Section R.9.
(3) Drain the oil from the engine and gearbox.
(4) Drain the coolant from the engine and radiator.
(5) Disconnect the oil cooler and oil pressure gauge pipes from the engine.
(6) Disconnect the top and bottom radiator hoses from the engine.
(7) Remove the oil cooler securing bolts.
(8) Remove the radiator and diaphragm assembly—Section C.6—complete with oil cooler and pipes.
(9) Disconnect the wiring from the dynamo or alternator and the distributor.
(10) Detach the leads from the sparking plugs and remove the distributor cap.
(11) Disconnect the heater hoses and control cable.
(12) Disconnect the thermometer thermal transmitter from the engine.
(13) Disconnect the wiring from the starter.
(14) Disconnect the petrol pipe and choke cable and remove the air cleaners from the carburetters.
(15) Disconnect the tachometer (if drive type).

*Synchromesh gearbox and overdrive*

(16) Remove the gear lever surround, raise the rubber boot, unscrew and remove the lever retaining bolts. Remove the gear lever.
(17) Disconnect the wiring from the reverse light and overdrive isolation switches.
(18) Detach the clutch slave cylinder from the gearbox casing and move it clear of the assembly.

*Automatic gearbox*

(19) Disconnect the down-shift cable from the carburetters.
(20) Disconnect the manual control lever from the gearbox shaft.
(21) Disconnect the wiring from the inhibitor and reverse light switch.

*All models*

(22) Disconnect the speedometer cable from the gearbox.
(23) Take the weight of the engine on a crane and support the gearbox
(24) Disconnect the exhaust pipe from the manifold and release the pipe clip from the gearbox.
(25) Disconnect and remove the propeller shaft—Section G.2.
(26) Remove the bolts securing the engine front mountings to the frame.

*Early cars*

(27) Remove the bolts securing the rear cross-member to the chassis frame and allow the gearbox to rest on the fixed cross-member.
(28) Remove the stay rod from the gearbox.
(29) Unscrew the bolts securing the rear mountings to the gearbox and remove the cross-member complete with stay-rod.

*Later cars*

(30) Remove the four bolts securing the cross-member to the chassis frame.
(31) Remove the two bolts securing the bottom tie bracket to the cross-member.
(32) Lower the gearbox so that it rests on the fixed cross-member, remove the rear mounting securing nuts and remove the cross-member.

*All models*

(33) Ease the assembly forward until the gearbox is clear of the cross-member, tilt the assembly and lift from the car.
(34) Remove the gearbox from the engine as described in Section F.1, Section Fb.1, or Section Fd.10.

**Refitting the engine**

(35) Reverse the removing procedure, refill the engine and gearbox with a recommended lubricant, and refill the cooling system with coolant.

## Section A.22

### ENGINE CONTROLS

The engine controls comprise the accelerator pedal and the choke control. The accelerator pedal assembly pivots about a bush and distance piece that is supported by a bracket at the top of the pedal box. An accelerator pedal stop is bolted on the front of the pedal box and a pedal return stop on the pedal lever abuts a return stop screw in the top of the pedal box.

The upper end of the lever is slotted to accommodate a nipple on the inner end of the accelerator inner cable. The cable passes through a guide bolted to the bulkhead, through an outer cable, and then through a guide on the top of the heat shield. The outer end of the cable is clamped to a pin attached to the spindle and lever assembly that operates the throttles. The outer cable ends abut the bulkhead guide and the heat shield guide.

One end of an accelerator return spring is attached to the spindle and lever assembly and the other end to a clip at the bottom of the heat shield.

The choke control is mounted on the fascia adjacent to the ignition/starter switch and comprises a knob, inner cable, and outer cable assembly. The cables pass through a rubber grommet in the bulkhead.

The end of the inner cable is clamped to a choke control bracket secured to the front carburetter flange by the air cleaner securing bolts. The outer cable terminates in an abutment that is pinned to the carburetter choke operating lever. When the control is operated the inner cable length is reduced between the bulkhead guide and clamp pin. The outer cable therefore moves along the inner cable and operates the choke lever.

The upper ends of the choke return springs are fitted to the cam levers of the forward and rear carburetters and the lower ends to clips on the bottom of the heat shield.

To remove the accelerator control take off the air cleaners as detailed in Section D, release the inner cable from the spindle and lever assembly and withdraw the inner cable from the outer cable at the pedal end.

Release the accelerator cable clip in the engine compartment and remove the outer cable.

Unscrew the pedal assembly fulcrum bolt from the pedal bracket and remove the pedal, bush, and distance piece.

Replacement is a reversal of the removal sequence.

After replacement ensure that the control is adjusted correctly as detailed in Section D.

To remove the choke control release the inner cable from the pin on the choke control bracket and retain the cable abutment.

Free the grommet from the bulkhead.

From the rear of the fascia panel unscrew the securing nut and withdraw the assembly from the fascia.

Replacement is a reversal of the removal sequence.

After replacing the accelerator and choke controls ensure that they are correctly adjusted as detailed in Section D.

## Section A.23

### EXHAUST SYSTEM

The exhaust system is made up of a front pipe and junction assembly, a front intermediate pipe, a front silencer, a rear intermediate pipe, and a rear silencer and tail pipe. The assembly is welded together.

The front pipe and junction assembly mate with the exhaust manifold ports; joint washers are fitted to each pipe.

The front intermediate pipe is supported by a clip which is attached to a strap and bracket. The bracket is bolted to the lower flange of the gearbox.

A rubber mounting bolted to the rear frame member supports a bushed housing to which is attached the front silencer mounting bracket.

A second rubber mounting bolted to the rear frame embodies a split clamp that secures the tail pipe.

To remove the exhaust assembly remove the six nuts from the manifold port studs, allow the twin pipes to drop, and retain the joint washers.

Loosen the rear mounting tail pipe securing bolt, remove the front silencer support clip bolt, and draw the assembly forward and downward.

To refit the assembly reverse the removal sequence but leave the front silencer and tail pipe securing bolts slack until the twin down pipes and joint washers have been fitted and tightened. Tighten the silencer clip bolt and finally the split clamp bolt on the tail pipe.

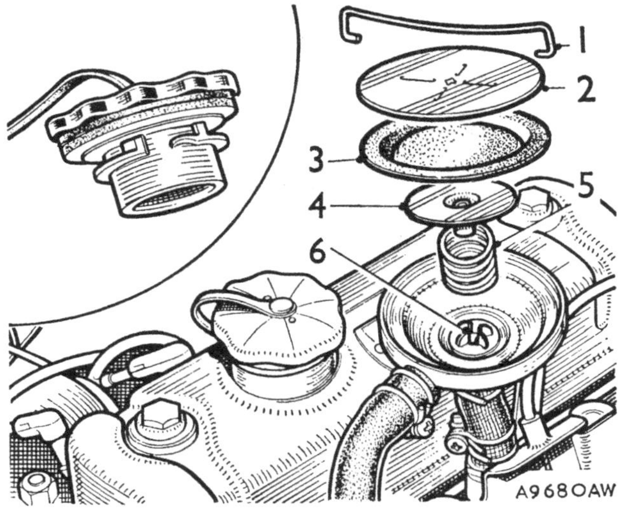


*Fig. A.28*
*(2nd-type)*

*Closed-circuit breathing arrangement. (Inset) Oil filler cap with combined air filter and the breather control valve*

1. Spring clip.
2. Cover.
3. Diaphragm.
4. Metering lever.
5. Spring.
6. Cruciform guides.

## Section A.24

### CLOSED-CIRCUIT BREATHING SYSTEM

(See Supplementary notes at end of section.)

**Oil filler cap**

An air filter is incorporated in the oil filler cap and the two are renewed as an assembly, on engines other than types 18GH, 18GJ or 18GK.

**Breather control valve (18GA, 18GB and 18GD engines only)**

*Testing*

Run the engine at idling speed at normal running temperature. Remove the oil filler cap and if the valve is functioning correctly the engine speed will be heard to rise by approximately 200 r.p.m. If no change in engine speed occurs the valve must be serviced or renewed.

*Servicing*

Remove the spring clip and dismantle the valve.

Clean all metal parts with a solvent (trichlorethylene, fuel, etc.). Do not use an abrasive. If deposits are difficult to remove, immerse the parts in boiling water before applying the solvent.

Clean the diagram with detergent or methylated spirits.

Replace components showing signs of wear or damage.

Reassemble the valve, making sure the metering needle is in the cruciform guides and the diaphragm is seated correctly.

**NOTE.—The 1st-type valve assembly (without the cruciform guides) is serviced as an assembly.**

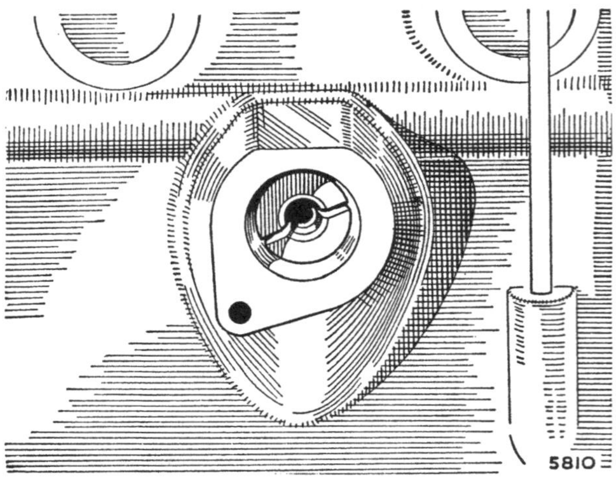


*Fig. A.29*
*The distributor drive with the slot in the correct position and the large offset uppermost*

## Section A.25

### DISTRIBUTOR DRIVING SPINDLE (18G/18GA)

**Removing**

Remove the distributor as detailed in Section B.

Take out the screw securing the distributor housing to the cylinder block and withdraw the housing.

Screw a $\frac{5}{16}$ in. UNF. bolt approximately $3\frac{1}{2}$ in. (89 mm.) long into the threaded end of the distributor drive spindle and withdraw the spindle.

**Refitting**

Turn the crankshaft until No. 1 piston is at T.D.C. on its compression stroke. When the valves on No. 4 cylinder are 'rocking' (i.e. exhaust just closing and inlet just opening) No. 1 piston is at the top of its compression stroke. If the engine is set so that the groove in the crankshaft pulley is in line with the largest pointer on the timing chain cover, or the dimples in the crankshaft and camshaft gears are in line, the piston is exactly at T.D.C.

Screw the $\frac{5}{16}$ in. by $3\frac{1}{2}$ in. UNF. bolt into the threaded end of the distributor drive gear and, holding the drive gear with the slot just below the horizontal and the large offset uppermost, enter the gear. As the gear engages with the camshaft the slot will turn in an anti-clockwise direction until it is approximately in the two o'clock position.

Remove the bolt from the gear, insert the distributor housing, and secure it with the special bolt and washer.

Ensure that the correct bolt is used and that the head does not protrude above the face of the housing.

Refit the distributor, referring to Section B.5 if the clamp plate has been released.

## Section A.26

### DISTRIBUTOR DRIVING SPINDLE (18GB)

**Removing**

Remove the distributor as detailed in Section B.

Take out the screw securing the distributor housing to the cylinder block and withdraw the housing.

Screw a $\frac{5}{16}$ in. UNF. bolt approximately $3\frac{1}{2}$ in. (89 mm.) long into the threaded end of the distributor drive spindle and with the crankshaft at 90° or A.T.D.C. (pistons halfway up the bores) withdraw the spindle.

**Refitting**

Screw the $\frac{5}{16}$ in. UNF. bolt into the threaded end of the distributor drive spindle and with the crankshaft in the 90° B. or A.T.D.C. position, enter the spindle.

Turn the crankshaft until No. 1 piston is at T.D.C. on its compression stroke. When the valves on No. 4 cylinder are 'rocking' (i.e. exhaust just closing and inlet just opening) No. 1 piston is at the top of its compression

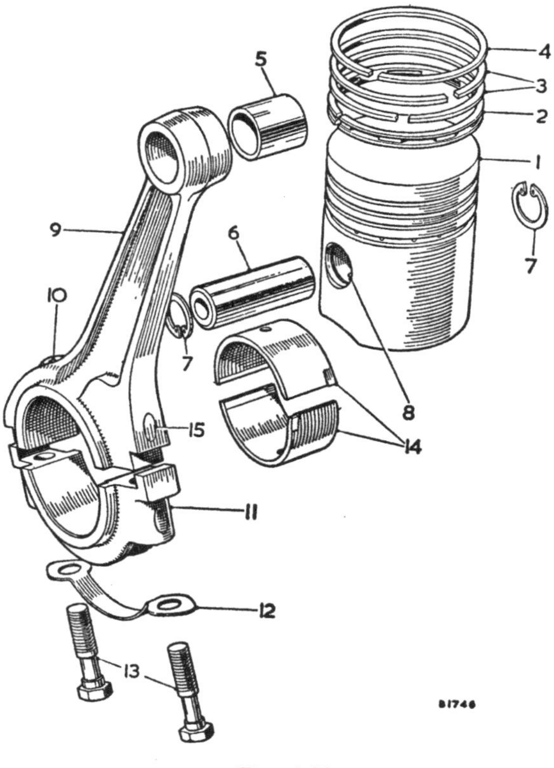


*Fig. A.30*
*Piston and connecting rod*

1. Piston.
2. Piston ring—scraper.
3. Piston rings—taper.
4. Piston ring—parallel.
5. Small-end bush.
6. Gudgeon pin.
7. Circlip.
8. Gudgeon pin lubricating hole.
9. Connecting rod.
10. Cylinder wall lubricating jet.
11. Connecting rod cap.
12. Lock washer.
13. Bolts.
14. Connecting rod bearings.
15. Connecting rod and cap marking.

stroke. If the engine is set so that the groove in the crankshaft pulley is in line with the largest pointer on the timing chain cover, or the dimples in the crankshaft and camshaft gears are in line, the piston is exactly at T.D.C.

Withdraw the spindle to clear the camshaft gear and holding the drive gear with the slot just below the horizontal and the large offset uppermost, re-enter the gear. As the gear engages with the camshaft the slot will turn in an anti-clockwise direction until it is approximately in the two o'clock position.

Remove the bolt from the gear, insert the distributor housing, and secure it with the special bolt and washer.

Ensure that the correct bolt is used and that the head does not protrude above the face of the housing.

Refit the distributor, referring to Section B if the clamp plate has been released.

## Section A.27

### PISTONS AND CONNECTING RODS (18GB)

For removing and refitting pistons, piston rings and connecting rods see Section A.14.

**Dismantling**

The gudgeon pins are fully floating; remove the two circlips locating each pin and press the pins out. It is essential that the piston assemblies should be replaced in their own bores and fitted the same way round: they should be marked to facilitate this.

**Reassembling**

Assemble the pistons to the connecting rods with the gudgeon pin, which should be a hand push fit at a room temperature of 20° C. (68° F.). Secure each pin in its piston with two circlips, ensuring that they fit well into their grooves.

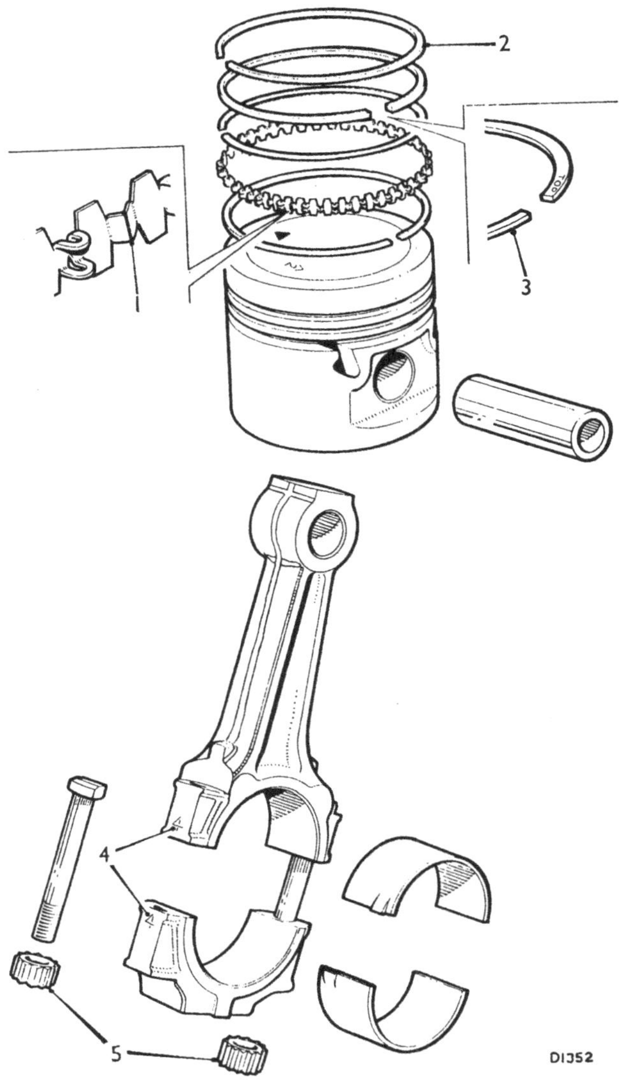


*Fig. A.31*

*A piston and connecting rod assembly with a press fit gudgeon pin*

1. Expander rail must butt.
2. Top compression ring.
3. Second compression ring.
4. Connecting rod and cap identification.
5. Multi-sided nut.

## Section A.28

### OIL COOLER

**Removing**

(1) Disconnect the oil cooler pipes from the oil filter, cylinder block and oil cooler connections.

(2) Remove the four oil cooler securing bolts and remove the cooler.

(3) Withdraw the pipes from the radiator diaphragm grommets.

**Refitting**

(4) Reverse the removing procedure.

## Section A.29

### PISTONS AND CONNECTING RODS

**(Press fit gudgeon pin)**

Pistons and connecting rod assemblies having a press fit gudgeon pin are fully interchangeable in complete sets with earlier assemblies having a fully floating gudgeon pin.

**Removing**

(1) Remove the power unit (Section A.21).

(2) Remove the engine sump.

(3) Remove the cylinder head assembly (Section A.6).

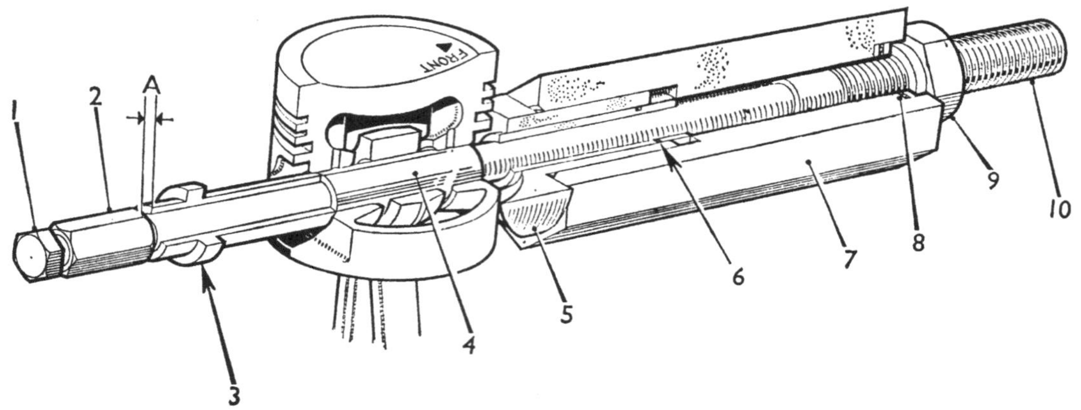


*Fig. A.32*

*Removing a gudgeon pin using Service tool 18G 1150 and adaptor 18G 1150 C*

'A' = $\frac{1}{32}$ in. (·8 mm.).
1. Lock screw.
2. Stop nut.
3. Flange away from gudgeon pin remover/replacer bush.
4. Gudgeon pin.
5. Piston support adaptor.
6. Groove in sleeve away from gudgeon pin.
7. Service tool body.
8. Thrust race.
9. Large nut.
10. Centre screw.

(4) Mark the bearing caps and connecting rods for reassembly, remove the big-end nuts and withdraw the bearing caps.
(5) Withdraw the piston and connecting rod assemblies from the top of the cylinder block and refit the bearing caps.

**Dismantling**

(6) Hold the hexagon body of 18G 1150 in a vice.
(7) Screw the large nut back until it is flush with the end of the centre screw, ensure the screw and large nut are well lubricated, and push the screw in until the nut contacts the thrust race.
(8) Fit adaptor 18G 1150 C on to the centre screw with the piston ring cut-away uppermost.
(9) Slide the parallel sleeve, groove end first, onto the centre screw.
(10) Fit the piston, 'FRONT' or ▲ towards the adaptor, onto the centre screw.
**IMPORTANT.**—The gudgeon pin bore is offset and it is essential that the 'FRONT' or ▲ (marked on the piston crown) of the piston is fitted against the adaptor face.
(11) Fit the remover/replacer bush on the centre screw, flange end towards the gudgeon pin.
(12) Screw the stop nut onto the centre screw and adjust it until approximately $\frac{1}{32}$ in. (·8 mm.) end-play ('A' Fig. A.32) exists on the assembly and lock the stop nut securely with the lock screw.
(13) Check that the remover/replacer bush and parallel sleeve are correctly positioned in the bore on both sides of the piston.
(14) Check that the curved face of the adaptor is clean and slide the piston on the tool so that it fits into the curved face of the adaptor with the piston rings over the cut-away.
(15) Screw the large nut up to the thrust race.
(16) Hold the lock screw, and turn the large nut (use 18G 587) until the gudgeon pin is withdrawn from the piston.
(17) Remove the piston rings over the crown of the piston.

**Reassembling**

*Piston rings*

(18) Enter the piston rings squarely into the top of the cylinder bore and check the gaps (see **'GENERAL DATA'**).
(19) Check the ring to piston groove clearances.
(20) Fit the bottom rail of the oil control ring to the piston and position it below the bottom groove.
Fit the oil control expander into the bottom groove and move the bottom oil control ring rail up into the bottom groove. Fit the top oil control rail into the bottom groove.
(21) Check that the ends of the expander are butting but not overlapping (see Fig. A.31) and set the gaps of the rails and the expander at 90 degrees to each other.
(22) Fit the thinner of the two compression rings into the second groove with the face marked 'TOP' uppermost.
(23) Fit the top compression ring.

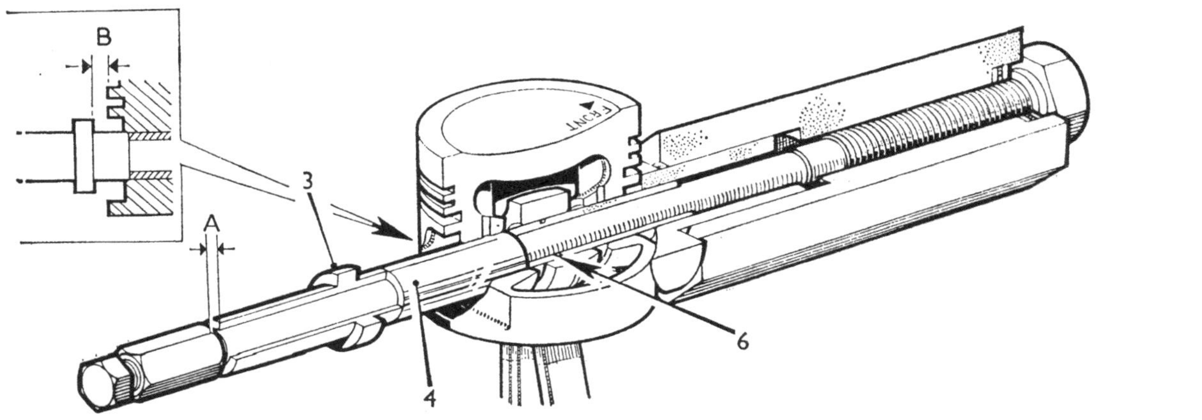


*Fig. A.33*

*Fitting a gudgeon pin using Service tool* 18G 1150C

'A' = $\frac{1}{32}$ in. (·8 mm.) end-play.
'B' = ·04 in. (1 mm.) from piston.

3 Flange towards gudgeon pin remover/replacer bush.
4. Gudgeon pin.
6. Groove in sleeve towards gudgeon pin.

*Gudgeon pin*

(24) Remove the large nut of 18G 1150 and pull the centre screw out of the body a few inches. Ensure the nut and screw are well lubricated, and the piston support adaptor is in place.

(25) Slide the parallel sleeve groove end last onto the centre screw and up to the shoulder.

(26) Lubricate the gudgeon pin and bores of the connecting rod and piston with graphited oil (Achesons Colloids 'Oildag').

(27) Fit the connecting rod and piston 'FRONT' to the tool, with the connecting rod entered on the sleeve up to the groove.

(28) Fit the gudgeon pin into the piston bore up to the connecting rod.

(29) Fit the remover/replacer bush flange end towards the gudgeon pin.

(30) Screw the stop nut onto the centre screw; adjust the nut to give a $\frac{1}{32}$ in. (·8 mm.) end play ('A' Fig. A.33) and lock the nut securely with the lock screw.

(31) Check that the curved face of the adaptor is **clean** and slide the piston on the tool so that it fits into the curved face of the adaptor with the piston rings over the adaptor cut-away.

(32) Screw the large nut up to the thrust race.

(33) Set the torque wrench 18G 537 to 12 lb. ft. (1·64 kg. m.). This represents the **minimum load for an acceptable fit.**

(34) Using the torque wrench and 18G 587 on the large nut, and holding the lock screw, pull the gudgeon pin in until the flange of the remover/replacer bush is ·04 in. (1 mm.) from the piston skirt. Under no circumstances must the flange be allowed to contact the piston. Remove the service tool.

**NOTE.**—If the torque wrench **has not broken** throughout the pull, the fit of the gudgeon pin to the connecting rod is not acceptable and necessitates the renewal of components. The large nut and centre screw of the tool must be kept well oiled.

(35) Check that the piston pivots freely on the pin, and is free to slide sideways. If stiffness exists, wash the assembly in fuel or paraffin (kerosene), lubricate the gudgeon pin with graphited oil (Achesons Colloids 'Oildag') and re-check. If stiffness persists, dismantle and check for ingrained dirt or damage.

(36) Check the piston and connecting rod for alignment.

**Refitting**

(37) Reverse the removal procedure noting the following points.

(*a*) Lubricate the piston rings with graphited oil and stagger the compression ring gaps at 90° to each other.

(*b*) Compress the piston rings using Service tool 18G 55 A.

(*c*) Ensure that each connecting rod and piston is refitted into its original bore, the correct way round.

(*d*) Check that the big-end bearings are correctly located in the connecting rods and caps. Tighten the nuts to the torque figure given in **'GENERAL DATA'.**

## Section A.30

### VALVE SEAT INSERTS

Valve seats that cannot be restored to their original standard by normal cutting and refacing should have valve seat inserts fitted.

To fit an insert, machine the cylinder head to the dimensions given (see Fig. A.34). Press in the insert, which will be an interference fit of ·0025 to ·0045 in. (·06 to ·11 mm.). Grind or machine the seats to the dimensions given. Normal valve 'grinding-in' may also be necessary to ensure efficient valve seating.

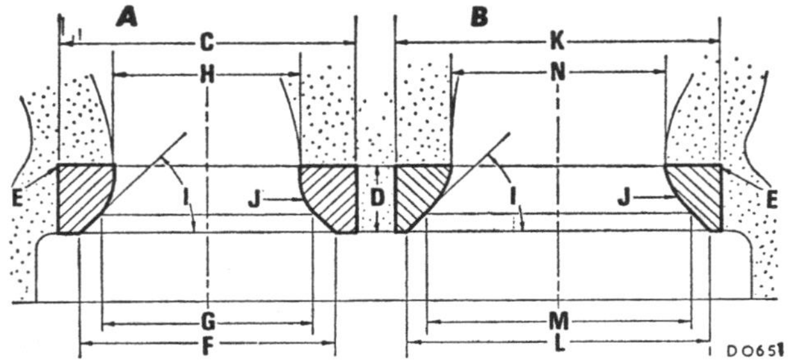


*Fig. A.34*

*Valve seat machining dimensions*

| | Exhaust (A) | | Inlet (B) |
|---|---|---|---|
| C. | 1·437 to 1·438 in. (36·50 to 36·53 mm.). | K. | 1·592 to 1·593 in. (40·43 to 40·46 mm.). |
| D. | ·186 to ·188 in. (4·72 to 4·77 mm.). | D. | ·186 to ·188 in. (4·72 to 4·77 mm.). |
| E. | Maximum radius ·015 in. (·38 mm.). | E. | Maximum radius ·015 in. (·38 mm.). |
| F. | 1·33 to 1·35 in. (33·78 to 34·29 mm.). | L. | 1·552 to 1·572 in. (39·42 to 39·93 mm.). |
| G. | 1·218 to 1·228 in. (30·94 to 31·19 mm.). | M. | 1·427 to 1·447 in. (36·25 to 36·75 mm.). |
| H. | Throat diameter 1·146 to 1·166 in. (29·11 to 29·62 mm.). | N. | Throat diameter 1·302 to 1·322 in. (33·07 to 33·58 mm.). |
| I. | 45°. | I. | 45°. |
| J. | Blend from seating to throat diameter. | J. | Blend from seating to throat diameter. |

## SUPPLEMENTARY NOTES

### CLOSED-CIRCUIT BREATHING SYSTEM

In order to conform to regulations in some countries concerning air pollution, later engines (from serial numbers prefixed 18GA) are not vented to atmosphere by a simple vent pipe. Instead, crankcase fumes enter a modified cylinder side cover which incorporates a mesh filter element and which acts as an oil separator returning fluid oil to the engine sump. A rubber hose connects the cylinder side cover to a breather control valve which is in turn connected by a short rubber hose to an adaptor fitted to the induction manifold.

Crankcase fumes are thus breathed to the induction manifold via the breather control valve which is of the diaphragm type and acts as a non-return valve.

## EDITOR'S NOTES

### A. The Engine

*Removing and replacing valves*
Any commercially available "C-type" valve spring compressor of proper size may be substituted for service tool 18G45,

*Removing and replacing valve guides*
A better method for removing and replacing guides involves the use of an arbor press and an adaptor similar to the above punch.

*Grinding and testing valves and seatings*
Badly worn or pitted valve seats and faces are best resurfaced by a competent automotive machine shop using power driven grinders. Lightly worn or newly ground valves should be lapped in with a suction hand grinder (similar to service tool 18G29) and valve lapping compound.

*Adjusting valve rocker clearances*
The number one rocker is at the front of the engine, number eight is at the rear.

*Figure A.10*
In the absence of the spanner 18G98, a sturdy box end wrench or socket and flex-handle and hammer may be used.

*Timing chain tensioner*
The check for ovality should be made with a suitable telescoping gauge and micrometer. A less elegant but acceptable check is merely to inspect for excessive wobble of the adjuster body. If "slop" can be easily felt, replacement is in order.

*Main and big-end bearings*
Although the bearing inserts are carefully machined to provide the correct clearance, it is advisable to check the clearance when fitting. A convenient method is to use Plastigage, a thin plastic string. A piece of Plastigage is placed on the journal, the bearing cap is installed and torqued down and then removed. The width of the flattened Plastigage indicates the actual clearance. The crankshaft must not be turned while doing this and all traces of Plastigage should be removed before reinstalling the bearings and cap. An incorrect clearance reading calls for further attention to correct assembly or worn parts.
Any commercially available liquid gasket compound may be substituted for the jointing compound referred to.

*Pistons and Connecting Rods*
If there is a ridge at the top of the cylinder bore, it should be removed with a ridge reamer before removing the piston and rod. Failure to do this may result in broken rings and ring bands on the piston.
When installing the piston and rings it is essential that a piston ring compressor be used (See Service Tools at end of this section). Failure to do so will result in damage to the piston and rings.

*Examination*
Automotive machine shops are equipped to check connecting rod alignment.

*Renewing the Camshaft bearings*
Unless the proper equipment is available, this job is best left to a machine shop.

*Figure A-27*
The front main bearing cap may be removed by very careful topping with a soft-faced hammer in the absence of Service tool 18G42A.

# SECTION B

## THE IGNITION SYSTEM



**† These operations must be followed by an exhaust emission check.** 

## Section B.1

### LUBRICATION

**Distributor**

*Cam bearing*

Lift the rotor off the top of the spindle by pulling it squarely, and add a few drops of thin oil to the cam bearing. Do not remove the cam securing screw which is exposed: there is a clearance between the screw and the inner face of the spindle for the oil to pass.

Replace the rotor with its drive lug correctly engaging the spindle slot and push it onto the shaft as far as it will go.

*Cam*

Lightly smear the cam with a small very amount of grease or, if this is not available, clean engine oil may be used.

*Automatic timing control*

Add a few drops of thin oil through the hole in the contact breaker base through which the cam passes. Do not allow oil to get on or near the contacts. Do not over-oil.

## Section B.2

### IGNITION COIL

**Removing**

Disconnect the switch, contact breaker and H.T. cables, release the coil securing bolts and remove the coil.

**Testing**

An indication of the state of serviceability of the coil may be obtained by assessing the quality of spark

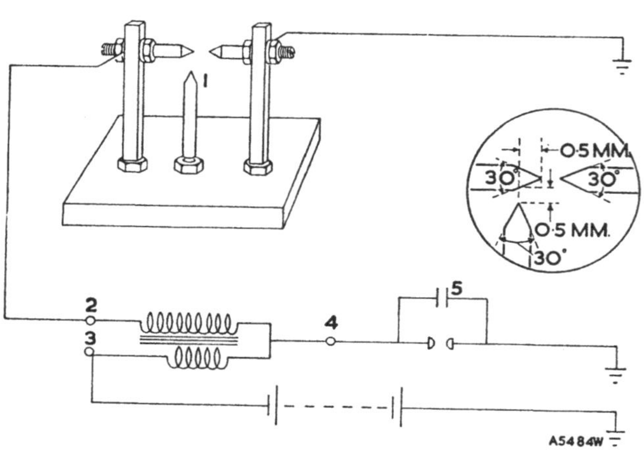


*Fig. B.1*

*Test circuit and gap measurement*

1. Auxiliary electrode.
2. H.T. contact.
3. S.W. contact.
4. C.B. contact.
5. Distributor.

produced by the coil under test conditions. The test circuit and equipment are illustrated in Fig. B.1.

The primary circuit make and break is obtained by using a six-lobe contact breaker cam and graduated scale and adjusting the contact breaker gap until a closed period of between 18 and 20° is obtained. The gap must be accurately set by using feeler gauges on each lobe of the cam in turn. The contact breaker lever spring tension must be between 18 and 20 oz. (510 and 566 gm.) measured at the contacts and the condenser capacity ·18 to ·25 mf.

The three-point test gap equipment is to be connected so that the auxiliary electrode is on the H.T. side of the circuit. This electrode causes the gap to break down at low voltage, a condition used during the setting of the gap for the high-speed performance test.

The coil must be mounted at an angle of 45° on the test rig with the C.B. terminal uppermost and the coil case earthed.

The mounting angle ensures that the primary winding take-off to the L.T. terminal is not covered with fluid and any internal tracking between the iron core and the primary lead will be shown up during test.

Earthing the coil case detects whether or not short circuits to earth are present.

**High-speed performance test**

Connect a peak voltmeter across the spark gap, run the distributor at any speed and then produce irregular sparking by tapping down the battery or by inserting a resistance in the primary circuit.

Adjust the gap until the peak reading on the voltmeter is 8 kV. Remove the voltmeter, and revert to 12 volts or remove the resistance from the circuit.

Run the distributor at 3,750 r.p.m. and check the spark. No missing should occur at this speed.

**Low-speed performance test**

Set the spark gap to ·44 in. (11 mm.) and retain the auxiliary electrode in the circuit. The coil should be at 20° C. (68° F.).

Run the distributor at 100 r.p.m. and note the number of misses. These should not exceed 30 per minute.

**Resistance test**

Remove the coil and connect a Megger across the C.B. and H.T. contacts and note the reading and then across the S.W. and H.T. contacts and note the reading.

The readings must be within the limits given in the **'GENERAL DATA'.**

**(See Editor's notes at end of section.)**

## Section B.3

### DISTRIBUTOR

The Type 25D distributor incorporates a one-piece body-shank casting which houses the distributor drive shaft, automatic advance mechanism, contact breaker plate assembly, capacitor, cam, rotor arm and their associated electrical connections. The direction of rotation is stamped on the body.

The upper end of the body is closed by a bakelite cap which houses the plug lead segments and the H.T. lead brush and spring.

The automatic advance mechanism comprises a centrifugally operated rolling weight mechanism and a vacuum-operated mechanism which together advance the ignition point in proportion to the engine speed and load. The vacuum-operated mechanism has a vernier adjustment screw to enable small adjustments to be made to suit varying grades of fuel or to satisfy tuning requirements. A double-headed arrow marked 'A' and 'R' is stamped on the body adjacent to the vernier screw. A fuel trap is incorporated in the vacuum line from the induction manifold.

The contact breaker assembly is made up of a base plate secured to the distributor body, a moving contact breaker plate, a fixed contact breaker plate, the contact breaker points, and a capacitor.

The moving plate is supported on the base plate by two nylon pads, which minimizes friction when the automatic advance moves the plate. A 'C' spring anchored to a vacuum control spring post bears against the under side of the base plate and so pre-tilts the moving plate. The pressure of the cam on the heel of the contact breaker points supplements this action and so minimizes rocking of the moving plate at high cam speeds.

A stud fixed to the under side of the moving contact plate engages a slot in the base plate and so limits the horizontal movement of the moving plate; the stud also limits the angle of tilt of the moving plate.

The fixed contact breaker plate is secured by a screw which passes through a slot in the fixed plate. A notch cut in the free end of the plate permits the engagement of a screwdriver for adjustment purposes when setting the contact breaker points gap.

The distributor is secured to the engine by a split housing plate and clamp bolt. On some distributors the clamp bolt is trapped and the nut free and on others the nut is trapped and the bolt is free.

## Section B.4

### SERVICING THE DISTRIBUTOR

**Contact breaker points**

Examine the contact breaker points and if they are found to be burnt or blackened remove them as described in Section B.6 and clean them with a fine carborundum stone or emery-cloth.

After cleaning, remove all traces of dust and grease with a petrol-moistened cloth. Lightly smear the contact breaker pivot pin with molybdenized non-creep oil a or suitable grease and refit the points.

It is important that no oil or grease is allowed to contaminate the contact breaker points.

Reset the gap by rotating the crankshaft until the contacts are at their maximum open position, slacken the fixed plate securing screw, insert a screwdriver between the notches in the fixed plate and adjust the gap until it is between ·014 and ·016 in. (·36 and ·40 mm.). Tighten the securing screw and re-check the gap setting.

**Distributor cover**

Thoroughly clean the moulded cover with a petrol-moistened cloth, paying particular attention to the spaces between the electrodes. Examine the cover for cracks and signs of tracking.

Inspect the carbon brush for serviceability and ensure that it moves freely in its holder.

Check the H.T. leads for security and examine them for signs of deterioration. Cracked or perished leads must be renewed.

To fit new H.T. leads, remove the old leads, fill the holes in the distributor cap with Silicone grease, and cut the new leads off to the required length. Push the ends of the leads fully home in the cap and secure them with the securing screws.

Remove the suppressors from the old leads and fit them to the new ones.

**Vacuum advance**

To check the vacuum advance, fit a modified cap having a window cut in the side. Start the engine, operate the throttle sharply, and observe the movement of the moving contact breaker plate.

**Centrifugal advance**

Remove the distributor cover, grasp the rotor firmly, and turn it in the direction of rotation. Release the rotor, when it should return to its original position without showing any tendency to stick.

**Capacitor**

If the capacitor is suspect it may be tested by substitution. Disconnect the suspect capacitor and connect a new one between the L.T. terminal and earth.

Should a new capacitor be necessary, remove the old one and fit the new as described in Section B.6.

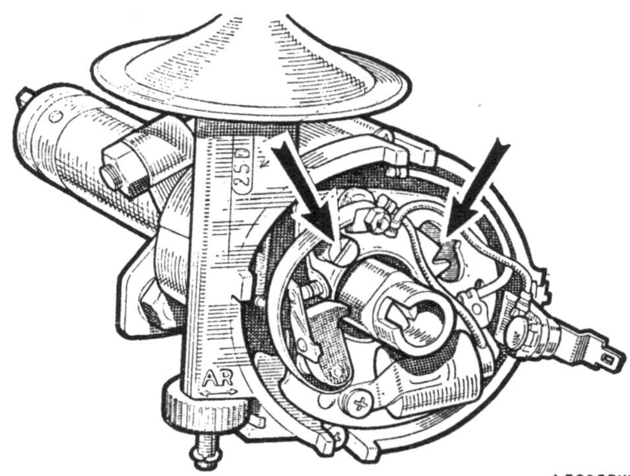


A5325BW

*Fig. B.2*
*The distributor fixed plate securing screw and screw-driver notches*

## Section B.5

### REMOVING AND REPLACING THE DISTRIBUTOR

To remove the distributor from the engine disconnect the suction advance pipe union from the distributor, the H.T. leads from the sparking plugs, and the L.T. lead from the distributor. Remove the two screws securing the split housing plate to the engine and lift the distributor from its housing.

Replacement is a reversal of these instructions, after which the ignition timing must be set or re-checked as described in Section B.9.

If it is desired to remove and replace the distributor without disturbing the timing proceed as follows:

(1) Rotate the crankshaft until the rotor arm is pointing to the segment in the cover for No. 1 cylinder plug lead. This will provide a datum for replacement.

(2) Remove the distributor cover, L.T. lead, and the suction advance pipe union from the distributor.

(3) Remove the two screws securing the split housing plate to the distributor housing. **Do not remove or disturb the split housing plate clamp bolt.**

(4) Withdraw the distributor from its housing.

(5) Insert the distributor into its housing until the driving dogs on the distributor drive shaft rest on the distributor driving spindle.

(6) Slowly rotate the rotor arm until the driving dogs engage the slots in the drive spindle. Both the dogs and the slots are offset to ensure correct replacement.

(7) Turn the distributor body to align the holes in the housing plate with those in the housing and secure the plate with the two screws.

(8) Provided the crankshaft has not been rotated the rotor arm will be opposite the segment for No. 1 plug lead.

(9) Refit the cap, plug leads, L.T. lead, and vacuum pipe union.

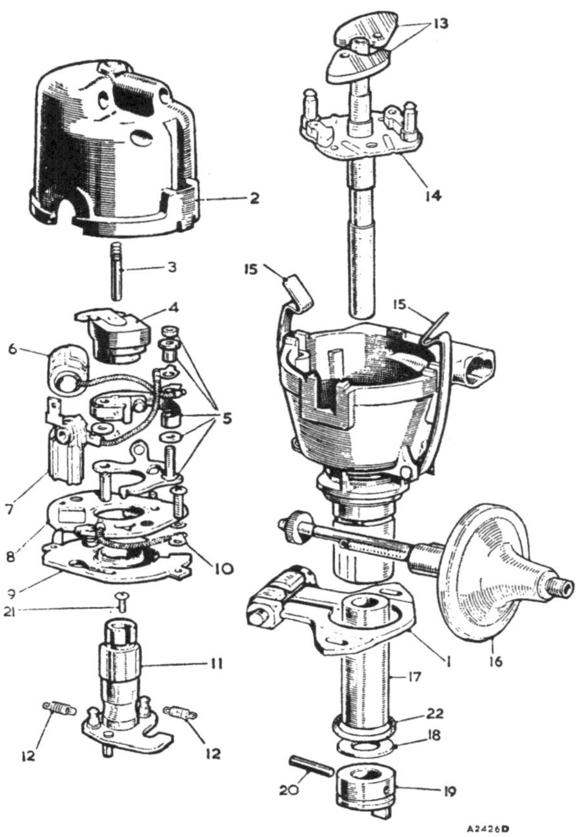


*Fig. B.3*

*The distributor components*

1. Clamping plate.
2. Moulded cap.
3. Brush and spring.
4. Rotor arm.
5. Contacts (set).
6. Capacitor.
7. Terminal and lead (low-tension).
8. Moving contact breaker plate.
9. Contact breaker base plate.
10. Earth lead.
11. Cam.
12. Automatic advance springs.
13. Weight assembly.
14. Shaft and action plate.
15. Cap retaining clips.
16. Vacuum unit.
17. Bush.
18. Thrust washer.
19. Driving dog.
20. Parallel pin.
21. Cam screw.
22. 'O' ring oil seal.

## Section B.6

### DISMANTLING THE DISTRIBUTOR

To dismantle the distributor for overhaul remove it from the engine as described in Section B.5 and then proceed as follows:

(1) Remove the distributor cap and from it withdraw the carbon brush and its spring. Remove the rotor arm.

(2) Lift the L.T. connector housing from the body, unscrew the two screws securing the contact breaker plates assembly, and disconnect the vacuum advance spring from its anchorage. Lift off the contact breaker plates assembly.

(3) Unscrew the contact breaker spring anchorage nut and remove the nut, collar, L.T. lead and capacitor lead tags, moving contact breaker point, and the fibre washer.

(4) Unscrew the fixed contact breaker plate securing screw and lift off the plate.

(5) Remove the capacitor securing screw and the capacitor.

(6) Rotate the moving plate and disengage the stud on it from the base plate. Disengage the base plate from the 'C' spring on the moving plate.

(7) Remove the cam securing screw, the rolling weight springs, cam and action plate, and the weights.

(8) Check the driving shaft end-float and clearance in the body bush.

(9) Drive out the driving dog securing pin and remove the driving dog and thrust washer from the drive shaft. Withdraw the drive shaft from the body.

(10) Remove the micro adjuster spring clip, unscrew the nut, and remove the spring and vacuum control from the body.

## Section B.7

### EXAMINING THE DISTRIBUTOR

In addition to carrying out the examinations detailed in Section B.5 examine all parts for wear, deterioration, and thread damage, and the driving dog and rolling weights for indentation. Examine the 'O' ring oil seal (if fitted) on the shank for deterioration. Worn or unserviceable parts should be renewed.

If the clearance between the action plate shaft and the body bush is excessive the bush may be renewed as follows:

(1) Soak the new bush in S.A.E. 30 or 40 oil for 24 hours. Alternatively, place the bush in oil and heat to 100° C. (212° F.) for 2 hours. Allow the oil to cool and remove bush.

(2) Using a shouldered mandrel, press the old bush out from the inside of the body.

(3) Insert the new bush into the shank end of the body smallest diameter of the bush leading. The bush will be a push fit until the large diameter contacts the body.

(4) Use a mandrel, vice, press, or bush drawing tool to press the bush fully home with an applied steady pressure.

(5) When pressed fully home the bush should be flush with the end of the body shank and should protrude slightly inside the body.

(6) Using the shank drain hole as a guide, drill through the bush and then remove all fragments of metal.

(7) Lubricate the action shaft with clean engine oil, ensure that no fraze exists around the hole in the shaft through which the driving dog pin passes, and then insert the shaft into the body shank.

(8) If the shaft is a tight fit in the body repeat operation (7) until the shaft is free to rotate without binding. This is most important.

(9) Run the shaft in the body with a lathe or test rig for 15 minutes, re-lubricate the shaft, and assemble the distributor.

(10) Under no circumstances must the bush be over-bored since this will impair the porosity and thereby the effective lubricating qualities of the bush.

## Section B.8

### ASSEMBLING THE DISTRIBUTOR

Assembly is a reversal of the dismantling sequence, but attention must be given to the following points during assembly:

(1) Lubricate the components.

(2) Set the micro-adjuster to the mid-way position.

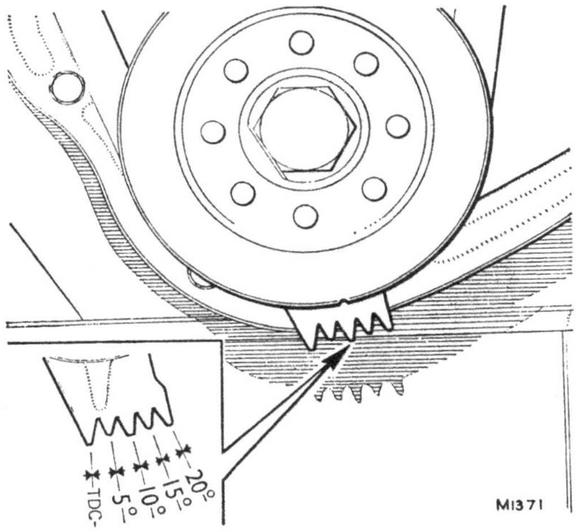


*Fig. B.4*

*The notch in the pulley approaching the T.D.C. position for pistons 1 and 4. The inset shows the timing scale 5° marks*

(3) Adjust the contact breaker points to the correct gap.

(4) Leave the clamp plate securing bolt slack.

(5) After fitting the distributor to its housing rotate the rotor arm until the driving dog engages the recess in the distributor drive from the camshaft. The lugs on the dog are offset to ensure correct replacement.

## Section B.9

### STATIC IGNITION TIMING

To set the ignition timing proceed as follows:

(1) Check the contact breaker gap setting.

(2) Rotate the crankshaft until the valves of No. 4 cylinder are 'rocking' (i.e. exhaust valve just closing and inlet valve just opening). In this position No. 1 cylinder is approximately at T.D.C. compression stroke.

(3) Refer to the **'GENERAL DATA'** for the correct ignition setting and rotate the crankshaft until the notch on the crankshaft pulley is opposite the correct pointer on the timing cover case for the required setting. The long pointer indicates T.D.C.

(4) Rotate the distributor body anti-clockwise until the points are fully closed, then slowly rotate it in a clockwise direction until the points just commence to open.

Alternatively, connect a 12-volt lamp in parallel with the contact breaker points (i.e. one lead to the L.T. terminal and the other to earth), switch on the ignition and turn the distributor as above until the lamp lights. This indicates that the points have just opened.

(5) Slacken the clamp plate set bolts and secure the distributor body in this position by tightening the

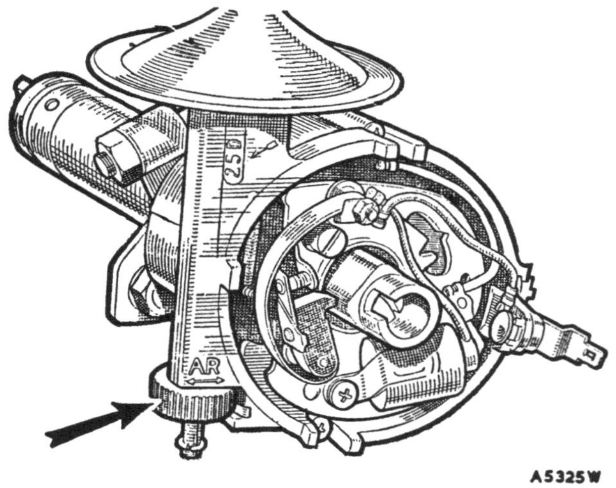


*Fig. B.5*
*The vernier adjustment nut*

clamp plate nut or bolt. The correct torque tightening figures are given in the **'GENERAL DATA'**. Tighten the clamp plate set bolts.

(6) If the static ignition timing figure is between T.D.C. and 5° B.T.D.C., or between 5° B.T.D.C. and 10° B.T.D.C., adjust the vacuum control adjusting screw to obtain the correct setting. Approximately 55 clicks of the screw will move the vacuum control barrel one graduation on the scale, and each graduation is equal to approximately 5° of movement on the moving plate.

(7) Check that the rotor is opposite the correct electrode in the distributor cap for No. 1 cylinder.

Should a stroboscopic lamp be used for timing the ignition, care must be taken that with the engine running the speed is low enough to ensure that the centrifugal advance weights are not in operation. If the vacuum advance take-off is direct from the induction manifold the take-off must be disconnected before attempting the timing, otherwise the timing will be retarded.

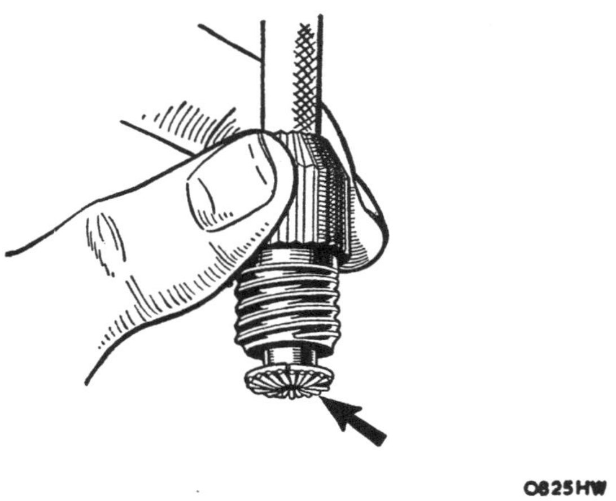


*Fig. B.6*
*The correct method of fitting a high-tension cable to the ignition coil terminal nut*

## Section B.10

### HIGH-TENSION CABLES

The high-tension cables must be examined carefully and any which have the insulation cracked, perished, or damaged in any way must be renewed.

To fit the cables to the terminal of the ignition coil thread the knurled moulded terminal nut over the lead, bare the end of the cable for about ¼ in. (6 mm.), thread the wire through the brass washer removed from the original cable, and bend back the strands over the washer. Finally, screw the terminal into the coil.

To make the connections to the terminals in the distributor moulded cap first remove the cap and slacken the screws on the inside of the moulding till they are clear of the cables.

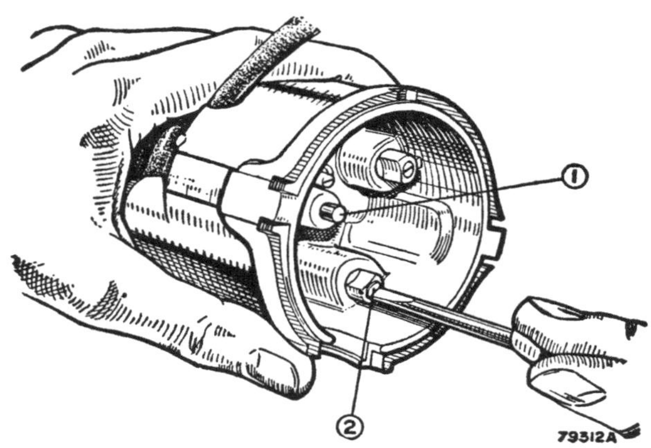


*Fig. B.7*
*The correct method of connecting high-tension leads*

1. Carbon brush. 2. Cable securing screw.

Remove the cables. Fill the holes in the distributor cap with Silicone grease, then cut the new cables off to the required length, push them completely home, and tighten the securing screws.

The cables from the distributor to the sparking plugs must be connected up in the correct firing order, which is 1, 3, 4, 2. Secure them firmly to the connectors.

Radio-frequency-suppressed high-tension cables are used on later cars. In this cable, a graphite-impregnated core replaces the metallic conductor of the earlier type of cable. Connectors used with the metal-cored cable are unsuitable for use with the later cable and new-type cable connectors have been introduced; these are a push-fit to the coil and distributor covers.

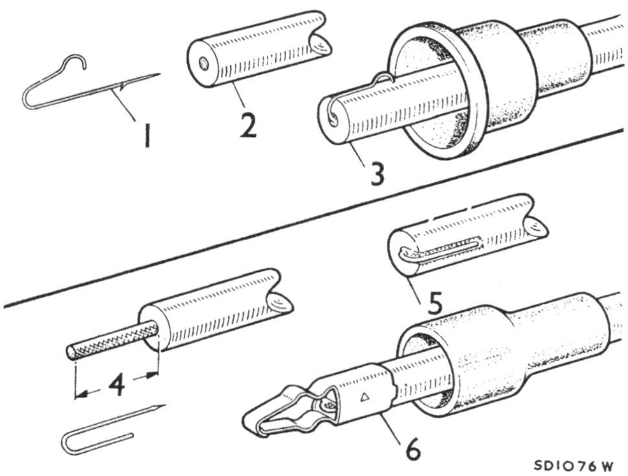


*Fig. B.8*

*The correct assembly method for later-type suppressed high tension cables*

| *Coil lead* | *Plug leads* |
|---|---|
| 1. Fish-hook connector. | 4. Insulation removed for ½ in. (12·7 mm.). |
| 2. Flush cable end. | 5. Inner cord folded onto cable, staple pushed into the centre of the cord as far as possible. |
| 3. Assembly of fish-hook and lead cover. | 6. Cord and staple must make a good contact with body of connector. |

## Section B.11

### SPARKING PLUGS

Inspect, clean, adjust and renew sparking plugs at the recommended mileage intervals.

Before replacing the plug in the engine, test it for correct functioning under air pressure in a plug tester, following out the instructions issued by the makers of the plug tester. Generally speaking, a plug may be considered satisfactory for further service if it sparks continuously under a pressure of 100 lb. per sq. in. (7 kg./cm.$^2$) with the gap between the points set at ·022 in. (·56 mm.). It is essential that the plug points should be reset to the correct gap (see **'GENERAL DATA'**) before the plug is refitted to the engine.

When replacing a plug, always screw it down by hand as far as possible and use the spanner for final tightening only. Whenever possible use a box spanner to avoid possible fracture of the insulator. Where a torque spanner is available 14-mm. plugs should be tightened to between 27 and 30 lb. ft. (3·7 and 4·1 kg. m.).

## Section B.12

### LOCATING THE CAUSE OF UNEVEN FIRING

Start the engine and set it to run at a fairly fast idling speed.

Short-circuit each plug in turn by placing a hammer head or the blade of a screwdriver with a wooden or insulated handle between the terminal and the cylinder head. No difference in the engine performance will be noted when short-circuiting the plug in the defective cylinder. Shorting the other plugs will make uneven running more pronounced.

Having located the cylinder which is at fault, stop the engine and remove the cable from the terminal of the sparking plug. Restart the engine and hold the end of the cable about $\frac{3}{16}$ in. (4·8 mm.) from the cylinder head.

If the sparking is strong and regular, the fault probably lies in the sparking plug. Remove the plug, clean, and adjust the gap to the correct setting, or alternatively, fit a new plug (see Section B.11).

If there is no spark or if it is weak and irregular, examine the cable from the sparking plug to the distributor. After a long period of service, the insulation may be cracked or perished, in which case the cable should be renewed.

Finally, examine the distributor moulded cap, wipe the inside and outside with a clean dry cloth, see that the carbon brush moves freely in its holder and examine the moulding closely for signs of breakdown. After long service it may become tracked, that is, a conducting path may have formed between two or more of the electrodes or between one of the electrodes and some part of the distributor in contact with the cap. Evidence of a tracked cap is shown by the presence of a thin black line in the places indicated. A replacement distributor cap must be fitted in place of one that has become tracked.

## Section B.13

### TESTING THE LOW-TENSION CIRCUIT

Spring back the securing clips on the distributor and remove the moulded cap and rotor. If the rotor is a tight fit, it can be levered off carefully with a screwdriver.

Check that the contacts are clean and free from pits, burns, oil, or grease. Turn the engine and check that the contacts are opening and closing correctly and that the clearance is correct when the contacts are fully opened to between ·014 and ·016 in. (·36 and ·40 mm.).

Correct the gap if necessary.

Disconnect the cable at the contact breaker terminal of the coil and at the low-tension terminal of the distributor, and connect a test lamp between these terminals. If the lamp lights when the contacts close and goes out when the contacts open, the low-tension circuit is in order. Should the lamp fail to light, the contacts are dirty or there is a broken or loose connection in the low-tension wiring.

## Section B.14

### LOCATING A LOW-TENSION CIRCUIT FAULT

Having determined, by testing as previously described, that the fault lies in the low-tension circuit, switch on the ignition, and turn the engine until the contact breaker points are fully opened.

Refer to the wiring diagram and check the circuit with a voltmeter (0–20 volts) as follows.

**NOTE.—If the circuit is in order, the reading on the voltmeter should be approximately 12 volts.**

(1) **Battery to control box terminal 'B'.** Connect a voltmeter between control box terminal 'B' and earth. No reading indicates a damaged cable or loose connections.

(2) **Control box terminal 'B' to ignition switch terminal (brown lead).** Connect a voltmeter between the ignition terminal and earth. No reading indicates a damaged cable or loose connections.

(3) **Ignition switch (white lead).** Connect a voltmeter between the ignition switch terminal and earth. Turn the ignition key to the ignition position. No reading indicates a fault in the ignition switch.

(4) **Ignition switch (white with red lead).** Connect a voltmeter between the ignition switch terminal and earth. Turn the ignition key to the start position. No reading indicates a fault in the ignition switch.

(5) **Ignition switch to fusebox terminal 'A3' (white lead).** Connect a voltmeter between the fuse unit terminal 'A' and earth. No reading indicates a damaged cable or loose connections.

(6) **Fusebox terminal 'A3' to ignition coil terminal 'SW'.** Connect a voltmeter to the ignition coil terminal 'SW' and to earth. No reading indicates a fault in the primary winding of the coil and a new coil must be fitted.

### EDITOR'S NOTES

**B. The Ignition System**
*Ignition Coil*
Most service stations and automobile repair shops are equipped to test the output of the coil in operation on the car's engine. These "ignition analyzers" are not only more convenient to use but can pinpoint problems elsewhere in the system, such as with points, wiring, or condenser.

# SECTION C

## THE COOLING SYSTEM

## KEY TO THE RADIATOR, GRILLE, AND DIAPHRAGM COMPONENTS

(Illustrated on pages C.3 and C.4)

| No. | Description |
|---|---|
| 1. | Core assembly—radiator. |
| 2. | Cap—filler. |
| 3. | Tap—drain. } Early cars. |
| 4. | Washer for tap. } Early cars. |
| 5. | Tube—drain. |
| 6. | Clip—drain tube. |
| 7. | Packing. |
| 8. | Diaphragm—radiator. |
| 9. | Screw—long—radiator to diaphragm. |
| 10. | Screw—short—radiator to diaphragm. |
| 11. | Washer—spring—for screw. |
| 12. | Screw—diaphragm to body. |
| 13. | Washer—plain—for screw. |
| 14. | Washer—spring—for screw. |
| 15. | Rubber—radiator air seal. |
| 16. | Tie—radiator—R.H. |
| 17. | Tie—radiator—L.H. |
| 18. | Washer—plain—radiator tie. |
| 19. | Washer—spring—radiator tie. |
| 20. | Nut—tie. |
| 21. | Pipe—water pump connector. |
| 22. | Plug—connector pipe. |
| 23. | Washer—plug. |
| 24. | Hose—connector. |

| No. | Description |
|---|---|
| 25. | Clip—hose. |
| 26. | Hose—top and bottom. |
| 27. | Clip—hose. |
| 28. | Case and grille assembly—radiator. |
| 29. | Case assembly. |
| 30. | Grille assembly. |
| 31. | Bar—grille fixing—top. |
| 32. | Bar—grille fixing—bottom. |
| 33. | Slats—grille. |
| 34. | Rivet—slat fixing. |
| 35. | Bar and badge housing—centre. |
| 36. | Badge. |
| 37. | Fixing—blind badge. |
| 38. | Screw—grille to case. |
| 39. | Nut—spring. |
| 40. | Fix—push-on—badge housing to case. |
| 41. | Washer—spring. |
| 42. | Washer—plain. |
| 43. | Screw—grille to steady bracket. |
| 44. | Bracket—steady grille centre assembly. |
| 45. | Bracket—steady grille side assembly—R.H. |
| 46. | Bracket—steady grille side assembly—L.H. |
| 47. | Screw—steady bracket to bonnet lock platform. |
| 48. | Washer—plain. |

| No. | Description |
|---|---|
| 49. | Washer—spring. |
| 50. | Nut. |
| 51. | Buffer—grille top rail. |
| 52. | Screw—grille to radiator duct panel. |
| | *Grille assembly (GHN5/GHD5 cars only)* |
| 53. | Rubber—radiator air seal. |
| 54. | Screw—radiator grille to bonnet lock platform. |
| 55. | Plug—sealing. |
| 56. | Grille assembly. |
| 57. | Washer. |
| 58. | Washer—spring. |
| 59. | Screw. |
| 60. | Heater hose connection. |
| 61. | Sealing washer. |
| 62. | Self-tapping screw. |
| | *Grille assembly (GHN5/GHD5 cars)* |
| 63. | Case assembly. |
| 64. | Grille—R.H. |
| 65. | Grille—L.H. |
| 66. | Bracket—retaining grille to case. |
| 67. | Screw. |
| 68. | Motif. |

## THE RADIATOR, GRILLE, AND DIAPHRAGM COMPONENTS (GHN5/GHD5 CARS)

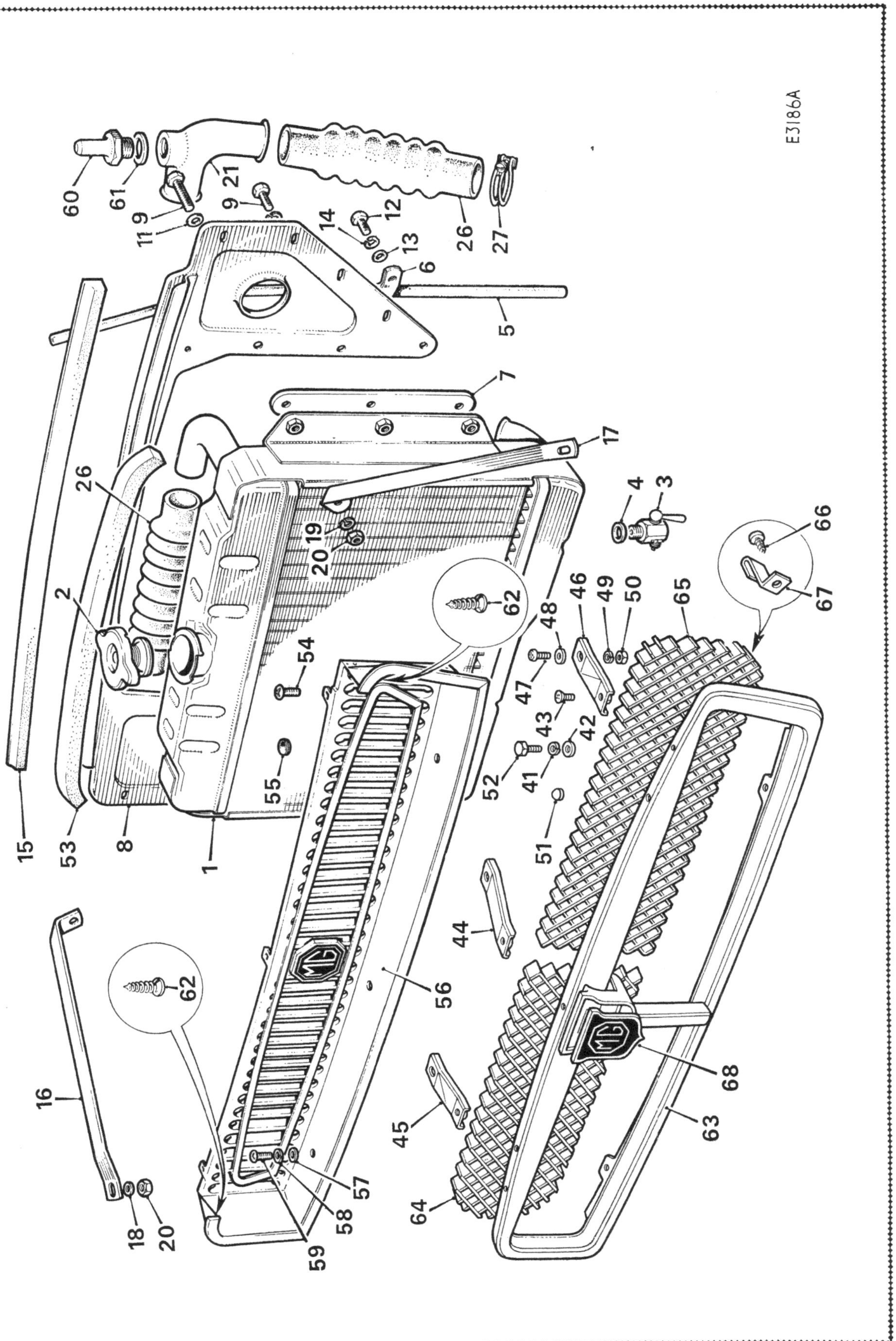

## THE RADIATOR, GRILLE, AND DIAPHRAGM COMPONENTS

(Refer to page C.2 for the key)

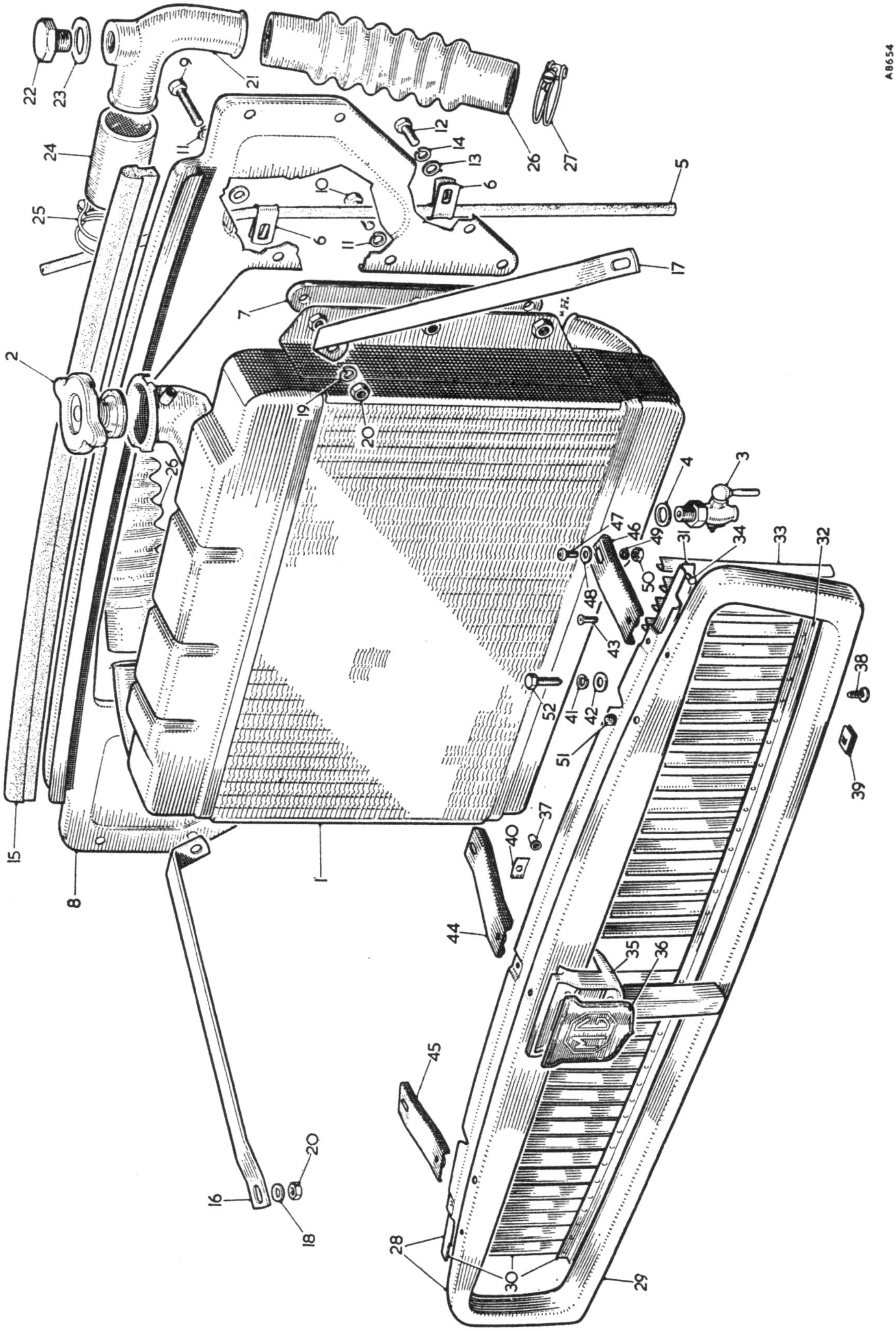

## GENERAL DESCRIPTION

The cooling system is pressurized and the water circulation is assisted by a pump attached to the front of the engine and driven by a belt from the crankshaft pulley. A relief valve is incorporated in the radiator filler cap which controls the pressure.

The water circulates from the base of the radiator and passes round the cylinders and cylinder head, reaching the header tank of the radiator via the thermostat and top hose. From the header tank is passes down the radiator core to the base tank of the radiator. Air is drawn through the radiator by a fan attached to the water pump pulley.

## Section C.1

### LUBRICATION

**Early cars**

Remove the plug from the water pump body and add a small quantity of one of the recommended greases. Do not pressure-lubricate the water pump or the seal may be damaged.

**Later cars**

The greasing plug has been deleted from the water pump as no lubrication is necessary.

## Section C.2

### THERMOSTAT

**IMPORTANT.—In hot climates where the thermostat is removed as a permanent measure, it is essential that a thermostat blanking sleeve is fitted. Failure to fit a sleeve will result in higher operating temperatures and possible damage.**

**Removing**

(1) Drain the cooling system—Section C.4.

(2) Disconnect the radiator top hose from the water outlet elbow. Unscrew the three retaining nuts, detach the water outlet elbow and joint washer and withdraw the thermostat.

**Inspection**

(3) Examine the thermostat for damage and check that the valve is in the closed position. Renew the thermostat if damaged or if the valve is open.

(4) Immerse the thermostat in water heated to the temperature marked on the thermostat; the valve will start to open if it is functioning correctly. If the valve fails to open renew the thermostat.

**Refitting**

(5) Reverse the removing procedure using a new joint washer, then refill the cooling system.

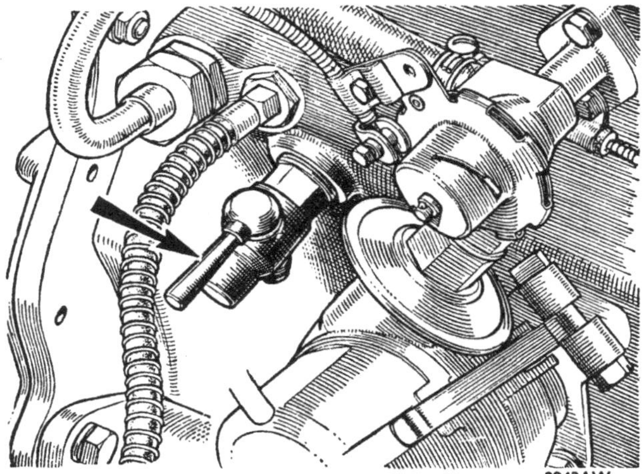


*Fig. C.1*
*The engine drain tap on the right-hand side of the engine, later cars have a drain plug*

## Section C.3

### TEMPERATURE GAUGE

A temperature gauge unit, consisting of a thermal transmitter and dial indicator, is fitted to the vehicle. The thermal transmitter is held in the cylinder head at the base of the thermostat housing by a gland nut. The dial indicator is situated in the instrument panel and is connected to the transmitter by a capillary tube.

If the unit fails to register the water temperature, check the thermal transmitter and dial indicator by substitution.

Cars built to certain specifications are fitted with electrical temperature gauges.

## Section C.4

### DRAINING, FLUSHING, AND REFILLING THE COOLING SYSTEM

**Removing the filler cap**

The cooling system is under pressure while the engine is hot, and the radiator filler cap must be removed very carefully or left in position until the water has cooled.

If it is necessary to remove the filler cap when the engine is hot it is absolutely essential to release the cap gradually, and the filler spout is provided with a specially shaped cam to enable this to be done easily.

Unscrew the cap slowly until the retaining tongues are felt to engage the small lobes on the end of the filler spout cam, and wait until the pressure in the radiator is fully released before finally removing the cap.

It is advisable to protect the hand against escaping steam while removing the cap.

**Draining the cooling system**

*Early cars*

Open the two drain taps, one at the rear of the engine cylinder block on the right-hand side, the other at the

## THE FAN AND WATER PUMP COMPONENTS

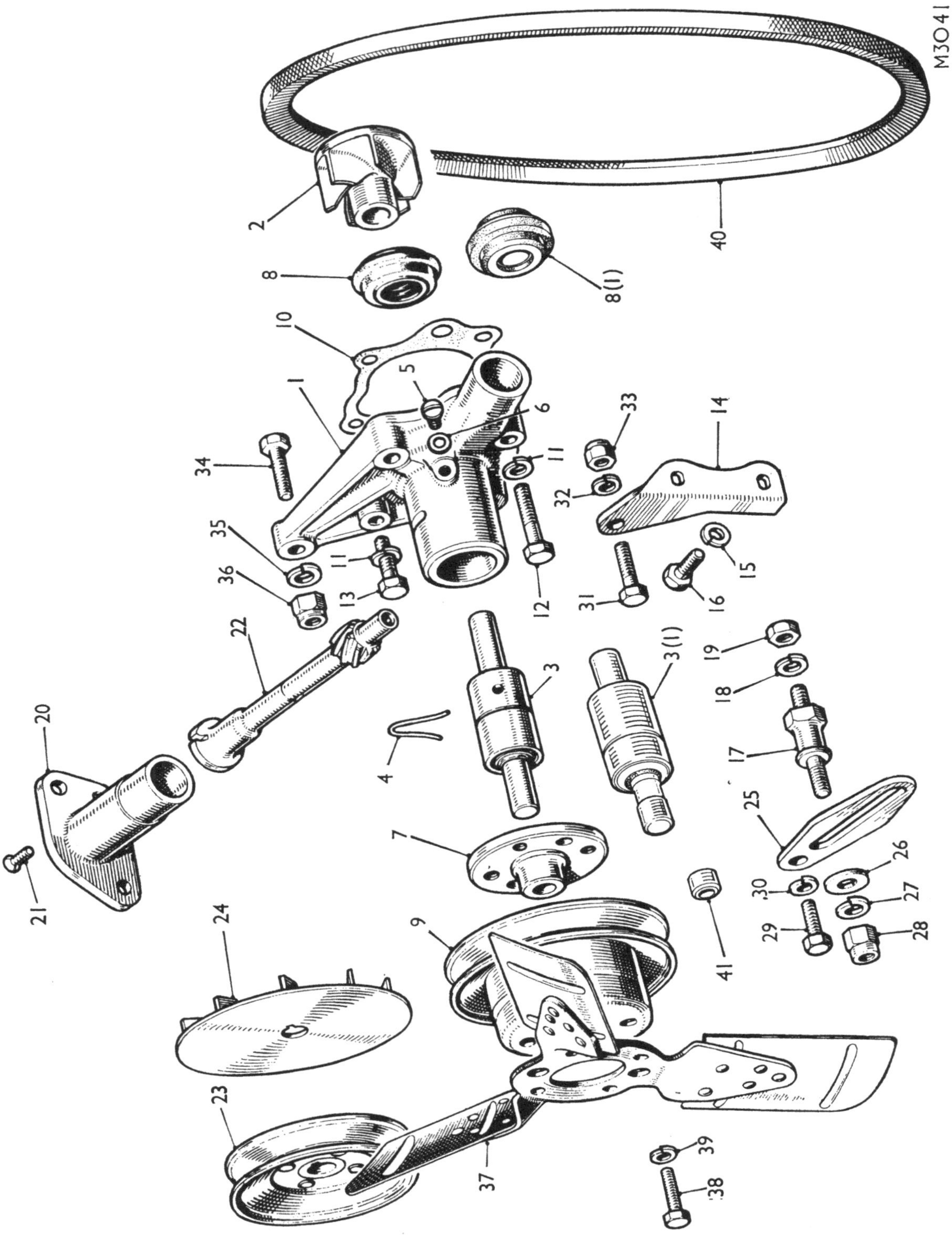

## KEY TO THE FAN AND WATER PUMP COMPONENTS

| No. | Description | No. | Description | No. | Description |
|---|---|---|---|---|---|
| 1. | Pump assembly—water (with hub less pulley). | 13. | Screw—short—water pump to crankcase. | 27. | Washer—spring. |
| 2. | Vane—water pump. | 14. | Bracket—rear—dynamo. | 28. | Nut—adjusting link to pillar. |
| 3. | Bearing assembly complete with spindle. | 15. | Washer—spring—screw. | 29. | Screw—adjusting link to pillar. |
| 3. | (1) Bearing assembly (later type). | 16. | Screw—dynamo bracket to crankcase. | 30. | Washer—spring—screw. |
| 4. | Wire—bearing locating. | 17. | Pillar—dynamo adjusting link. | 31. | Bolt—dynamo to mounting bracket. |
| 5. | Screw—lubricating point. } (early pumps only) | 18. | Washer—spring—pillar. | 32. | Washer—spring—bolt. |
| 6. | Washer—fibre—screw. } (early pumps only) | 19. | Nut—dynamo pillar. | 33. | Nut—bolt. |
| 7. | Hub—pulley. | 20. | Housing—distributor. | 34. | Bolt—dynamo to water pump body. |
| 8. | Seal—water pump. | 21. | Screw—housing to crankcase. | 35. | Washer—spring—bolt. |
| 8. | (1) Pressure balanced steel (later pumps). | 22. | Spindle—distributor driving. | 36. | Nut—bolt. |
| 9. | Pulley—fan and water pump. | 23. | Pulley—dynamo. | 37. | Fan—water pump (six-bladed). |
| 10. | Joint washer—pump to crankcase. | 24. | Fan—dynamo. | 38. | Screw—fan to pulley. |
| 11. | Washer—spring—screw. | 25. | Link—dynamo adjusting. | 39. | Washer—spring—screw. |
| 12. | Screw—long—water pump to crankcase. | 26. | Washer—adjusting link to pillar. | 40. | Belt—fan. |

base of the radiator. On 18GD and 18GG engines the engine cylinder block drain tap is replaced by a removable plug and sealing washer.

*Later cars*

Later cars have no provision for draining the radiator and the use of anti-freeze all-the-year round is recommended. To drain the coolant from the radiator, slacken the radiator bottom hose clip and remove the hose at its connection to the radiator.

When the system is completely drained and refilling is to be deferred until some later date a suitable notice should be fixed to the radiator filler cap, indicating that the coolant has been drained. As an alternative, place the radiator filler cap on the driver's seat or leave the filler cap access panel open as a reminder to fill the cooling system before the vehicle is used again.

NOTE.—**If a heater is fitted, under no circumstances should draining of the cooling system be resorted to as an alternative to the use of anti-freeze mixture, due to the fact that complete draining of the heater unit by means of the cooling system drain taps is not possible.**

To ensure sufficient circulation of the coolant and to reduce the formation of scale and sediment in the radiator the system should be periodically flushed out with clear running water, preferably before putting in anti-freeze solution and again after taking it out.

The water should be allowed to run through until it comes out clear from the taps.

This method is adequate under normal conditions, but in extreme cases where excessive 'furring up' is experienced a more efficient method is to completely remove the radiator and flush in the reverse way to the flow, i.e. turn the radiator upside-down and let the water flow in through the bottom hose and out of the top connection. (See Editor's notes at end of section.)

**Filling the cooling system**

Close the radiator and cylinder block drain taps.

*Fig. C.2*

*The location of the radiator drain tap (Early cars)*

Ensure that all hose connections are tight.

Fill up the system through the filler in the radiator header tank until the level of water can just be seen. Run the engine until it is hot and then add sufficient water to raise the level to within 1 in. (25·4 mm.) of the bottom of the filler neck.

When possible, soft water, such as clean rain-water, should be used to fill the system.

When using anti-freeze solution avoid overfilling to prevent loss from expansion.

Anti-freeze solutions having an alcohol base are unsuitable for use in the cooling system owing to the high temperatures attained in the top radiator tank. Only anti-freeze solutions of the ethylene glycol type incorporating the correct type of corrosion inhibitor should be employed. (See Section C.8.)

## Section C.5

### FAN BELT

To fit a new fan belt slacken slightly the two bolts on which the dynamo pivots and release the bolt securing it to the slotted link and the nut securing the slotted link to the engine. Move the dynamo to the engine as far as possible. Slide the belt over the fan and onto the fan pulley; ease the belt onto the crankshaft pulley and dynamo pulley. It may be found helpful to turn the engine with the starting-handle whilst easing the belt over the dynamo pulley.

Adjustment is made by raising the dynamo upwards away from the engine. **A gentle hand pull only must be exerted on the dynamo, or the belt tension will be excessive and undue strain will be thrown onto the dynamo bearings.** Tighten up the bolts with the dynamo in this position. The belt should be sufficiently tight to prevent slip, yet it should be possible to move the belt laterally 1 in. (2·5 cm.) in the centre of its longest run.

On cars fitted with alternators the fan belt is renewed and adjusted in a similar manner.

## Section C.6

### RADIATOR AND DIAPHRAGM

The radiator block assembly is supported in a metal diaphragm which is secured to the body sides by screws and washers. The diaphragm has two holes in the right-hand side to permit the oil cooler pipes to pass from the cooler to the engine. A rubber air seal is fitted across the top channel of the diaphragm.

The radiator is supported in the diaphragm by three screws and washers each side and a tie-bar running forward from each top screw to the body sides.

The overflow pipe is secured by two clips that are retained by a radiator fixing screw and a diaphragm fixing screw respectively.

**Removing**

To remove the radiator drain the coolant—Section C.4, release the top and bottom hose clips, and detach the hoses from their connectors.

Where an oil cooler is not fitted the radiator and diaphragm may be removed as a complete assembly by removing the top radiator to diaphragm screws to release the stays and then removing the screws securing each side of the diaphragm to the body.

If an oil cooler is fitted the pipe connections must be disconnected from the cooler and the engine before removing the radiator and diaphragm assembly.

To remove the radiator without the diaphragm undo the radiator to diaphragm securing screws, slacken the diaphragm to body securing screws, remove the screw retaining the overflow pipe clip, and then lift the diaphragm sufficiently to allow the radiator to be pulled forward and lifted from the car.

**Replacing**

Replacement is a reversal of the removal sequence, but ensure that the two packing pieces are correctly positioned either side of the radiator and that the overflow pipe is secured.

Close the drain taps, fill with coolant, and check for leaks.

## Section C.7

### WATER PUMP

**Removing**

(1) Remove the radiator—Section C.6.
(2) Remove the dynamo or alternator.
(3) Unscrew the fan and pulley retaining screws and remove the fan and pulley.
(4) Remove the water pump securing bolts and withdraw the pump.

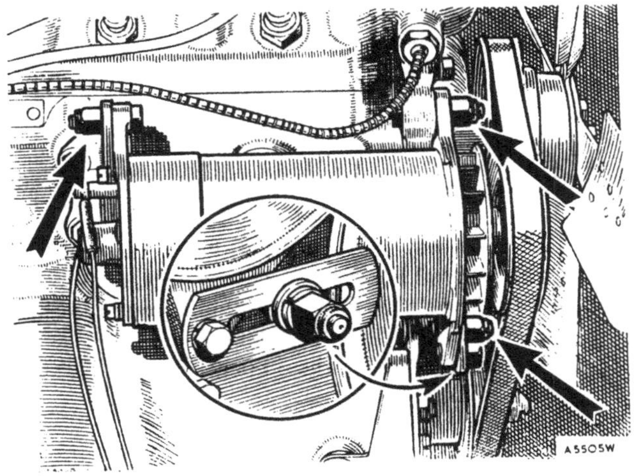


*Fig. C.4*

*The dynamo mounting bolts which must be slackened for fan belt tension adjustment*

**Dismantling**

(5) Withdraw the pulley hub from the spindle.
(6) Withdraw the bearing locating wire from the body aperture.
(7) Tap the spindle rearwards and remove the spindle complete with the bearing. The bearing cannot be removed from the spindle.
(8) Withdraw the impeller from the spindle and remove the seal.

**Reassembling**

(9) If the interference fit of the fan hub was impaired when the hub was withdrawn from the spindle a new nut must be fitted. The hub must be fitted with its face flush with the end of the spindle (B, Fig. C.3).
(10) Press the spindle bearing assembly into the pump body and take measurement E, Fig. C.5; adjust (See Editor's notes at end of section.)

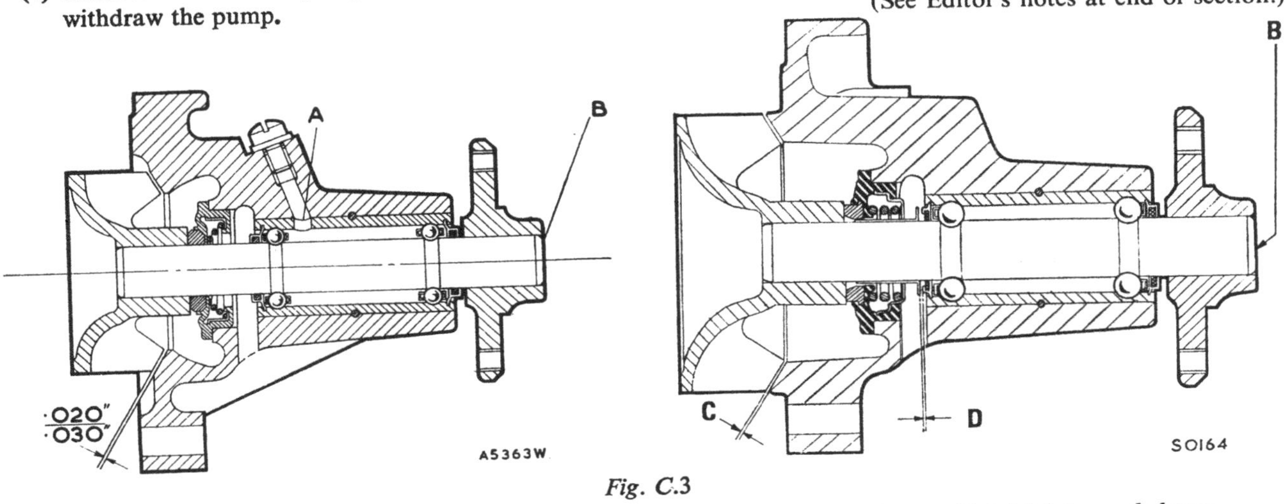


*Fig. C.3*

*A section through the water pump showing: (left) early type with grease plug, and (right) later sealed type*

A. Lubricating hole.
B. Hub face flush with spindle.
C. ·020 to ·030 in. (·51 to ·76 mm.) clearance.
D. ·042 to ·062 in. (1·1 to 1·6 mm.) clearance.

the position of the spindle bearing assembly to obtain the correct measurement.

(*a*) On pumps fitted with the early-type seal and thrower, check the bearing to thrower clearance before fitting the spindle (D, Fig. C.3).

(*b*) On pumps fitted with a greaser plug, align the holes in the bearing and body (A, Fig. C.3).

(11) Fit a new seal. Smear the jointing face of the seal with a mineral base oil to ensure a watertight joint.

(12) Press the impeller onto the spindle, ensuring that the correct running clearance (C, Fig. C.3) is maintained between the impeller vanes and the pump body.

**Refitting**

(10) Reverse the removing procedure in (1) to (4).

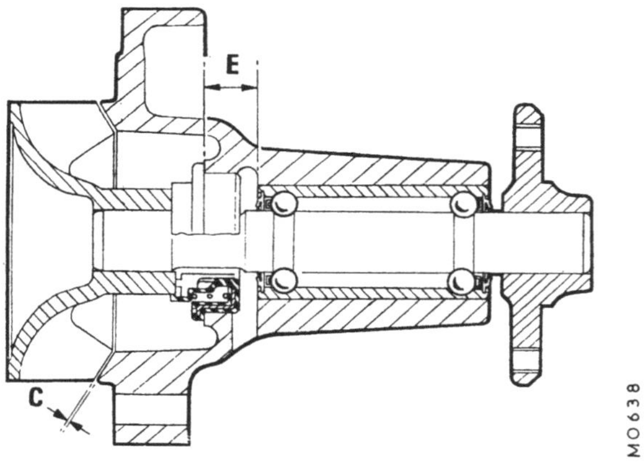


*Fig. C.5*

*Pump with later-type seal, thrower deleted*

E. Distance from rear face of spindle bearing (outer track) to seal housing shoulder should be ·534 in. ± ·005 in. (13·56 mm. ± ·13 mm.) (pumps fitted with later-type seal only).

## Section C.8

### COLD WEATHER PRECAUTIONS

As the cooling system is pressurized, relatively high temperatures are developed in the radiator upper tank. For this reason anti-freeze solutions having an alcohol base are unsuitable owing to their high evaporation rate producing rapid loss of coolant and a consequent interruption of the circulation of coolant.

Only anti-freeze solution of the ethylene glycol type incorporating the correct type of corrosion inhibitor is suitable for use in the cooling system, and the recommended make is Bluecol, or any anti-freeze solution which conforms to Specification B.S.3151 or B.S.3152. The anti-freeze solution should be made up in the proportions given in the table below.

It is advisable for vehicles with an anti-freeze solution in the cooling system to have an identification mark on the header tank of the radiator.

The following precautions are necessary on vehicles so marked.

Make sure that the strength of the mixture is, in fact up to that instructed on the container of the particular anti-freeze solution used.

The strength of the mixture must be maintained by topping up with the anti-freeze solution as necessary when the system is hot. Topping up with water alone will reduce the degree of protection provided.

If the cooling system has to be emptied run the contents into clean containers, strain, and use again.

Anti-freeze can remain in the cooling system for two years provided that the specific gravity of the coolant is checked periodically and anti-freeze added as required. Specialized equipment, which can be obtained from the anti-freeze manufacturer, is necessary to check the specific gravity.

After the second winter, drain the system and flush out. Refill with fresh water or the recommended anti-freeze solution.

If for any reason the coolant is lost and the system is filled with water **remove the identification mark from the radiator header tank.**

**(See Editor's notes at end of section.)**

| *Anti-freeze* | *Commences to freeze* | | *Frozen solid* | | *Amount of anti-freeze* | | |
|---|---|---|---|---|---|---|---|
| % | °*C.* | °*F.* | °*C.* | °*F.* | *Pts.* | *U.S. Pts.* | *Litres* |
| 25 | −13 | 9 | −26 | −15 | 2½ | 3 | 1·4 |
| 33⅓ | −19 | −2 | −36 | −33 | 3½ | 4 | 2·0 |
| 50 | −36 | −33 | −48 | −53 | 5 | 6 | 2·8 |

### EDITOR'S NOTES

**C. The Cooling System**

*Draining the cooling system*

For best results, the thermostat should be removed when flushing. This allows water to circulate freely in the block.

*Dismantling and assembling*

The fan hub and impeller should be removed with a suitably sized gear puller.

*Cold weather precautions*

An anti-freeze hydrometer is used to check the specific gravity of the coolant. With this device, a small quantity of the coolant solution is sucked into a cylinder and the position of a glass float indicates the specific gravity, usually in terms of the temperature at which ice will begin to form.

# SECTION D

## THE FUEL SYSTEM



**† These operations must be followed by an exhaust emission check** 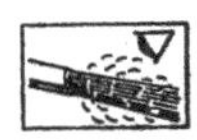

## Section D.1

### LUBRICATION

**Carburetters**

Unscrew the oil cap at the top of each suction chamber, top up with oil to within ½ in. (13 mm.) from the top of the hollow piston rod. Under no circumstances should a heavy-bodied lubricant be used.

Failure to lubricate the piston dampers will cause the pistons to flutter and reduce acceleration.

## Section D.2

### FUEL TANK

**Removing**

*Early tourer models* (10 *gal.* [12 *U.S. gal.*, 45·4 *litres*])

Disconnect the battery and raise the car to a workable height.

Remove the hexagon drain plug and empty the tank.

Slacken the two clips on the filler neck hose, withdraw the filler extension and seal.

Disconnect the fuel pipe at the union and the fuel gauge cable from the tank unit, each on the right-hand side of the tank.

Support the tank and remove the two nuts from the bolts securing the straps at the rear of the tank to the anchorage brackets.

Swing away the straps and lower the tank.

*Later tourer and GT models* (12 *gal.* [14 *U.S. gal.*, 54·5 *litres*]).

Disconnect the battery and raise the car to a workable height.

Remove the hexagon drain plug and empty the tank.

Slacken the two clips on the filler neck hose, withdraw the filler extension and seal. Disconnect the fuel pipe at the union and the fuel gauge cable from the tank unit, each on the right-hand side of the tank.

Support the tank and from inside the luggage compartment remove the set screws securing the tank to the luggage compartment floor.

Lower the tank.

Replacement is a reversal of the above instructions.

**Fuel gauge tank unit**

*Early tourer models*

Remove the six set screws securing the unit to the tank and withdraw the tank unit, taking care not to strain or bend the float lever. When replacing, a new joint washer must be fitted and a sealing compound used to make a fuel-tight joint.

*Later tourer and GT models*

Remove the tank gauge locking ring with Service tool 18G 1001 and lift out the gauge assembly and rubber sealing ring. When refitting the gauge unit a new rubber sealing ring should be fitted if necessary to ensure a fuel-tight joint.

## Section D.3

### FUEL PUMP

**Removing and refitting**

The fuel pump is mounted on a bracket secured to the heelboard adjacent to the front mounting right-hand rear spring, and is accessible from beneath the car.

Disconnect the earth lead from the batteries and the earth and supply leads from the terminals on the pump.

Disconnect the inlet and outlet pipe unions and the breather pipes (later cars).

Remove the two bolts securing the pump bracket to the heelboard.

When refitting, reverse the removing procedure ensuring that the breather pipes are connected to the pump and correctly retained by the 'T' pieces in the harness grommets in the boot floor.

**Dismantling**

*Contact breaker*

(1) Remove the insulated sleeve or knob, terminal nut. and connector, together with its shakeproof washer. Remove the tape seal and take off the end cover,

(2) Unscrew the 5 B.A. screw which holds the contact blade to the pedestal and remove the condenser from its clip. This will allow the washer, the long coil lead, and the contact blade to be removed.

*Coil housing and diaphragm*

(3) Unscrew the coil housing securing screws, using a thick-bladed screwdriver to avoid damaging the screw heads.

(4) Remove the earthing screw.

(5) *Roller type.* Hold the coil housing over a bench or receptacle to prevent the 11 brass rollers from being damaged or lost as they come free, unscrew the diaphragm assembly anti-clockwise until the armature spring pushes it free of the housing.
*Guide plate type.* Turn back the edge of the diaphragm and carefully lever the two end lobes of the armature guide plate from the recess in the coil housing, unscrew the diaphragm assembly anti-clockwise until the armature spring pushes it free of the housing. Remove the armature guide from the diaphragm assembly.

*Pedestal and rocker*

(6) Remove the end cover seal washer, unscrew the terminal nut, and remove the lead washer. This will have flattened on the terminal tag and thread and is best cut away with cutting pliers or a knife. Remove the terminal tag spring washer. Unscrew the two 2 B.A. screws holding the pedestal to the coil housing, remove the earth terminal tag together with the condenser clip. Tip the pedestal and withdraw the terminal stud from the terminal tag. The pedestal may now be removed with the rocker mechanism attached.

# THE AUF 300 AND HP FUEL PUMP COMPONENTS

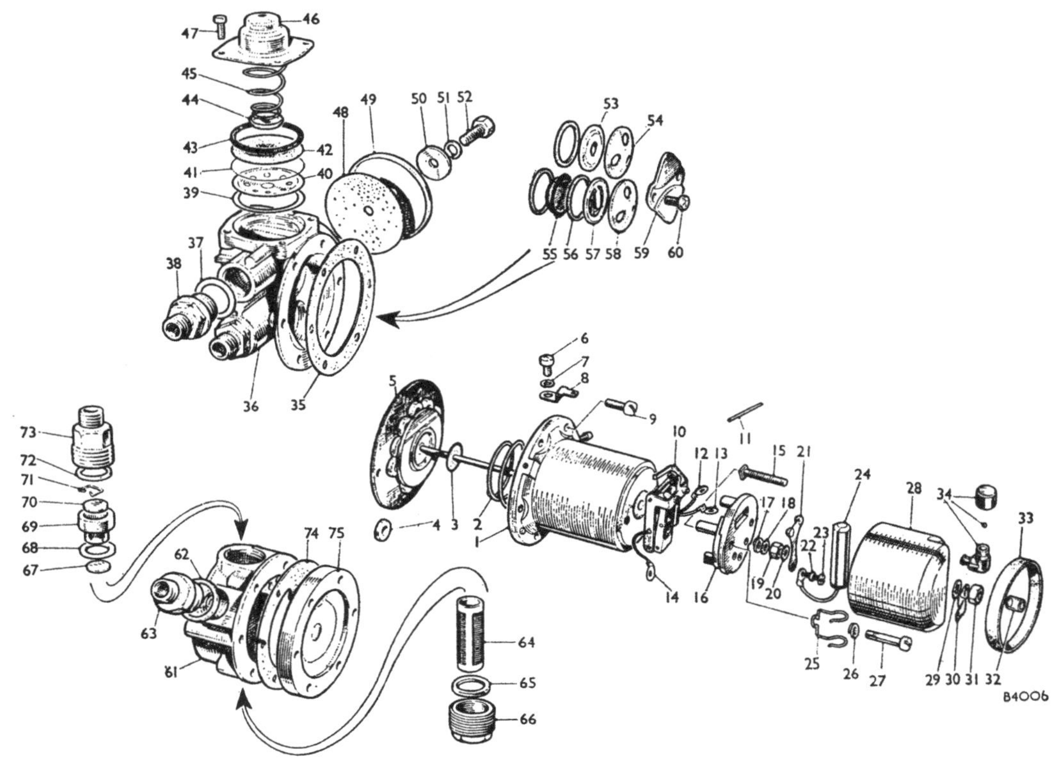


| No. | Description | No. | Description | | No. | Description | |
|---|---|---|---|---|---|---|---|
| 1. | Coil housing. | 26. | Spring washer. | | 51. | Spring washer. | AUF 300 Type. |
| 2. | Armature spring. | 27. | Pedestal screw. | | 52. | Set screw. | AUF 300 Type. |
| 3. | Impact washer. | 28. | End cover. | | 53. | Outlet valve. | AUF 300 Type. |
| 4. | Armature centralizing roller. | 29. | Shakeproof washer. | | 54. | Valve cap. | AUF 300 Type. |
| 5. | Diaphragm and spindle assembly. | 30. | Lucar connector. | | 55. | Filter. | AUF 300 Type. |
| 6. | Set screw. | 31. | Nut. | | 56. | Sealing washer. | AUF 300 Type. |
| 7. | Spring washer. | 32. | Insulating sleeve. | | 57. | Inlet valve. | AUF 300 Type. |
| 8. | Earth connector. | 33. | Sealing band. | | 58. | Valve cap. | AUF 300 Type. |
| 9. | Set screw. | 34. | Vent valve. | | 59. | Clamp plate. | AUF 300 Type. |
| 10. | Rocker mechanism. | 35. | Gasket. | AUF 300 Type. | 60. | Set screw. | AUF 300 Type. |
| 11. | Rocker pivot pin. | 36. | Pump body. | AUF 300 Type. | 61. | Pump body. | HP Type. |
| 12. | Terminal tag. | 37. | Fibre washer. | AUF 300 Type. | 62. | Fibre washer. | HP Type. |
| 13. | Terminal tag. | 38. | Outlet connection. | AUF 300 Type. | 63. | Outlet connection. | HP Type. |
| 14. | Earth tag. | 39. | Sealing washer. | AUF 300 Type. | 64. | Filter. | HP Type. |
| 15. | Terminal stud. | *40. | Diaphragm plate. | AUF 300 Type. | 65. | Washer. | HP Type. |
| 16. | Pedestal. | 41. | Plastic diaphragm barrier. | AUF 300 Type. | 66. | Plug. | HP Type. |
| 17. | Spring washer. | *42. | Rubber diaphragm. | AUF 300 Type. | 67. | Inlet valve. | HP Type. |
| 18. | Lead washer. | 43. | Rubber 'O' ring. | AUF 300 Type. | 68. | Thin fibre washer. | HP Type. |
| 19. | Terminal nut. | *44. | Spring end cap. | AUF 300 Type. | 69. | Outlet valve cage. | HP Type. |
| 20. | End cover seal washer. | *45. | Diaphragm spring. | AUF 300 Type. | 70. | Outlet valve. | HP Type. |
| 21. | Contact blade. | †46. | Delivery flow smoothing device cover. | AUF 300 Type. | 71. | Spring clip. | HP Type. |
| 22. | Washer. | 47. | Set screw. | AUF 300 Type. | 72. | Medium fibre washer. | HP Type. |
| 23. | Contact blade screw. | 48. | Gasket | AUF 300 Type. | 73. | Outlet connection. | HP Type. |
| 24. | Condenser. | 49. | Inlet air bottle cover. | AUF 300 Type. | 74. | Gasket. | HP Type. |
| 25. | Condenser clip. | 50. | Dished washer. | AUF 300 Type. | 75. | Sandwich plate. | HP Type. |

* Early pumps. † Delivery air bottle (later pumps).

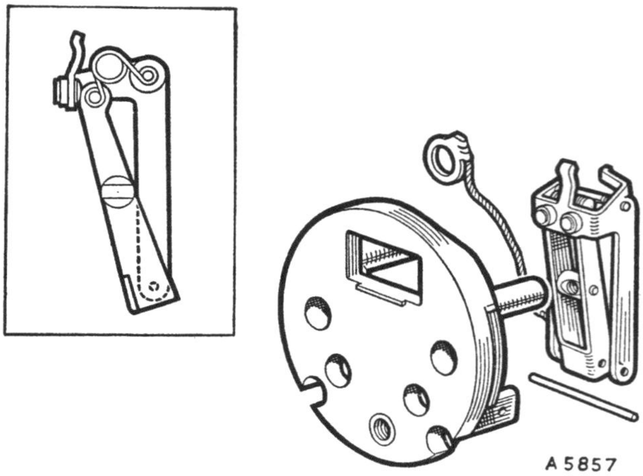


*Fig. D.1*

*Fitting the rocker assembly to the pedestal. (Inset) the correct position of the centre toggle spring*

(7) Push out the hardened steel pin which holds the rocker mechanism to the pedestal.

*Body and valves (type AUF 300)*

(8) Unscrew the two Phillips screws securing the valve clamp plate, remove the valve covers, valves, sealing washers, and filter.

NOTE.—**Dismantling of the delivery flow smoothing device should only be undertaken if the operation of it is faulty, and if the necessary equipment for pressure-testing after assembly is available. On this understanding proceed as follows:**

(9) Remove the four 4 B.A. screws securing the delivery flow smoothing device vent cover, remove the cover, the diaphragm spring, rubber 'O' rings, spring cap, diaphragm, barrier, diaphragm plate, and sealing washer.

(10) Remove the single 2 B.A. screw securing the inlet air bottle cover, remove the cover and gasket.

(11) Unscrew the inlet and outlet connections.

*Body and valves (type HP)*

(12) Remove the inlet union, the outlet union, the outlet valve cage, and the inlet valve disc. Remove the base plug and filter.

**Inspecting**

*General*

If gum formation has occurred in the fuel used in the pump, the parts in contact with the fuel will have become coated with a substance similar to varnish. This has a strong, stale smell and will attack the neoprene diaphragm. Brass and steel parts so affected can be cleaned by being boiled in a 20 per cent. solution of caustic soda, dipped in a strong nitric acid solution, and finally washed in boiling water. Light alloy parts must be well soaked in methylated spirits and then cleaned.

(1) Clean the pump and inspect for cracks, damaged joint faces, and threads.

(2) Clean the filter with a brush and examine for fractures. Renew if necessary.

(3) Examine the coil lead tags for security and the lead insulation for damage.

(4) Examine the contact breaker points for signs of burning and pitting. If this is evident the rockers and spring blade must be renewed.

(5) Examine the pedestal for cracks or other damage particularly to the narrow ridge on the edge of the rectangular hole on which the contact blade rests.

(6) Examine the diaphragm for signs of deterioration.

(7) Examine the non-return vent valve in the end cover for damage, and ensure that the small ball valve is free to move.

*Type AUF 300*

(8) Examine the plastic valve assemblies for kinks or damage to the valve plates. They can best be checked by blowing or sucking with the mouth.

(9) Check that the narrow tongue on the valve cage which prevents the valve being forced out of position, has not been distorted but allows a valve lift of approximately $\frac{1}{16}$ in. (1·6 mm.).

(10) *Early pumps:* Examine the delivery flow smoothing device diaphragm, barrier, plate, spring, and spring cap for damage. If in doubt, renew the diaphragm. *Later pumps:* Examine the delivery air bottle diaphragm.

(11) Examine the inlet air bottle cover and gasket for damage.

(12) Examine the valve recesses in the body for damage and corrosion. If it is impossible to remove the corrosion or if the recesses are badly pitted the body must be discarded.

*Type HP*

(13) Remove the circlip in the outlet valve cage and examine the inlet and outlet valve discs for signs of wear. Scrap, if worn.

(14) Examine the valve seat in the body and outlet valve cage for damage and corrosion. If it is impossible to remove the corrosion or if the seat is pitted the body or cage must be discarded.

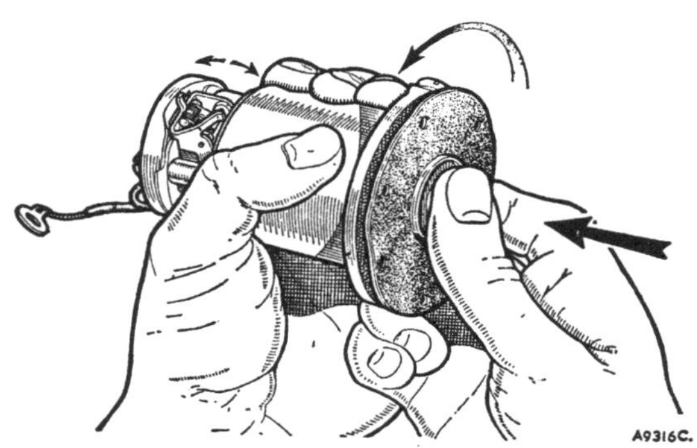


*Fig. D.2*

*Setting the diaphragm. Unscrew until the rocker just 'throws over'*

**Reassembling**

*Pedestal and rocker*

**NOTE.—The steel pin which secures the rocker mechanism to the pedestal is specially hardened and must not be replaced by other than a genuine S.U. part.**

(1) Invert the pedestal and fit the rocker assembly to it by pushing the steel pin through the small holes in the rockers and pedestal struts. Then position the centre toggle so that with the inner rocker spindle in tension against the rear of the contact point the centre toggle spring is above the spindle on which the white rollers run. This positioning is important to obtain the correct 'throw-over' action. It is also essential that the rockers are perfectly free to swing on the pivot pin and that the arms are not binding on the legs of the pedestal. If necessary, rockers can be squared up with a pair of thin-nosed pliers.

(2) Assemble the square-headed 2 B.A. terminal stud to the pedestal, the back of which is recessed to take the square head.

(3) Assemble the 2 B.A. spring washer and put the terminal stud through the 2 B.A. terminal tag then fit the lead washer, and the coned nut with its coned face to the lead washer (this makes better contact than an ordinary flat washer and nut). Tighten the 2 B.A. nut and finally add the end cover seal washer.

(4) Assemble the pedestal to the coil housing by fitting the two 2 B.A. pedestal screws, ensuring that the condenser wire clip on the left-hand screw (9 o'clock position) is between the pedestal and the earthing tag. The spring washer is not fitted when a condenser is used.

(5) Tighten the screws, taking care to prevent the earthing tag from turning as this will strain or break the earthing flex. Do not overtighten the screws or the pedestal will crack. **Do not fit the contact blade at this stage.**

*Diaphragm assembly*

(6) Place the armature spring into the coil housing with its large diameter towards the coil.

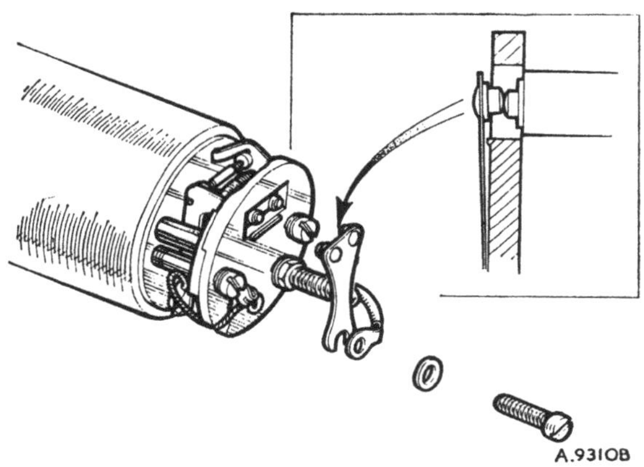


*Fig. D.3*

*Setting the correct relative position of blade and rocker contact points*

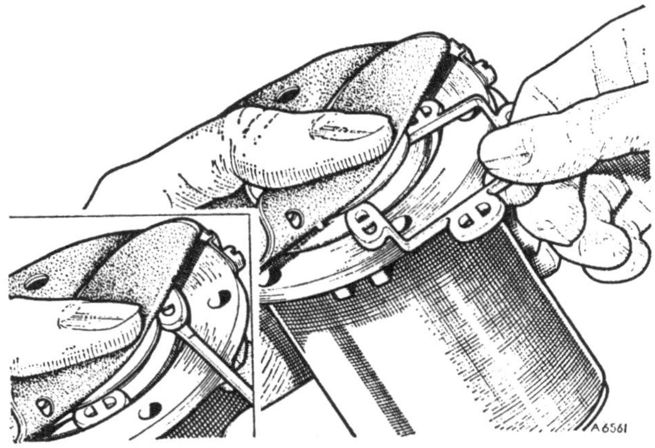


*Fig. D.4*

*Fitting the armature guide plate. (Inset), levering one of the end lobes from the recess*

(7) Before fitting the diaphragm make sure that the impact washer (a small neoprene washer that fits in the armature recess) is fitted to the armature. Do not use jointing compound or dope on the diaphragm.

(8) Fit the diaphragm by inserting the spindle in the coil and screwing it onto the threaded trunnion in the centre of the rocker assembly.

(9) Screw in the diaphragm until the rocker will not 'throw over'. This must not be confused with jamming the armature on the coil housing internal steps.

(10) *Roller type.* With the pump held with the rocker end downwards, turn back the edge of the diaphragm and fit the 11 brass rollers into the recess in the coil housing.

**On later-type rocker mechanisms with adjustable fingers fit the contact blade and adjust the finger settings as described under those headings, then carefully remove the contact blade.**

(11) With the pump held horizontally, slowly unscrew the diaphragm while at the same time actuating it, until the rocker just throws over. Unscrew the diaphragm until the holes are aligned, then unscrew it a further quarter of a turn (four holes).

(12) *Roller type.* Press the centre of the armature and fit the retaining fork at the back of the rocker assembly.

*Body components (type AUF 300)*

(13) *Guide plate type.* Turn back the edge of the diaphragm and insert one end lobe of the armature guide plate into the recess between the armature and the coil housing. Progressively position all four lobes, then commencing in the centre and finishing with the two end ones, press the lobes firmly into the recess.

(14) Note that the inlet valve recess in the body is deeper than the outlet recess to allow for the filter and extra washer. Screw in the inlet and outlet connections with their sealing rings. Assemble the

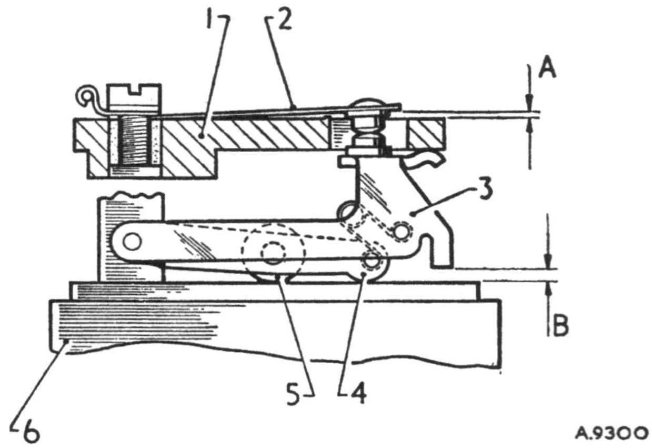


*Fig. D.5*
*The rocker finger settings on modified rocker assemblies*

1. Pedestal.
2. Contact blade.
3. Outer rocker.
4. Inner rocker.
5. Trunnion.
6. Coil housing.

A = ·035 in. (·9 mm.). B = ·070 in. (1·8 mm.).

outlet valve components into the outlet recess in the following order: first a joint washer, then the valve (tongue side downwards), then the valve cover.

(15) Assemble the inlet valve into the recess as follows: first a joint washer, then the filter (dome side downwards), then another joint washer, followed by the valve assembly (tongue side uppermost), then the valve cover.

(16) Take care that both valve assemblies nest down into their respective recesses, place the clamp plate on top of the valve covers and tighten down firmly on to the body with the two screws. Replace the inlet air bottle cover with its joint washer and tighten down the central screw.

(17) *Early pumps:* Place the sealing washer in the bottom of the delivery flow smoothing device recess, follow this with the perforated diaphragm plate (dome side downwards), then the plastic barrier followed by the rubber diaphragm. Insert the 'O' section sealing ring into the recess and ensure that it seats evenly. Place the diaphragm spring (large end towards the vented cover) into the cover, place the spring end cap on to the small end of the spring, pass the assembly tool through the cover, spring, and end cap, turn the tool through 90°, and tension the spring. Finally, fit the cap and spring assembly on to the diaphragm, tighten the four retaining screws and release the assembly tool. *Later pumps:* Place the sealing washer in the bottom of the delivery air bottle recess, place the plastic diaphragm, dome side downwards, then add the 'O' section sealing ring and tighten down the cap with its four screws.

*Body components (type HP)*

(18) Assemble the brass outlet valve disc to the outlet valve cage, making sure that the smooth face of the disc faces the valve seat, retain it in position with the circlip, which must be located in its groove. The valve must rattle freely when the valve cage is shaken.

(19) Drop the inlet valve disc, smooth face downwards, on to the inlet valve seat in the body of the pump, insert the thin fibre washer, drop the outlet valve cage in position and insert the medium fibre washer, then screw in the outlet union and tighten with a spanner. Fit the inlet union and filter.

*Body attachment (type AUF 300)*

(20) Fit the joint washer to the body, aligning the screw holes, and offer up the coil housing to the body, ensuring correct seating between them.

(21) Line up the six securing screw holes, making sure that the cast lugs on the coil housing are at the bottom, and insert the six 2 B.A. screws finger tight. Fit the earthing screw with its Lucar connector.

(22) *Roller type.* Carefully remove the retaining fork from the rocker assembly and check that the rollers are correctly positioned.

(23) Tighten the securing screws in diagonal sequence.

*Body attachment (type HP)*

(24) Place the sandwich plate joint gasket on the face of the body, lining up the holes in the body and washer, fit the sandwich plate (concave face to diaphragm) to the gasket on the body, again lining up the holes.

(25) Offer up the coil housing to the body and sandwich plate and ensure correct seating between them, with the connections at the top and filter at the bottom.

(26) Line up the six securing screw holes, making sure that the two cast lugs on the coil housing are at the bottom, and insert the six 2 B.A. screws finger tight. Fit the earthing screw with its Lucar connector.

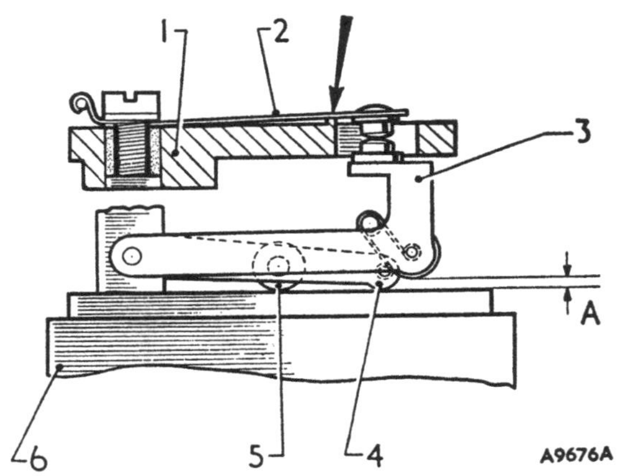


*Fig. D.6*
*The contact gap setting on earlier-type rocker assemblies*

1. Pedestal.
2. Contact blade.
3. Outer rocker.
4. Inner rocker.
5. Trunnion.
6. Coil housing.

A = ·030 in. (·8 mm.).

(27) **Remove the roller retaining fork carefully,** making sure that the rollers retain their position; a displaced roller will cut the diaphragm.
(28) Tighten the securing screws in sequence as they appear diametrically opposite each other.

*Contact blade*

(29) Fit the contact blade and coil lead to the pedestal with the 5 B.A. washer and screw. Place the condenser tag beneath the coil lead tag.
(30) Adjust the contact blade so that the points on it are a little above the contact points on the rocker when the points are closed, also that when the contact points make or break one pair of points wipes over the centre line of the other in a symmetrical manner.

The contact blade attachment screw slot allows a degree of adjustment.

(31) Tighten the contact blade attachment screw when the correct setting is obtained.

*Contact gap settings*

(32) Check that when the outer rocker is pressed on to the coil housing the contact blade rests on the narrow rib which projects above the main face of the pedestal. If it does not, slacken the contact blade attachment screw, swing the blade clear of the pedestal, and bend it downwards a sufficient amount so that, when re-positioned, it rests against the rib.

*(a) Modified rocker assemblies*

(33) Check the lift of the contact blade above the top of the pedestal (A) (Fig. D.5) with a feeler gauge, bending the stop finger beneath the pedestal, if necessary, to obtain a lift of ·035±·005 in. (·9±·13 mm.).

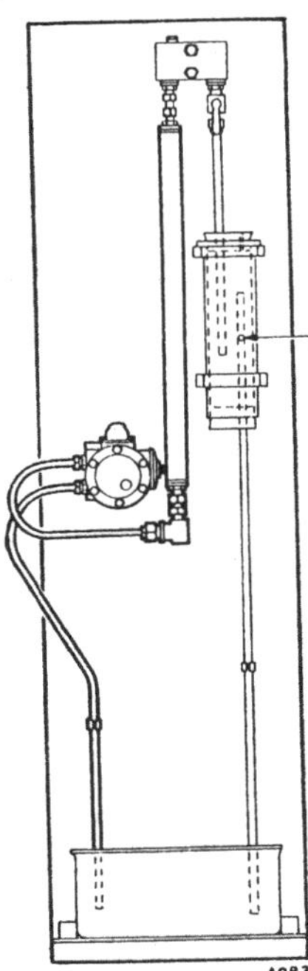


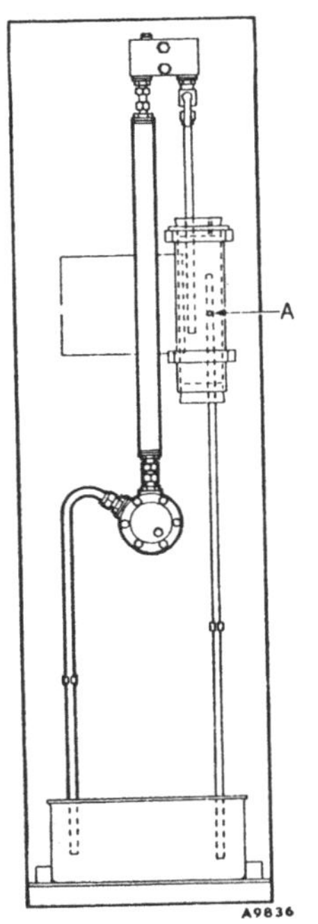


*Fig. D.7*

(1) *The AUF 300 type pump mounted on the S.U. test stand*
(A) Hole ·187 to ·192 in. (4·74 to 4·9 mm.) dia.

(2) *The HP type pump mounted on the S.U. test stand.*
(A) Hole ·156 in. (4 mm.) dia.

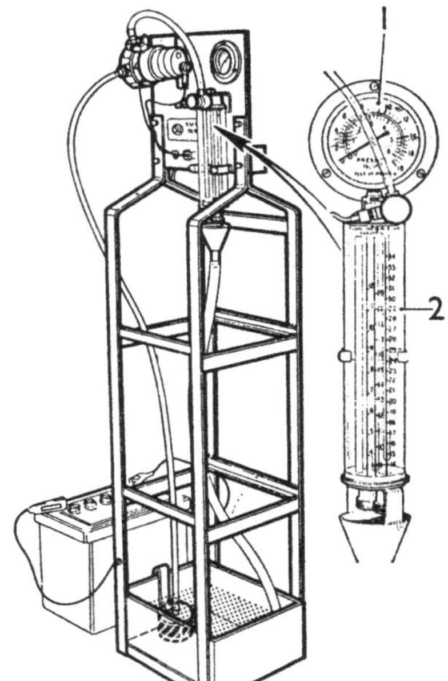


*Fig. D.8*

*A checking rig for S.U. fuel pumps available from V. L. Churchill and Co. Ltd. The rig measures output in gallons of paraffin (kerosene) per hour, against required suction and delivery heads*

1. Pressure gauge. 2. Flow glass.

(34) Check the gap between rocker finger and coil housing (B) (Fig. D.5) with a feeler gauge, bending the stop finger, if necessary, to obtain a gap of ·070±·005 in. (1·8±·13 mm.).

*(b) Earlier-type rocker assemblies*

(35) Check the gap between the points indirectly by carefully holding the contact blade against the rib on the pedestal, without pressing against the tip (Fig. D.6), then check if a ·030 in. (·8 mm.) feeler will pass between the fibre rollers and the face of the coil housing. If necessary, the tip of the blade can be set to correct the gap.

*End cover*

(36) Ensure that the end cover seal washer is in position on the terminal stud, fit the bakelite end cover and shakeproof washer, and secure with the brass nut. Fit the terminal tag or connector and the insulated sleeve. After test, replace the rubber sealing band over the end cover gap and seal with adhesive tape. This must be retained when the pump is not mounted internally in a moisture-free region.

**Testing on a test stand**

*Preparation*

(1) Churchill test rig:

Secure the pump in the clamping ring, with the outlet connection uppermost. Connect to a 12-volt battery, and with the switch in the 'OFF' position, clip the connector to the pump. Connect the delivery and return of the correct bore to the pump.

S.U. test rig:

Mount the pump on the test stand, using the appropriate adaptor set according to the type of pump. Connect the feed and earth terminals to the test battery and check the contact gap setting as described under that heading. Replace the end cover with a cut-away one which allows observation of the rocker assembly while retaining the pivot pin.

Use paraffin (kerosene) in the test tank. Ensure an adequate supply.

*Priming*

(2) Unscrew the regulator valve (Churchill rig only) and switch on: the pump should prime from dry in 10 to 15 seconds. Allow the pump to run for a minute to stabilize the flow.

*Air leak check*

(3) When the pump is first started air bubbles will be mixed with the liquid discharged from the pipe projecting downwards into the flow-meter; these bubbles should cease after a minute or so. If they do not, an air leak is indicated either in the pump or the connecting unions, and this must be rectified.

*Valve seat check*

(4) Let the pump run for about 10 minutes and then test as follows:

With the regulator valve (delivery tap) turned completely off the pump should stand without repeating for a minimum of 20 seconds at the correct delivery head. If it repeats, the inlet valve is not seating correctly. On AUF 300 type pumps malfunction of the inlet valve must be investigated. On HP-type pumps with metal valve discs this may be remedied by removing the discs and rubbing down the smooth face, using fine lapping paste on a smooth surface, or by dressing the valve seat.

*Delivery check*

(5) Churchill: Obtain a delivery head reading of 4 feet (1220 mm.) on the gauge by adjusting the regulator valve on top of the flow glass. When correct, the pump flow rate may be read directly from the appropriate colour scale on the flow glass.

S.U.: The paraffin (kerosene) should rise in the glass tube until it flows over the top of the pipe in which a side hole is drilled; if the output is not up to specification, the side hole will carry off all paraffin (kerosene) pumped and none will flow over the top.

The maximum delivery should be timed as follows:

| | |
|---|---|
| AUF 300 type pump .. | .. 1 pint in 30 sec. |
| HP-type pump .. | .. 1 pint in 1 min. 5 sec. |

*Minimum delivery check*

(6) Check with the tap turned on only slightly, and also by pressing gradually inwards on the tip of the contact blade, so as to reduce the effective stroke, that the pump continues to work with an increasing frequency until it eventually stops because there is no gap left between the points.

*Reduced voltage*

(7) Connect a resistance and voltmeter in circuit and test the pump at 9·5 volts with regulator valve open (tap full on); the pump should work satisfactorily although with reduced output.

*Sparking check*

(8) Check for excessive sparking at the contact points. A moderate degree is permissible; excessive sparking would indicate that the special leak wire incorporated in the coil winding has fractured, necessitating a new coil unit, or that the condenser is faulty.

**Fuel pump faults**

1. *Suspected fuel feed failure*

Disconnect the fuel line at the carburetter and check for flow.

(*a*) If normal, examine for obstructed float-chamber needle seating or gummed needle.

(*b*) If normal initially, but diminishing rapidly and accompanied by slow pump operation, check for correct tank venting by removing the filler cap. Inadequate venting causes a slow power stroke with resultant excessive burning of contact points.

(*c*) If reduced flow is accompanied by slow operation of the pump, check for any restriction on the inlet side of the pump, such as a clogged filter, which should be removed and cleaned.

In the case of reduced flow with rapid operation of the pump, check for an air leak on the suction side, dirt under the valve, or faulty valve sealing washers.

(*d*) If no flow, check for:

(i) *Electrical supply*

Disconnect the lead from the terminal and check if current is available.

(ii) *Faulty contact points*

If electrical supply is satisfactory, the bakelite cover should be removed to check that the tungsten points are in contact. The lead should then be replaced on the terminal and a short piece of bared wire put across the contacts. If the pump then performs a stroke, the fault is due to dirt or corrosion, or maladjustment of the tungsten points. The points may be cleaned by folding a small piece of fine emery-paper and inserting it between them and sliding it to and fro. To re-adjust the contact points follow the procedure laid down under the appropriate heading.

(iii) *Obstructed pipeline between fuel tank and pump*

The inlet pipe should be disconnected. If the pump then operates, the trouble is due to a restriction in the pipeline between the pump and the tank. This may be cleared by the use of compressed air after removing the fuel tank filler cap. It should be noted, however, that compressed air should not be passed through the pump as this will cause serious damage to the valves.

(iv) *Faulty diaphragm action*

In the event of the previous operations failing to locate the trouble, it may be due to a stiffening of the diaphragm fabric or the presence of abnormal friction in the rocker 'throw-over' mechanism, or a combination of both. To remedy these faults the coil housing should be removed and the diaphragm flexed a few times, taking care not to lose any of the 11 rollers under the diaphragm. Prior to this resetting it is advisable to apply, very sparingly, a little thin oil to the throw-over spring spindles at the point where they pivot in the brass rockers. The diaphragm/armature assembly should then be reassembled in accordance with instructions given under that heading.

2. *Noisy pump*

If the pump is noisy in operation, an air leak at the suction line may be the cause. Such a leak may be checked by disconnecting the fuel pipe from the carburetter and allowing the pump to discharge into a suitable container with the end of the pipe submerged. The emission of continuous bubbles at this point will confirm the existence of an air leak. The fault should be rectified by carrying out the following procedure:

(*a*) Check that all connections from the fuel tank to the pump are in good order.

(*b*) Check that the inlet union is tight and that the sealing 'O' ring is not damaged.

(*c*) Check that the coil housing securing screws are well and evenly tightened.

Air leaks on the suction side cause rapid operation of the pump and are the most frequent cause of premature failure.

3. *Pump operates without delivering fuel*

If the pump operates continuously without delivery of fuel the most likely causes are:

(*a*) A very serious air leak on the suction side or,

(*b*) Foreign matter lodged under one of the valves, particularly the inlet valve.

To remedy (*a*) see Section 2 (above)

In order to remove any foreign matter lodged under the valves, these should be removed for cleaning and great care taken that the plastic material of the valve disc is not scratched or damaged during this operation.

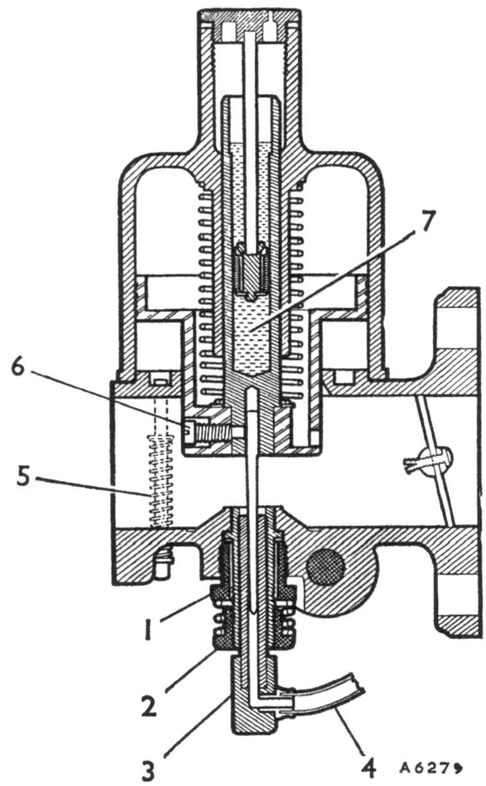


*Fig. D.*10

*A sectional view of the HS4 carburetter*

1. Jet locking nut.
2. Jet adjusting nut.
3. Jet head.
4. Feed tube from float-chamber.
5. Piston lifting pin.
6. Needle securing screw.
7. Oil damper reservoir.

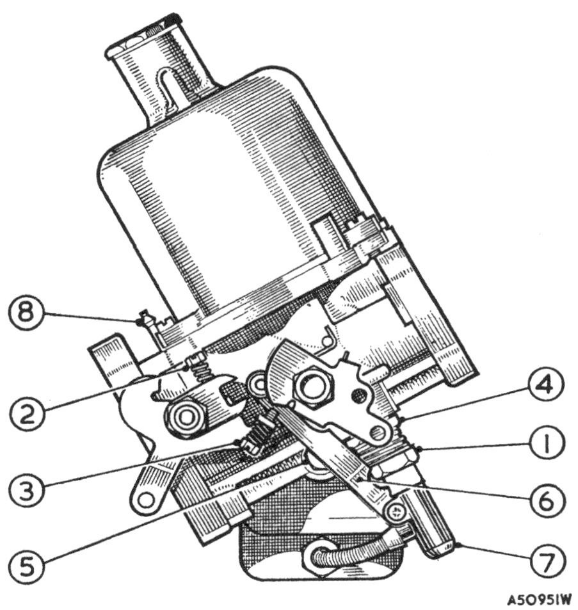


*Fig. D.9*

*The HS4 carburetter*

1. Jet adjusting nut.
2. Throttle stop screw.
3. Choke or fast-idle screw.
4. Jet locking nut.
5. Float-chamber securing nut.
6. Jet link.
7. Jet head.
8. Vacuum ignition take-off.

# THE CARBURETTER COMPONENTS

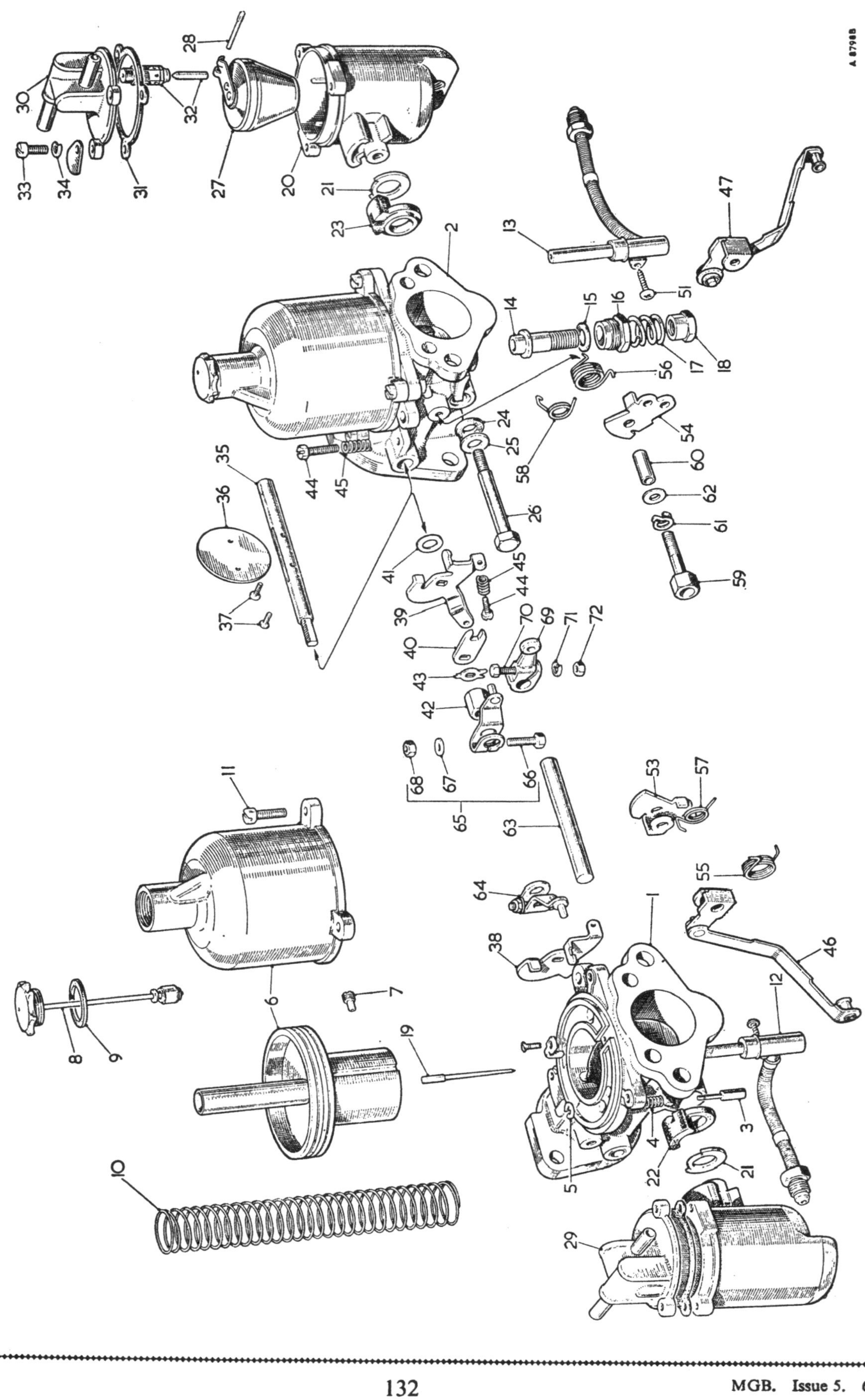

## KEY TO THE CARBURETTER COMPONENTS

| No. | Description |
|---|---|
| 1. | Body—front carburetter. |
| 2. | Body—rear carburetter. |
| 3. | Pin—piston lifting. |
| 4. | Spring—pin. |
| 5. | Circlip—pin. |
| 6. | Chamber and piston assembly. |
| 7. | Screw—needle locking. |
| 8. | Cap and dampers assembly. |
| 9. | Washer—cap (fibre). |
| 10. | Spring—piston (red). |
| 11. | Screw—chamber to body. |
| 12. | Jet assembly—front carburetter. |
| 13. | Jet assembly—rear carburetter. |
| 14. | Bearing—jet. |
| 15. | Washer—jet bearing (brass). |
| 16. | Screw—jet locking. |
| 17. | Spring—jet locking. |
| 18. | Screw—jet adjusting. |
| 19. | Needle. |
| 20. | Chamber—float. |
| 21. | Washer—support. |
| 22. | Grommet—front carburetter (rubber). |
| 23. | Grommet—rear carburetter (rubber). |
| 24. | Washer (rubber). |

| No. | Description |
|---|---|
| 25. | Washer—plain. |
| 26. | Bolt—float-chamber fixing. |
| 27. | Float assembly. |
| 28. | Pin—hinged lever. |
| 29. | Lid—float-chamber—front carburetter. |
| 30. | Lid—float-chamber—rear carburetter. |
| 31. | Washer—lid. |
| 32. | Needle and seat assembly. |
| 33. | Screw—lid. |
| 34. | Washer—spring—screw. |
| 35. | Spindle—throttle. |
| 36. | Disc—throttle. |
| 37. | Screw—disc. |
| 38. | Lever—throttle return—front carburetter. |
| 39. | Lever—throttle return—rear carburetter. |
| 40. | Lever—lost motion. |
| 41. | Washer—spacing. |
| 42. | Nut—lever. |
| 43. | Washer—tab—nut. |
| 44. | Screw—throttle stop. |
| 45. | Spring—throttle stop screw. |
| 46. | Lever and link assembly—pick-up—front carburetter. |

| No. | Description |
|---|---|
| 47. | Lever and link assembly—pick-up—rear carburetter. |
| 51. | Screw—link to jet. |
| 53. | Lever—cam—front carburetter. |
| 54. | Lever—cam—rear carburetter. |
| 55. | Spring—pick-up lever—front carburetter. |
| 56. | Spring—pick-up lever—rear carburetter. |
| 57. | Spring—cam lever—front carburetter. |
| 58. | Spring—cam lever—rear carburetter. |
| 59. | Bolt—pivot. |
| 60. | Tube—pivot bolt. |
| 61. | Washer—spring—pivot bolt. |
| 62. | Washer—distance. |
| 63. | Rod—jet connecting. |
| 64. | Lever and pin assembly—front carburetter. |
| 65. | Lever and pin assembly—rear carburetter. |
| 66. | Bolt—lever. |
| 67. | Washer—bolt. |
| 68. | Nut—bolt. |
| 69. | Lever—choke operating. |
| 70. | Bolt—lever. |
| 71. | Washer—spring. |
| 72. | Nut. |

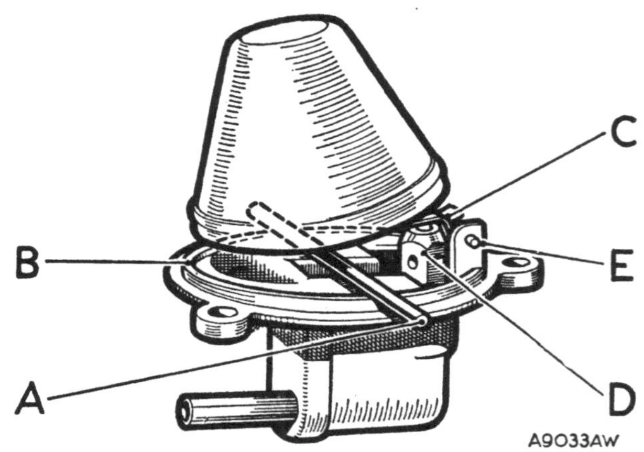


*Fig. D.*11

*Checking the nylon float level*

A. $\frac{1}{8}$ to $\frac{3}{16}$ in. (3·18 to 4·76 mm.) dia. bar.
B. Machined lip.
C. Float lever reseting point.
D. Needle valve assembly.
E. Hinge pin.

## Section D.4

### CARBURETTERS

IMPORTANT.—The instructions given in **this section** for adjusting, dismantling and reassembling the carburetters applies only to **cars not fitted with exhaust emission control equipment.** Carburetters fitted to **cars with exhaust emission control equipment** must be tuned and serviced in accordance with the instructions given in **The Engine Emission Control Supplement to this manual.**

**Removing and refitting**

Release the spring clips and detach the breather hose between the air cleaner and rocker cover (early cars).

Disconnect the fuel supply pipe at the rear carburetter or at each carburetter float-chamber. Disconnect the overflow pipes.

Slacken off the bolt and nut securing the mixture and throttle cables respectively and withdraw the cables complete. Detach the throttle return springs. Pull off the rubber connection for the vacuum ignition control pipe from the top of the rear carburetter body.

Remove the four nuts securing the carburetter flanges and withdraw the carburetters and air cleaners as an assembly.

When refitting, the centre throttle return spring end eye is located between the flat washer and the throttle lever. The throttle linkage must be checked, and re-adjusted if necessary, after refitting.

**Slow-running adjustments and synchronization**

As the needle size is determined during engine development, tuning of the carburetters is confined to correct idling setting.

Run the engine until it attains its normal running temperature.

Slacken off the pinch-bolt of one of the coupling levers locating the carburetter interconnecting shaft to the carburetter throttle spindles so that each carburetter can be operated independently.

Disconnect the mixture control cable and slacken off the two pinch-bolts to free the choke actuating lever and unscrew the fast-idling screws.

Remove the pistons and suction chambers. Screw the jet adjusting nuts until each jet is flush with the bridge of its carburetter, or as near to this as possible (each jet being in the same relative position to the bridge of its respective carburetter). Replace the pistons and suction chamber assemblies, and check that the pistons fall freely onto the bridge of the carburetters (by use of the piston lifting pins). Turn down the jet adjusting nut two complete turns (12 flats).

Restart the engine, and adjust the throttle adjusting screws to give the desired idling speed by moving each throttle adjusting screw an equal amount. By listening to the hiss in the intake, adjust the throttle adjusting screws until the intensity of the hiss is similar on each intake. This will synchronize the throttle setting.

When this is satisfactory the mixture should be adjusted by screwing each jet adjusting nut up or down by the same amount until the fastest idling speed is obtained consistent with even firing. During this adjusting it is necessary that the jets are pressed upwards to ensure that they are in contact with the adjusting nuts.

As the mixture is adjusted the engine will probably run faster, and it may therefore be necessary to unscrew the throttle adjusting screws a little, each by the same amount, to reduce the speed.

Now check the mixture strength by lifting the piston of the rear carburetter by approximately $\frac{1}{32}$ in. (·75 mm.), when:

(1) If the engine speed increases, the mixture strength of the rear carburetter is too rich.
(2) If the engine speed immediately decreases, the mixture strength of the rear carburetter is too weak.

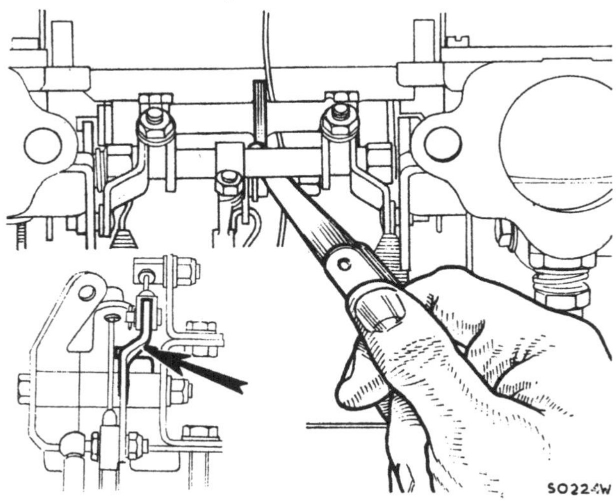

*Fig. D.*12

***A feeler gauge between the throttle shaft stop and the choke interconnecting spindle. Inset shows the position for the feeler gauge on the cars fitted with automatic transmission***

(3) If the engine speed momentarily increases very slightly, the mixture strength of the rear carburetter is correct.

Repeat the operation at the front carburetter, and after adjustment re-check the rear carburetter, since the carburetters are interdependent.

When the mixture is correct the exhaust note should be regular and even. If it is irregular, with a splashy type of misfire and colourless exhaust, the mixture is too weak. If there is a regular or rhythmical type of misfire in the exhaust beat, together with a blackish exhaust, then the mixture is too rich.

The carburetter throttle on each carburetter is operated by a lever and pin, with the pin working in a forked lever attached to the throttle spindle. A clearance exists between the pin and the fork which must be maintained when the throttle is closed and the engine idling, to prevent any load from the accelerator linkage being transferred to the throttle butterfly and spindle.

To set the clearance on synchromesh cars, with the throttle shaft levers free on the throttle shaft, place a ·12 in. (·3 mm.) feeler gauge between the throttle shaft stop and the choke interconnecting rod (see Fig. D.12). On cars fitted with automatic transmission place a ·020 in. (·5 mm.) feeler gauge between the throttle lever and the throttle lever stop (inset, Fig. 12). Move each throttle shaft lever downwards in turn until the lever pin rests lightly on the lower arm of the fork in the carburetter throttle lever. Tighten the clamp bolt of the throttle shaft lever at this position. When both carburetters have been dealt with, remove the feeler gauge. The pins on the throttle shafts should then have clearance in the forks.

Reconnect the choke cable, ensuring that the jet heads return against the lower face of the jet adjusting nuts and that the choke actuating levers are fully returned when the choke control is pushed fully in.

Pull out the mixture control knob on the dash panel until the linkage is about to move the carburetter jets (a minimum of ¼ in. or 6 mm.) and adjust the fast-idle adjusting screws to give an engine speed of about 1,000 r.p.m. when hot.

A small clearance must be maintained between the fast-idle cams and the abutment screws when the mixture control is pushed fully in.

**Jet-centring**

To check the jet for concentricity with the jet needle set the jet head and the jet adjusting nut in the uppermost position, lift the suction piston with the piston lifting pin, and allow the piston to fall. It should fall freely, and a definite soft, metallic click will be heard as the base of the piston strikes the jet bridge.

If this does not happen with the jet raised, but does occur when the jet is lowered, the jet bearing and jet must be recentred as follows.

Disconnect the lever between the interconnecting lever and the jet head.

Unscrew the union holding the nylon feed tube into the base of the float-chamber and withdraw the tube and jet together.

Unscrew the jet adjusting nut and remove the lock spring; screw up the nut to its fullest extent and refit the jet head and feed tube.

Slacken off the jet locking nut until the jet bearing is just free to rotate with finger pressure.

Remove the piston damper from the top of the suction chamber and gently press the piston down onto the jet bridge.

Tighten the jet locking nut, at the same time ensuring that the jet head is still in its correct angular position. Lift the piston and check that it falls freely and evenly, hitting the jet bridge with a soft, metallic click, with the jet in the raised and lowered position. If the result is not satisfactory and the piston is not sticking, the recentring operation must be repeated until the correct result is obtained.

When the operation is completed replace the adjusting nut lock spring and the jet operating lever.

This adjustment is best effected with the carburetters removed from the engine.

**Float-chamber**

The position of the hinged float lever must be such that the level of the float (and therefore the height of the fuel at the jet) is correct.

To check the float level, hold the float-chamber lid and float assembly upside-down and place a ⅛ in. (3·18 mm.) diameter bar across the diameter of the machined lip of the float-chamber lid, parallel to the float lever hinge pin and under the float lever (see Fig. D.11). The face of the float lever should just rest on the bar when the float needle is held fully on its seating. If this is not so, carefully reset the angle made between the straight portion of the float lever and its hinge until the correct position is obtained.

**Float-chamber flooding**

This is indicated by the fuel dripping from the drain pipe, and is generally caused by grit between the float-chamber needle and its guide. This is cured by removing the float-chamber, washing the valve and float-chamber components, and reassembling.

**Float needle sticking**

If the engine stops, apparently through lack of fuel, when there is plenty in the tank and the pump is working properly, the probable cause is a sticking float needle. An easy test for this is to disconnect the pipe from the electric pump to the carburetters and switch the ignition on and off quickly while the end of the pipe is directed onto a pad of cloth or into a container.

If fuel is delivered, starvation is almost certainly being caused by the float needle sticking to its seating, and the float-chamber lid should therefore be removed and the needle and seating cleaned and refitted. At the same time it will be advisable to clean out the entire fuel feed system as this trouble is caused by foreign matter in the fuel, and unless this is removed it is likely to recur. It is of no use whatever renewing any of the component parts of either carburetter, and the only cure is to make sure that the fuel tank and pipe lines are entirely free from any kind of foreign matter or sticky substance capable of causing this trouble.

**Needles**

Remove the piston and suction chamber assembly. Slacken the needle clamping screw, extract the needle, and check its identifying mark (see **'GENERAL DATA'**). Refit the correct needle, ensuring that the shoulder on the shank is flush with the piston base.

**Piston sticking**

The piston assembly comprises the suction disc and the piston forming the choke, into which is inserted the hardened and ground piston rod which engages in a bearing in the centre of the suction chamber and in which is, in turn, inserted the jet needle. The piston rod running in the bearing is the only part which is in actual contact with any other part, the suction disc, piston, and needle all having suitable clearances to prevent sticking. If sticking does occur the whole assembly should be cleaned carefully and the piston rod lubricated with a spot of thin oil. No oil must be applied to any other part except the piston rod. A sticking piston can be ascertained by removing the piston damper and lifting the piston by pressing the piston lifting pin; the piston should come up quite freely and fall back smartly onto its seating when released. On no account should the piston return spring be stretched or its tension altered in an attempt to improve its rate of return.

## Section D.5

### AIR CLEANERS

The air intake filters are of the dry paper element type requiring no attention between filter replacement.

The air filter casings should be removed and cleaned out periodically to remove any dust deposit.

**Filter replacement**

Disconnect the engine rocker cover breather pipe from the front filter (early cars). Unscrew the bolts securing each air filter to the choke bracket and fixing plate respectively and remove the filters from the vehicle complete.

Remove the base plate and throw away the old paper element. Clean the inside casing and intake tubes thoroughly and reassemble, using new paper elements.

# Section Da

## THE FUEL SYSTEM (EVAPORATIVE LOSS)

**The information given in this Section refers specifically to service operations on, or affected by, equipment fitted to the MGB (GHN5 and GHD5) in conformity with local and territorial requirements.**

## Section Da.1

### FUEL TANK

**Removing**

(1) Remove the spare wheel.

(2) Disconnect the battery (negative earth) and drain the fuel tank.

(3) Slacken the two hose clips securing the rubber hose to the tank fuel inlet pipe and filler tube.

(4) Remove the screw, large plain washers and sealing washers securing the tank inside the boot.

(5) Underneath the car, disconnect the petrol feed pipe and gauge unit electrical connections.

(6) Remove the nuts and washer retaining the tank lip to the four studs. Release the two vapour pipe clips and move the pipe to one side.

(7) Remove the tank.

**Refitting**

(8) Reverse the removing procedure (1) to (7).

## Section Da.2

### VAPOUR SEPARATOR TANK

**Removing**

(1) Disconnect the battery.

(2) Remove the two flexible pipe connections from the tank. Remove the screw, nut and washer retaining the tank.

(3) Remove the tank. **Caution.**—Under extreme operating conditions the tank may contain fuel.

**Refitting**

(4) Reverse the removing procedure (1) to (3).

## Section Da.3

### LEAK TESTING

If a fault in the operation of the Evaporative Loss Control System is suspected or components of the system other than the filters or canister have been removed and refitted, the system must be pressure-tested in accordance with the test procedure given **in The Engine Emission Control Supplement to this manual.**

# SECTION E

## THE CLUTCH

# THE CLUTCH COMPONENTS

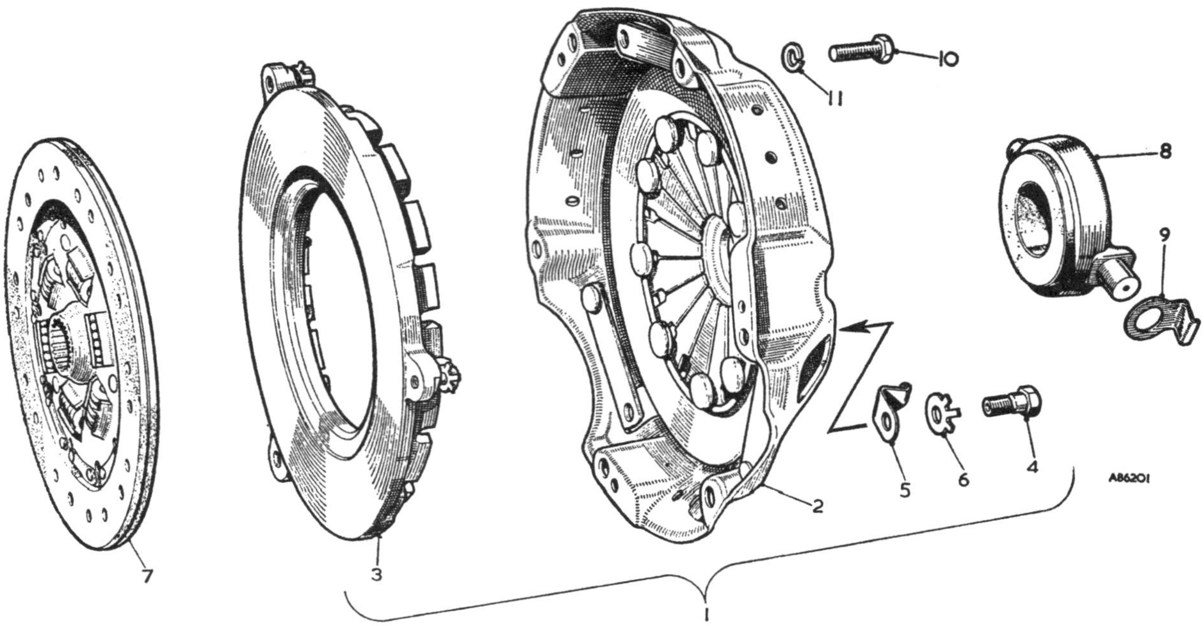


| *No.* | *Description* |
|---|---|
| 1. | Cover assembly. |
| 2. | Cover with straps, diaphragm spring and release plate. |
| 3. | Plate—pressure. |
| 4. | Bolt—strap. |
| 5. | Clip—pressure plate. |
| 6. | Washer—tab. |
| 7. | Plate assembly—driven. |
| 8. | Bearing assembly—release. |
| 9. | Retainer—bearing. |
| 10. | Screw—clutch to flywheel. |
| 11. | Washer for screw—spring. |

## GENERAL DESCRIPTION

The clutch mechanism is hydraulically operated and consists of a driven plate, a pressure plate, and a diaphragm spring and cover assembly. The cover is bolted to the flywheel and encloses the driven plate, pressure plate, and diaphragm spring.

The hydraulic system comprises a master cylinder coupled to a slave cylinder which operates the clutch release mechanism.

## Section E.1

### CLUTCH ASSEMBLY

The driven plate comprises a splined hub connected to a flexible steel plate by a spring mounting. The annular friction facings are riveted to the plate and damper springs are assembled around the hub to absorb power shocks and torsional vibration.

The diaphragm spring is interposed between two annular rings which provide fulcrum points for the diaphragm when it is flexed. The rings and the diaphragm are located and secured to the cover by nine equally spaced rivets. Three clips that engage the outer edge of the diaphragm are bolted to the pressure plate. The bolts pass through three straps which are riveted to the inside of the cover; the straps prevent the diaphragm and the pressure plate from rotating in relation to the cover.

The release plate is secured directly to the diaphragm and is an integral part of the clutch cover assembly. On early cars the release plate is retained with a circlip and can be removed when the circlip is released.

The release bearing is graphite and is mounted in a cup which fits into the fork of the clutch withdrawal lever. The cup is held in position by two spring retainers.

**Removing**

Remove the engine as described in Section A.21.

Loosen each of the bolts securing the clutch assembly to the flywheel by slackening them a turn at a time until spring pressure is released. The clutch cover can now be disengaged from the dowels on the flywheel and the assembly removed.

**Dismantling**

On earlier cars remove the circlip securing the release plate to the diaphragm and lift the plate from the diaphragm.

Unscrew the three screws securing the clips to the pressure plate, a turn at a time, until the diaphragm contacts the cover. Remove the screws, clips, and washers and the pressure plate.

Rotate the release bearing spring retainers through 90° and withdraw the bearing from the withdrawal lever fork.

**Assembling**

Assembly is a reversal of the dismantling sequence, but ensure that the release bearing retainers are correctly located.

**Replacing**

Position the driven plate assembly on the flywheel with the large end of the hub away from the flywheel.

Centralize the plate by using Service tool 18G 680, which fits the splined hub of the driven plate and the pivot bearing in the flywheel. As an alternative, a spare first motion shaft can be used.

Locate the cover assembly on the flywheel dowels and secure it with the bolts; tighten the bolts down a turn at a time by diametrical selection. Do not remove the centralizer until all bolts are securely tightened.

Remove the clutch centralizer and refit the engine. The weight of the gearbox must be supported during refitting in order to avoid strain on the first motion shaft and distortion or displacement of the release plate and straps, or driven plate assembly.

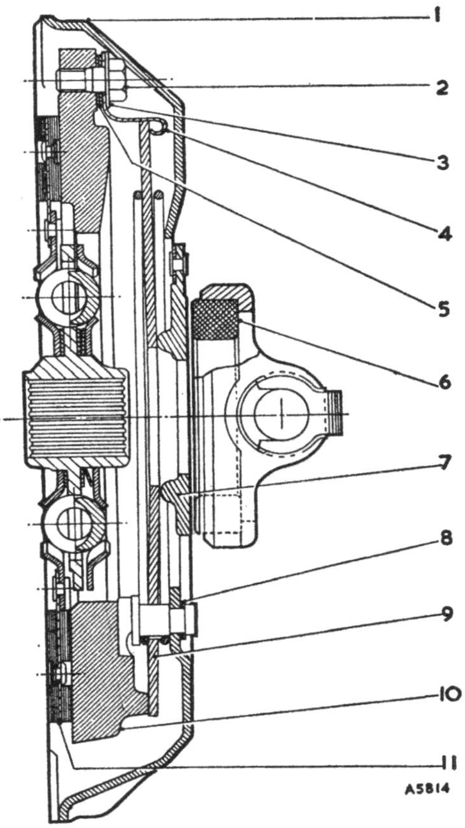


*Fig. E.1*

*A section through the clutch*

1. Cover.
2. Strap bolt.
3. Washer—tab.
4. Clip.
5. Strap—diaphragm cover.
6. Release bearing.
7. Release plate.
8. Strap—release plate/cover.
9. Diaphragm spring.
10. Pressure plate.
11. Driven plate.

## Section E.2

### SERVICING THE CLUTCH

**Driven plates**

It is important that neither oil nor grease should contact the clutch facings.

It is essential to install a complete driven plate assembly when the renewal of the friction surfaces is required. If the facings have worn to such an extent as to warrant renewal, then slight wear will have taken place on the splines, and also on the torque reaction springs and their seatings. The question of balance and concentricity is also involved. Under no circumstances is it satisfactory to repair or rectify faults in clutch driven plate centres.

**Tolerances**

Wear on the working faces of the driven plate is about ·001 in. (·02 mm.) per 1,000 miles (1600 km.) under normal running conditions. The accuracy of the alignment of the face of the driven plate must be within ·015 in. (·38 mm.).

**Condition of clutch facings in service**

It is natural to assume that a rough surface will give a higher frictional value against slipping than a polished one, but this is not necessarily correct. A roughened surface consists of small hills and dales, only the 'high-spots' of which make contact. As the amount of useful friction for the purpose of taking up the drive is dependent upon the area in actual contact, it is obvious that a perfectly smooth face is required to transmit the maximum amount of power for a given surface area.

Since non-metallic facings of the moulded asbestos type have been introduced in service the polished surface is common, but it must not be confused with the glazed surface which is sometimes encountered due to conditions to be detailed subsequently. The ideally smooth or polished condition will therefore provide proper surface contact, but a glazed surface entirely alters the frictional value of the facing and will result in excessive clutch slip. These two conditions might be simply illustrated by comparison between a piece of smoothly finished wood and one with a varnished surface; in the former the contact is made directly by the original material, whereas in the latter instance a film of dry varnish is interposed between the contact surfaces and actual contact is made by the varnish.

If the clutch has been in use for some time under satisfactory conditions, the surface of the facings assumes a high polish through which the grain of the material can be seen clearly. This polished facing is of light colour when in perfect condition.

Should oil in small quantities gain access to the clutch and find its way onto the facings, it will be burnt off as a result of the heat generated by the slipping occurring under normal starting conditions. The burning of this small quantity of lubricant has the effect of gradually darkening the facings, but provided the polish of the facing remains such that the grain of the material can be distinguished clearly, it has little effect on clutch performance.

Should increased quantities of oil obtain access to the facing, then one of two conditions, or a combination of these, may arise, depending upon the nature of the oil.

(1) The oil may burn off and leave a carbon deposit on the surface of the facings, which assume a high glaze, producing further slip. This is a very definite, though very thin, deposit, and in general it hides the grain of the material.

(2) The oil may partially burn and leave a resinous deposit on the facings. This has a tendency to produce a fierce clutch, and may also cause excessive 'spinning' due to the tendency of the face of the linings to adhere to the surface of the flywheel or pressure plate.

(3) There may be a combination of conditions (1) and (2) which produces a tendency to 'judder' on such engagement.

Still greater quantities of oil produce a dark and soaked appearance of the facings, and the result will be further slip, accompanied by fierceness or 'juddering'.

If the conditions enumerated above are experienced, the clutch driven plate should be replaced by a new one. **The cause of the presence of the oil must be traced and removed.** It is, of course, necessary for the clutch and flywheel to be cleaned out thoroughly before assembly.

Where the graphite release bearing ring is badly worn in service a complete replacement assembly should be fitted, returning the old assembly for salvage of the metal cup. These graphite rings are inserted into their metal cup by heating the metal cup to a cherry red, then forcing the graphite ring into position. Immediately the ring is forced into position the whole should be quenched in oil. Alignment of the thrust pad in relation to its face and the trunnions should be within ·005 in. (·12 mm.).

In almost every case of rapid wear on the splines of the clutch driven plate misalignment is responsible.

Looseness of the driven plate on the splined shaft results in noticeable backlash in the clutch. Misalignment also puts undue stress on the driven member, and may result in the hub breaking loose from the plate, with consequent total failure of the clutch.

It may also be responsible for a fierce chattering or dragging of the clutch, which makes gear-changing difficult. In cases of persistent difficulty it is advisable to check the flywheel for truth with a dial indicator. The dial reading should not vary more than ·003 in. (·07 mm.) anywhere on the flywheel face.

## Section E.3

### MASTER CYLINDER

The master cylinder has an integral-type supply tank in which the barrel passes through the tank. A piston contained within the barrel has a rubber main cup and is spring-loaded against its inner end; between the piston and cup is a thin washer which prevents the cup being drawn into the feed holes drilled around the head of the piston. The outer end of the piston carries a secondary cup and is formed with a depression to receive the spherical end of the push-rod assembly. The push-rod has a piston stop that is retained in the body by a circlip. A rubber boot through which the push-rod passes is fitted to the end of the body.

At the opposite end of the barrel to the push-rod an end plug screws down against a gasket. This plug forms the outlet connection for the pipe line to the slave cylinder.

**Removing**

Remove the screws securing the brake and clutch master cylinder cover and take off the cover.

Drain the fluid from the supply tank by attaching a rubber tube to the bleed screws in the clutch slave cylinder, opening the screw one full turn and then depressing the clutch pedal. Hold the pedal down and tighten the screw and then let the pedal return unassisted. Repeat this operation until the tank is empty.

Remove the split pin, washer, and clevis pin from the push-rod and disengage the clutch pedal lever.

Clean the pipe connection, disconnect the pipe line, and fit a plug to the end of the cylinder to prevent the entry of dirt.

Unscrew the fixing bolts and detach the master cylinder from the box assembly.

**Dismantling**

Detach the rubber boot from the barrel.

Depress the piston to relieve the load on the circlip, then remove the circlip and the push-rod assembly.

Withdraw the piston, piston washer, main cup, spring retainer, and spring.

Remove the secondary cup by carefully stretching it over the end of the piston.

**Examination**

Place all metal parts in a tray of clean Clutch and Brake Fluid to soak. Dry them with a clean, non-fluffy cloth. Rubber components are to be examined for swollen or perished cups or other signs of deterioration. Any suspect parts must be renewed.

Swill the main castings in industrial methylated spirit and thoroughly dry out before assembly.

Ensure that the by-pass ports are free of obstruction. The port is drilled with a $\frac{1}{8}$ in. (3·17 mm.) drill for half its length and then finished with a ·028 in. (·711 mm.) drill.

**Assembling**

Dip all components in Clutch and Brake Fluid and assemble when wet.

Stretch the secondary cup over the piston with the lip of the cup facing the head of the piston. When the cup is in its groove work it round gently with the fingers to ensure that it is correctly seated.

Insert the return spring, largest diameter first, into the barrel and position the spring seat on the small-diameter end of the spring.

Assemble the main cup, piston washer, piston, and push-rod. When assembling the cups carefully enter the lip edge of the cups into the barrel first.

Depress the piston, position the piston stop, and retain it in the barrel with the circlip.

Place the rubber boot in position and fit the dust excluder.

**Replacing**

Refit the master cylinder to the master cylinder box and secure it with the bolts. The long bolt passes through the stiffener plate.

Remove the dust excluder and fit the pipe connection to the master cylinder.

Refit the clutch pedal lever to the push-rod and secure it with the clevis pin, washer, and a new split pin.

Refit the master cylinder cover.

Fill the master cylinder and then prime and bleed the system.

## Section E.4

### SLAVE CYLINDER

The slave cylinder incorporates two threaded connections for the feed hose and the bleed screw and accommodates in the body a piston, a cup, and a spring.

A rubber boot through which passes a push-rod is fitted to the body and is retained by two clips. The push-rod has an eye end which connects with the clutch withdrawal fork.

**Removing**

Drain the system as described in Section E.3.

Release the feed pipe from the cylinder and remove the two screws securing the cylinder to the clutch housing. The cylinder may be withdrawn, leaving the push-rod attached to the clutch withdrawal fork, or the rod may be detached from the fork.

**Dismantling**

Remove the rubber dust cover and with an air line blow out the piston and seal. Extract the spring and cup filler.

Examine all components and renew any that are suspect.

**Assembling**

Place the spring in the cylinder, followed by the filler, cup, and piston. Depress the piston with the push-rod and refit the rubber boot. Secure the boot to the rod with the small clip and then fit the boot to the cylinder and secure it with the large circlip.

## THE CLUTCH CONTROL COMPONENTS

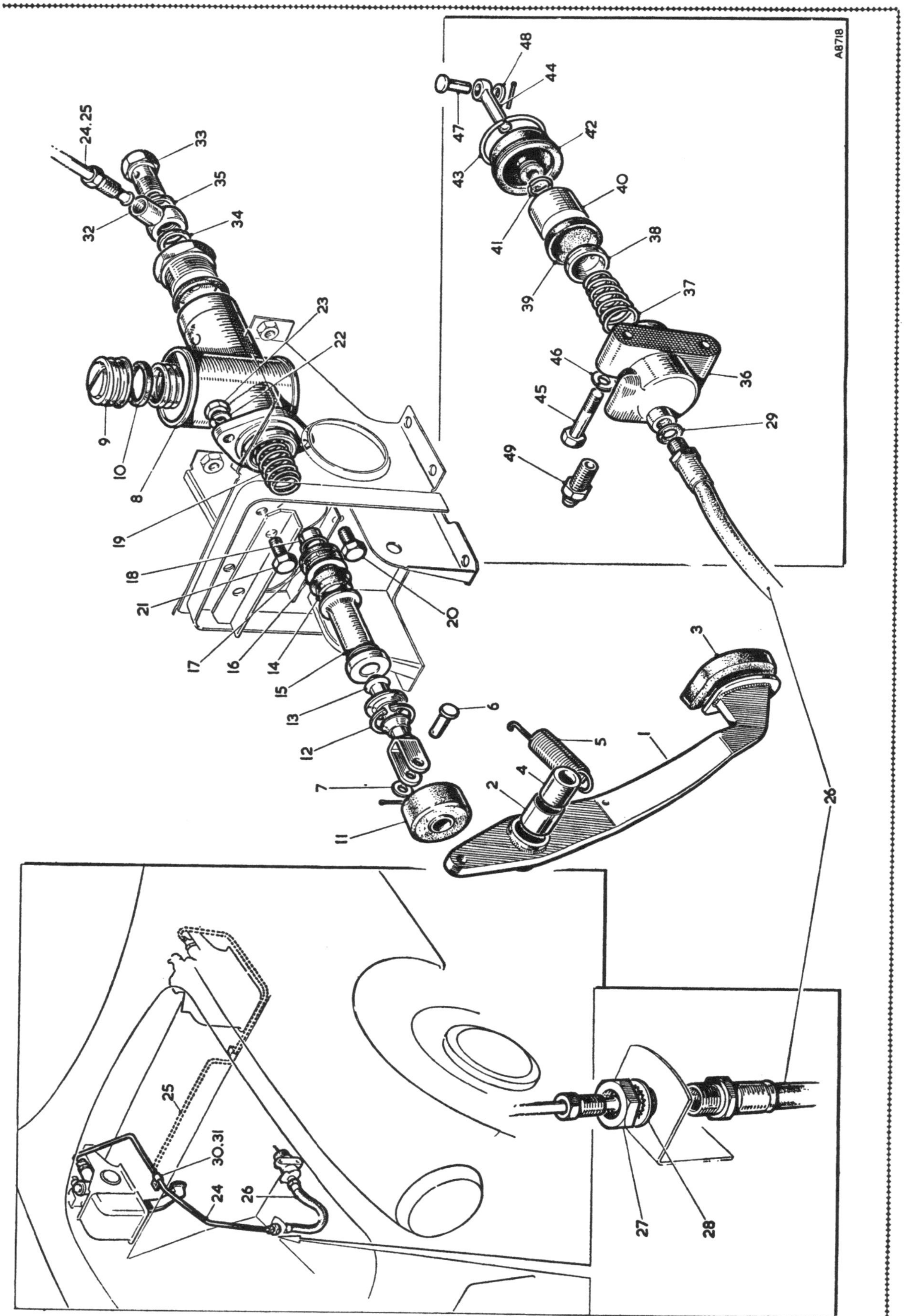

## KEY TO THE CLUTCH CONTROL COMPONENTS

| No. | Description | No. | Description | No. | Description |
|---|---|---|---|---|---|
| 1. | Pedal—clutch. | 18. | Retainer—spring. | 34. | Gasket. |
| 2. | Bush. | 19. | Spring. | 35. | Gasket. |
| 3. | Pad—pedal (rubber). | 20. | Screw—cylinder to box. | 36. | Body. |
| 4. | Tube—distance—clutch pedal. | 21. | Screw—cylinder and stiffener to box. | 37. | Spring—cup filler. |
| 5. | Spring—pedal pull-off. | 22. | Washer—spring—for screw. | 38. | Filler—piston cup. |
| 6. | Pin—clevis—fork end to pedal. | 23. | Nut for screw. | 39. | Cup piston. |
| 7. | Washer—plain—for pin. | 24. | Pipe—master cylinder to hose. | 40. | Piston. |
| 8. | Barrel and tank. | 25. | Pipe—master cylinder to hose. | 41. | Clip—boot (small). |
| 9. | Cap—filler. | 26. | Hose—clutch. | 42. | Boot. |
| 10. | Seal—cap. | 27. | Locknut for hose. | 43. | Clip—boot (large). |
| 11. | Boot. | 28. | Washer—shakeproof. | 44. | Rod—push. |
| 12. | Circlip. | 29. | Gasket—hose to body. | 45. | Bolt—cylinder to gearbox. |
| 13. | Rod—push. | 30. | Clip—clutch pipe to bulkhead. | 46. | Washer—spring—for bolt. |
| 14. | Cup—secondary. | 31. | Clip—clutch pipe to bulkhead. | 47. | Pin—clevis—cylinder to clutch fork. |
| 15. | Piston. | 32. | Connection—banjo. | 48. | Washer—plain—for pin. |
| 16. | Washer—piston. | 33. | Bolt for banjo connection. | 49. | Screw—bleeder. |
| 17. | Cup—main. | | | | |

**Replacing**

Fit the cylinder to the clutch housing and secure it with the two screws. Assemble the push-rod to the clutch withdrawal fork. Connect the feed line and fill, prime, and bleed the system as described in Section E.5.

## Section E.5

### BLEEDING THE CLUTCH SYSTEM

Open the bleed screw on the slave cylinder three-quarters of a turn and attach a tube, immersing the open end in a clean receptacle containing a small quantity of the recommended hydraulic fluid. Fill the master cylinder reservoir with the recommended fluid (see **'GENERAL DATA'**). Using slow, full strokes, pump the clutch pedal until the fluid entering the container is completely free from air bubbles. On a downstroke of the pedal tighten the bleed screw and remove the bleed tube.

*Fig. E.2*
*Bleeding the clutch hydraulic system at the slave cylinder*

# SECTION F

## THE GEARBOX

## THE GEARBOX COMPONENTS

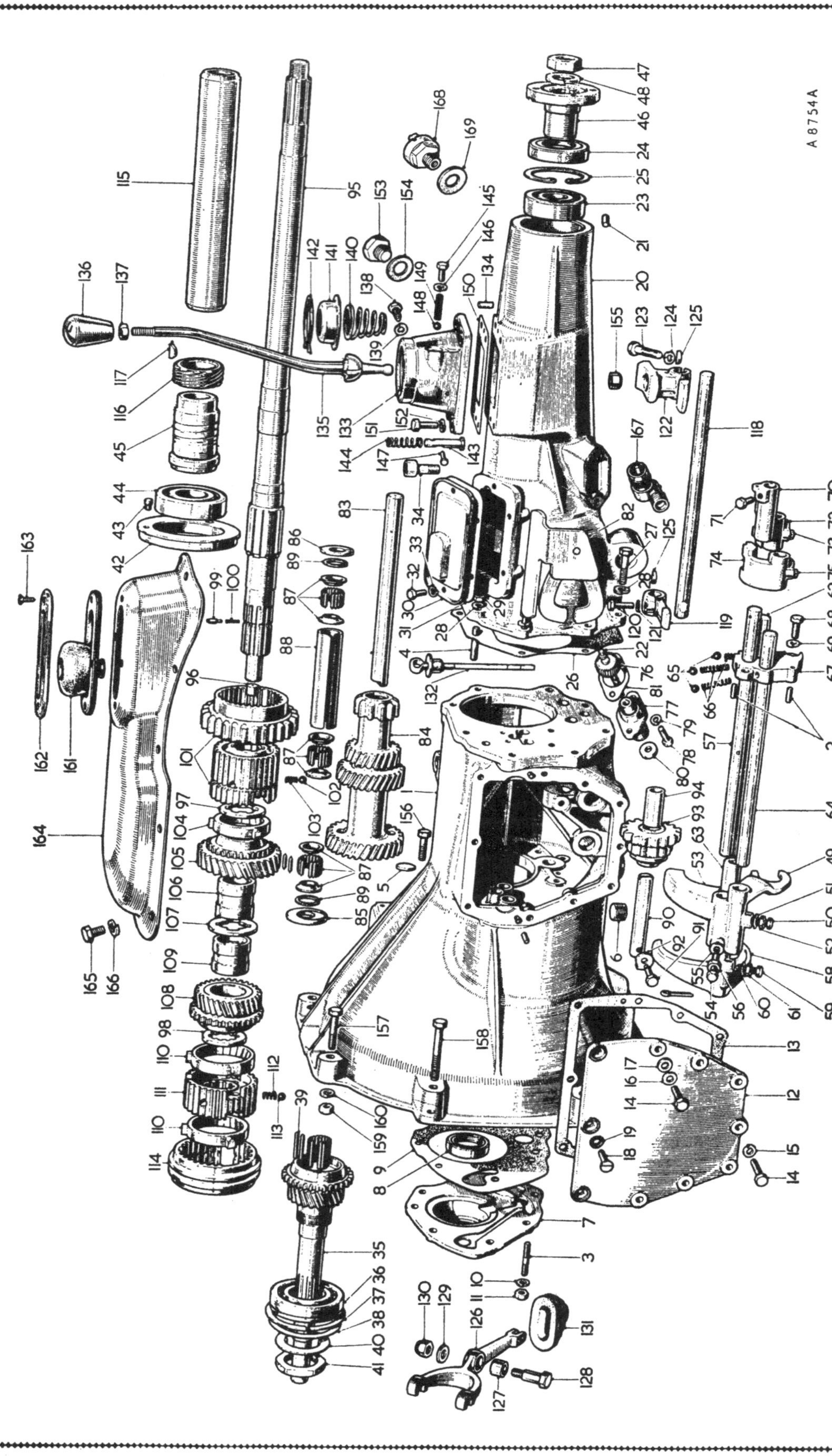

## KEY TO THE GEARBOX COMPONENTS

| No. | Description |
|---|---|
| 1. | Casing assembly. |
| 2. | Dowel—locating block to gearbox. |
| 3. | Stud—front cover. |
| 4. | Stud—rear extension. |
| 5. | Plug—welch—casing. |
| 6. | Plug—drain. |
| 7. | Cover assembly—front. |
| 8. | Seal—oil. |
| 9. | Joint washer—front cover to casing. |
| 10. | Washer—spring—stub—front cover to casing. |
| 11. | Nut—stud. |
| 12. | Cover—gearbox side. |
| 13. | Joint washer—side cover to casing. |
| 14. | Screw—gearbox side cover. |
| 15. | Washer—spring—screw. |
| 16. | Washer—plain—screw. |
| 17. | Washer—fibre—screw. |
| 18. | Screw—countersunk—gearbox side cover. |
| 19. | Washer—shakeproof—screw. |
| 20. | Extension—rear. |
| 21. | Plug—taper—rear extension. |
| 22. | Button—thrust—speedometer. |
| 23. | Bearing—rear extension. |
| 24. | Seal—oil—rear extension. |
| 25. | Circlip—oil seal. |
| 26. | Joint washer—extension to casing. |
| 27. | Screw—extension to casing. |
| 28. | Washer—spring—screw. |
| 29. | Nut—stud—rear extension to casing. |
| 30. | Cover—side—rear extension. |
| 31. | Joint washer—side cover to extension. |
| 32. | Screw—side cover to extension. |
| 33. | Washer—spring—screw. |
| 34. | Breather assembly. |
| 35. | Shaft—first motion. |
| 36. | Bearing. |
| 37. | Ring—spring—bearing. |
| 38. | Shim—first motion shaft—·002 in. (·051 mm.). |
| 39. | Roller—needle—first motion shaft. |
| 40. | Washer—locking—first motion shaft nut. |
| 41. | Nut—first motion shaft. |
| 42. | Housing—rear bearing. |
| 43. | Peg—locating—rear bearing housing. |
| 44. | Bearing—rear—third motion shaft. |
| 45. | Distance piece assembly—speedometer gear to rear bearing. |
| 46. | Flange—third motion shaft. |
| 47. | Nut—third motion shaft flange. |
| 48. | Washer—spring—nut. |
| 49. | Fork—reverse. |
| 50. | Screw—reverse fork locating. |
| 51. | Washer—shakeproof—screw. |
| 52. | Nut—screw—fork locating. |
| 53. | Fork—first and second speed. |
| 54. | Screw—fork locating. |
| 55. | Washer—shakeproof—screw. |
| 56. | Nut—screw—fork locating. |
| 57. | Rod—first and second speed fork. |
| 58. | Fork—third and fourth speed. |
| 59. | Screw—fork locating. |
| 60. | Washer—shakeproof—screw. |
| 61. | Nut—screw—fork locating. |
| 62. | Rod—third and fourth speed fork. |
| 63. | Distance piece—third and fourth speed fork rod. |
| 64. | Rod—reverse fork. |
| 65. | Ball—fork locating. |
| 66. | Spring—locating ball. |
| 67. | Block—sliding shaft locating. |
| 68. | Screw—locating block to gearbox. |
| 69. | Washer—spring—screw. |
| 70. | Selector—first and second gear. |
| 71. | Screw—selector locating. |
| 72. | Selector—third and fourth gear. |
| 73. | Screw—selector locating. |
| 74. | Selector—reverse gear. |
| 75. | Screw—selector locating. |
| 76. | Pinion—speedometer. |
| 77. | Bush—speedometer pinion. |
| 78. | Screw—speedometer pinion bush. |
| 79. | Washer—lock—screw. |
| 80. | Seal—oil—speedometer pinion. |
| 81. | Joint washer—speedometer pinion bush. |
| 82. | Arm assembly—interlocking. |
| 83. | Layshaft. |
| 84. | Gear unit—layshaft. |
| 85. | Washer—thrust—front—laygear. |
| 86. | Washer—thrust—rear—·154 to ·156 in. (3·912 to 3·962 mm.). |
| 87. | Roller—needle bearing—layshaft. |
| 88. | Tube—distance—laygear bearing. |
| 89. | Ring—spring—layshaft. |
| 90. | Shaft—reverse. |
| 91. | Screw—reverse shaft. |
| 92. | Washer—locking screw. |
| 93. | Gear assembly—reverse. |
| 94. | Bush—reverse gear. |
| 95. | Shaft—third motion. |
| 96. | Restrictor—oil. |
| 97. | Washer—thrust—rear—third motion shaft. |
| 98. | Washer—thrust—front—·1565 to ·1575 in. (3·962 to 3·9887 mm.). |
| 99. | Peg—thrust washer. |
| 100. | Spring—peg. |
| 101. | First speed wheel and synchronizer assembly. |
| 102. | Ball—synchronizer. |
| 103. | Spring—synchronizer ball. |
| 104. | Baulk ring—second speed gear. |
| 105. | Gear—second speed. |
| 106. | Bush—second speed mainshaft gear. |
| 107. | Ring—interlocking—second and third gear bushes. |
| 108. | Gear—third speed. |
| 109. | Bush—third speed gear. |
| 110. | Baulk ring—third and fourth speed gear. |
| 111. | Synchronizer—third and fourth speed. |
| 112. | Spring—synchronizer ball. |
| 113. | Ball—synchronizer. |
| 114. | Coupling—sliding—third and fourth speed. |
| 115. | Distance piece—third motion shaft. |
| 116. | Gear—speedometer. |
| 117. | Key—speedometer gear. |
| 118. | Shaft—remote-control (rear extension). |
| 119. | Lever—selector—front. |
| 120. | Screw—selector lever—front. |
| 121. | Washer—shakeproof—screw. |
| 122. | Lever—selector—rear. |
| 123. | Screw—selector lever—rear. |
| 124. | Washer—spring—screw. |
| 125. | Key—selector lever. |
| 126. | Lever—clutch withdrawal. |
| 127. | Bush—withdrawal lever. |
| 128. | Bolt—clutch withdrawal lever. |
| 129. | Washer—bolt. |
| 130. | Nut—stiff—bolt. |
| 131. | Cover—dust—clutch withdrawal lever. |
| 132. | Indicator—oil level—gearbox. |
| 133. | Tower—remote-control. |
| 134. | Dowel—remote-control tower. |
| 135. | Lever—change speed. |
| 136. | Knob—change speed lever. |
| 137. | Locknut—change speed knob. |
| 138. | Pin—locating—change speed lever. |
| 139. | Washer—spring. |
| 140. | Spring—change speed lever. |
| 141. | Cover—ball spring. |
| 142. | Circlip—ball spring cover. |
| 143. | Plunger—reverse selector. |
| 144. | Spring—reverse plunger. |
| 145. | Screw—reverse plunger spring. |
| 146. | Washer—spring—screw. |
| 147. | Pin—locating—reverse plunger. |
| 148. | Ball—reverse plunger. |
| 149. | Spring—reverse plunger detent. |
| 150. | Gasket—control tower. |
| 151. | Screw—tower to extension. |
| 152. | Washer—spring—screw. |
| 153. | Plug—reverse light switch hole. |
| 154. | Joint washer—plug. |
| 155. | Bush—change speed lever. |
| 156. | Bolt—gearbox to mounting plate. |
| 157. | Bolt—gearbox to mounting plate. |
| 158. | Bolt—gearbox to mounting plate. |
| 159. | Nut—bolt. |
| 160. | Washer—spring—bolt. |
| 161. | Grommet—gear lever. |
| 162. | Retainer—gear lever grommet. |
| 163. | Screw—retainer to cover. |
| 164. | Cover—gearbox remote-control. |
| 165. | Screw—cover to tunnel. |
| 166. | Washer—spring. |
| 167. | Box—speedometer drive adaptor. |
| 168. | Reverse light switch (later cars). |
| 169. | Washer for switch (later cars). |

## GENERAL DESCRIPTION

The gearbox has four forward gears and a reverse gear. Top, third, and second gear engagement are effected by synchromesh hubs with direct drive for top gear and constant-mesh gears on the layshaft and mainshaft for third and second gears. First and reverse gears are sliding spur gears.

A combined dipstick and filler plug is housed in the top of the gearbox and a drain plug in the bottom.

The bell housing is integral with the gearbox and accommodates the clutch release mechanism and the starter motor driving pinion.

An extension bolted to the rear end of the gearbox contains the gear selector mechanism.

The gearbox is bolted at the front end to the engine mounting plate and is supported at the rear by two rubber mountings that locate on the rear cross-member mounting brackets. An engine stay-rod attached to the under side of the rear extension is anchored to the rear cross-member.

## LUBRICATION

**Oil level**

When topping up the gearbox ensure that it is not filled above the 'HIGH' mark on the dipstick. If the level is too high, oil may get into the clutch case and cause clutch slip.

The combined filler plug and dipstick for the gearbox and overdrive is located beneath the rubber plug on the gearbox cover.

**Draining**

Remove the gearbox drain plug and drain off the oil.

When the gearbox has been drained, completely refill with fresh oil.

# Section F.1

## REMOVING THE GEARBOX

(1) Prepare the engine for removal from the car as detailed in Section A and take the weight of the engine.

(2) Drain the oil from the gearbox.

(3) Disconnect and remove the propeller shaft as detailed in Section G and disconnect the speedometer pinion drive.

(4) Disconnect the clutch slave cylinder push-rod from the clutch withdrawal lever and the slave cylinder from the clutch housing. Tie the cylinder clear of the gearbox.

(5) Remove the screws securing the rear cross-member to the body and lower the engine and gearbox until the gearbox rests on the fixed body cross-member.

(6) Release the engine stay-rod bracket from the rear cross-member and remove the four nuts and washers securing the gearbox rear rubber mountings to the cross-member. Withdraw the rear cross member from the body.

(7) Remove the gear lever knob and the rubber cover from the tunnel. Remove the remote-control tower complete with gear lever from the gearbox.

(8) Ease the assembly forward until it is clear of the fixed body cross-member and then tilt the assembly and lift it from the car.

(9) Remove the screws securing the gearbox to the engine rear mounting plate and separate the gearbox from the engine.

**NOTE.—Great care must be taken during this operation to ensure that no load is placed on the clutch release plate drive straps. Even slight damage to these may result in the clutch breaking up at high speeds.**

(10) Remove the stiffnut and washer from the clutch withdrawal lever bolt and take the dust cover and lever from the clutch housing.

(11) Remove the nut, spring washer, and pin from the engine stay-rod and the stay-rod from the gearbox.

(12) If the stay-rod and bracket are to be dismantled remove the rear nut and washer and slide off the plates, buffers, brackets, and distance tube. Separate the plates, buffers, and bracket from the distance tube.

# Section F.2

## DISMANTLING THE GEARBOX

(1) Remove the dipstick, drain plug, and speedometer drive pinion.

(2) Using Service tool 18G 2, remove the propeller shaft flange.

(3) Remove the remote-control tower, the gearbox extension side cover, and the interlock plate and bracket.

(4) Slacken the locating screw on the remote-control front selector lever, unscrew the screws and nuts securing the extension to the gearbox, and remove the extension.

(5) Retain the remote-control selector lever, which will fall free as the extension is withdrawn. Withdraw the shaft and rear selector lever from the rear extension and remove the selector lever from the shaft; withdraw the split bush and circlip from the selector.

(6) Remove the three countersunk screws and seven hexagon-headed screws from the gearbox side cover and lift off the cover and gasket.

(7) Cut the locking wire, unscrew the three selector locating screws, and remove the selectors.

(8) Unscrew the shift shaft locating block screws and remove the block from the rear face of the gearbox. Retain the three balls and springs which will be released when the block is withdrawn from the shafts. Two dowels on the block locate it on the rear face of the gearbox.

(9) Release the locknuts and remove the fork locating screws and star washers from the change speed

forks. Slide out the shafts and remove the forks in the following order: reverse, fourth and third speed, second and first speed.

(10) If difficulty is experienced in removing the shafts remove the front cover and use a soft drift to tap the shafts out from the front of the box.

(11) Remove the nuts and washers securing the front cover and then remove the cover, gasket, and the first motion shaft bearing shims. Do not remove the oil seal from the cover unless it has to be renewed.

(12) Unscrew the reverse shaft locating screw and remove the shaft and gear.

(13) Using a soft drift, tap out the layshaft from the front of the box and allow the laygear to rest in the bottom of the box.

(14) Ease the rear bearing housing from the rear of the gearbox and withdraw the third motion shaft assembly.

(15) Withdraw the first motion shaft from the front of the gearbox. A soft drift may be used to tap the shaft from the inside of the box.

(16) Reassemble the laygear on its shaft and check the end-float of the gear (see **'GENERAL DATA'**).

(17) Remove the layshaft, laygear, and thrust washers.

**(See Editor's notes at end of section.)**

## Section F.3

### DISMANTLING AND ASSEMBLING THE CONTROL TOWER

Do not dismantle the control tower unless worn or broken parts are to be renewed.

(1) Remove the ball spring cover circlip to ease the tension on the spring and then remove the two change speed lever locating pins. Withdraw the lever, and from it remove the knob, cover, and spring.

(2) From the rear of the control tower unscrew the reverse plunger detent screw and remove the reverse plunger detent spring and ball.

(3) From the side of the tower remove the reverse plunger locating pin and retain the plunger and spring that will be released from inside the tower.

Assembly is a reversal of the dismantling sequence, but care must be taken to ensure that the front and rear selectors are correctly lined up and that the clinch bolts pass across the flats on the remote-control shaft.

When fitting a new pin to retain the reverse selector plunger ensure that the plunger is depressed sufficiently enough to permit the pin to engage the flats on the plunger.

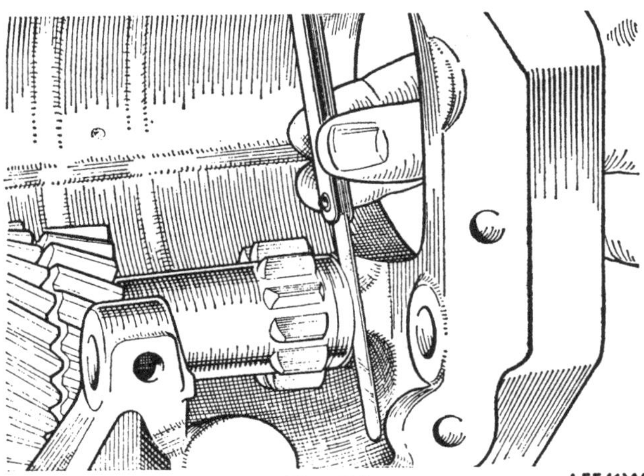

A5541W

*Fig. F.1*

*Checking the laygear end-float*

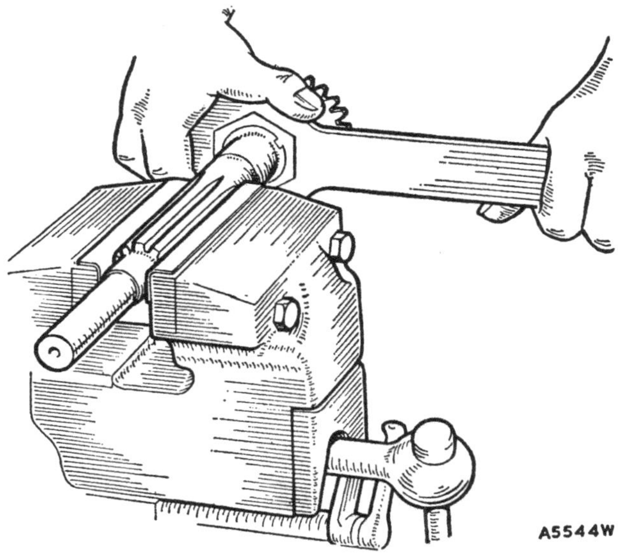

A5544W

*Fig. F.2*

*Using Service tool 18G 5 to unscrew a first motion shaft nut. This nut has a left-hand thread*

## Section F.4

### DISMANTLING AND ASSEMBLING THE REAR EXTENSION

(1) Remove the rear selector lever clamp bolt, lever, and the Woodruff key and withdraw the shaft from the rear extension. Do not dismantle the rear extension any further unless the oil seal and/or bearing has to be renewed, in which case proceed as follows.

(2) Using Service tool 18G 389 and 18G 389 C, remove the rear oil seal.

(3) Remove the rear extension bearing circlip and press out the bearing.

Assembly is a reversal of the dismantling sequence. If the bearing and/or oil seal have/has been removed, press in the new bearing and secure it with the circlip, and, using Service tool 18G 134 and adaptor 18G 134 N, fit the new oil seal.

**(See Editor's notes at end of section.)**

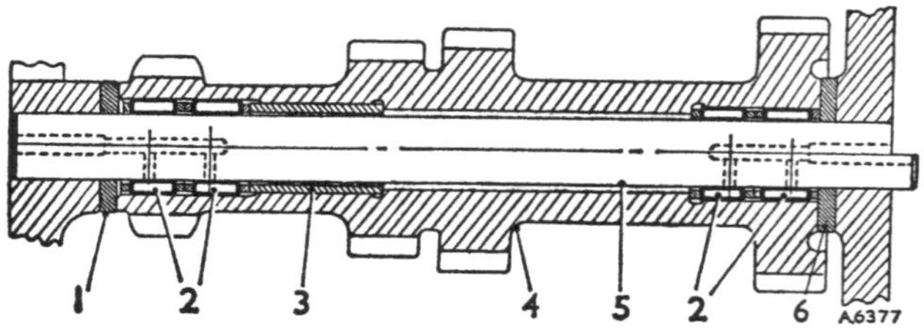


*Fig. F.3*

*Laygear assembly—second type*

1. Thrust washer—small.
2. Needle-roller bearing (pair).
3. Distance tube.
4. Laygear.
5. Layshaft.
6. Thrust washer—large.

## Section F.5

### DISMANTLING AND ASSEMBLING THE FIRST MOTION SHAFT

(1) Remove the needle rollers from the spigot bearing housing.

(2) Secure the shaft in a soft-jawed vice, release the lock washer, and, using Service tool 18G 5, unscrew the locknut.

**NOTE.—The locknut has a left-hand thread.**

(3) Press the bearing from the shaft and remove the spring ring from the bearing.

Assembly is a reversal of the dismantling sequence.

(See Editor's notes at end of section.)

## Section F.6

### DISMANTLING AND ASSEMBLING THE LAYGEAR

*First type*

Remove the circlips from the ends of the laygear and withdraw the needle-roller bearing assemblies and the distance tube.

To assemble the laygear proceed as follows:

(1) Hold the layshaft in a vice, stepped end downwards.

(2) Smear the shaft with grease and assemble a bearing assembly, the distance tube, and the remaining two bearing assemblies to the shaft.

(3) Fit a circlip to the front end of the laygear and place the laygear on the shaft.

(4) Remove the laygear and shaft from the vice, fit the remaining circlip, and remove the shaft from the gear.

*Second type*

The second-type laygear has a larger diameter shaft and a pair of caged needle-roller bearings at each end.

Extract the two pairs of needle-roller bearings from the ends of the laygear and the distance tube from the small end.

To assemble the laygear proceed as follows:

(1) Dip the bearings in oil.

(2) Hold the laygear in a vice and insert the distance tube and a pair of bearings at the small end.

(3) Insert a pair of bearings at the large end.

## Section F.7

### DISMANTLING THE THIRD MOTION SHAFT

Before dismantling the third motion shaft check the end-float on the third speed mainshaft gear to ensure the fitting of the correct thrust washer during reassembly of the third motion shaft.

(1) Remove the third and fourth speed gear front baulk ring, the synchromesh sleeve and hub, and the rear baulk ring. The synchromesh sleeve may be withdrawn from the hub, but care must be taken to retain the balls and springs that will be released when the sleeve is withdrawn from the hub.

(2) Depress the front thrust washer retaining peg, rotate the washer to align its splines with those on the shaft, and remove the washer.

(3) Withdraw the third speed gear and its bush and then remove the bush interlock ring to free the second speed gear. Remove the gear, baulk ring, bush, and thrust washer.

(4) Withdraw the first and second speed synchromesh and gear. The first gear may be disengaged from the hub, but care must be taken to retain the balls and springs that will be released when the gear is withdrawn from the hub.

(5) Remove the speedometer drive gear and its Woodruff key and then slide the distance piece from the shaft.

(6) Press the bearing from the shaft and then from its housing.

## Section F.8

### EXAMINATION OF COMPONENTS

Thoroughly clean, dry, and examine all components for wear, distortion, deterioration, and thread damage, paying particular attention to the following:

(1) Bushes fitted to the clutch withdrawal fork, third motion shaft, and reverse speed gear. Worn or damaged bushes should be renewed (Section F.11).

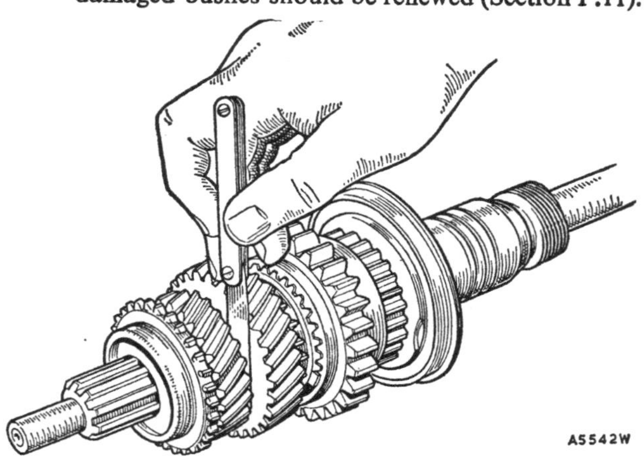


*Fig. F.4*

*Checking the mainshaft second and third speed gear end-float*

(2) Bearings fitted to the first motion shaft, third motion shaft, and rear extension.
Examine the bearings for looseness, pitting, security of cages, and fit in their housings and on their journals.
Examine also the needle rollers in the first motion shaft and the laygear.
Unserviceable bearings should be renewed.
(3) Spring fitted to the synchromesh hubs, shaft locating block, reverse shaft, and reverse plunger detent.
Comparisons may be made with new springs, but where facilities for test are available refer to the weights and lengths given under **'GENERAL DATA'**.
(4) The rubber gaiters for the clutch withdrawal fork and gear lever for cracks or signs of deterioration.
(5) Ensure that all oilways are clear of obstruction, including the oil restrictor in the front of the third motion shaft.
(6) Examine the interlock arm for burrs and for security of the rivet.

## Section F.9

### ASSEMBLING THE THIRD MOTION SHAFT

Lubricate all contact surfaces with oil, and smear the bore of the gears with Duckham's Moly Disulphide Grease (max. size 5 microns).
(1) Use Service tools 18G 222 and 18G 223 to assemble the second speed synchromesh hub and third and top gear synchromesh hub respectively.
(2) Press the rear bearing into its housing and the bearing onto the shaft.

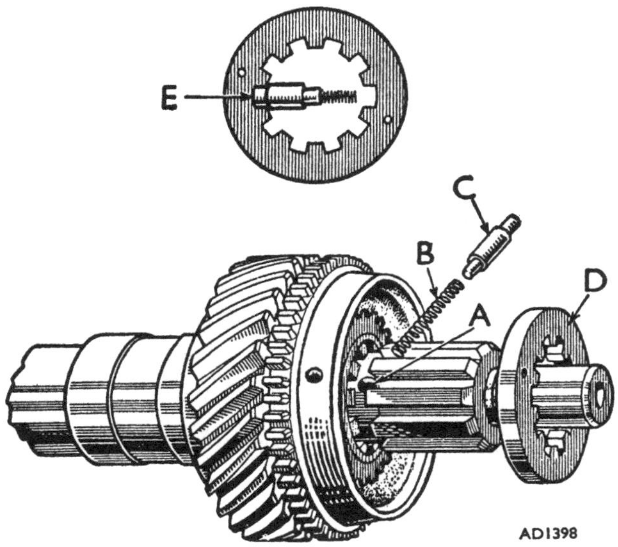


*Fig. F.5*
*Securing the mainshaft gears*

A. Hole for spring. C. Locating peg.
B. Spring. D. Locking washer.
E. Peg located in the washer.

*Fig. F.6*
*Assembling a synchromesh hub, using Service tool* 18G 222

(3) Fit the speedometer gear drive to the shaft with its key.
(4) Fit the first gear and second gear synchromesh to the shaft, followed by the baulk ring and rear thrust washer.
(5) Fit the second speed gear bush to the shaft and ensure that the lugs face forward and that the oil hole in the bush is in alignment with the oil hole in the shaft.
(6) Assemble the second speed gear and the interlock washer so that the washer engages the lugs on the bush.
(7) Fit the third speed gear bush, lugs first, and ensure that the lugs engage the interlock washer and that the oil hole and cut-away in the bush are in alignment with the holes in the shaft.
(8) Place the retaining pin spring and pin in the shaft and the third speed gear on the bush with the cone facing forward.
(9) Position the gear so that the hole in the cone is in line with the retaining peg, depress the peg with a thin drift, fit the thrust washer to the shaft, then turn the washer to allow the peg to lock the washer in position.

Check the end-float of the second and third speed gear (see **'GENERAL DATA'** and Fig. F.4).
Thrust washers are available in four thicknesses as follows:
·1565 to ·1575 in. (3·96 to 3·98 mm.).
·1585 to ·1595 in. (4·01 to 4·03 mm.).
·1605 to ·1615 in. (4·06 to 4·08 mm.).
(10) Assemble the third and fourth speed gear rear baulk ring, synchromesh, and front baulk ring.

**(See Editor's notes at end of section.)**

## Section F.10

### ASSEMBLING THE GEARBOX

Any gaskets or locking devices that are unserviceable or suspect should be renewed during assembly of the gearbox.
(1) Place the laygear in the bottom of the gearbox together with the thrust washers (see Section F.2, paragraph [16]) and thread a piece of stiff wire through the laygear to allow Service tool 18G 471 to pick up the thrust washers and laygear. The

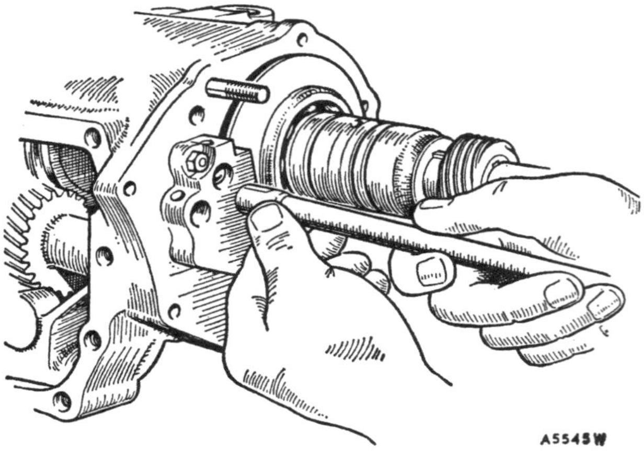


*Fig. F.7*
*Assembling a selector shaft, using Service tool* 18G 4

washers are available in four thicknesses as follows:

·154 to ·156 in. (3·81 to 3·96 mm.).
·157 to ·159 in. (3·95 to 4·03 mm.).
·160 to ·161 in. (4·06 to 4·08 mm.).
·163 to ·164 in. (4·13 to 4·16 mm.).

(2) Fit the first motion shaft assembly to the gearbox.
(3) Insert the third motion shaft from the rear of the box, use the gearbox extension gasket to align the dowel in the bearing housing, enter the spigot in the first motion shaft, and push home the shaft and bearing housing.
(4) Using Service tool 18G 471, fit the layshaft to the laygear and ensure that the cut-away end of the shaft faces forward.
(5) Assemble the reverse gear and shaft and secure the shaft with the locating screw and a new tab washer. Lock the screw with the washer.
(6) Refit the first motion shaft bearing shims, align the step on the end of the layshaft with the inside edge of the front cover, and, using Service tool 18G 598, fit the front cover as described in Section F.12.
(7) Assemble the clutch withdrawal lever with its bolt and stiffnut.
(8) Bolt the shaft locating block to the rear face of the gearbox and insert the three springs and balls into the block.
(9) Assemble the selectors to the shafts and secure them with their locating screws. Lock the screw heads to the selectors with new locking wire.
(10) Using Service tool 18G 41, depress the springs and balls in the locating block and pass the shafts through the block.
(11) Position the gear change forks in the box in the following order: reverse, first and second speed, third and fourth speed.
(12) Fit the distance piece to the third and fourth speed shaft, push the shafts through the forks, align the holes in the shafts and forks, and secure them with the fork locating screws. Tighten the locknut on each screw.
(13) Replace the gearbox side cover and gasket.
(14) Place the third motion shaft distance piece on the shaft and the extension gasket over the studs on the rear face of the gearbox.
(15) Fit the rear extension to the gearbox and engage the rear extension remote-control shaft in the front selector and the dowel on the third motion shaft bearing housing with the hole in the extension.
(16) Secure the selector with its Woodruff key and pinch-bolt and replace the extension side cover and gasket.
(17) Refit the speedometer drive pinion, drain plug, dipstick, and rear joint flange.
(18) After refitting the gearbox to the vehicle replace the control tower and its gasket and fill the box with clean oil.

(See Editor's notes at end of section.)

## Section F.11

### FITTING REPLACEMENT BUSHES

When fitting replacement bushes to the reverse gear, the clutch withdrawal fork, and the third motion shaft remove the old bushes and fit the new ones as follows.

**Reverse gear**

Press the new bush into the gear from the small end of the gear until the end of the bush is flush with the face of the small end.

Finish to an internal diameter of between ·6255 and ·6265 in. (15·8 and 15·9 mm.) concentric with the gear teeth.

**Clutch withdrawal fork**

Press in the new bush and finish to an inside diameter of between ·4996 and ·5004 in. (12·6 and 12·7 mm.).

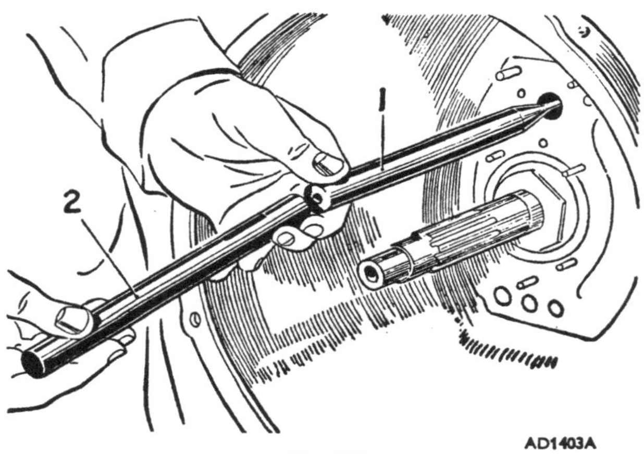


*Fig. F.8*
*Using tool* 18G 471 *as a pilot when installing the layshaft*

1. Pilot. 2. Layshaft.

**Third motion shaft bushes**

Heat the bushes to between 180 and 200° C. (356 and 392° F.), ensure that the locating tongues on the bushes are in line with the splines on the shaft and that the oil holes are in line, then assemble the second speed bush, interlock washer, and third speed bush.

## Section F.12

### FITTING THE GEARBOX FRONT COVER

To prevent oil leaking past the gearbox front cover oil seal the cover must be correctly fitted to ensure that the seal is concentric with the first motion shaft. When refitting the cover proceed as follows.

Clean off the cover and examine it carefully for burrs and bruising, particularly around the bore, stud holes, and machined surfaces. Check the flat surfaces for twist and warp, and correct it if necessary. If the condition is too bad to correct, fit a new cover.

Remove and discard the front cover to gearbox gasket and clean off the flat surfaces around the base of all studs.

Offer the front cover (less oil seal) to the gearbox and push it fully home on the studs. The cover should be free to move in all directions, and points at which the holes may be binding on the studs must be relieved until the cover is free to 'float'.

Remove the cover and, using Service tool 18G 134 with adaptor 18G 134 N, fit the oil seal so that its lip faces inwards towards the gearbox. Lightly grease and fit a new gasket to the gearbox front face, then fit the centralizer (Service tool 18G 598) to the bore of the front cover and push it in until it is tight. Lightly oil the seal and pass the cover over the first motion shaft, taking particular care not to cut or damage the knife edge of the seal.

Keep the centralizer firmly in position, push the cover onto the studs, and fit the spring washers and nuts, tightening the nuts finger tight only. Use a suitable socket spanner and long extension to tighten the nuts a half-turn at a time by diametrical selection until all nuts are fully tightened.

Remove the centralizer and refit the clutch-operating components.

**(See Editor's notes at end of section.)**

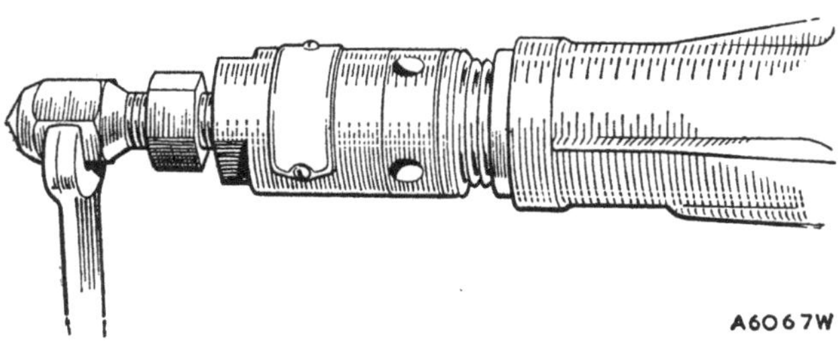


*Fig. F.9*

*Removing a rear oil seal, using Service tools* 18G 389 *and* 18G 389 *B*

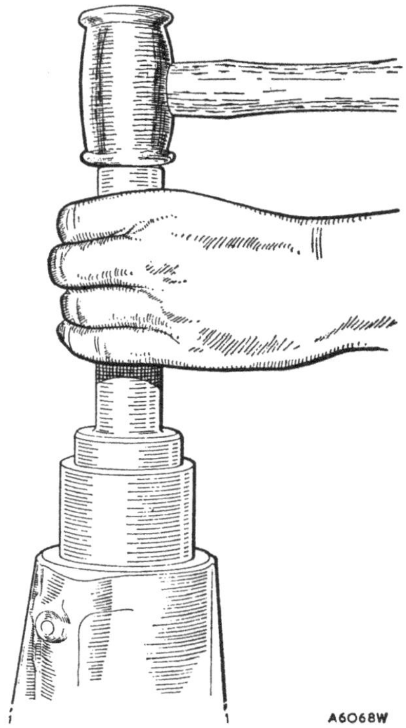


*Fig. F.*10

*Installing a new rear oil seal, using Service tools* 18G 134 *and* 18G 134 *N*

### EDITOR'S NOTES

**F. The Gearbox**

*Dismantling the gearbox*

Service tool 18G2 is a gear puller. Any suitably sized puller will work.

*Dismantling and assembling the rear extension*

The rear oil seal can be removed and installed through the careful use of a suitable drift punch and hammer. Care must be taken to avoid damaging the seal and housing.

*Dismantling and assembling the first motion shaft*

A suitably sized wrench may be substituted for Service tool 18G5.

*Assembling the third motion shaft*

Service tools 18G222 and 18G223 make the assembly of the synchromesh hubs easier. The assembly can be done without these tools, but generally requires more than one pair of hands.

*Assembling the gearbox*

Service tool 18G471 is not strictly necessary, but if it is not used, great care must be exercised in order to avoid damaging the thrust washers on either end of the laygear.

Service tool 18G598 centers the front cover on the first motion shaft. If this tool is not available, the job may be done by carefully centering the cover by eye and then cautiously tightening the bolts so as not to shift the cover.

Service tool 18G41 is not needed if the balls and springs are carefully depressed with a small flat-bladed screwdriver as the shafts are installed.

After assembling the transmission, check for correct operation in all gears.

*Fitting the Gearbox front cover*

Careful use of a drift punch and hammer may be substituted for the use of Service tools 18G134 and 18G134N.

# Section Fa

## THE OVERDRIVE (TYPE 'D')

## THE OVERDRIVE COMPONENTS

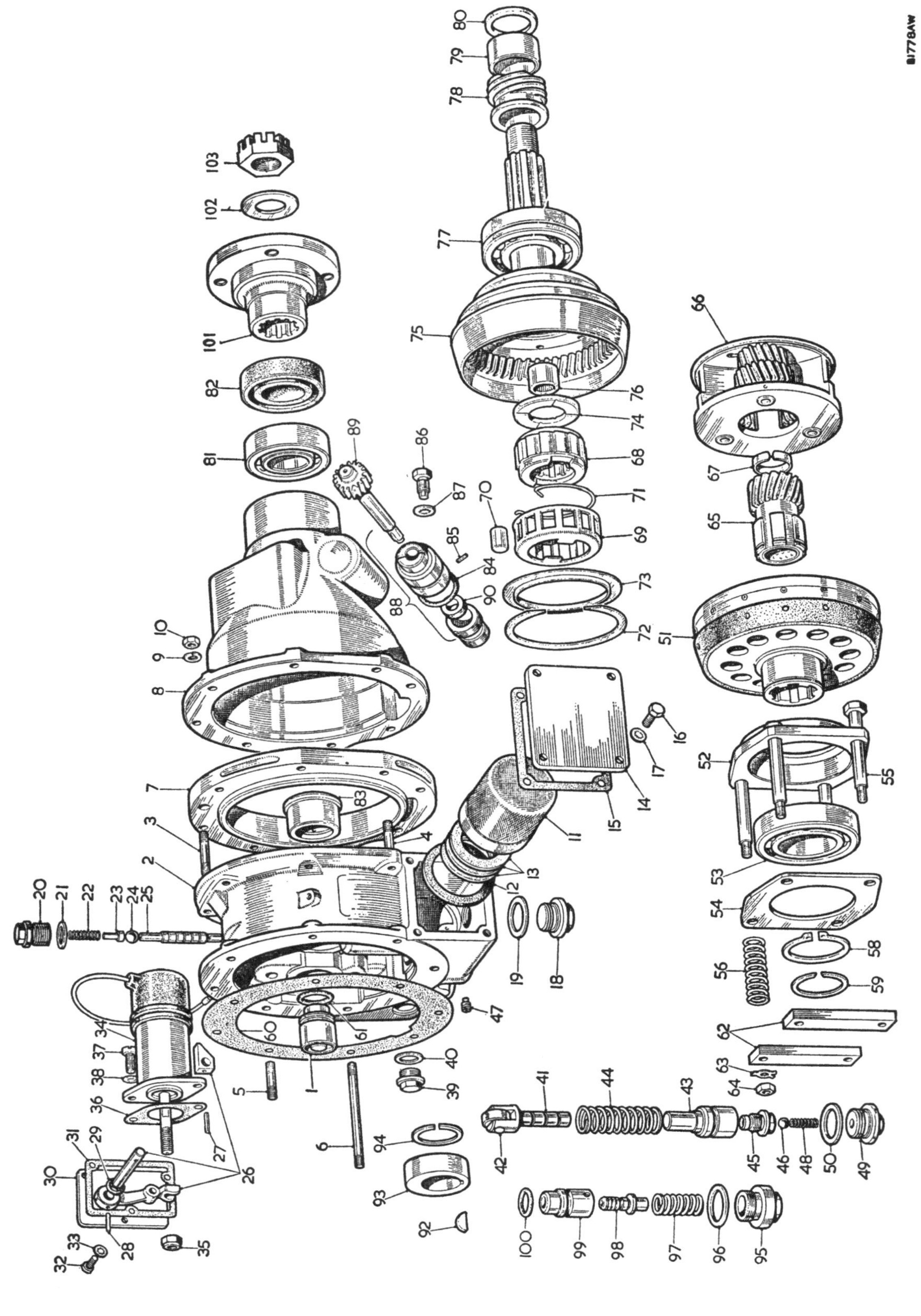

Fa.2

## KEY TO THE OVERDRIVE COMPONENTS

| No. | Description |
|---|---|
| 1. | Operating piston. |
| 2. | Main casing assembly. |
| 3. | Stud—main casing to rear casing. |
| 4. | Stud—main casing to rear casing. |
| 5. | Stud—main casing to gearbox adaptor. |
| 6. | Stud—main casing to gearbox adaptor. |
| 7. | Brake ring. |
| 8. | Intermediate casing. |
| 9. | Spring washer. |
| 10. | Nut. |
| 11. | Filter. |
| 12. | Sealing plate. |
| 13. | Magnetic rings. |
| 14. | Side cover-plate. |
| 15. | Joint for cover-plate. |
| 16. | Set screw. |
| 17. | Shakeproof washer. |
| 18. | Drain plug. |
| 19. | Washer for plug. |
| 20. | Plug for operating valve. |
| 21. | Washer for plug. |
| 22. | Spring—operating valve. |
| 23. | Plunger—operating valve. |
| 24. | Steel ball. |
| 25. | Operating valve. |
| 26. | Operating valve lever assembly. |
| 27. | Mills pin. |
| 28. | Mills pin. |
| 29. | 'O' ring. |
| 30. | Cover—solenoid. |
| 31. | Joint for cover-plate. |
| 32. | Set screw. |
| 33. | Shakeproof washer. |
| 34. | Solenoid. |

| No. | Description |
|---|---|
| 35. | Self-locking nut. |
| 36. | Joint for solenoid. |
| 37. | Screw for solenoid. |
| 38. | Shakeproof washer. |
| 39. | Plug for solenoid bracket. |
| 40. | Washer for plug. |
| 41. | Pump plunger. |
| 42. | Pin. |
| 43. | Pump body. |
| 44. | Pump plunger spring. |
| 45. | Non-return valve body. |
| 46. | Steel ball. |
| 47. | Screw locating pump body. |
| 48. | Spring—non-return valve. |
| 49. | Plug—non-return valve. |
| 50. | Washer for plug. |
| 51. | Clutch assembly. |
| 52. | Bearing housing. |
| 53. | Bearing—thrust. |
| 54. | Plate—retainer. |
| 55. | Bolt for bearing housing. |
| 56. | Spring—clutch return. |
| 58. | Circlip—bearing retaining. |
| 59. | Snap ring—sun wheel retaining. |
| 60. | Joint for front cover. |
| 61. | 'O' ring. |
| 62. | Bridge piece. |
| 63. | Lock washer. |
| 64. | Nut. |
| 65. | Sun wheel assembly. |
| 66. | Planet carrier. |
| 67. | Locating ring—third motion shaft. |
| 68. | Inner member—uni-directional clutch. |
| 69. | Cage—uni-directional clutch. |

| No. | Description |
|---|---|
| 70. | Roller—uni-directional clutch. |
| 71. | Spring for clutch. |
| 72. | Snap ring. |
| 73. | Oil thrower. |
| 74. | Thrust bearing. |
| 75. | Annulus assembly. |
| 76. | Bearing—needle-roller. |
| 77. | Bearing—inner. |
| 78. | Speedometer driving gear. |
| 79. | Bush. |
| 80. | Thrust washer. |
| 81. | Bearing. |
| 82. | Oil seal. |
| 83. | Steady bush—third motion shaft. |
| 84. | 'O' ring. |
| 85. | Pin. |
| 86. | Locking screw—speedometer bearing. |
| 87. | Washer for locking screw. |
| 88. | Speedometer drive bearing assembly. |
| 89. | Speedometer driven gear. |
| 90. | Oil seal—speedometer bearing. |
| 92. | Key—third motion shaft cam. |
| 93. | Cam—third motion shaft. |
| 94. | Snap ring—third motion shaft. |
| 95. | Plug—relief valve. |
| 96. | Washer for plug. |
| 97. | Spring—relief valve. |
| 98. | Plunger—relief valve. |
| 99. | Body—relief valve. |
| 100. | 'O' ring—relief valve. |
| 101. | Flange—coupling. |
| 102. | Washer—flange. |
| 103. | Nut—flange. |

## Section Fa.1

### LUBRICATION

The overdrive shares a common oil supply with the gearbox; it is filled and topped up through the gearbox dipstick hole and the oil level is checked with the gearbox dipstick.

**Draining**

When an overdrive is fitted, remove the overdrive drain plug marked 'DRAIN' and the gearbox drain plug, and drain off the old oil.

Remove the overdrive filter cover-plate and gasket from the left-hand side of the unit and withdraw the filter gauze, together with the filter seal and magnetic rings. Wash the gauze in fuel and clean out the filter housing. Clean the filter seal and remove all deposits from the magnetic rings.

Replace the filter seal (metal surface inwards) and place the magnetic rings in the filter location. Refit the filter, gasket, and cover-plate, replace the drain plugs, and refill with fresh oil. Check the filter cover-plate and drain plugs for leaks.

## Section Fa.2

### DESCRIPTION

The overdrive unit comprises a hydraulically controlled epicyclic gear housed in a casing which is directly attached to an extension at the rear of the gearbox.

The gearbox third motion shaft is extended and carries at its end the inner member of a uni-directional clutch. The outer ring of the clutch is carried in the combined annulus and output shaft.

Also mounted on the third motion shaft are the planet carrier and a freely rotatable sun wheel, and sliding thereon is a cone clutch member the inner lining of which engages the outside of the annulus while the outer lining engages a cast-iron brake ring sandwiched between the main and tail casings.

The cone clutch is held in contact with the annulus by compression springs, thus locking the sun wheel to the annulus so that the entire gear train rotates as a solid unit, giving direct drive. In this condition the drive is taken through the uni-directional clutch. Reverse torque or overrun is taken by the planet wheels, which being locked by the sun wheel, transmit the drive via the planet carrier and third motion shaft.

Hydraulic pressure produced when overdrive is selected forces two pistons in the unit housing against the cone clutch and overcomes the spring pressure, causing the cone clutch to engage with the stationary brake ring and bring the sun wheel to rest, allowing the annulus to overrun the uni-directional clutch and give an increased speed to the output shaft, i.e. 'overdrive'.

When changing from overdrive to direct gear, if the accelerator pedal is released the vacuum switch holds 'overdrive' until the engine takes up the drive. If the accelerator pedal is not released, when contact between the cone clutch and brake is broken, the unit still operates momentarily in its overdrive ratio as engine speed and road speed remain unchanged. When the load on the engine is released it begins to accelerate, speeding up the sun wheel from rest until, just at the instant when its speed synchronizes with the speed of the annulus, the whole unit revolves solidly and the uni-directional clutch takes up the drive once more. The movement of the cone clutch is deliberately slowed down so that the uni-directional clutch is driving before the cone clutch contacts, ensuring a perfectly self-synchronized change.

## Section Fa.3

### CONSTRUCTION

The third motion shaft of the gearbox is extended to carry first a cam operating the oil pump and then a steady bearing with a plain phosphor-bronze bush carried in the main housing. Next is the sun wheel of the epicyclic gear carried on a Clevite bush, and behind this the shaft is splined to take the planet carrier and uni-directional clutch. The end of the shaft is reduced and is carried in a needle-roller bearing in the output shaft. The latter is supported in the rear housing by two roller bearings. The clutch member slides on the splines of the

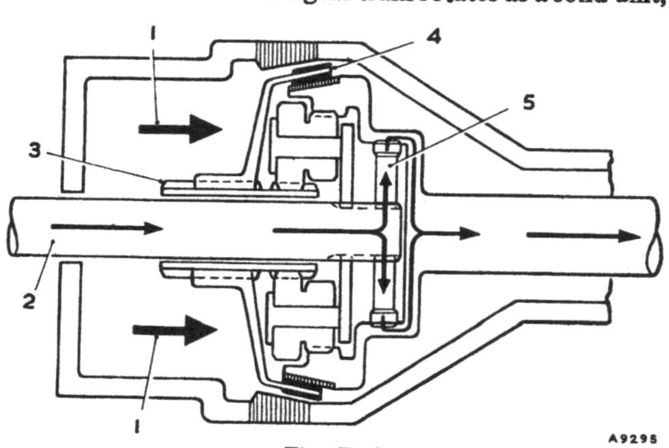


*Fig. Fa.1*
*Direct drive*

1. Spring pressure.
2. Third motion shaft.
3. Sun wheel.
4. Cone clutch.
5. Uni-directional clutch.

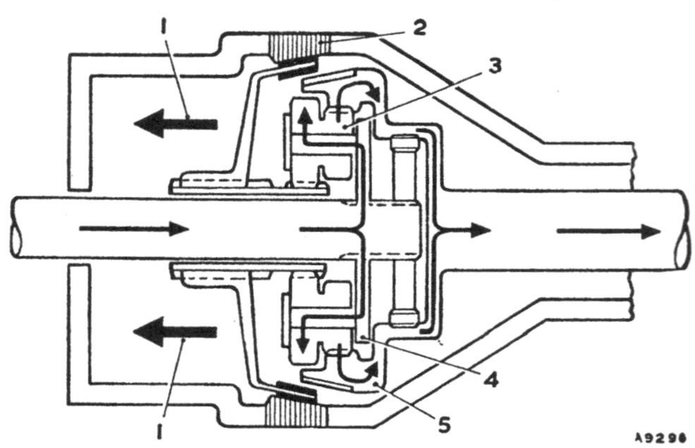


*Fig. Fa.2*
*Overdrive engaged*

1. Hydraulic pressure.
2. Brake ring.
3. Planet wheel.
4. Planet carrier.
5. Annulus.

sun wheel extension to contact either the annulus or the cast-iron braking ring forming part of the unit housing.

A ball bearing housed in a flanged ring is secured to the cone clutch member. This carries four bolts which act as guides to compression springs by which the ring, and with it the clutch member, is held against the annulus. The action of the springs prevent free-wheeling on overrun and they are of sufficient strength to handle reverse torque. Also secured to the four bolts are two bridge pieces against which bear two pistons operating in cylinders formed in the unit housing. The cylinders are connected by ports in the hydraulic system to a relief valve which regulates the pressure developed by the oil pump. When the valve is open (overdrive selected), oil under pressure is admitted to the cylinders and pushes the pistons forward to engage the overdrive clutch. Closing the valve cuts off the supply of oil to the cylinders and allows it to escape. Under the influence of the springs the clutch member moves back to engage direct drive position. The escape of oil from the cylinders is deliberately restricted so that the clutch takes up slowly.

The sun wheel and pinions are case-hardened and the annulus heat-treated. Gear teeth are helical. The pinions have needle-roller bearings and run on case-hardened pins.

The outer ring of the uni-directional clutch is pressed into the annulus member. The clutch itself is of the caged roller type, loaded by a lock-type spring made of round wire.

The hydraulic system is supplied with oil by a plunger-type pump operated by a cam on the gearbox third motion shaft. The pump body is pressed into the main housing and is located by a grub screw. The pump delivers oil through a non-return valve to a relief valve, in which a piston moves back against a compression spring until the required pressure is reached, at this point a hole in the relief valve is uncovered. The spill oil from the relief valve is led through drilled passages to an annular groove in the steady bush on the third motion shaft. Radial holes in the shaft collect oil and deliver it through an axial drilling to other radial holes in the shaft from which it is fed to the needle-roller bearing, thrust washer, and uni-directional clutch, and is led to the planet gear pins by a thrower attached to the planet carrier.

When 'Overdrive' is selected the valve is lifted, thus holding the ball off its seat against the spring pressure of the plunger spring. Oil passes to the operating cylinders, forcing the pistons forward to engage the overdrive clutch. With the selection of 'Normal' drive the spring plunger pushes the ball onto its seat and the valve falls away from the ball. Oil from the cylinders then returns through the centre of the valve to the sump. Near the bottom of the valve is a small jet which slows down the emptying of the cylinders, to provide smooth re-engagement of direct drive.

For direct drive the oil is returned to the sump via the operating valve and flows in open circuit (Fig. Fa.3).

## Section Fa.4

### GUIDE TO SERVICE DIAGNOSIS

**Overdrive does not engage**

(1) Insufficient oil in the unit.
(2) Solenoid not operating due to fault in electrical system.
(3) Solenoid operating lever out of adjustment.
(4) Insufficient hydraulic pressure due to pump non-return valve incorrectly seating (probably dirt on ball seat).

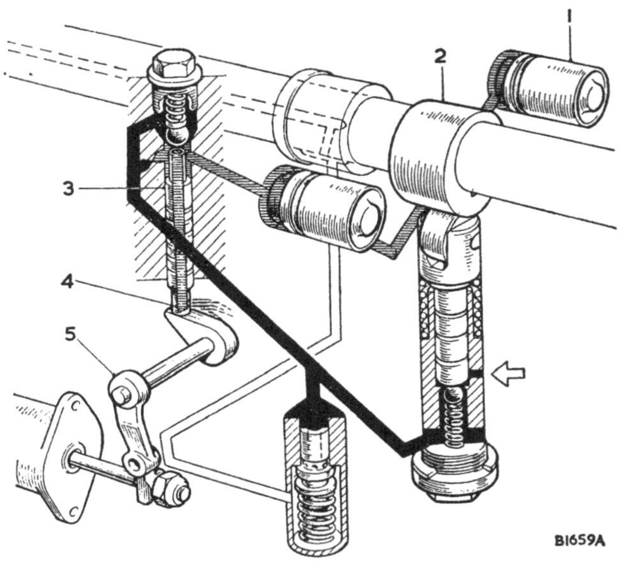


*Fig. Fa.3*

*Hydraulic system—direct drive*

1. Operating cylinders.
2. Cam.
3. Operating valve.
4. Jet.
5. Operating linkage.

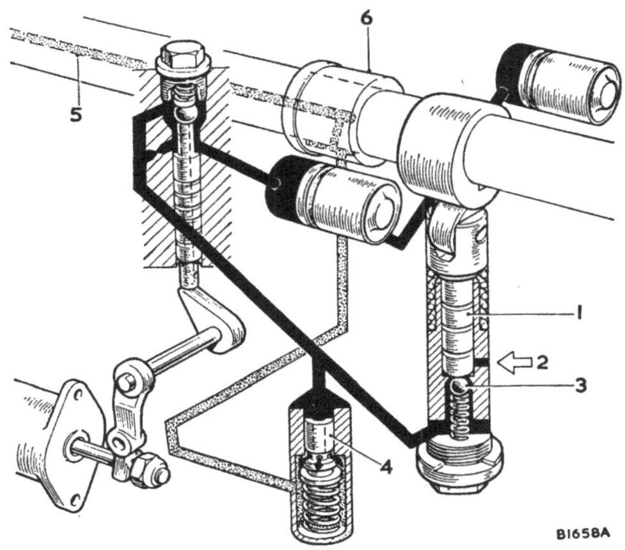


*Fig. Fa.4*

*Hydraulic system—overdrive*

1. Oil pump.
2. Oil inlet.
3. Non-return valve.
4. Oil relief valve.
5. Axial drilling—third motion shaft.
6. Third motion shaft steady bush.

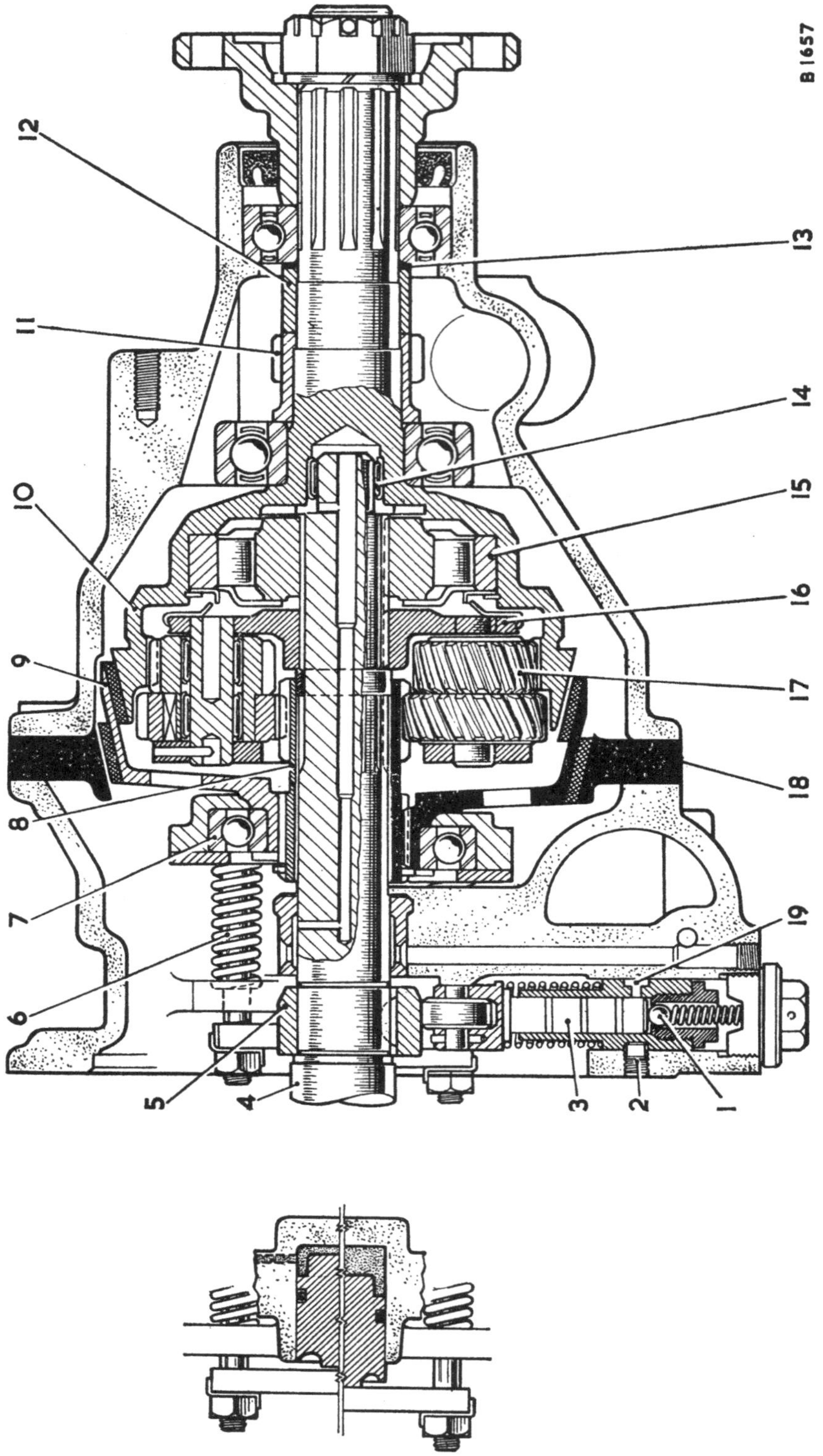


*Fig. Fa.5*

***Sectional view of the overdrive unit with the upper half showing direct drive and the lower half showing overdrive engaged. Operating cylinder detailed left***

1. Non-return valve.
2. Grub-screw—pump body.
3. Pump plunger.
4. Third motion shaft.
5. Cam.
6. Clutch spring.
7. Thrust bearing.
8. Sun wheel.
9. Cone clutch assembly.
10. Annulus.
11. Speedometer driving gear.
12. Spacer.
13. Shim.
14. Needle-roller bearing.
15. Uni-directional clutch.
16. Planet carrier.
17. Planet wheel.
18. Brake ring.
19. Oil inlet.

(5) Insufficient hydraulic pressure due to sticking or worn relief valve.
(6) Damaged gears, bearings, or moving parts within the unit requiring removal and inspection of the assembly.

**Overdrive does not release**

**(NOTE.—Do not attempt to reverse the car or damage may be caused within the overdrive unit.)**

(1) Fault in electrical control system.
(2) Blocked restrictor jet in operating valve.
(3) Solenoid operating lever adjustment.
(4) Sticking clutch.
(5) Damaged parts within the unit necessitating removal and inspection of the assembly.

**Clutch slip in overdrive**

As (1), (3), (4), and (5) under **'Overdrive does not engage'.**

**Clutch slip in reverse or free-wheel condition on overrun**

(1) Solenoid operating lever out of adjustment.
(2) Partially blocked restrictor jet in operating valve.
(3) Worn clutch linings.

## Section Fa.5

### PUMP NON-RETURN VALVE

Access to the pump non-return valve is gained through the centre cap in the bottom of the main casing (see [2], Fig. Fa.6).

**Removing**

(1) Remove the engine stay rod from the gearbox adaptor and the stay rod bracket from the rear cross-member.
(2) Remove the drain plugs and drain off the oil.
(3) Unclip the locking wire. Unscrew the valve cap and take out the spring and ball and unscrew the non-return valve body from the pump body.

Clean the components and examine the seating for pits, etc. If the ball valve is not seating correctly the ball should be tapped sharply onto its seat in the non-return valve body, using a soft-metal drift.

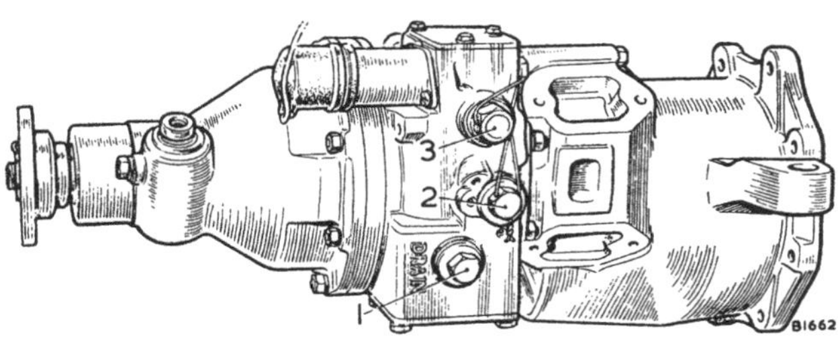


*Fig. Fa.6*

*Bottom view of overdrive unit*

1. Drain plug.
2. Non-return valve plug.
3. Relief valve plug.

Reassembly is the reverse of the above sequence. Position the ball on the spring with petroleum jelly. Ensure that the copper washer between the valve cap and main casing is fitted correctly over the step on the cap.

**NOTE.—The valve body must be removed from the pump body to reseat the ball valve, otherwise damage may be caused to the pump.**

## Section Fa.6

### RELIEF VALVE

Access to the relief valve is gained through the right-hand cap in the bottom of the main casing (see [3], Fig. Fa.6).

**Removing**

(1) Remove the engine stay rod from the gearbox adaptor and the stay rod bracket from the rear cross-member.
(2) Remove the drain plugs and drain off the oil.
(3) Unclip the locking wire. Unscrew the valve cap and remove the spring and piston.
(4) Remove the relief valve body with circlip pliers, taking care not to damage the piston bore.

Reassembly is the reverse of the above sequence. Ensure that the copper washer between the cap and main casing is fitted correctly over the step on the cap.

## Section Fa.7

### HYDRAULIC LEVER SETTING

The setting of the operating lever can be checked by means of a hole in the lever accessible from under the car after removal of the cover-plate on the right-hand side of the unit.

The controls are set correctly when a $\frac{3}{16}$ in. (4·76 mm.) diameter rod can be passed through the hole in the lever into the hole in the casing with the solenoid energized, i.e. with the ignition switched on, top gear engaged, and the fascia switch in the 'Overdrive' position.

If the solenoid operates but does not move the setting lever far enough to allow the rod to be inserted, the solenoid plunger must be adjusted as follows.

Screw the self-locking nut on the plunger in or out, with the plunger pushed into the solenoid as far as it will go. The plunger must be held against rotation with a spanner on two flats on the spindle. The operating lever fork should just contact the nut with the $\frac{3}{16}$ in. (4·74 mm.) rod in position.

Operate the overdrive several times and check that the setting rod can be inserted and that the solenoid current does not exceed 2 amperes. If the current is maintained at 17 amperes (approx.), it indicates that the solenoid plunger is not moving far enough to switch from the operating to the holding coil; the plunger must therefore be readjusted.

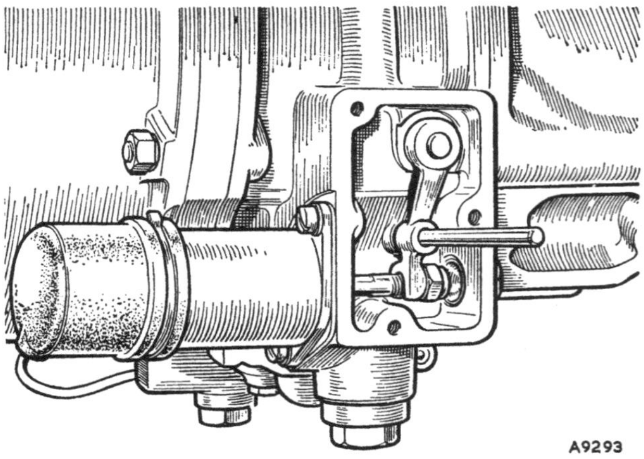


*Fig. Fa.7*
*Align the operating lever to check the adjustment of the operating plunger*

A new plunger must be fitted with a new solenoid. It will be necessary to check the setting for correct valve operation after fitment.

## Section Fa.8

### OPERATING VALVE

Access may be gained to the operating valve after removing the gear selector remote control assembly from inside the car (see Section Fa.9).

Having gained access, unscrew the plug and remove the spring and plunger. The ball valve will then be seen inside the valve chamber. The ball should be lifted $\frac{1}{32}$ in. (·8 mm.) off its seat when the solenoid is operated.

If the ball does not lift, the fault lies in the control mechanism (see Section Fa.7).

Remove the ball with a magnet and pull out the valve, using a piece of $\frac{1}{8}$ in. (3·175 mm.) dia. wire, taking care not to damage the seating at the top of the valve. Near the bottom of the valve will be seen a small hole breaking through the centre drilling; this is the jet for restricting the exhaust of oil from the operating cylinders. Check the bore of the operating valve with a $\frac{1}{8}$ in. (3·175 mm.) diameter drill and blow out the jet.

If the ball valve is not seating correctly the ball should be tapped sharply onto its seat in the casing, using a soft-metal drift.

If the unit fails to operate and the ball valve is seating and lifting correctly check that the pump is functioning, as follows.

Jack up the rear wheels, then with the engine idling and the valve plug removed engage top gear. Watch for oil being pumped into the valve chamber. If none appears, then the pump is not functioning.

Possible causes of trouble are: (1) Failure of the non-return valve due to foreign matter on the seat or to a broken valve spring, and (2) Breakage of the spring holding the pump plunger in contact with the cam.

**Testing oil pressure**

Remove the operating valve plug, fit an adaptor (Service tool 18G251D) in its place, and connect up a pressure gauge. Jack up the rear wheels and then with the engine idling engage top gear. Operate the overdrive solenoid; a pressure of 540 to 560 lb./sq. in. (37·97 to 39·37 kg./cm.$^2$) should be recorded.

## Section Fa.9

### REMOVING AND REPLACING

To dismantle the unit further than already described necessitates the removal of the unit from the car.

It is not possible to lift out the overdrive alone from the car; the engine, gearbox, and overdrive must be removed complete.

**Removing**

(1) Prepare the engine for removal from the car as detailed in Section A, and take the weight of the engine.
(2) Remove the retainer and pull the gear lever grommet clear of the remote control cover. Release the circlip and withdraw the gear lever from its housing.
(3) Withdraw the four screws securing the speaker panel and remove the panel. Pull back the tunnel carpet and remove the gearbox remote control cover.
(4) Pull the solenoid wire from the snap connector and disconnect the gear switch wires.
(5) Drain the oil from the gearbox and overdrive.
(6) Disconnect and remove the propeller shaft as detailed in Section G, and disconnect the speedometer pinion drive.
(7) Disconnect the clutch slave cylinder push-rod from the clutch withdrawal lever and the slave cylinder from the clutch housing. Tie the cylinder clear of the gearbox.
(8) Remove the screws securing the rear cross-member to the body and lower the engine and gearbox until the gearbox rests on the fixed body cross-member.
(9) Release the engine stay-rod bracket from the rear cross-member and remove the four nuts and washers securing the gearbox rear rubber mountings to the cross-member. Withdraw the rear cross-member from the body.
(10) Ease the assembly forward until it is clear of the fixed body cross-member and then tilt the assembly and lift it from the car.

Replacement is a reversal of the removal sequence.

After refitting the unit refill with fresh oil.

## Section Fa.10

### DISMANTLING AND REASSEMBLING

**Dismantling**

Remove the remote control assembly after withdrawing the six bolts and spring washers securing the control to the adaptor and the two screws from the rear support bracket. Take care not to damage the joint washer.

The unit is split at the gearbox adaptor, which is attached to the main casing by eight ¼ in. (6·35 mm.) diameter studs, one of which is extra long. After removing the eight nuts the unit can then be drawn off the third motion shaft.

Remove the operating valve plug (see Section Fa.8). This will allow air to enter the cylinders of the operating pistons and thus facilitate their removal. Release the lock washers securing the nuts on the two operating bridge pieces and remove the nuts, lock washers, and bridge pieces. Withdraw the operating pistons by gripping their spigots with pliers.

As the adaptor is separated from the unit the grub screw securing the pump body can be withdrawn and the body pulled from the casing after removing the non-return valve (see Section Fa.5).

The main and rear casings are secured by eight ¼ in. (6·35 mm.) diameter studs: unscrew the nuts a little at a time, gradually releasing the four clutch springs. Remove the main casing complete with brake ring. Take the four clutch springs off their guide bolts on the thrust plate and remove the clutch assembly complete with sun wheel.

The brake ring is spigoted into each casing and will come away with the main casing. A few light taps with a mallet around its flange will release the ring.

Remove the sun wheel from the cone clutch boss by withdrawing the snap ring in the forward end of the sun wheel hub. Remove the large circlip and withdraw the thrust bearing and housing complete.

Lift out the planet carrier assembly from the annulus. If it is necessary to remove the uni-directional roller clutch, first remove the circlip and brass retaining washer which are in front of the clutch.

Place Service tool 18G178 centrally over the clutch and lift the inner member into it. This will ensure that the rollers do not fall out of the inner member. Remove the thrust washer. **DO NOT remove the outer bearing ring,** as it is expanded into the annulus.

A caged needle-roller bearing is fitted in the annulus centre spigot. If it is necessary to remove this use an extractor.

To remove the annulus from the rear casing withdraw the speedometer pinion and bush located by one dowel screw, take off the coupling flange at the rear of the unit, and drive out the annulus from the back. The front bearing will come away on the shaft, leaving the rear bearing in the casing.

(See Editor's notes at end of section.)

**Inspection**

Each part should be thoroughly cleaned and inspected to determine which parts need renewing. It is important to appreciate the difference between parts which are worn sufficiently to affect operation of the unit and those which are merely 'worn in'.

(1) Inspect the main casing for cracks, damage, etc. Examine the bores of the operating cylinders for scores or wear. Check for leaks at the plugged ends of the oil passages. Inspect the support bush in the centre bore for wear or damage.

(2) Examine the clutch sliding member assembly. Ensure that the clutch linings are not burned or worn. Inspect the clutch spring locating bolts and bridge pieces and see that they are not distorted. Ensure that the ball bearing is in good condition and rotates freely. See that the sliding member slides easily on the splines of the sun wheel.

(3) Check the springs for distortion and collapse (see **'GENERAL DATA'**).

(4) Inspect the teeth of the gear train for damage. If the gears are damaged or the needle rollers worn, renew the planet carrier assembly. The sun wheel bush is bored to the pitch line of the teeth, and if worn, the gear will have to be renewed.

(5) Examine the uni-directional clutch thrust washer.

(6) See that the rollers of the uni-directional clutch are not chipped and that the outer ring is tight in the annulus. Ensure that the spring is free from distortion.

(7) Inspect the output shaft ball bearings and see that there is no roughness when they are rotated slowly.

(8) Ensure that there are no nicks or burrs on the third motion shaft splines and that the oil holes are open and clean.

(9) Inspect the oil pump for wear on the pump plunger and roller pin. Ensure that the plunger spring is not distorted (see **'GENERAL DATA'**). Inspect the valve seat and ball and make sure that they are free from nicks and scratches.

(10) Check the operating valve for distortion and damage and see that it slides easily in its bore. Inspect the relief valve body, 'O' ring, and ball, etc.

**Reassembling the unit**

Assemble the unit after all the parts have been thoroughly cleaned and checked to ensure that none is damaged or worn.

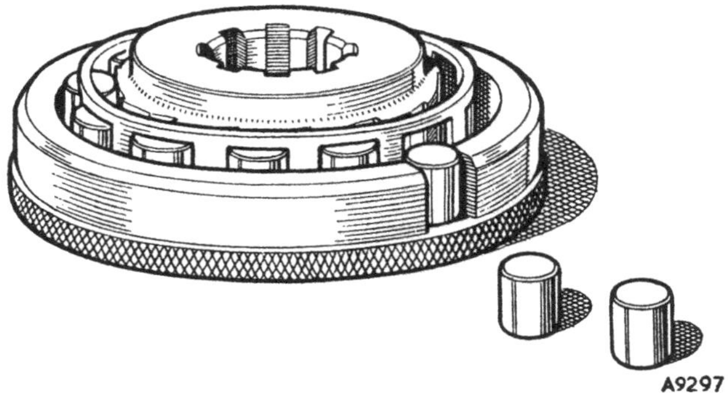


*Fig. Fa.8*

*Use Service tool 18G178 when removing or replacing the uni-directional clutch inner member*

Fit the annulus assembly with the speedometer drive gear, spacing tube, and shim which fit on the shaft between the front and rear ball bearings in the rear casing.

If a new bearing is fitted it becomes necessary to assess the thickness of shim required. To do this replace the annulus in the rear casing with the speedometer drive gear and distance tube in position, and with a depth gauge measure the distance between the rear ball bearing seating and the distance tube. To this figure add ·005 to ·010 in. (·13 to ·25 mm.): this will give the actual shimming required.

The shims are available in the following thicknesses for selective assembly, which should allow an end-float of ·005 to ·010 in. (·13 to ·25 mm.) on the output shaft and no preloading of the bearings:

·090 in.±·0005 in. (2·28 mm.±·01 mm.)
·095 in.±·0005 in. (2·41 mm.±·01 mm.)
·100 in.±·0005 in. (2·54 mm.±·01 mm.)
·105 in.±·0005 in. (2·67 mm.±·01 mm.)

Replace the coupling flange and tighten the nut to a torque reading of 100 to 130 lb. ft. (13·83 to 17·97 kg. m.), holding the flange against rotation with the aid of a wrench (Service tool 18G34A).

Fit the speedometer drive bush and pinion and lock in position with the dowel screw.

Replace the thrust washer and uni-directional clutch, using Service tool 18G178 to enter the rollers in the outer ring. Ensure that the spring is fitted correctly so that the cage urges the rollers up the ramp on the inner member.

Fit the brass retaining washer and locate it in position with the circlip.

To ensure that the second set of teeth on each planet wheel will mesh with the annulus teeth turn each planet wheel until its etched line coincides with an etched line on the carrier. Insert the sun wheel to hold the planet wheel in these relative positions. Fit the assembly to the annulus and withdraw the sun wheel.

**As long as the planet carrier remains in the annulus the sun wheel can be removed and reinserted at any time. Should the planet carrier be removed from the annulus, the whole lining-up procedure must be repeated.**

Pass the sun wheel into the cone clutch member and fit the small circlip at the forward end of the sun wheel. Press the thrust bearing into its housing, fit the four bolts to the thrust bearing housing, and then fit the assembly onto the forward end of the cone clutch hub, securing in place with the large circlip. Fit the clutch assembly to the annulus, engaging the sun and planet wheels. Place the thrust plate and springs on the bolts.

Assemble the brake ring to the main casing (large end of taper towards the rear casing) with a jointing compound. Position the clutch bolts through the holes in the main casing. Start the nuts on the casing studs and gradually tighten to secure the two casings together. Ensure that the clutch spring bolts do not bind in their holes and that the casings go together easily.

Fit the two operating pistons, carefully easing their rubber sealing rings into the cylinder bores (the centre spigots of the pistons face towards the front of the unit). Fit the two bridge pieces, lock washers, and nuts.

If the pump body has been removed, insert its small end into the casing in the middle hole at the bottom of the casing, ensuring that the oil inlet faces to the rear. Gently tap into position until the groove lines up with the grub screw hole at the bottom of the casing front face. Fit the grub screw and tighten. Refit the non-return valve (see Section Fa.5). Insert the pump plunger and spring from inside the casing.

Replace the operating valve plug.

Support the unit upright and insert a dummy shaft or a spare third motion shaft so that the planet carrier and uni-directional clutch line up with each other; a long, thin screwdriver should be used to line by eye the splines (turn anti-clockwise only) in the planet carrier and the clutch before inserting the dummy shaft. Gently turn the dummy shaft to assist in feeling it into the splines, making sure that it goes fully home.

Make sure that the lowest part of the oil pump cam will contact the pump plunger and that the third motion shaft clip is seated in its groove; then with top gear engaged, carefully thread the third motion shaft into the centre bushing in the unit. Gently turn the first motion shaft to and fro to assist in engaging the splines of the planet carrier. As the adaptor and overdrive come together watch carefully to see that the oil pump engages the cam properly.

**NOTE.—The gearbox third motion shaft should enter the overdrive easily, provided that the lining-up procedure previously described is carried out and the unit is not disturbed. If any difficulty is experienced it is probable that one of the components has been misaligned, and the gearbox should be removed and the overdrive realigned with the dummy shaft.**

Replace the remote control tower, ensuring that the change speed lever engages the selector.

**(See Editor's notes at end of section.)**

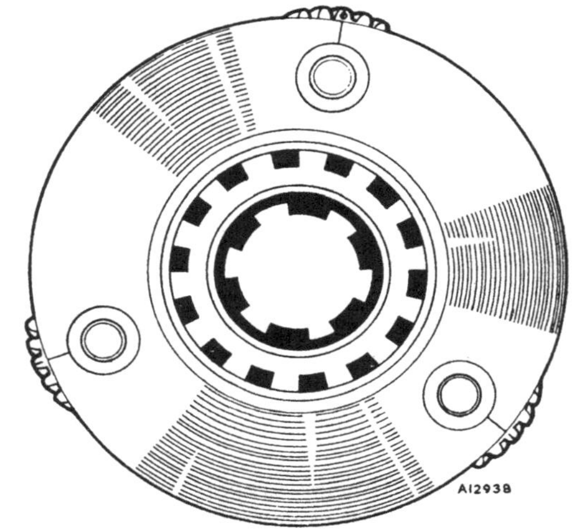


*Fig. Fa.9*

*Before assembling the planet carrier to the sun wheel rotate the planet wheels until the etched line on the planet wheel lines up with the etched line on the planet carrier*

## Section Fa.11

### DRIVE SHAFT BEARINGS AND SEAL

**Removing**

The drive shaft front bearing must be drawn from the shaft with a suitable puller.

The oil seal may be removed with the overdrive in position in the car after removing the propeller shaft and drive flange.

The rear bearing should be pressed from its seating with a suitable spigot.

**Replacing**

A press should be used when replacing the bearings. Fit a new oil seal.

## Section Fa.12

### OVERDRIVE RELAY SYSTEM

**Description**

Engagement of overdrive is controlled electrically through a manually operated toggle switch. The circuit is shown in Fig. Fa.10 and includes the following components:

(1) *Relay.* An electro-magnetic switch used with item (2) to enable an interlocking safeguard to be incorporated against changing out of overdrive with the throttle closed.

(2) *Throttle switch.* A vacuum-operated switch to over-ride the toggle switch under closed throttle conditions.

(3) *Gear switch.* A small plunger-operated switch allowing overdrive to be engaged only in the two highest forward gear positions.

(4) *Solenoid unit.* An electro-magnetic actuator to engage the overdrive mechanism by opening the hydraulic control valve.

**Operation**

When the driver engages 'overdrive' by closing the contacts of the toggle switch, current is fed by way of the ignition switch and fuse unit supply terminal 'A3' to energize the relay operating coil. Closure of the relay contacts connects terminal 'A3' to the gear switch and, provided third or top gear is engaged, will energize the solenoid unit and effect a change from direct drive to overdrive.

Change from overdrive to direct drive is effected by selecting a low gear (when the gear switch contact will open) or by moving the toggle switch to 'NORMAL' with the throttle open (when the vacuum switch will open) with the throttle open (when the vacuum switch will open).

If effected with the throttle closed (high manifold depression) the vacuum switch will over-ride the toggle switch, delaying the change until the engine takes up the drive.

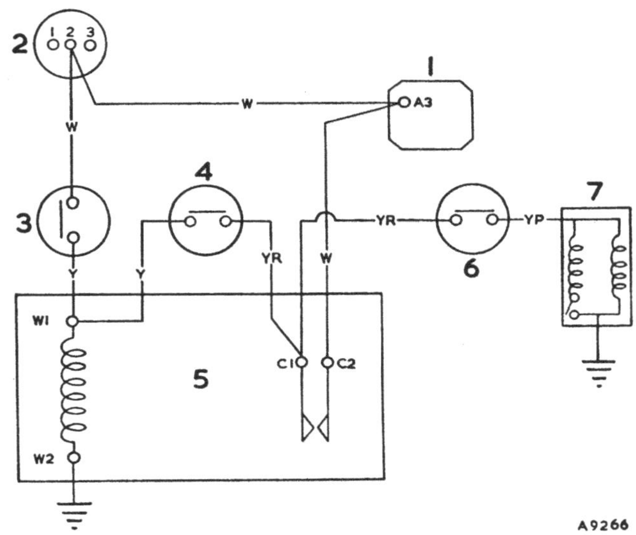


*Fig. Fa.*10

*Diagram of the overdrive electrical circuit*

1. Fuse block.
2. Ignition switch.
3. Driver's switch.
4. Vacuum switch.
5. Relay.
6. Gearbox switch.
7. Solenoid.

CABLE COLOUR CODE

When a cable has two colour code letters the first denotes the main colour and the second denotes the tracer colour.

P. Purple. R. Red. W. White. Y. Yellow.

### EDITOR'S NOTES

**Fa. The Overdrive**

*Dismantling*

In lieu of Service tool 18G178, a piece of string may be used to hold the rollers in the uni-directional clutch cage.

*Reassembling the unit*

As the rollers enter the outer ring, slip the string off.

In the absence of a dummy shaft, the alignment may be carried out by eye, using a long thin screwdriver.

# Section Fb

## THE GEARBOX (ALL SYNCHROMESH)

## Section Fb.1

### GEARBOX ASSEMBLY

**Removing**

(1) Remove the engine and gearbox from the car—Section A.21.

(2) Remove the starter.

(3) Unscrew the bolts securing the gearbox to the engine and withdraw the gearbox.

**Refitting**

(4) Reverse the removing procedure in (1) to (3) and refill the gearbox with the correct quantity (see **'GENERAL DATA'**) of a recommended lubricant.

## Section Fb.2

### REAR EXTENSION

**Removing**

(1) Drain the gearbox.

(2) Using tool 18G 34 A to prevent rotation, unscrew the propeller shaft flange retaining nut and withdraw the flange.

(3) Unscrew the bolts securing the remote-control housing to the gearbox and remove the remote control.

(4) Withdraw the selector interlocking arm and plate assembly.

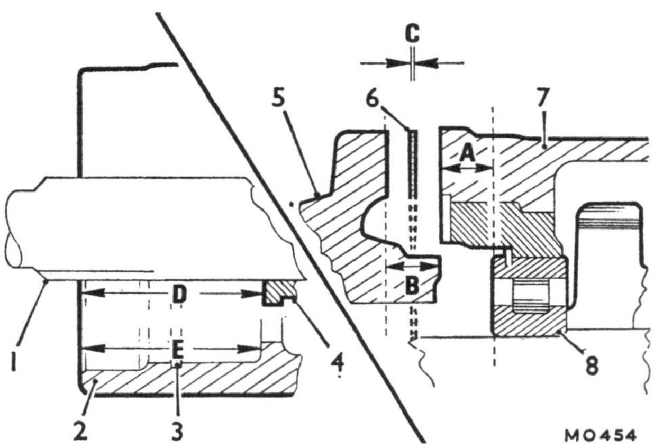


*Fig. Fb.1*

*Rear extension setting up*

1. Third motion shaft.
2. Rear extension.
3. Rear bearing circlip groove.
4. Distance tube.
5. Rear extension.
6. Gasket for rear extension.
7. Gearbox main casing.
8. Bearing.

A. Depth between bearing face and gearbox main casing face.
B. Depth between rear extension face and bearing flange.
C. Thickness of joint washer.
D. Depth from rear extension face to distance tube.
E. Depth from rear extension face to bearing register.

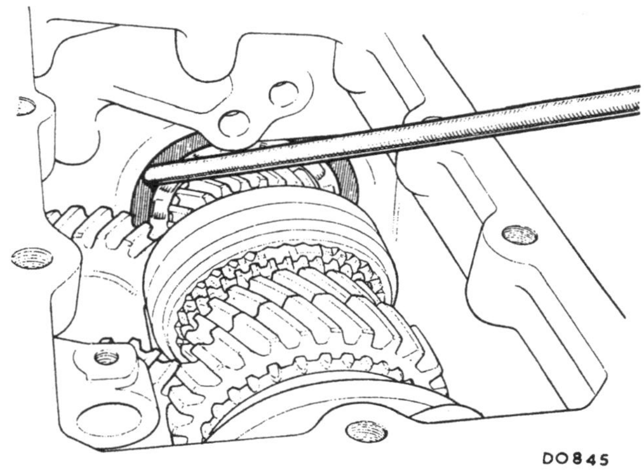


*Fig. Fb.2*

*Drifting the first motion shaft from the casing*

(5) Unscrew the nuts securing the rear extension to the gearbox and withdraw the extension, ensuring that the shims fitted to the gearbox third motion shaft are not mislaid.

**Dismantling**

The following operations are only necessary if the bearing or oil seal are to be renewed.

(6) Remove and discard the oil seal.

(7) Remove the circlip retaining the bearing.

(8) Press the bearing from the extension.

**Reassembling**

**The following procedure must be carried out when fitting a new gearbox extension or if any of the third motion shaft components have been changed.**

(9) Calculate the thickness of shims required between the third motion shaft front bearing, and the gearbox rear extension case, as follows:

(*a*) Measure the depth (A) between the face of the front bearing and its housing, and add the thickness (C) of the joint washer.

(*b*) Measure the depth (B) from the joint face to the bearing register face of the rear extension.

(*c*) Fit shims between the extension and the face of the front bearing to bring the dimension measured in (*b*) ·000 to ·001 in. (·000 to ·025 mm.) less than the dimension in (*a*). Shims of ·002 in. (·05 mm.) and ·004 in. (·10 mm.) thickness are available.

(*d*) Fit the rear extension.

(10) Calculate the thickness of shims required between the distance tube, and the third motion shaft rear bearing, as follows:

(*a*) Measure the depth (D) between the rear face of the extension and the rear face of the distance tube.

(*b*) Measure the depth (E) of the rear bearing register from the rear face of the extension.

(*c*) Fit shims between the distance tube and the bearing to bring the dimension measured in (*a*) from ·000 to ·001 in. (·000 to ·025 mm.) less than the dimension measured in (*b*). Shims of ·002 in. (·05 mm.), ·005 in. (·13 mm.) and ·010 in. (·25 mm.) thickness are available.

(*d*) Press the bearing into the extension using tool 18G 186.

(*e*) Select and fit the thickest circlip that will enter the rear bearing circlip groove from the following range:

·096 to ·098 in. (2·43 to 2·49 mm.)
·098 to ·100 in. (2·49 to 2·54 mm.)
·100 to ·102 in. (2·54 to 2·59 mm.)

(11) Using tool 18G 134 with adaptor 18G 134 BK, fit a new oil seal.

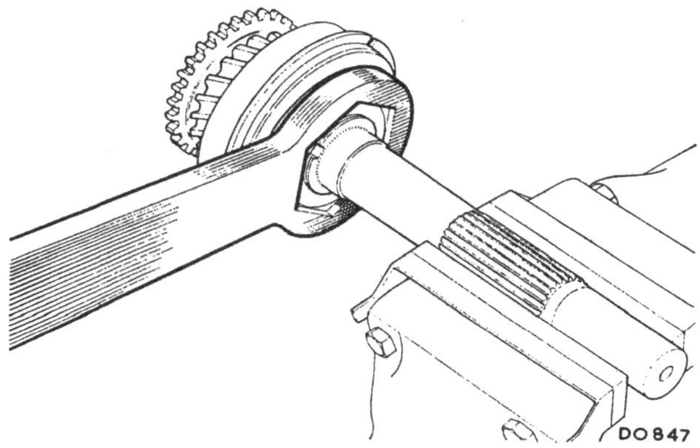


*Fig. Fb.3*
*Removing the left-handed-threaded first motion shaft nut using tool* 18G 48

**Refitting**

(12) Check that the shims on the third motion shaft are correctly positioned against the distance tube shoulder, and fit the extension.

(13) Fit and tighten the extension securing nuts.

(14) Reverse the removing procedure in (1) to (4).

## Section Fb.3

### FRONT COVER

**Removing**

(1) Turn the clutch release bearing retaining clips and withdraw the bearing from the lever.

(2) Remove the rubber boot from the clutch lever.

(3) Unscrew the front cover retaining nuts and remove the cover, collecting the shims fitted in front of the first motion shaft bearing.

**Dismantling**

(4) Remove the oil seal and gasket.

(5) Unscrew the nut retaining the clutch lever pivot bolt, remove the bolt, and withdraw the lever.

**Reassembling**

**The following procedure must be carried out when fitting a new front cover or if any of the first motion shaft components have been changed.**

(6) Calculate the thickness of shims required between the front cover and the bearing, as follows:

(*a*) Measure the depth (A) from the face of the cover to the bearing register and add ·012 in. (·31 mm.) for the compressed thickness (C) of the joint washer.

(*b*) Measure the distance (B) that the first motion shaft bearing protrudes from the casing.

(*c*) Fit shims between the bearing and cover to bring the dimension measured in (*b*) from ·000 to ·001 in. (·000 to ·025 mm.) greater than the dimension in (*a*). Shims of ·002 in. (·05 mm.) and ·004 in. (·10 mm.) thickness are available.

(7) Using tool 18G 134 with adaptor 18G 134 Q, fit a new oil seal to the cover.

(8) Position the clutch lever, fit the pivot bolt, screw on and tighten the bolt retaining nut.

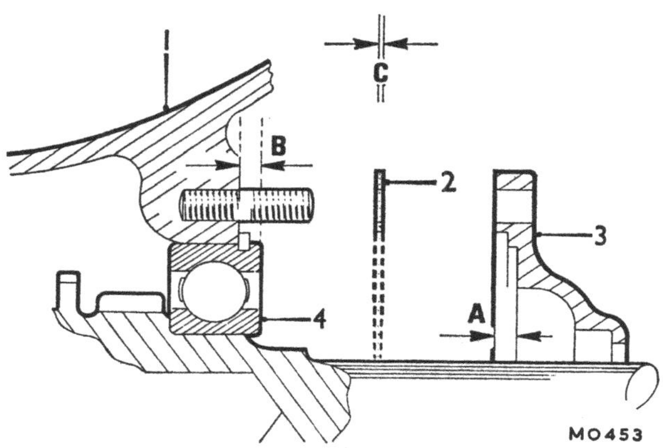


*Fig. Fb.4*
*Front cover setting up*

1. Gearbox main casing.
2. Gasket for front cover.
3. Front cover.
4. Bearing.

A. Depth of front cover to bearing register.
B. First motion shaft bearing protrusion.
C. Compressed thickness of joint washer.

**Refitting**

(9) Reverse the removing procedure noting the following:

(*a*) Use a new gasket when refitting the cover; ensure that all traces of the old gasket are removed from the cover and gearbox jointing faces before fitting the new gasket.

(*b*) Check that the first motion shaft bearing shims are correctly positioned.

## Section Fb.4

### SELECTOR RODS AND FORKS

**Removing**

(1) Remove the rear extension—Section Fb.3, (1) to (3)—or the overdrive adaptor—Section Fc.1.

(2) Unscrew the side cover retaining bolts and remove the cover.

(3) Remove the selector detent plunger plugs and springs.

(4) Slacken the locknuts on each of the selector fork retaining bolts and remove the bolts.

(5) Withdraw the selector rods and remove the selector forks.

**Refitting**

(6) Reverse the removing procedure, noting the following points:

(*a*) Clean any burrs from the rods at the fork attachment points before refitting the rods.

(*b*) Check that the detent plungers are clear of the holes before fitting the rods.

## Section Fb.5

### REVERSE IDLER GEAR

**Removing**

(1) Remove the rear extension—Section Fb.3, (1) to (3)—or the overdrive adaptor—Section Fc.1.

(2) Remove the selector rods and forks—Section Fb.4, (2) to (5).

(3) Knock back the locking tab on the shaft retaining bolt and remove the bolt.

(4) Remove the shaft and withdraw the gear.

**Inspection**

(5) Inspect the shaft for signs of excessive wear; renew the shaft if necessary.

(6) Examine the gear teeth for wear and damage; renew the gear if necessary.

(7) Inspect the gear bushes for signs of excessive wear; worn bushes may be removed and new bushes pressed in.

**Refitting**

(8) Reverse the removing procedure in (1) to (4) using a new locking washer on the shaft retaining bolt.

## Section Fb.6

### FIRST MOTION SHAFT

**Removing**

(1) Remove the rear extension—Section Fb.2, (1) to (5).

(2) Remove the front cover—Section Fb.3, (1) to (3).

(3) Remove the selector rods and forks—Section Fb.4, (2) to (5).

(4) Remove the reverse idler gear—Section Fb.5, (2) to (4).

(5) Carefully drift the layshaft from the gearbox.

(6) Check that the laygear teeth are clear of the first motion shaft gear teeth.

(7) Using a soft metal drift registering against the bearing outer track, carefully drift the shaft assembly forward from the gearbox.

**Dismantling**

(8) Withdraw the needle-roller bearing from inside the rear end of the shaft.

(9) Tap back the locking tab on the bearing retaining nut.

(10) Fit tool 18G 49 onto the nut and clamp the shaft in a soft-jawed vice.

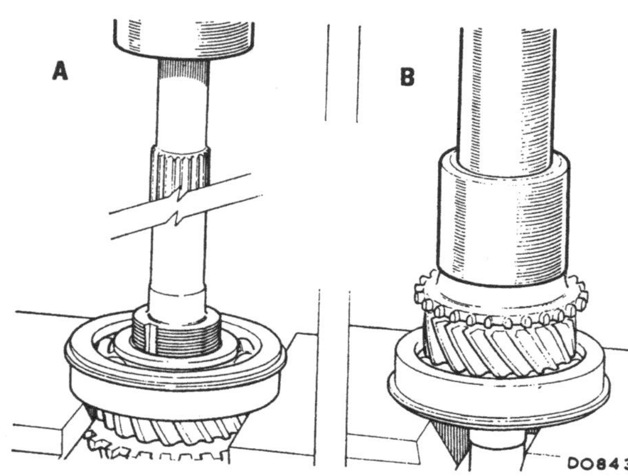


*Fig. Fb.5*

*First motion shaft bearing*

A. Removing the bearing from the shaft.
B. Refitting the bearing.

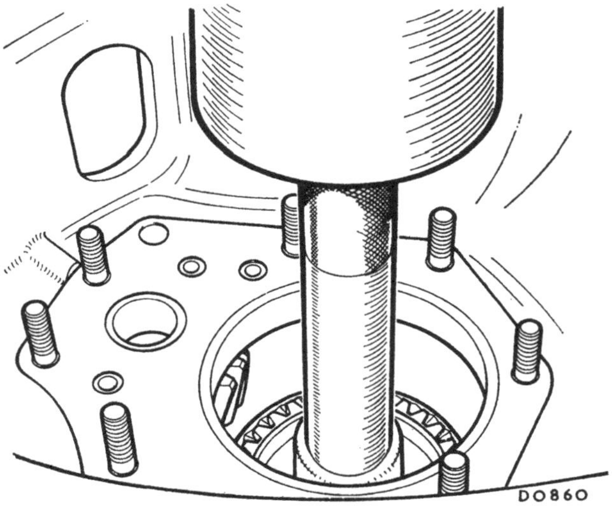


*Fig. Fb.6*
*Pressing out the third motion shaft using tool* 18G 1045

(11) Unscrew the **left-hand-threaded** nut; remove the nut and lock washer.
(12) Press the bearing from the shaft.

**Inspection**

(13) Inspect the shaft for signs of wear and damage to the gear or splines.
(14) Examine the bearings for wear and damage; renew the bearing as necessary.

**Reassembling**

(15) Reverse the dismantling procedure in (8) to (12), using a new shaft nut locking washer.

**Refitting**

(16) Carefully drift the assembled shaft into the casing ensuring that the laygear teeth do not foul the shaft gear teeth.
(17) Reverse the removing procedure in (1) to (6).

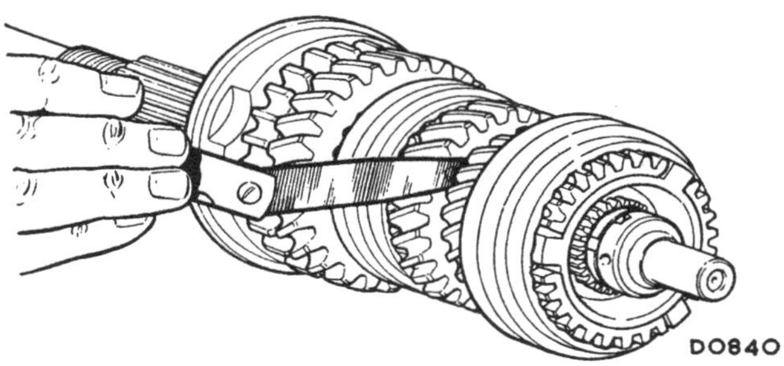


*Fig. Fb.7*
*Checking the first and second speed gear end-float*

## Section Fb.7

### THIRD MOTION SHAFT—NON-OVERDRIVE

**Removing**

(1) Remove the rear extension—Section Fb.2, (1) to (5).
(2) Remove the front cover—Section Fb.3, (1) to (3).
(3) Remove the selector rods and forks—Section Fb.4, (2) to (5).
(4) Remove the reverse idler gear—Section Fb.5, (2) to (4).
(5) Remove the first motion shaft—Section Fb.6, (5) to (7).
(6) Remove the shims and distance tube from the third motion shaft.
(7) Unscrew the rear extension securing studs from the gearbox casing.
(8) Check that the laygear teeth are clear of the gears and synchronizers.
(9) Using tool 18G 1045, press the third motion shaft rearwards from the gearbox.

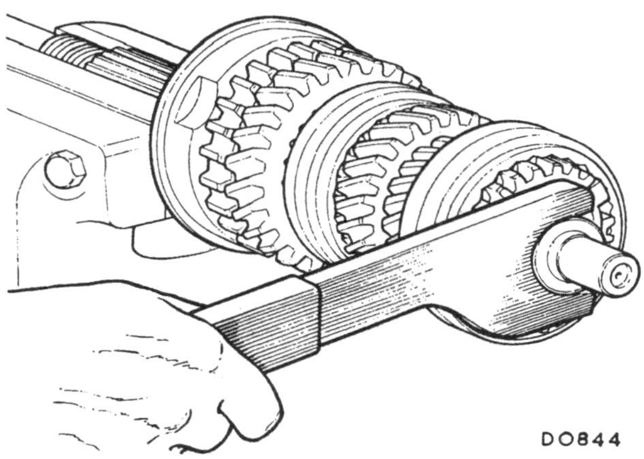


*Fig. Fb.8*
*Using tool* 18G 1024 *to remove the front gear locking nut for the third motion shaft*

**Dismantling**

*Shaft assembly*

(10) Check the end-float on the first, second, and third speed gears against the figures given in **'GENERAL DATA'**.
(11) Withdraw the third and fourth speed synchronizer assembly complete with baulk rings.
(12) Tap back the locking tabs on front gear locking nut; unscrew the nut using tool 18G 1024 and remove the lock washer.
(13) Withdraw the shaft sleeve, third speed gear, and interlocking thrust washer.
(14) Withdraw the second speed gear and thrust washer.

(15) Remove the first and second speed synchronizer assembly complete with baulk rings.
(16) Withdraw the speedometer gear and remove the gear driving key.
(17) Remove the shaft distance piece.
(18) Press the first speed gear, reverse gear, and the bearing complete with housing from the shaft (see Fig. Fb.9).
(19) Press the bearing from its housing (see Fig. Fb.10).

**Synchronizers**

(20) Remove the baulk rings.
(21) Wrap a cloth loosely around the assembly to retain the balls and springs and push the synchronizer hub from the sliding coupling.

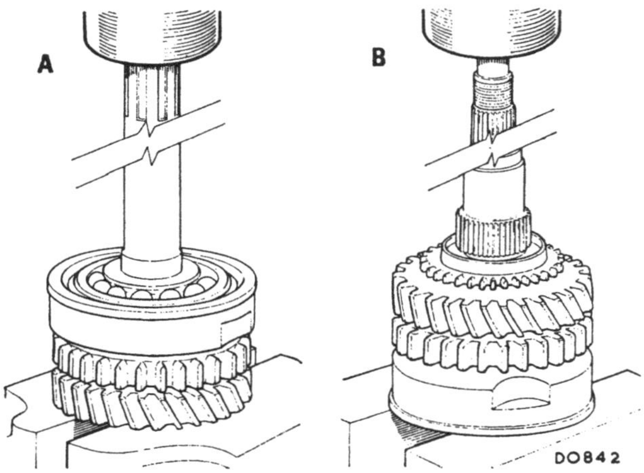


*Fig. Fb.9*
*Third motion shaft*

A. Pressing off the first speed and reverse gears and bearing with housing.
B. Refitting gears and bearing.

**Inspection**

(22) If the end-float on the gears when checked in (10) exceeds the figures given, inspect the relevant gear and thrust washer bearing faces for wear. Renew worn gears and thrust washers as necessary.
(23) Examine the gears, hubs, and couplings for wear and damaged or worn parts; replace as necessary.
(24) Inspect the shaft for wear and damage to splines and bearing surfaces.
(25) Examine the bearing for wear and damage.
(26) Check the synchronizer springs against the dimension and pressure given in **'GENERAL DATA'**.

**Reassembling**

*Synchronizers*

(27) Fit tool 18G 1026 over the synchronizer hub.

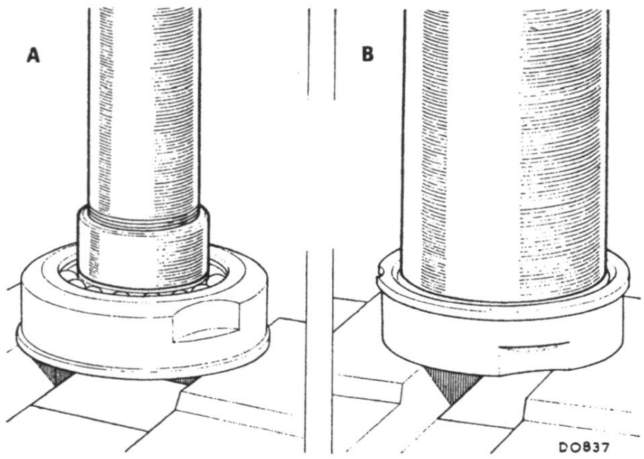


*Fig. Fb.10*
*Third motion shaft bearing*

A. Removing the bearing from its housing.
B. Refitting the bearing.

(28) Turn the hub until one of the spring pockets is in line with the hole in the tool.
(29) Fit the spring and ball into the pocket; depress the ball and turn the hub so that the ball is retained by the tool.
(30) Repeat operations (28) and (29) for the remaining balls and springs.
(31) Line up the cut-outs in the coupling with those in the hub and enter the hub into the coupling.
(32) Press the hub from the tool into the coupling so that the springs and balls are retained by the coupling as the tool is removed.
(33) Fit the baulk rings to the hub.

**NOTE.**—The baulk rings of the two synchronizer assemblies are not interchangeable. The first and second speed synchronizer baulk rings are identified (see Fig. Fb.14) by a drill point on one of the baulk ring lugs or by the fillets at the base of the lugs.

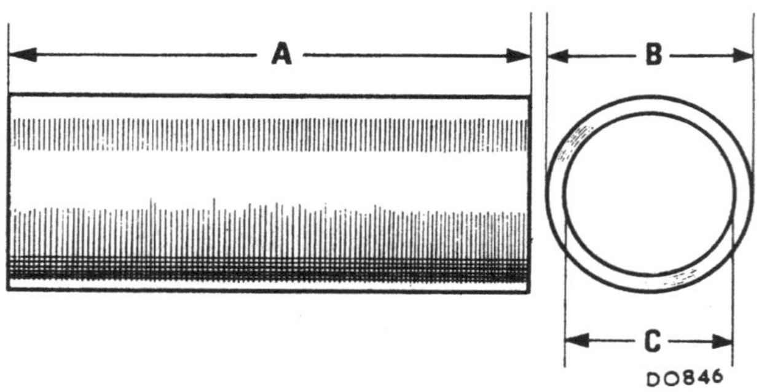


*Fig. Fb.11*
*Dimensions of tool for refitting the third motion shaft*

A=19 in. (483 mm.) B=3 in. (76 mm.)
C=2·75 in. (69 mm.)

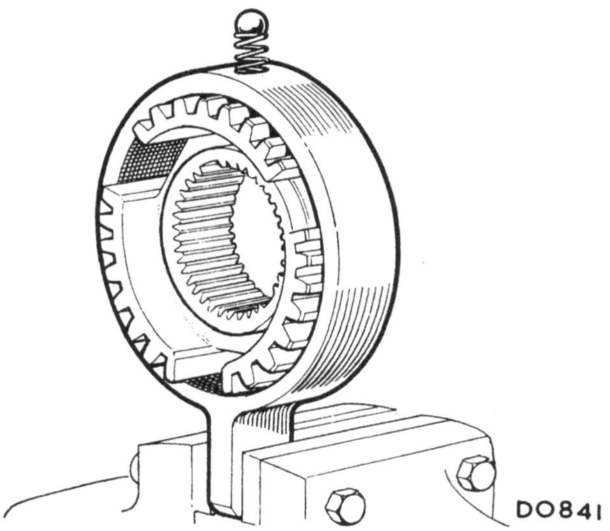


*Fig. Fb.12*
*Reassembling the synchronizers using tool 18G 1026*

*Shaft assembly*

(34) Reverse the dismantling procedure in (10) to (19), noting the following points:

(*a*) Ensure that the lugs on the shaft sleeve are aligned with the cut-outs in the interlocking thrust washers.

(*b*) The synchronizer assemblies may be fitted either way round, but their baulk rings or the complete assemblies must not be interchanged with each other.

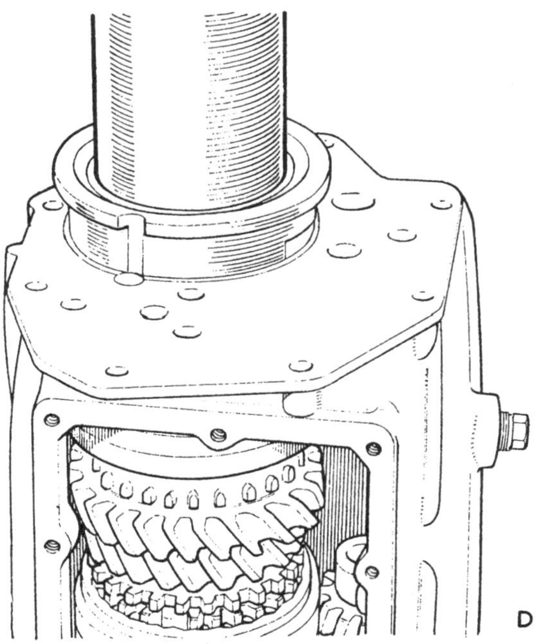


*Fig. Fb.13*
*Refitting the assembled third motion shaft using the fabricated tool*

(*c*) After reassembly re-check the end-float on the first, second, and third speed gears.

**Refitting**

(35) Enter the assembled shaft into the gearbox from the rear; ensure that the third and fourth speed synchronizer assembly is fitted to the shaft.

(36) Check that the laygear teeth are clear of the shaft gears.

(37) Using a fabricated tool to the dimensions shown in Fig. Fb.11 registering on the outer track of the bearing, press the shaft assembly into the gearbox.

(38) Reverse the removing procedure in (1) to (5).

## Section Fb.8

### THIRD MOTION SHAFT—WITH OVERDRIVE

**Removing**

(1) Remove the overdrive unit and adaptor—Section Fc.1, (2) to (10).

(2) Carry out the operations for removing the non-overdrive third motion shaft given in Section Fb.7, (2) to (9).

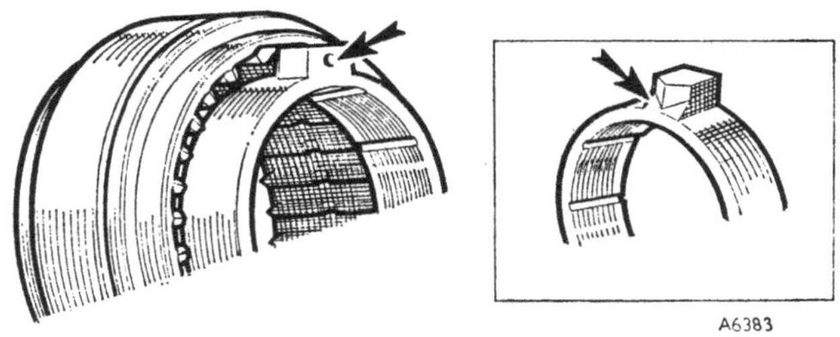


*Fig. Fb.14*
*The identification drill point or fillets on the lugs of the first and second speed synchronizer baulk rings*

**Dismantling**

(3) Carry out operations in Section Fb.7, (10) to (15).

(4) Tap back the locking tab on rear shaft nut.

(5) Using tool 18G 391, unscrew the nut and remove the locking washer.

(6) Withdraw the distance piece.

(7) Carry out operations in Section Fb.7, (18) to (21).

**Inspection**

(8) Carry out operations in Section Fb.7, (22) to (26).

**Reassembling**

(9) Carry out operations in Section Fb.7, (27) to (33).

(10) Reverse the dismantling procedure in (3) to (7).

**Refitting**

(11) Reverse the removing procedure in Section Fb.7, (2) to (9).

(12) Refit the overdrive unit and adaptor—Section Fc.9.

## Section Fb.9

### LAYGEAR

**Removing**

(1) Remove the rear extension—Section Fb.2, (1) to (5).

(2) Remove the front cover—Section Fb.3, (1) to (3).

(3) Remove the selector rods and forks—Section Fb.4, (2) to (5).

(4) Remove the reverse idler gear—Section Fb.5, (2) to (4).

(5) Remove the first motion shaft—Section Fb.6, (5) to (7).

(6) Remove the third motion shaft—Section Fb.7, (6) to (9).

(7) Temporarily refit the layshaft and check the laygear end-float.

(8) Withdraw the layshaft and remove the laygear and its thrust washers.

**Dismantling**

(9) Withdraw the needle-roller bearings from inside the laygear.

**Inspection**

(10) Inspect the layshaft for wear and damage to the bearing surfaces.

(11) Examine the roller bearings for wear and damage.

(12) Inspect the laygear teeth for damage and the bearing faces and bore for wear.

**Reassembling**

(13) Fit the needle-roller bearings into the laygear.

**Refitting**

(14) Position the laygear and front (large) thrust washer; enter the layshaft through the housing and thrust washer, and into the laygear.

(15) If the laygear end-float measured in (7) is outside the limits given, select a rear thrust washer to give the correct end-float.

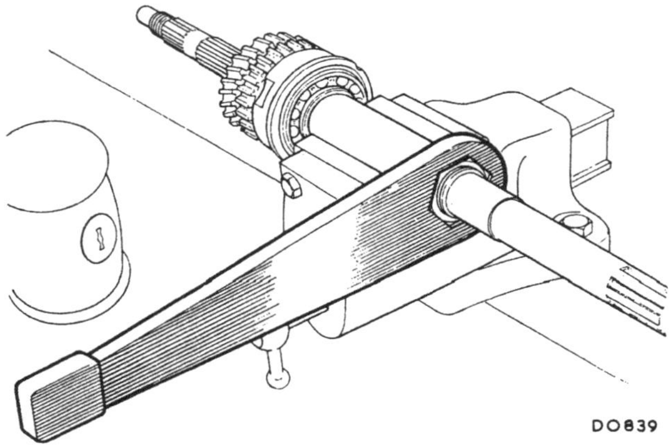


*Fig. Fb.15*
*Using tool 18G 391 to unscrew the rear nut on the overdrive gearbox third motion shaft*

Rear thrust washers are available in the following thicknesses:

·154 to ·156 in. (3·91 to 3·96 mm.)
·157 to ·158 in. (3·99 to 4·01 mm.)
·160 to ·161 in. (4·06 to 4·08 mm.)
·163 to ·164 in. (4·14 to 4·16 mm.)

(16) Fit the rear thrust washer and thread a piece of soft wire through the housing, thrust washer, and into the laygear.

(17) Withdraw the layshaft, at the same time thread the wire through the laygear and front thrust washer. This will enable the laygear and thrust washers to be retained ready for refitting the layshaft, and at the same time allow sufficient movement of the laygear to prevent the gear teeth obstructing the gears on the first and third motion shafts during refitting.

(18) Refit the third motion shaft—Section Fb.7, (35) to (37).

(19) Refit the first motion shaft—Section Fb.6, (16).

(20) Use the wire to align the laygear and thrust washers.

(21) Enter the layshaft, plain end first, into the front of the casing; line up the cut-away in the end of the shaft with the front cover recess and press the shaft home.

(22) Reverse the removing procedure in (1) to (4).

# Section Fc

## THE OVERDRIVE (TYPE L.H.)

## THE OVERDRIVE COMPONENTS (TYPE L.H.)

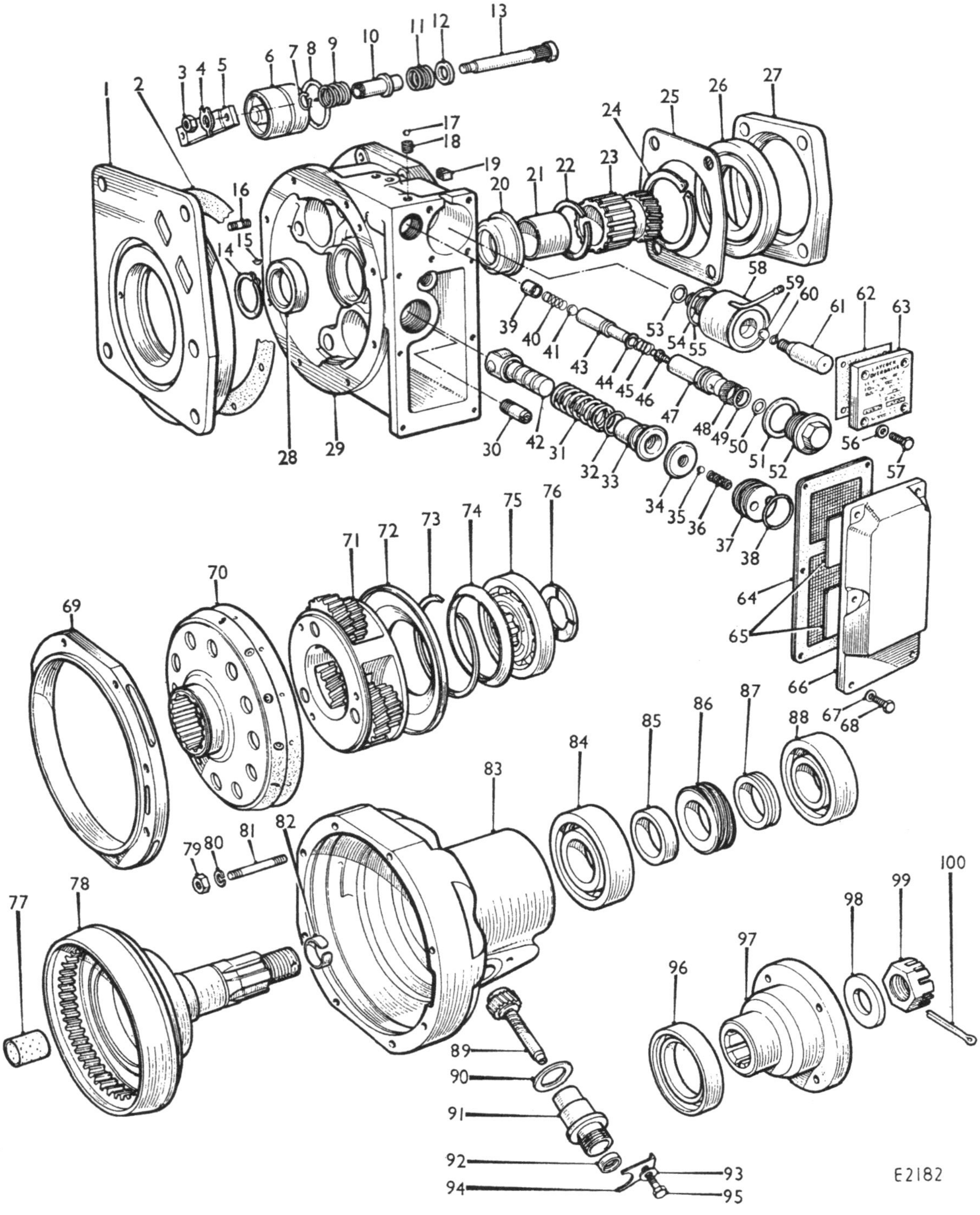

## KEY TO THE OVERDRIVE COMPONENTS (TYPE L.H.)

| *No.* | *Description* |
|---|---|
| 1. | Adaptor plate. |
| 2. | Gasket. |
| 3. | Nut. |
| 4. | Tab washer. |
| 5. | Bridge-piece. |
| 6. | Operating piston. |
| 7. | Circlip. |
| 8. | 'O' ring. |
| 9. | Spring. |
| 10. | Thrust rod. |
| 11. | Spring. |
| 12. | Washer. |
| 13. | Thrust housing pin. |
| 14. | Circlip. |
| 15. | Key. |
| 16. | Stud. |
| 17. | Steel ball. |
| 18. | Plug. |
| 19. | Grommet. |
| 20. | Sun wheel thrust bush. |
| 21. | Sun wheel bush. |
| 22. | Circlip. |
| 23. | Sun wheel. |
| 24. | Circlip. |
| 25. | Retainer plate. |
| 26. | Thrust ball-race. |
| 27. | Thrust ring. |
| 28. | Pump cam. |
| 29. | Main casing. |
| 30. | Pump suction tube. |
| 31. | Spring. |
| 32. | 'O' ring. |
| 33. | Pump body. |
| 34. | Non-return valve seat. |
| 35. | Valve ball. |
| 36. | Valve spring. |
| 37. | Pump plug. |
| 38. | 'O' ring. |
| 39. | Low pressure valve plug. |
| 40. | Valve spring. |
| 41. | Valve ball. |
| 42. | Pump plunger. |
| 43. | Low pressure valve body. |
| 44. | Washer. |
| 45. | Relief valve spring. |
| 46. | Valve plunger. |
| 47. | Valve body. |
| 48. | Filter. |
| 49. | 'O' ring. |
| 50. | 'O' ring. |
| 41. | Washer. |
| 52. | Plug. |
| 53. | 'O' ring. |
| 54. | Solenoid valve body. |
| 55. | 'O' ring. |
| 56. | Washer. |
| 57. | Screw. |
| 58. | Solenoid coil. |
| 59. | Valve ball. |
| 60. | 'O' ring. |
| 61. | Solenoid plunger. |
| 62. | Gasket. |
| 63. | Solenoid cover. |
| 64. | Sump filter and gasket. |
| 65. | Filter magnets. |
| 66. | Sump. |
| 67. | Washer. |
| 68. | Screw. |
| 69. | Brake-ring. |
| 70. | Clutch sliding member. |
| 71. | Planet carrier assembly. |
| 72. | Oil catcher. |
| 73. | Circlip. |
| 74. | Oil thrower. |
| 75. | Uni-directional clutch. |
| 76. | Thrust washer. |
| 77. | Bus. |
| 78. | Annulus. |
| 79. | Nut. |
| 80. | Washer. |
| 81. | Stud. |
| 82. | Spring ring. |
| 83. | Rear casing. |
| 84. | Annulus front bearing. |
| 85. | Spacer. |
| 86. | Speedometer drive gear. |
| 87. | Selective spacer. |
| 88. | Annulus rear-bearing. |
| 89. | Speedometer driven gear. |
| 90. | Sealing washer. |
| 91. | Speedometer bearing. |
| 92. | Oil seal. |
| 93. | Retaining clip. |
| 94. | Washer. |
| 95. | Bolt. |
| 96. | Oil seal. |
| 97. | Drive flange. |
| 98. | Washer. |
| 99. | Nut. |
| 100. | Split pin. |

## Section Fc.1

### GENERAL DESCRIPTION

The Laycock Type L.H. Overdrive fitted between the gearbox and propeller shaft is a self-contained gear unit which provides a higher overall gear ratio than that given by the final drive.

The overdrive gears consist of a central sun wheel and three planet gears which mesh with an internally toothed annulus. Fitted inside the annulus is a uni-directional clutch. A sliding clutch-member is secured to the sun wheel and is free to move forward and rearward on the sun wheel splines. Attached to a ball bearing, secured on the sliding clutch by a circlip, is a static thrust ring. The thrust ring is actuated by two hydraulic pistons and returned by primary and secondary return springs.

An electrically operated solenoid valve, mechanical pump, relief valve, and low pressure valve comprise the main components of the hydraulic system.

#### Overdrive engaged

With the overdrive switch selected the solenoid is energized and the ball valve is held in the closed position by the solenoid rod.

The input shaft to the overdrive unit carries a cam which operates the overdrive oil pump. The pump draws

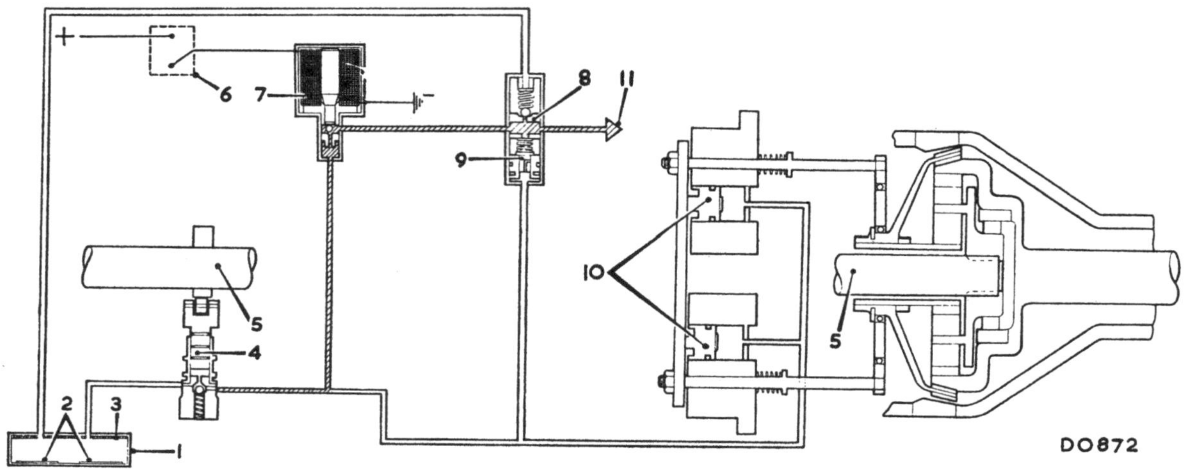


*Fig. Fc.1*

*Direct drive*

1. Sump.
2. Magnet filters.
3. Gauze filter.
4. Pump.
5. Third motion shaft (gearbox).
6. Control switch.
7. Solenoid operating valve.
8. Low pressure valve.
9. Relief valve.
10. Operating pistons.
11. Oil return to sump.

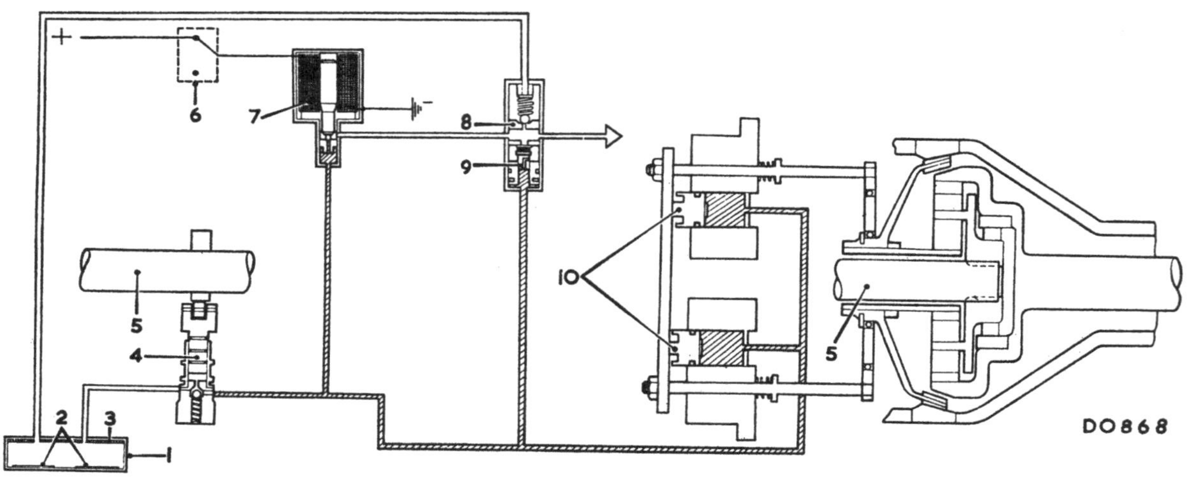


*Fig. Fc.2*

*Overdrive engaged*

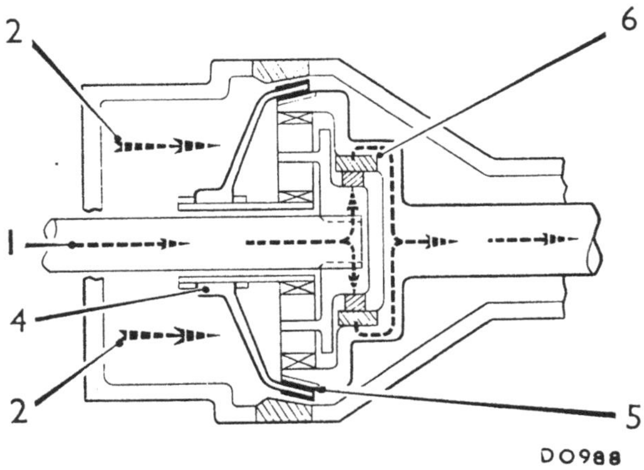


*Fig. Fc.3*

*Direct drive*

1. Third motion shaft.
2. Spring pressure.
4. Sun wheel.
5. Cone clutch.
6. Uni-directional clutch.

oil from the sump and the oil from the pump discharge is ducted to the two operating pistons of the clutch sliding member. A build-up of oil pressure operates the two pistons and moves the sliding clutch-member, its outer brake surface contacts the stationary brake ring and the complete sliding member and sun wheel cease to rotate.

At a predetermined spring pressure the relief valve operates and relieves any pressure, in excess of the pressure required to keep the pistons operative, into the low pressure lubricating system. Build-up of pressure in the low pressure lubricating system is relieved by operation of the low pressure valve and the oil relieved returns to the sump.

The planet gear carrier is splined to, and rotates with, the input shaft. The planet gears, forced to turn about their own axis by movement of the carrier, while in mesh with the stationary sun wheel, impart driving force to turn the annulus. Because of the gearing arrangement the annulus turns faster than the input shaft. The propeller shaft, coupled to a flange rigidly secured to the shaft of the annulus, revolves at the same speed as the annulus.

In overdrive the outer bearing surface of the uni-directional clutch inside the annulus, over-rides the rollers, cage, and clutch, attached to the slower-moving input shaft. In this condition the uni-directional clutch is in the unlocked or free condition.

The solenoid valve also acts as a safety valve. If the pressure becomes excessive the ball would be blown off its seat against the load of the solenoid.

**Overdrive disengaged**

**CAUTION.—IF OVERDRIVE DOES NOT DISENGAGE, DO NOT REVERSE THE CAR OTHERWISE EXTENSIVE DAMAGE MAY RESULT.**

With the overdrive switch in the 'off' position the operating solenoid is de-energized and the valve ball is free to move away from the valve seat. Oil from the pump discharge lifts the valve from its seat, flows to the low pressure valve, and is used for lubrication. This action relieves the oil pressure maintaining the pistons operative. The return springs act to close the pistons and the oil returning from the piston chambers is forced to mix with the pump flow, thus causing a restriction and damping the return movement of the sliding member. Action of the secondary return springs force the clutch sliding member rearward and its inner brake surface contacts the annulus brake ring. The sliding member and annulus commence to revolve in unison. As the sun wheel is splined to the sliding clutch member the complete gear train is locked.

Thrust from the input shaft locks the uni-directional clutch against the outer bearing surface, inside the annulus, and direct drive is applied to the propeller shaft through the annulus extension on which the propeller shaft coupling flange is mounted.

During over-run and reverse, additional load is imparted to the clutch sliding member by the helix thrust of the sun wheel, thus helping to retain direct drive.

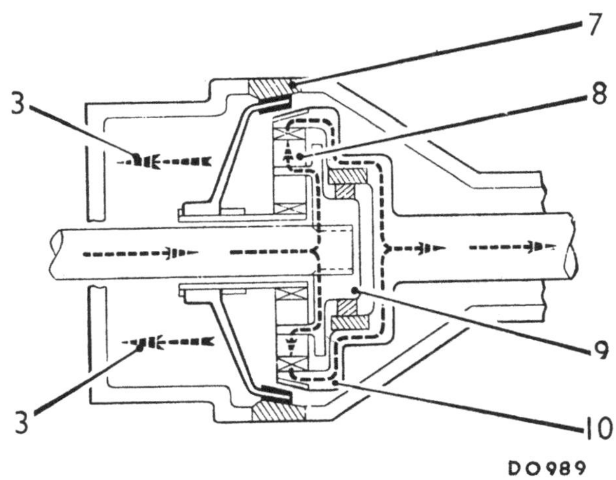


*Fig. Fc. 4*

*Overdrive engaged*

3. Hydraulic pressure.
7. Brake ring.
8. Planet wheel.
9. Planet carrier.
10. Annulus.

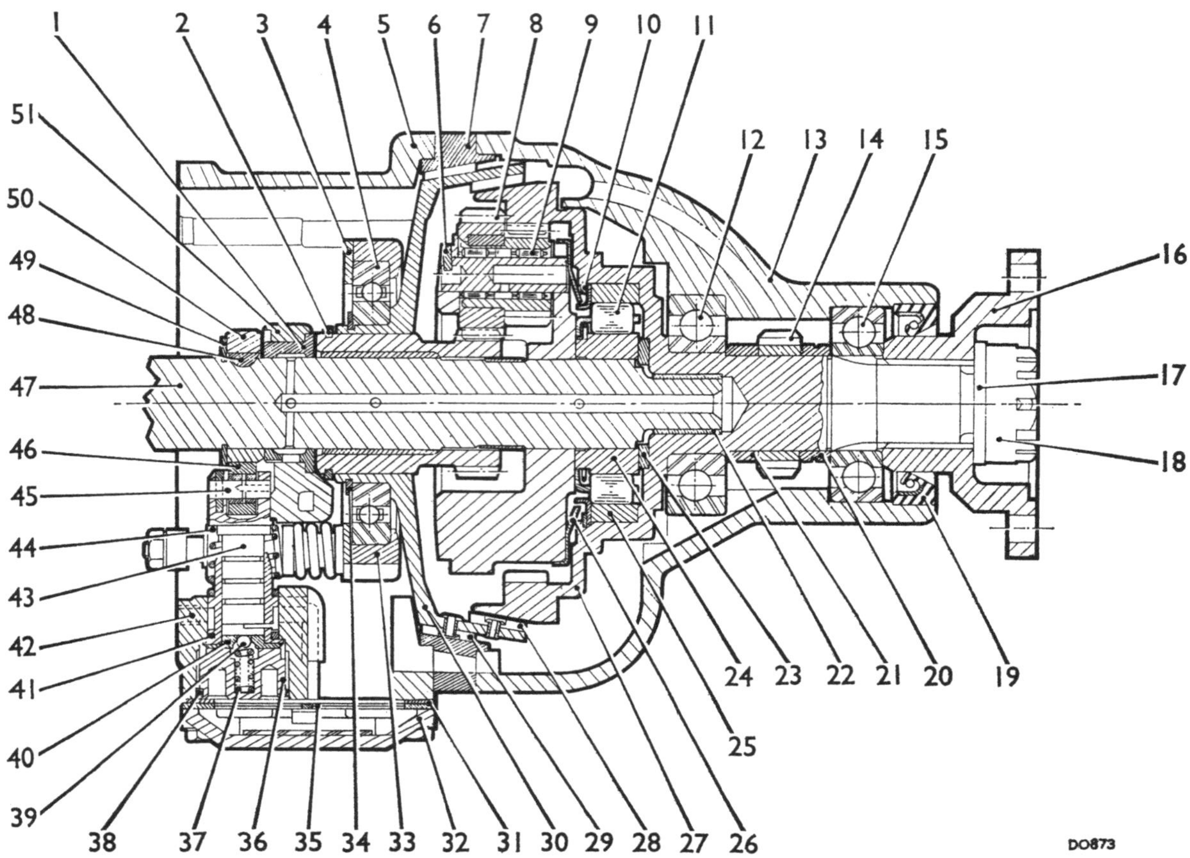


*Fig. Fc.5*

*A section through the overdrive unit*

1. Sun wheel thrust bush.
2. Snap ring for sun wheel.
3. Bearing retainer plate.
4. Thrust bearing.
5. Main casing.
6. Grooved pin.
7. Brake ring.
8. Planet wheel.
9. Planet bearing.
10. Circlip.
11. Rollers for uni-directional clutch.
12. Annulus front bearing.
13. Rear casing.
14. Speedometer driving gear.
15. Annulus rear bearing.
16. Drive flange.
17. Washer.
18. Nut.
19. Oil seal.
20. Selective spacer.
21. Spacer.
22. Bush.
23. Thrust washer
24. Uni-directional clutch, inner member.
25. Uni-directional clutch, outer member.
26. Oil thrower.
27. Annulus.
28. Inner lining of clutch sliding member.
29. Outer lining of clutch sliding member.
30. Clutch sliding member.
31. Sump filter and gasket assembly.
32. Sump.
33. Thrust bearing housing.
34. Circlip.
35. Filter.
36. Pump plug.
37. Non-return valve spring.
38. 'O' ring.
39. Non-return valve ball.
40. Non-return valve seat.
41. Pump body.
42. Main case housing for pump.
43. Pump plunger.
44. Spring for pump plunger.
45. Pin for pump roller.
46. Pump roller.
47. Gearbox third motion shaft.
48. Key for pump cam.
59. Circlip.
50. Pump cam.
51. Main case housing for sun wheel and oil transfer bush.

## Section Fc.2

### FAULT DIAGNOSIS

**OVERDRIVE DOES NOT ENGAGE**

| | |
|---|---|
| (1) Insufficient oil in gearbox .. .. .. .. | Top up to correct level. |
| (2) Electrical system fault .. .. .. .. | Check and rectify fault; check operation of solenoid. |
| (3) Low hydraulic pressure: | |
| (*a*) Pump non-return valve not seating .. .. | Remove and check the valve; clean or renew. |
| (*b*) Solenoid ball valve not seating .. .. | Remove and check the valve; clean or renew. |
| (*c*) Pump filter choked .. .. .. .. | Remove and clean filter. |
| (4) Damaged parts within unit .. .. .. .. | Remove, dismantle, and inspect unit. |

**OVERDRIVE DOES NOT DISENGAGE**

**NOTE.—If the overdrive does not disengage, do not reverse the car as extensive damage may result.**

| | |
|---|---|
| (5) Electrical control system fault .. .. .. | Check and rectify fault. |
| (6) Sticking clutch .. .. .. .. .. | On a new unit this may be due to insufficient bedding-in of the clutch. The clutch can usually be freed by giving the brake-ring, accessible from beneath the car, several sharp blows with a hide mallet. |
| (7) Damaged parts within unit .. .. .. .. | Remove, dismantle, and inspect unit. |

**OVERDRIVE SLIPS WHEN ENGAGED**

| | |
|---|---|
| (8) Insufficient oil in gearbox .. .. .. .. | Top up to correct level. |
| (9) Low hydraulic pressure: | |
| (*a*) Pump non-return valve not seating .. .. | Remove and check the valve; clean or renew. |
| (*b*) Relief valve not seating .. .. .. | Remove and check the valve; clean or renew. |
| (*c*) Solenoid valve not seating .. .. .. | Remove and check the valve; clean or renew. |
| (*d*) Partially blocked pump or relief valve filter .. | Remove and clean filters. |
| (10) Worn or glazed clutch lining .. .. .. | Remove, dismantle, and replace faulty parts. |

**SLIP IN REVERSE OR FREE WHEEL ON OVER-RUN**

| | |
|---|---|
| (11) Worn or glazed clutch lining .. .. .. | Remove, dismantle, and replace faulty parts. |
| (12) Broken circlip on sun wheel .. .. .. | Remove, dismantle, and replace faulty parts. |

## Section Fc.3

### HYDRAULIC PRESSURE TEST

(1) Check the gearbox oil level, and top up as necessary.
(2) Raise and support the rear of the car so that the rear wheels are clear of the ground.
(3) Securely chock the front wheels.
(4) Remove the relief valve plug and its sealing washer.
(5) Fit adaptor 18G 251 E, using the relief valve plug sealing washer, into the relief valve orifice.
(6) Connect pressure gauge 18G 251.
(7) Start the engine, engage top gear and select overdrive.
(8) Set the speed to a speedometer reading of 30 m.p.h. (48 km.p.h.) and note the pressure gauge reading; the correct reading is 400 to 420 lb./sq. in. (28 to 29·5 kg./cm.²).
**NOTE.—No pressure reading will be recorded with the overdrive disengaged.**

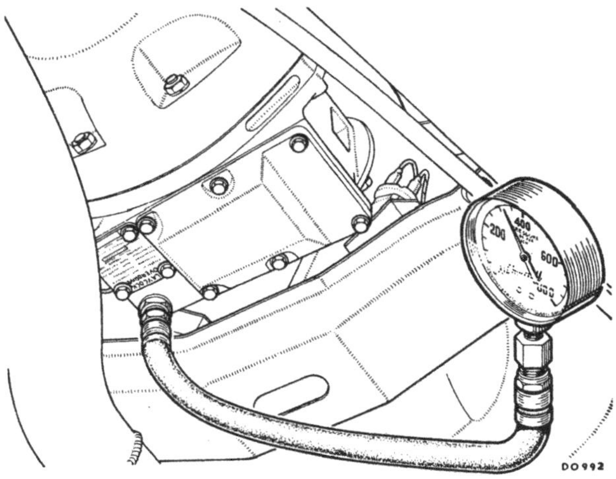


*Fig. Fc.6*

*Pressure gauge* 18G 251 *and adaptor* 18G 251 *E connected to check the hydraulic pressure*

## Section Fc.4

### LUBRICATION

**IMPORTANT.—Anti-friction additives must not be used in the gearbox and overdrive.**

**Draining**

(1) Remove the gearbox drain plug to drain the oil from the gearbox and overdrive unit.

**Filling**

(2) Remove the gearbox dipstick and add the correct quantity (see **'GENERAL DATA'**) of one of the recommended lubricants.
(3) Run the car for a short distance, allow to stand for a few minutes and check the level with the dipstick.

## Section Fc.5

### FILTERS

**Sump**

(1) Drain the gearbox and overdrive unit.
(2) Clean the sump cover and its surroundings.
(3) Unscrew the sump cover securing screws and remove the cover and sump filter.
(4) Clean all metallic particles from the two magnets fitted inside the cover; wash the cover and filter in petrol (gasoline).
(5) Refit the filter and sump cover.
(6) Refill the gearbox and overdrive unit with the correct quantity (see **'GENERAL DATA'**) of one of the recommended lubricants.

**Relief valve**

(7) Drain the gearbox and overdrive unit.
(8) Remove the relief valve plug and sealing washer.
(9) Withdraw the relief valve body and remove the filter.
(10) Clean the filter in petrol (gasoline).
(11) Refit the filter to the valve body.
(12) Refit the valve body assembly, sealing washer, and plug.
(13) Refill the gearbox and overdrive unit with a recommended lubricant.

## Section Fc.6

### RELIEF AND LOW PRESSURE VALVE

**Removing**

(1) Drain the gearbox and overdrive unit.
(2) Remove the relief valve plug and sealing washer.
(3) Withdraw the relief valve assembly.

**Dismantling**

(4) Remove the filter, spacer tube, low pressure valve assembly, and relief valve spring.
(5) Remove the relief valve plunger.

**Inspection**

(6) Examine the relief valve plunger and seat for pitting, scoring, and wear. Renew worn or damage parts as necessary.
(7) Inspect the relief valve body 'O' rings for signs of deterioration, and renew if necessary.
(8) Check the relief valve spring for signs of collapse or weakness against the figures given in **'GENERAL DATA'.**
(9) Test the low pressure valve for correct setting. The complete valve must be renewed if faulty.

**Reassembling**

(10) Reverse the dismantling procedure in (4) and (5).

**Refitting**

(11) Reverse the removing procedure in (2) and (3), noting that the spacer tube must be fitted with its slotted end farthest from the filter with the slots lining up respectively with the oil outlet hole and locating stud.

## Section Fc.7

### SOLENOID VALVE

**Removing**

(1) Drain the gearbox and overdrive unit.
(2) Unscrew the four screws securing the solenoid cover (name plate) and remove the cover and gasket.
(3) Remove the solenoid and valve assembly by carefully pulling on the solenoid lead.

**Dismantling**

(4) Withdraw the solenoid rod and operating valve assembly from the solenoid housing.

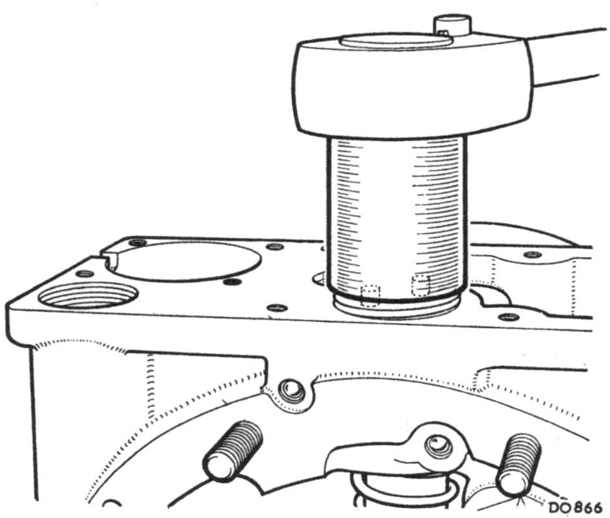


*Fig. Fc.7*
*Removing the pump retaining plug using tool* 18G 1118

(5) Press the solenoid coil and base cap from the housing.
(6) Remove the operating valve plunger and ball by shaking from the solenoid rod.

**Inspection**

(7) Examine the valve ball and seat for pitting and scoring and renew damaged parts; the ball may be reseated by light tapping onto the seat using a suitable drift.
(8) Inspect the 'O' ring seals for signs of deterioration, and renew as necessary.

**Reassembling**

(9) Reverse the dismantling procedure in (4) to (6).

**Refitting**

(10) Reverse the removing procedure in (2) and (3), ensuring that the lead grommet is pressed fully into its slot, and using a new gasket under the solenoid cover.

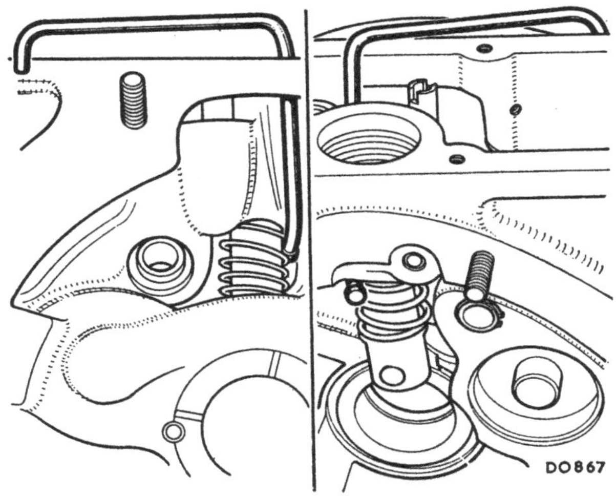


*Fig. Fc.8*
*Using tool* 18G 1117 *to retain the pump components*

## Section Fc.8

### PUMP AND NON-RETURN VALVE

**Removing**

(1) Drain the gearbox and overdrive unit.
(2) Remove the sump filter—Section Fc.5, (2) and (3).
(3) Using tool 18G 1118, unscrew the pump retaining plug.
(4) Remove the non-return valve spring and ball.
(5) Remove the pump body, the pump plunger spring and plunger.

**Dismantling**

(6) Using a suitable drift, separate the non-return valve seat from the pump body taking care not to damage the bore of the pump body.

**Inspection**

(7) Examine the 'O' ring seals for signs of deterioration, and renew where necessary.
(8) Examine the non-return valve ball and seat for scoring and pitting, and renew damaged parts; the ball may be reseated by lightly tapping onto its seat using a suitable drift.

**Reassembling**

(9) Carefully refit the non-return valve seat to the pump body.

**Refitting**

(10) Insert the pump plunger into the casing, ensuring that the flat side of the plunger is towards the rear of the overdrive unit and retain in position with tool 18G 1117.
(11) Reverse the removing procedure in (2) to (5).

## Section Fc.9

### OVERDRIVE ASSEMBLY

**Removing**

*Overdrive unit*

(1) Remove the engine and gearbox from the car as described in Section A.
(2) Remove the gearbox drain plug and drain the oil from the gearbox and overdrive.
(3) Unscrew the six bolts securing the remote control assembly to the gearbox and remove the remote control.
(4) Remove the eight nuts from the studs securing the overdrive unit to the gearbox adaptor.
(5) Withdraw the overdrive unit from the gearbox adaptor.

*Overdrive adaptor*

(6) Remove the overdrive unit as in (1) to (5).
(7) Slide the overdrive pump driving cam from the gearbox shaft and remove the cam locking ball from its pocket in the shaft.
(8) Using tool 18G 1004, remove the pump cam circlip from the shaft.
(9) Withdraw the selector interlocking arm and plate assembly.
(10) Unscrew the eight nuts securing the adaptor to the gearbox and withdraw the adaptor.

**Dismantling**

(11) Remove the solenoid cover and gasket.
(12) Withdraw the solenoid and plunger.
(13) Remove the relief valve assembly.
(14) Remove the sump and filter.
(15) Using tool 18G 1118, unscrew the pump retaining plug and remove the pump components.
(16) Remove the speedometer drive retaining clip, drive gear, and sealing washer.

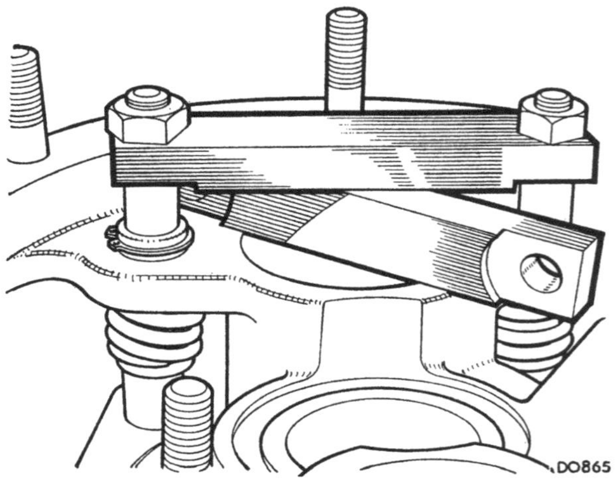


*Fig. Fc.9*

*Using the bridge-pieces to remove the thrust rods*

(17) Unscrew the six rear retaining stud nuts and the two front nuts securing the rear casing to the main casing assembly.

(18) Separate the rear casing from the main casing and remove the planet gear assembly.

(19) Unlock and remove the nuts securing the bridge pieces and pistons in the main casing.

(20) Remove the bridge piece, pistons, clutch sliding member, brake-ring and sun wheel.

(21) Remove the four clutch springs and selective washers.

(22) Lever the sun wheel split-ring from the retaining groove and push out the sun wheel.

(23) Lift off the bearing retaining plate.

(24) Using tool 18G 257, remove the bearing circlip.

(25) Drift the clutch sliding member from the bearing.

(26) Press the bearing from its housing.

(27) Fit the four studs in the thrust ring and insert the studs into the thrust rods from the rear of the main casing.

(28) Lay one bridge-piece on its side across one of the piston chambers and fit the remaining bridge-piece, inverted to its normal assembly position, on the two adjacent thrust ring studs and secure with two nuts.

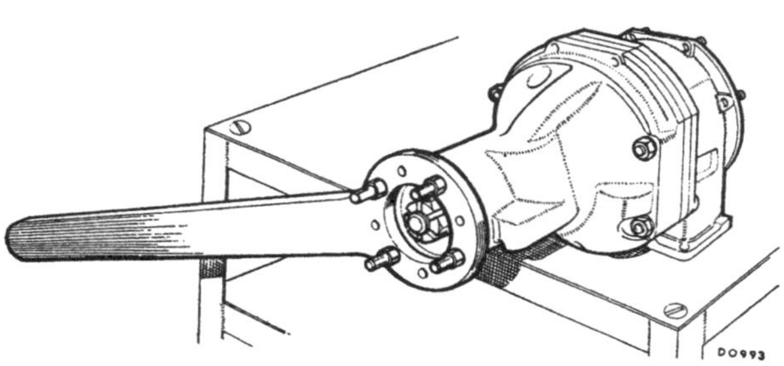

*Fig. Fc.10*

*Using tool 18G 34A to prevent the drive flange from rotating*

(29) Tighten both nuts evenly until the springs are compressed.

(30) Remove the circlips securing the thrust rods and springs.

(31) Slacken the two nuts evenly until the spring pressure is relieved and remove the nuts.

(32) Repeat the operations in (28) to (31) for the remaining thrust rods.

(33) Unlock the flange nut, using tool 18G 34A to prevent the shaft turning; remove the nut and washer and withdraw the flange.

(34) Press the annulus forward out of the rear casing complete with the front bearing. If the bearing remains in the housing, carefully drift it out with a drift registering on the inner bearing track.

(35) Remove the spacer, speedometer driving gear and selective spacer, noting the position of the selective spacer which is identified by a groove cut in its periphery.

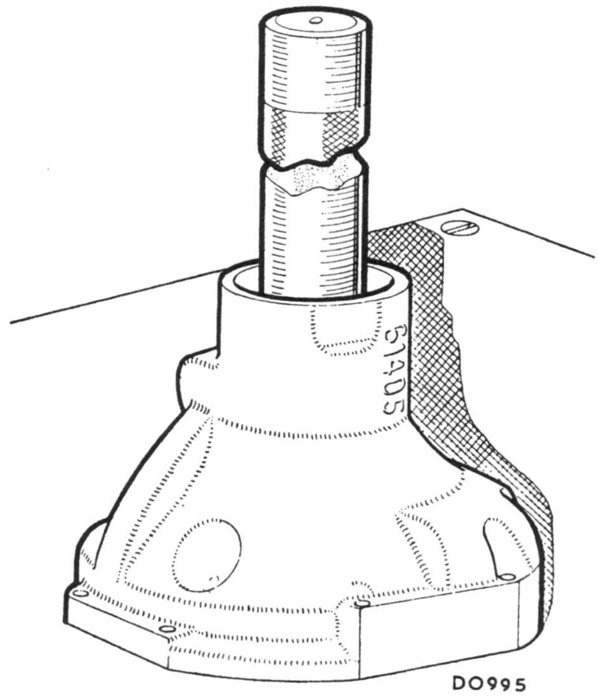


*Fig. Fc.11*

*Fitting the rear bearing using tool 18G 186*

(36) Press out the rear bearing and oil seal.

(37) Remove the front bearing from the annulus if it has been retaining to the shaft.

(38) Remove the split ring, oil thrower ring, and uni-directional clutch, collecting the rollers as they come free.

(39) Remove the thrust washer.

**Inspection**

(40) Thoroughly clean the components in a cleaning solvent and dry them using lint-free or nylon material or compressed air.

(41) Thoroughly examine each component for damage and excessive wear.

(42) Examine the bearings for pitting or scoring of the tracks and balls.

(43) Check the springs for weakness, collapse or distortion against the following:

(44) Inspect the pistons and bores for signs of scoring.

(45) Examine the rubber components for signs of deterioration.

(46) Inspect the clutch for worn or charred linings and loose rivets. If the clutch is defective a complete new clutch sliding member must be fitted.

(47) Examine the shaft splines for signs of wear and damage.

(48) Check the condition of the sun wheel bush. If the bush is defective a complete new sun wheel assembly must be fitted.

(49) Examine the selective washers for wear. If worn, a new set (four) of washers must be fitted.

**Reassembling**

(50) Press the front bearing into its housing in the rear casing, ensuring that the outer track of the bearing is firmly against the shoulder of the casing.

(51) Press the annulus into the front bearing.

(52) Fit the spacer, speedometer driving gear, and selective washer.

(53) If the rear casing, annulus, speedometer gear, or spacer have been renewed, proceed as follows:

(*a*) Using dial gauge 18G 191, take a reading from the rear face of the selective washer (Fig. Fc.12).

(*b*) Take another reading from the shoulder of the rear bearing housing.

(*c*) The dimension in (*a*) should be ·010 in. (·254 mm.) $\pm^{0.000}_{0.005}$ in. (·127 mm.) greater than the dimension taken in (*b*). If the correct reading is not obtained, check that the front bearing and annulus are firmly in contact with the shoulders of their respective housings.

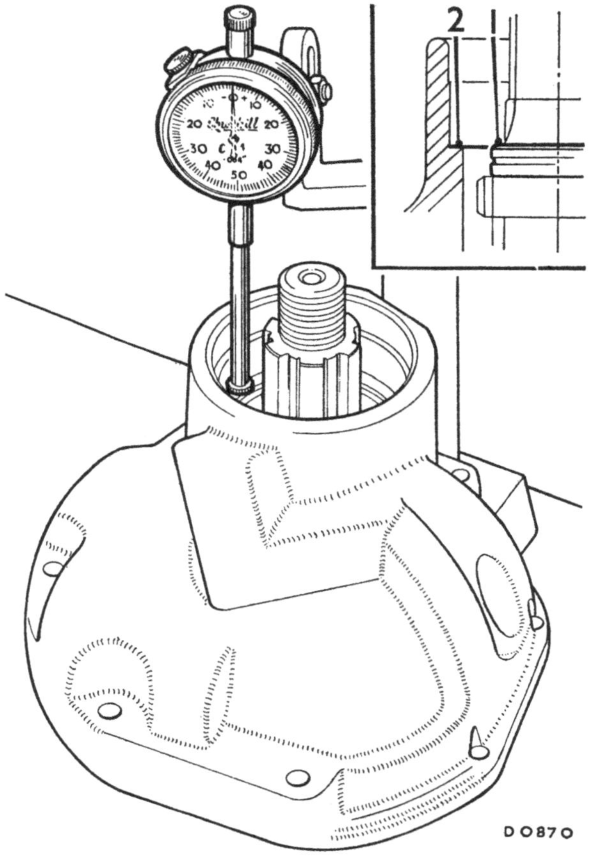


*Fig. Fc.*12

*Using dial gauge* 18G 191 *to take measurements for the selective washer. Inset shows:*

1. Rear face of the selective washer.
2. Shoulder of the bearing housing.

| Spring | No. of coils | Wire diameter | Free length | Fitted length | Load at fitted length |
|---|---|---|---|---|---|
| Clutch engagement .. | 3¾ | ·098 in. (2·5 mm.) | ·78 in. (19·8 mm.) | ·52 in. (13·2 mm.) | 65 lb. (29·5 kg.) |
| Clutch release .. .. | 4½ | ·128 in. (3·3 mm.) | ·85 in. (21·6 mm.) | ·625 in. (15·9 mm.) | 150 lb. (68 kg.) |
| Relief valve .. .. | 8 | ·092 in. (2·3 mm.) | 1·182 in. (30 mm.) | ·995 in. (25·3 mm.) | 38·35 lb. (17·5 kg.) |
| Pump .. .. .. | 7 | ·064 in. (1·6 mm.) | 1·75 in. (44·5 mm.) | ·896 in. (22·8 mm.) | 10·99 lb. (5 kg.) |

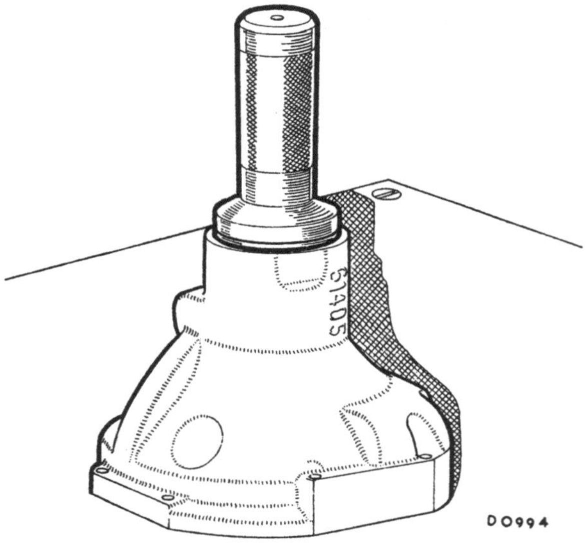


*Fig. Fc.*13

*Fitting the rear oil seal using tool* 18*G* 177

(*d*) If the correct reading is still unobtainable, select a new selective washer to suit. Washers are available in the following thicknesses:

| | |
|---|---|
| ·36 in. (9·1 mm.) | ·380 in. (9·7 mm.) |
| ·365 in. (9·3 mm.) | ·385 in. (9·8 mm.) |
| ·370 in. (9·4 mm.) | ·39 in. (9·9 mm.) |
| ·375 in. (9·5 mm.) | |

(54) Using tool 18G 186, fit the rear bearing.

(55) Smear the outer case of the new rear oil seal with grease, and fit the seal fully into its housing using tool 18G 177.

(56) Fit the rear flange, tighten the nut to the torque figure given in **'GENERAL DATA'** and lock the nut using a new split pin.

(57) Fit the thrust ring.

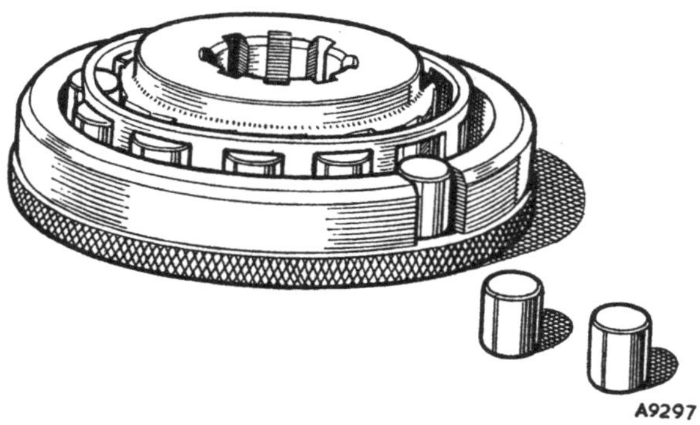


*Fig. Fc.*14

*Assembling the uni-directional clutch using tool* 18*G* 178

(58) Fit the spring of the uni-directional clutch to the inner clutch member, and the inner member and spring into the clutch cage, then insert the assembly into tool 18G 178 and fit the clutch rollers.

(59) Fit the uni-directional clutch assembly and remove tool 18G 178.

(60) Fit the oil thrower and split ring and check that the clutch rotates in an anti-clockwise direction.

(61) Rotate each planet wheel until the centre punch mark is radially outwards and in line with the wheel shaft locking pin.

(62) Insert the sun wheel into the planet carrier and re-check the alignment of the planet wheel markings.

(63) Install the assembled planet carried into the annulus.

(64) Remove the sun wheel.

(65) Using tool 18G 185, align the splines in the planet carrier and uni-directional clutch.

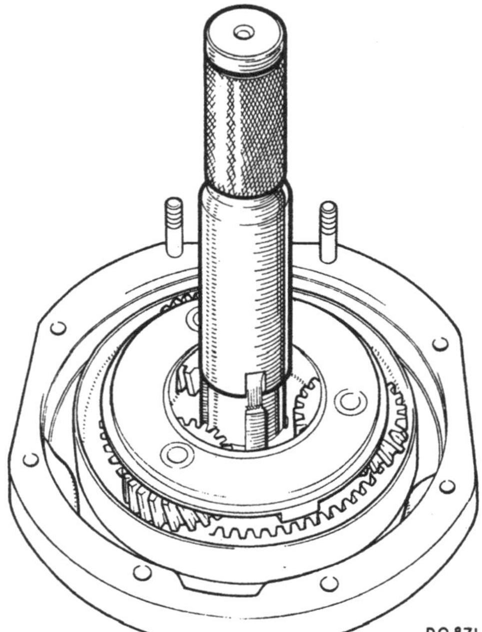


*Fig. Fc.*15

*Using tool* 18*G* 185 *to align the splines*

(66) Refit the four clutch release thrust rods using the dismantling method in (27) to (32); fit new circlips if required.

(67) Refit the pistons.

(68) Reassemble the clutch sliding member and components by reversing the procedure in (21) to (26), ensuring that the four selective washers fitted are a complete set.

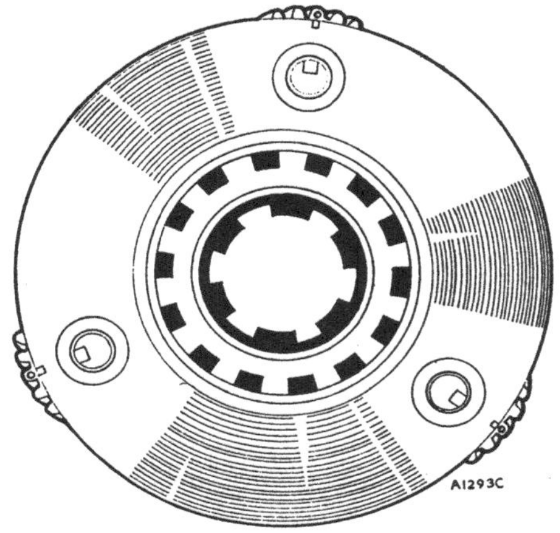


*Fig. Fc.16*

*The planet wheels correctly positioned for fitting the sun wheel and assembling in the annulus*

(69) Fit the sun wheel, brake ring and clutch sliding member reversing the procedure in (19) to (21) and noting that the jointing faces of the brake ring must be smeared with a liquid jointing compound and the nuts locked using new tab washers.

(70) Smear the jointing faces of the two casings with liquid jointing compound and assemble the casings ensuring that the sun wheel meshes with the planet gears.

(71) Fit and tighten the retaining nuts and spring washers.

(72) Reverse the dismantling procedure in (11) to (16).

**Refitting**

*Overdrive adaptor*

(73) Reverse the removing procedure in (6) to (10).

*Overdrive unit*

(74) Reverse the removing procedure in (1) to (5).

# SECTION G

## THE PROPELLER SHAFT

THE PROPELLER SHAFT COMPONENTS

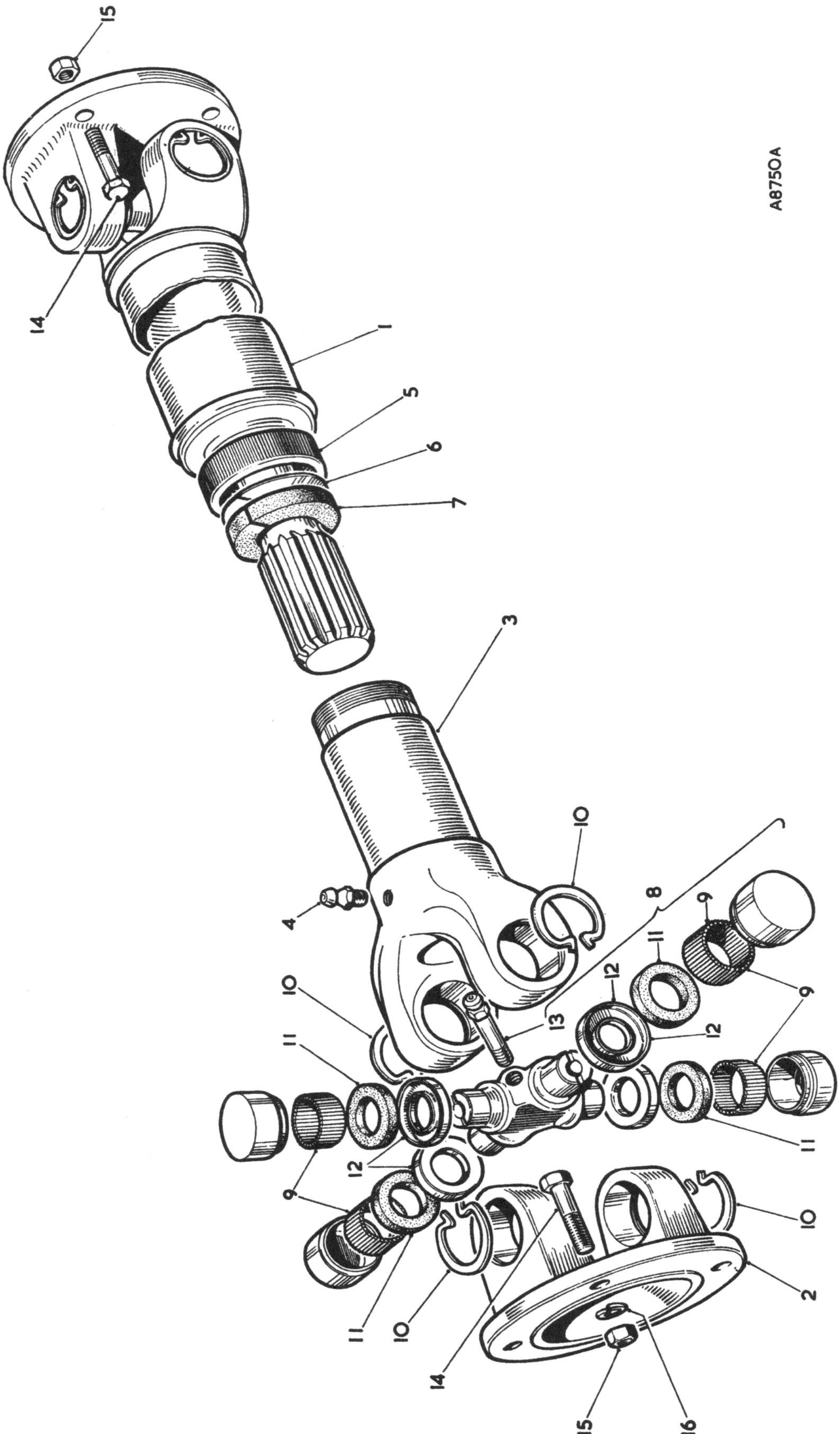

## KEY TO THE PROPELLER SHAFT COMPONENTS

| *No.* | *Description* | *No.* | *Description* |
|---|---|---|---|
| 1. | Shaft assembly—propeller. | 9. | Bearing assembly—needle. |
| 2. | Flange yoke. | 10. | Circlip. |
| 3. | Sleeve assembly—yoke. | 11. | Gasket. |
| 4. | Lubricator. | 12. | Retainer—gasket. |
| 5. | Cap—dust. | 13. | Lubricator—journal. |
| 6. | Washer—dust cap (steel). | 14. | Bolt—shaft to pinion flange (front and rear). |
| 7. | Washer—dust cap (cork). | 15. | Nut for bolt. |
| 8. | Journal assembly. | 16. | Washer—spring—for front bolt. |

## GENERAL DESCRIPTION

The propeller shaft and universal joints are of the Hardy Spicer type with needle-roller bearings in the universal joints.

The rear end of the shaft is flanged and carries the rear universal joint flange yoke. The front end of the shaft is splined and engages a sleeve and yoke assembly. In the assembled condition a dust cap, steel washer, and cork washer seal the end of the sleeve and the sliding joint.

Each universal joint is made up of a spider and four gaskets, retainers, and needle bearing assemblies. Each needle bearing assembly is retained in its yoke by a circlip.

The yoke flanges are secured to the pinion and gearbox flanges respectively with eight bolts, spring washers, and Aerotight or Nyloc nuts.

## Section G.1

### LUBRICATION

A nipple is provided for lubricating the sliding yoke at the front end of the propeller shaft. On non-sealed type propeller shafts lubrication nipples are also provided on the front and rear universal joints. Three or four strokes of a grease gun are required when lubricating.

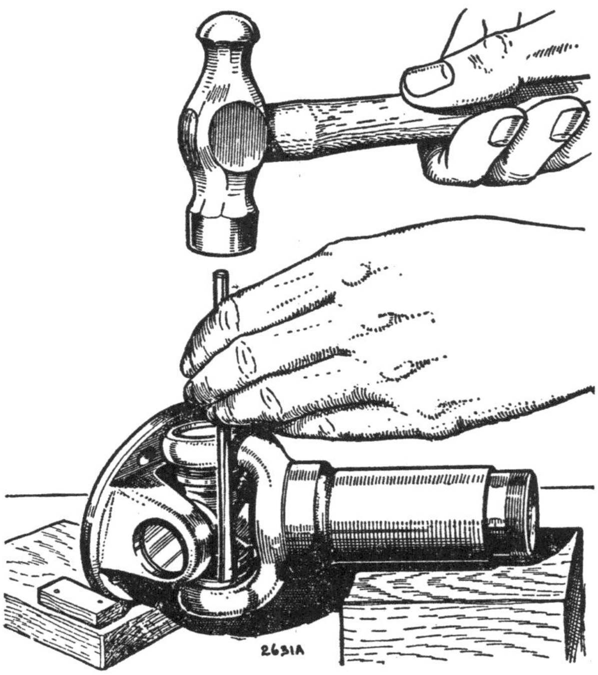


*Fig. G.1*

*When dismantling a universal joint the bearings may be tapped out with a small-diameter rod from the inside as shown. Take care not to damage the roller races*

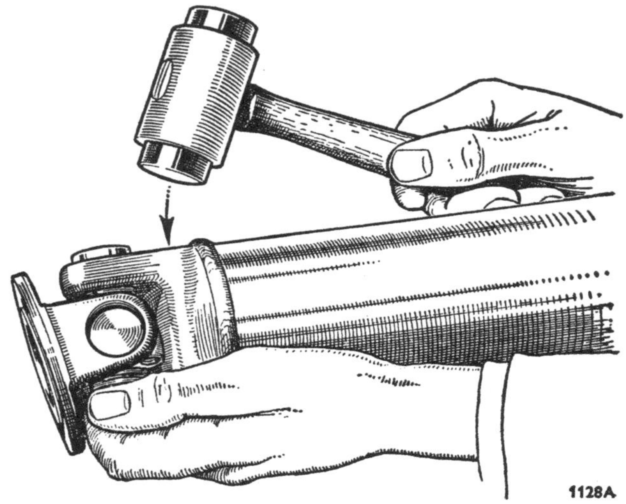


*Fig. G.2*

*Where to apply light blows to the yoke after removing the retaining circlip*

## Section G.2

### REMOVING AND REPLACING THE PROPELLER SHAFT

To remove the propeller shaft proceed as follows:

(1) Mark the yoke flanges and the gearbox and rear axle flanges to assist in refitting them in their original positions. **This is most important.**

(2) Remove the nuts, washers, and bolts securing the flanges and lower the propeller shaft.

Replacement is a reversal of the removal sequence, but ensure that the joint faces of the flanges are perfectly smooth and clean and that they are correctly aligned with the gearbox and rear axle flanges.

## Section G.3

### DISMANTLING THE PROPELLER SHAFT

(1) Unscrew the dust cap from the sleeve and slide the sleeve off the shaft. Remove the steel washer and the cork washer.

(2) Remove all circlips from the universal joints. If a circlip appears to be tight in its groove lightly tap the end of the bearing race to relieve the pressure against the ring.

(3) Remove the lubricators from the journals and the sleeve.

(4) Hold the shaft yoke in one hand and tap the radius of the yoke with a hammer. The bearing should begin to emerge. Turn the yoke over and remove the bearing with the fingers.

(5) If necessary, tap the bearing race from inside with a small diameter bar taking care not to damage the bearing face.

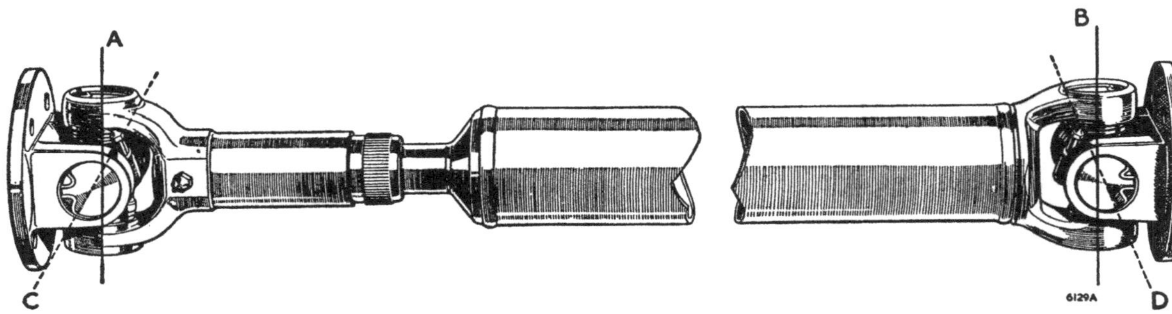


*Fig. G.3*

*When the splined shaft is assembled to the drive shaft it is essential to see that the forked yokes on both shafts have their axes parallel to each other. In other words, the yoke* (A) *must be in alignment with the yoke* (B), *and the flange yoke* (C) *must be in alignment with the flange yoke* (D)

(6) Hold the bearing in a vertical position and remove the race from the bottom side to avoid dropping the needle rollers.

(7) Repeat operations (4), (5), and (6) for the opposite bearing.

(8) Rest the exposed trunnions on wood blocks and tap the top lug of the flange yoke to remove the bearing race. Turn the yoke over and repeat the operation to remove the remaining bearing.

(9) Repeat operations (4) to (8) to remove the bearings from the sleeve and yoke assembly.

(10) Remove the gaskets and their retainers from the spider journals.

## Section G.4

### EXAMINING THE PROPELLER SHAFT

Thoroughly clean, wash, and dry all components prior to examination, paying particular attention to the lubrication passages.

(1) Check the splines on the sleeve and shaft for indentation.

(2) Examine the bearing races and journals for wear.

(3) Examine the holes in the yokes and flanges for ovality.

(4) Ensure that the bearing races are a light driving fit in their yokes.

(5) Check the flange yoke faces for burrs, cracks, or fractures.

All worn or unserviceable parts must be renewed.

## Section G.5

### ASSEMBLING THE PROPELLER SHAFT

(1) To provide a good seal, apply a thin coat of shellac to the gasket retaining shoulders on the spiders.

(2) Use a hollow drift to refit the retainers and then fit new gaskets.

(3) Smear the walls of the races with grease (page P.2), fit the needle rollers to the races and then fill the races with grease.

(4) Insert the spiders in the flange yokes and **ensure that the lubricating nipple bosses are facing away from the yoke flanges.**

(5) Using a soft drift $\frac{1}{32}$ in. (·79 mm.) smaller than the diameter of the races fit the bearings to the spider journals and the holes in the yokes.

(6) Place the dust cap, steel washer, and a new cork gasket over the spined portion of the shaft.

(7) Grease the inside of the sleeve, line up the arrows on the sleeve and splined portion of the shaft, and fit the sleeve to the shaft. Fit the washers and screw the dust cap to the sleeve.

(8) Wipe off any excess grease and fit the lubricators to the sleeve and spiders.

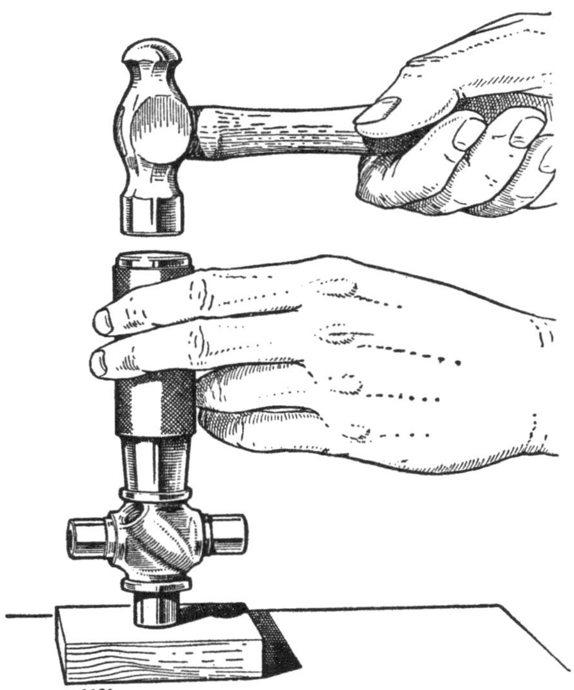


*Fig. G.4*

*When replacing the gasket retainer use should be made of a hollow drift to tap it into place without damage*

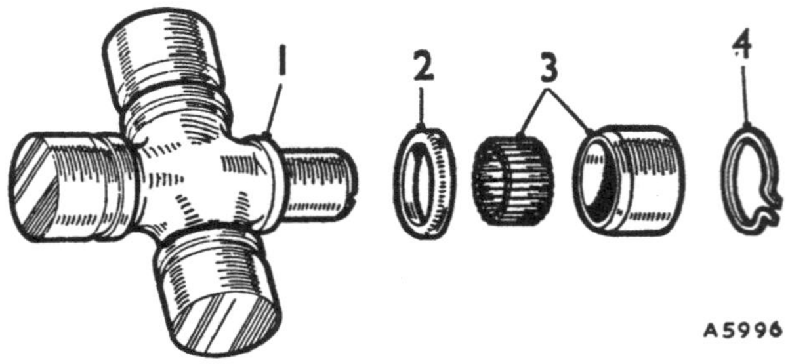


*Fig. G.5*

*A universal joint bearing—sealed type*

1. Journal spider.
2. Rubber seal.
3. Needle rollers and bearing.
4. Circlip.

## Section G.6

### PROPELLER SHAFT—SEALED TYPE

**Dismantling**

Carry out operations (1), (2), (4), (5), (6), (7), (8), and (9) of Section G.3.

**Inspection**

As detailed in Section G.4.

**Reassembling**

It is of extreme importance that the assembly of the journals be carried out under absolutely clean, dust-free conditions.

(1) Fill the reservoir holes in the journal spider with the recommended grease taking care to exclude all air pockets. Fill each bearing assembly with grease to a depth of $\frac{1}{8}$ in. (3 mm.).

(2) Fit new seals to the spider journals and insert the spider into the flange yoke, tilting it to engage in the yoke bores.

(3) Fit a bearing assembly into the yoke bore in the bottom position, and using a soft-nosed drift slightly smaller in diameter than the hole in the yoke, tap it into the yoke bore until it is possible to fit the circlip. Repeat this operation for the other three bearings starting opposite the bearing first fitted.

(4) After assembly, carefully remove all surplus grease with a soft cloth. If the bearing appears to bind, tap lightly with a wooden mallet; this will relieve any pressure of the bearing on the ends of the journals.

# SECTION H

## THE REAR AXLE AND SUSPENSION

### THREE-QUARTER FLOATING

## THE REAR AXLE COMPONENTS

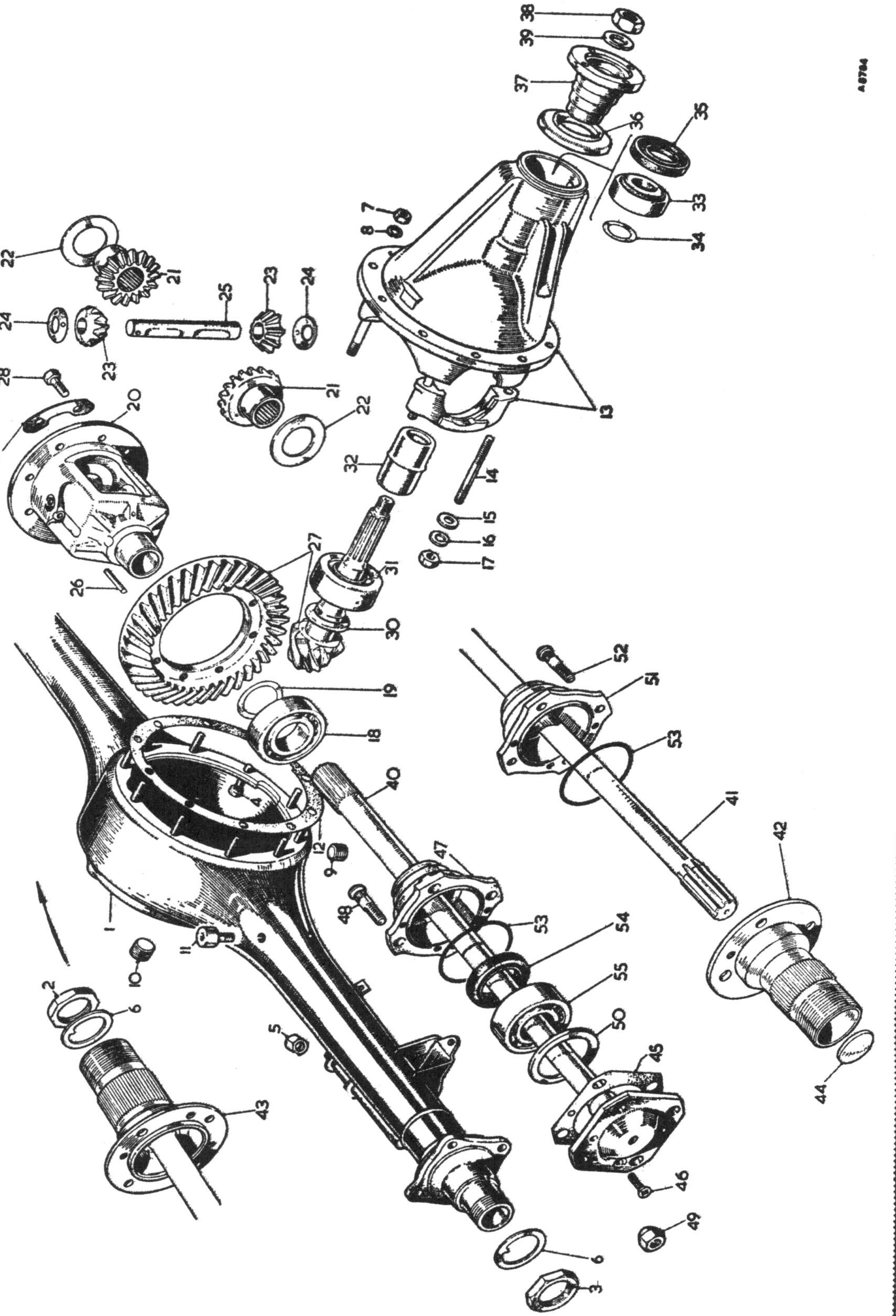

## KEY TO THE REAR AXLE COMPONENTS

| No. | Description |
|---|---|
| 1. | Case assembly. |
| 2. | Nut—bearing retaining—R.H.T. |
| 3. | Nut—bearing retaining—L.H.T. |
| 4. | Stud—gear carrier. |
| 5. | Nut—rebound spindle. |
| 6. | Washer—bearing retaining nut. |
| 7. | Nut—gear carrier to axle case. |
| 8. | Washer—spring—nut. |
| 9. | Plug—drain. |
| 10. | Plug—filler. |
| 11. | Breather assembly. |
| 12. | Joint—gear carrier to axle case. |
| 13. | Carrier assembly. |
| 14. | Stud—bearing cap. |
| 15. | Washer—plain—bearing cap. |
| 16. | Washer—spring—bearing cap. |
| 17. | Nut—stud. |
| 18. | Bearing—differential. |
| 19. | Washer—bearing—packing—·002 to ·010 in. (·051 to ·254 mm.). |
| 20. | Cage—differential. |
| 21. | Wheel—differential. |
| 22. | Washer—thrust—differential wheel. |
| 23. | Pinion—differential. |
| 24. | Washer—thrust—differential pinion. |
| 25. | Pin—pinion. |
| 26. | Peg—pinion pin. |
| 27. | Crown wheel and pinion—11/43. |
| 28. | Bolt—crown wheel to differential cage. |
| 29. | Lock washer—bolt. |
| 30. | Washer—pinion—thrust—·112 to ·126 in. (2·85 to 3·20 mm.). |
| 31. | Bearing—pinion—inner. |
| 32. | Spacer—bearing. |
| 33. | Bearing—pinion—outer. |
| 34. | Shim—outer bearing—·004 to ·030 in. (·102 to ·762 mm.). |
| 35. | Seal—oil. |
| 36. | Cover—dust. |
| 37. | Flange—universal joint. |
| 38. | Nut—pinion. |
| 39. | Washer—spring—nut. |
| 40. | Shaft—axle (disc wheels). |
| 41. | Shaft—axle (wire wheels). |
| 42. | Hub extension—R.H. (wire wheels). |
| 43. | Hub extension—L.H. (wire wheels). |
| 44. | Plug—welch—hub extension (wire wheels). |
| 45. | Joint—shaft to hub. |
| 46. | Screw—countersunk—shaft to hub. |
| 47. | Hub assembly (disc wheels). |
| 48. | Stud—wheel (disc wheels). |
| 49. | Nut—wheel stud (disc wheels). |
| 50. | Spacer—bearing (disc wheels). |
| 51. | Hub assembly (wire wheels). |
| 52. | Stud—wheel (wire wheels). |
| 53. | Ring—oil seal. |
| 54. | Seal—hub. |
| 55. | Bearing—hub. |

## GENERAL DESCRIPTION

The rear axle is of the three-quarter-floating type, incorporating hypoid final reduction gears. The axle shafts, pinion, and differential assemblies can be withdrawn without removing the axle from the vehicle. Wire wheels are available as optional equipment.

The rear axle wheel bearing outer races are located in the hubs; the inner races are mounted on the axle tube and secured by nuts and lock washers. Wheel studs in the hubs pass through the brake-drums and axle shaft driving flanges.

The differential and pinion shaft bearings are preloaded, the amount of preload being adjustable by shims. The position of the pinion in relation to the crown wheel is determined by a spacing washer. The backlash between the gears is adjustable by shims.

Suspension is by semi-elliptic leaf springs, rubber-mounted, and the shackles are fitted with rubber bushes of the flexing type.

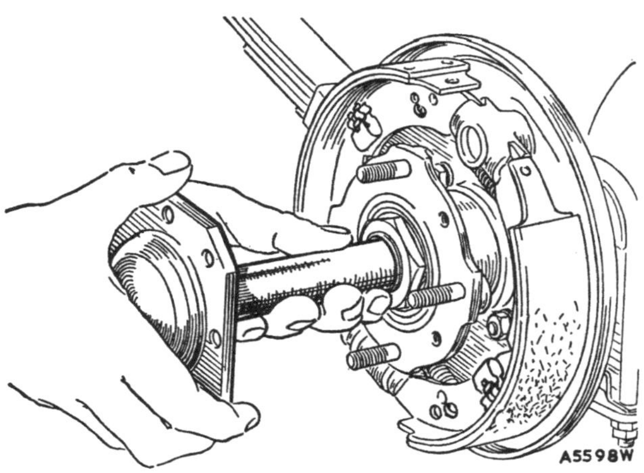


*Fig. H.1*
*Remove the brake drum and drive shaft flange locating screws and then withdraw the axle shaft*

## LUBRICATION

**Oil level**

Check the level, and top up if necessary. The filler plug is located on the rear of the axle and also serves as an oil level indicator. After topping up allow time for any surplus oil to run out. This is most important; if the axle is overfilled, the oil may leak through to the brake linings and lessen their efficiency.

**NOTE.—It is essential that only Hypoid oil be used in the rear axle.**

**Draining**

The most suitable time for draining is after a long journey whilst the oil is still warm. Clean the drain plug before it is replaced and tightened.

Refill the axle with fresh oil.

# Section H.1

### REMOVING AND REPLACING A BRAKE-DRUM AND AXLE SHAFT

Jack up the car and place blocks under the spring as close as possible to the axle.

Remove the wheel.

Release the hand brake.

**Disc wheels**

Unscrew and remove the two countersunk Phillips screws locating the drum and tap it from the hub. It may be necessary to slacken off the brake adjustment slightly if the shoes hold the drum.

Unscrew the countersunk Phillips locating screw in the axle shaft driving flange.

Withdraw the axle shaft by gripping the flange or carefully prising it with a screwdriver. Note the gasket and bearing spacer between the axle flange and the hub.

**Wire wheels**

Unscrew the four nuts securing the drum to the hub and tap the drum from the hub. It may be necessary to slacken off the brake adjustment to free the drum.

Unscrew the countersunk Phillips screw securing the hub extension flange to the hub. Withdraw the hub extension and axle shaft. The extension flange has two threaded holes so that the shaft can be pulled out if it is tight. Note the rubber ring oil seal between the hub extension flange and the hub.

To replace the shaft and drum reverse the above sequence of operations.

# Section H.2

### REMOVING AND REPLACING A HUB

Remove the drum and axle shaft as detailed in Section H.1.

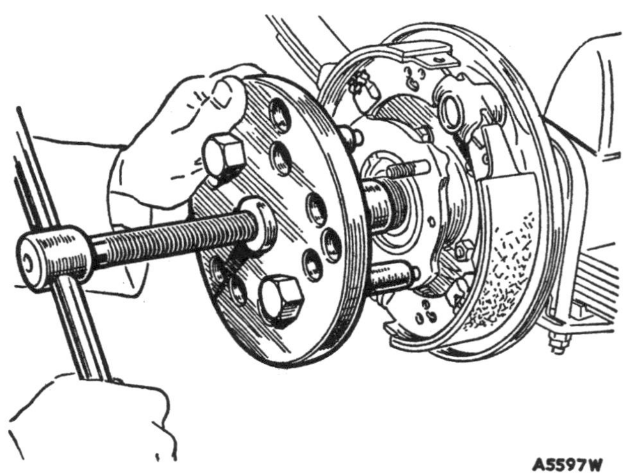


*Fig H.2*
*Use Service tool 18G 304 with adaptor 18G 304 A to withdraw the hub*

Remove the bearing spacer.

Knock back the tab of the locking washer and unscrew the nut with a suitable spanner.

Tilt the lock washer to disengage the key from the slot in the threaded portion of the axle casing; remove the washer.

The hub can then be withdrawn with a suitable puller such as Service tools 18G 304 and 18G 304 A. The bearing and oil seal will be withdrawn with the hub.

Before assembly, repack the hub bearings with grease.

The bearing is not adjustable and is replaced in one straightforward operation.

When reassembling it is essential that the outer face of the bearing spacer should protrude from ·001 in. (·025 mm.) to ·004 in. (·102 mm.) beyond the outer face of the hub when the bearing is pressed into position. This ensures that the bearing is gripped between the abutment shoulder in the hub and the driving flange of the axle shaft.

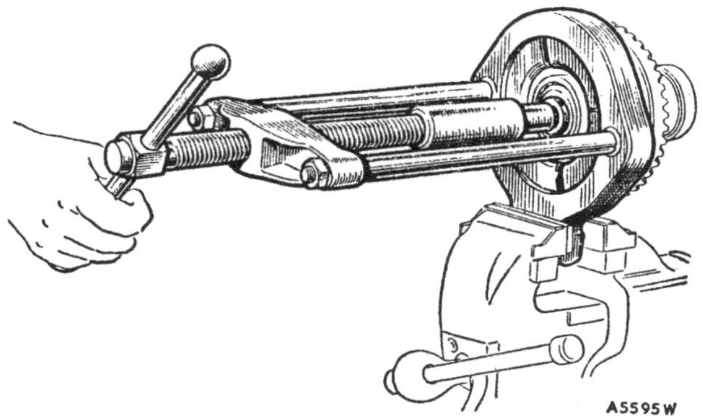


*Fig. H.4*
*Use Service tool 18G 47 C and adaptor 18G 47 T to remove the differential bearings*

Reconnect the propeller shaft, taking care to fit the two flanges with the locating marks in alignment.

## Section H.3

### RENEWING THE PINION OIL SEAL

Mark the propeller shaft and the pinion driving flanges so that they may be replaced in the same relative positions. Disconnect the propeller shaft.

Knock back the lock washer and unscrew the nut in the centre of the driving flange. Remove the nut and washer and withdraw the flange and pressed-on end cover from the pinion shaft.

Extract the oil seal from the casing.

Press a new oil seal into the casing with the edge of the sealing ring facing inwards.

Replace the driving flange end cover, taking care not to damage the edge of the oil seal. Tighten the nut with a torque wrench to a reading of 140 lb. ft. (19·34 kg. m.).

*Fig.* H.3
*Use Service tools* 18G 134 *and* 18G 134 *P to drift the hub into position*

## Section H.4

### REMOVING THE DIFFERENTIAL PINIONS

Drain the oil from the axle casing.

Remove the axle shafts as detailed in Section H.1.

Mark the propeller shaft and pinion shaft driving flanges so that they may be replaced in the same relative positions; unscrew the self-locking nuts and disconnect the joint.

Unscrew the 10 nuts securing the bevel pinion and gear carrier to the axle casing; withdraw the gear carrier complete with the pinion shaft and differential assembly.

Make sure that the differential bearing housing caps are marked so that they can be replaced in their original positions, then remove the four nuts and spring washers. Withdraw the bearing caps and differential assembly.

Tap out the dowel pin locating the differential pinion shaft. The diameter of the pin is $\frac{3}{16}$ in. (4·8 mm.) and it must be tapped out from the crown wheel side as the hole into which it fits has a slightly smaller diameter at the crown wheel end to prevent the pin from passing right through. It may be necessary to clean out the metal peened over the entry hole with a $\frac{3}{16}$ in. (4·8 mm.) drill in order to facilitate removal of the dowel pin. Drive out the differential pinion shaft. The pinions and thrust washers can then be removed from the cage.

(See Editor's notes at end of section.)

## Section H.5

### REPLACING THE DIFFERENTIAL PINIONS

Examine the pinions and thrust washers and renew as required.

Replace the pinions, thrust washers, and pinion shaft in the differential cage and insert the dowel pin. Peen over the entry hole.

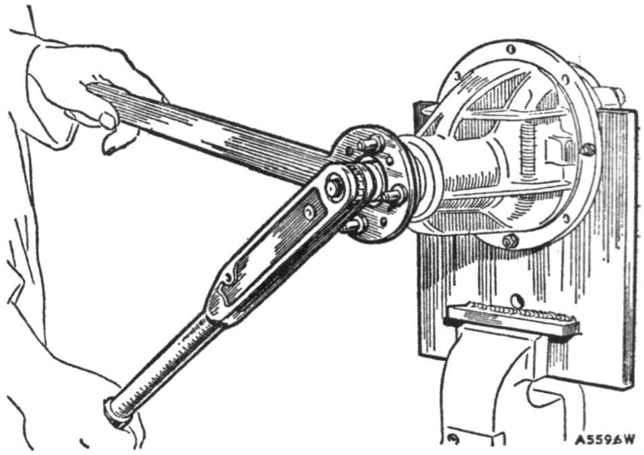


*Fig. H.5*

*Use Service tool 18G 34 A to hold the bevel pinion flange while loosening or tightening the bevel pinion nut*

Reassembly is now a reversal of the instructions given in Section H.4. Refill the axle with the recommended oil.

If it proves necessary to fit any new parts other than those detailed in Sections H.2, H.3, or H.5 the axle assembly must be set up as in Section H.7.

## Section H.6

### DISMANTLING THE CROWN WHEEL AND PINION

Remove the differential assembly as detailed in Section H.4.

Remove the differential bearings from the differential cage, using Service tool 18G 47 C with adaptors 18G 47 T. Note that the word 'THRUST' is stamped on the thrust face of each bearing and that shims are fitted between the inner ring of each bearing and the differential cage.

Knock back the tabs of the locking washers, unscrew the nuts from the bolts securing the crown wheel to the differential cage, and remove the crown wheel.

Knock back the tab of the locking washer and unscrew the pinion nut; remove the driving flange and the pressed end cover.

Drive the pinion shaft towards the rear; it will carry with it the inner race and the rollers of the rear bearing, leaving the outer race and the complete front bearing in position.

The inner race of the front bearing may be removed with the fingers and the outer races of both bearings withdrawn with Service tool 18G 264, using adaptors 18G 264 E and 18G 264 F.

Slide off the pinion sleeve and shims; withdraw the rear bearing inner race from the pinion shaft using Service tool 18G 47 C with adaptors 18G 47 AH (see Fig. H.8), noting the spacing washer against the pinion head.

Assembly and adjustment procedure are detailed in Section H.7.

## Section H.7

### ASSEMBLING AND SETTING THE CROWN WHEEL AND PINION

Apart from the fitting of components as detailed in Sections H.2, H.3, and H.5 it is not permissible to fit any new parts (e.g. crown wheel and pinion, pinion bearings, differential bearings, etc.) to the axle assembly without working through the procedure given in this Section. Furthermore, if a new crown wheel or a new pinion is needed, **a mated pair—crown wheel and pinion—must be fitted.**

Fitting a new crown wheel and pinion involves four distinct operations:

(1) Setting the position of the pinion.
(2) Adjusting the pinion bearing preload.
(3) Setting the crown wheel position.
(4) Adjusting the backlash between the gears.

The following Service tools are required to enable these operations to be carried out correctly:

Bevel pinion and differential setting gauge.
Bevel pinion inner race remover and replacer.
Bevel pinion outer race remover and replacer.
Bevel pinion preload gauge.

#### 1. SETTING THE PINION POSITION

(1) Fit the bearing outer races to the gear carrier, using the special pinion race replacing tool.
(2) Smooth off the pinion head with an oil-stone, but do not erase any markings that may be etched on the pinion head.
(3) Assemble the pinion and rear bearings with a washer of known thickness behind the pinion head.

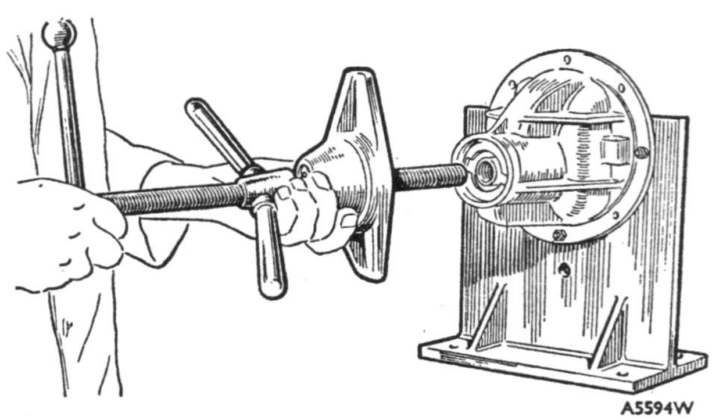


*Fig. H.6*

*Use Service tool 18G 264 with adaptor 18G 264 E and 18G 264 F to remove the pinion front and rear bearing outer races*

(4) Position the pinion in the gear carrier without the shims, bearing spacer, and oil seal.

(5) Fit the inner ring of the front bearing and the universal joint driving flange and tighten the nut gradually until a bearing preload of 10 to 12 lb. in. (·12 to ·14 kg. m.) is obtained.

(6) Remove the keep disc from the base of the magnet. Adjust the dial indicator to zero on the machined step 'B' of the setting block.

(7) Clean the pinion head and place the magnet and dial indicator in position (Fig. H.9). Move the indicator arm until the foot of the gauge rests on the centre of the differential bearing bore at one side and tighten the knurled locking screw. Obtain the maximum depth reading and note any variation from the zero setting. Repeat the check in the opposite bearing bore. Add the two variations together and divide by two to obtain a mean reading.

(8) Take into consideration any variation in pinion head thickness. This will be shown as an unbracketed figure etched on the pinion head and will always be minus (—). If no unbracketed figure is shown the pinion head is of nominal thickness. Using the mean clock gauge reading obtained and the unbracketed pinion head figure (if any), the following calculation can be made:

(*a*) **If the clock reading is minus** add the clock reading to the pinion head marking, the resulting sum being minus. **Reduce** the washer thickness by this amount.

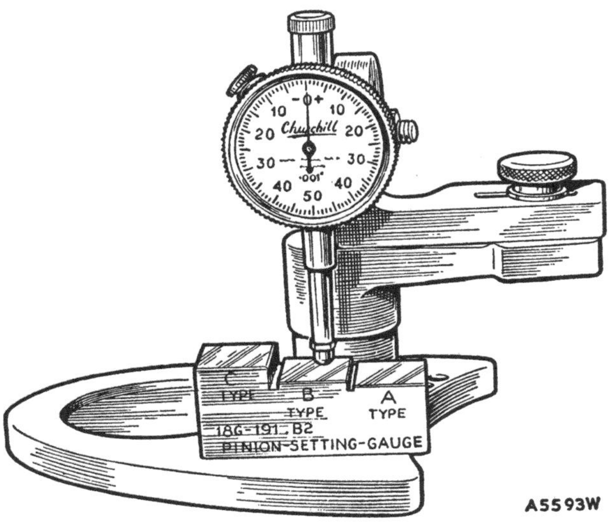


*Fig. H.7*

*Setting the gauge to zero on the special block for determination of the pinion position*

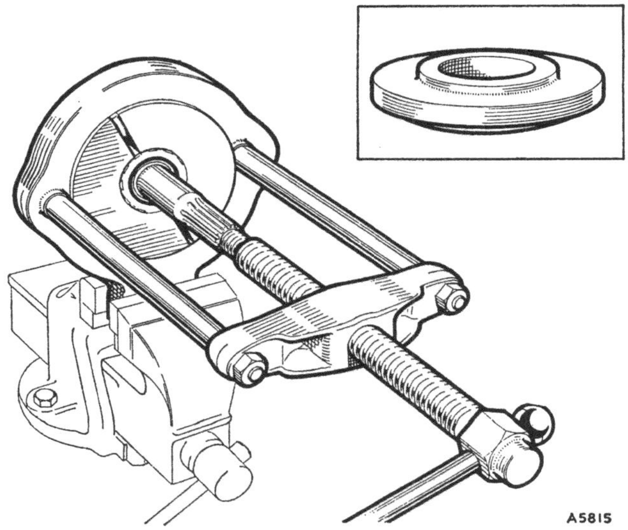


*Fig. H.8*

*Use Service tool* 18G47C *and adaptors* 18G47AH *to remove the pinion bearing inner race. (Inset) Part of the adaptor set used to replace the bearing race*

*Example:*

| | | |
|---|---|---|
| Clock reading | .. .. | —·002 in. |
| Pinion marking | .. .. | —·005 in. |
| Variation from nominal | .. | —·007 in. |

**Reduce the washer thickness by this amount.**

(*b*) **If the clock reading is plus and numerically less** than the pinion marking **reduce** the washer thickness by the difference.

*Example:*

| | | |
|---|---|---|
| Pinion marking | .. .. | —·005 in. |
| Clock reading | .. .. | +·003 in. |
| Variation from nominal | .. | —·002 in. |

**Reduce the washer thickness by this amount.**

(*c*) **If the clock reading is plus and numerically greater** than the pinion marking **increase** the washer thickness by the difference.

*Example:*

| | | |
|---|---|---|
| Clock reading | .. .. | +·008 in. |
| Pinion marking | .. .. | —·003 in. |
| Variation from nominal | .. | +·005 in. |

**Increase the washer thickness by this amount.**

The only cases where no alterations are required to the washer thickness are when the clock reading is **plus** and **numerically equal** to the unbracketed pinion marking, or when the clock reading is zero and there is no unbracketed marking on the pinion head.

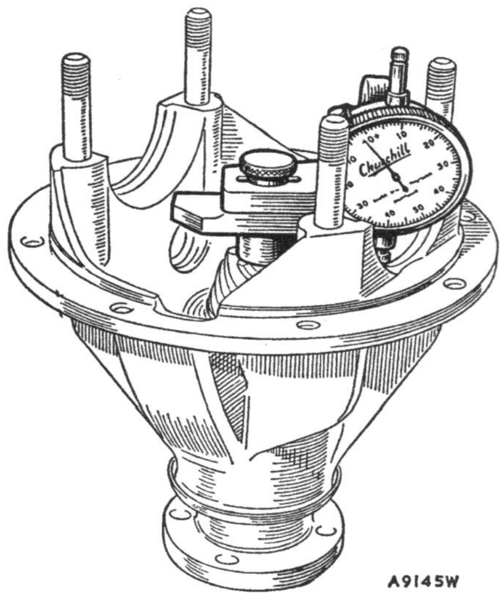


*Fig. H.9*

*The gauge in position on the pinion with the dial indicating a variation from the standard setting*

(9) Allowance should then finally be made as follows for the mounting distance marked on the pinion head in a rectangular bracket:

If the marking is a **plus** figure **reduce** the washer thickness by an equal amount.

If the marking is a **minus** figure **increase** the washer thickness by an equal amount.

A tolerance of ·001 in. is allowed in the thickness of the washer finally fitted.

*Table of washer and shim thicknesses*

| | |
|---|---|
| Pinion head washer thickness | ·112 to ·126 in. in steps of ·002 in. |
| Pinion bearing preload shims .. .. .. | ·004 to ·012 in. in steps of ·002 in., plus ·020 in. and ·030 in. |
| Crown wheel bearing shims | ·002 in., ·004 in., and ·006 in. |
| Pinion bearing preload .. | 10 to 12 lb. in. without oil seal; 13 to 15 lb. in. with oil seal |
| Crown wheel bearing pinch | ·002 in. each side |

## 2. ADJUSTING PINION BEARING PRELOAD

Fit the appropriate washer to the pinion head.

Assemble the pinion shaft, bearings, distance tube, and shims to the gear carrier; fit the oil seal and driving flange. Shims to a thickness of ·008 to ·011 in. (·2 to ·28 mm.) should be used as a starting-point for adjustment of the bearing preload.

Tighten the driving flange nut gradually with a torque wrench to 140 lb. ft. (19·34 kg. m.) and check the preload on the bearings during tightening to ensure that it does not exceed 13 to 15 lb. in. (·15 to ·173 kg. m.), i.e. 3 lb. in. (·034 kg. m.) greater than the recommended figure since the oil seal is now fitted. If the preload is too great more shims must be added. If the preload is too small when the nut is tightened correctly the shim thickness must be reduced.

## 3. SETTING THE CROWN WHEEL POSITION

(1) Before fitting the crown wheel and differential assembly to the gear carrier it is necessary to calculate the shim thickness required behind each bearing to give the required pinch. To facilitate the calculation, machining variations are indicated by stamped numbers on the carrier adjacent to the bearing bores. The dimensions to be considered are shown in Fig. H.10, (A) being the distance from the centre-line to the bearing register of the carrier on the left-hand side and (B) the distance from the centre-line to the bearing register of the carrier on the right-hand side. The (C) dimension is from the bearing register on one side of the cage to the register on the other side, while the (D) dimension is from the rear face of the crown wheel to the bearing register on the opposite side. Any variation from nominal on the (A) dimension will be found stamped on the carrier adjacent to the bearing bore, and similarly with the (B) dimension. The variations from nominal on (C) and (D) dimensions are stamped on the machined face of the differential cage.

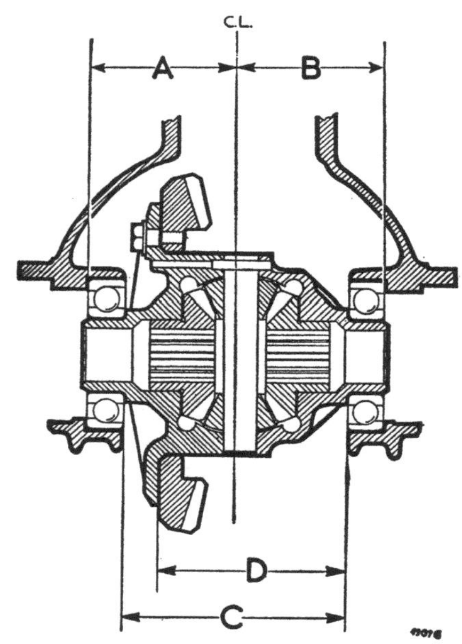


*Fig. H.10*

*The dimensions referred to in the instructions for differential setting*

It is possible to calculate the shim thickness required on the **left-hand side** by the use of the following formula:

A+D−C+·007 in.

Substituting the actual variations shown, this formula gives the shim thickness required to compensate for the variations in machining plus the extra ·002 in. (·05 mm.) to give the necessary bearing pinch. In addition, allowance must be made for variations in bearing thickness in the following manner.

Rest the bearing, with the inner race over the recess and the outer ring thrust face downwards, on the small surface plate of tool 18G191A. Drop the magnet onto the surface plate and set the clock gauge to zero on the small gauge block on the step marked 'B'. (See Fig. H.11.) This is the thickness of the standard bearing. Swing over the indicator until it rests on the plain surface of the inner race and, holding the inner race down against the balls, take a reading (Fig. H.12). Normally the bearing will be standard to −·003 in., though in some cases the tolerance may be from standard to −·005 in. A negative variation shown by this test indicates the additional thickness of shimming to be added to that side of the differential.

The formula for the **right-hand side** is:

B−D+·006 in.

and here again final allowance must be made for variation in bearing thickness.

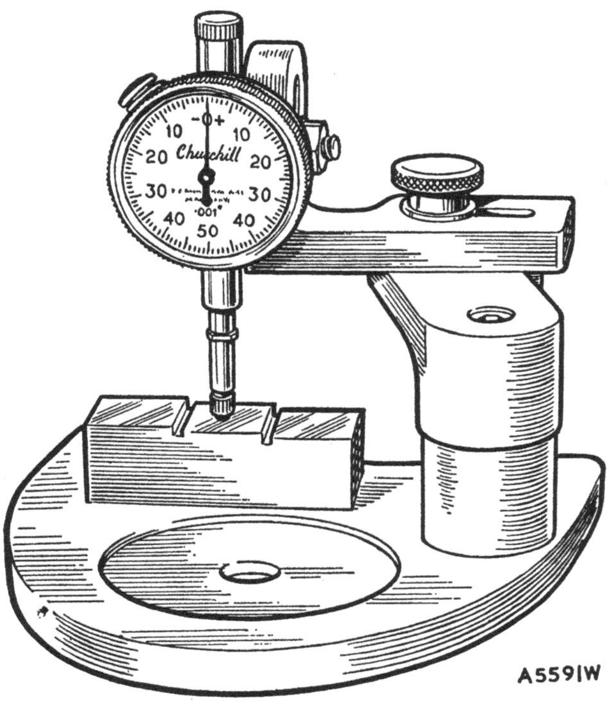


*Fig. H.11*

*To measure variations in bearing thickness first zero the gauge on the appropriate portion of the gauge block*

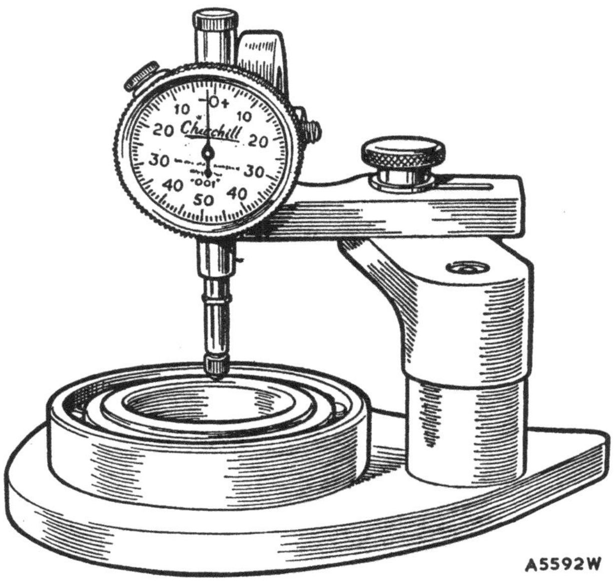


*Fig. H.12*

*With the gauge set to zero, place the bearing on the surface plate with the outer ring thrust face down, and take a reading while the indicator foot contacts the inner ring*

(2) When a framed number is marked on the back of the crown wheel, e.g. −2, it must be taken into account before assembling the shims and bearings to the differential cage. This mark assists in relating the crown wheel with the pinion.

If, for example, the mark is +2, then shims to the value of ·002 in. (·05 mm.) must be transferred from the left-hand side (the crown wheel side) to the right-hand side. If the marking is −2, then shims to the value of ·002 in. (·05 mm.) must be moved from the right-hand side to the left-hand side.

## 4. ADJUSTING THE BACKLASH

(1) Assemble the bearings and shims as calculated to the differential cage.

(2) Bolt the crown wheel to the differential case, but do not knock over the locking tabs. Tighten the nuts to a torque wrench reading of 60 lb. ft. (8·3 kg. m.).

Fit the shims and differential bearings with the thrust faces outwards.

Mount the assembly on two 'V' blocks and check the amount of run-out of the crown wheel, as it is rotated, by means of a suitably mounted dial indicator.

The maximum permissible run-out is ·002 in. (·05 mm.) and any greater irregularity must be corrected. Detach the crown wheel and examine the joint faces on the flange of the differential case and crown wheel for any particles of dirt.

When the parts are thoroughly cleaned it is unlikely that the crown wheel will not run true.

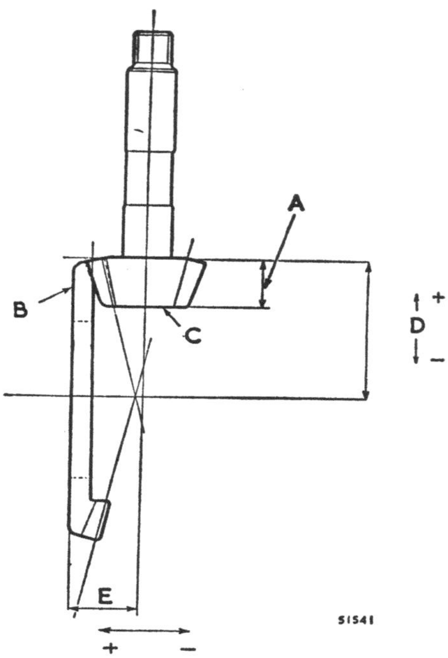


*Fig. H.13*

*Crown wheel and pinion markings*

A. Pinion head thickness.
B. Crown wheel marking.
C. Pinion marking.
D. Pinion mounting distance.
E. Crown wheel mounting distance.

Tighten the bolts to the correct torque wrench reading and knock-over the locking tabs.

(3) Fit the differential to the gear carrier. Replace the bearing caps and tighten the nuts to a torque wrench reading of 65 lb. ft. (8·99 kg. m.). Bolt the Service tool surface plate to the gear carrier flange and mount the clock gauge on the magnet bracket in such a way that an accurate backlash figure may be obtained. (See Fig. H.14.) The correct figure for the backlash to be used with any particular crown wheel and pinion is etched on the rear face of the crown wheel concerned and must be adhered to strictly.

A movement of ·002 in. (·05 mm.) shim thickness from one side of the differential to the other will produce a variation in backlash of approximately ·002 in. (·05 mm.). Thus it should be possible to set up the differential, even though the backlash is incorrect, by removing the bearings on one occasion only.

Great care must be taken to ensure absolute cleanliness during the above operations, as any discrepancies resulting from dirty assembly would affect the setting position of the crown wheel or pinion.

## Section H.8

### REMOVING AND REFITTING THE AXLE

Raise the rear of the car.

Mark the propeller shaft coupling flanges so that they may be replaced in the original relative positions. Remove the four bolts and self-locking nuts and release the rear end of the propeller shaft from the axle. Remove the nuts and spring and flat washers securing each end of each check strap to the anchor pins and remove the check straps.

Remove the split pin and clevis pin securing the brake cables to each brake operating lever. Remove the small nut and Phillips recessed-head screw securing the hand brake cable clip to the axle casing. Remove the self-locking nut and large flat washer securing the brake balance lever to the pivot on the axle casing.

Remove the nut and spring washer securing the lower end of each damper link to the rear spring clamp plate.

Unscrew the brake fluid supply pipe union and release the flexible pipe from the battery box support bracket.

Release the exhaust pipe from the exhaust manifold and the two supporting brackets and lower the exhaust pipe assembly.

Remove the nut and spring washer from the spring front anchor pin.

Support the axle casing and remove the rear shackle plates, brackets, and rubbers. Lower the axle support until the axle and spring assembly rests on the road wheels. Withdraw the front anchor pins and roll the assembly from beneath the car.

The axle is now free to be withdrawn on the stand rearwards from the car.

Replacement is the reverse of the above sequence of operations.

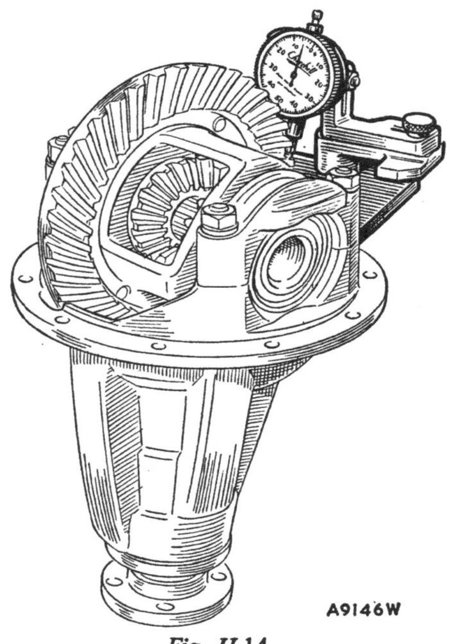


*Fig. H.14*

*Measuring the crown wheel backlash*

## Section H.9

### REAR ROAD SPRINGS

**To remove a rear spring**

(1) Remove the road wheel adjacent to the spring to be removed.
(2) Raise and support the body and support the axle with a hydraulic jack to enable the axle to be lowered to relieve the tension in the spring.
(3) Disconnect the shock absorber link from its bracket and the rebound strap from the rebound spindle.
(4) Remove the nuts and spring washers from the eyebolt and shackle plate pins and take off the outer shackle plate.
(5) Using a suitable drift, tap each shackle plate pin alternately until the plate and pins are free of the spring and the mounting bracket.
(6) Withdraw the eyebolt from the front of the spring.
(7) From the two 'U' clips remove the locknuts and nuts. Retain the shock absorber bracket, locating plate, and pad which will fall from the under side of the spring.
(8) Remove the spring and retain the upper locating plate, pad, pedestal, and 'U' clips.

**To refit a rear spring**

(1) Offer the front of the spring to its bracket and locate it with the front eyebolt.
(2) Fit the shackle plate and pins to the rear eye and the body bracket.
(3) Fit the upper locating plate and pad to the spring and locate the hole in the axle spring seat over the head of the centre-bolt.
(4) Place the pedestal over the axle, fit the lower pad, locating plate, and shock absorber bracket, and pass the 'U' clips over the axle and through the locating plates. Secure the 'U' clips with the nuts and locknuts.
(5) Refit the road wheel, remove the axle and body supports, and refit the washers and nuts to the front eyebolt and the shackle plate, nuts, and washers to the rear shackle plate pins.
(6) Refit the shock absorber link and rebound strap.

## Section H.10

### DISMANTLING AND ASSEMBLING A REAR SPRING

**Dismantling**

(1) Remove the rear spring as detailed in Section H.9.
(2) Straighten the ends of the clips on the third and fourth spring leaves.
(3) Support the spring in a vice with the top and bottom leaves against the vice jaws and the centre-bolt just clear of the jaws.
(4) Remove the locknut and nut from the centre-bolt and withdraw the centre-bolt and distance piece from the spring.
(5) Slowly open the vice to relieve the tension in the spring leaves and remove the leaves from the vice.
(6) Remove the two rubber bushes from the rear eye of the main leaf and then press out the Silentbloc bush from the front eye.
(7) Remove and discard the spring clip rivets and clips.

Clean and examine each leaf for signs of cracks or fractures, particularly around the centre-bolt holes.

Before assembly, cover each leaf with Shell Ensis 260 Fluid.

**Assembling**

(1) Press a new Silentbloc bush into the main leaf front eye and ensure that the outer bush is perfectly central in the eye.
(2) Rivet new spring clips to the third and fourth leaves.
(3) Use a tapered mandrel having the same maximum diameter as the centre-bolt to align the holes in the spring leaves and assemble the leaves. The longest half of the spring faces the rear of the car.
(4) Keep the leaves aligned and slowly compress them in a vice.
(5) Remove the mandrel and assemble the distance piece and centre-bolt. Secure the centre-bolt with its nut and locknut.
(6) Align the side of the spring and bend the spring clips over the main leaf.

After assembly, refit the spring to the car as detailed in Section H.9.

### EDITOR'S NOTES

**H. Rear Axle and Suspension**

*Removing and replacing a hub*

The pullers described in this and following sections may be replaced by any suitable commercially available pullers.

*Removing the differential pinions*

Before starting any work which involves the assembly of the differential unit, be advised that the correct reassembly procedure involves several very specialized tools and operations. It is imperative that these operations are carried out with extreme accuracy, otherwise the differential unit will be noisy and wear out quickly. Unless the proper tools are available, it is recommended that this work be left to qualified personnel.

## THE REAR SUSPENSION COMPONENTS

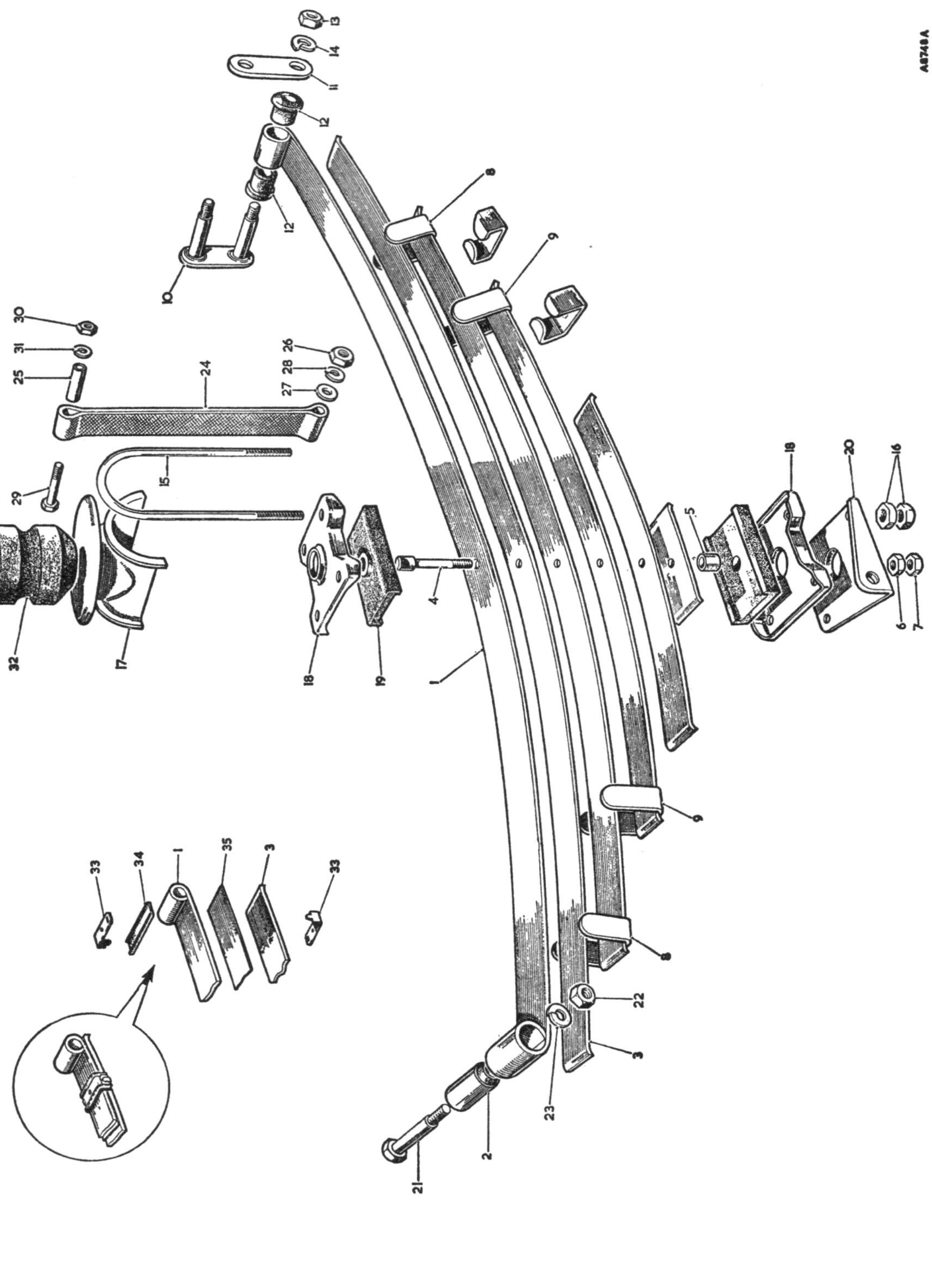

## KEY TO THE REAR SUSPENSION COMPONENTS

| *No.* | *Description* | *No.* | *Description* | *No.* | *Description* |
|---|---|---|---|---|---|
| 1. | Leaf assembly—main. | 13. | Nut—shackle plate pin. | 25. | Distance tube—rebound strap. |
| 2. | Bush. | 14. | Washer—spring—nut. | 26. | Nut—rebound strap to axle. |
| 3. | Leaf—second. | 15. | Clip—'U'—rear spring. | 27. | Washer—plain—nut. |
| 4. | Bolt—locating. | 16. | Nut—'U' clip. | 28. | Washer—spring—nut. |
| 5. | Distance piece. | 17. | Pedestal—bump rubber. | 29. | Bolt—strap to bracket. |
| 6. | Nut—bolt. | 18. | Plate—spring locating. | 30. | Nut—bolt. |
| 7. | Locknut—bolt. | 19. | Pad—spring seating. | 31. | Washer—spring—nut. |
| 8. | Clip—third leaf. | 20. | Bracket—shock absorber to rear spring—R.H. | 32. | Bump—rubber. |
| 9. | Clip—fourth leaf. | 21. | Bolt—spring—front end. | 33. | Clip—second leaf } later type only. |
| 10. | Plate—shackle and pins. | 22. | Nut—bolt. | 34. | Pad—second leaf clip } later type only. |
| 11. | Plate—shackle. | 23. | Washer—spring—nut. | 35. | Strip—interleaf, 1 2, 2 3, 3 4 } later type only. |
| 12. | Bush—shackle plate (rubber). | 24. | Strap—rebound. | | |

# Section Ha

## THE REAR AXLE AND REAR SUSPENSION
### (SEMI-FLOATING TYPE)

### GENERAL DESCRIPTION

The rear axle assembly is of the semi-floating type Adjustment to the bearing is by means of spacers, as also is the position of the pinion in relation to the crown wheel and the backlash between the gears.

Suspension is by semi-elliptic leaf springs, rubber-mounted, and the shackles are fitted with rubber bushes.

## Section Ha.1

### LUBRICATION

**Oil level**

Check the level, and top up if necessary. The filler plug is located on the rear of the axle and also serves as an oil level indicator. After topping up allow time for any surplus oil to run out. This is most important; if the axle is overfilled, the oil may leak through to the brake linings and lessen their efficiency.

NOTE.—**It is essential that only Hypoid oil be used.**

**Draining**

The most suitable time for draining is after a long journey while the oil is still warm. Clean the drain plug before it is replaced and tightened.

Refill the axle with fresh oil.

## Section Ha.2

### AXLE UNIT

**Removing**

Raise the rear of the car.

Mark the propeller shaft driving flanges so that they may be replaced in their original relative positions. Remove the four bolts and self-locking nuts and release the rear end of the propeller shaft from the axle. Remove the nuts and spring and flat washers securing each end of each check strap to the anchor pins on the axle and remove the check straps.

Remove the split pin and clevis pin securing the brake cable to each brake operating lever. Remove the set screw securing the hand brake cable clip to the axle casing. Remove the self-locking nut and large flat washer securing the brake balance lever to the pivot on the axle casing.

Remove the nut and spring washer securing the lower end of each damper link to the rear spring clamp plate.

Unscrew the brake fluid supply pipe union and release the flexible pipe from the battery box support bracket.

Release the exhaust pipe from the exhaust manifold and the supporting brackets and lower the exhaust pipe assembly.

Remove the nut and spring washer from each of the spring front anchor pins.

Support the axle casing and remove the rear shackle plates, brackets, and rubbers. Lower the axle support until the axle and spring assembly rests on the road wheels. Withdraw the front anchor pins and roll the assembly from beneath the car.

**Refitting**

Reverse the removal procedure when refitting.

Bleed and adjust the brakes (see Section M).

## Section Ha.3

### HUB AND AXLE SHAFT

**Removing**

Jack up the car and place blocks under the spring as close as possible to the axle. Remove the wheel. Release the hand brake.

Remove the two countersunk screws (disc wheels) or the four nuts (wire wheels) locating the brake-drum and tap it from the hub. It may be necessary to slacken off the brake adjusters slightly if the shoes hold the drum.

Remove the split pin, unscrew the slotted axle shaft nuts and withdraw the hub. Remove the clevis pin securing the brake cable to the operating lever, and disconnect the hydraulic pipe from the wheel cylinder. Remove the backplate. Remove the oil seal collar, bearing hub cap and oil seal from the axle shaft. Remove the axle shaft using impulse extractor 18G 284 with adaptor 18G 284 D and press the bearing from the shaft.

**Refitting**

Repack the bearings with grease before refitting.

Reverse the removal procedure, using tool 18G 1067 to drift the axle shaft into position. Lubricate and fit a new oil seal, lip facing inwards. Refit the oil seal collar and bearing hub cap. Tighten the axle shaft nut to the torque figure given in **'GENERAL DATA'**. Bleed and adjust the brakes as in Section M.

## Section Ha.4

### RENEWING THE PINION OIL SEAL

**Removing**

(1) Raise and support the rear of the car.
(2) Remove the rear road wheels and brake drums.
(3) Drain the rear axle.
(4) Mark the propeller shaft and pinion flanges to ensure correct reassembly and disconnect the propeller shaft.
(5) Using 18G 207 (see Fig. Ha.5) measure and record the torque required to rotate the pinion.
(6) Using 18G 34A to prevent the pinion from rotating, remove the flange retaining nut and washer, and withdraw the pinion flange.
(7) Extract and discard the oil seal.

**Inspection**

(8) Examine the pinion flange for damage, paying particular attention to the oil seal track area.

**Refitting**

(9) Grease the periphery and sealing lip of the new oil seal and fit the seal flush with the axle casing.

## THE REAR AXLE COMPONENTS

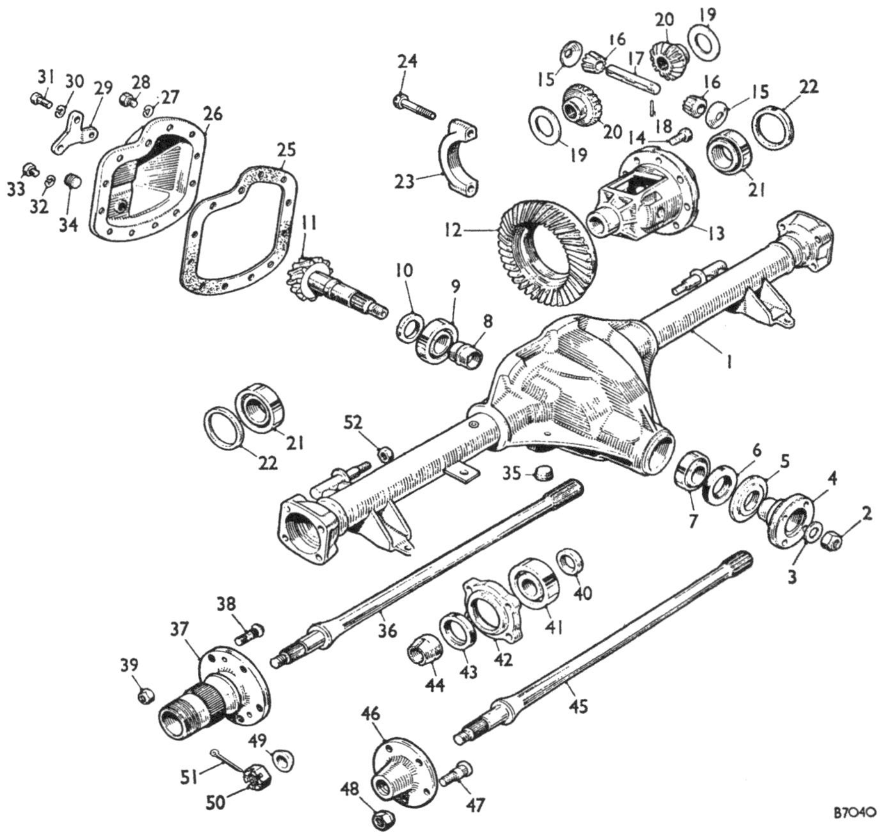


| No. | Description | No. | Description | No. | Description | |
|---|---|---|---|---|---|---|
| 1. | Case assembly | 18. | Roll-pin. | 35. | Drain plug. | |
| 2. | Nut. | 19. | Thrust washer. | 36. | Axle shaft. | Wire wheels only. |
| 3. | Plain washer. | 20. | Differential wheels. | 37. | Driving flange. | Wire wheels only. |
| 4. | Universal joint flange. | 21. | Differential bearing. | 38. | Stud. | Wire wheels only. |
| 5. | Dust cover. | 22. | Distance collars. | 39. | Nut. | Wire wheels only. |
| 6. | Oil seal. | 23. | Bearing cap. | 40. | Bearing spacer. | |
| 7. | Outer pinion bearing. | 24. | Bolt. | 41. | Bearing. | |
| 8. | Bearing spacer. | 25. | Joint washer. | 42. | Bearing hub cap. | |
| 9. | Inner pinion bearing. | 26. | Axle case cover. | 43. | Oil seal. | |
| 10. | Pinion thrust washer. | 27. | Spring washer. | 44. | Oil seal collar. | |
| 11. | Pinion. | 28. | Set screws. | 45. | Axle shaft. | Disc wheels only. |
| 12. | Crown wheel. | 29. | Compensating lever bracket. | 46. | Driving flange. | Disc wheels only. |
| 13. | Differential cage. | 30. | Spring washer. | 47. | Wheel stud. | Disc wheels only. |
| 14. | Bolt. | 31. | Set screw. | 48. | Wheel nut. | Disc wheels only. |
| 15. | Thrust washer. | 32. | Spring washer. | 49. | Axle shaft collar. | |
| 16. | Differential pinions. | 33. | Set screw. | 50. | Axle shaft nut. | |
| 17. | Pinion pin. | 34. | Filler and level plug. | 51. | Split pin. | |

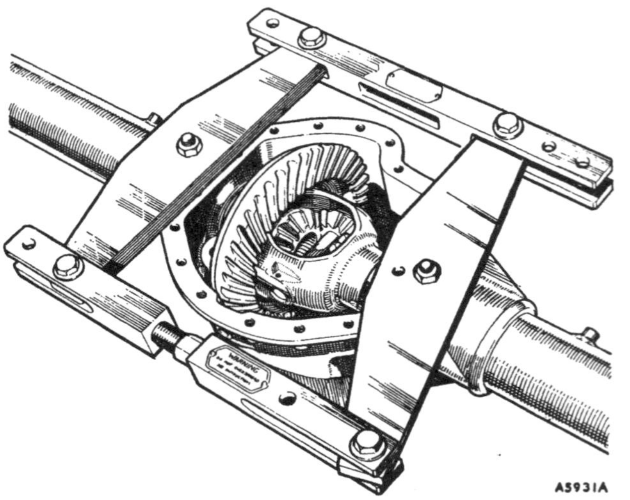


*Fig. Ha.*1
*Axle stretcher* 18*G* 131 *C*

(10) Refit the pinion flange and washer.
(11) Screw on the retaining nut, **tightening the nut gradually** until resistance is felt.
(12) Rotate the pinion to settle the bearings and measure the torque required to rotate the pinion.
If the reading obtained is less than that recorded in Operation 5, before the seal was removed; tighten the nut **A VERY SMALL AMOUNT,** resettle the bearings and recheck the torque reading. Repeat this procedure until a reading equal to that recorded in Operation 5 **BUT NOT LESS THAN 4 to 6 lb. in. (0·04 to 0·07 kg. m.)** is obtained.
e.g. Reading in Operation 5 = 9 lb. in. (0·10 kg. m.).
Adjust torque to this figure
or
Reading in Operation 5 = 0 lb. in.
Adjust torque to 4 to 6 lb. in. (0·04 to 0·07 kg. m.).

**CAUTION. Preload build up is rapid, tighten the nut with extreme care. IF AN ORIGINAL TORQUE READING, in excess of 6 lb. in. (0·7 kg. m.) is exceeded, THE AXLE MUST BE DISMANTLED AND A NEW COLLAPSIBLE SPACER FITTED.**

## Section Ha.5

### DIFFERENTIAL AND PINION

**Removing the differential**

Drain and remove the axle from the car (see Section Ha.2).

Remove the axle shafts and wheel hubs (see Section Ha.3).

Remove the differential cover. Mark each differential bearing cap before removal to ensure correct replacement unscrew the two bolts and withdraw both caps.

Before the differential assembly can be withdrawn the axle case must be stretched with Service tool 18G 131 C. When using this tool tighten the turnbuckle one flat at a time until the differential unit can be prised out. Each flat on the turnbuckle is numbered to provide a check on the amount turned on the buckle. Prise the differential assembly out with two levers, one on each side of the differential case opening, using suitable packing between the levers and the gear carrier, and ensure that no leverage is placed upon the axle stretcher.

NOTE.—**To prevent the axle case being permanently damaged, it must not be stretched any more than is absolutely necessary.**

**Maximum stretch is 0·012 in. (·31 mm.) 9 flats.**

Release the stretcher.

**Dismantling**

Bend back the locking tabs and remove the crown wheel securing screws. Remove the crown wheel from the differential cage.

Drive out the pinion pin peg and remove the pin.

Turn the differential wheels by hand until the differential pinions are opposite the openings in the differential cage, remove the differential pinions and the thrust washers fitted behind them.

Remove the differential wheels and their thrust washers.

Should it be necessary to withdraw the differential inner bearing races, Service tool 18G 47 C, adaptor 18G 47 AK and adaptor plug 18G 47 AR must be used.

**Reassembling**

Use Service tool 18G 134 and adaptor 18G 134 CM to replace the differential bearings.

Fit the differential wheels with the thrust washers in position.

Insert the differential pinions through the opening in the differential cage and mesh them with the differential wheels. Hold the pinion thrust washers in position and install the pinion pin.

Line up the pinion pin with the hole in the differential cage and fit the pinion pin peg.

Bolt the crown wheel to the differential cage but do not knock over the locking tabs. Tighten the bolts to a torque wrench reading of 60 to 65 lb. ft. (8·3 to 8·9 kg. m.).

Mount the assembly on two 'V' blocks and check the amount of run-out of the crown wheel, as it is rotated, by means of a suitably mounted dial indicator. The maximum permissible run-out is ·002 in. (·05 mm.) and any greater irregularity must be corrected. If there is excessive run-out, detach the crown wheel and examine the joint faces on the flange of the differential cage and on the crown wheel for any particles of dirt.

When the parts are thoroughly cleaned it is unlikely that the crown wheel will not run true.

Change the position of the crown wheel on the differential cage to correct any misalignment.

Tighten the bolts to the correct torque wrench reading and knock over the locking washers.

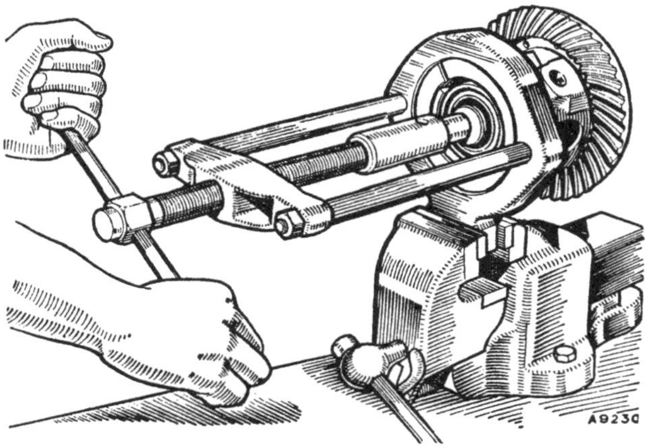


*Fig. Ha.2*

*Use 19G 47 C, adaptor 18G 47 AK, and adaptor plug 18G 47 AR to remove the differential bearings*

**Removing the pinion**

Before removing the pinion measure the preload: if this is zero, fit new pinion bearings.

Hold the flange with wrench 18G 34 A and remove the nut and washer.

Withdraw the flange by tapping it with a hide hammer.

Remove the bevel pinion. To prevent damaging the outer bearing the pinion must be pressed and not driven out.

**Dismantling**

To remove the bevel pinion front and rear outer races from within the gear carrier casing use Service tool 18G 264, with adaptors 18G 264 AA and 18G 264 AD for the front race, and adaptors 18G 264 AA and 18G 264 AB for the rear race. As the front outer race is withdrawn it will carry the pinion oil seal with it.

Use service tool 18G 47 C and adaptor 18G 47 AS to remove the rear bearing inner race from the bevel pinion.

**Reassembling**

Refit the pinion front inner race using Service tool 18G 47C with adaptor 18G 47 AS. With the same tool assemble the rear bearing inner race to the bevel pinion, ensuring that the pressure is exerted on the inner race only.

**Refitting the differential and pinion using original parts**

Where it is only necessary to fit a new oil seal the differential can be refitted in the reverse order, assuming that the original shim thicknesses are retained, but ensure that the pinion preload figure noted before removing the pinion is achieved when tightening the flange nut.

If any new parts are fitted, i.e. crown wheel and pinion, pinion bearings, etc. the setting of the pinion (its position relative to the crown wheel) must be checked.

Examine the crown wheel teeth. If a new crown wheel is needed, a mated pair—crown wheel and pinion—must be fitted.

Fitting a new crown wheel and pinion involves four distinct operations.

(1) Setting the pinion position.
(2) Setting the pinion bearing preload.
(3) Setting the crown wheel.
(4) Setting the backlash.

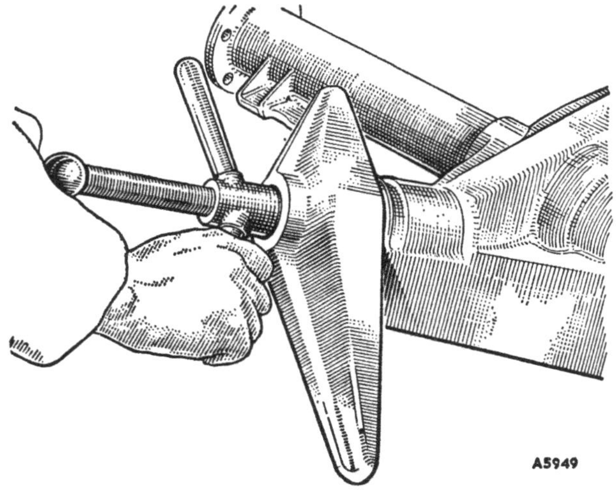


*Fig. Ha.3*

*Use 18G 264 and adaptors 18G 264 AA, and 18G 264 AD, or 18G 264 AB to remove front and rear bevel pinion outer races*

**Refitting**

1. SETTING THE PINION POSITION

Fit the bearing outer races to the axle casing using tool 18G 264 and 18G 264 AA with adaptors 18G 264 AD or 18G 264 AB.

The variation in the pinion head thickness is etched on the pinion head and is to be ignored for this setting.

Using the dummy pinion 18G 191 H, fit the inner race of the bearing to the pinion in the gear carrier without

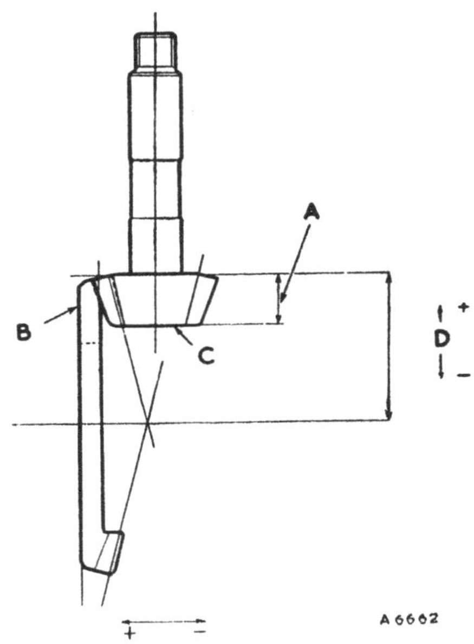


*Fig. Ha.4*

*Crown wheel and pinion markings*

A. Pinion head thickness. Max. −·007 in. (−·178 mm.).
B. Crown wheel marked here.
C. Pinion marked here.
D. Pinion mounting distance. Max. ±·004 in. (±·102 mm.).

the collapsible spacer and oil seal. Fit the inner race of the front bearing.

NOTE.—**The standard pinion head spacer (0·208 in. (5·3 mm.)) is incorporated in the dummy pinion.**

Refit the universal joint driving flange and tighten the nut gradually until a bearing preload figure of 10 to 20 lb. in. (·11 to ·2 kg. m.) is obtained.

Clean the dummy pinion head and remove the keep disc from the base of the magnet. Position the dial indicator foot on the pinion head and adjust the dial indicator to zero (see Fig. Ha.6).

Move the indicator arm until the foot of the gauge rests on the centre of the differential bore at one side and tighten the knurled locking screws. Obtain the maximum depth reading and note any variation from the zero setting. Repeat this check on the opposite bearing bore. Add the two variations together and divide by two to obtain a mean reading.

The reading shown on the clock will be the amount of correction necessary to the spacer which is 0·208 in. (5·3 mm.) thick.

(*a*) If the clock reading is **negative** (—) reduce the spacer thickness by this amount.

(*b*) If the clock reading is **positive** (+) increase the spacer thickness by this amount.

Allowance must finally be made for the mounting distance marked on the pinion head in a rectangular bracket. Proceed as follows:

If the marking is **positive** (+) reduce the washer thickness by an equal amount.

If the marking is **negative** (—) increase the washer thickness by an equal amount.

A tolerance of ·001 in. is allowed in the thickness of the washer finally fitted.

Remove the dummy pinion.

**Available washers for pinion position**

Eight washers are available in ·002 in. (·051 mm.) steps from 0·208 in. (5·3 mm.) to ·222 in. (5·64 mm.).

## 2. PINION BEARING PRELOAD

A washer of the thickness indicated by the use of the tools and by the calculations should be fitted under the pinion head. Fit the pinion inner bearing race. Using 18G 47 C, and adaptor 18G 47 AS. Insert the pinion into the case and fit the collapsible spacer with the small diameter towards the shoulder of the pinion head.

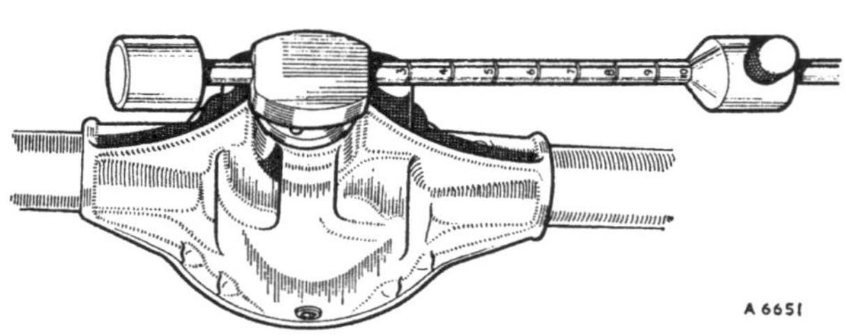


*Fig. Ha.5*

*Checking the bevel pinion bearing preload* (18*G* 207)

Support the head of the pinion and press the outer bearing onto the pinion; **do not compress the spacer.** Grease the periphery and sealing lip of the seal and fit the seal flush with the axle casing. Fit the universal joint flange, washer and nut.

Tighten the flange nut to 140 lb. ft. (20·7 kg. m.) using 18G 592 and flange wrench 18G 34 A, progressively increase the torque load by 10 lb. ft. (1 kg. m.) until the spacer starts to collapse. Rotate the pinion to settle the bearings and using 18G 207 B check the pinion pre-load. **BEARING PRELOAD IS RAPID. IF THE TORQUE OF 24 LB. IN. (0·3 KG. M.) IS EXCEEDED THE PINION MUST BE DISMANTLED AND A NEW COLLAPSIBLE SPACER FITTED.** In practice the pinion nut will require tightening **A VERY SMALL AMOUNT BETWEEN EACH PINION TORQUE CHECK.** When the nut is correctly tightened it should provide a pre-load of 14 to 18 lb. in. (0·10 to 0·16 kg. m.).

## 3. SETTING THE CROWN WHEEL POSITION

Assemble each inner and outer bearing race to the differential assembly.

Using crown wheel setting tools (18G 191, 18G 191 F, and 18G 191 J) place the differential assembly onto the jig and load the crown wheel assembly. Spin the unit to settle the bearings.

The standard measurement of the bearing bores is 7·243 in. (183·98 mm.). Any excess machined from the bores will be marked A and B on the axle casing. The A and B tolerances added to the standard measurement will determine the overall dimension (see **Example A**).

*Fig. Ha.6*

*Taking a reading of the bearing bore and (inset) the dial gauge set at zero on the dummy pinion*

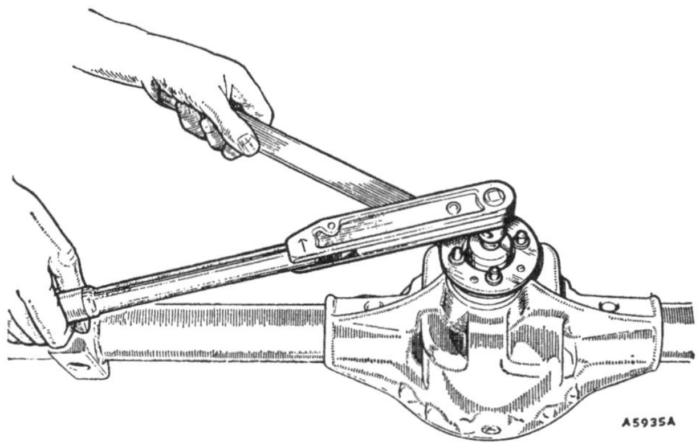


*Fig. Ha.7*

*Use flange wrench* 18G 34 *A to hold the flange against rotation when tightening the pinion nut with torque wrench* 18G 592

Clean the head of the jig pillar (18G 191 J) and remove the keep disc from the base of the magnet. Position the magnet on the pillar head and adjust the dial gauge to zero when the dial gauge foot is positioned on the pillar head (see Fig. Ha.9).

**NOTE.—The pillar is the standard height of the differential assembly of 6·972 in. (177·10 mm.).**

Spin the differential to settle the bearings, move the arm of the dial indicator to the machined face of the jig (see Fig. Ha.9), and take a reading. Add the variation to the standard height to obtain the total width of the differential assembly (see **Example B**).

Subtract the total width of the differential assembly, and the total distance collar thickness will be found. To this figure a further ·004 in. (·102 mm.) must be added to give bearing 'pinch' (see **Example C**). Divide this total by two to obtain the distance collars of equal thickness (see **Example D**).

**Example A**

| | |
|---|---|
| Differential case standard .. | 7·243 in. (183·98 mm.) |
| Stamped on case 'A' .. | ·001 in. (·025 mm.) |
| Stamped on case 'B' .. .. | ·002 in. (·038 mm.) |
| Total distance between bores | 7·246 in. (184·05 mm.) |

**Example B**

| | |
|---|---|
| Differential assembly standard | 6·972 in. (177·10 mm.) |
| Clock reading (positive) .. | ·008 in. (·203 mm.) |
| Total differential assembly .. | 6·980 in. (177·30 mm.) |

**Example C**

| | |
|---|---|
| Total distance between bores | 7·246 in. (184·05 mm.) |
| Minus total differential assembly .. .. .. .. | 6·980 in. (177·30 mm.) |
| | ·266 in. (6·75 mm.) |
| Plus bearing pre-load .. | ·004 in. (·102 mm.) |
| | ·270 in. (6·86 mm.) |

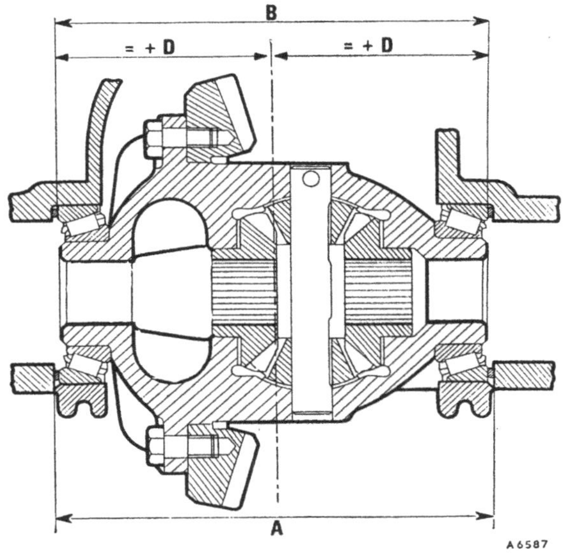


*Fig. Ha.8*

*Illustrates the points from which the calculations must be made to determine the spacer thickness for the bearings on each side of the carrier*

**Example D**

| | | |
|---|---|---|
| Spacers each side .. | ·270 in.÷2 .. | ·135 in. |
| | (6·86 mm.÷2) .. | (3·43 mm.) |

**Available spacers for differential assembly**

Eighteen spacers are available in ·002 in. (·051 mm.) steps from 0·115 in. (2·92 mm.) to ·149 in. (3·77 mm.).

**Refitting**

The correct figure for the backlash to be used with any particular crown wheel and pinion is etched on the rear face of the crown wheel concerned. Increase the spacer thickness on the opposite side of the crown wheel by this amount and decrease the spacer thickness on the crown wheel side by the same amount (see **Example E**).

**Example E**

| | |
|---|---|
| Backlash ·008 | |
| Spacer calculated .. .. | ·135 in. (3·43 mm.) |
| Minus backlash .. .. | ·008 in. (·203 mm.) |
| Crown wheel side spacer .. | ·127 in. (3·23 mm.) |
| Spacer calculated .. .. | ·135 in. (3·43 mm.) |
| Plus backlash .. .. .. | ·008 in. (·203 mm.) |
| Opposite crown wheel side spacer .. .. .. | ·143 in. (3·62 mm.) |

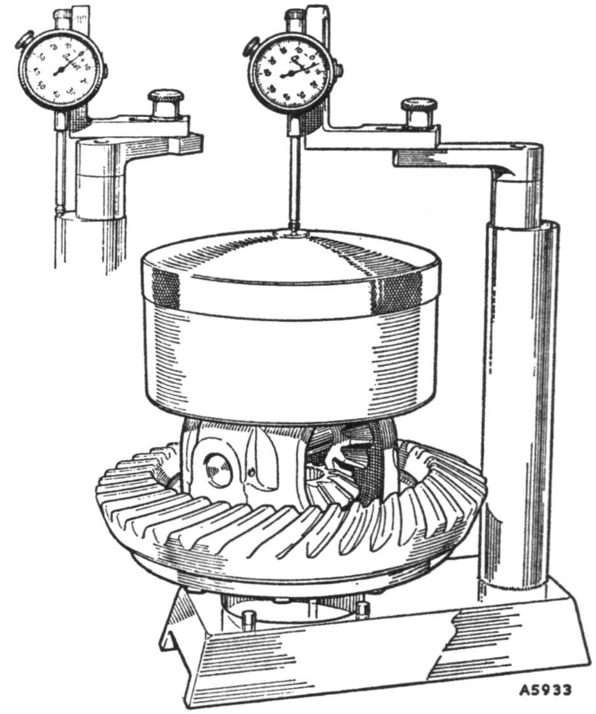


*Fig. Ha.9*

*Taking a reading on the differential assembly and (inset) the dial gauge set at zero on the jig pillar*

### 4. ADJUSTING THE BACKLASH

Fit the differential to the gear carrier. Replace the bearing caps and tighten the bolts to the correct torque wrench reading as given in ('GENERAL DATA'). Bolt the special tool surface plate to the gear carrier flange and mount the clock gauge in such a way that an accurate backlash figure may be obtained (see Fig. Ha.10).

A movement of ·002 in. (·051 mm.) shim thickness from one side of the differential cage to the other will produce a variation in backlash of approximately ·002 in. (·051 mm.).

*Fig. Ha.10*

*Checking crown wheel backlash*

Ensure absolute cleanliness during the above operations, as any discrepancies resulting from dirty assembly would affect the setting of the crown wheel or pinion.

Continue refitting by reversing the removing procedure. Finally, bend back the locking tabs around the bolt heads.

NOTE.—**Ensure that the axle case jointing washer is coated with Hylomar Jointing Compound to obtain a perfect oil seal.**

## Section Ha.6

### ROAD SPRINGS

For removal and dismantling see Section H.9 and H.10.

# SECTION J

## THE STEERING GEAR

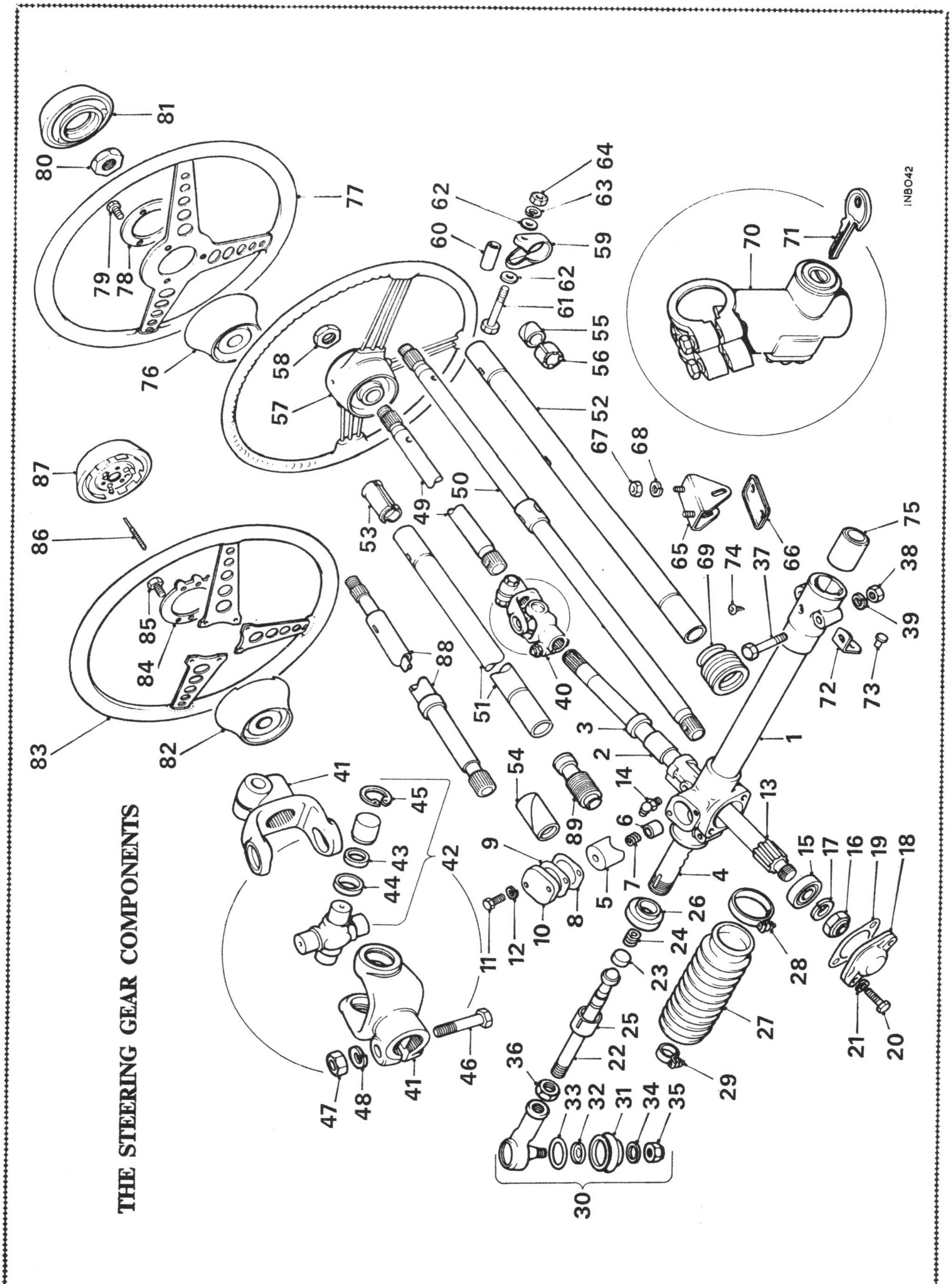


THE STEERING GEAR COMPONENTS

## KEY TO THE STEERING GEAR COMPONENTS

| No. | Description |
|---|---|
| 1. | Housing assembly. |
| 2. | Bush—pinion. |
| 3. | Seal—oil. |
| 4. | Rack. |
| 5. | Yoke—rack support. |
| 6. | Pad—damper. |
| 7. | Spring—damper pad. |
| 8. | Shim—cover-plate. |
| 9. | Joint—cover-plate. |
| 10. | Plate—cover—yoke. |
| 11. | Bolt—cover-plate. |
| 12. | Washer—spring—bolt. |
| 13. | Pinion. |
| 14. | Lubricator—pinion shaft. |
| 15. | Bearing—ball—pinion. |
| 16. | Nut—pinion bearing. |
| 17. | Washer—spring—nut. |
| 18. | Cover—end. |
| 19. | Joint—end cover. |
| 20. | Bolt—end cover. |
| 21. | Washer—spring—bolt. |
| 22. | Rod—tie. |
| 23. | Seat—ball. |
| 24. | Spring—thrust—ball seat. |
| 25. | Housing—ball. |
| 26. | Locknut—ball housing. |
| 27. | Seal—rack. |
| 28. | Clip—inner—seal. |
| 29. | Clip—outer—seal. |

| No. | Description |
|---|---|
| 30. | Socket assembly—ball. |
| 31. | Boot. |
| 32. | Retainer—boot. |
| 33. | Spring—garter. |
| 34. | Washer—boot. |
| 35. | Nut—ball pin. |
| 36. | Locknut. |
| 37. | Bolt—track to bracket. |
| 38. | Nut—bolt. |
| 39. | Washer—spring. |
| 40. | Joint—universal—steering-column. |
| 41. | Yoke. |
| 42. | Journal—assembly. |
| 43. | Joint—journal. |
| 44. | Retainer—joint. |
| 45. | Circlip. |
| 46. | Bolt—universal joint. |
| 47. | Nut—bolt. |
| 48. | Washer—spring—nut. |
| 49. | Column assembly—inner. |
| 50. | Column assembly—inner—R.H.D. |
| 51. | Tube—column—outer. |
| 52. | Tube—column—outer—R.H.D. |
| 53. | Bearing—upper—column. |
| 54. | Bearing—lower—column. |
| 55. | Bush (felt) } When steering lock is fitted. |
| 56. | Clip—retaining } When steering lock is fitted. |
| 57. | Wheel—steering. |
| 58. | Nut—steering-wheel. |
| 59. | Clamp—steering-column. |

| No. | Description |
|---|---|
| 60. | Distance piece. |
| 61. | Bolt—clamp. |
| 62. | Washer—plain—clamp bolt. |
| 63. | Washer—spring—clamp bolt. |
| 64. | Nut—bolt. |
| 65. | Bracket—lower—steering-column. |
| 66. | Plate—blanking—bracket. |
| 67. | Nut—bracket to body. |
| 68. | Washer—spring—nut. |
| 69. | Excluder—draught—column to bulkhead. |
| 70. | Lock assembly—steering and ignition. |
| 71. | Key. |
| 72. | Shim—cross-member bracket. |
| 73. | Rivet. |
| 74. | Screw—retaining bush. |
| 75. | Bush—rack housing. |
| | *GHN5/GHD5 car only* |
| 76. | Wheel hub. |
| 77. | Steering-wheel. |
| 78. | Lock ring. |
| 79. | Screw. |
| 80. | Nut. |
| 81. | Motif and housing. |
| 82. | Wheel hub. |
| 83. | Steering wheel. |
| 84. | Lock ring. |
| 85. | Bolt. |
| 86. | Horn push contact. |
| 87. | Horn push. |
| 88. | Steering column } From Car No. 258001. |
| 89. | Sealing tube } From Car No. 258001. |

## GENERAL DESCRIPTION

The steering gear is of the direct-acting rack-and-pinion type, providing light and accurate control under all conditions.

It consists of a rack bar and toothed pinion mounted on the front suspension cross-member.

No adjustment for bearing wear in the box is provided, except by the fitting of the necessary new parts.

The steering inner column is attached to the pinion by a universal coupling.

CAUTION.—If the vehicle is hoisted with its front wheels clear of the ground care should be taken to avoid forceful movement of the wheels from lock to lock, as damage may occur within the steering mechanism.

## LUBRICATION

**(Early cars)**

Give the lubricating nipple on the steering-gear housing up to 10 strokes of the oil gun, but no more.

On R.H.D. cars the nipple is accessible from above the steering gearbox and on L.H.D. cars from below the car under the radiator.

# Section J.1

## FRONT WHEEL ALIGNMENT

The wheels should toe in $\frac{1}{16}$ to $\frac{3}{32}$ in. (1·6 to 2·4 mm.).

See that the tyres are inflated to the correct pressures. Set the wheels in the straight-ahead position.

Set the arms of a suitable trammel to the height of the hub centre on the outside of the wheels.

Place the trammel to the rear of the wheels and adjust the pointers to register with the wheel rims. Chalk the position of the pointers in each wheel rim and push the car forward one half-turn of the wheels. Take the front reading from the same marks on the rims.

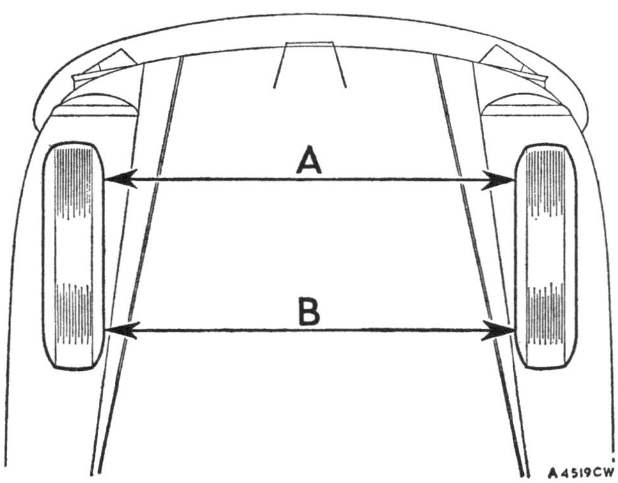


*Fig. J.1*

*The front wheel alignment check must be taken with the front wheels in the straight-ahead position. Dimension* (B) *is* $\frac{1}{16}$ *to* $\frac{3}{32}$ *in.* (1·6 *to* 2·4 *mm.*) *greater than* (A)

If adjustment is necessary, proceed as follows.

Slacken the locknuts at the ends of the short tie-rods and the clips securing the rubber gaiters to the tie-rods.

Use a wrench to rotate each of the tie-rods **equally** in the desired direction. These both have right-hand threads.

NOTE.—**To ensure that the steering gearbox is in the central position and that the steering geometry is correct, it is important that the tie-rods are adjusted to exactly equal lengths. This can be ascertained by measuring the amount of thread visible behind each locknut, which should be equal.**

After adjustment retighten the ball joint locknuts and rubber gaiter clips and ensure that the machined under sides of the ball joints are in the same plane.

# Section J.2

## STEERING-WHEEL

Release the three grub screws and remove the horn-push hub centre. Unscrew the steering-wheel nut and mark the wheel hub and column to ensure replacement in the original position. Pull off the wheel with a suitable tool. When replacing the wheel position it on the column splines in the original position to place the spokes equally about a horizontal datum line. Tighten the nut to the torque wrench setting given in **'GENERAL DATA'**.

**GHN and GHD5 cars**

(1) Remove the steering-wheel motif assembly; it is a press-fit. Later type, remove the horn contact plunger.

(2) Turn back the lock ring tabs and remove the bolts, lock ring, and steering-wheel.

NOTE: Early cars only, have lock tabs integral with the lock ring.

*Hub*

(3) Slacken the steering-wheel nut and fit Service tool 18G 1181 to the hub using the special bolts.* Mark the hub and column to assist correct re-alignment and pull the hub until it is a loose fit on the steering-column. Remove 18G 1181, the steering-wheel nut and hub.

(4) When refitting the hub, position it on the column splines in the original position. Fit the nut and tighten to the torque wrench setting given in **'GENERAL DATA'**.

* *Later type steering wheel use two* $\frac{5}{16}$ *UNF,* $\frac{7}{8}$ *in. threaded,* $\frac{7}{8}$ *in. plain shank, bolts to fit service tool* 18G 1181.

# Section J.3

## STEERING-COLUMN

**Removing and replacing**

Withdraw the clamping bolt and nut securing the universal joint to the inner steering-column. Unscrew the four set screws from the direction indicator cowling and remove the indicator. Withdraw the clamping bolts, nuts, and spring and plain washers from the two brackets which support the column and remove the column complete.

**NOTE.—Disconnect the wiring from the column lock when fitted and remove the ignition and auxiliary circuit fuses.**

**Bushes**

Remove the steering-wheel and column and withdraw the inner column. Prise out the felt bush and pull out the polythene bush.

Soak a new felt bush in graphite oil before fitting.

## Section J.4

### STEERING-COLUMN UNIVERSAL JOINT

**Removing**

Bolts and nuts clamp the universal joint splines on the steering inner column and steering pinion, and the bolts must be withdrawn completely to release the universal joint assembly.

Slacken the bolts supporting the steering-column below the dash panel.

Withdraw the clamping bolts from the universal joint.

Move the steering-column assembly upwards to withdraw the steering inner column from the universal joint.

Withdraw the universal joint from the steering pinion.

**Dismantling and reassembling**

The Hardy Spicer joint has four needle-roller bearings retained on a centre spider by circlips. The joints are packed with grease on assembly and there is no further provision for lubrication.

Remove any enamel and dirt from the circlips and bearing races. Remove the circlips by pinching the ears together and prising them out with a screwdriver.

If a ring does not slide readily from its groove, tap the end of the bearing race lightly to relieve the pressure against the bearing.

Hold the joint in one hand with the side of a yoke at the top and tap the radius of the yoke lightly with a copper hammer. The bearing should begin to emerge; turn the joint over and remove the bearing and needle rollers with the fingers. If necessary, tap the bearing race from the inside with a small-diameter bar, taking care not to damage the bearing face, or grip the needle bearing race in a vice and tap the yoke clear.

Repeat this operation for the opposite bearing.

One yoke can now be removed. Rest the two exposed trunnions on wood or lead blocks to protect their ground faces, and tap the top lug of the flange yoke to remove the bearing race.

Turn the yoke over and repeat the operation.

When reassembling, replace the cork gaskets and gasket retainers on the spider journals, using a tubular drift. The spider journal shoulders should be shellacked prior to fitting the retainers to ensure a good oil seal.

Smear the walls of the races with grease and assemble the needle rollers to the bearing races and pack with grease.

Insert the spider in one yoke and, using a soft-nosed drift slightly smaller in diameter than the hole in the yoke, tap the bearings into position. It is essential that the bearing races are a light drive fit in the yoke trunnions.

Repeat this operation for the other bearings and replace the circlips, making sure that they are firmly located in their grooves. If the joint appears to bind, tap lightly with a wooden mallet to relieve any pressure by the bearings on the ends of the journals.

## Section J.5

### STEERING GEAR

**Removing**

The procedure detailed here will remove the steering rack and pinion from a completely assembled car.

Support the front end of the car by placing jacks beneath the lower suspension arm spring pans and remove the road wheels.

Remove the locknuts and drive the tie-rod ball pins from the steering-arms. Turn the steering onto the left lock (R.H.D. cars) or right lock (L.H.D. cars). Withdraw the clamping nut and bolt from the universal joint on the pinion shaft. Remove the nuts and bolts securing the steering rack to the front suspension cross-member, noting that the front bolts are fitted with self-locking nuts, and packing shims may be found between the rack and the frame brackets.

The steering assembly can now be withdrawn downwards.

**Replacing**

The steering gear is assembled to the car by reversing the above procedure although special attention should be given to the instructions in Section J.7.

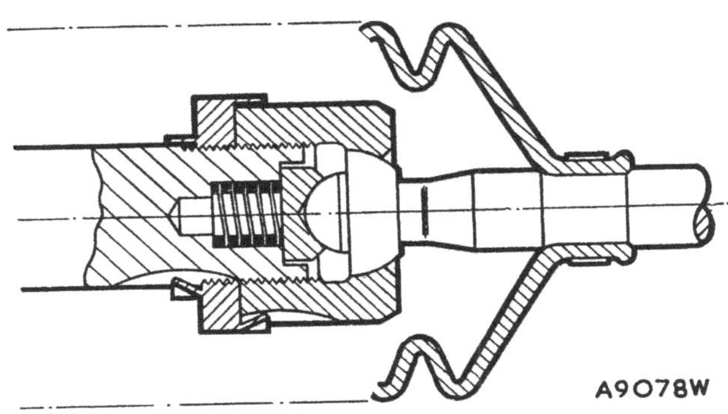


*Fig. J.2*
*The assembly of a tie-rod ball joint*

## Section J.6

### RACK AND PINION

**Dismantling**

Hold the rack housing between suitable clamps in a vice. Remove the pinion end cover and joint washer, placing a container to catch the oil that may drain from the housing. Remove the damper cover and shims, exposing the yoke, damper pad, and spring. After removal of these the pinion, complete with ball race and locknut, may be withdrawn.

Unlock the tie-rod ball end locknuts and disconnect the ball end assemblies. Release the rubber gaiter seal clips and withdraw the seals.

Prise up the indentations in the locking rings clear of the slots in the rack and ball housing. Slacken back the locking ring and unscrew the housing to release the tie-rod, ball seat, and seat tension spring.

Withdraw the rack from the pinion end of the housing; if removed from the other end the teeth may damage the rack housing bush.

To remove the rack housing bush unscrew the self-tapping screw retaining it and carefully drive the bush out.

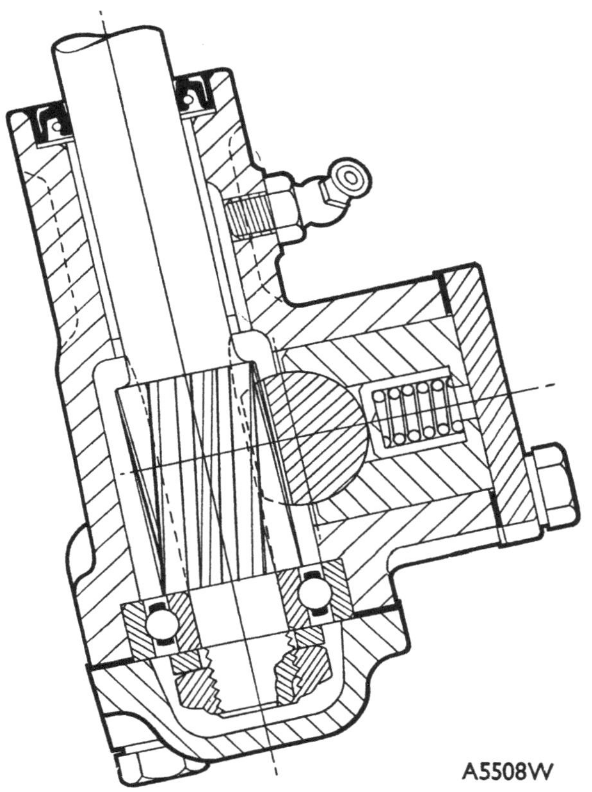


*Fig. J.3*

*A section through the rack and pinion housing, showing the damper and seals*

**Examining for wear**

Thoroughly clean and examine all parts of the assembly; components showing signs of wear must be replaced with new parts. Fractures, hollows, or roughness in the surfaces of the rack or pinion teeth will render them unserviceable.

Take particular note of the rubber gaiters; should they be damaged or show the slightest sign of deterioration, they must be replaced with new ones. The tie-rod ball housing and ball seat should also be subjected to a careful check and new parts fitted if excessive wear is evident.

The outer ball socket assembly cannot be dismantled, and must therefore be renewed complete if it is worn or damaged.

Examine the bush fitted in the end of the rack housing and fit a new one if it shows signs of damage or wear.

**Reassembling**

Insert the rack housing bush and carefully drive or press it in until the bush is flush with the housing end. Enter a $\frac{7}{64}$ in. (2·78 mm.) diameter drill through the retaining screw hole and drill the outer shell of the bush to receive the retaining screw. A depth of ·24 in. (6·3 mm.) will achieve this.

NOTE.—**Bushes of sintered iron with an outer shell of pressed steel and rubber injected between the two have superseded the lead-bronze type in earlier use.**

Smear the head of the screw with jointing compound before tightening so as to ensure an oil-tight joint.

Replace the rack from the pinion end. Refit the seat spring, seat, tie-rod, and ball housings, smearing the ball seats liberally with S.A.E. 90 oil. Tighten the ball housings until the tie-rod is held firm and without free play. When correctly adjusted a torque of 32 to 52 lb. in. (·359 to ·594 kg. m.) must be required on the rod to produce articulation. Relock the housing by tightening up the locking ring to a torque of 33 to 37 lb. ft. (4·60 to 5·63 kg. m.) and secure in the locked position by punching the lips of the locking ring into the slots in the ball housing and rack. It is recommended that new locking rings are fitted whenever these have been disturbed.

Insert the pinion complete with ball race and locking nut into its housing. Replace the pinion end cover and seal, using a sealing compound to make an oil-tight joint. The outer edge of the ball race locknut must be peened into the slot in the pinion shaft if removed and refitted.

To adjust the rack damper replace the plunger in the housing and tighten the cover down without the spring or shims until it is just possible to rotate the pinion shaft by drawing the rack through its housing. With a feeler gauge measure the clearance between the cover and its seating on the rack housing. To this dimension must be added an additional clearance of ·0005 to ·003 in. (·013 to ·076 mm.) to arrive at the correct thickness of shims which must be placed beneath the damper cover-plate.

Remove the damper cover-plate and plunger and replace the assembly, using a sealing compound to make an oil-tight joint.

Refit the rubber gaiters to the housing and the tie-rods. Before securing the gaiter clip on the tie-rod at the pinion end, stand the assembly upright, and pour in

¼ pint (·4 U.S. pint, ·2 litre) of Extreme Pressure S.A.E. 90 oil through the end of the gaiter, or pump the oil into the rack housing through the nipple provided.

Refit and tighten the gaiter clip.

Refit the ball ends and locknuts.

## Section J.7

### STEERING-COLUMN ALIGNMENT

Set the steering in the straight-ahead position and check that the trafficator cancellation stud at the top of the column is in the correct angular position when the serrations on the coupling and column are aligned.

When assembling the steering-column or steering gearbox assembly to the car care must be taken to ensure a free condition at the universal joint before the column or gearbox securing bolts are tightened. For the universal joint to be completely unloaded the centre-line of the steering-column and the centre-line of the steering rack pinion must pass through the centre of the universal joint spider when the assembly is viewed from above and from the side. Failure to ensure complete freedom at the universal joint will load the steering pinion upper bearing and cause extreme wear and steering stiffness.

To enable the assembly to be secured in the correct position the steering-column support bracket and plate may be moved up and down and sideways, and packing shims may be fitted between the steering rack mounting bosses and the brackets on the front suspension member.

Tighten the universal joint clamp bolts.

With all column and rack securing bolts slack, position the universal joint and tighten the support bracket clamp bolts at the lower end of the column.

Should there be a gap between the gearbox bosses and mounting brackets, remove the bolts, pack with shims as required, and replace and tighten the securing bolts.

To ensure complete alignment again slacken and re-tighten the steering-column lower support bolt.

Tighten the upper support bracket bolt.

## Section J.8

### NYLON-SEATED BALL JOINTS

Nylon-seated ball joints, which are sealed in manufacture and require no lubrication, are fitted at the steering-arm tie-rod ball joints.

It is essential that no dirt or abrasive matter should enter the nylon ball joint; in the event of a rubber boot being torn or damaged in service it is probable that the ball joint has been left exposed, and it is therefore important to renew both the ball joint and the boot.

If damage to the boot occurs whilst the steering side- or cross-rod is being removed in the workshop, only a new rubber boot need be fitted, provided the ball joint is clean. Smear the area adjacent to the joint with a little Dextragrease Super G.P. prior to assembling the boot.

Early production cars are fitted with a plain washer under the steering ball joint nut.

Later production cars are fitted with steering levers having a reduced diameter taper hole. These levers have a longer ball joint pin, are fitted with a thicker nut and have no plain washer.

When early production cars with a plain washer under the steering ball joint nut are fitted with service steering levers the plain washer must not be replaced.

## Section J.9

### STEERING LOCK IGNITION SWITCH

Cars exported to certain markets are fitted with a combined ignition/starter switch and steering-column lock mounted on the steering-column.

On cars fitted with the lock a sleeve integral with the inner column is slotted to permit engagement of the lock tongue; the outer column is also slotted to allow the lock tongue to pass through. A hole drilled in the upper surface of the outer column locates the steering lock bracket. The bracket is secured by two bolts each waisted below the head to permit removal of the heads by shear action during assembly.

To remove the lock, disconnect the battery and the ignition/starter switch connections and turn the lock setting to 'GARAGE' to unlock the steering. Free the steering-column assembly as described in Section J.3 and remove the lock securing bolts with an easy-out.

## Section J.10

### STEERING-COLUMN
### (GHN5 and HGD5 cars from Car No. 258001)

**Removing**

(1) Disconnect the battery.

(2) Turn the steering to the straight-ahead position.

(3) Mark the inner steering-column and the universal joint to ensure correct alignment when refitting.

(4) Remove the pinch bolt and nut securing the universal joint to the steering-column.

(5) Disconnect the multi connector block.

(6) Disconnect the wiring from the ignition switch.

(7) Remove the bolts retaining the steering-column upper and lower support clamp brackets.

(8) Remove the steering-column complete with the sealing tube, steering-wheel, direction indicator and ignition/steering lock switch.

**Refitting**

(9) Check that the steering rack is in the straight ahead position.

(10) Check that the steering-column is in the straight-ahead position.

(11) Reverse the removing procedure in (1) and (4) to (8) noting.

  (*a*) That the sealing tube is fitted to the steering-column with the gaitered end towards the toe-board.

  (*b*) Ensure that the marks made in (3) align when fitting the inner column to the universal joint.

## Section J.11

### STEERING LOCK AND IGNITION STARTER SWITCH
### (GHN5 and GHD5 cars from Car No. 258001)

**Removing**

(1) Remove the steering-column—Section J.10.
(2) Remove the steering-wheel and hub—Section J.2.
(3) Remove the direction indicator/headlight flasher/low-high beam switch—Section N.30.
(4) Remove the direction indicator switch trip from the inner steering-column.
(5) Slide the top clamp bracket off the steering-column.
(6) Turn the ignition key to position '1' to ensure that the steering lock is disengaged.
(7) Drill out or remove with a suitable proprietary tool, the retaining shear bolts.
(8) Unscrew the steering lock locating grub screw.
(9) Remove the steering lock and ignition starter switch.

**Refitting**

(10) Reverse the removing procedure in (1) to (9), using new shear bolts and ensuring that the shear bolts are tightened until the bolt heads shear at the waisted point giving a torque tightness of 12 lbf. ft. (1·66 kgf. m.).

# Section Ja

## THE STEERING GEAR (ENERGY ABSORBING COLUMN)

**The information given in this Section refers specifically to service operations on, or affected by, equipment fitted to the MGB (GHN5/GHN4 and and GHD5/GHD4) in conformity with local and territorial requirements, and must be used in conjunction with Section J.**

## Section Ja.1

### STEERING-COLUMN

**Removing**

(1) Disconnect the batteries.
(2) Remove the carburetter air cleaners.
(3) Remove the upper pinch-bolt from the steering-column universal joint.
(4) Remove lower panel from under the left-hand side of the fascia panel.
(5) Disconnect the steering-column switch wiring at the snap connectors below the fascia.
(6) Unscrew the three bolts securing the steering-column to the toe-plate.
(7) Note the location, quantity, and thickness of the packing washers between the column upper fixing flanges and the body brackets, remove the securing bolts and collect the packing washers. **If the packing washers are mislaid or their fitting positions are not recorded, the steering must be aligned as described in Section Ja.3 when refitted.**
(8) Withdraw the steering-column complete with steering-wheel and switches.

**Dismantling**

(9) Withdraw the motif disc from the centre of the steering-wheel; unscrew the steering-wheel retaining nut and, using a suitable puller-type extractor, remove the wheel.
(10) Remove the three left-hand switch cowl retaining screws.
(11) Unscrew the panel light switch retaining screws and remove the left-hand switch cowl.
(12) Remove the screw retaining the right-hand switch cowl.
(13) Unscrew the two screws securing the combined overdrive and windshield wiper/washer switch and remove the switch complete with wiring.
(14) Unscrew the two screws securing the combined headlamp flasher, horn and direction indicator switch, and remove the switch complete with wiring.
(15) Unscrew the four ignition switch retaining screws and remove the switch complete with wiring.

**Reassembling**

(16) Reverse the dismantling procedure in (9) to (15).

**Refitting**

NOTE.—**If a new steering-column is being fitted it must be aligned as described in Section Ja.3.**

(17) Fit the column assembly into the car and enter the inner column into the universal joint.
(18) Fit the packing washers in their original positions between the column fixing flanges and the body brackets; fit the three securing bolts, tightening them by hand until the packing washers are just pinched.
(19) Fit and tighten the three column to toe-plate fixing bolts.
(20) Tighten the three upper fixing bolts to the torque figure given in **'GENERAL DATA'**.
(21) Fit and tighten the universal joint pinch-bolt to the torque figure given in **'GENERAL DATA'**.

## Section Ja.2

### RACK ASSEMBLY

**Removing**

(1) Raise and support the front of the car and remove both front road wheels.
(2) Remove the tie-rod ball pin locknuts.
(3) Using tool 18G 1063, detach the tie-rod ball pins from the steering-arms.
(4) Turn the steering onto the right lock and remove the universal joint lower pinch-bolt.
(5) Remove the nuts and bolts securing the rack assembly to the front cross-member, noting that the front bolts are fitted with self-locking nuts.
(6) Withdraw the rack assembly downwards from the car.

**Dismantling**

(7) Carry out the operations detailed in Section J.6.

**Refitting**

NOTE.—**If a new rack assembly is being fitted it must be aligned as described in Section Ja.3.**

(8) Reverse the removing procedure in (1) to (6) ensuring that when entering the pinion shaft into the universal joint both the steering-column and rack are in the straight-ahead position.

## Section Ja.3

### ALIGNMENT

(1) Fit the toe-plate seal to the steering-column and position the column in the car.
(2) Fit one packing washer between each of the two top column fixing flanges and the body brackets.
(3) Screw in the two top fixing bolts by hand until the packing washers are just pinched.
(4) Check that the column is free to move forwards, backwards and sideways.
(5) Screw in the three toe-plate fixings bolts, tightening them by hand only.
(6) Check that the lower end of the column is free to move and position it centrally in the aperture.
(7) Fit one of the two larger point gauges of tool 18G 1140 on the lower end of the column, positioning it so that the retaining screw tightens onto the flat machined in the splines.
(8) Fit the remaining large points gauge of 18G 1140 to the pinion shaft so that the retaining screw tightens onto the radial groove machined in the splines.
(9) Fit the rack assembly and tighten the rack fixing bolts.

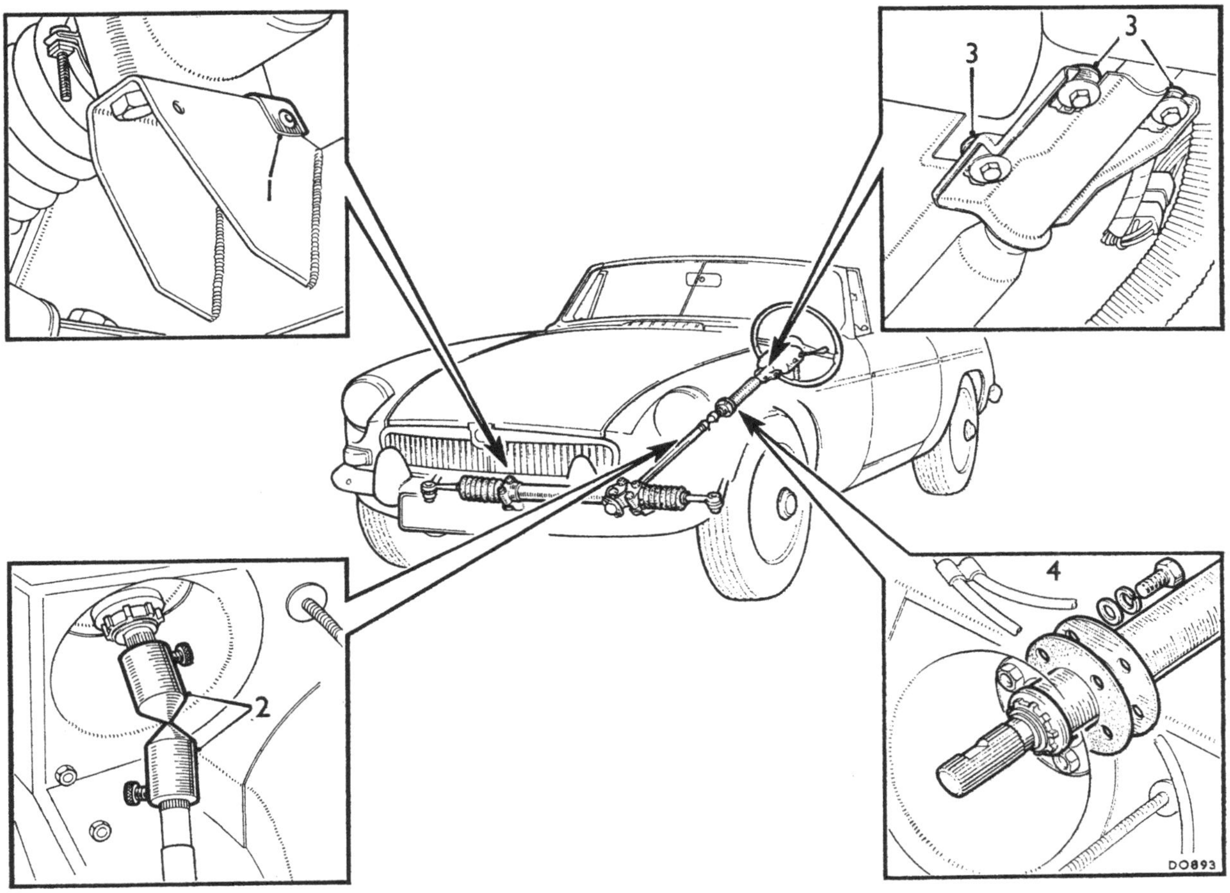


*Fig. Ja.*1
*Steering alignment*

1. Rack shims.
2. Alignment gauge, 18G 1140.
3. Packing washers fitted between column top fixing brackets.
4. Toe-plate fixings.

(10) Slacken the screw on the column point gauge and slide the gauge down until the points of both gauges are on the same plane but not overlapping.
(11) The steering-column and rack assembly are correctly aligned when the gauges meet exactly at their points (see Fig. Ja.1).
(12) Correct any horizontal misalignment by moving the end of the steering-column.
(13) If any vertical misalignment exists between the gauge points, proceed as follows:
  (*a*) Remove the rack fixing bolts.
  (*b*) Add sufficient shims (shim thickness ·020 in. [·508 mm.]) to correct the misalignment, between the left-hand and right-hand mounting brackets and the cross-member mounting bosses. Do not rivet the shims to the bosses at this stage.
  (*c*) Refit and tighten the rack mounting bolts.
  (*d*) Recheck the alignment of the gauge points.
  (*e*) If the alignment is still unsatisfactory, adjust the shim thickness.
  (*f*) When correct alignment of the gauge points has been achieved, rivet the shims to the mounting bosses through the holes provided.
(14) Remove the rack assembly.
(15) Remove the gauges from the pinion and steering-column.
(16) Check that the steering-column is in the straight-ahead position, and fit the universal joint on to its splines with the pinch-bolt hole aligned with the machined flat.
(17) Check that the rack is in the straight-ahead position
(18) Fit the rack assembly and fully tighten the rack mounting bolts.
(19) Tighten the three toe-plate to steering-column securing bolts.
(20) Tighten the two steering-column upper fixing bolts to the torque figure given in **'GENERAL DATA'**.
(21) Measure the gap between the column upper mounting flange and the body bracket at the third bolt position. Fit packing washers to the thickness of the gap, then fit and tighten the bolt to the torque figure given in **'GENERAL DATA'**.
(22) Fit the two universal joint pinch-bolts and tighten them to the torque figure given in **'GENERAL DATA'**.

# SECTION K

## THE FRONT SUSPENSION

## GENERAL DESCRIPTION

The independent front suspension is of the wishbone and coil type in which the suspension units are mounted one on each end of the cross-member assembly. The cross-member is rubber-mounted and bolted to the body side-members.

This design allows the front suspension to be removed as a complete unit.

The cross-member embodies two mounting brackets for the steering rack, a bracket at each end for the bump rubbers, a towing-eye plate, two front brake hose brackets, and two brake pipe clips.

Each suspension unit comprises a coil spring, a swivel axle unit, a lower wishbone assembly, and an upper assembly formed by the arms of the double-acting dampers mounted on top of the cross-member.

The lower arms are rubber-mounted on a pivot that is bolted to the cross-member; the outer ends of the arms are bolted to the lower end of the swivel pin. The outer ends of the shock absorber arms are secured to the swivel pin upper trunnion link by a fulcrum pin and tapered rubber bushes.

A spring pan secured between the lower wishbone arms supports the coil spring; the upper end of the spring is located by a spigot bolted to the under side of the cross-member.

Both suspension units may be interconnected by a rubber-mounted anti-roll bar.

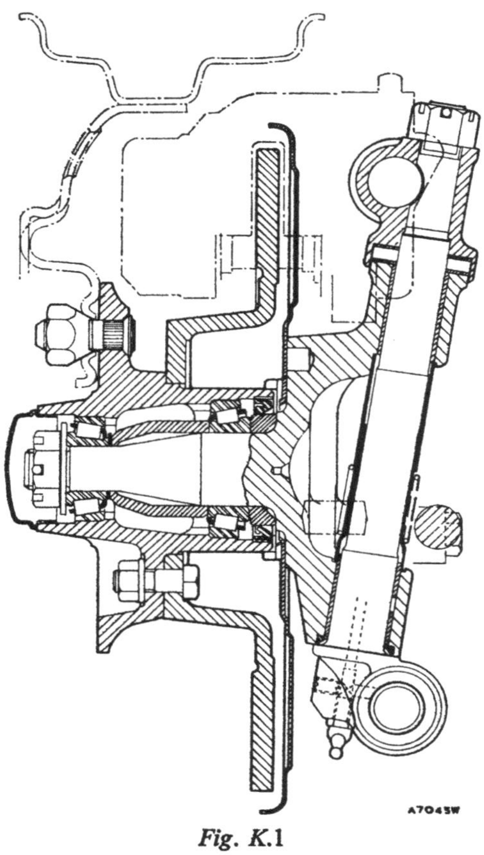


*Fig. K.1*
*Cross-section of a swivel axle and hub*

## Section K.1

### LUBRICATION

Lubrication nipples are provided on the top and bottom swivel pin bushes and in the base of the swivel pin. Each nipple should be charged with grease.

## Section K.2

### FRONT HUBS

**Disc wheels**

The front hubs are supported on taper-roller bearings that are mounted on the swivel axles. Four wheel disc studs are pressed into the hub, and the brake disc is secured to the hub by nuts and bolts.

An oil seal is fitted to the inner end of the hub and a spacer and shims are interposed between the inner and outer bearing. The assembly is retained on the swivel axle by a washer, nut, and split pin, and the outer end of the hub is closed by a grease-retaining cap.

**Wire wheels**

The construction of the wire wheel hubs is similar to that of the disc wheel hubs. Externally the outer end of the hub takes the form of splines, and the grease retainer is fitted inside the hub to close an access hole to the split pin. The outer edge of the hub is threaded externally to permit the fitting of a hub cap.

**NOTE.—The right-hand hub is threaded right-hand and the left-hand hub is threaded left-hand.**

**Removing a hub**

To remove a hub apply the hand brake, lift and support the front of the car, and remove the road wheel.

Remove the two studs securing the brake calliper to the swivel axle and support the calliper clear of the hub assembly.

Withdraw the grease retainer and remove the split pin from the stub axle nut. Unscrew the nut.

Using Service tool 18G 363 (wire wheels) or 18G 304 with adaptors 18G 304 B and 18G 304 J (disc wheels), withdraw the hub and disc assembly.

From the hub withdraw the bearing retaining washer, outer bearing, shims, spacer, inner bearing, oil seal collar, and oil seal.

The outer bearing races should be left in the hub unless they are to be renewed.

**(See Editor's notes at end of section.)**

**Inspection of bearings**

Wash the bearings in paraffin and thoroughly dry them in an air blast or with a non-fluffy cloth.

Examine the rollers for chips, pitting, or other damage and for security in their cages. Examine also the inner and outer races.

Bearings damaged or suspect must be renewed.

After examination immerse the bearings in mineral oil.

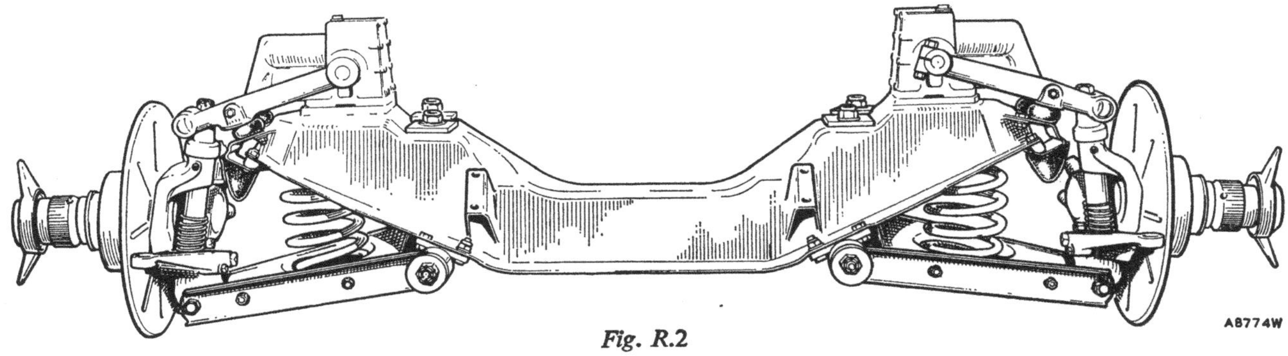


*Fig. R.2*

*The front suspension cross-member and the suspension units*

**Refitting a hub**

If the bearing outer races have been removed refit the new ones by pressing them into the hub.

Fill each bearing with grease ensuring a small protrusion either side of the bearing. Fill the cavity between the bearing and the oil seal and lightly smear the spacer with grease. Do not fill the cavity between the bearings or the grease-retaining cap with grease.

Fit the inner bearing to its race and the collar and seal to the hub, position the spacer and outer bearing, and assemble the hub to the axle.

Adjust the hub bearings as follows to obtain the correct end-float of between ·002 and ·004 in. (·05 and ·10 mm.).

(1) Assemble the hub without the shims and mount the assembly on the axle. Fit the retaining washer and nut and tighten the nut until the bearings bind. This will pull the outer races fully against their locating flanges inside the hub.

(2) Remove the nut and washer and pull out the roller race of the outer bearing. Insert sufficient shims to produce an excessive end-float and note the thickness of shims used. Refit the bearing, washer, and nut and tighten the nut.

(3) Using a clock gauge, measure accurately the end-float in the bearings. Remove the nut, washer, and outer bearing and reduce the number of shims to produce the required end-float.

The shims are available in thicknesses of ·003 in. (·076 mm.), ·005 in. (·127 mm.), and ·010 in. (·254 mm.).

(4) Replace the bearing, washer, and nut and tighten the nut to a torque loading of between 40 and 70 lb. ft. (5·3 and 9·6 kg. m.). Latitude for the torque wrench reading is given so that the nut can be tightened sufficiently to align a slot in the nut with the hole in the axle. Fit a new split pin.

(5) Refit the hub caps.

**(See Editor's notes at end of section.)**

# Section K.3

## COIL SPRINGS

Each coil spring is located between a spring pan bolted to the lower wishbone arms and a spigot bolted to the under side of the cross-member assembly.

To remove a spring fit a spring compressor (Service tool 18G 693) to the lower wishbone arms and take the weight of the spring.

Remove the anti-roll bar link, if fitted, and the bolts securing the spring pan to the wishbone arms.

Unscrew the spring compressor to release the tension in the spring and then remove the spring pan and spring.

Replacement is a reversal of the removal sequence.

**(See Editor's notes at end of section.)**

# Section K.4

## SWIVEL AXLES

The swivel axles work on the swivel pins, which are supported at their upper ends by the trunnion links, which are bolted to the shock absorber arms, and at their lower ends by the fulcrum pins, which connect with the lower wishbone arms.

The centre portion of the pins are protected by upper and lower spring-loaded dust shields.

The steering levers and the disc brake dust plates are bolted to the inside of the swivel axles.

**Removing and dismantling**

To remove the swivel axles jack up and support the car and, dealing with each axle in turn, proceed as follows:

(1) Remove the front road wheel.

(2) Detach the brake calliper and support it clear of the hub.

(3) Remove the hub and brake disc assembly as described in Section K.2.

(4) Remove the steering lever bolts and detach the lever and then the disc cover bolts and the cover.

(5) Remove the coil spring as described in Section K.3.

(6) Extract the split pins from the upper trunnion pin and fulcrum pin and remove the nuts.

(7) Unscrew the clamp bolt and centre-bolt of the shock absorber arm, ease the arm outwards, and remove the swivel axle.

## THE FRONT SUSPENSION COMPONENTS

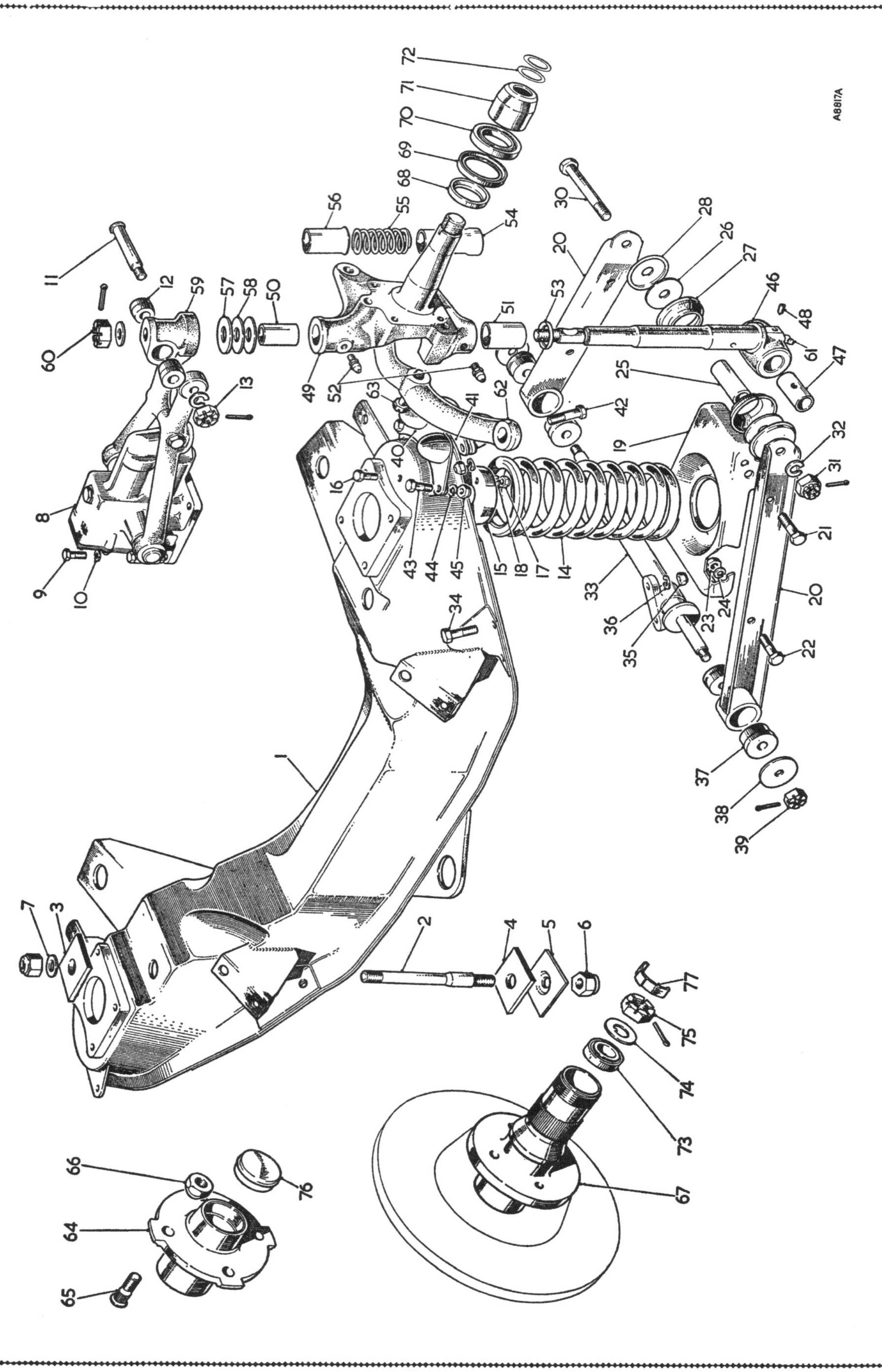

## KEY TO THE FRONT SUSPENSION COMPONENTS

| No. | Description | No. | Description | No. | Description |
|---|---|---|---|---|---|
| 1. | Cross-member. | 27. | Seal—link. | 53. | Ring—sw vel axle pin (cork). |
| 2. | Bolt—cross-member to body. | 28. | Support—link seal. | 54. | Tube—dust excluder—bottom. |
| 3. | Pad—mounting—upper (rubber). | 29. | Nut—wishbone pivot. | 55. | Spring—dust excluder. |
| 4. | Pad—mounting—lower (rubber). | 30. | Bolt—wishbone to link. | 56. | Tube—dust excluder—top. |
| 5. | Plate—clamp. | 31. | Nut—bolt. | 57. | Washer—thrust. |
| 6. | Nut—mounting bolt. | 32. | Washer—spring—nut. | 58. | Washer—floating thrust—·052 to ·057 in. (1·32 to 1·44 mm.). |
| 7. | Washer—plain—nut. | 33. | Pivot—wishbone. | 59. | Trunnion—suspension link. |
| 8. | Absorber—shock. | 34. | Bolt—pivot to member. | 60. | Nut—swivel axle pin. |
| 9. | Screw—shock absorber to cross-member. | 35. | Nut—bolt. | 61. | Lubricator—swivel pin. |
| 10. | Washer-screw—spring. | 36. | Washer—spring—nut. | 62. | Lever—steering—R.H. |
| 11. | Pin—fulcrum—top link to shock absorber arm. | 37. | Bush—wishbone. | 63. | Bolt—steering lever to swivel axle. |
| 12. | Bearing—link. | 38. | Washer—wishbone pivot. | 64. | Hub assembly. |
| 13. | Nut—fulcrum pin. | 39. | Nut—wishbone pivot. | 65. | Stud—wheel. |
| 14. | Spring—coil. | 40. | Buffer—rebound. | 66. | Nut—wheel stud. |
| 15. | Spigot—spring. | 41. | Distance piece. | 67. | Hub assembly—R.H. |
| 16. | Screw—spigot to cross-member. | 42. | Bolt—rebound buffer to cross-member. | 68. | Collar—oil seal. |
| 17. | Nut—screw. | 43. | Screw—rebound buffer to cross-member. | 69. | Seal—oil. |
| 18. | Washer—spring—nut. | 44. | Washer—spring. | 70. | Bearing for hub—inner. |
| 19. | Pan assembly—spring. | 45. | Nut. | 71. | Spacer—bearing. |
| 20. | Wishbone assembly—bottom. | 46. | Pin—swivel. | 72. | Shim—·003 in. (·76 mm.). |
| 21. | Screw—spring pan to wishbone. | 47. | Bush—swivel pin. | 73. | Bearing—hub—outer. |
| 22. | Screw—spring pan to wishbone. | 48. | Screw—grub—swivel pin. | 74. | Washer—bearing retaining. |
| 23. | Nut—screw. | 49. | Axle assembly—swivel—R.H. | 75. | Nut—bearing retaining. |
| 24. | Washer—spring—nut. | 50. | Bush—swivel—top. | 76. | Cup—grease-retaining. |
| 25. | Tube—distance—link. | 51. | Bush—swivel—bottom. | 77. | Cup—grease-retaining. |
| 26. | Washer—thrust—link. | 52. | Lubricator—swivel bush. | | |

(8) Extract the split pin from the swivel axle and remove the nut, upper trunnion suspension link, steel and bronze thrust washers, swivel pin, and dust covers and spring. From the swivel pin remove the cork washer.

**Examination**

Wash all parts and thoroughly dry them with a non-fluffy cloth. Examine all parts for wear or damage, paying particular attention to the swivel pins, lower fulcrum pins, and all bushes. Check the pins for ovality. Worn or suspect pins or bushes must be renewed.

**Assembling and replacing**

Reverse the dismantling sequence to reassemble the axles, and, if necessary, renew the thrust washers by selective assembly to produce a condition that will permit the swivel axle to rotate freely on the pin with a minimum amount of end-play. The maximum permissible end-play is ·002 in. (·05 mm.).

The thrust washers are available in the following sizes: ·052 to ·057 in. (1·32 to 1·44 mm.), ·058 to ·063 in. (1·47 to 1·60 mm.), ·064 to ·069 in. (1·62 to 1·75 mm.).

After assembly reverse the removal sequence to refit the assembly to the hub.

## Section K.5

### REMOVING AND REFITTING BUSHES

**Swivel pin bush**

(1) Press out the old bush and ensure that the lubrication channels are clean and free from obstruction.
(2) Position the new bush with the split in the bush adjacent to the outer face of the base. This will ensure that the grease channel in the bush is in line with the channel in the pin.

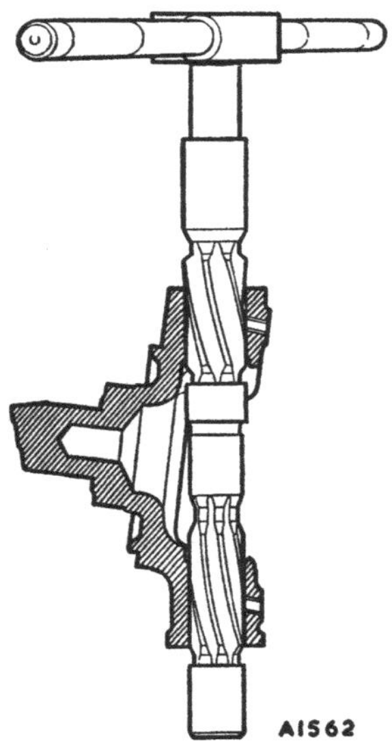


*Fig. K.3*
*Line-reaming a swivel axle*

(3) Press the bush into the pin squarely and evenly and ream to between ·7495 and ·7505 in. (19·02 and 19·05 mm.).

**Swivel axle bushes**

(1) Press out the old bushes from the bottom of the axle.
(2) Press in the new upper bushes from the bottom of the axle and ensure that the open end of the oil groove enters the axle first and that the hole in the bush is in line with the lubrication channel in the axle. Press the bush in until its top face is flush with the top face of the axle.
(3) Assemble the bottom bush in a similar manner but press it in until the lower face of the bush is flush with the counterbore in the under side of the axle.
(4) Line-bore the bushes to the following dimensions:

| | |
|---|---|
| Top bush .. | ·7815 to ·7820 in. (19·83 to 19·86 mm.). |
| Bottom bush .. | ·9075 to ·9080 in. (23·03 to 23·06 mm.). |

## Section K.6

### FRONT SUSPENSION UNIT

**Removing**

(1) Jack up and support the front of the car and remove the road wheels.
(2) Disconnect the anti-roll bar links from the spring pans.
(3) Remove the steering-rack as described in Section J.5.
(4) Drain the hydraulic fluid from the braking system.
(5) Disconnect the brake pipes from the flexible hoses and from the clips on the front suspension cross-member.
(6) Support the cross-member with a trolley jack positioned under its centre point.
(7) Remove the nuts and washers from the tops of the support bolts, lower the cross-member, and remove it from the under side of the car.
(8) From the under side of the cross-member remove the bolts, mounting plates, and the upper and lower rubber mounting pads.

**Dismantling**

(1) Remove the coil springs as described in Section K.3, the hubs as described in Section K.2, and the swivel axles as described in Section K.4.
(2) Retain the rubber bushes from the upper trunnion link and the distance tube, thrust washers, seals, and supports from the lower end of the swivel pin.
(3) Unscrew the shock absorber bolts and remove the shock absorbers.
(4) Unscrew the pivot to cross-member securing bolts and remove the pivot and wishbone arms.
(5) Unscrew the pivot nuts and from each end of the pivot remove the washer, bush, arm, and inner bush.

(6) Remove the two bolts and two screws securing the rebound buffer brackets to the cross-member and remove the brackets.

**Examination of components**

Thoroughly clean and dry all components and inspect them for cracks, fractures, deterioration, and thread damage. Additionally, carry out the following examinations:

(1) Check the hub bearings as described in Section K.2 and the swivel axles and pins as described in Section K.4.
(2) Renew rubber bushes or seals that are perished, split, eccentric, or oil-soaked.
(3) Examine the holes in the wishbone arms and spring pans for elongation.
(4) Examine the coil spring for correct length and weight as given under **'GENERAL DATA'.**
(5) Check the fulcrum pin distance tubes for scoring or wear. The tubes are 2·337 in. (59·36 mm.) long by ·748 in. (19·0 mm.) diameter.
(6) Examine the fulcrum pin thrust washers for wear or ridging. The faces should be flat and parallel to within ·0005 in. (·01 mm.). The washers are between ·065 and ·068 in. (1·68 and 1·73 mm.) thick.

**Assembling and replacing**

Assembly and replacement is a reversal of the dismantling and removal sequence, but attention must be given to the following:

(1) The swivel axles must be free to rotate on the pins as described in Section K.4.
(2) The fulcrum pin distance tubes, thrust washers, seal supports, and seals should be temporarily assembled and the swivel pin end-float checked. This should be between ·008 and ·013 in. (·20 and ·32 mm.).
(3) When assembling the lower wishbone arms the inner pivot must be fully tightened before the coil spring is fitted. This may be done in either the static or rebound position.

## Section K.7

### ADDITIONAL SWIVEL PIN GREASE NIPPLE

On later production cars an additional grease nipple is provided on the swivel pin lower bush.

Earlier cars can be modified as follows:

(1) Remove the swivel axles as described previously.
(2) Using a No. 3 (·213 in. [·541 mm.]) drill, carefully drill each swivel axle in the position shown at 'X' in Fig. K.4.

**NOTE.—Drilling is in the same position for both hands, i.e. facing forward on L.H. side, rearward on R.H. side.**

(3) Tap the ¼ in. UNF. drilled holes ¼ in. (6·35 mm.) deep.

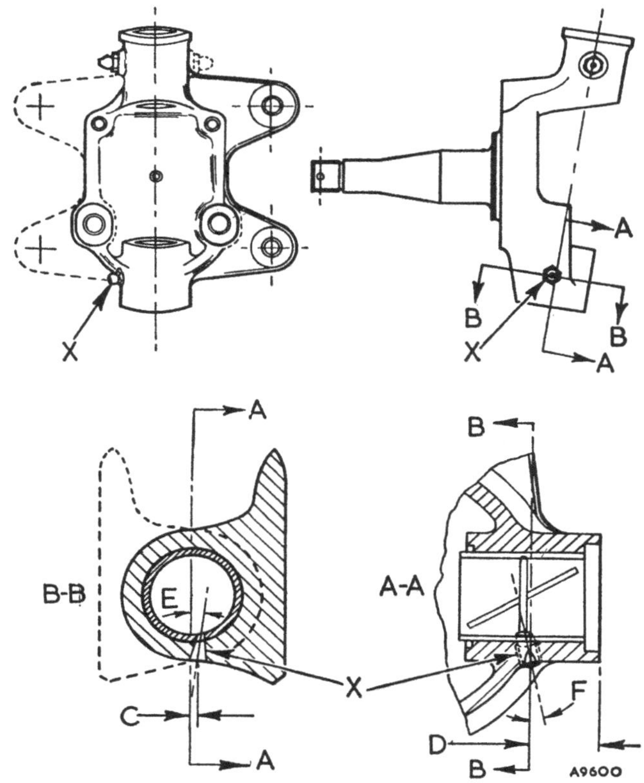


*Fig. K.4*

*Fitting additional grease nipple to swivel pin lower bush (earlier cars)*

C. $\frac{1}{16}$ in. (1·6 mm.). E. 6°.
D. ¾ in. (19 mm.). F. 12°.

(4) Remove drilling swarf and any fraze from inside bush bore.
(5) Fit grease nipple (UHN 400) to both swivel axles.
(6) Coat swivel pins with grease before assembly.

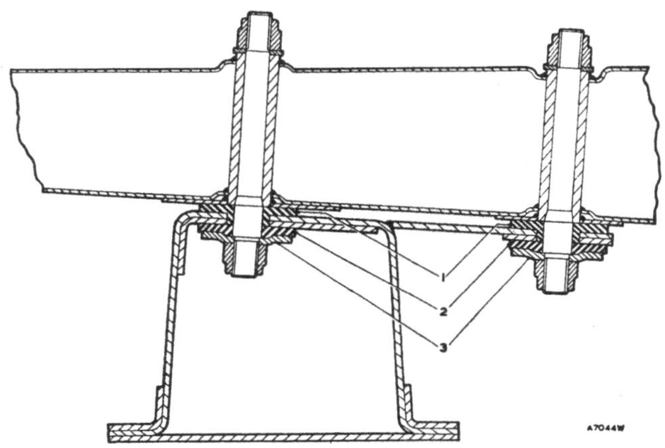


*Fig. K.5*

*Assembly of front suspension cross-member to body side-member, showing:*

1. Upper mounting pad. 2. Lower mounting pad.
3. Clamp plate.

# THE ANTI-ROLL BAR COMPONENTS

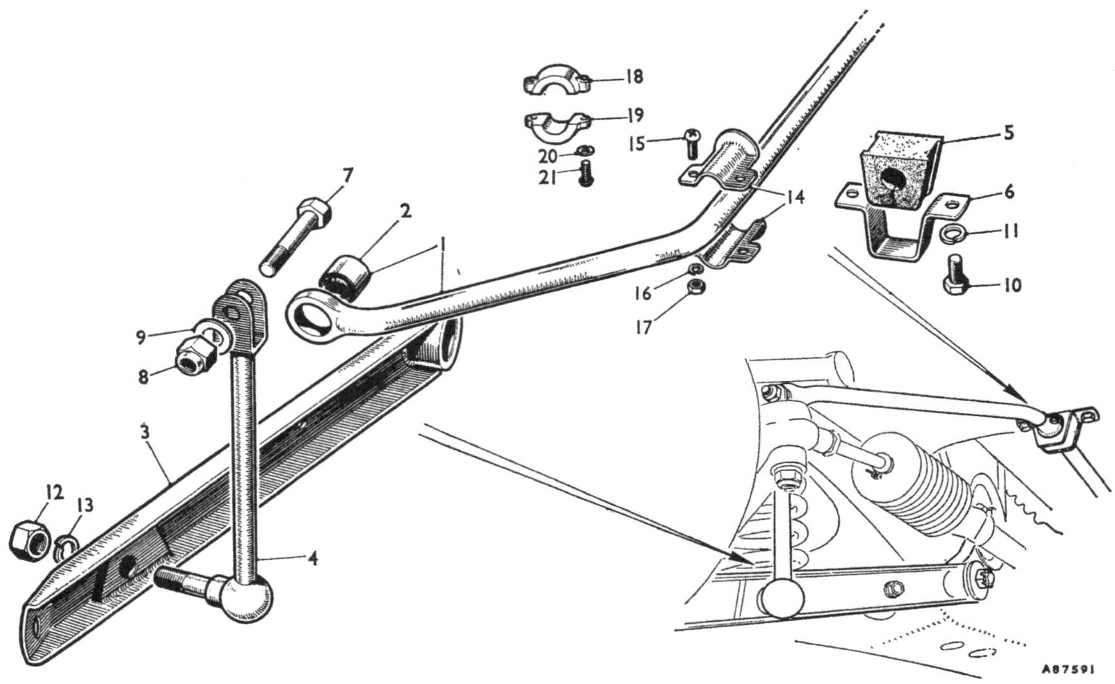


| No. | Description |
|---|---|
| 1. | Anti-roll bar assembly. |
| 2. | Bush. |
| 3. | Bottom wishbone assembly—RH. |
| 4. | Anti-roll bar link—R.H. |
| 5. | Anti-roll bar bearing. |
| 6. | Bearing strap. |
| 7. | Clamping bolt. |
| 8. | Nut. |
| 9. | Plain washer. |
| 10. | Screw. |
| 11. | Spring washer. |

| No. | Description | |
|---|---|---|
| 12. | Nut. | |
| 13. | Spring washer. | |
| 14. | Location stop end. | |
| 15. | Screw. | |
| 16. | Spring washer. | |
| 17. | Nut. | |
| 18. | Upper locator. | GT only. |
| 19. | Lower locator. | GT only. |
| 20. | Spring washer. | GT only. |
| 21. | Set screws. | GT only. |

## Section K.8

### ANTI-ROLL BAR

**Tourer**

The front suspension units on later cars are interconnected by a rubber mounted anti-roll bar; this can also be fitted to early cars.

To fit the anti-roll bar proceed as follows:

(1) Remove the coil springs as described in Section K.3 and the front lower wishbone arms as described in Section K.6.

(2) Assemble the new right-hand and left-hand wishbone arms to the pivot and swivel pins.

(3) Refit the coil spring and then fit the lower ends of the anti-roll bar links to the wishbone and spring pan.

(4) Fit the two rubber bearings to the bar and loosely assemble the end location stops outboard of the bearings.

(5) Fit the straps to the bearings and then secure the straps to the right-hand and left-hand body side-members respectively with the spring weshers and screws.

(6) Position the location stops so that their inner faces are $11\frac{1}{16}$ in. (28·1 cm.) from the centre-line of the bar. Tighten the screw and nuts to secure the stops.

(7) Place the ends of the bar into the fork ends of the links and secure them with the clamping bolts, washers, and stiff nuts. The Metalastik bushes are pressed into the ends of the anti-roll bar.

**(GT) Removing**

Raise the car to a workable height. Remove the nuts and washers securing the anti-roll bar link to the bottom wishbones. Remove the set screws securing the bearing straps to the front longitudinal members and lift away the anti-roll bar assembly.

**Refitting**

Reverse the removal procedure.

### EDITOR'S NOTES

**K. The Front Suspension**

*Removing a hub*

Service tool 18G304 may be replaced by a universal wheel puller. Service tool 18G363 is required if wire wheels are fitted. Often, no puller is needed as the hubs can be pulled off by hand or by light tapping with a soft-faced hammer.

*Refitting a hub*

A clock gauge is a dial indicater.

*Coil springs*

A hydraulic or scissors type jack between the spring pan and the ground may be used instead of the spring compressor (Service tool 18G693). Care must be taken that the jack will not slip out when the pan bolts are removed.

# SECTION L

## THE HYDRAULIC DAMPERS

## THE REAR HYDRAULIC DAMPER

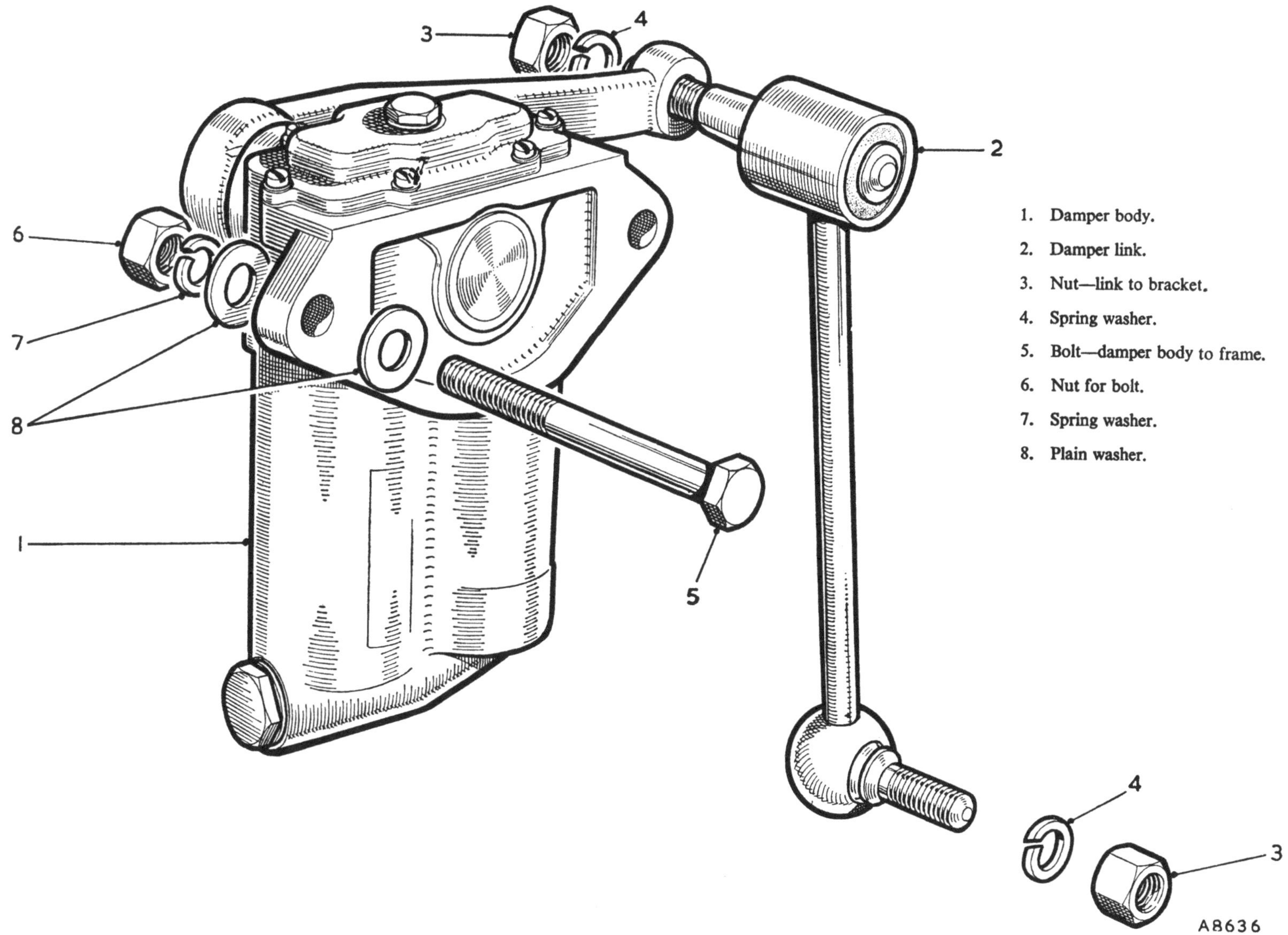


1. Damper body.
2. Damper link.
3. Nut—link to bracket.
4. Spring washer.
5. Bolt—damper body to frame.
6. Nut for bolt.
7. Spring washer.
8. Plain washer.

### GENERAL DESCRIPTION

The hydraulic dampers are Armstrong double-acting-type in which resistance is offered to the compression and to the recoil of the front suspension coil and rear leaf springs.

The front dampers are bolted to the front suspension cross-member above the coil spring seats. The damper arms are connected to the upper swivel pin link.

The rear dampers are bolted to the body rear side-members and are connected by a link to the rear axle.

## Section L.1

### MAINTENANCE

The maintenance of the hydraulic dampers, when in position on the vehicle, is confined to examination for leakage and examination of the anchorage to the chassis to ensure that the fixing bolts are tight. No adjustment of the hydraulic dampers is required or provided. They are accurately set before leaving the manufacturer to give the amount of damping most suitable for the car. **Any attempt to dismantle the assembly will seriously affect the operation and performance.**

The dampers are topped up by removing the filler plug and filling up to the bottom of the filler plug hole. **Use Armstrong Super (Thin) Shock Absorber Fluid No. 624.** (If this fluid is not available, any good-quality mineral oil to specification S.A.E. 20W should be used, but this alternative is not suitable for low-temperature operation.)

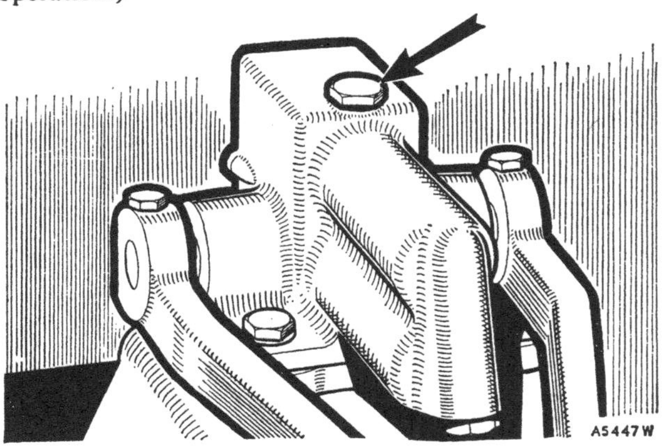


*Fig. L.1*
*The front hydraulic damper filler plug*

**Before removing the filler cap, which is located on the top of the damper, carefully wipe the exterior, as it is of utmost importance that no dirt whatever enters through the filler hole.**

On no account neglect the operation of topping up the damper fluid because if the low-pressure chamber of the unit is allowed to become empty, air will enter the pressure cylinders and the action of the damper will be impaired.

Access to the rear dampers is obtained by removing the rubber plugs in the rear floor panel adjacent to the battery cover.

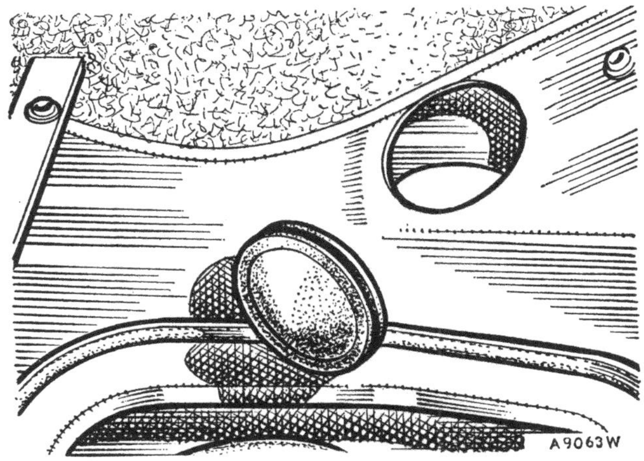


*Fig. L.2*
*A rubber plug removed from the rear floor panel to give access to the rear hydraulic damper filler plug*

## Section L.2

### REMOVING AND REPLACING THE DAMPERS

**Front**

Jack up the car under the lower wishbone spring pan until the wheel is clear of the ground.

Remove the wheel and take out the swivel pin top pivot bolt. Swing out the hub unit clear of the upper wishbone and support it on a suitable stand to prevent straining the brake hose. Unscrew the four nuts holding the damper to the cross-member.

**Rear**

Jack up the rear of the car below the axle or rear springs and remove the rear wheel.

Remove the nut and spring washers securing the damper arm to the bracket on the rear spring.

Remove the nuts, spring and flat washers from the two bolts securing the damper to the side-member, and withdraw the damper.

NOTE.—**When handling hydraulic dampers that have been removed for any purpose, it is important to keep the assemblies upright as far as possible, otherwise air may enter the operating chamber and result in free movement.**

Replacement is a reversal of the above procedure; ensure that the securing bolts are tightened to the torque figure given in **'GENERAL DATA'**.

## Section L.3

### TESTING THE DAMPERS

If the hydraulic dampers do not appear to function satisfactorily, the resistance may be roughly checked by bouncing each corner of the car up and down. A uniform movement indicates that no attention is required, but if the resistance is erratic or free movement of the car is felt, the damper should be removed for checking and topping up.

Indication of their resistance can be obtained by carrying out the following check.

Bolt the damper, in an upright position, to a plate held in a vice.

Move the lever arm up and down through its complete stroke. A moderate resistance throughout the full stroke should be felt. If the resistance is erratic, and free movement in the lever arm is noted, it may indicate lack of fluid.

While adding fluid the lever arm must be worked throughout its full stroke to expel any air that may be present in the operating chamber.

If the addition of fluid gives no improvement a new damper should be fitted.

Too much resistance, i.e. when it is not possible to move the lever arm by hand, indicates a broken internal part or a seized piston.

The arms should not be removed from the dampers at any time as it is essential that they should be assembled to the damper shaft in the right relation to the damper cam lever so that there is the full range of movement on either side of the centre-line.

It must be clearly understood that there is no provision for adjusting the setting of the dampers, and if they are in any way defective they must be returned to the manufacturers for attention.

Leakage from the damper lid may be rectified by fitting a new lid gasket and fully tightening the screws.

A weep from the valve screw may be stopped by gently tightening the screw.

# SECTION M

## THE BRAKING SYSTEM

## GENERAL DESCRIPTION

The Lockheed hydraulic brake system comprises a master cylinder and fluid supply reservoir, automatically adjusted disc brake assemblies at the front of the vehicle, and manually adjusted drum-type brakes at the rear.

The master cylinder and fluid supply reservoir assembly is mounted on the engine side of the bulkhead above the brake pedal.

The front brake units are of the rotating disc and rigidly mounted calliper type, each calliper carrying two friction pad assemblies between which the disc revolves. The rear brake units are of the internal-expanding shoe and drum type, each unit being operated by a single wheel cylinder which is actuated hydraulically by the foot pedal and mechanically by the hand brake.

## Section M.1

### MAINTENANCE

Periodically examine the quantity of brake fluid in the master cylinder. The fluid level should never be higher than ¼ in. (6·35 mm.) below the bottom of the filler neck, and the cylinder must never be less than half-full.

A considerable drop in the level of the fluid in the reservoir is an indication of a leak in the system which should at once be traced and rectified.

The grease nipple on the hand brake cable should be charged with grease.

**Brake adjustment**

*Front*

Wear on the front pads is automatically compensated during braking, and manual adjustment therefore is not provided. In order to maintain peak braking efficiency and to obtain the maximum life from the friction pads they should be examined at the recommended periods, and if the wear on one pad is greater than that on the other their operating positions must be changed over.

*Fig. M.1*

*The square head of the adjuster spindle*

*Rear*

As the brake-shoe linings wear, the movement of the brake pedal will increase, indicating that adjustment is necessary.

Block the front wheels and jack up each rear wheel in turn. Fully release the hand brake. Turn the expander in a clockwise direction until the wheel is locked, then turn back one notch only. The wheel should be free to rotate without the shoe rubbing. Adjust the other rear brake in a similar way. Adjustment of the brake-shoes automatically adjusts the hand brake mechanism.

**Preventive maintenance**

To safeguard against the possible effects of wear, or deterioration, it is recommended that:

(1) Disc brake pads, drum brake linings, hoses, and pipes should be examined at intervals no greater than those laid down in the Passport to Service.

(2) Brake fluid should be changed completely every 18 months or 24,000 miles (40000 km.) whichever is the sooner.

(3) All fluid seals in the hydraulic system and all flexible hoses should be examined and renewed if necessary every 3 years or 40,000 miles (65000 km.) whichever is the sooner. At the same time the working surface of the pistons and of the bores of the master cylinder, wheel cylinders, and other slave cylinders should be examined and new parts fitted where necessary.

Care must be taken always to observe the following points

(*a*) At all times use the recommended brake fluid.

(*b*) Never leave fluid in unsealed containers. It absorbs moisture quickly and this can be dangerous.

(*c*) Fluid drained from the system or used for bleeding is best discarded.

(*d*) The necessity for absolute cleanliness throughout cannot be over-emphasized.

## Section M.2

### BRAKE PEDAL

The brake and clutch pedals are mounted at a common fulcrum point inside the master cylinder cover.

**Removing**

Remove the screws securing the brake and clutch master cylinder cover and take off the cover.

Extract the split pin and withdraw the clevis pin to release the master cylinder push-rod from the brake pedal. Detach the brake pedal return spring and remove the nut and spring washer from the pedal fulcrum bolt The fulcrum bolt may now be withdrawn together with the brake pedal and internal distance sleeve. The clutch pedal will remain in position held by its return spring and master cylinder push-rod.

Reverse the above procedure to replace the brake pedal, noting that the head of the fulcrum bolt is at the clutch pedal side of the assembly.

## Section M.3

### HAND BRAKE

The hand brake, which is of the lever type, incorporating a thumb-operated ratchet release in the handle, is located on the right-hand side of the floor tunnel between the seats.

Pulling the handle operates the rear shoes only by means of cable-operated levers. The action of the inner and outer cables is employed through the action of a balance lever mounted on the rear axle to ensure even braking on the rear wheels.

Adjustment of the rear brake-shoes automatically adjusts the hand brake, but excessive movement of the hand brake lever due to cable stretch should be taken up as follows.

Adjust the inner cable length by turning the brass adjusting nut at the lower end of the hand brake lever below the car floor. The adjustment is correct if the hand brake is applied fully when the lever is pulled up three or four notches.

The rear brake-shoes should be adjusted as described in Section M.1 before taking up the hand brake cable stretch.

**To remove (lever and cable assembly)**

Unscrew and remove the adjusting nut, withdraw the end of the cable from the lower end of the lever, and remove the spring and flat washers.

Remove the nut securing the brake lever to the hand brake spindle and withdraw the spring washer, forked lever, and plain washer. The hand lever may now be withdrawn from inside the car.

Remove the right-hand seat (see Section R). Unscrew the three screws securing the ratchet plate to the floor tunnel. Lift the carpet and remove the nut and spring washer securing the outer cable front abutment. Disconnect the clips securing the cable assembly to the body and rear axle.

Remove the bolt, nut, and spring washer connecting the two halves of the brake compensating lever to each other, slacken fully the self-locking nut securing the lever to the axle bracket, and release the cable abutment trunnion from the lever. Extract the split pins and withdraw the clevis pins to release the cable yokes from the levers on the brake backplates.

On replacement, check the hand brake lever operation as detailed above.

## Section M.4

### BLEEDING THE BRAKE SYSTEM

The following procedure should be adopted either for initial priming of the system or to bleed in service if air has been permitted to enter the system. Air may enter the system if pipe connections become loose or if the level of fluid in the reservoir is allowed to fall below the recommended level. During the bleeding operation it is important to keep the reservoir at least half-full to avoid drawing air into the system.

Check that all connections are tightened and all bleed screws closed. Fill the fluid reservoir with the recommended fluid (see **'GENERAL DATA'**). Keep it at least half-full throughout the bleeding operation, otherwise air will be drawn into the system, necessitating a fresh start.

Attach the bleeder tube to the bleed screw on the near-side rear brake and immerse the open end of the tube in a small quantity of brake fluid contained in a clean glass jar. Slacken the bleed screw and depress the brake pedal slowly through its full stroke and allow it to return without assistance. Repeat this pumping action with a slight pause before each depression of the pedal. When fluid entering the jar is completely free of air bubbles hold the pedal firmly against the floorboards and tighten the bleeder screw.

This process must be repeated at each of the three remaining brake assemblies.

Top up the fluid reservoir to its correct level, ¼ in. (6·35 mm.) below the filler neck.

If the disc brake callipers have been disturbed it will be necessary to pump the brake pedal several times to restore the automatic adjustment of the friction pads.

Apply a normal working load on the brake pedal for a period of two or three minutes and examine the entire system for leaks.

**(See Editor's notes at end of section.)**

## Section M.5

### MASTER CYLINDER

**Description**

The master cylinder piston is backed by a rubber cup normally held in the 'off' position by a piston return spring. Immediately in front of the cup, when it is in the 'off' position, is a compensating orifice connecting the cylinder with the fluid supply. Pressure is applied to the piston by means of the push-rod attached to the brake pedal.

The reduced skirt of the piston forms an annular space

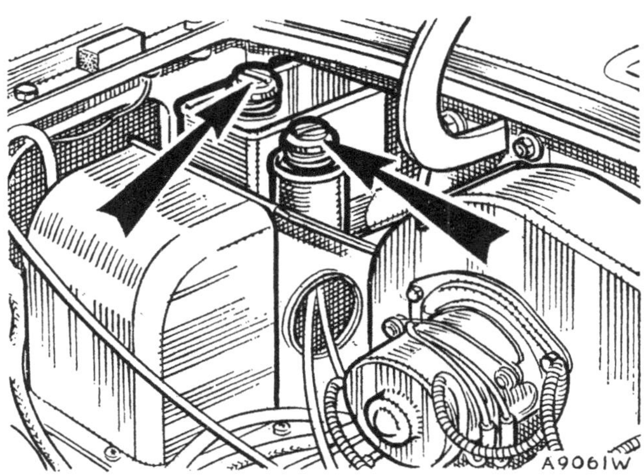


*Fig. M.2*

*The upper arrow shows the brake master cylinder filler cap*

which is filled with fluid from the supply tank. On releasing the brake pedal after application the piston is returned quickly to its stop by the return spring. A small bleed hole is drilled in the side of the check valve body to prevent any fluid pressure being retained in the system.

**Removing**

Remove the screws securing the brake and clutch master cylinder cover and take off the cover.

Drain the fluid from the supply tank reservoir by attaching a rubber tube to a brake calliper bleed screw, open the screw one full turn, and then depress the brake pedal. Hold the pedal down, tighten the bleed screw, and then let the pedal return unassisted. Repeat this operation until the tank is empty.

Remove the split pin, washer, and clevis pin from the push-rod and disengage the brake pedal lever.

Clean the pipe connection, disconnect the pipe line, and fit a plug to the end of the cylinder to prevent the entry of dirt.

Unscrew the fixing bolts and detach the master cylinder from the box assembly.

**Dismantling**

Detach the rubber dust cover from the cylinder barrel and move it along the push-rod.

Depress the push-rod to relieve the load on the circlip, then remove the circlip and the push-rod assembly.

Withdraw the piston, piston washer, main cup, spring retainer, and valve assembly.

Remove the secondary cup by carefully stretching it over the end of the piston.

**Assembling**

Clean all parts thoroughly, using brake fluid or methylated spirit. If spirit is used the rubber should not be allowed to soak but should be dried and dipped in clean brake fluid. The main casing, if cleaned with spirit, must be dried out before assembly. Ensure that the compensating port in the cylinder barrel is clear by probing with a piece of fine wire.

Examine all the rubber parts for damage or distortion. It is usually advisable to renew the rubbers when rebuilding the cylinder. Dip all the internal parts in brake fluid and assemble when wet.

Stretch the secondary cup over the piston with the lip of the cup facing the head of the piston. When the cup is in its groove work it round gently with the fingers to ensure that it is correctly seated.

Fit the check valve assembly onto the large end and the spring retainer onto the small end of the return spring and insert the spring, check valve first, into the cylinder.

Insert the main cup, lip first, and press it down onto the spring retainer, taking care not to damage or turn back the lip.

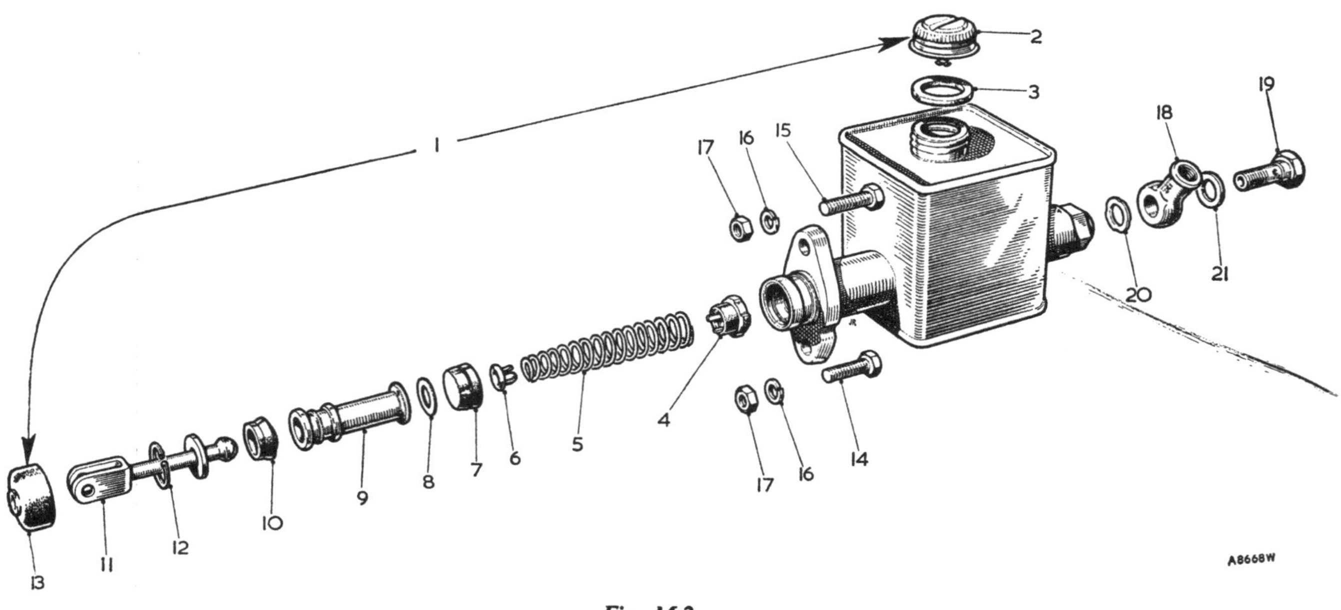


*Fig. M.3*
*The brake master cylinder*

1. Cylinder and supply tank assembly.
2. Cap—filler.
3. Seal—cap.
4. Valve assembly.
5. Spring—piston return.
6. Retainer—return spring.
7. Cup—main.
8. Washer—piston.
9. Piston.
10. Cup—secondary.
11. Rod—push.
12. Circlip.
13. Boot.
14. Screw—cylinder to box.
15. Screw—cylinder and stiffener to box.
16. Washer—spring—screw.
17. Nut—screw.
18. Banjo—master cylinder.
19. Bolt—banjo.
20. Gasket—banjo connection.
21. Gasket—banjo connection.

Place the main cup washer in position with its concave face next to the main cup and insert the piston, taking care not to damage or turn back the lip of the secondary cup.

Place the push-rod in position, push the piston down the bore, and with the piston stop washer in the bore refit the retaining circlip. Refit the dust cover.

**Replacing**

Refit the master cylinder to the master cylinder box and secure it with bolts. The long bolt passes through the stiffener plate.

Remove the dust excluder and fit the pipe connection to the master cylinder.

Refit the clutch pedal lever to the push-rod and secure it with the clevis pin, washer, and a new split pin.

Refit the master cylinder cover.

Fill the master cylinder and then prime and bleed the system.

## Section M.6

### DISC BRAKE UNITS

**Description**

Each front wheel brake unit comprises a hub-mounted disc rotating with the wheel and a braking unit rigidly attached to the swivel axle. The brake unit consists of a calliper manufactured in two halves—the mounting half and the rim half—which are held together by three bolts. A cylinder in each calliper half houses a self-adjusting hydraulic piston, a fluid seal, a dust seal, and a seal retainer. The pistons are interchangeable side for side.

The friction pad assemblies are fitted adjacent to the pistons and are retained in position by a retainer spring and pin.

Fluid pressure generated in the master cylinder enters the mounting half of each calliper and passes through the internal fluid ports into the rim half. An even pressure is therefore exerted on both hydraulic pistons, moving them along the cylinder bores until the friction pad assemblies contact the disc. In order to compensate for wear of the pads the pistons move progressively along each cylinder. The movement of the piston deflects the fluid seal in the cylinder bore, and on releasing the pressure the piston moves back into its original position thus providing the required clearance for the friction pads.

**Removing the disc friction pads**

Apply the hand brake, jack up the car, and remove the road wheel.

Depress the pad retaining springs and remove the split pins and the retaining springs; lift the pads out of the calliper.

Before the lining material has worn down to a minimum permissible thickness of $\frac{1}{16}$ in. (1·59 mm.) the friction pads must be renewed.

Thoroughly clean the exposed end of each piston and ensure that the recesses which are provided in the calliper to receive the friction pads are free from rust and grit.

Before fitting new friction pads the calliper pistons which will be at their maximum adjustment, must be returned to the base of the bores, using a suitable clamp.

**NOTE.—The level of the fluid in the master cylinder supply tank will rise during this operation and it may be necessary to siphon off any surplus fluid to prevent it from overflowing.**

Check that the portion that has been machined away from the face of each piston is correctly positioned at the inner end of the calliper (see Fig. M.6). Insert the friction pads (which are interchangeable side for side), replace the retaining springs, and fit the split pins. Ensure that the pad assemblies are free to move easily in the calliper recesses. Remove any high-spots from the pad pressure plate by filing carefully.

Pump the brake pedal several times to readjust the pistons and top up the fluid supply reservoir.

**Removing a calliper unit**

Apply the hand brake, jack up the car, and remove the road wheel. Withdraw the brake friction pads.

Attach a bleeder tube to the bleed screw and drain the fluid by pumping the brake pedal. Disconnect the flexible hose on the mounting half of the calliper (see Section M.4) and plug the end of the hose to prevent the entry of foreign matter.

Press back the ears of the locking washer, unscrew the two bolts securing the calliper to the swivel axle and withdraw the calliper complete.

**Replacing**

Reverse the above instructions. Tighten the calliper securing bolts to a torque spanner reading of 40 to 45 lb. ft. (5·6 to 6·2 kg. m.). Finally, bleed the system as described in Section M.4.

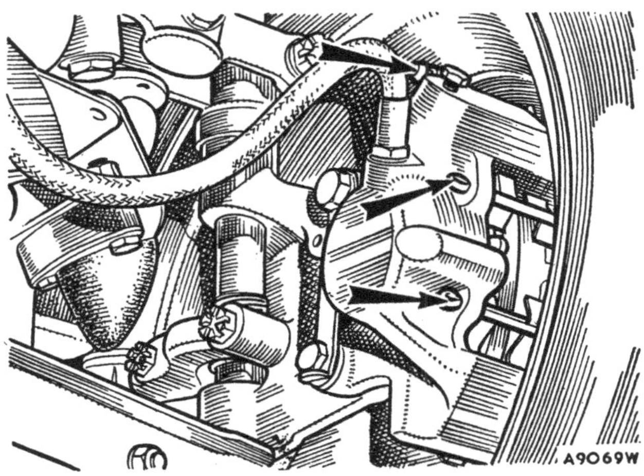


*Fig. M.4*

*Friction pad retaining pins and bleed screw location*

## THE DISC BRAKE COMPONENTS

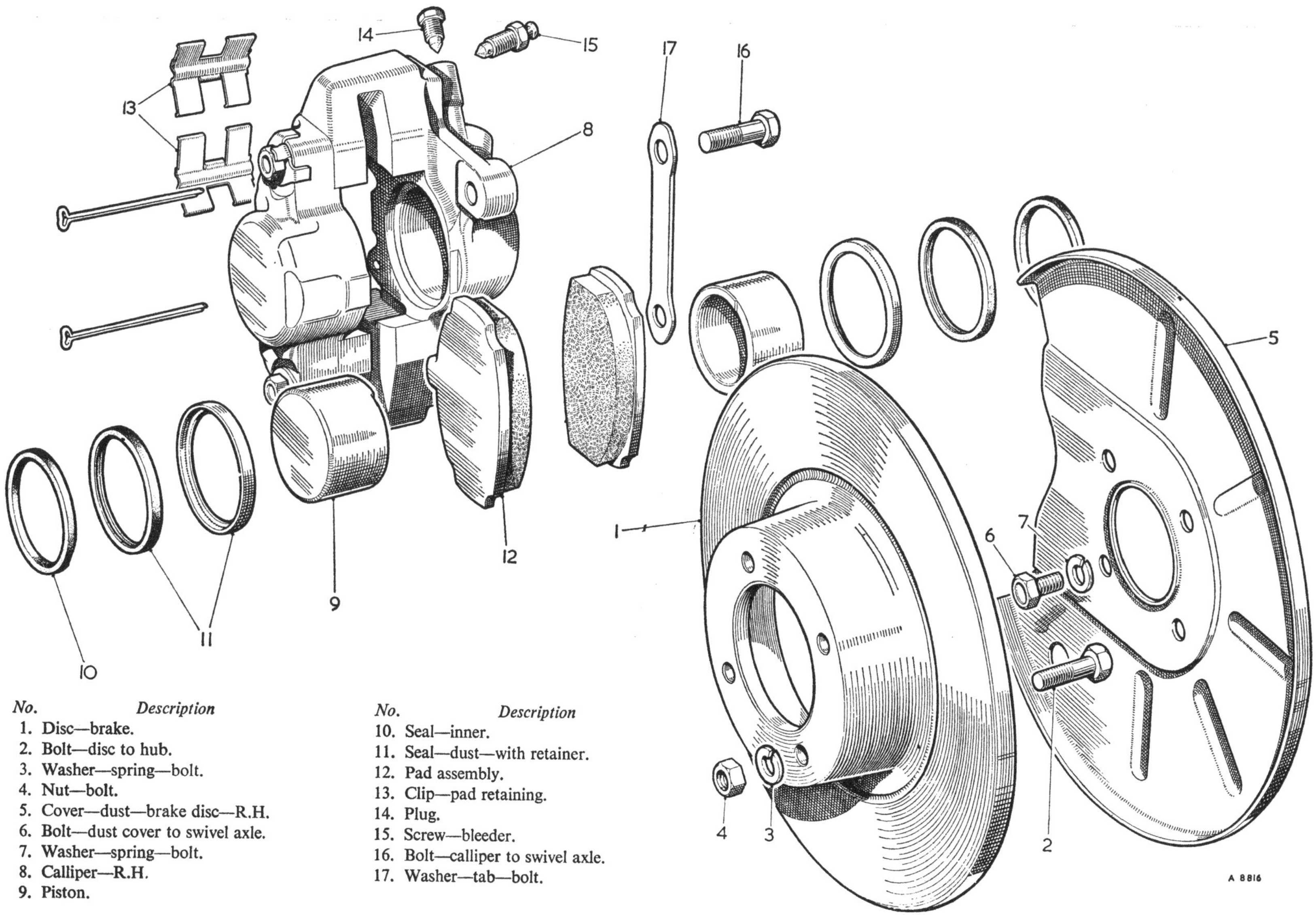


| No. | Description |
|---|---|
| 1. | Disc—brake. |
| 2. | Bolt—disc to hub. |
| 3. | Washer—spring—bolt. |
| 4. | Nut—bolt. |
| 5. | Cover—dust—brake disc—R.H. |
| 6. | Bolt—dust cover to swivel axle. |
| 7. | Washer—spring—bolt. |
| 8. | Calliper—R.H. |
| 9. | Piston. |
| 10. | Seal—inner. |
| 11. | Seal—dust—with retainer. |
| 12. | Pad assembly. |
| 13. | Clip—pad retaining. |
| 14. | Plug. |
| 15. | Screw—bleeder. |
| 16. | Bolt—calliper to swivel axle. |
| 17. | Washer—tab—bolt. |

**Removing the calliper pistons**

Unscrew and remove the two bolts securing the calliper to the front hub and withdraw the calliper from the disc and hub. Do not remove the rubber hose, and support the calliper to avoid straining the hose. Remove the friction pads and clean the outside of the calliper, making sure that all dirt and traces of cleaning fluid are completely removed.

Clamp the piston in the mounting half of the calliper and gently apply the foot brake. This operation will force the piston in the rim half of the calliper to move outwards. Continue with gentle pressure on the foot pedal until the piston has emerged sufficiently for it to be removed by hand. Have a clean receptacle ready to catch the fluid as the piston is removed.

With a suitable blunt-nosed tool remove the fluid seal from its groove in the bore of the calliper, taking great care not to damage the bore of the calliper or the seal retaining groove.

The dust seal retainer can be removed by inserting a screwdriver between the retainer and the seal and gently prising the retainer from the mouth of the calliper bore. The rubber seal can then be detached.

Remove the clamp from the mounting-half piston. To remove the mounting-half piston from the calliper it is necessary first to refit the rim-half piston, and thereafter the procedure is as previously detailed.

When cleaning out the calliper it is essential that only methylated spirit or Lockheed Brake Fluid be used as a cleaning medium. Other types of cleaning fluid may damage the internal rubber seal between the two halves of the calliper.

(See Editor's notes at end of section.)

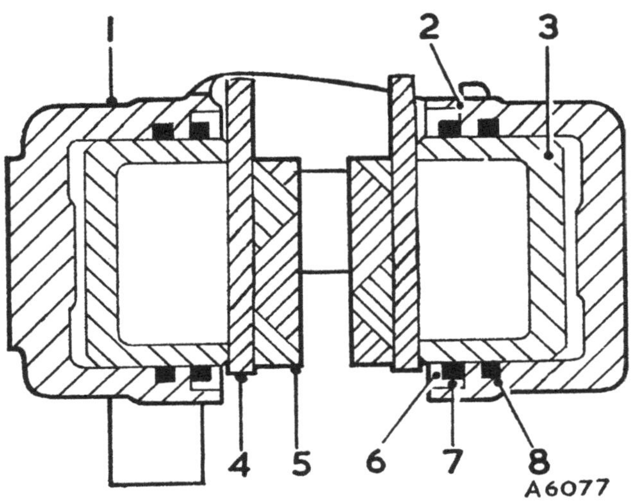


*Fig. M.5*

*A disc brake in section*

1. Calliper—mounting half.
2. Calliper—rim half.
3. Hydraulic piston.
4. Pad backing plate.
5. Friction pad.
6. Dust seal retainer.
7. Dust seal.
8. Fluid seal.

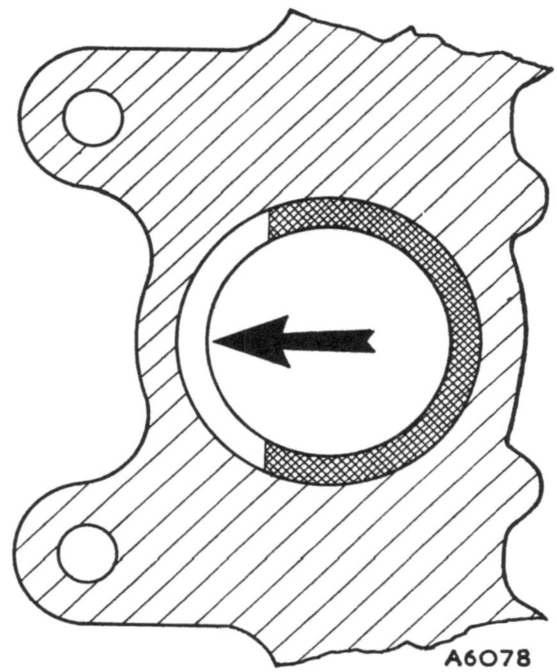


*Fig. M.6*

*The cut-away portion of the piston (arrowed) must be located at the inner edge of the calliper, i.e. towards the hub*

**Reassembling**

Coat a new fluid seal with Lockheed Disc Brake Lubricant, making sure that the seal is absolutely dry before so doing, and ease the seal into its groove with the fingers until it is seating correctly in the groove.

Slacken the bleeder screw in the rim half of the calliper one complete turn. Coat the piston with Lockheed Disc Brake Lubricant and locate the piston squarely in the mouth of the bore with the cutaway portion of the piston face correctly positioned (see Fig. M.6).

Press in the piston until approximately $\frac{5}{16}$ in. (7·94 mm.) of the piston is protruding from the bore. Take great care to prevent the piston tilting during this operation. If the dust seal and retainer have been previously removed, take a new, perfectly dry dust seal, coat it with Lockheed Disc Brake Lubricant, and fit the seal into its retainer. Position the seal assembly on the protruding portion of the piston with the seal innermost, ensuring that the assembly is square with the piston. Press home the piston and seal assembly with clamp. Retighten the bleeder screw.

The mounting-half piston is dealt with in the same manner as described for the rim-half piston. The rubber hose must be disconnected to allow the clamp to be used and the bleeder screw must be slackened.

Reconnect the hose and bolt the calliper to the hub. Do not depress the brake pedal. Fit the friction pad assemblies, together with their retaining springs and split pins, and bleed the system.

After bleeding operate the brake pedal several times to adjust the brake.

**Dismantling the calliper**

Further servicing of the calliper should be confined to removing the bleeder screw and the fluid pipe line and blowing the fluid passages clear with compressed air.

**Unless it is absolutely unavoidable the calliper should not be separated into two halves. In the event of separation becoming essential, the fluid channel seal, clamping bolts, and lock plates must be renewed when reassembling. Only bolts supplied by BMC Service Ltd. may be used. On assembly these must be tightened with a torque wrench set at between 35·5 and 37 lb. ft. (4·9 and 5·1 kg. m.).**

**Ensure that the calliper faces are clean and that the threaded bolt holes are thoroughly dry. Make certain that the new fluid channel seal is correctly located in the recessed face before assembling the two calliper halves.**

## Section M.7

### BRAKE DISCS

**Removing**

Remove the brake calliper as detailed in Section M.6 without disconnecting the fluid supply and withdraw the hub by the method described in Section K.

Separate the disc from the hub by removing the four securing nuts and washers.

**Replacing**

Assemble the brake disc to the hub and refit the assembly to the swivel hub.

Check the disc for true rotation by clamping a dial indicator to a suitable fixed point on the vehicle with the needle pad bearing on the face of the hub. Run-out must not exceed ·003 in. (·076 mm.), and in the event of this dimension being exceeded the components should be examined for damage and, if necessary, renewed. Damaged disc faces may be rectified by grinding (see below).

Replace the brake calliper as detailed in Section M.6.

A certain amount of concentric and even scoring of the disc faces is not detrimental to the satisfactory operation of the brakes.

If it is found necessary to regrind the disc faces they can be ground up to a maximum of only ·040 in. (1·016 mm.) off the original thickness of ·350 to ·340 in. (8·89 to 8·63 mm.). This may be ground off equally each side, or more on one side than the other, provided that the total reduction does not exceed the maximum limit of ·040 in. (1·016 mm.). The reground surface must not exceed 63 micro-in.

After grinding, the faces must run true to within a total clock reading of ·002 in. (·05 mm.) and the thickness must be parallel to within ·001 in. (·0254 mm.) clock reading.

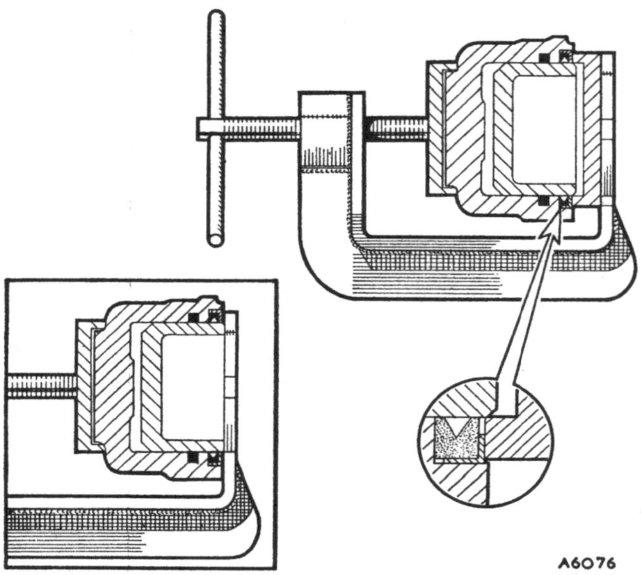


*Fig. M.7*

*Using Service tool 18G590 to install the dust seal and retainer in the recessed mouth of the calliper cylinder. Shown inset is the tool, less the adaptor, being used to reset a piston*

## Section M.8

### FLEXIBLE HOSES

Do not attempt to release a flexible hose by turning either end with a spanner. It should be removed as follows.

Unscrew the metal pipe line union nut from its connection to the hose.

Remove the locknut securing the flexible hose union to the bracket and unscrew the hose from the wheel cylinder.

## Section M.9

### REAR BRAKE ASSEMBLY

The rear brakes are of the leading and trailing shoe type, giving the advantage of equal braking action whether the car is travelling forwards or backwards.

The hand brake lever operates the brakes mechanically through linked levers which apply a force to each shoe. When the foot brake pedal is depressed the master cylinder piston applies pressure to the fluid, thus causing the pistons in the wheel cylinder to operate on the tip of the leading and trailing shoes.

When pressure on the brake pedal is released the brake-shoe springs return the shoes, thrust the pistons back into the wheel cylinders, and the fluid passes back to the master cylinder.

**Dismantling**

Jack up the car and remove the road wheel.

Remove the brake-drum as described in Section H.1.

Slacken fully the brake-shoe adjuster.

Depress each shoe steady spring retaining washer and turn to release them from the anchor brackets on the backplate. Pull the trailing shoe against the load of the springs and disengage at each end; on releasing the tension on the springs the other shoe will fall away.

# THE REAR BRAKE COMPONENTS

| No. | Description |
|---|---|
| 1. | Backplate. |
| 2. | Backplate to axle case bolt. |
| 3. | Nut. |
| 4. | Spring washer. |
| 5. | Shoe assembly. |
| 6. | Pull-off spring—cylinder end. |
| 7. | Pull-off spring—adjustment end. |
| 8. | Brake-shoe steady pin. |
| 9. | Brake-shoe steady spring. |
| 10. | Retainer washer. |
| 11. | Adjuster assembly. |
| 12. | Tappet. |
| 13. | Wedge spindle. |
| 14. | Adjuster to backplate nut. |

| No. | Description |
|---|---|
| 15. | Spring washer. |
| 16. | Wheel cylinder assembly. |
| 17. | Piston. |
| 18. | Piston seal. |
| 19. | Piston boot. |
| 20. | Wheel cylinder retaining clip. |
| 21. | Bleeder screw. |
| 22. | Hand brake lever. |
| 23. | Hand brake lever boot. |
| 24. | Brake-drum. |
| 25. | Drum to hub screw. |
| 26. | Drum retaining nut (wire wheels). |

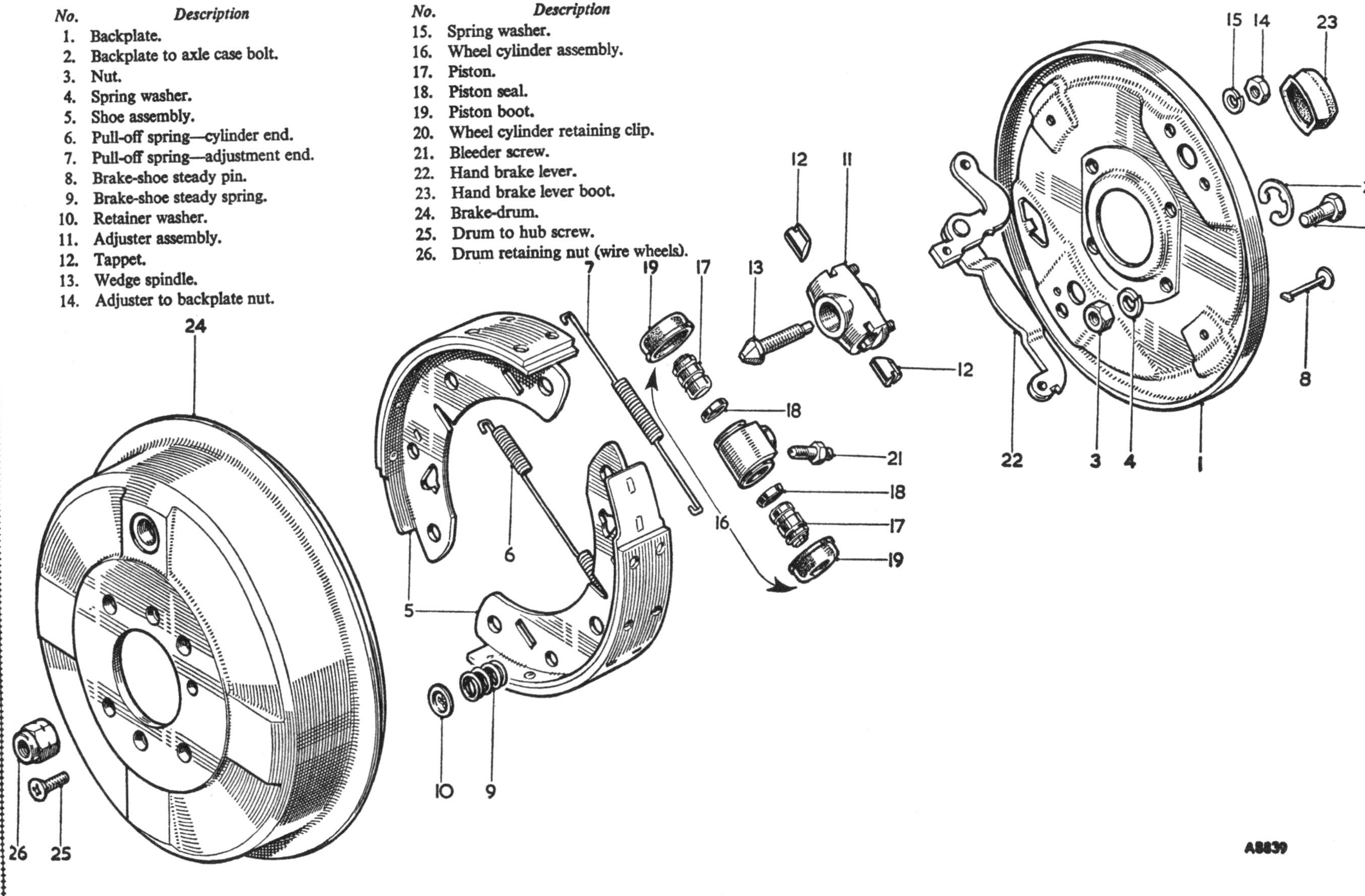

To remove the wheel cylinder disconnect the brake fluid supply pipe, placing a container to catch the fluid. Withdraw the circlip and retaining washer and remove the cylinder.

Extract the split pin and withdraw the clevis pin to release the hand brake cable from the lever. Detach the rubber dust cover from the brake lever at the rear of the backplate and withdraw the lever.

Withdraw the tappets from the spindle adjuster and screw the adjusting spindle inwards until clear of the threads. Remove the two nuts and spring washers from the rear of the backplate to release the adjuster body.

**Assembling**

Thoroughly clean the adjuster body, tappets, and adjuster. Smear the adjuster threads and tappets with Lockheed Expander Lubricant. Screw the adjuster fully home and slide the tappets into the body, ensuring that the tapered porton on each is facing inwards.

Examine the rubber seals on both pistons and renew them should they appear damaged or distorted. It is usually advisable to renew the rubbers when rebuilding the cylinders. Smear all internal parts with fluid and reassemble. Replace the dust covers. Hold the cylinder up against the backplate and replace the flat washer and circlip. Reconnect the fluid supply pipe.

Hold the brake lever against the backplate and replace the rubber boot. Reconnect the hand brake cable.

The brake-shoes are interchangeable, but when replacing, the pull-off springs must be on the backplate side of the shoes and located in the shoes as shown in Fig. M.8.

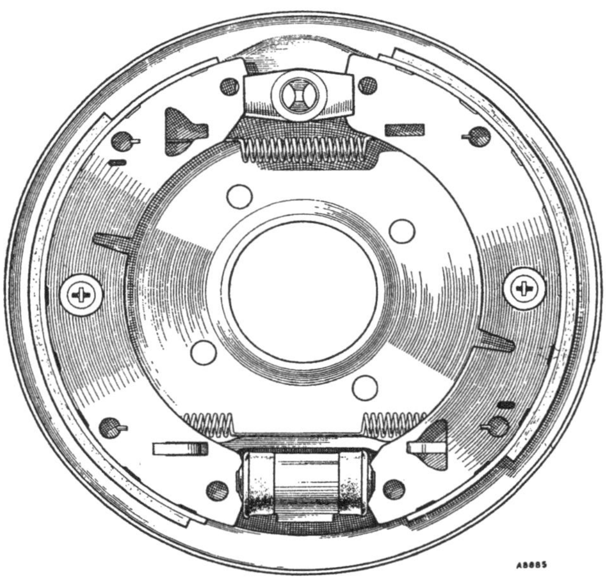

*Fig. M.8*

***The rear brake assembly***

Replace the steady springs on the shoe web and locate them with the brackets on the backplate by depressing and turning the retaining washer.

Ensure that all adjustments are off and that the shoes are centralized. Fit the drum and the road wheel, bleed the system, and adjust the brakes.

# Section M.10

## RELINING THE BRAKE-SHOES

Owing to the need for the brake linings to be finished so that they are perfectly concentric with the brake-drums, special precautions must be taken when relining the shoes.

**It is imperative that all brake linings should be of the same make, grade, and condition to ensure even braking.**

**When brake linings are in need of renewal they must always be replaced in axle sets, and the relining of the shoes in one brake-drum must be avoided.**

Any variations from this will give an unequal and unsatisfactory braking performance.

After riveting the new brake linings to the brake-shoes it is essential that any high-spots should be removed before replacement of the backplate assembly.

When new linings are fitted it is necessary to return the spindle adjuster to the fully off position. The hand brake must also be in the fully released position.

Do not allow grease, paint, oil, or brake fluid to come into contact with the brake linings.

(See Editor's notes at end of section.)

# Section M.11

## REAR BRAKE BACKPLATE

**Removing**

Jack up the car and remove the road wheel. Remove the brake-drum and withdraw the axle half-shaft (see Section H.1 or Ha.3) and the rear hub (see Section H.2 or Ha.3).

Disconnect the brake pipe from the back of the wheel cylinder. Extract the split pin and remove the clevis pin to detach the hand brake cable from the brake lever.

Unscrew the four nuts securing the backplate to the flange on the axle and withdraw the backplate complete with shoes, wheel cylinder, and shoe adjuster.

**Replacing**

Reverse the above instructions; readjust the brakes (see Section M.1) and finally, bleed the hydraulic system as detailed in Section M.4.

## EDITOR'S NOTES

**M. The Braking System**

*Bleeding the brake system*

When bleeding the brake system, always start with the bleed screw farthest from the master cylinder (in terms of tubing length) and finish with the one nearest the master cylinder.

*Removing the calliper pistons, Reassembling*

Service tool 18G590 is quite useful for performing these operations.

*Relining the brake shoes*

It is common practice to purchase a relined set of brake shoes on an exchange basis. If the drums are severely worn, a limited amount of material may be cut to trim the braking service. The relined shoes should be ground to match the brake drums before installation.

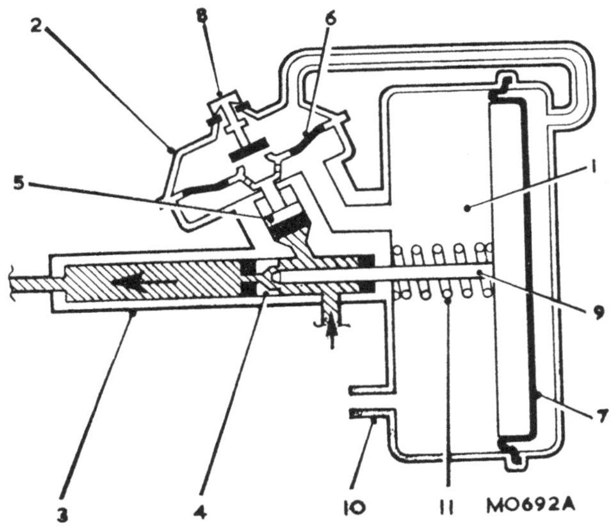


*Fig. M.9*

*A schematic diagram showing the principle of operation and the main components of the vacuum servo unit. The shaded area represents brake fluid*

## Section M.12

### SERVO UNIT
### (Lockheed Type 6)

**Operation**

Refer to Fig. M.9. The vacuum-operated servo unit consists of three main components, namely the vacuum cylinder (1), the air valve assembly (2), and the slave cylinder (3) which is connected in the hydraulic circuit between the main master cylinder and the wheel cylinders. Under light braking, fluid is allowed to pass directly to the wheel cylinders via the hollow centre of the slave piston (4) and no braking assistance is obtained; fluid pressure acting on the air valve piston (5) closes the diaphragm (6), thus separating the chamber behind the main servo diaphragm (7) from the one in front. Under heavier braking, further movement of the air valve piston opens the air valve and allows air to enter the chamber behind the main diaphragm, destroying the vacuum. The central rod (9) is thus pushed to the left, sealing the hollow centre of the slave piston and pushing it down its bore, so increasing the fluid pressure at the wheel cylinders. When the brake pedal is released, the pressure beneath the air valve piston is destroyed, the diaphragm (6) re-opens and the air valve closes. Via the non-return valve (10), a suspended vacuum is re-created around the main diaphragm. Under the action of the spring (11), the diaphragm and push-rod, and thus the slave piston, are returned to their original positions, and the pressure in the wheel cylinders is lost.

**Removing**

(1) Disconnect the battery.

(2) Remove the two captive screws securing the base of the servo cylinder bracket (LHD cars), one screw (RHD cars).

(3) Remove the bolt securing the vacuum hose clip to the servo cylinder bracket.

(4) Disconnect the master cylinder feed outlet hydraulic connections at the servo. Ensure the feed pipe from the master cylinder is blanked to prevent fluid loss.

*R.H.D. Cars.* Remove the three nuts and spring washers retaining the servo to the rear bracket. Remove the servo unit and extract the vacuum hose connection from the servo.

*L.H.D. Cars.* From inside the car, remove the bulkhead grommet and remove the three nuts and spring washers retaining the servo to the rear bracket. Remove the servo unit and extract the vacuum hose connection from the servo.

**Dismantling**

*Air valve assembly (Fig. M.10)*

(5) Grip the slave cylinder in a soft jawed vice with the air valve uppermost and disconnect the rubber pipe from the connection on the end cover.

(6) Remove the screws securing the plastic air valve cover to the valve housing, lift off the cover complete with the air valve sub-assembly. Suspect functioning of the air valve must be remedied by fitting a replacement air valve cover assembly comprising cover, filter, and air valve as an assembled part of the relevant repair kit.

(7) Remove the rubber diaphragm and its plastic support to obtain access to valve housing securing screws. Remove the three screws and take off the housing and joint washer.

(8) Seal one of the slave cylinder fluid ports with a finger, apply a low-pressure air-line to the remaining port and blow the air control valve piston from its bore. Remove the rubber cup from the valve piston.

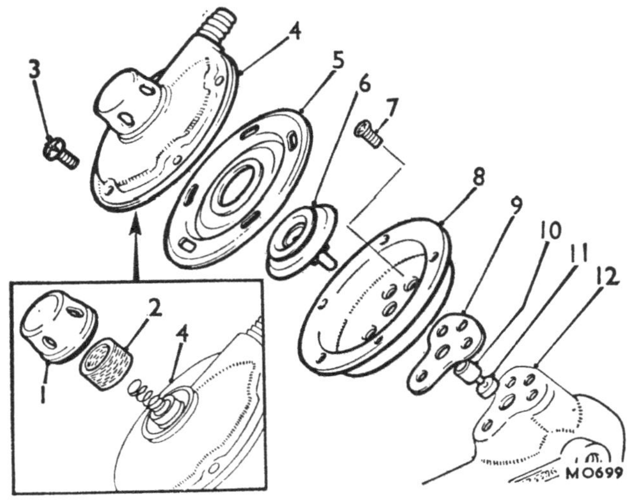


*Fig. M.10*

*The air valve assembly components and piston*

1. Domed cover for filter.
2. Air filter.
3. Air valve cover securing screws.
4. Air valve cover.
5. Diaphragm.
6. Diaphragm support.
7. Valve housing securing screws.
8. Valve housing.
9. Joint washer.
10. Piston.
11. Piston cup.
12. Slave cylinder.

*Servo cylinder assembly*

(9) Use thumb pressure to prise the non-return valve from the servo shell and extract the rubber valve mounting.

(10) To remove the end cover from the servo shell, locate Service tool C 2030* over the three studs and secure tightly with suitable nuts (see Fig. M.11). Turn the removal tool anti-clockwise with a ½ in. (13 mm.) square drive socket 'T' bar as far as the stops on the cover will allow and lift off the cover.

(11) Turn the diaphragm support so that the push-rod retaining key faces downwards. Apply light fluctuating hand pressure on the support into the shell; this will allow the retaining key to drop out, releasing the support from the push-rod under pressure of the main return spring. Extract the spring from the servo shell.

(12) Bend back the locking tabs of the servo shell to slave cylinder securing bolts; remove the bolts, together with the locking and abutment plates. Lift off the servo shell and retrieve the joint washer from the mounting face of the slave cylinder.

*Slave cylinder assembly*

(13) Pull the servo push-rod and the piston assembly attached to it from the slave cylinder bore. Slide off the plastic bearing, rubber cup and plastic spacer, noting their relative positions for refitting.

(14) Prise off the rubber seal from the head of the slave piston. Open the retaining clip with a small screw-driver to expose the connecting pin, which may then be driven out to separate the piston from the rod (see Fig. M.12).

**Inspecting**

Examine all parts for faults and wear and be prepared to fit new rubber parts throughout. If the air valve is faulty, a replacement kit must be fitted. Dust deposits on the air filter, which is otherwise in good condition, can be removed by blowing through with a low-pressure air-line. Do not use a cleaning fluid or lubricant of any description on the filter.

Wash all original components of the slave cylinder assembly and remove light deposits from the cylinder bore with clean brake fluid. If the slave cylinder bore is scored, a replacement unit will be required.

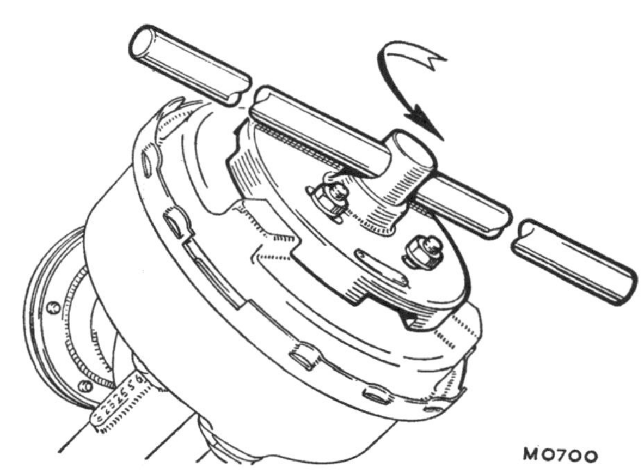


*Fig. M.11*

*Using Service tool C 2030* to remove and refit the end cover*

* Obtainable from: V. L. Churchill & Co. Ltd

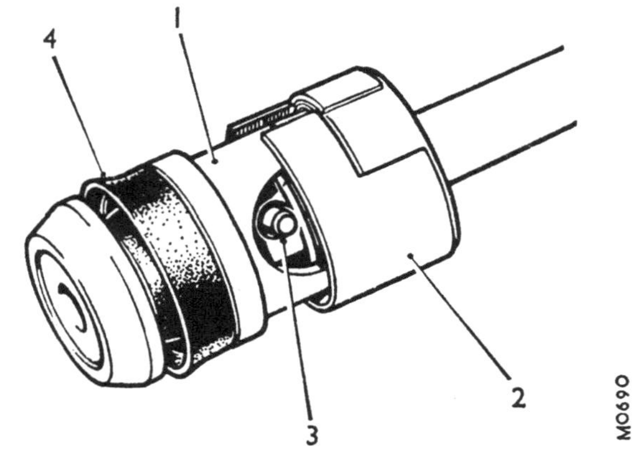


*Fig. M.12*

*The slave piston, with the retaining clip withdrawn to expose the connecting pin*

1. Piston body.
2. Retaining clip.
3. Connecting pin.
4. Piston seal.

**Reassembling**

Scrupulous cleanliness of all parts of the servo unit is essential. Lay out all parts to be assembled on a clean sheet of paper. Use clean brake fluid as a lubricant when reassembling the hydraulic components of the servo unit.

*Slave cylinder assembly*

(15) If the piston and push-rod were separated in item (10), a new retaining clip and connecting pin will be required. Insert the chamfered end of the push-rod into the rear of the piston and depress the spring inside it to uncover the hole in the end of the rod. Fit the pin, followed by its retaining clip; see that this is a snug fit, and does not protrude beyond its groove, otherwise the bore will become scored.

Using only the fingers, bed the rubber seal evenly into the groove on the head of the piston, with the lips of the seal pointing away from the push-rod.

(16) Insert the piston into the lubricated bore of the slave cylinder, then, one by one, slide the spacer, the rubber cup, and the bearing over the push-rod into the mouth of the bore. Take care not to bend back the lips of either the seal or the cup, and bed in each part individually.

## THE VACUUM SERVO COMPONENTS

(Lockheed Type 6)

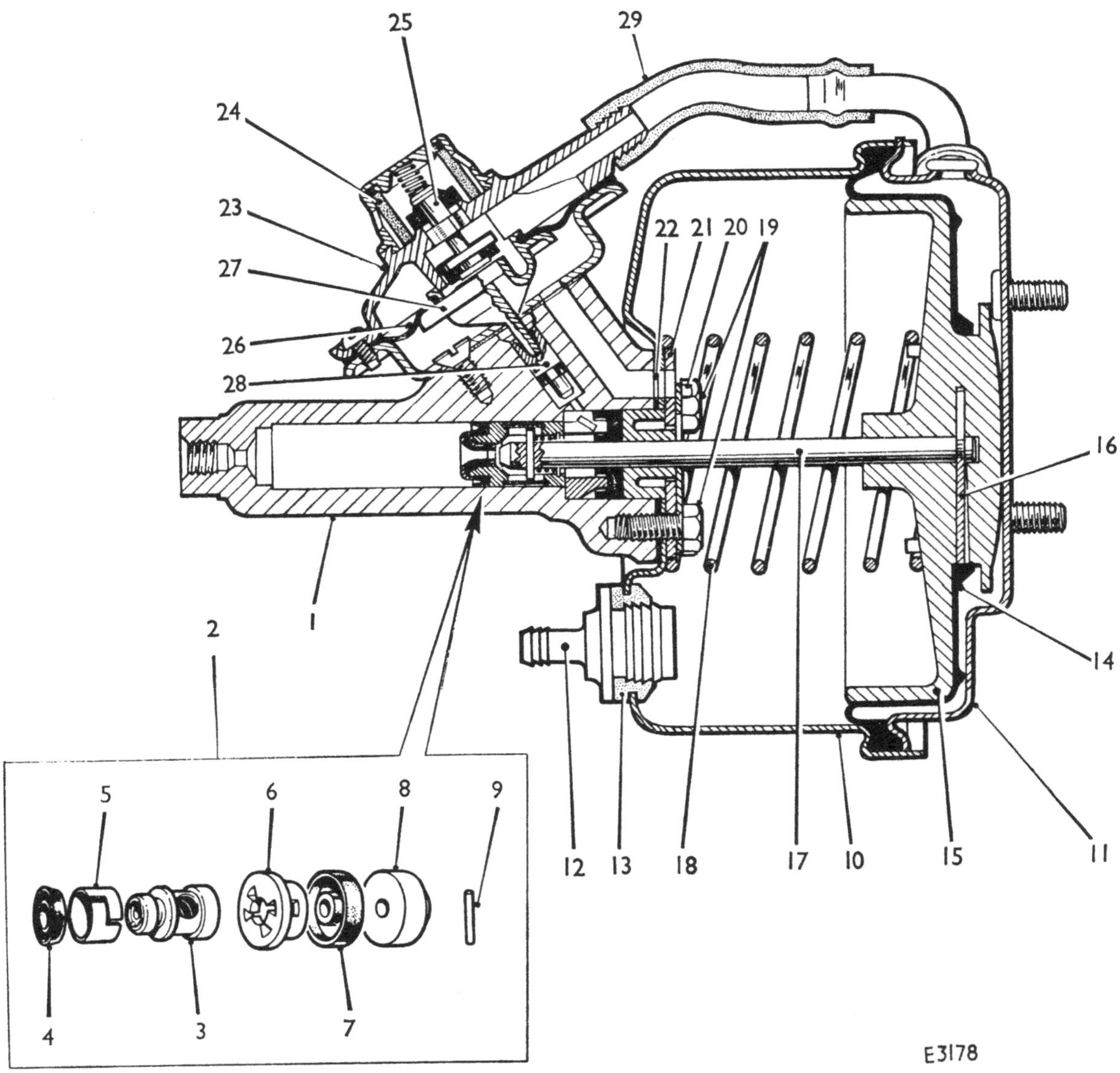


| No. | Description | No. | Description | No. | Description |
|---|---|---|---|---|---|
| 1. | Slave cylinder. | 11. | End cover. | 21. | Abutment plate. |
| 2. | Slave piston components. | 12. | Non-return valve. | 22. | Joint washer. |
| 3. | Slave piston. (Shown inset) | 13. | Rubber mounting | 23. | Air valve cover. |
| 4. | Piston seal. (Shown inset) | 14. | Main servo diaphragm. | 24. | Air filter. |
| 5. | Retaining clip. (Shown inset) | 15. | Diaphragm support. | 25. | Air valve. |
| 6. | Spacer. (Shown inset) | 16. | Retaining key. | 26. | Air valve diaphragm. |
| 7. | Cup. (Shown inset) | 17. | Push-rod. | 27. | Diaphragm support. |
| 8. | Bearing. (Shown inset) | 18. | Main return spring.. | 28. | Air valve piston. |
| 9. | Connecting pin. (Shown inset) | 19. | Servo shell retaining bolts. | 29. | Rubber pipe. |
| 10. | Servo shell. | 20. | Locking plate. | | |

*Servo cylinder assembly*

(17) Grip the slave cylinder in a soft-jawed vice, position the mounting face joint washer and refit the servo shell. After correctly positioning the abutment plate and locking plate, which must be renewed if it has been used more than once previously, tighten the three bolts evenly to the torque figure given in **'GENERAL DATA'** and tap up the locking plate tabs.

(18) Pull out the push-rod to its limit, fit the main return spring followed by the diaphragm support, with its key slot facing upwards. Make sure that the two end coils are located round the abutment plate and support boss respectively. Press the support into the shell until the groove in the end of the push-rod aligns with the key slot, and insert the key.

Ensure that both the rubber diaphragm and its support are perfectly dry and then fit the diaphragm to the support. Gently stretch the diaphragm to seat its inner edge in the groove of the support.

(19) Smear the outer edge of the rubber diaphragm with Lockheed Disc Brake Lubricant where it will contact the rim of the end cover and of the shell, and position the diaphragm evenly around the rim of the shell.

(20) Refit and secure Service tool C2030* to the end cover (see Fig. M.11). Position the end cover with the shell so that the elbow will be in alignment with the air valve. Whilst maintaining a downward pressure on the cover, turn it clockwise as far as the stops will allow, taking care not to trap the edge of the rubber diaphragm. Remove the Service tool.

(21) Push in the non-return valve with its rubber mounting.

*Air valve assembly* (*Fig. M.*10)

(22) Using only the fingers, fit the rubber cup to the spigot of the air valve piston, with the lips pointing away from the drilled head, and insert the piston into its bore, spigot end first. Do not bend over the lips of the cup.

(23) Fit the valve housing, with its joint washer, to the mounting face of the slave cylinder and tighten the three securing screws to the torque figure given in **'GENERAL DATA'**.

(24) Insert the spigot of the diaphragm support into the drilled head of the air valve piston, and then fit the inner edge of the air valve diaphragm into the groove of its support and align the screw hole slots. Do not use any lubricant.

(25) Refit or replace the filter and snap-fit the dome cover if these have been removed in item (6). Place the valve cover over the diaphragm making sure that the projections on the under surface of the cover engage in the slots of the diaphragm. Tighten the five securing screws firmly, progressively and diametrically; do not overtighten, since the smallest air leakage into the air valve assembly will impair the action of the servo.

(26) Refit the rubber pipe to join the end cover elbow to the valve cover port.

**Refitting**

(27) Reverse the removing procedure, items (1) to (4).

(28) Bleed the braking system using clean Lockheed Brake Fluid.

Fluid drained from the system or that used for bleeding should be discarded.

## EDITOR'S NOTES

**M. The Braking System**

*Bleeding the brake system*

When bleeding the brake system, always start with the bleed screw farthest from the master cylinder (in terms of tubing length) and finish with the one nearest the master cylinder.

*Removing the calliper pistons, Reassembling*

Service tool 18G590 is quite useful for performing these operations.

*Relining the brake shoes*

It is common practice to purchase a relined set of brake shoes on an exchange basis. If the drums are severely worn, a limited amount of material may be cut to trim the braking service. The relined shoes should be ground to match the brake drums before installation.

# Section Ma

## THE BRAKING SYSTEM (TANDEM MASTER CYLINDER)

**The information given in this Section refers specifically to service operations on, or affected by, equipment fitted to the MGB (GHD5/GHD4 and GHN5/GHN4) in conformity with local and territorial requirements, and must be used in conjunction with Section M.**

## Section Ma.1

### PEDAL FREE MOVEMENT

A free movement of $\frac{1}{8}$ in. (3·2 mm.) measured at pedal pad must be maintained on the brake pedal.

To adjust the free movement, slacken the stop light switch locknut and turn the switch clockwise to decrease or anti-clockwise to increase the clearance. Tighten the stop light switch locknut.

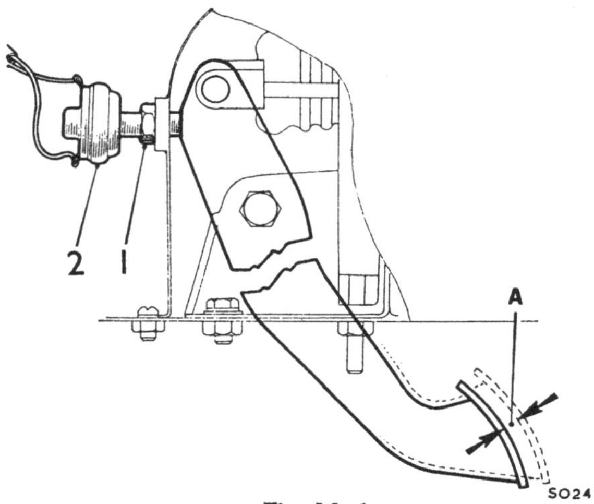


*Fig. Ma.1*
*Brake pedal adjustment*
1. Switch locknut. 2. Stop light switch.
A=$\frac{1}{8}$ in. (3·2 mm.).

## Section Ma.2

### TANDEM MASTER CYLINDER

**Removing**

(1) Remove the four screws securing the master cylinder cover and remove the cover.

(2) Drain the fluid from the reservoir as follows:

(*a*) Attach a rubber tube to one of the front brake calliper bleed screws.

(*b*) Attach a rubber tube to one of the rear wheel brake cylinder bleed screws.

(*c*) Slacken both bleed screws one complete turn.

(*d*) Depress the brake pedal, hold it in the down position and tighten the bleed screws.

(*e*) Repeat the operations in (*d*) until the reservoir is empty.

(3) Drain the clutch master cylinder reservoir by the same method on the brakes from the slave cylinder bleed screw.

(4) Disconnect the pipes from the brake master cylinder and plug the pipe ends.

(5) Disconnect the pipe from the clutch master cylinder and plug the pipe and connection.

(6) Remove the split pin and clevis pin from the push-rod and disconnect the pedal lever.

(7) Disengage the clutch and brake pedal return springs.

(8) Remove the left-hand lower fascia panel.

(9) Unscrew the two bolts securing the pedal box to the toe-plate.

(10) Unscrew the nut retaining the pedal pivot bolt, withdraw the bolt, remove the centre spacer and temporarily refit the bolt.

(11) Unscrew the six bolts securing the pedal box to the bulkhead and remove the pedal box complete with master cylinders and pedals.

(12) Unscrew the two bolts securing the brake master cylinder and remove the master cylinder.

**Dismantling**

(13) Ensure that the pipe connections are securely plugged and the reservoir cap is fitted and thoroughly clean the exterior of the assembly.

(14) Detach the rubber boot and withdraw the push-rod.

(15) Grip the cylinder body in a soft-jawed vice with the mouth of the bore uppermost.

(16) Compress the return spring and remove the Spirolox ring from its groove in the primary piston, taking care not to distort the coils of the ring or score the bore of the cylinder.

(17) Using tool 18G 1112, remove the piston retaining circlip.

(18) Move the piston up and down in the bore to free the nylon guide bearing and cap seal; remove the guide bearing and seal.

(19) Remove the plain washer.

(20) Using tool 18G 1112, remove the inner circlip.

(21) Withdraw the primary and secondary piston assembly complete with stop washer.

(22) Remove the stop washer.

(23) Compress the spring separating the two pistons and drive out the roll-pin retaining the piston link.

(24) Note the positions of the rubber cups by their moulded identations and remove the cups and washers from the pistons.

(25) Unscrew the four bolts securing the plastic reservoir to the body and remove the reservoir.

(26) Remove the two reservoir sealing rings.

(27) Unscrew the front brake pipe connection adaptor and discard the copper gasket.

(28) Withdraw the springs and trap valves from the connection ports.

**Inspection**

(29) Clean all the parts thoroughly using the recommended brake fluid and dry them with clean lint-free cloth.

(30) Examine the metal parts for wear and damage; inspect the rubber components for swelling, perishing, distortion or any other signs of deterioration. Renew all worn, damaged, or suspect parts.

**Reassembling**

(31) Dip all the internal components in the recommended brake fluid and assemble them while wet.

(32) Locate the piston washer on the head of the secondary piston, convex surface first.

(33) Carefully ease the secondary main cap, lip last, over the end of the piston, using the fingers, and seat it correctly in the groove adjacent to the washer.

(34) Carry out opreations (32) and (33) with the washer and main cup of the primary piston.

(35) Reverse the dismantling procedure in (13) to (23) and (25) to (28).

**Refitting**

(36) Reverse the removing procedure in (1) and (4) to (12).

(37) Fill the brake and clutch master cylinders with the recommended fluid (see **'GENERAL DATA'**).

(38) Bleed the brake system (see Section M.4) commencing with the rear brakes.

(39) Bleed the clutch system—Section E.5.

(40) Check and, if necessary, adjust the brake pedal free movement—Section Ma.1.

## Section Ma.3

### PRESSURE FAILURE SWITCH ASSEMBLY

**Removing**

(1) Disconnect the wiring from the switch.

(2) Clean the switch assembly and its surroundings, particularly the pipe connections.

(3) Disconnect and plug the hydraulic pipes.

(4) Unscrew the retaining bolt and remove the assembly.

**Dismantling**

(5) Remove the end plug and discard the copper washer.

(6) Unscrew the nylon switch.

(7) Withdraw the shuttle valve piston assembly from the bore; use a low pressure air-line to free the piston if necessary.

(8) Remove and discard the two piston seals.

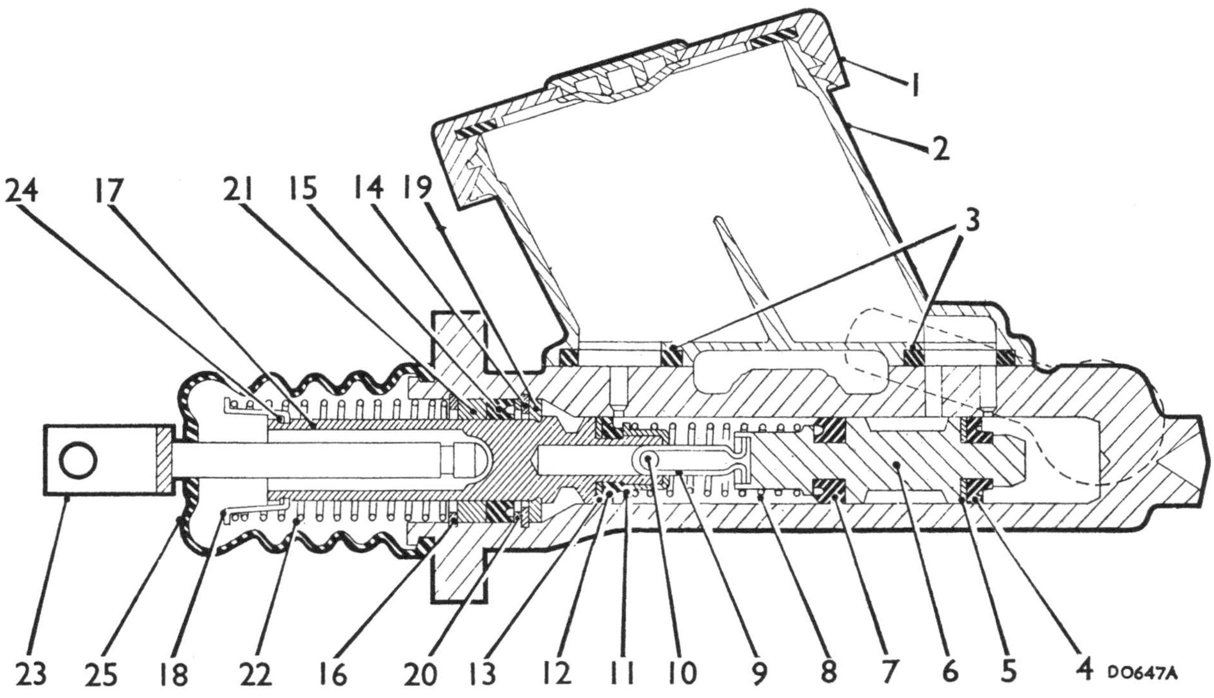


*Fig. Ma.2*

*A section through the master cylinder*

1. Filler cap.
2. Plastic reservoir.
3. Reservoir seals.
4. Main cup.
5. Piston washer.
6. Piston.
7. Main cup.
8. Spring.
9. Piston link.
10. Pin.
11. Pin retainer.
12. Main cup.
13. Piston washer.
14. Circlip.
15. Cup.
16. Circlip.
17. Piston.
18. Spring retainer.
19. Stop washer.
20. Washer.
21. Bearing.
22. Spring.
23. Push-rod.
24. Spirolox ring.
25. Rubber boot.

**Inspection**

(9) Thoroughly clean all the components using methylated spirit (denatured alcohol) or the recommended brake fluid, and dry with lint-free cloth.

(10) Inspect the bore of the casing for scoring and damage; the complete assembly must be renewed if the bore is not in a perfect condition.

(11) Reconnect the wiring to the switch and actuate the switch plunger, with the ignition switched on, to test the switch operation and warning light circuit.

**Reassembling**

(12) Fit two new seals, lip facing outwards, to the piston.

(13) Lubricate the piston assembly with Lockheed Disc Brake Lubricant and fit the piston into the bore, taking care that the lip of the leading seal is not turned back.

(14) Fit a new copper washer to the end plug, screw in and tighten the plug to the torque figure given in 'GENERAL DATA'.

(15) Screw in the switch and carefully tighten it to the torque figure given in 'GENERAL DATA'.

**Refitting**

(16) Reverse the removing procedure in (1) to (4), fill the master cylinder with the recommended fluid (see 'GENERAL DATA') and bleed the system (see Section M.4) commencing with the rear brakes.

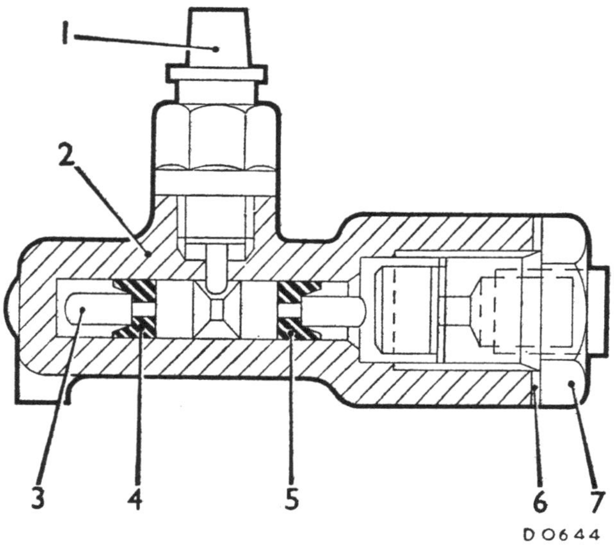


*Fig. Ma.3*

*A section through the pressure failure switch assembly*

1. Nylon switch.
2. Switch body.
3. Shuttle valve piston.
4. Piston seal.
5. Piston seal.
6. Copper washer.
7. End plug.

# SECTION N

## THE ELECTRICAL SYSTEM

### GENERAL DESCRIPTION

The electrical system is 12-volt NEGATIVE earth return. The electrical system on early cars is 12-volt POSITIVE earth.

Two 6-volt batteries wired in series are carried in trays mounted beneath the rear cockpit floor; each battery is retained by a clamp plate and two fixing bolts. A battery access panel is provided in the rear cockpit floor and is secured to the floor by five quick-release fasteners.

## Section N.1

### BATTERY

The batteries are 6-volt lead-acid type, having six cells in each consisting of a group of positive and negative plates immersed in a solution of sulphuric acid (electrolyte).

The Type CA batteries fitted to later cars have a special filling device which automatically ensures the correct electrolyte level when topping-up.

Remove the battery vent cover; use the grip at the centre of the cover, this will ensure that the filling valves are operated correctly. If no electrolyte is visible inside the battery, pour distilled water into the filling trough until the three tubes and the connecting trough are filled. Refit the vent cover. Check the second battery.

Except when topping-up or taking hydrometer readings the vent cover must be kept fitted. The electrolyte will flood if the cover is removed for a long period during or within 30 minutes of the battery being normal (5 amp.) or fast charged (30–40 amps.).

**Caution.**—Single-cell discharge testers cannot be used on the Type CA battery. Operation of the filling device will be destroyed if the battery case is drilled or punctured.

#### Fast charging

During fast charging the vent cover may be removed, providing all filling tubes are pressed downwards to break the air lock.

The battery is 'fast charged' at 30–40 amps., when the on-charge voltage across the battery terminals is less that 15·5 volts. Above 15·5 volts, charge the battery at normal rate until the specific gravity readings indicate that the battery is fully-charged.

The battery has three functions; to supply current for starting, ignition, and lighting; to provide a constant supply of current to the electrical equipment under normal operating conditions and when the consumption of the electrical equipment exceeds the output of the dynamo; and to control the voltage of the electrical supply system.

#### Adjustments in the vehicle

The purpose of the following operations is to maintain the performance of the battery at its maximum.

The battery and its surrounding parts should be kept dry and clean, particularly the tops of the cells, as any dampness could cause a leakage between the securing strap and the battery negative terminal and result in a partially discharged battery. Clean off any corrosion from the battery bolts, clamps, and trays with diluted ammonia, afterwards painting the affected parts with anti-sulphuric paint.

The electrolyte levels should be maintained just level with the tops of the separator guards by adding distilled water. Never add acid.

Check the terminal posts. If they are corroded remove the cables and clean with diluted ammonia. Smear the posts with petroleum jelly before remaking the connections and ensure that the cable terminal screws are secure.

Check the condition of the battery cells using a hydrometer. If the level of the electrolyte is too low for hydrometer readings to be taken, top up with distilled water and recharge the battery for at least 30 minutes before taking readings. The hydrometer readings and their indications are as follows:

*For climates below* 27° *C.* (80° *F.*)

| | |
|---|---|
| Cell fully charged .. .. .. | 1·270 to 1·290 |
| Cell about half-charged .. .. | 1·190 to 1·210 |
| Cell completely discharged.. .. | 1·110 to 1·130 |

*For climates above* 27° *C.* (80° *F.*)

| | |
|---|---|
| Cell fully charged .. .. .. | 1·210 to 1·230 |
| Cell about half-charged .. .. | 1·130 to 1·150 |
| Cell completely discharged.. .. | 1·050 to 1·070 |

The figures given in the table are corrected to an electrolyte temperature of 16° C. (60° F.) and the hydrometer readings obtained must also be corrected to suit the temperature of the electrolyte.

For every 3° C. (5° F.) above 16° C. (60° F.) add ·002.

For every 3° C. (5° F.) below 16° C. (60° F.) subtract ·002.

The readings of all the cells should be approximately the same. If one cell gives a reading which differs from the remainder by 40 points (·040 S.G.) or more, an internal fault in that cell is indicated. Should the battery be in a low state of charge, it should be recharged by taking the car for a long daytime run or by charging from an external source of D.C. supply at a current rate of 5·0 amps. until the cells are gassing freely.

#### Removing

Remove the moulded carpet and release the five quick-release screws in the panel. Take off the panel, disconnect the electrical leads to the batteries, remove the four fixing bolts and two clamp plates, and lift out the batteries.

#### Viewing

Place the battery on a lead-covered bench or on a wooden bench treated with anti-sulphuric paint.

Check the electrolyte levels.

Inspect the container for cracks, which may be indicated by external corrosion or extreme variation in the electrolyte levels.

**Recharging from an external source**

The length of time for a used battery to remain on charge before it can be accepted as fully charged depends entirely on the specific gravity before charging commences and the charging rate. The charging should continue at 5 amps. until all cells are gassing freely and evenly and the specific gravity in each of the six cells has reached a maximum, i.e. has shown no further rise in four hours. The specific gravity at the end of charging should be within the limits given and should not vary ·005 from the values given.

Do not allow the temperature of the electrolyte to exceed the maximum permissible temperature, i.e.:
For climates below 27° C. (80° F.) .. 38° C. (100° F.).
For climates above 27° C. (80° F.) .. 49° C. (120° F.).

If this temperature is reached the charge should be suspended to allow the temperature to fall at least 6° C. (10° F.), otherwise the life of the battery will tend to be shortened.

**Refitting**

The installation of the battery is a reversal of the procedure **'Removing'**. Smear the terminal posts and cable connections with petroleum jelly and tighten the retaining screws sufficiently to prevent the cables from moving on the terminal posts when tested by hand, but do not overtighten.

**NOTE.—Whenever booster charging of the battery or electric welding of the body is carried out, the battery earth lead must be disconnected to prevent damage to the electrical system.**

## Section N.2

### PREPARING DRY-CHARGED BATTERIES FOR SERVICE

Dry-charged batteries are supplied without electrolyte but with the plates in a charged condition. When they are required for service it is only necessary to fill each cell with sulphuric acid of the correct specific gravity. No initial charging is required.

**Preparing electrolyte**

The electrolyte is prepared by mixing together distilled water and concentrated sulphuric acid, taking the precautions given in Section N.3. The specific gravity of the filling electrolyte depends on the climate in which the battery is to be used.

**Filling the battery**

Carefully break the seals in the cell filling holes and fill each cell with electrolyte to the top of the separator guards **in one operation.** The temperature of the filling room, battery, and electrolyte should be maintained between 16 and 38° C. (60 and 100° F.). If the battery has been stored in a cool place it should be allowed to warm up to room temperature before filling.

**Putting into use**

Measure the temperature and specific gravity of the electrolyte in each of the cells. Allow to stand for 20 minutes and then re-check. The battery is ready for service unless the electrolyte temperature has risen by more than 5·5° C. (10° F.), or the specific gravity has fallen by more than 10 points (·010 S.G.). In this event, re-charge the battery at the normal re-charge rate until the specific gravity remains constant for three successive hourly readings and all cells are gassing freely. During the charge the electrolyte must be kept level with the top of the separator guards by the addition of distilled water.

## Section N.3

### PREPARING NEW, UNFILLED, UNCHARGED BATTERIES FOR SERVICE

**Preparing electrolyte**

Batteries should not be filled with acid until required for initial charging. Electrolyte is prepared by mixing distilled water and concentrated sulphuric acid, usually of 1·840 S.G. The mixing must be carried out either in a lead-lined tank or in suitable glass or earthenware vessels. Slowly add the acid to the water, stirring with a glass rod. **Never add the water to the acid,** as the resulting chemical reaction causes violent and dangerous spurting of the concentrated acid. The approximate proportions of acid and water are indicated in the following table:

| *For climates* | *To obtain specific gravity (corrected to 16° C. [60° F.]) of* | *Add 1 vol. of 1·840 S.G. acid to distilled water as below* |
|---|---|---|
| Below 27° C. (80° F.) | 1·260 | 3.2 volumes of water |
| Above 27° C. (80° F.) | 1·210 | 4·3 volumes of water |

Heat is produced by mixing acid and water, and the electrolyte should be allowed to cool before taking hydrometer readings—unless a thermometer is used to measure the actual temperature and a correction applied to the readings before pouring the electrolyte into the battery.

**Filling the battery**

The temperature of the electrolyte, battery, and filling room should be maintained between 16° and 38° C. (60° and 100° F.).

Carefully break the seals in the filling holes and **half-fill** each cell with electrolyte of the appropriate specific gravity. Allow the battery to stand for at least 12 hours in order to dissipate the heat generated by the chemical action of the acid on the plates and separators, and then add sufficient electrolyte to fill each cell to the top of the separators. Allow to stand for a further two hours and then proceed with the initial charge.

**Initial charge**

The initial charging rate is 3·5 amps. Charge at this rate until the voltage and specific gravity readings show no increase over five successive hourly readings. This will take from 48 to 80 hours, depending on the length of time the battery has been stored before charging.

Keep the current constant by varying the series resistance of the circuit or the generator output. **This charge should not be broken by long rest periods.** If, however, the temperature of any cell rises above the permissible maximum, i.e.:

For climates below 27° C. (80° F.) .. 38° C. (100° F.),

For climates above 27° C. (80° F.) .. 49° C. (120° F.),

the charge must be interrupted until the temperature has fallen at least 5·5° C. (10° F.) below that figure. Throughout the charge the electrolyte must be kept level with the top of the separators by the addition of acid solution of the same specific gravity as the original filling-in acid until specific gravity and charge readings have remained constant for five successive hourly readings. If the charge is continued beyond that point, top up with distilled water.

At the end of the charge carefully check the specific gravity in each cell to ensure that, when corrected to 16° C. (60° F.) it lies between the specified limits. If any cell requires adjustment some of the electrolyte must be siphoned off and replaced either by distilled water or by acid of strength originally used for filling in, depending on whether the specific gravity is too high or too low. Continue the charge for an hour or so to ensure adequate mixing of the electrolyte and again check the specific gravity readings. If necessary, repeat the adjustment process until the desired reading is obtained in each cell. Finally, allow the battery to cool, and siphon off any electrolyte over the top of the separators.

## Section N.4

### DYNAMO

**Testing on vehicle when dynamo is not charging**

Make sure that belt slip is not the cause of the trouble. It should be possible to deflect the belt approximately ½ in. (13 mm.) with moderate hand pressure at the centre of its longest run between two pulleys. If the belt is too slack, tightening is effected by slackening the two dynamo suspension bolts and then the bolt of the slotted adjustment link. A gentle pull on the dynamo outwards will enable the correct tension to be applied to the belt and all three bolts should then be tightened firmly.

Check that the dynamo and control box are connected correctly. The dynamo terminal 'D' should be connected to the control box terminal 'D' and the dynamo terminal 'F' connected to the control box terminal 'F'.

After switching off all lights and accessories disconnect the cables from the dynamo terminals marked 'D' and 'F' respectively.

Connect the two terminals with a short length of wire.

Start the engine and set to run at normal idling speed.

Clip the negative lead of a moving-coil-type voltmeter calibrated 0–20 volts to one dynamo terminal and the other lead to a good earthing point on the dynamo yoke.

Gradually increase the engine speed: the voltmeter reading should rise rapidly and without fluctuation. Do not allow the voltmeter reading to reach 20 volts. Do not race the engine in an attempt to increase the voltage. It is sufficient to run the dynamo up to a speed of 1,000 r.p.m.

If there is no reading check the brush gear.

If the reading is low (approximately ½ to 1 volt) the field winding may be faulty.

If the reading is approximately 4 to 5 volts the armature winding may be faulty.

If the dynamo is in good order leave the temporary link in position between the terminals and restore the original connections, taking care to connect the dynamo terminal 'D' to the control box terminal 'D' and the dynamo terminal 'F' to the control box terminal 'F'. Remove the lead from the 'D' terminal on the control box and connect the voltmeter between this cable and a good earthing point on the vehicle. Run the engine as before. The reading should be the same as that measured directly on the dynamo. No reading on the voltmeter indicates a break in the cable to the dynamo. Carry out the same procedure for the 'F' terminal, connecting the voltmeter between cable and earth. Finally, remove the link from the dynamo. If the reading is correct test the control box.

**Removing**

To remove the dynamo disconnect the dynamo leads from the dynamo terminals.

Slacken all the attachment bolts and pivot the dynamo towards the cylinder block to enable the fan belt to be removed from the dynamo pulley. The dynamo can then be removed by withdrawing the two upper and one lower attachment bolts.

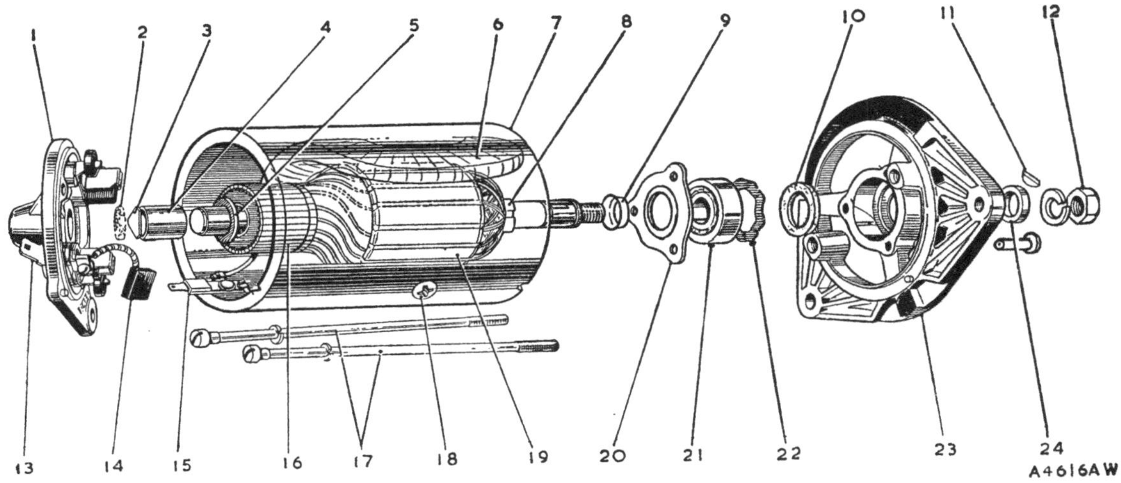


*Fig. N.1*

*The C40/1 type dynamo*

1. Commutator end bracket.
2. Felt ring.
3. Felt ring retainer.
4. Bronze bush.
5. Thrust washer.
6. Field coils.
7. Yoke.
8. Shaft collar.
9. Shaft collar retaining cup.
10. Felt ring.
11. Shaft key.
12. Shaft nut.
13. Output terminal 'D'.
14. Brushes.
15. Field terminal 'F'.
16. Commutator.
17. Through-bolts.
18. Pole-shoe securing screws.
19. Armature.
20. Bearing retaining plate.
21. Ball bearing.
22. Corrugated washer.
23. Driving end bracket.
24. Pulley spacer.

**Dismantling**

Remove the securing nut and take off the drive pulley.

Remove the Woodruff key from the commutator shaft.

Unscrew and remove the two through-bolts and take off the commutator end bracket. The driving end bracket, together with the armature and its ball bearing, can now be lifted out of the yoke. Unless the ball bearing is damaged or requires attention it need not be removed from the armature. Should it be necessary to remove the bearing, the armature must be separated from the end bracket by means of a hand press.

**Servicing**

*Brushes*

Lift the brushes up in the brush boxes and secure them in that position by positioning each brush spring at the side of the brush. Fit the commutator end bracket over the commutator and release the brushes. Hold back each of the brush springs and move the brush by pulling gently on its flexible connector. If the movement is sluggish, remove the brush from its holder and ease the sides by lightly polishing it on a smooth file. Always refit the brushes in their original positions. If the brushes are badly worn, new brushes must be fitted and bedded to the commutator. The minimum permissible length of brush is $\frac{9}{32}$ in. (7·14 mm.).

Test the brush spring tension, using a spring scale. The tension of the springs when new is 18 to 24 oz. (510·7 to 681 gm.). In service it is permissible for this value to fall to 15 oz. (425 gm.) before performance may be affected. Fit new springs if the tension is low.

*Commutator*

A commutator in good condition will be smooth and free from pits or burned spots. Clean the commutator with a cloth moistened with fuel. If this is ineffective carefully polish with a strip of fine glass-paper while rotating the armature. To remedy a badly worn commutator mount the armature (with or without the drive end bracket) in a lathe, rotate at high speed, and take a light cut with a very sharp tool. Do not remove more metal than is necessary. Polish the commutator with very fine glass-paper. Undercut the mica insulation between the segments to a depth of $\frac{1}{32}$ in. (·8 mm.) with a hacksaw blade ground down to the thickness of the mica. (See Editor's notes at end of section.)

Some commutators are of the moulded type and may be skimmed to a minimum diameter of 1·45 in. (36·8 mm.). The undercut must conform to the following dimensions:

| | |
|---|---|
| Width .. .. .. | ·040 in. (1·02 mm.). |
| Depth .. .. .. | ·020 in. (·51 mm.). |

It is important that the sides of the undercut clear the moulding material by a minimum of ·015 in. (·38 mm.).

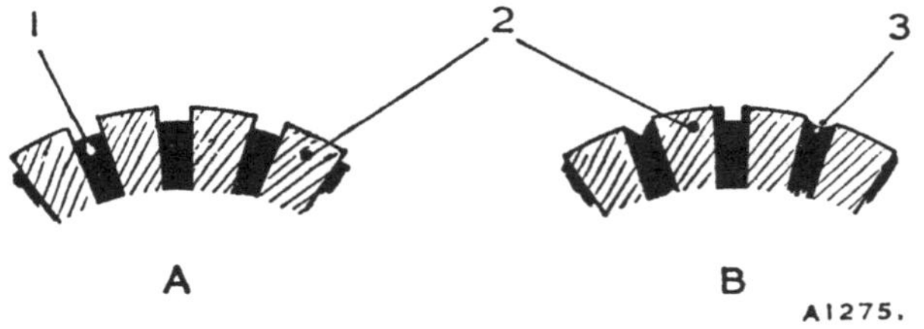


*Fig. N.2*

*Undercutting the commutator* (A) *the right way and* (B) *the wrong way*

1. Insulator. 2. Segments. 3. Insulator.

The most common armature faults are usually confined to open- or short-circuited windings. Indication of an open-circuited armature winding is given by burnt commutator segments. A short-circuited armature winding is easily identified by discoloration of the overheated windings and badly burnt commutator segments.

*Field coils*

Test the field coils with an ohmmeter without removing them from the dynamo yoke. The reading on the ohmmeter should be between 6·0 and 6·3 ohms. If this is not available connect a 12-volt D.C. supply with an ammeter in series between the field terminal and the dynamo yoke. The ammeter reading should be approximately 2 amps. If no reading is indicated the field coils are open-circuited and must be renewed.

If the current reading is much more than 2 amps. or the ohmmeter reading much below 6 ohms, it is an indication that the insulation of one of the field coils has broken down.

In either case, unless a substitute dynamo is available, the field coils must be renewed. To do this carry out the procedure outlined below.

Drill out the rivet securing the field coil terminal assembly to the yoke and unsolder the field coil connections.

Remove the insulation piece which is provided to prevent the junction of the field coils from contacting the yoke.

Mark the yoke and pole-shoes in order that they can be refitted in their original positions.

Unscrew the two pole-shoe retaining screws by means of a wheel-operated screwdriver.

Draw the pole-shoes and coils out of the yoke and lift off the coils.

Fit the new field coils over the pole-shoes and place them in position inside the yoke. Take care that the taping of the field coils is not trapped between the pole-shoes and the yoke.

Locate the pole-shoes and field coils by lightly tightening the fixing screw.

Fully tighten the screws by means of a wheel-operated screwdriver and lock them by caulking.

Replace the insulation piece between the field coil connections and the yoke.

Resolder the field coil connections to the field coil terminal tags and re-rivet the terminal assembly to the yoke.

*Armature*

The testing of the armature winding requires the use of a voltage drop-test and growler. If these are not available the armatures should be checked by substitution. No attempt should be made to machine the armature core or to true a distorted armature shaft.

*Bearings*

Bearings which are worn to such an extent that they will allow side-movement of the armature shaft must be renewed.

To renew the bearing bush in a commutator end bracket proceed as follows.

Remove the old bearing bush from the end bracket. The bearing can be withdrawn with a suitable extractor or by screwing a $\frac{5}{8}$ in. tap into the bush for a few turns and pulling out the bush with the tap. Screw the tap squarely into the bush to avoid damaging the bracket.

Press the new bearing bush into the end bracket, using a shouldered, highly polished mandrel of the same diameter as the shaft which is to be fitted in the bearing, until the visible end of the bearing is flush with the inner face of the bracket. Porous bronze bushes should not be opened out after fitting or the porosity of the bush may be impaired.

**NOTE.—Before fitting the new bearing bush it should be allowed to stand for 24 hours completely immersed in thin (S.A.E. 20) engine oil; this will allow the pores of the bush to be filled with lubricant. In cases of extreme urgency this period may be shortened by heating the oil to 100° C.**

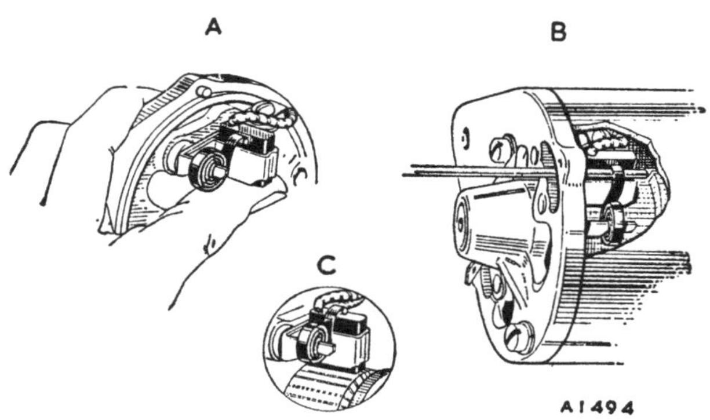


*Fig. N.3*

*Fitting the commutator end bracket*

A. Brush trapped by spring in raised position.
B. Releasing the brush onto the commutator.
C. Normal position of brush.

**(212° F.) for two hours, then allowing it to cool before removing the bearing bush.**

The ball bearing is renewed as follows.

Drill out the rivets which secure the bearing retaining plate to the end bracket and remove the plate. Press the bearing out of the end bracket and remove the corrugated washer, felt washer, and oil-retaining washer.

Before fitting the replacement bearing see that it is clean and pack it with high-melting-point grease.

Place the oil-retaining washer, felt washer, and corrugated washer in the bearing housing in the end bracket.

Press the bearing into the housing. The outer bearing journal is a light push-fit in the bearing housing.

Refit the bearing retaining plate, using rivets having the same dimensions as those originally fitted.

NOTE.—**When fitting a drive end bracket to the armature shaft the inner journal of the bearing MUST be supported by a mild-steel tube—do not use the drive end bracket.**

**Reassembling**

The reassembly of the dynamo is a reversal of the dismantling sequence.

If the end bracket has been removed from the armature in dismantling, press the bearing end bracket onto the armature shaft, taking care to avoid damaging the end plate and armature winding. When assembling the commutator end bracket the brushes must first be held clear of the commutator by partially withdrawing them from their boxes until each brush is trapped in position by the side pressure of its spring. The brushes can be released onto the commutator by a small screwdriver or similar tool when the end bracket is assembled to within about ½ in. (13 mm.) of the yoke. Before closing the gap between the end bracket and the yoke see that the springs are in correct contact with the brushes.

Add a few drops of oil through the hole in the armature end cover.

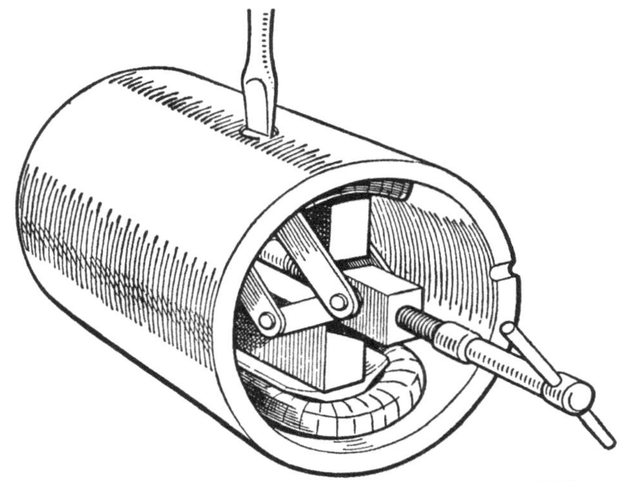


*Fig. N.5*

*Expanding the pole-shoes in the yoke*

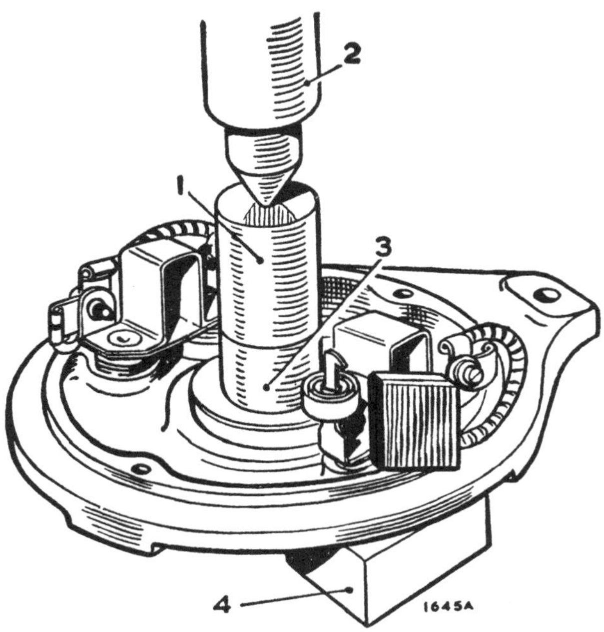


*Fig. N.4*

*The method of pressing in the commutator end bracket bush is shown in this illustration*

1. Shouldered mandrel.
2. Hand press.
3. Bearing bush.
4. Support block.

**Refitting**

Reverse the removal procedure, noting that on later models plain washers are fitted under the heads of the two upper fixing bolts.

## Section N.5

### CONTROL BOX MODEL RB340

**General description**

The RB340 control box operates on the current-voltage system of dynamo output regulation. Three units are housed in the control box: two separate vibrating armature-type single-contact regulators and a cut-out relay. One regulator is responsive to changes in current and the other to changes in voltage.

The voltage regulator and cut-out relay are temperature-compensated to allow for operating changes in the circuit resistance and for climatic variations in battery voltage. The effect of temperature fluctuation on control box settings is further minimized by the use of a swamp resistor connected in series with the shunt coils of the voltage regulator and cut-out relay.

For adjustment purposes toothed cams are carried on the front limb of each magnet frame to enable voltage and current settings to be made with a special tool (see Fig. N.7).

The control box settings are accurately adjusted during manufacture and the cover should not be removed unnecessarily.

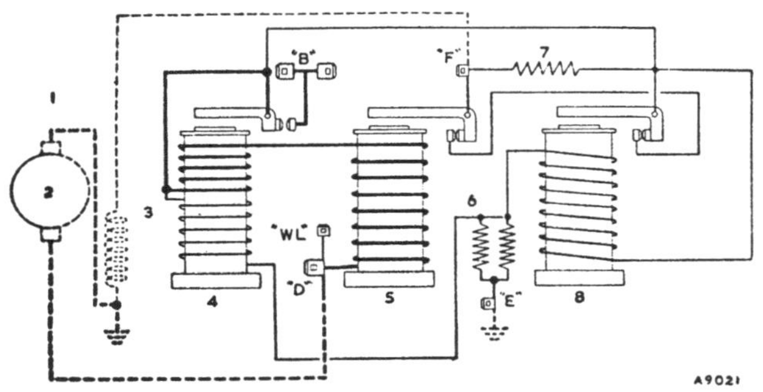


*Fig. N.6*

*The charging circuit*

1. Dynamo.
2. Armature.
3. Field.
4. Cut-out relay.
5. Current regulator.
6. Swamp resistor.
7. Field resistor.
8. Voltage regulator.

**Preliminary checking of charging circuit**

Before disturbing any electrical adjustments examine the items mentioned below to ensure that the fault does not lie outside the control box.

Check the battery by substitution or with a hydrometer and a heavy-discharge tester. Inspect the dynamo driving belt This should be just taut enough to drive without slipping. Check the dynamo by substitution, or by disconnecting the cables from the dynamo terminals and linking the large terminal 'D' to the small terminal 'F', connecting a voltmeter between this link and earth, and running the dynamo up to about 1,000 r.p.m., when a rising voltage should be shown.

Inspect the wiring of the charging circuit and carry out continuity tests between the dynamo, control box, and battery.

Check the earth connections, particularly that of the control box.

In the event of reported undercharging, ascertain that this is not due to low mileage.

**Voltage regulator (open circuit)**

*Method of adjustment*

Checking and adjusting should be completed as rapidly as possible to avoid errors due to heating of the operating coil. Withdraw the cables from the control box terminal blades 'B'. To enable the engine to be started it will be necessary to join the ignition and battery feeds together with a suitable lead. Connect a first-grade 0–20 moving-coil voltmeter between control box terminal 'D' and a good earthing point. A convenient method of making this connection is to withdraw the ignition warning light feed from the control box terminal 'WL' and to clip the voltmeter lead of appropriate polarity to the small terminal blade thus exposed, this terminal being electrically common with terminal 'D'. Start the engine and run the dynamo at 3,000 r.p.m. The voltmeter reading should be steady and lie between the appropriate limits according to the temperature (see **'GENERAL DATA'**).

An unsteady reading may be due to dirty contacts. If the reading is steady but occurs outside the appropriate limits an adjustment must be made. Proceed as follows.

Stop the engine and remove the control box cover. Restart the engine and run the dynamo at 3,000 r.p.m. Using a suitable tool (see Fig. N.7), turn the voltage adjustment cam until the correct setting is obtained. Turn the tool clockwise to raise the setting or anti-clockwise to lower it. Check the setting by stopping the engine and then again raising the generator speed to 3,000 r.p.m. Restore the original connections and refit the cover.

**Cut-out relay**

*Cut-in adjustment*

Checking and adjusting should be completed as rapidly as possible to avoid errors due to heating of the operating coil.

Connect a first-grade 0–20 moving-coil voltmeter between control box terminal 'D' and a good earthing point. A convenient method of making this connection is to withdraw the ignition warning light feed from control box terminal 'WL' and to clip the voltmeter lead of appropriate polarity to the small terminal blade thus exposed, this terminal being electrically common with terminal 'D'. Switch on an electrical load such as the headlamps, start the engine, and gradually increase the engine speed. Observe the voltmeter pointer. The voltage should rise steadily and then drop slightly at the instant of contact closure. The cut-in voltage is that which is indicated immediately before the pointer drops back and should be within the limits 12·7 to 13·3 volts. If the

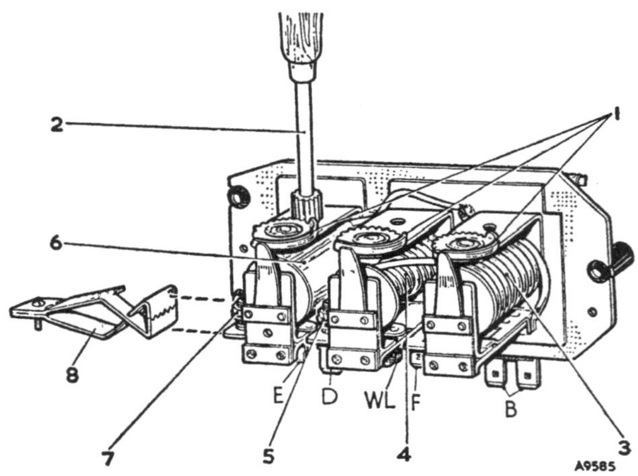


*Fig. N.7*

*The control box with cover removed*

1. Adjustment cams.
2. Setting tool.
3. Cut-out relay.
4. Current regulator.
5. Current regulator contacts.
6. Voltage regulator.
7. Voltage regulator contacts.
8. Clip.

cut-in occurs outside these limits an adjustment must be made. In this event proceed as follows.

Remove the control box cover. Using a suitable tool (see Fig. N.7), turn the cut-out relay adjustment cam until the correct setting is obtained. Turn the tool clockwise to raise the setting or anti-clockwise to lower it. Stop the engine, restore the original connections, and refit the cover.

*Drop-off adjustment*

Withdraw the cables from control box terminal blades 'B'. To enable the engine to be started it will be necessary to join the ignition and battery feeds together with a suitable lead. Connect a first-grade 0–20 moving-coil voltmeter between control box terminal 'B' and a good earthing point. Start the engine and run up to approximately 3,000 r.p.m.

Slowly decelerate and observe the voltmeter pointer. Opening of the contacts, indicated by the voltmeter pointer dropping to zero, should occur between 9·5 and 11 volts. If the drop-off occurs outside these limits an adjustment must be made. Proceed as follows.

Stop the engine and remove the control box cover. Adjust the drop-off voltage by carefully bending the fixed contact bracket. Reducing the contact gap will raise the drop-off voltage and increasing the gap will lower it. Retest, and if necessary readjust until the correct drop-off setting is obtained. This should result in a contact 'follow through' or blade deflection of ·010 to ·020 in. (·254 to ·508 mm.). Restore the original connections and refit the cover.

**Current regulator**

*On-load setting*

The current regulator on-load setting is equal to the maximum rated output of the dynamo, which is 22 amps.

*Method of adjustment*

The dynamo must be made to develop its maximum rated output whatever the state of charge of the battery might be at the time of setting. The voltage regulator must therefore be rendered inoperative, and to do this the bulldog clip shown in Fig. N.7 is used to keep the voltage regulator contacts together.

Remove the control box cover and, using a bulldog clip, short out the contacts of the voltage regulator.

Withdraw the cables from the control box terminal blades 'B' and connect a first-grade 0–40 moving-coil ammeter between these cables and one of the terminal blades 'B'. It is important that terminal 'B' should carry only this one connection. All other load connections, including the ignition coil feed, must be made to the battery. Switch on all lights and accessories, start the engine, and run the dynamo at 3,000 r.p.m. The ammeter pointer should be steady and indicate a current of 19–22 amps. An unsteady reading may be due to dirty contacts. If the reading is too high or too low an adjustment must be made. Proceed as follows.

Using a suitable tool (see Fig. N.7), turn the current adjustment cam until the correct setting is obtained. Turn the tool clockwise to raise the setting or anti-clockwise to lower it. Stop the engine, restore the original connections, and refit the control box cover.

**Adjustment of air gap settings**

Air gap settings are accurately adjusted during manufacture and should require no further attention. If the original settings have been disturbed, it will be necessary to make adjustments in the manner described below.

*Armature-to-bobbin core gaps of voltage and current regulators*

Disconnect the battery. Using a suitable tool (see Fig. N.7), turn the adjustment cam of the regulator being adjusted to the point giving minimum lift to the armature tensioning spring (by turning the tool to the fullest extent anti-clockwise). Slacken the appropriate contact locking nut and unscrew the contact. Insert a feeler gauge of ·056 in. (1·42 mm.) thickness between the armature and the regulator head as far back as the two rivet heads on the underside of the armature. With the gauge in position, press squarely down on the armature, and screw in the contact until it just touches the armature contact. Tighten the locknut and withdraw the gauge. Repeat this procedure on the remaining regulator.

**NOTE.—On earlier type regulators having a copper shim on the regulator head the air gap setting is ·045 in. (1·15 mm.) and care must be taken not to damage the copper shim.**

Carry out the electrical setting procedure.

*Contact 'follow through' and armature-to-bobbin core gap of cut-out relay*

Press the armature squarely down against the copper separation on the core face. Adjust the fixed contact bracket to give a 'follow through' or blade deflection of the moving contact of ·010 to ·020 in. (·254 to ·508 mm.).

Adjust the armature back stop to give a core gap of ·035 to ·045 in. (·889 to 1·147 mm.).

Check the cut-in and drop-off voltage settings.

**Cleaning contacts**

*Regulator contacts*

To clean the voltage or current regulator contacts use fine carborundum stone or silicon-carbide paper, followed by methylated spirit (denatured alcohol).

*Cut-out relay contacts*

To clean the cut-out relay contacts use a strip of fine glass-paper—never carborundum stone or emery-cloth.

## Section N.6

### STARTER

**Testing on vehicle when starter is not operating**

In the following test it is assumed that the battery is in a charged condition.

Switch on the lamps and operate the starter control. If the lights go dim but the starter is not heard to operate, an indication is given that the current is flowing through the starter motor windings, but that for some reason the armature is not rotating; possibly the starter pinion is

## THE STARTER MOTOR COMPONENTS

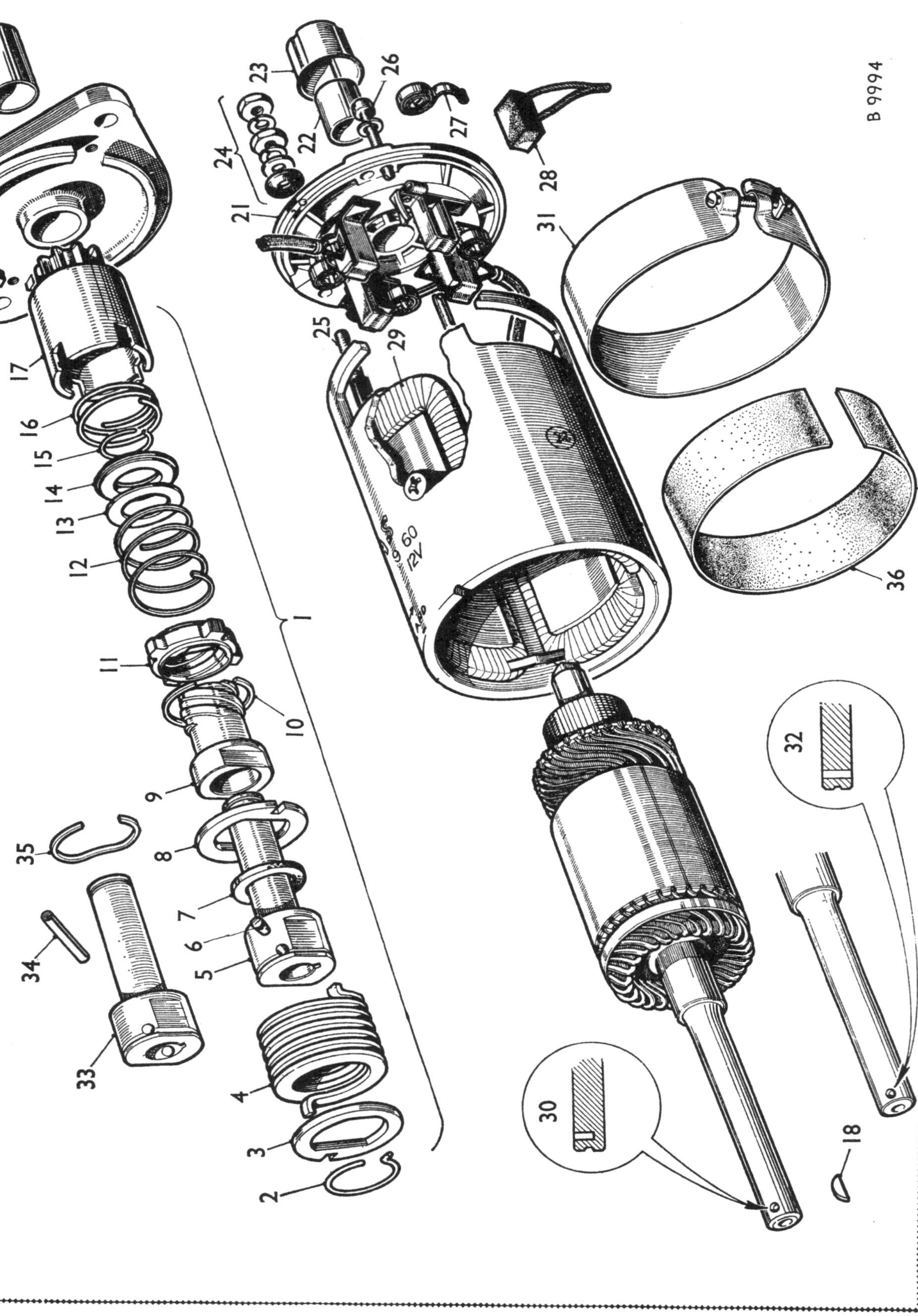

## KEY TO THE STARTER MOTOR COMPONENTS

| *No.* | *Description* | *No.* | *Description* | *No.* | *Description* |
|---|---|---|---|---|---|
| 1. | Drive assembly. | 13. | Thrust washer—control nut. | 25. | Terminal post. |
| 2. | Retaining ring. | 14. | Locating collar. | 26. | Through-bolt. |
| 3. | Anchor plate—front. | 15. | Retaining ring. | 27. | Brush tension spring. |
| 4. | Main spring. | 16. | Spring. | 28. | Brush. |
| 5. | Centre sleeve.* | 17. | Pinion and barrel. | 29. | Field coils. |
| 6. | Retaining pin.* | 18. | Woodruff key.* | 30. | Armature.* |
| 7. | Thrust washer (fibre). | 19. | Drive end bracket. | 31. | Cover band. |
| 8. | Anchor plate—rear. | 20. | Bush—driving end. | 32. | Armature.† |
| 9. | Screwed sleeve. | 21. | Commutator end bracket. | 33. | Centre sleeve.† |
| 10. | Retaining ring.* | 22. | Bush—commutator end. | 34. | Spiral pin.† |
| 11. | Control nut. | 23. | Shaft cap. | 35. | Waved circlip.† |
| 12. | Restraining spring. | 24. | Terminal nuts and washers. | 36. | Cover band seal.† |

* For starter motors Serial No. 25555.

† For starter motors Serial No. 25598.

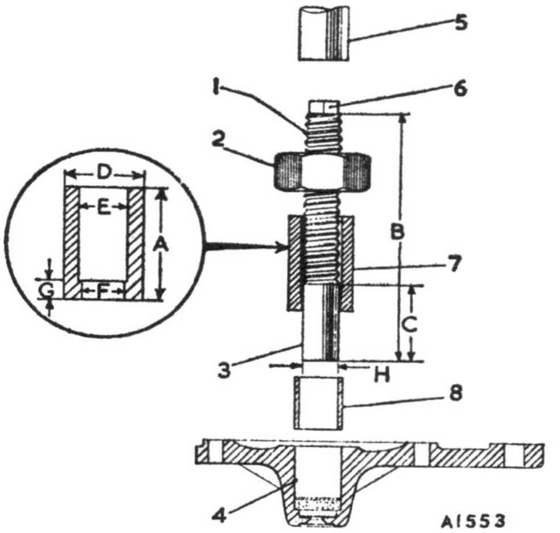


*Fig. N.8*

*Using a self-extracting-type tool to fit an end bracket bush*

| | | | |
|---|---|---|---|
| 1. | ⅝ in. B.S.F. truncated thread. | A. | 1½ in. (38 mm.). |
| 2. | Extracting nut. | B. | 4 in. (10·2 cm.). |
| 3. | Fitting pin. | C. | 1¼ in. (31·8 mm.). |
| 4. | Bearing housing. | D. | 1 5/16 in. (33·3 mm.). |
| 5. | Hand press. | E. | ·625 in. (15·87 mm.). |
| 6. | Squared end. | F. | ·605 in. (15·37 mm.). |
| 7. | Sleeve. | G. | ¼ in. (6·35 mm.). |
| 8. | Bush. | H. | ·5924 in. (15·05 mm.). |

meshed permanently with the geared ring on the flywheel. This could be caused by the starter being operated while the engine is still moving. In this case the starter motor must be removed from the engine for examination.

Should the lamps retain their full brilliance when the starter switch is operated, check that the switch is functioning. Next, if the switch is in order examine the connections at the battery and starter switch, and also examine the wiring joining these units. Continued failure of the starter to operate indicates an internal fault in the starter, which must be removed for examination.

Sluggish action of the starter is usually caused by a poor connection in the wiring which causes a high resistance in the starter circuit. Check the wiring as described above.

**Removing**

Disconnect the cable from the earth terminal on the battery and the cable from the terminal on the starter. Remove the external oil filter, the ignition coil and bracket, and the distributor. Unscrew the two bolts securing the starter to the flywheel housing and engine rear plate and withdraw the starter upwards from the engine compartment.

**Servicing**

*Examination of commutator and brush gear*

Remove the starter cover band and examine the brushes and the commutator. Hold back each of the brush springs and move the brush by pulling gently on its flexible connector. If the movement is sluggish remove the brush from its holder and ease the sides by lightly polishing on a smooth file. Always replace brushes in their original positions. If the brushes are worn so that they no longer bear on the commutator or if the brush flexible lead has become exposed on the running face, they must be renewed.

If the commutator is blackened or dirty clean it by holding a cloth moistened with fuel against it while the armature is rotated.

Secure the body of the starter in a vice and test by connecting it with heavy-gauge cables to a 12-volt battery. One cable must be connected to the starter terminal and the other held against the starter body or end bracket. Under these light load conditions the starter should run at a very high speed.

If the operation of the starter is still unsatisfactory it should be dismantled for detailed inspection and testing.

*Dismantling*

Take off the cover band at the commutator end, hold back the brush springs, and take out the brushes.

Remove the circlip from the outer end of the drive head sleeve and take off the front spring anchor plate, the main spring and the rear spring anchor plate.

Withdraw the pin securing the drive head sleeve to the armature shaft, push the sleeve assembly down the shaft and remove the woodruff key. Slide the complete drive assembly from the armature shaft.

Remove the barrel retaining ring from the inside of the barrel and pinion assembly and withdraw the barrel and anti-drift spring from the screwed sleeve.

From the inner end of the drive head sleeve remove the circlip, locating collar, control nut thrust washer, cushioning spring, control nut, screwed sleeve and the drive head thrust washer.

Remove the terminal nuts and washers from the terminal post and screw out the two through-bolts.

Remove the commutator end bracket, the drive end bracket, and the armature.

*Brushes*

(1) Test the brush springs with a spring balance. The correct tension is 30 to 40 oz. (850 to 1134 gm.). Fit a new spring if the tension is low.

(2) If the brushes are worn so that they no longer bear on the commutator, or if the flexible connector has become exposed on the running face, they must be renewed. Two of the brushes are connected to terminal eyelets attached to the brush boxes on the commutator end bracket. The other two brushes (Fig. N.9) are connected to tappings on the field coils.

The flexible connectors must be removed by unsoldering and the connectors of the new brushes secured in their place by soldering. The brushes are preformed so that bedding of the working face to the commutator is unnecessary.

*Drive*

(1) If the pinion is tight on the screwed sleeve wash away any dirt with paraffin (kerosene).

(2) If any parts are worn or damaged they must be renewed.

*Commutator*

A commutator in good condition will be smooth and free from pits and burned spots. Clean the commutator with a cloth moistened with petrol (gasoline). If this is ineffective carefully polish with a strip of fine glass-paper while rotating the armature. To remedy a badly worn commutator dismantle the starter drive as described above and remove the armature from the end bracket. Now mount the armature in a lathe, rotate it at a high speed, and take a light cut with a very sharp tool. Do not remove any more metal than is absolutely necessary, and finally polish with very fine glass-paper.

The mica on the starter commutator **must not be undercut.**

(See Editor's notes at end of section.)

*Field coils*

The field coils can be tested for an open circuit by connecting a 12-volt battery, having a 12-volt bulb in one of the leads, to the tapping-point of the field coils to which the brushes are connected and the field terminal post. If the lamp does not light there is an open circuit in the wiring of the field coils.

Lighting of the lamp does not necessarily mean that the field coils are in order as it is possible that one of them may be earthed to a pole-shoe or to the yoke. This may be checked by disconnecting the battery lead from the tapping-point of the field coils and holding it on a clean part of the starter yoke. Should the bulb now light, it indicates that the field coils are earthed.

NOTE.—**When carrying out this test the brushes connected to the field coils must not be in contact with the starter yoke.**

Should the above tests indicate that the fault lies in the field coils, they must be renewed. When renewing field coils carry out the procedure detailed in Section N.4.

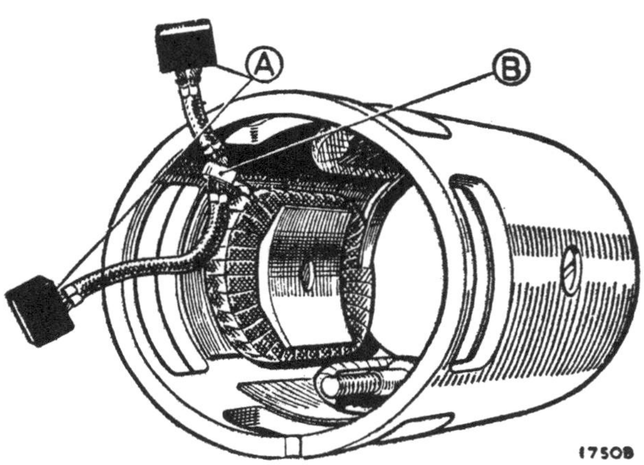


*Fig. N.9*

*The starter brush connections*

A. Brushes. B. Tapping on field coils.

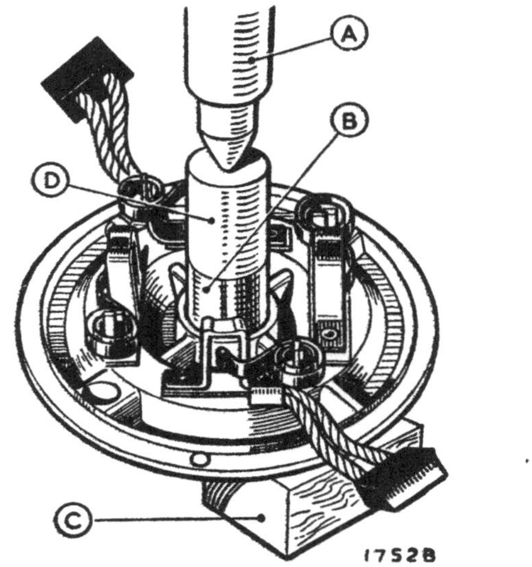


*Fig. N.10*

*Fitting a new bush, using a shouldered mandrel*

A. Hand press. B. Bearing bush. C. Supporting block. D. Shouldered mandrel.

*Armature*

Examination of the armature will in many cases reveal the cause of failure, e.g. conductors lifted from the commutator due to the starter being engaged while the engine is running and causing the armature to be rotated at an excessive speed. A damaged armature must in all cases be renewed—no attempt should be made to machine the armature core or to true a distorted armature shaft.

*Bearings (commutator end)*

Bearings which are worn to such an extent that they will allow excessive side-play of the armature shaft must be renewed. To renew the bearing bush proceed as follows.

Press the new bearing bush into the end bracket, using a shouldered mandrel of the same diameter as the shaft which is to fit in the bearing.

NOTE.—**The bearing bush is of the porous phosphor-bronze type, and before fitting, new bushes should be allowed to stand completely immersed for 24 hours in thin engine oil in order to fill the pores of the bush with lubricant.**

*Reassembling*

The reassembly of the starter is a reversal of the operations described in this Section.

## Section N.7

### FUSES

The fuses are mounted in a separate holder located under a large plastic cover on the right-hand wing valance.

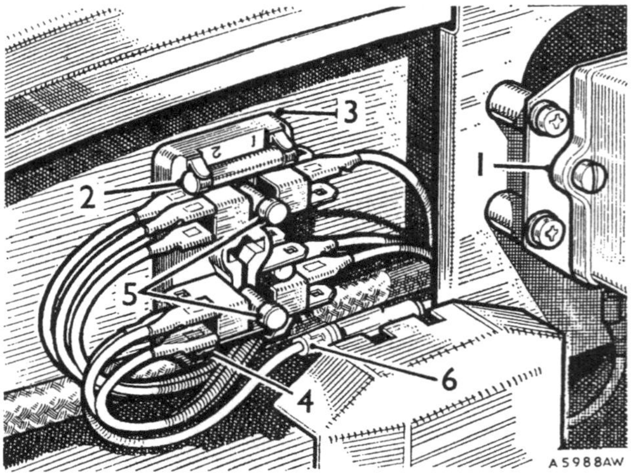


*Fig. N.*11

*Regulator and fuse block*

1. Regulator.
2. AUX. IGN. fuse (35-amp.).
3. Fuse block.
4. AUX. fuse (35-amp.).
5. Spare fuses.
6. Line fuse—heated back-light light (GT).

The fuse between terminals '1' and '2' protects the accessories which operate irrespective of whether the ignition is on or off and the fuse between terminals '3' and '4' protects the accessories which operate only when the ignition is switched on.

Two spare fuses are provided, and it is important that only fuses of the correct value, marked inside the fuse tube, should be used.

**Blown fuses**

The units which are protected by each fuse can readily be identified by referring to the wiring diagram.

A blown fuse is indicated by the failure of all the units protected by it and is confirmed by examination of the fuse, which can easily be withdrawn from the spring clips. If it has blown the fused state of the wire will be visible inside the glass tube. Before renewing a blown fuse inspect the wiring of the units that have failed for evidence of a short circuit or other faults which may have caused the fuse to blow, and remedy the cause of the trouble.

## Section N.8

### WINDSHIELD WIPER

Apart from the renewal of perished wiper blades, the windshield wiper requires no periodic maintenance.

Inspect the rubber wiping elements, which after long service become worn and should be renewed.

The gearbox and cable rack are packed with grease during manufacture and need no further lubrication.

(See supplementary notes at end of section.)

**Resetting the limit switch**

If the wiper fails to park or parks unsatisfactorily, the limit switch in the gearbox cover should be checked. Unless the limit switch is correctly set, it is possible for the wiper motor to overrun the open-circuit position and continue to draw current.

Slacken the four screws securing the gearbox cover and observe the projection near the rim of the limit switch. Position the projection in line with the groove in the gearbox cover. Turn the limit switch 25° in an anti-clockwise direction and tighten the four securing screws. If the wiper blades are required to park on the opposite side of the windshield, the limit switch should be turned back 180° in a clockwise direction.

**Testing on a vehicle**

If the wiper fails to operate, or operates unsatisfactorily, connect a 0–15 moving-coil ammeter in the wiper circuit, switch on the wiper, and note the current being supplied to the motor. The normal running current should be 2·7 to 3·4 amps.

*Wiper takes no current*

Examine the fuse protecting the wiper circuit. If the fuse has blown examine the wiring of the motor circuit and of all other circuits protected by that fuse. Renew, if necessary, any cables which are badly worn or chafed, fitting protective sleeving over the cables to prevent a recurrence of the fault.

If the external wiring is found to be in order, replace the fuse with one of the recommended rating. Then proceed as for the wiper taking an abnormally high current.

If the fuse is intact, examine the wiring of the motor circuit for breaks and ensure that the wiper control switch is operating correctly.

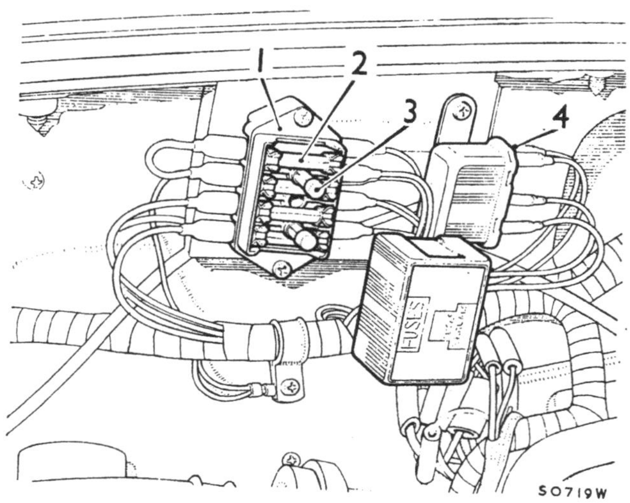


*Fig. N.*12

*Lucas type 7FJ fuse block (GHN5 and GHD5 cars)*

1. Four-way fuse block.
2. Current rated 17-amp. (35-amp. blow rated) fuses.
3. Spare fuses.
4. Starter solenoid.

*Wiper takes abnormally low current*

Check that the battery is fully charged. The performance of the motor is dependent on the condition of the battery.

Remove the commutator end bracket and examine the brush gear, ensuring that it bears firmly on the commutator. The tension spring must be renewed if the brushes do not bear firmly on the commutator. Brush levers must move freely on the pivots. If these levers are stiff they should be freed by working them backwards and forwards by hand.

Examine the commutator and, if necessary, clean with a fuel-moistened cloth. A suspected armature should be checked by substitution.

*Wiper takes abnormally high current*

If an abnormally high current is shown on the ammeter, this may be due to excessive load on the driving shaft. The stall current of the motor when cold is 14 amps. and when hot is 8 amps.

If there is no obvious reason for this, such as a sticking wiper blade, a check should be made at the gearbox.

Remove the gearbox cover and examine the gear assembly, checking that a blow on the gearbox end bracket has not reduced the armature end-float. The armature end-float adjusting screw must be set to give an armature end-play of ·008 to ·012 in. (·20 to ·30 mm.).

Sluggish operation with excessive current consumption may be caused through frictional losses in badly positioned or defective connecting tubes. The connecting tubes can be checked, using a cable gauge. (Details of this gauge can be obtained from any Lucas Agent.) The gauge cable is similar in appearance to the driving rack but is ·010 in. (·25 mm.) larger in diameter and is less flexible. The gauge will not easily pass through connecting tubes having less than the minimum permissible curvature.

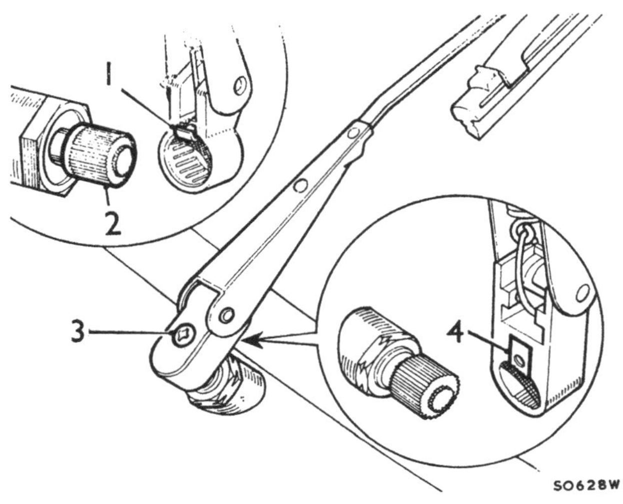


*Fig. N.*12

*Wiper arm fixing details*

1. Retaining clip.
2. Splined drive.
3. Locking screw.
4. Locking wedge.

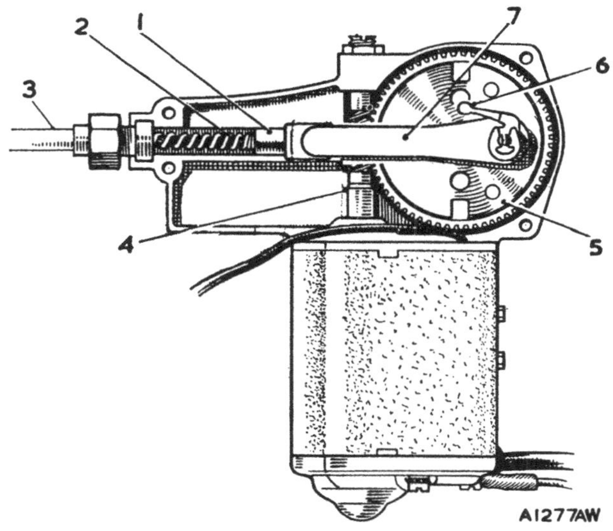


*Fig. N.*13

*Windshield wiper drive components*

1. Cross-head.
2. Cable rack.
3. Outer casing.
4. Armature.
5. Final gear.
6. Park switch.
7. Connecting rod.

To check the tubing remove the motor and inner rack. Insert the gauge into the connecting tube as far as the first wheelbox and then withdraw it. Remove the tubing connecting the wheelboxes. Insert and withdraw the gauge. If the gauge moves freely the tubing is correctly installed. If the gauge does not move freely the tubing must be checked for sharp bends and obstructions. Check the wheelboxes for alignment and then reassemble.

**Removing the motor, gearbox, and wheelboxes**

The motor and gearbox is located beneath the passenger's side of the fascia panel and is mounted on a bracket secured to the bulkhead panel by three screws.

The cable rack connected to the cross-head in the gearbox passes through outer casings which connect the gearbox to the first wheelbox and the first wheelbox to the second wheelbox.

Disconnect the wiper arms, the electrical connections from the motor, and the outer cable from the gearbox housing. Remove the three screws securing the bracket to the bulkhead panel and withdraw the motor, bracket, and cable rack from beneath the fascia panel.

Slacken the cover screws in each wheelbox and remove the cable rack outer casings.

Remove the nut, front bush, and washer from the front of each wheelbox and remove the wheelbox together with the rear bush and spindle tube from beneath the fascia panel.

Replacement is a reversal of the removal sequence, but care must be taken to ensure that the wheelboxes are correctly lined up and that the cable rack engages the gear and spindle assemblies.

## THE WINDSHIELD WIPER COMPONENTS

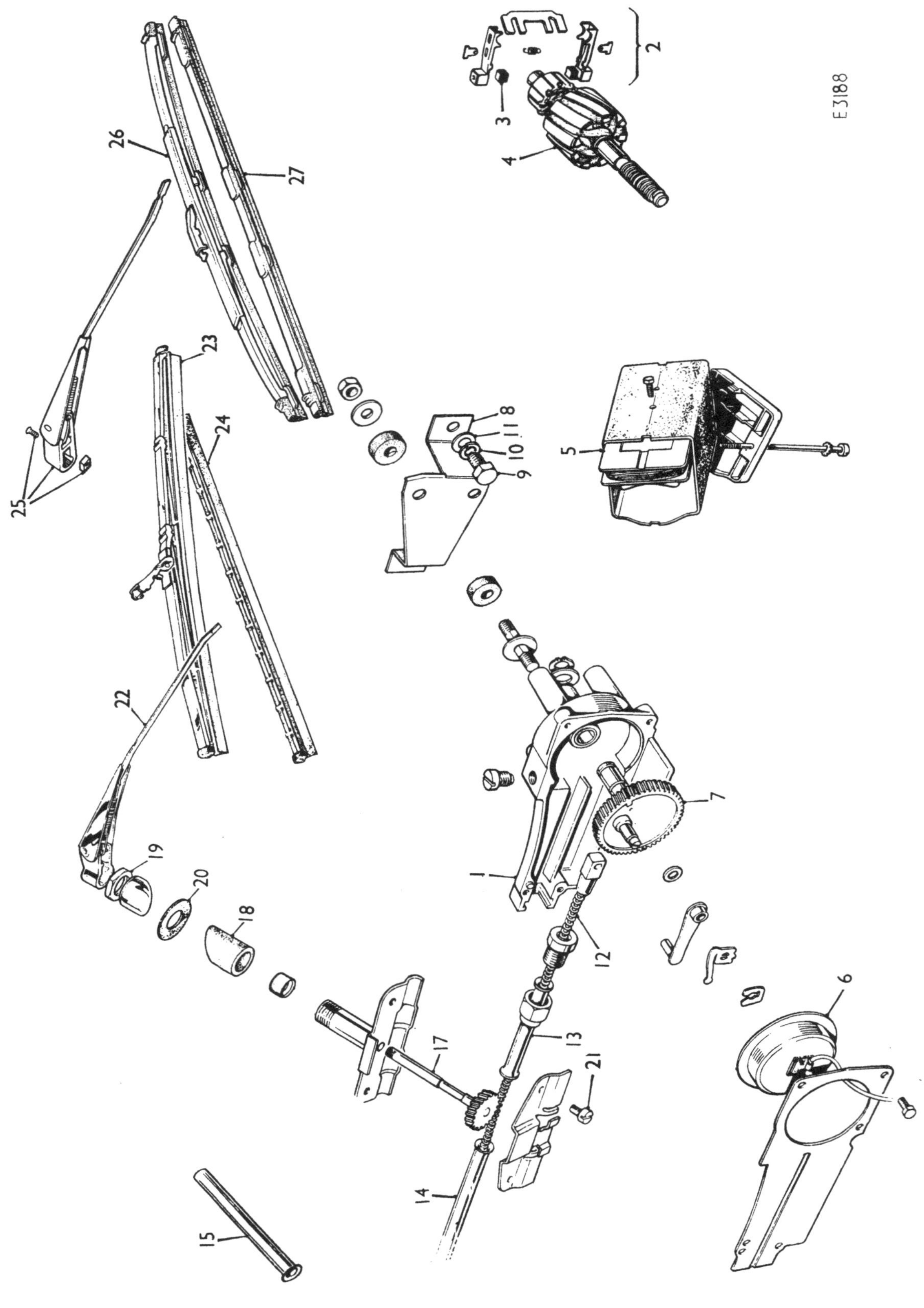

## KEY TO THE WINDSHIELD WIPER COMPONENTS

| No. | Description | No. | Description | No. | Description |
|---|---|---|---|---|---|
| 1. | Motor—windshield wiper. | 9. | Screw—bracket to bulkhead. | 17. | Spindle and gear. |
| 2. | Brush gear. | 10. | Washer—spring. | 18. | Tube—spindle—rubber. |
| 3. | Brush. | 11. | Washer—plain. | 19. | Nut—wheelbox. |
| 4. | Armature. | 12. | Crosshead and rack. | 20. | Washer (rubber). |
| 5. | Coil—field. | 13. | Casing—motor to wheelbox. | 21. | Screw—cover. |
| 6. | Switch—parking. | 14. | Casing—wheelbox to wheelbox. | 22. | Arm—wiper. |
| 7. | Shaft and gear. | 15. | Casing—wheelbox to extension. | 23. | Blade—wiper. |
| 8. | Bracket—wiper motor. | 16. | Wheelbox. | 24. | Rubber. |

Later-type wiper arms (item 22), are retained to the wheelbox spindle by a screw and locking wedge. The end of the wiper arm and blade connector are also slightly modified (page N.16).

**Dismantling the motor**

Withdraw the four screws securing the gearbox cover and remove the cover.

Withdraw the connectors and through-bolts at the commutator end bracket.

Remove the commutator end bracket clear of the yoke.

The brush gear can be removed by lifting it clear of the commutator and withdrawing it as a unit. Care should be taken at this point to note the particular side occupied by each brush so that each may be replaced in its original setting on the commutator.

Access to the armature and field coils can be gained by withdrawing the yoke.

If it is necessary to remove the field coil, unscrew the two screws securing the pole-piece to the yoke. These screws should be marked so that they can be returned to their original holes.

Press out the pole-piece complete with field coil, marking the pole-piece so that it can be replaced in its correct position inside the yoke. The pole-piece can now be pressed out of the field coil.

Pieces of carbon short-circuiting adjacent segments of the commutator will also cause excessive current consumption. The resistance between adjacent commutator segments should be ·34 to ·41 ohm. Cleaning the commutator and brush gear removes this fault. When dismantling, check the internal wiring of the motor for evidence of short-circuiting due to chafed or charred insulation. Slip a new piece of sleeving over any charred connections, and arrange them so that they do not rub against sharp edges.

While the motor is dismantled check the value of the field resistance. If it is found to be lower than 12·8 to 14 ohms a short circuit in the windings is indicated and a new field coil must be fitted. Other evidence of a short circuit will be given by charred leads from the field coil.

**Dismantling the gearbox unit**

Remove the circlip and washer from the cross-head connecting link pin and lift off the cross-head and cable rack assembly. Then remove the circlip and washer from the final gear shaft located underneath the gearbox unit. Remove any burr from the circlip groove before lifting out the final gear. The armature and worm drive can now be withdrawn from the gearbox. All gear teeth should be examined for signs of damage or wear and, if necessary, new gears fitted.

**Reassembling**

Reassembly is a reversal of the above procedures. When reassembling, the following components should be lubricated, using the lubricants recommended.

*Armature bearings*
These should be lubricated with S.A.E. 20 engine oil, the self-aligning bearing being immersed in this for 24 hours before assembly.

*Armature shaft (commutator end)*
Apply S.A.E. 20 engine oil.

*Felt lubricator in gearbox*
Apply S.A.E. 20 engine oil.

*Worm wheel bearings, cross-head, guide channel, connecting rod, crankpin, worm, and final gear shaft*
Grease liberally.

*Cable rack and wheelboxes*
Grease liberally.

**Testing**

Switch on the ignition and the wiper control. The two wiper areas should be approximately symmetrical on the windshield.

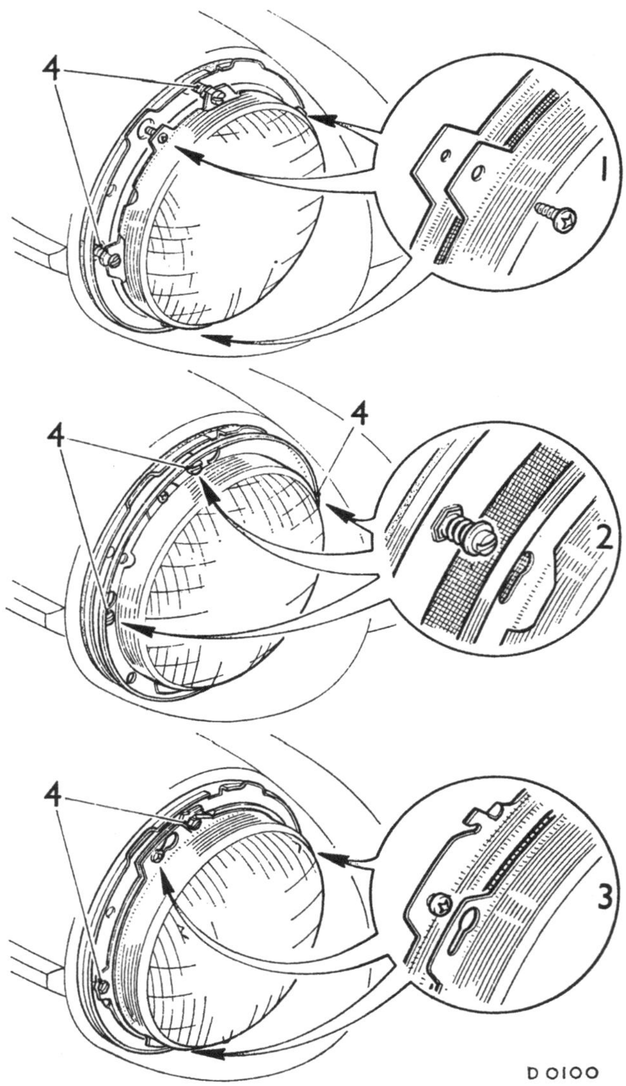


*Fig. N.14*
*Light unit retaining screws*

1. Removable screw.
2. Combined adjusting/retaining screw.
3. Captive retaining screw.
4. Beam-setting adjusting screws.

## Section N.9

### HEADLAMPS

The two types of headlamp fitted have sealed-beam units or, alternatively, replaceable bulb light units.

Variations within the two basic types cater for the local lighting regulations existing in the country for which the car was produced. The method of retaining and adjusting the light unit together with the type of lens and bulb used are subject to territorial variation. Fig. N.14 shows the alternative methods of retaining and adjusting the light unit and Fig. N.15 the different types of bulb fittings used.

**Removing a light unit**

*Removable-type retaining screw* (1) *Fig. N.*14

(1) Ease the bottom of the outer rim forward away from the lamp, and lift it off the retaining lugs at the top of the lamp.

(2) Remove the rubber dust excluder (if fitted).

(3) Remove the three screws securing the light unit retaining plate and withdraw the light unit from the lamp body.

*Combined adjusting/retaining screw* (2) *Fig. N.*14

(4) Carry out the operations detailed in (1) and (2).

(5) Press the light unit inwards against the tension of the springs and turn it in an anti-clockwise direction until the heads of the screws can pass through the enlarged ends of the slots in the retaining plate, then withdraw the light unit.

*Captive-type retaining screw* (3) *Fig. N.*14

(6) Carry out the operations detailed in (1) and (2).

(7) Slacken the three retaining screws and turn the light unit retaining plate in an anti-clockwise direction until the heads of the screws can pass through the enlarged ends of the slots in the retaining plate, then withdraw the retaining plate and light unit.

**Bulb replacement**

*Cap-type holder* (1) *Fig. N.*15

(8) Press and turn the cap anti-clockwise, lift off the cap and withdraw the bulb.

(9) Fit the bulb into the reflector ensuring that the notch in the bulb flange locates on the ridge in the reflector.

(10) Engage the cap lugs in the reflector slots, press and turn the cap clockwise.

*Spring clip type* (2) *Fig. N.*15

(11) Withdraw the three-pin socket.

(12) Disengage the spring clip from the reflector lugs, swing the clip up and withdraw the bulb.

(13) Fit the bulb into the reflector ensuring that the pip on the bulb flange engages the slot in the reflector.

(14) Swing the spring clip back, ensure that the coils in the clip are resting on the base of the bulb, and engage the legs of the spring clip under the reflector lugs.

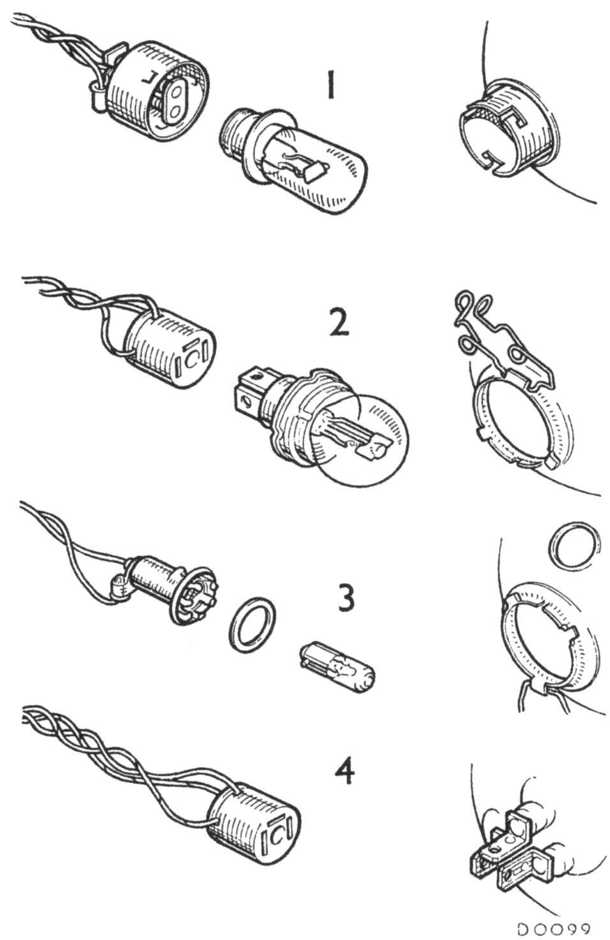


*Fig. N.*15

*Bulb holders and fixings*

1. Cap-type holder.
2. Spring clip type.
3. Headlamp pilot lamp.
4. Sealed-beam unit.

*Headlamp pilot lamp* (3) *Fig. N.*15

(15) Withdraw the holder from the reflector.

(16) Press and turn the bulb anti-clockwise and withdraw the bulb.

(17) Locate the pins of the bulb in the grooves in the holder, press and turn the bulb clockwise.

(18) Press the holder into its hole in the reflector.

**Refitting a light unit**

(19) Reverse the removing procedure, noting that the outer rim is refitted as follows.

(*a*) Position the rim on the retaining lugs with the cut-away portion of the rim at the bottom of the lamp.

(*b*) Press the rim downwards and inwards.

**Beam setting**

The headlamps must be set so that the main driving beams are parallel with the road surface or in accordance with local regulations.

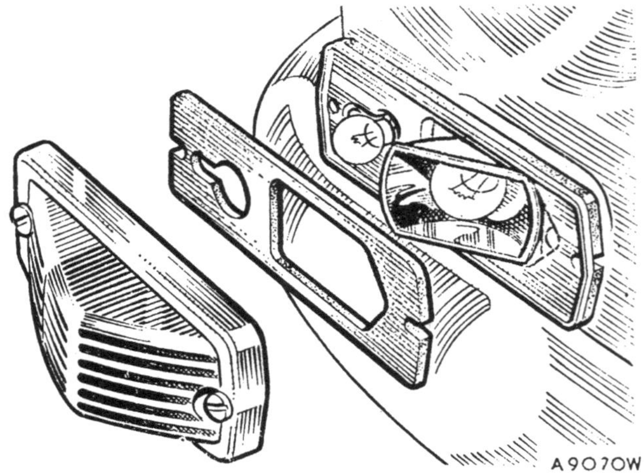


*Fig. N. 16*
*The side and direction indicator lamps*

*Adjusting*

(20) Carry out the operations detailed in (1) and (2).

(21) Adjust the beam vertically by turning the adjusting screw, located at the top of the lamp, in the required direction.

(22) Set the beam horizontally by turning the adjusting screw located on the right-hand side or, on some models, the adjusting screws on both sides of the lamp (see Fig. N.14) in the required direction.

## Section N.10

### SIDE AND DIRECTION INDICATOR LAMPS

To gain access to the side and direction lamps remove the two screws securing the lamp lens and take off the lens. Both bulbs are of the bayonet-fixing type and may be fitted either way round.

The lamp bodies are secured to the front of each wing by two studs and nuts.

## Section N.11

### TAIL AND STOP AND DIRECTION INDICATOR LAMPS

The tail lamp bulbs are of the double-filament type, the second filament giving a marked increase in brilliance when the brakes are applied.

Access to the bulbs is gained by extracting the securing screws from the outer face of the lamp lens to release the lens.

The tail and stop lamp bulbs must be fitted one way only; offset retaining pegs ensure that they are replaced correctly.

The lamp body can be removed when the lens is taken off as indicated in Fig. N.17 and the three screws located in the lamp body withdrawn. When refitting the glass to the body make certain that it is seating correctly over the sealing rubber.

## Section N.12

### NUMBER-PLATE LAMPS

The rear number-plate lamps are fitted one to each inside face of the rear bumper over-riders.

To dismantle a lamp remove the two securing screws and retain the nuts and distance pieces from the inside of the over-rider. Take off the lamp hood and remove the lens, washer, and plinth.

When refitting the screws and nuts ensure that the thick edge of the wedge-shaped distance piece is fitted so that it faces the front of the car.

## Section N.13

### PANEL, MAP, AND WARNING LIGHTS

The location of the panel and warning lights are shown in the accompanying illustration.

The panel and warning light bulb holders are a push fit in their housings and are accessible from below the fascia panel.

To change a map lamp bulb remove the lamp cover screws and the cover.

## Section N.14

### RELAYS

Relays are electromagnetically operated switches that enable large currents to be controlled by smaller pilot currents. They are adjusted and sealed and the only maintenance required is an occasional check of the terminal connections for looseness.

**Fog lamp relay (Early cars only)**

The relay is attached by two screws and washers to the engine side of the bulkhead. When the switch is closed

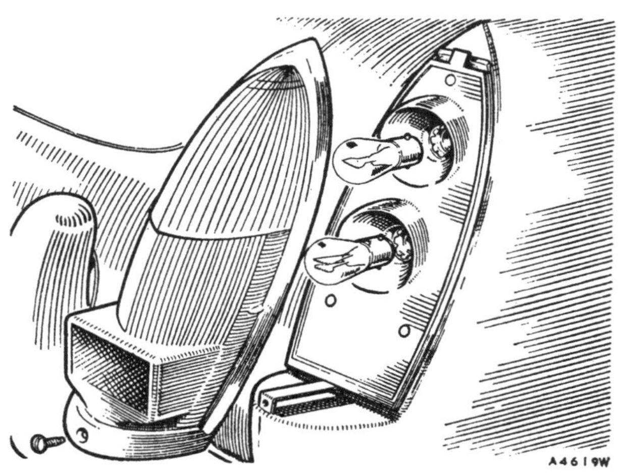


*Fig. N.17*
*The tail, stop, and direction indicator lamps*

the relay contacts close and permit current to flow to the lamps. When the switch is opened the relay contacts open and interrupt the circuit.

**Starter motor relay (Car Number prefix GHN5 and GHD5 cars)**

The non-adjustable relay is mounted next to the four-way fuse block.

**Removing**

Disconnect the battery. Remove the electrical connections and securing screws. Remove the relay.

## Section N.15

### HORN

**Removing the horn**

Disconnect the electrical connections, remove the screws and washers securing the horn to the bonnet lock platform or from the wing valance and withdraw the horn.

**Maintenance**

If the horn fails to operate, or operates unsatisfactorily, first carry out the following external checks.

Check the current consumption, which should be 3 to 3½ amps. when the horn is operating correctly.

After making a thorough external check remove the horn cover and examine the cable connections inside the horn. Examine the contact breaker contacts. If they are burnt or blackened clean them with a fine file, then wipe with a petrol-moistened cloth.

**Adjusting**

Withdraw the cover securing screw and remove the cover and its retaining strap.

Slacken the locking nut on the fixed contact stem and rotate the adjusting nut clockwise until the contacts just separate (indicated by the horn failing to sound when the horn-push is operated). Rotate the adjusting nut one half-turn in the opposite direction and secure it in this position with the locknut. Refit the cover and strap.

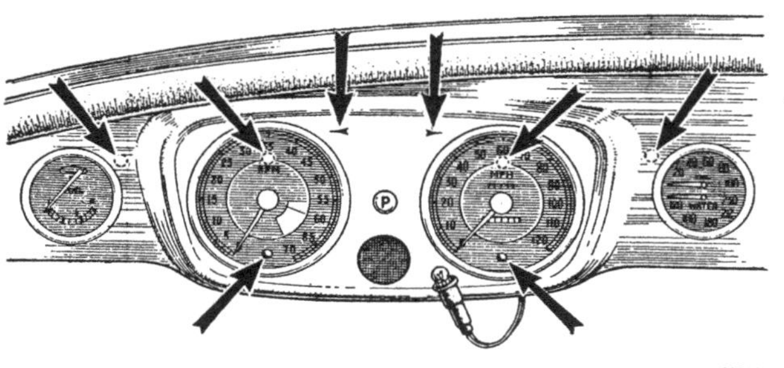


*Fig. N.18*
*The location of the panel and warning lights*

*Fig. N.19*
*A rear number-plate lamp (early cars)*

## Section N.16

### FLASHER UNIT

**Description**

The unit is contained in a small cylindrical metal container, one end of which is rolled over onto an insulated plate carrying the mechanism and three terminals. The unit depends for its operation on the linear expansion of a length of wire which becomes heated by an electric current flowing through it. This actuating wire controls the movement of a spring-loaded armature attached to a central steel core and carrying a moving contact—the sequence of operation being as follows.

When the direction indicator switch is turned either to the left or right, current flows through the actuating wire, ballast resistor, and coil wound on the central core and hence to earth via the flasher lamp filaments. This current is limited by the ballast resistor to a value which will ensure that the flasher lamp filaments do not light at this stage. The actuating wire increases in length under the heating influence of the current and allows the armature to move inwards to its alternative position, thereby closing a pair of contacts in the supply circuit to the flasher lamps and at the same time short-circuiting the actuating wire. The increased electromagnetic attraction of the armature to the core, due to the full lamp current now flowing through the coils, serves to hold the closed contacts firmly together. At the same time a secondary spring-loaded armature is attracted to the core and closes a pilot warning lamp circuit so that now both flasher lamps and warning lamp are illuminated.

Since, however, heating current no longer flows through the short-circuited actuating wire, the latter cools and consequently contracts in length. The main armature is therefore pulled away from the core, the contacts opened,

## THE SWITCHES

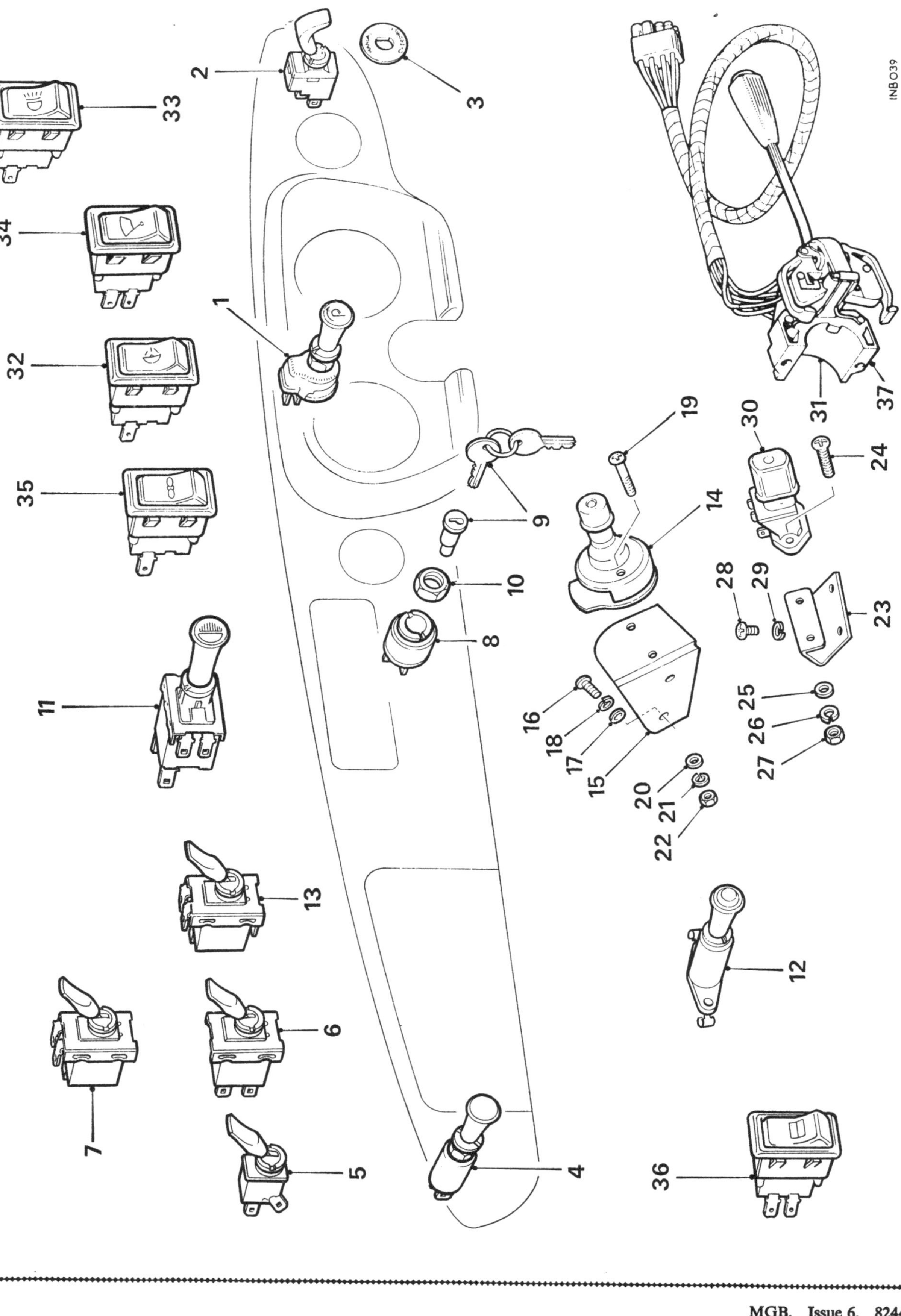

## KEY TO THE SWITCHES

| No. | *Description* |
|---|---|
| 1. | Panel light switch. |
| 2. | Overdrive switch. |
| 3. | Bezel for overdrive switch. |
| | **Up to and including early GHN5 and GHD5 cars** |
| 4. | Map light switch. |
| 5. | Heater blower switch. |
| 6. | Windscreen wiper switch (GT). |
| 7. | Windscreen wiper switch (Tourer). |
| 8. | Ignition/starter switch. |
| 9. | Ignition switch lock and key. |
| 10. | Locknut for ignition/starter switch. |
| 11. | Light switch (later cars). |
| 12. | Electrically heated back-light switch (GT). |
| | **GHN3 and GHD3 cars** |
| 13. | Light switch. |
| 14. | Headlight dip switch. |
| | **Up to and including GHN4 and GHD4 cars—R.H.D.** |
| 15. | Bracket for dip switch (R.H.D.). |
| 16. | Screw—bracket to pedal box. |
| 17. | Plain washer. |
| 18. | Spring washer. |
| 19. | Screw—dip switch to bracket. |
| 20. | Plain washer. |
| 21. | Spring washer. |
| 22. | Nut. |
| | **Up to and including GHN4 and GHD4 cars—L.H.D.** |
| 23. | Bracket for dip switch (L.H.D.). |
| 24. | Screw—dip switch to bracket. |
| 25. | Plain washer. |
| 26. | Spring washer. |
| 27. | Nut. |
| 28. | Screw—bracket to pedal box. |
| 29. | Spring washer. |
| | **GHN4 and GHD4 cars** |
| 30. | Headlight dip switch. |
| | **Early GHN5 and GHD5 cars** |
| 31. | Combined direction indicator, headlamp, high-low beam, headlamp flasher, horn-push switch. |
| | **GHN5 and GHD5 cars from Car No. 258001** |
| 32. | Fog and spot light switch. |
| 33. | Light switch. |
| 34. | Windscreen wiper switch. |
| 35. | Heater blower switch. |
| 36. | Electrically heated back-light switch (GT). |
| 37. | Combined direction indicator, headlamp high-low beam, headlamp flasher switch. |

and the light signals extinguished. The consequent reduction of electromagnetism in the core allows the secondary armature to return to its original position and so extinguish the pilot warning light. The above sequence of operations continues to be repeated until the indicator switch is returned to the 'off' position.

**Functions of warning lamp**

The warning lamp not only serves to indicate that the flasher unit is functioning correctly, but also gives a warning of any bulb failure occurring in the external direction indicator lamps—since a reduction in bulb current flowing through the coil reduces the electromagnetic effect acting on the secondary armature and so prevents closure of the pilot light contacts.

**Checking faulty operation**

In the event of trouble occurring with a flashing light direction indicator system the following procedure should be followed.

Check with a voltmeter between the flasher unit terminal 'B' (or '+') and earth that battery voltage is present.

Check the bulbs for broken filaments.

Refer to the vehicle wiring diagram and check all flasher circuit connections.

Check the appropriate fuse.

Switch on the ignition.

Connect together flasher unit terminals 'B' (or '+') and 'L' and operate the direction indicator switch. If the flasher lamps now light, the flasher unit is defective and must be renewed.

**Maintenance**

Flasher units cannot be dismantled for subsequent reassembly. A defective unit must therefore be renewed.

When renewing a flasher unit or installing a flashing light system it is advisable to test the circuits before connections to flasher terminals are made. When testing join the cables normally connected to those terminals (green, green with brown, and light green) together and operate the direction indicator switch. In the event of a wrong connection having been made, the ignition auxiliaries fuse will blow but no damage will be done to the flasher unit.

## Section N.17

### SWITCHES

Always disconnect the batteries before removing any switches and refer to the wiring diagram and switch contact numbers when refitting switches.

**Ignition/starter switch**

Disconnect the Lucar connectors and, using Service tool 18G 671, remove the bezel lock ring. Withdraw the switch assembly from the rear of the fascia.

To remove the locking barrel from the switch body insert the ignition key and turn the switch to 'ON' to align the barrel retaining plunger with the small hole in the switch body. Using an awl, depress the plunger and withdraw the barrel and key.

Reverse the removal and dismantling procedure to assemble and refit the switch.

**Lighting, wiper, and blower switches**

Disconnect the Lucar connectors and, using Service tool 18G 671, unscrew the bezel lock rings. Remove the switches from the rear of the fascia.

To refit the switches reverse the removal sequence.

**Panel and map-reading lamp switches**

The panel light switch is controlled by a rheostat-type rotary switch and the map-reading light by a push-pull-type switch.

To remove the switches depress the spring-loaded plunger in the switch knobs and withdraw the knobs. Use Service tool 18G 671 to remove the bezel lock rings and remove the switches from the rear of the fascia panel.

Reverse this sequence to refit the switches.

**Direction indicator and headlight flasher switch (GHN4 and GHD4 cars)**

Unscrew the four screws securing the two halves of the cowling and remove the cowling. Disconnect the switch lead connectors and remove the two switch bracket fixing screws.

Reverse this sequence to refit the switch and cowling but ensure that the locating pin on the switch and the locating peg on the cowling are correctly aligned.

NOTE.—**When replacing an early-type switch with one using spring-loaded metal paws and spring leaves, a square-headed trip stud must be used. Set the height of the stud to between 1·177 and 1·187 in. (29·9 and 30·45 mm.) from the underside of the column.**

**Dip switch (GHN4 and GHD4 cars)**

The dip switch is mounted on a bracket secured to the front pedal box by two screws. Remove the screws to free the switch and bracket, disconnect the electrical connections, and remove the switch.

To refit the switch reverse the removal sequence.

**Direction indicator/headlamp flasher/high–low beam/horn switch (GHN5 and GHD5 cars)**

Refer to the instructions given for removing and refitting the direction indicator and headlamp flasher switch (GHN4 and GHD4 cars).

## Section N.18

### LOCATION AND REMEDY OF FAULTS

Although every precaution is taken to eliminate possible causes of trouble, failure may occasionally develop through lack of attention to the equipment or damage to the wiring. The following set-out is the recommended procedure for a systematic examination to locate and remedy the causes of some of the more usual faults encountered.

The sources of trouble are by no means always obvious, and in some cases a considerable amount of deduction from the symptoms is needed before the cause is disclosed.

For instance, the engine might not respond to the starter switch; a hasty inference would be that the starter motor is at fault. However, as the motor is dependent on the battery it may be that the battery is exhausted.

This in turn may be due to the dynamo failing to charge the battery, and the final cause of the trouble may be, perhaps, a loose connection in some part of the charging circuit.

If, after carrying out an examination, the cause of the trouble is not found the equipment should be checked.

### CHARGING CIRCUIT

**1. Battery in low state of charge**

(*a*) This state will be shown by lack of power when starting, poor light from the lamps, and the hydrometer readings below 1·200. It may be due to the dynamo not charging or giving low or intermittent output. The ignition warning light will not go out if the dynamo fails to charge, or will flicker on and off in the event of intermittent output.

(*b*) Examine the charging and field circuit wiring. tighten any loose connections, or renew any broken cables. Pay particular attention to the battery connections.

(*c*) Examine the dynamo driving belt; take up any undue slackness by swinging the dynamo outwards on its mounting after slackening the attachment bolts.

(*d*) Check the regulator setting, and adjust if necessary.

(*e*) If, after carrying out the above, the trouble is still not cured, have the equipment examined.

**2. Battery overcharged**

This will be indicated by burnt-out bulbs, very frequent need for topping up the battery, and high hydrometer readings. Check the charge reading with an ammeter when the car is running. It should be of the order of only 3 to 4 amps.

If the ammeter reading is in excess of this value it is advisable to check the regulator setting, and adjust if necessary.

### STARTER MOTOR

**1. Starter motor lacks power or fails to turn engine**

(*a*) See if the engine can be turned over by hand. If not, the cause of the stiffness in the engine must be located and remedied.

(*b*) If the engine can be turned by hand first check that the trouble is not due to a discharged battery.

(*c*) Examine the connections to the battery, starter, and starter switch, making sure that they are tight and that the cables connecting these units are not damaged.

(*d*) It is also possible that the starter pinion may have jammed in mesh with the flywheel. To disengage the pinion rotate the squared end of the starter shaft by means of a spanner.

**2. Starter operates but does not crank the engine**

This fault will occur if the pinion of the starter drive is not allowed to move along the screwed sleeve into engagement with the flywheel, due to dirt having collected on the screwed sleeve. Remove the starter and clean the sleeve carefully with paraffin (kerosene).

**3. Starter pinion will not disengage from flywheel when engine is running**

Stop the engine and see if the starter pinion is jammed in mesh with the flywheel, releasing it if necessary by rotation of the squared end of the starter shaft. If the pinion persists in sticking in mesh have the equipment examined. Serious damage may result to the starter if it is driven by the flywheel.

### LIGHTING CIRCUITS

**1. Lamps give insufficient illumination**

(*a*) Test the state of charge of the battery, recharging it if necessary from an independent electrical supply.

(*b*) Check the setting of the lamps.

(*c*) If the bulbs are discoloured as the result of long service they should be renewed.

**2. Lamps light when switched on but gradually fade out**

As paragraph 1 (*a*).

**3. Brilliance varies with speed of car**

(*a*) As paragraph 1 (*a*).

(*b*) Examine the battery connections, making sure that they are tight, and renew any faulty cables.

## Section N.19

### BI-METAL RESISTANCE INSTRUMENTATION

**General description**

The bi-metal resistance equipment for fuel and temperature gauges consists of an indicator head and transmitter unit connected to a common voltage stabilizer The system by which the equipment functions is voltage-sensitive and the voltage stabilizer is necessary to ensure a constant supply of a predetermined voltage to the equipment.

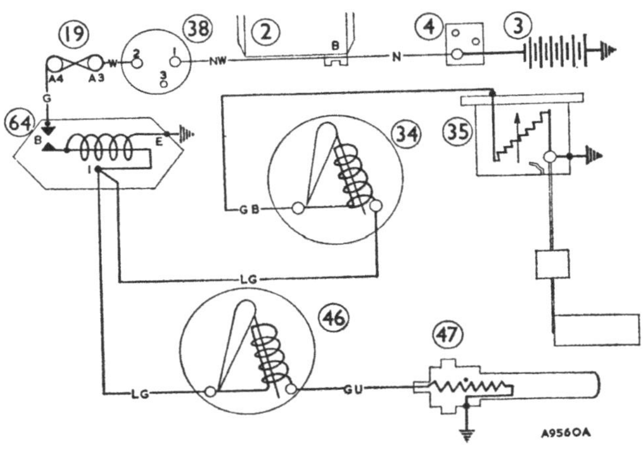


*Fig. N.20*

*The bi-metal resistance instrumentation circuit*

| | | | |
|---|---|---|---|
| 2. | Control box. | 38. | Ignition switch. |
| 3. | Batteries (12-volt). | 46. | Coolant temperature. gauge. |
| 4. | Starter solenoid. | 47. | Coolant temperature. transmitter. |
| 19. | Fuse (A3–A4). | 64. | Voltage stabilizer. |
| 34. | Fuel gauge. | | |
| 35. | Fuel tank unit. | | |

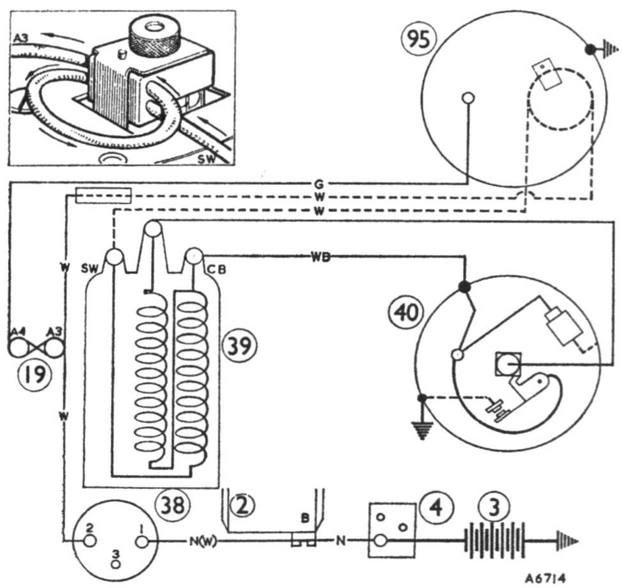


*Fig. N.21*

*Impulse tachometer circuit*

2. Control box.
3. Batteries (12-volt).
4. Starter solenoid.
19. Fuse—A3–A4.
38. Ignition switch.
39. Ignition coil.
40. Distributor.
95. Tachometer.

***Inset: symmetrical loop of pulse lead***

**Fault analysis**

For rapid diagnosis of a faulty unit use Smiths Automotive Instrument Tester (which incorporates a thermal sensitive voltmeter).

*Battery voltage*

Connect voltmeter to control box terminal 'B' and earth:

(*a*) Engine stationary: 12 volts, approx.

(*b*) Engine running at 1,000 r.p.m. and ignition warning light out, 12–13 volts approx.

*Wiring*

Check for continuity between each unit and check for leak to earth. Check for short-circuits in wiring to each transmitter. Check that the voltage stabilizer and transmitters are earthed.

*Voltage stabilizer*

Ignition on; after two minutes check the main voltage between the output terminal 'I' and earth, which should be 10 volts.

NOTE.—**If the voltage stabilizer is removed, it is essential to ensure that, when replacing, 'B' and 'E' are uppermost and not exceeding 20° from the vertical.**

Substitute voltage stabilizer if faulty.

*Gauges*

Check for continuity between terminals with wiring disconnected.

**The gauges must not be checked by short-circuiting to earth.**

Substitute gauge if faulty.

*Transmitter*

Check for continuity between terminal and case with lead disconnected.

Substitute transmitter if faulty.

## Section N.20

### TACHOMETER—IMPULSE TYPE

The equipment consists of an indicator head and pulse lead. The pulse lead is connected in series between the ignition switch and the ignition coil, and transmits voltage pulses to the indicator head.

**Faulty operation**

Check wiring connections to the indicator head and continuity of the circuit. Poor connections may result in faulty readings.

The pulse lead should form a symmetrical loop and not tight against the plastic forms (inset Fig. N.21).

## Section N.21

### REVERSE LAMPS

**Bulb replacement**

(1) Remove the screws securing the reverse lamp to the rear panel.

(2) Withdraw the lens.

(3) Press the bulb down towards the lower contact and withdraw if from the lamp.

(4) Fit one end of the bulb into the hole in the lower contact, then press the top of the bulb until the point on the bulb cap engages in the hole in the upper contact.

(5) Refit the lens and securing screws.

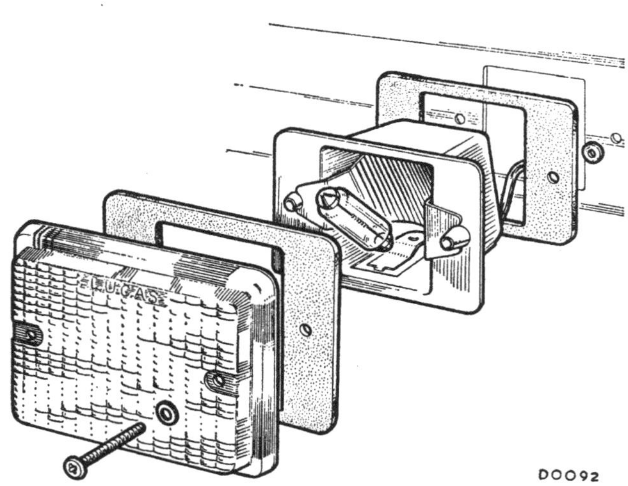


*Fig. N.22*

*The reverse lamp*

## Section N.22

### ELECTRICALLY HEATED BACK-LIGHT (GT)

**Removing and refitting**

For removing and refitting procedure refer to Section R.27.

**Fuse**

A line fuse to protect the heated back-light is located below the main fuse block (see Fig. N.11).

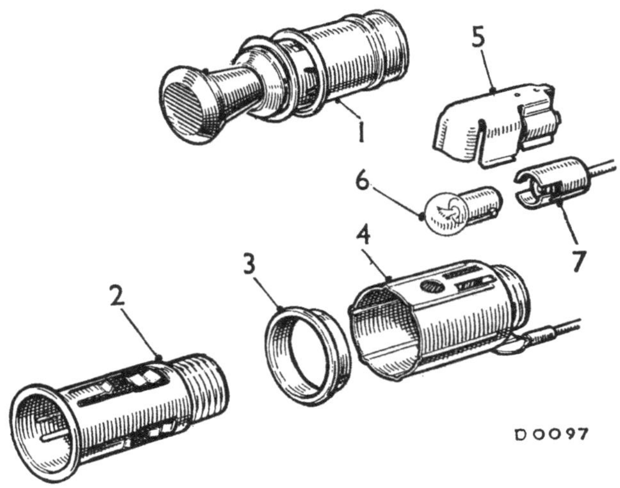


*Fig. N.23*

*The cigar lighter components*

1. Pop-out heater unit.
2. Switch base.
3. Glow ring.
4. Shell.
5. Lamp cover.
6. Bulb.
7. Bulb holder.

## Section N.23

### CIGAR-LIGHTER

**Removing**

(1) Disconnect the cigar-lighter feed wire from the ignition switch terminal.
(2) Disconnect the earth wire from the cigar-lighter.
(3) Press in the sides of the illumination lamp cover and withdraw the lamp unit from the lighter shell.
(4) Unscrew the lighter shell from the switch base.
(5) Withdraw the switch base and glow ring from the front of the fascia panel.

**Refitting**

(6) Reverse the removing procedure in (1) to (5).

## Section N. 24

### CHARGING SYSTEM (GHN4 and GHD4 Cars)

**Description**

The 12-volt NEGATIVE-earth charging system consists of an alternator and a voltage regulator supplying current to two 6-volt batteries connected in series.

Diodes contained within the alternator provide rectification of its output current for battery charging and also prevent reverse current flow. These and other design features overcome the necessity of a cut-out and current regulator.

**Precautions**

To prevent inadvertent damage to the charging system equipment the following points must be observed.

(1) Do not run the engine with the battery or any of the charging circuit cables disconnected.
(2) Ensure that all electrical connections in the generating and charging circuits are maintained tight at all times.
(3) Ensure that the correct battery polarity (NEGATIVE EARTH) is maintained at all times.
(4) If electric arc-welding equipment is to be used on the car, all electric cables must be disconnected from the alternator and control box terminals.

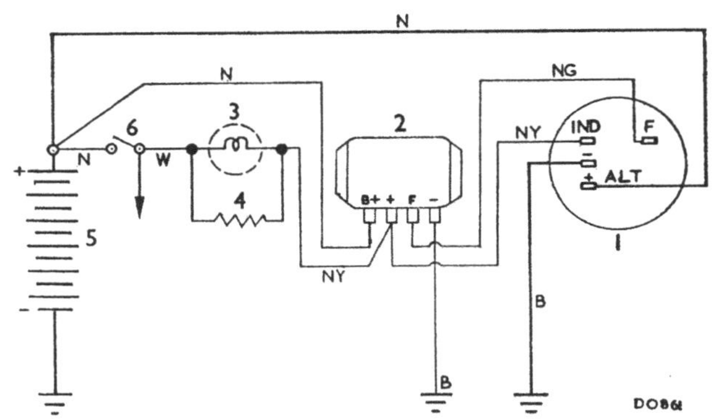


*Fig. N.24*

*16AC alternator charging circuit*

1. Alternator.
2. Control box.
3. Warning light.
4. Resistance.
5. Battery.
6. Ignition switch.

## Section N.25

### ALTERNATOR (Lucas Type 16AC)

**Testing in position**

(1) Check that the fan belt is correctly tensioned and that all charging circuit electrical connections are tight.

(2) Run the engine at fast idle speed until normal operating temperature is reached.

(3) Stop the engine and withdraw the moulded terminal connector block from the alternator.

*Fig. N.25*
*The 16AC alternator components*

1. Cover.
2. Rectifier pack.
3. Through-bolt.
4. Slip-ring end bracket.
5. Stator.
6. Slip-rings.
7. Slip-ring end bearing.
8. Rotor.
9. Field windings.
10. Drive end bearing.
11. Circlip.
12. Oil sealing ring.
13. Drive end bracket.
14. Shaft nut.
15. Brush box moulding.
16. Brush assembly.

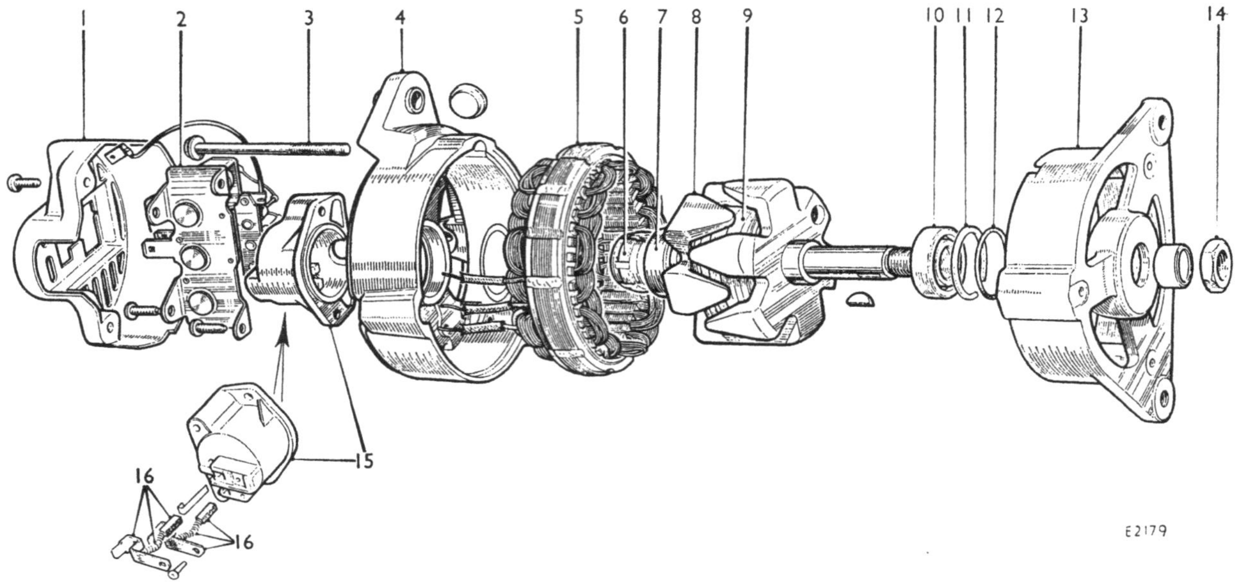


*Fig. N.26*
*16AC alternator output test circuit*

1. Alternator.
2. 0–40 amp. ammeter.
3. 12-volt, 2·2-watt bulb.
4. 0–15 ohm, 35-amp. variable resistance.
5. 0–20 volt voltmeter.
6. 12-volt battery.

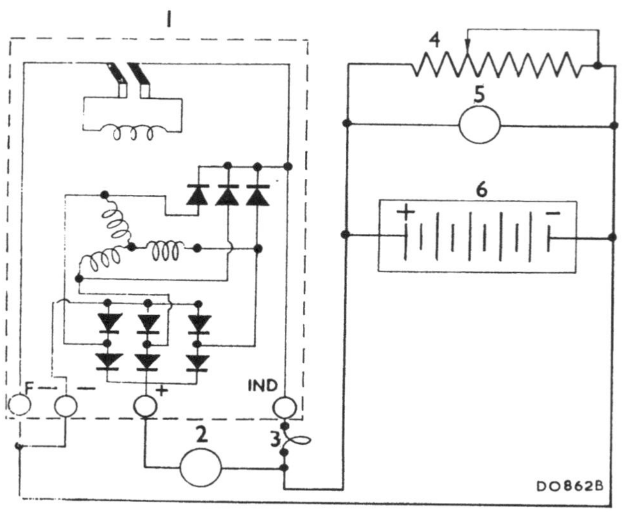


(4) Connect test equipment to the alternator terminals as shown in Fig. N.26. The variable resistance must only be connected across the battery for the amount of time necessary to carry out the tests.

**IMPORTANT.—Ensure that the correct battery and alternator polarity is preserved. Serious damage to the alternator will result if the polarity is reversed.**

(5) Start the engine and check the alternator output.

(*a*) Run the engine at 850 r.p.m. and observe the test bulb; the bulb should be extinguished.

(*b*) Increase the engine speed to 3,300 r.p.m. and adjust the variable resistance until a reading of 14 volts is registered on the voltmeter; at these settings the ammeter should register a reading of approximately 34 amps.

If the test bulb does not extinguish or its brilliance fluctuates, or if the ammeter reading varies considerably from the correct reading, a fault on the alternator is indicated and the alternator must be removed for further tests.

Failure of one or more of the diodes will effect the alternator output and in some instances raise the alternator temperature and noise level. The following table indicates the effects that diode failure may have on test results.

(6) If the tests in (5) show the alternator to be operating satisfactorily, disconnect the test circuit and reconnect the alternator terminal connector.

(7) Connect a low-range voltmeter between the positive terminals of the alternator and battery.

(8) Switch on the headlamps, start the engine and check the volt drop in the charging circuit as follows:

(*a*) Run the engine at 3,300 r.p.m. and note the voltmeter reading.

(*b*) Transfer the voltmeter connections to the negative terminals of the alternator and battery, and repeat the test in (*a*).

If the voltmeter reading in either of the tests exceeds ·2 volts a high resistance exists which must be traced and remedied.

**Removing**

(9) The following operations (*a*) to (*d*) are necessary on engines fitted with exhaust emission control air pumps.

(*a*) Disconnect the hoses from the air pump discharge connections.

(*b*) Slacken the air pump mounting bolt.

(*c*) Remove the air pump adjusting link bolt from the air pump.

(*d*) Slip the belt from the air pump pulley and raise the air pump.

(10) Withdraw the terminal block from the alternator.

(11) Remove the adjusting link bolt from the alternator.

(12) Slacken the alternator mounting bolts, lower the alternator and slip the fan belt from the alternator pulley.

(13) Unscrew the alternator mounting bolts and remove the alternator.

**Testing—alternator removed**

(14) Unscrew the cover retaining screws and remove the cover.

(15) Unsolder the three stator connections from the rectifier pack, noting the connection positions.

**IMPORTANT.—When soldering or unsoldering connections to the diodes great care must be taken not to overheat the diodes or bend the pins. During the soldering operations the diode pins should be gripped lightly with a pair of long-nosed pliers which will then act as a thermal shunt.**

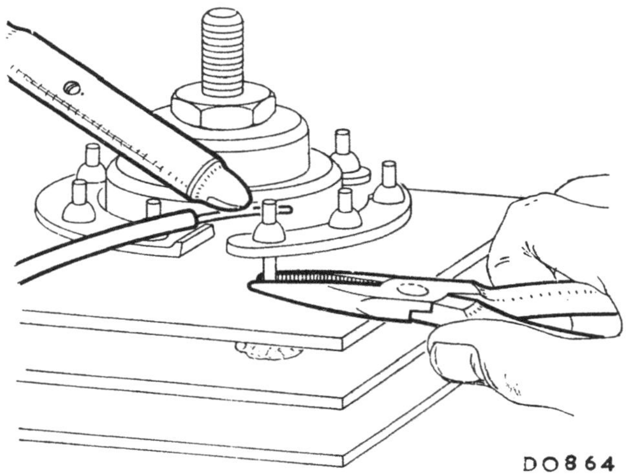


*Fig. N.27*

*Using pliers as a thermal shunt when soldering the alternator diodes*

| Test bulb | Alternator: Temperature | Alternator: Noise | Alternator: Output | Probable fault and associated damage |
|---|---|---|---|---|
| Glows at stand-still, goes out at 850 r.p.m., then glows progressively brighter as the speed is increased. | High | Normal | Higher than normal—40 amps. approximately at 3,300 r.p.m. | Live side output diode open-circuit. (May damage rotor windings and regulator output stage, overheat brush boxes, and blow warning light). |
| Light out under all conditions. | High | Excessive | Very low—10 amps. approximately at 3,300 r.p.m. | Live side output diode short-circuit. (May cause failure of associated field diode). |
| Glows at stand-still, dims appreciably at 850 r.p.m. and gets progressively dimmer as speed is increased. | Normal | Excessive | Poor at low speed, slightly below normal at 3,300 r.p.m. —32 amps. approximately. | Earth side output diode open-circuit. |
| Glows at a stand-still, dims slightly at 850 r.p.m. and remains so throughout speed range. | Normal | Excessive | Very low at all speeds above 850 r.p.m.—7 amps. approximately. | Earth side output diode short-circuit or one phase winding shorted to earth. |
| Glows at stand-still, dims slightly at 850 r.p.m. and remains so throughout speed range. | Normal | Normal | Lower than normal—29 amps. approximately at 3,300 r.p.m. | Field diode open-circuit. |
| Glows at stand-still, dims appreciably at 850 r.p.m. and remains so as speed is increased. | Normal | Excessive | Very low—7 amps. approximately at 3,300 r.p.m. | Field diode short-circuit. |

(16) Unscrew the two brush moulding securing screws.
(17) Slacken the rectifier pack retaining nuts and withdraw the brush moulding and rectifier pack.

*Brushes*

(18) Check the brushes for wear by measuring the length of brush protruding beyond the brushbox moulding. If the length protruding is ·2 in. (5 mm.) or less the brush must be renewed.
(19) Check that the brushes move freely in their holders. If a brush shows a tendency to stick, clean it with a petrol (gasoline) moistened cloth or, if necessary polish the sides of the brush with a fine file.
(20) Check the brush spring pressure using a push-type spring gauge. The gauge should register 7 to 10 oz. (198 to 283 gm.) when the brush is pushed back until its face is flush with the housing. If the gauge reading is outside the limits given, renew the brush assembly.

*Slip-rings*

(21) Clean the surfaces of the slip-rings using a petrol (gasoline) moistened cloth.
(22) Inspect the slip-ring surfaces for signs of burning; remove burn marks using very fine sand-paper. On no account must emery-cloth or similar abrasives be used or any attempt made to machine the slip-rings.

*Rotor*

(23) Connect an ohmmeter or a 12-volt battery and an ammeter to the slip-rings. An ohmmeter reading of 4·3 ohms or an ammeter reading of 3 amps. should be recorded.
(24) Using a 110-volt A.C. supply and a 15-watt test lamp, test for insulation between one of the slip-rings and one of the rotor poles. If the test lamp lights, the rotor must be renewed.

*Stator*

(25) Connect a 12-volt battery and a 36-watt test lamp to two of the stator connections. Repeat the test replacing one of the two stator connections with the third. If the test lamp fails to light in either of the tests the stator must be renewed.
(26) Using a 110-volt A.C. supply and a 15-watt test lamp, test for insulation between any one of the three stator connections and the stator laminations. If the test lamp lights, the stator must be renewed.

*Diodes*

(27) Connect a 12-volt battery and a 1·5-watt test lamp in turn to each of the nine diode pins and its corresponding heat sink on the rectifier pack, then reverse the connections. The lamp should light with the current flowing in one direction only. If the lamp lights in both directions or fails to light in either, the rectifier pack must be renewed.

**IMPORTANT.—See notes on soldering the diodes given in (2).**

*Dismantling*

(28) Carry out the operations detailed in (14) to (17).
(29) Remove the three through-bolts.
(30) Fit a tube of the dimensions given in Fig. N.28 over

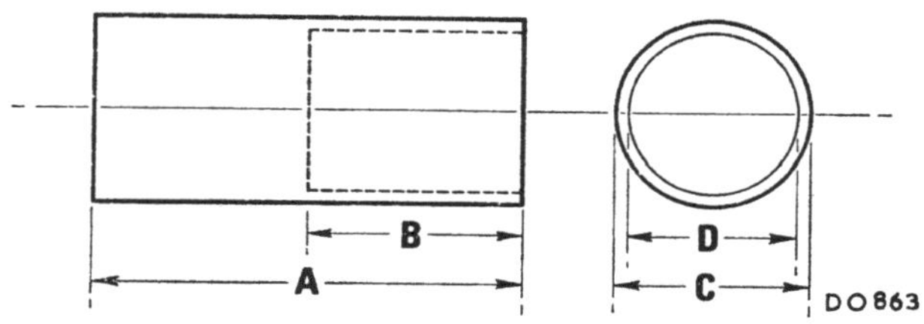


*Fig. N.28*
*Dimensions of the rotor removing tool*

A=3 in. (76 mm.) B=1·5 in. (38 mm.)
C=1·32 in. (33·5 mm.) D=1·24 in. (31·5 mm.)

the slip-ring moulding so that it registers against the outer track of the slip-ring end bearing and carefully drive the bearing from its housing.
(31) Remove the shaft nut, washer, pulley, fan, and shaft key.
(32) Press the rotor from the drive end bracket.
(33) Remove the circlip retaining the drive end bearing and remove the bearing.
(34) Unsolder the field connections from the slip-ring assembly and withdraw the assembly from the rotor shaft.
(35) Remove the slip-ring end bearing.

**Reassembling**

(36) Reverse the dismantling procedure in (29) to (35) and (14) to (17), noting the following points.
  (*a*) Use Shell Alvania 'RA' to lubricate the bearings where necessary.
  (*b*) When refitting the slip-ring end bearing, ensure that it is fitted with its open side facing towards the rotor and is pressed onto the rotor shaft as far as it will go.
  (*c*) Re-solder the field connections to the slip-rings using Fry's H.T.3 solder.
  (*d*) When refitting the rotor to the drive end bracket, support the inner track of the bearing with a suitable piece of tube. Do not use the drive end bracket as the only support for the bearing when fitting the rotor.
  (*e*) Tighten the through-bolts evenly.
  (*f*) Check that the brushes are entered in their housings before fitting the brush moulding.
  (*g*) Tighten the shaft nut to the torque figure given in '**GENERAL DATA**'.

## Section N.26

### ALTERNATOR (Lucas Type 16ACR)

The construction of the model 16ACR alternator is similar to the 16AC as illustrated in Fig. N.25, except that a voltage regulator of micro-circuit construction is incorporated on the slip-ring end casing (see Fig. N.29). The procedure given for removing, testing, and dismantling the 16AC alternator in Section N.25, operations (9) to (36), should be followed. For regulator removal see operations (17) to (18) of this Section.

**Testing in position**

*Output test*

(1) Check that the fan belt is correctly tensioned and that all charging circuit electrical connections are secure.

(2) Run the engine at fast idle speed until normal operating temperature is reached.

(3) Ensure that the battery is fully charged; if not, fit a slave battery.

(4) Disconnect both connector blocks from the alternator.

(5) Switch on the ignition and connect a voltmeter, negative to earth positive lead to each cable connector tag of the two connector blocks in turn. If battery voltage is not registered at any cable, locate and remedy the fault.

(6) Remove the alternator end cover.

(7) Bridge the regulator green connector 'F' to earth (see Fig. N.29).

(8) Refit the three-way connector block to the alternator. Connect an ammeter in series with the positive lead of the two-way connector block and the positive output tag of the alternator.

(9) Start the engine and run at 3,300 r.p.m. The ammeter should read 34 amps. If the correct alternator output cannot be obtained, replace or service the alternator.

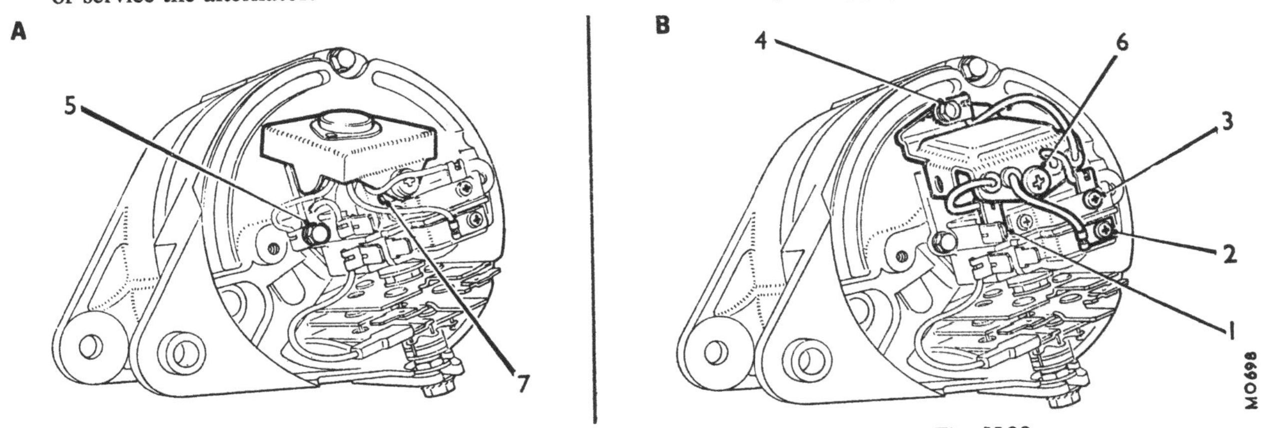


*Regulator test*

(10) Disconnect the bridge from the regulator green connector 'F' and earth.

(11) Connect a voltmeter across the battery terminals. Start the engine and run at 3,300 r.p.m. If the ammeter reads zero, the regulator pack must be replaced.

(12) Adjust the engine speed until the ammeter reading falls below 10 amps. The voltmeter should read 14·0 to 14·4 volts. If the voltmeter reading is not within the limits stated, a faulty regulator or high resistance in the charging circuit cables is indicated.

*Charging circuit resistance test*

(13) Connect a voltmeter between the positive terminal of the alternator and the positive terminal of the battery.

(14) Start the engine, switch on the headlamps, and run the engine at 3,300 r.p.m. The voltmeter reading should not exceed ·5 volts.

(15) Transfer the voltmeter connections to the negative terminals of the alternator and battery. Run the engine at 3,300 r.p.m. The voltmeter reading should not exceed ·25 volt.

If the readings in (14) and (15) are in excess of that stated, the charging circuit has developed a high-resistance fault which must be traced and remedied.

(16) If the charging circuit resistance test is satisfactory, the failure is due to the regulator pack and the alternator must be replaced or removed for overhaul.

**Removing the regulator**

(17) Remove the alternator.

(18) Remove the moulded end cover, and identify the type of regulator pack fitted; the 8TR-type has two short mounting screws at each end, while the 11TR type has a single longer one, with a spacer, screwed into the top lug of the brush box moulding only. Both types have two legs which locate in the brush box moulding.

Disconnect the coloured tag lead connectors from the brush box, (1, 2 and 3) and detach the black (earth) lead (4), after removing the lower mounting screw (8TR) or one of the brush box retaining screws (6) (11TR). Remove the remaining screw (5) securing the regulator pack. Collect the spacer (7), (11TR only).

*Fig. N.29*

*The in-built regulator connections, showing the 8TR* (B) *and* (A) *11TR type of regulator units*

1. Battery + (B+).
2. Positive (+).
3. Field (F).
4. Negative earth (−).
5. Brush box retaining screw.
6. Upper mounting screw.
7. Spacer (11TR only).

## Section N.27

### WINDSHIELD WIPERS—TWO-SPEED

**Removing**

*Motor and gearbox assembly*

(1) Disconnect the battery.
(2) Remove the wiper arms.
(3) Remove the right-hand fascia under-panel if fitted.
(4) Disconnect the wiring from the motor terminals.
(5) Unscrew the outer cable retaining nut from the motor housing.
(6) Remove the two motor securing bolts.
(7) Withdraw the motor and gearbox assembly complete with the inner drive cable. The inner cable will rotate the wiper spindles as it is withdrawn.

*Wheelboxes*

(8) Remove the motor and gearbox as in (1) to (7).
(9) Slacken the cover screws on each wheelbox and detach the cable outer casings.
(10) Unscrew the spindle housing retaining nut and remove the front bush and washer from each wheelbox.
(11) Remove the wheelboxes complete with rear bushes from beneath the fascia.

**Dismantling**

*Motor and gearbox assembly*

(12) Unscrew the four gearbox cover retaining screws and remove the cover.
(13) Remove the circlip and flat washer securing the connecting rod to the crankpin.
(14) Withdraw the connecting rod, taking care not to lose the flat washer fitted under it.
(15) Remove the circlip and washer securing the shaft and gear.

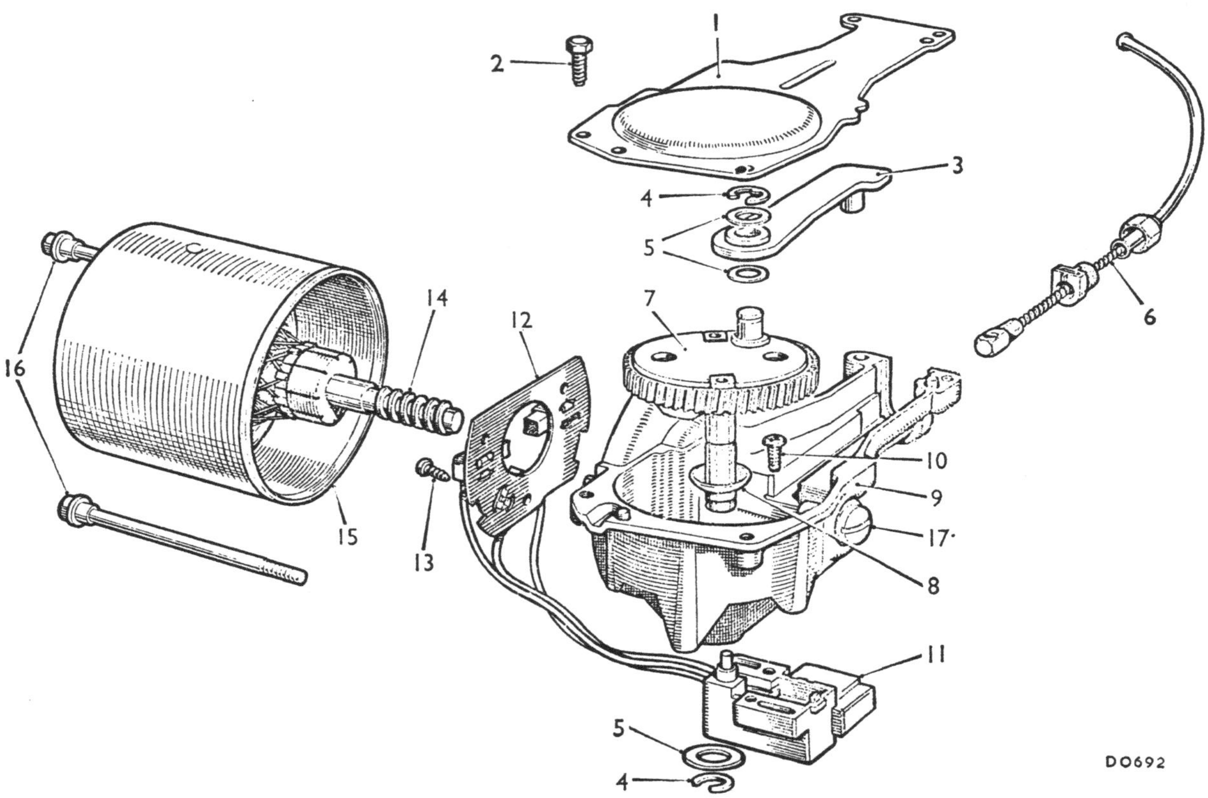


*Fig. N.30*
*The two-speed windshield wiper motor components*

1. Gearbox cover.
2. Screw for cover.
3. Connecting rod.
4. Circlip.
5. Plain washer.
6. Cable assembly.
7. Shaft and gear.
8. Dished washer.
9. Gearbox.
10. Screw for limit switch.
11. Limit switch assembly.
12. Brush gear.
13. Screw for brush gear.
14. Armature.
15. Yoke assembly.
16. Armature adjusting screw.

(16) Clean any burns from the gear shaft and withdraw the gear, taking care not to lose the dished washer fitted under it.
(17) Mark the yoke and gearbox for reassembly.
(18) Unscrew the two fixing bolts from the yoke and remove the yoke assembly and armature. The yoke must be kept clear of metallic particles which will be attracted to the pole-piece.
(19) Remove the screws securing the brush gear and the terminal end switch assembly. and remove both assemblies.

**Inspection**

*Motor and gearbox assembly*

(20) Examine the brushes for excessive wear. If the main bushes (diametrically opposite) are worn to $\frac{3}{16}$ in. (4·8 mm.) or if the narrow section of the third brush is worn to the full width of the brush the brush gear assembly must be renewed.
(21) Check the brush spring pressure with a push-type gauge. The gauge reading should be 5 to 7 oz. (140 to 200 gm.) when the bottom of the brush is level with the bottom of the slot in the brush box. The brush gear assembly must be renewed if the springs are not satisfactory.
(22) Test the armature for insulation and open- or short-circuits. Renew the armature if faulty.
(23) Examine the gear wheel for damage or excessive wear. Renew if necessary.

**Reassembling**

*Motor and gearbox assembly*

(24) Reverse the dismantling procedure in (12) to (19), noting the following points:
 (*a*) Use Ragosine Histate Grease to lubricate the gear wheel teeth and cam, armature shaft worm gear, connecting rod and pin, cross-head slide, cable rack, and wheelbase gear wheels.
 (*b*) Use Shell Turbo 41 oil to lubricate the bearing bushes, armature shaft bearing journals (sparingly), gear wheel shaft and crankpin, felt washer in the yoke bearing (thoroughly soak), and the wheelbox spindles.
 (*c*) Tighten the yoke fixing bolts to a torque figure of 20 lb. in. (·23 kg. m.).
 (*d*) If a replacement armature is being fitted, slacken the thrust screw to provide end-float for fitting the yoke.
 (*e*) Fit the thrust disc inside the yoke bearing with its concave side towards the end face of the bearing.
 (*f*) Fit the dished washer beneath the gear wheel with its concave side towards the gear wheel.
 (*g*) When fitting the connecting rod to the crank-pin ensure that the larger of the two flat washers is fitted under the connecting rod with the smaller one on top beneath the circlip.
 (*h*) With the thrust screw fully tightened against the gearbox casing, on end-float of ·004 to ·008 in. (·1 to ·21 mm.) should exist on the armature. Adjustment of the armature end-float can be achieved by adjustment of the thrust screw.

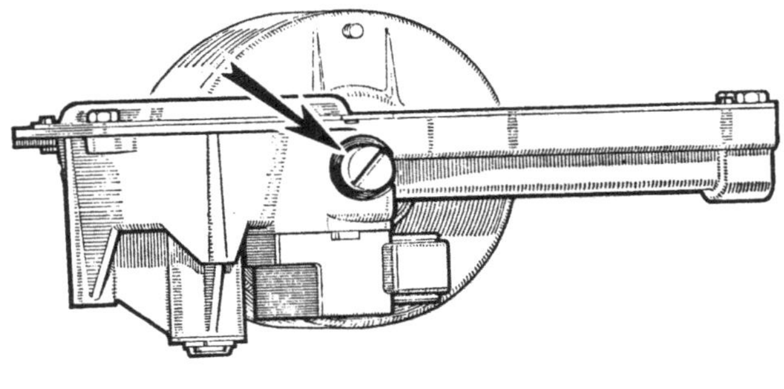

D0668

*Fig. N.31*
*The armature end-float adjusting screw*

**Refitting**

*Wheelboxes*

(25) Reverse the removing procedure in (8) to (11).

*Motor and gearbox assembly*

(26) Reverse the removing procedure in (1) to (7) ensuring that the inner cable engages correctly with the wheelbox gear teeth.

## Section N.28

### STARTER (Lucas Type M418G Pre-engaged)

**Removing**

(1) Disconnect the battery.
(2) Remove the distributor Section B.5.
(3) Remove the starter top securing bolt.
(4) Disconnect the wiring from the starter terminals.
(5) Remove the starter lower securing bolt and remove the starter.

**Dismantling**

(6) Disconnect the lower solenoid terminal marked 'STA'.
(7) Remove the solenoid unit securing nuts.
(8) Withdraw the solenoid from the drive end bracket, carefully disengaging the solenoid plunger from the drive engagement lever.
(9) Remove the cover band and lift the brushes from their holders.
(10) Unscrew the two through-bolts from the commutator end bracket.
(11) Slacken the drive engagement lever pivot pin securing nut; unscrew and withdraw the pin.

(12) Position the starter with the solenoid housing uppermost; separate the drive end bracket from the yoke and remove the drive engagement lever.

(13) Withdraw the armature and separate the commutator end bracket from the yoke.

(14) Remove the thrust washer from the armature shaft and the thrust collar from the end of the shaft extension.

(15) Prise the retaining ring from its groove and withdraw the drive assembly.

(16) Prise the retaining ring from the drive and remove the collar, operating plate, and thrust washer.

**Inspection**

*Brush gear*

(17) Check the length of the brushes. When they have worn to (or are approaching) $\frac{5}{16}$ in. (8 mm.) renew them by unsoldering their flexible connectors, two from the uninsulated brush holders and two from the field coils, then solder the new brush connectors in place. New brushes are pre-formed and bedding to the commutator is unnecessary.

(18) Check the brushes for freedom of movement in their holders and, if necessary, clean them with a petrol-moistened cloth.

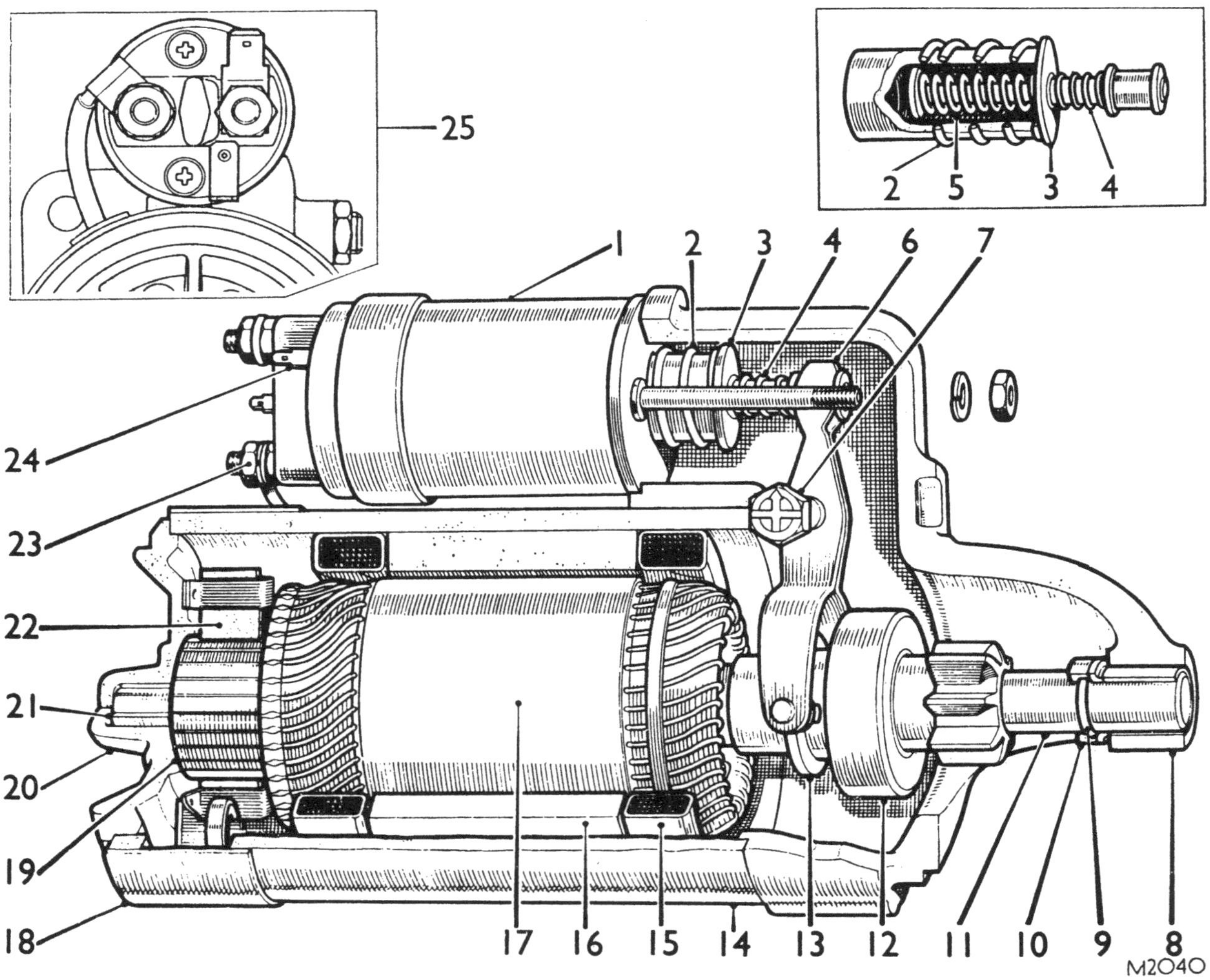


*Fig. N.32*

*The starter motor—Lucas type M418 pre-engaged*

1. Solenoid.
2. Plunger return spring.
3. Plunger.
4. Lost motion spring.
5. Drive engagement spring.
6. Engaging lever.
7. Pivot pin.
8. Fixing bracket.
9. Retaining ring
10. Thrust collar.
11. Armature shaft extension.
12. Clutch.
13. Drive operating plate.
14. Yoke.
15. Field coils.
16. Pole shoes.
17. Armature.
18. Cover band.
19. Commutator.
20. End bracket.
21. Bush bearing.
22. Brushes.
23. Solenoid 'STA' terminal.
24. Solenoid operating terminal.
25. Repositioned solenoid (later cars).

(19) Check the brush spring pressure. Each spring should exert a minimum of 36 oz. (1·02 kg.) with a new (full-length) brush in position.

*Commutator*

(20) The commutator should be burnished, free of pits or burned spots, and can be cleaned with a petrol-moistened cloth. If this cleaning is ineffective, spin the armature and polish the commutator with fine glass-paper, removing all abrasive dust with a dry air blast.

(21) In the case of a badly worn commutator, mount the armature between centres in a lathe and rotate it at high speed. Take a light cut with a very sharp tool and do not remove more metal than is necessary to clean the commutator. Finally, polish with very fine glass-paper. The insulators between the commutator segments **must not be undercut.**

*Armature*

(22) Examine the armature for conductors which have lifted from the commutator risers. This would indicate overspeeding and the drive clutch should be checked (see (28)).

(23) Check for signs of the armature fouling the pole shoes, indicating a distorted armature shaft or worn bearings (see (29) and (30)). Do not attempt to machine the armature core or to true a distorted shaft; a damaged armature must be renewed.

(24) To make a thorough check on the condition of the armature it should be tested on a 'growler'. The armature insulation can be checked by means of a 110-volt A.C. test lamp connected between each commutator segment in turn and the armature shaft. The test lamp should not light.

*Field coils*

(25) Test the field coils for continuity. Ensure that both brushes and their flexible connectors are clear of the yoke and connect a 12-volt test lamp and battery between the cable connection to 'STA' and each of the field coil brushes in turn. If the lamp does not light, an open circuit in the field coil is indicated and they should be renewed as follows:

- (*a*) Using a wheel-operated screwdriver, unscrew the four pole-shoe retaining screws.
- (*b*) Withdraw the inter-coil connector insulating piece and remove the pole shoes and coils from the yoke.
- (*c*) Fit new coils to the pole shoes and position them inside the yoke. Make sure that the coil taping is not trapped between the pole shoes and the yoke, then lightly tighten the pole-shoe retaining screws.
- (*d*) Fit the insulator to the field coil terminal and position the inter-coil connection insulator inside the yoke. Hold the pole shoes in position with a pole-shoe expander and tighten the retaining screws with a wheel-operated screwdriver.

(26) Test the field coils for insulation from the yoke. Ensure that both brushes and their flexible connectors are clear of the yoke and connect a 110-volt A.C. test-lamp between the terminal post and the yoke. If the lamp lights, earthing of the field coils is indicated and they should be renewed as described in (25) (*a*) to (*d*).

*Commutator end bracket*

(27) Clean all traces of brush deposit from the commutator end bracket and connect a 110-volt A.C. test lamp between the end bracket and each insulated brush box in turn. If the lamp lights, faulty insulation is indicated and the commutator end bracket should be renewed.

*Drive clutch*

(28) Check the operation of the drive clutch. It should provide instantaneous take-up of the drive in one direction, yet rotate easily and smoothly in the opposite direction. It should also move freely on the armature shaft splines without any roughness or tendency to bind.

*Bearings*

(29) Renew bearings which are worn to such an extent that they allow excessive side-play of the armature shaft. Remove the bearing from the commutator end bracket by screwing a tap squarely into the bearing and withdrawing the bearing with the tap. The drive end bracket bearing can be pressed out.

(30) Press the new bearings into position using a shouldered highly polished mandrel ·0005 in. (·013 mm.) larger in diameter than the shaft which will run in the bearing.

NOTE.—The commutator and drive end bracket bearings are porous bronze and should be immersed in clean engine oil for 24 hours before fitting. In an emergency this period may be reduced by maintaining the oil at a temperature of 100° C. (212° F.) for two hours and allowing the oil to cool before removing the bearing. Porous bronze bushes should not be reamed, as the porosity of the bush would be impaired.

**Reassembling**

(31) During reassembly lubricate all moving parts with Shell Retinax A grease, or an equivalent.

(32) Reassemble the starter motor by reversing the procedure in (6) to (16).

(33) Set the pinion travel as follows:

- (*a*) Connect a 6-volt supply, with a switch in circuit, between the small unmarked solenoid terminal and earth. The period of energization should be kept as brief as possible to avoid overheating the series winding.

(b) Close the switch (this throws the drive assembly into the engaged position) and measure the distance between the pinion and the thrust collar on the armature shaft extension. When making this measurement, take up the slack in the engagement linkage by lightly pressing the pinion towards the armature. The measurement should be ·005 to ·015 in. (·13 to ·4 mm.). To adjust this setting, slacken the locknut on the engagement lever pivot pin and rotate the pin until the measurement is correct. The arc of adjustment is 180 degrees and the head of the arrow marked on the pivot pin should be set only between the arrows on the arc marked on the drive end bracket.

(c) Tighten the locknut to secure the pivot pin.

(34) After setting the pinion travel, check the opening and closing of the starter switch contacts as follows:

(a) Connect a 10-volt supply, with a switch in circuit, between the small unmarked solenoid terminal and earth.

(b) Connect a separately energized test lamp circuit across the solenoid main terminals.

(c) Insert a stop in the drive end bracket to restrict the pinion travel to that of the normal out-of-mesh clearance.

(d) Close the switch (the period of energization should be as brief as possible to avoid overheating the series winding). The test lamp should now give a steady light indicating that the solenoid contact is remaining fully closed.

(e) Switch off and remove the stop.

(f) Switch on, hold the pinion assembly in the fully engaged position and switch off. The test lamp should now go out indicating that the solenoid contacts have opened.

**Bench testing**

*Light running test*

(35) Using heavy-gauge cables, connect the starter to a fully charged 12-volt 120-amp.-hour battery with a switch and ammeter in circuit. The starter motor should run freely at 5,800 to 6,500 r.p.m. with a current consumption of 70 amps.

*Torque test*

(36) Check the lock torque and running torque, against the figures given in 'GENERAL DATA'.

**Refitting**

(37) Reverse the procedure in (1) and (5).

## Section N.29

### CONTROL UNIT (Lucas Type 4TR)

**Testing**

(1) Check that the alternator is functioning correctly as described in Section N.25 (1) to (8).

(2) Check that the batteries are in a fully charged state; if not, temporarily connect fully charged replacements.

(3) Connect an accurate voltmeter across the battery terminals.

(4) Connect a 0–40 amp. ammeter in series with the alternator main output cable.

(5) Start the engine and run it at 2,750 r.p.m. until a steady ammeter reading of not more than 5 amps. is obtained. At this setting the voltmeter should give a steady reading of 14·3 to 14·7 volts. If the voltmeter reading is unsteady or is outside the limits given, the control unit must be renewed.

**Removing**

(6) Disconnect the positive connection from the battery terminal.

(7) Withdraw the terminal connector block from the control unit.

(8) Remove the control unit securing screws and remove the unit.

**Refitting**

(9) Reverse the removing procedure in (6) to (8).

## Section N.30

### SWITCHES

**(GHN5 and GHD5 cars from Car No. 258001)**

**Removing**

**IMPORTANT.** — Disconnect the batteries before attempting to remove any of the switches and refer to the wiring diagram and switch contact numbers when refitting the switches.

*Heater, wiper, and lighting switches*

(1) Remove the face level vents—Section R.35.

(2) Disconnect the wiring from the back of the switches.

(3) Using tool 18G 1202 press in the four retaining lugs and withdraw the heater and wiper switches.

(4) To remove the lighting switch, depress the retaining lugs on one side of the switch, ease one side of the switch through the aperture, depress the retaining lugs on the other side and withdraw the switch.

*Console switches*

(5) Remove the console—Section R.34.

(6) Using tool 18G 1202 press in the four retaining lugs and withdraw the switches.

*Ignition switch*
(7) Turn the key to position 'O'.
(8) Disconnect the wiring from the switch.
(9) Remove the two screws retaining the switch to the steering lock.
(10) Withdraw the switch.

*Direction indicator/headlight flasher/low-high beam switch*
(11) Unscrew the four switch cowl retaining screws and remove the cowl.
(12) Disconnect the switch wiring from the multi connector block and unclip it from the steering-column.
(13) Remove the two screws retaining the direction indicator/headlight flasher/low-high beam switch and remove the switch and retaining clip.

**Refitting**
*Heater, wiper and lighting switches*
(14) Reverse the removing procedure in (1) to (4).

*Console switches*
(15) Reverse the removing procedure in (5) to (6).

*Ignition switch*
(16) Reverse the removing procedure in (7) to (10).

*Direction indicator/headlight flasher/low-high beam switch*
(17) Reverse the removing procedure in (11) to (13) ensuring that the small tongue on the switch engages in the cut-away of the steering outer column.

## Section N.31

### STARTER (Lucas Type 2M100 Pre-engaged)

**Removing**
(1) Disconnect the battery.
(2) *Synchromesh gearbox only:* Detach the clutch slave cylinder from the gearbox and move the cylinder clear of the starter.
(3) Remove the starter top securing bolt.
(4) Disconnect the wiring from the solenoid terminals.
(5) Remove the starter motor lower securing bolt and withdraw the starter.

**Dismantling**
(6) Remove the nut securing the (solenoid-to-starter) connecting link to the solenoid 'STA' terminal.
(7) Pull back the connecting link from the solenoid terminal.
(8) Remove the nuts securing the solenoid to the drive end bracket. Refit torque 4·5 lb.ft. (0·63 kg.m.).
(9) Remove the solenoid unit from the drive-end bracket.
(10) Lift the solenoid plunger and return spring from the engagement lever.
(11) Remove the block-shaped grommet.

CAUTION.—Before carrying out instructions 7 to 15 ensure that new retaining rings are available for the armature shaft and engagement lever pivot pin are available for reassembly of the starter.
(12) Remove the end cap seal.
(13) Remove the armature shaft retaining ring (Spire nut).
(14) Remove the through-bolts. Refit torque 8 lb.ft. (1·11 kg.m.).
(15) Detach the commutator end cover from the yoke, disengaging the field brushes from the brush box moulding, and remove the end-cover.
(16) Remove the thrust washer from the armature shaft.
(17) Withdraw the yoke from the armature and drive-end assembly.
(18) Remove the retaining ring (Spire nuts) from the engagement lever pivot pin.
(19) Remove the pivot pin from the engagement lever and drive-end bracket.
(20) Move the thrust collar clear of the jump ring, and remove the jump ring from the armature shaft.
(21) Remove the thrust collar.
(22) Remove the roller clutch drive and engagement lever assembly from the armature shaft.

**Inspection**
*Brush gear, commutator and armature*
(23) Refer to Section N.28.

*Field windings*
(24) Connect a 12-volt battery-operated test lamp between each of the field brushes and a clean part of the yoke; the lamp will light if continuity is satisfactory between the brushes, windings, and yoke connection.
(25) Disconnect the field windings from their riveted connection with the yoke. Using a 110-volt A.C., 15-watt test lamp connected between each of the brushes in turn and the yoke, check the insulation of the field windings, if the lamp lights the windings must be renewed.
(26) The field windings may be renewed as follows:
*a.* Disconnect the end of the field winding where it is riveted to the yoke.
*b.* Remove the four pole-shoe retaining screws using a wheel-operated screwdriver.
*c.* Withdraw the field coil assembly from the yoke.
*d.* Clean the inside of the yoke and insulating pieces which the through-bolts locate.
*e.* Loosely fit the new field coil assembly with the pole-shoes into the yoke.
*f.* Fit the pole-shoe retaining screws but do not tighten.
*g.* Slide the insulating pieces between the field coils and the yoke, positioned 90° each side of the field coil brush connection point.
*h.* Tighten the pole-shoe screws evenly using a wheel-operated screwdriver. Torque 30 lb.ft. (4·15 kg.m.).
*j.* Reconnect the end of the winding to the yoke.

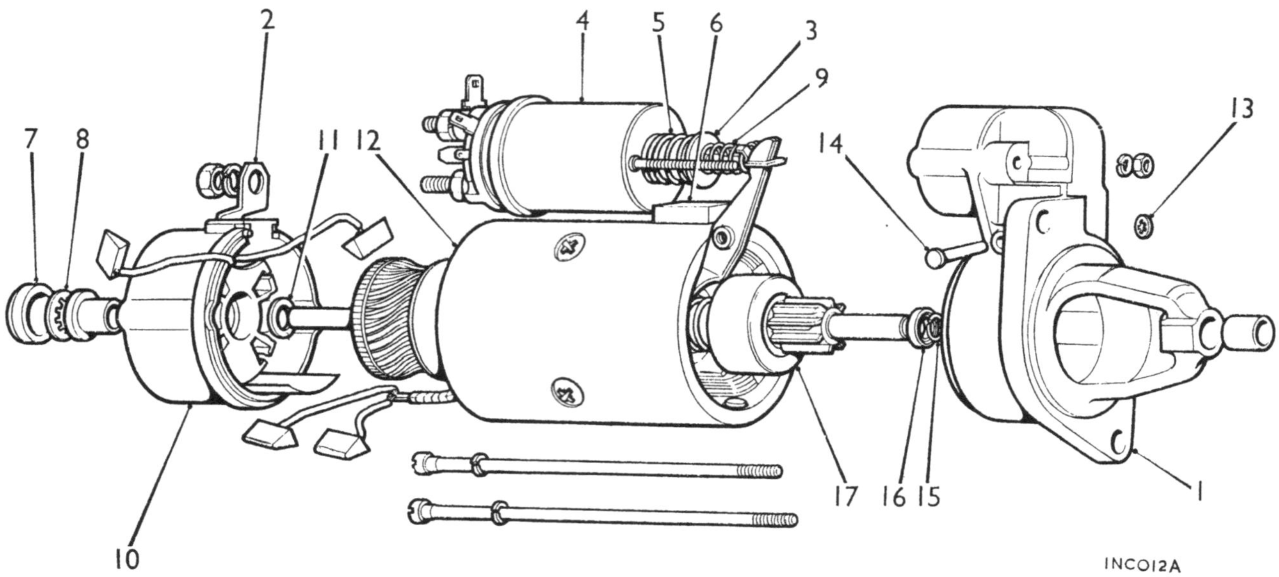


*Fig. N.33*

*The starter motor—Lucas type 2M100 pre-engaged*

1. Fixing bracket.
2. Solenoid 'STA' terminal.
3. Plunger.
4. Solenoid.
5. Plunder return spring.
6. Block shaped grommet.
7. End cap seal.
8. Armature shaft retaining ring.
9. Lost motion spring.
10. Commutator end cover.
11. Thrust washer.
12. Yoke.
13. Pivot pin retaining ring.
14. Pivot pin.
15. Jump ring.
16. Thrust collar.
17. Roller clutch.

*Commutator end bracket*

(27) Check the insulation of the brush springs by connecting a 110-volt A.C., 15-watt test lamp between a clean part of the commutator and cover each of the springs in turn; the lamp will light if the insulation is not satisfactory.

*Bearings*

(28) If the bearings in the commutator end cover and the drive-end bracket are worn sufficiently to allow excessive side-play of the armature shaft, the bearing bush must be renewed as follows:

*a.* Support the bracket and drive out the bush using a suitable sized mandrel.

*b.* Using a polished, shouldered mandrel, the same diameter as the shaft bearing journal, press the new bushes into the brackets.

**NOTE.**—New bushes must be immersed in S.A.E. 30/40 engine oil for 24 hours or in oil heated to 100° C. (212° F.) for two hours prior to fitting. The bushes must not be reamed after fitting.

*Roller clutch drive*

(29) Check the operation of the drive clutch, it should provide instantaneous take-up of the drive in one direction and rotate easily and smoothly in the other direction.

(30) Check that the drive moves freely on the armature splines without binding or sticking.

**Reassembling**

(31) Reverse 1 to 17, noting the following:

*a.* When assembling the drive-end bracket use a new retaining ring (Spire nut) to secure the engagement lever pivot pin.

*b.* Ensure that the internal thrust washer is fitted to the commutator end of the armature shaft before the armature end cover is fitted.

*c.* Fit a new retaining ring (Spire nut) onto the armature shaft to a maximum clearance of 0·010 in. (0·25 mm.) between the retaining ring and the bearing bush shoulder (Armature end-float).

**Bench testing**

*Lock torque and current*

(32) Clamp the starter firmly in a vice.

(33) Connect a starter switch, a 0–600 amp. ammeter and a 12-volt battery in series to the starter.

(34) Connect a voltmeter between the solenoid two main terminals.

(35) Secure an arm to the driving pinion.

(36) Connect a spring balance to the free end of the arm.

(37) Operate the switch and note the ammeter, voltmeter, and spring balance readings. Calculate the lock torque by multiplying the reading of the spring balance in pounds by the length of the arm in feet.

(38) Check the readings obtained in 37 against the figures given in **'GENERAL DATA'** for lock torque current and voltage.

**NOTE.**—If a constant-voltage supply is used for this test, a higher voltage may be registered on the voltmeter than the figure given. If this should occur, a variable resistance must be added to the circuit and adjusted to give the required reading and the test repeated.

*Solenoid—coil*

(39) Check the continuity of the coil windings between the terminal 'STA' and a good earth point on the solenoid body, using an ammeter (in series); a reading of 11·2 to 11·8 amps should be obtained. If no reading, renew the solenoid.

(40) Using an ohmmeter connected between the small unmarked Lucar terminal and terminal 'STA'; check the resistance of the closing (series) winding; if the reading differs considerably from the limits given the solenoid must be renewed.

(41) Using an ohmmeter connected between the small unmarked Lucar terminal and a good earth on the solenoid body, check the resistance of the hold-on (shunt) winding; if the reading differs considerably from the limits given, the solenoid must be renewed.

*Solenoid contacts*

(42) Check the contacts for correct opening and closing by carrying out the instructions 27 to 31 plus the following:
   (*a*) Operate the switch and note the voltmeter reading; it should register practically zero if the solenoid is satisfactory. Alternatively the solenoid should be checked by substitution.

(43) The contacts must only be renewed as a set, e.g. moving and fixed contacts. The fixed contacts are an integral part of the moulded cover. Renew a set of contacts as follows:
   (*a*) Remove the screws securing the terminal and base assembly to the solenoid.
   (*b*) Undersolder the coil connections from the cover terminals.
   (*c*) Clamp the solenoid in a vice.
   (*d*) Remove the terminal and base assembly.
   (*e*) Assemble the new terminal and base assembly.
   (*f*) Resolder the connections to the terminal and base assembly.
   (*g*) Refit the securing screws. Torque 1·8 lb. ft. (0·25 kg.m.).

**Refitting**

(44) Reverse the removing procedure in (1) to (5).

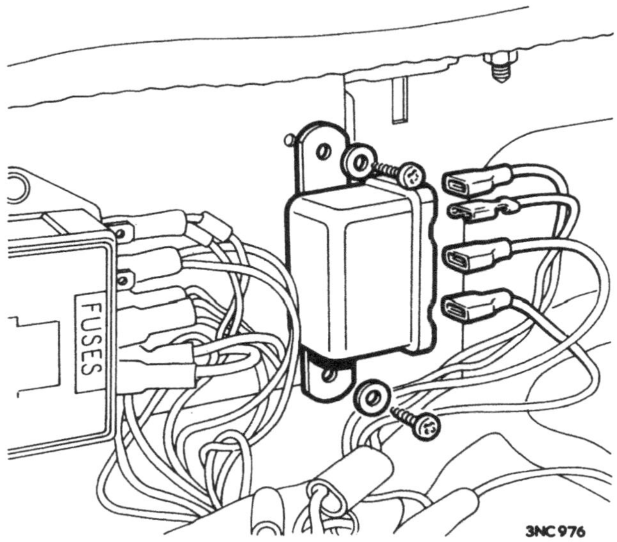


*Fig. N.34*
*Starter motor relay*

## Section N.32

### STARTER MOTOR RELAY

**Removing**

(1) Disconnect the batteries.
(2) Disconnect the wiring harness from the terminal blades on the starter motor relay.
(3) Remove the two screws to release the starter motor relay from the right-hand wing valance.

**Testing**

(4) Connect a 12-volt direct-current supply between the relay terminals 'W1' and 'W2', and a 12-volt, 2·2-watt test lamp in circuit with a 12-volt direct-current supply between terminals 'C1' and 'C2'.
   (*a*) If the test lamp fails to light check the relay winding resistance, using an ohmmeter connected between terminals 'W1' and 'W2'. Renew the relay if a reading of 76 ohms is not obtained.
   (*b*) If the winding resistance is correct, faulty contact adjustment is indicated which may be corrected as follows:
      (i) Uncrimp and remove the cover from the relay.
      (ii) Check the air gap between the relay bobbin core and the underside of the armature. The air gap should be 0·030± 0·005 in. (0·76±0·13 mm.) when the contact points are open, and 0·010± 0·003 in. (0·25±0·08 mm.) when the points are closed. Bend the fixed contact post as necessary.

(5) After any adjustment to the air gap, check the relay cut-in and drop-off voltages as follows:

(*a*) Connect a variable direct current supply between the relay terminals 'W1' and 'W2' and a 12-volt direct-current supply in circuit with a test lamp between terminals 'C1' and 'C2'.

(*b*) Raise the voltage slowly from zero to 15 volts and check that the test lamp lights at 4·0 to 7·5 volts.

(*c*) Reduce the voltage slowly from 15 to zero volts and check that the test lamp goes out at 5 volts maximum.

(*d*) Repeat operation 4 as necessary, and recheck the relay cut-in and drop-off voltages.

(6) Refit the relay cover and crimp the cover lip at the points provided.

**Refitting**

(7) Reverse the procedure in 1 to 3.

## SUPPLEMENTARY NOTES

### GENERAL DESCRIPTION

The dynamo is mounted on the right of the cylinder block and is belt-driven from the crankshaft pulley. A slotted link connecting the dynamo front plate and the cylinder block enables the belt tension to be adjusted.

The starter motor is mounted on the engine rear mounting plate and operates on the flywheel through a sliding pinion device. The starter motor solenoid is located on the right-hand wing valance.

An RB340 regulator unit is mounted on a bracket secured to the right-hand wing valance. A fusebox adjacent to the regulator carries two fuses in circuit and two spare fuses.

Instruments on later cars are voltage-controlled by a bi-metal resistance.

Lighting and warning equipment is fitted to conform to the regulations of the country or state in which the car is expected to be used.

The lighting system incorporates the double-filament dipping system with sealed-beam light units for the Home and U.S.A. markets.

**Fitting a wiper arm to the driving spindle**

First ensure that the wiper spindles are in the correct parking position by switching on the ignition and turning the wiper control on and then off.

To fit the arms, press the headpieces onto the spindles at the correct parking angle until the retaining clip is heard to snap over the end of the spindle drum.

Switch the wiper control on and off. The arms should come to rest in the correct parking position.

## EDITOR'S NOTES

**N. The Electrical System**
*Servicing — commutator*
Glass paper is sand paper. Do not use emery paper to polish a commutator.

# Section Na

## THE ELECTRICAL SYSTEM

**The information given in this Section refers specifically to service operations on, or affected by, equipment fitted to the MGB (GHN5/GHN4 and GHD5/GHD4) in conformity with local and territorial requirements, and must be used in conjunction with Section N.**

## THE SWITCHES

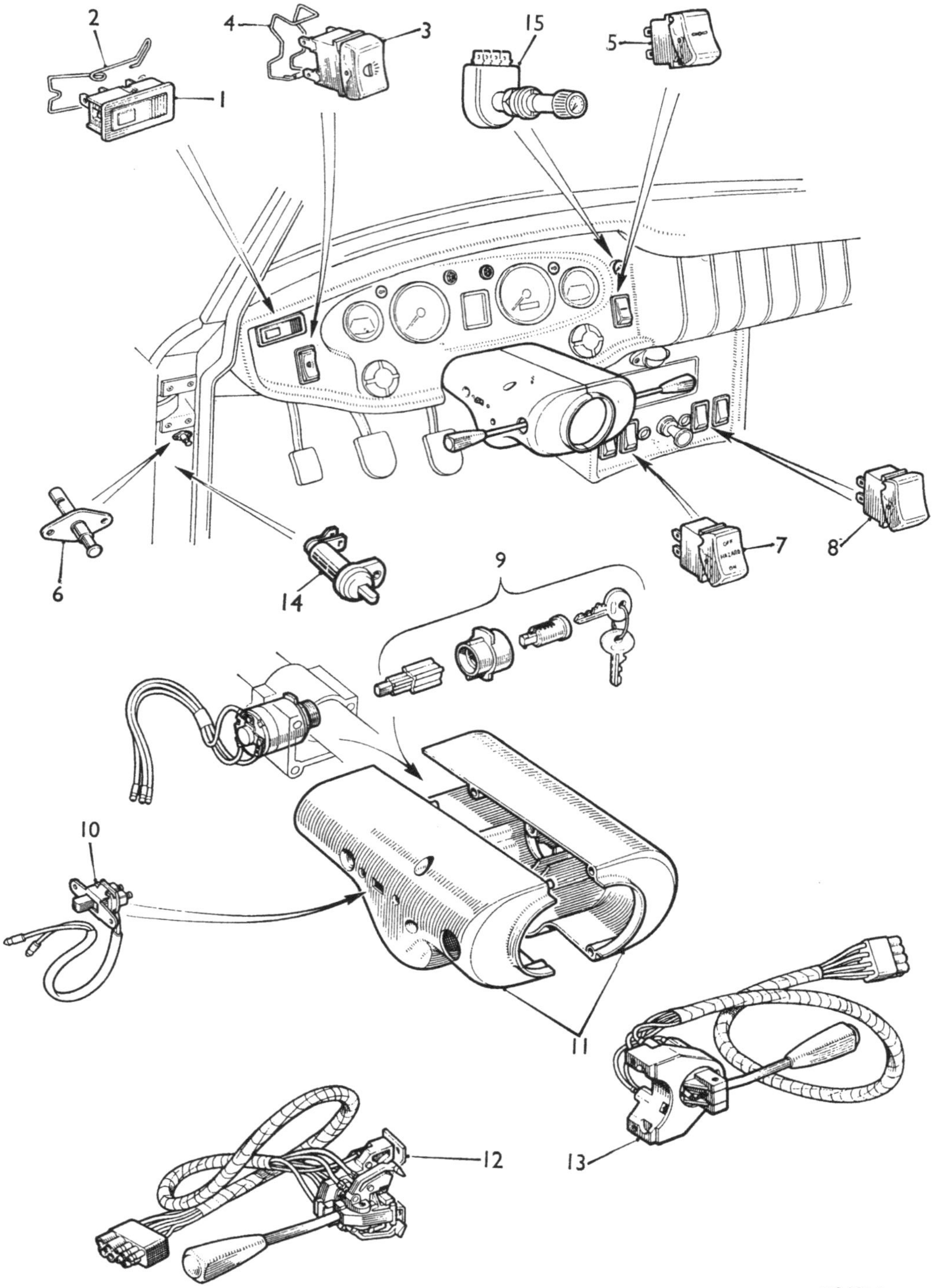


| No. | Description | No. | Description | No. | Description |
|---|---|---|---|---|---|
| 1. | Brake pressure—warning light/test push. | 6. | Door switch—interior light. | 12. | Direction indicator/headlight flasher—low/high beam/horn switch. |
| 2. | Retaining clip. | 7. | Hazard warning switch. | 13. | Windshield wiper/washer and overdrive switch. |
| 3. | Lighting switch. | 8. | Map light switch. | 14. | Audible warning door switch. |
| 4. | Retaining clip. | 9. | Ignition switch. | 15. | Panel lamp rheostat switch. |
| 5. | Heater blower switch. | 10. | Panel light switch. | | |
| | | 11. | Steering-column switch cowl. | | |

## Section Na.1

### INSTRUMENTS AND SWITCHES

**Removing**

**IMPORTANT.—Disconnect the battery before attempting to remove any of the switches or instruments.**

*Tachometer*

(1) Remove the left-hand lower fascia panel cover.
(2) Unscrew the two knurled retaining nuts, disconnect the earth cable and remove the retaining brackets.
(3) Withdraw the instrument from the fascia and disconnect the wiring.

*Oil pressure gauge*

(4) Remove the tachometer as in (1) to (3).
(5) Remove the ignition warning lamp and holder.
(6) Unscrew the two retaining nuts, withdraw the instrument from the fascia and disconnect the wiring.

*Speedometer*

(7) Remove the heater air control knob, unscrew the control retaining nut, and disengage the control from its mounting bracket.
(8) Unscrew the trip recorder reset knob retaining nut and disengage the reset from its bracket.
(9) Unscrew the two knurled retaining nuts remove the retaining brackets and withdraw the instrument from the fascia.
(10) Disconnect the drive cable from the back of the speedometer.

*Temperature gauge*

(11) Remove the heater air control as described in (7)
(12) Unscrew the knurled retaining nut and withdraw the instrument from the fascia.
(13) Disconnect the wiring from the back of the instrument.

*Mixture control (choke)*

(14) Remove the speedometer as in (7) to (10).
(15) Remove the temperature gauge as in (12) and (13).
(16) Disconnect the inner cable from the carburetter linkage.
(17) Release the cable from its clip on the left-hand side of the gearbox cover.
(18) Unscrew the control retaining nut from behind the fascia and withdraw the control.

*Heater blower switch*

(19) Remove the heater air control as described in (7).
(20) Disconnect the wiring from the back of the switch.
(21) Using service tool 18G 1145, depress the switch interior retaining lugs and withdraw the switch interior.

*Lighting switch*

(22) Disconnect the wiring from the back of the switch.
(23) Using service tool 18G 1145, depress the switch interior retaining lugs and withdraw the switch interior.

*Fuel gauge*

(24) Unscrew the knurled retaining nut from the back of the instrument.
(25) Withdraw the instrument from the fascia, disconnect the wiring and remove.

*Brake failure warning lamp assembly*

(26) Remove the speedometer as in (7) to (10).
(27) Remove the fuel gauge as in (24) to (25).
(28) Disconnect the wiring from the back of the lamp assembly.
(29) Disengage the spring retaining clip and withdraw the lamp assembly.

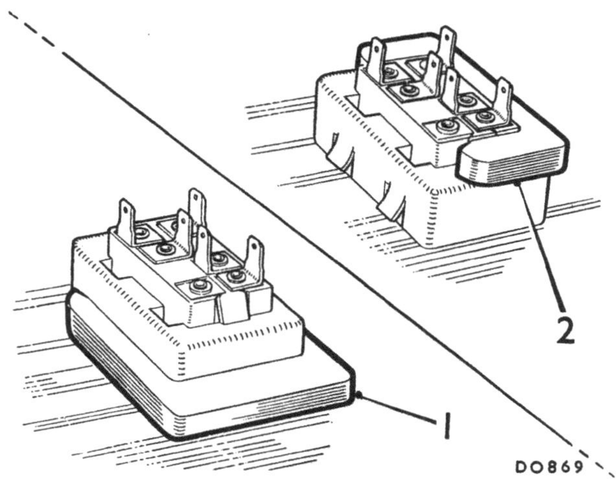


*Fig. Na.1*

*Using tool* 18G 1145 *to remove the switches*

1. Switch complete with bezel. 2. Switch interior only.

*Hazard warning switch*

(30) Remove the control console—Section Ra.1.
(31) Disconnect the wiring from the switch.
(32) Using tool 18G 1145 press in the four retaining lugs and withdraw the switch.

*Map light switch*

(33) Carry out operations (30) to (32) as for the hazard warning switch.

*Panel light switch*

(34) Withdraw the centre motif from the steering-wheel, unscrew the retaining nut and remove the steering-wheel using a pull-type extractor.
(35) Remove the three retaining screws from the left-hand switch cowl and remove the cowl half.

(36) Unscrew the two switch retaining screws, disconnect the switch wiring and remove the switch.

*Windshield wiper/washer and overdrive switch*

(37) Remove the left-hand switch cowl half as in (34) and (35).
(38) Remove the screw retaining the right-hand cowl half and remove the cowl.
(39) Remove the two switch retaining screws.
(40) Disconnect the switch wiring at the snap connectors and remove the switch complete with wiring.

*Direction indicator; horn and headlamp flasher switch*

(41) Detach the windshield wiper/washer and overdrive switch as in (37) to (39).
(42) Remove the two switch retaining screws.
(43) Disconnect the switch wiring at the snap connectors and remove the switch complete with wiring.

*Ignition switch*

(44) Remove both halves of the switch cowl as in (34), (35), and (38).
(45) Unscrew the four switch retaining screws.
(46) Disconnect the switch wiring at the snap connectors and remove the switch complete with wiring.

**Refitting**

*Tachometer*

(47) Reverse the removing procedure in (1) to (3).

*Oil pressure gauge*

(48) Reverse the removing procedure in (5) and (6).
(49) Refit the tachometer by reversing the removing procedure in (1) to (3).

*Speedometer*

(50) Reverse the removing procedure in (7) to (10).

*Temperature gauge*

(51) Reverse the removing procedure in (11) to (13).

*Mixture control (choke)*

(52) Reverse the removing procedure in (16) to (18).
(53) Refit the temperature gauge by reversing the removing procedure in (12) and (13).
(54) Refit the speedometer by reversing the removing procedure in (7) to (10).

*Heater blower switch*

(55) Press the switch firmly and evenly into the aperture until it is held by the retaining lugs.
(56) Re-connect the wiring to the switch.
(57) Refit the heater air control.

*Lighting switch*

(58) Press the switch firmly and evenly into the aperture until it is held by the retaining lugs.
(59) Re-connect the wiring to the switch.

*Fuel gauge*

(60) Reverse the removing procedure in (24) to (25).

*Brake failure warning lamp assembly*

(61) Fit the lamp assembly into the aperture and retain it with the spring clip.
(62) Reconnect the wiring.
(63) Refit the fuel gauge by reversing the removing procedure in (24) and (25).
(64) Refit the speedometer by reversing the removing procedure in (7) to (10).

*Hazard warning switch*

(65) Press the switch firmly and evenly into the aperture until it is held by the retaining lugs.
(66) Reconnect the wiring to the switch.
(67) Refit the control console—Section Ra.1.

*Map light switch*

(68) Carry out operations (65) to (67) as for the hazard warning switch.

*Panel light switch*

(69) Reverse the removing procedure in (34) to (36), tightening the steering-wheel nut to the torque figure given in **'GENERAL DATA'**.

*Windshield wiper/washer and overdrive switch*

(70) Reverse the removing procedure in (37) to (40), tightening the steering-wheel nut to the torque figure given in **'GENERAL DATA'**.

*Direction indicator, horn and headlamp flasher switch*

(71) Reverse the removing procedure in (41) to (43), tightening the steering-wheel nut to the torque figure given in **'GENERAL DATA'**.

*Ignition switch*

(72) Reverse the removing procedure in (44) to (46), tightening the steering-wheel to the torque figure given in **'GENERAL DATA'**.

**GHN5 and GHD5 cars only**

*Audible warning buzzer*

(73) Disconnect the electrical leads and remove the screw retaining the buzzer unit. No adjustment is possible and in the event of malfunction the buzzer unit must be replaced.

*Audible warning door switch*

(74) Remove the one self tapping screw, extract the switch unit and disconnect the leads from the rear of the switch.

## Section Na.2

### WINDSHIELD WASHER MOTOR

**Removing**

(1) Disconnect the battery.
(2) Disconnect the wiring from the motor terminals.
(3) Disconnect the water tubes.
(4) Unscrew the two pump mounting screws and remove the pump assembly.

**Refitting**

(5) Reverse the removing procedure in (1) to (4).

## Section Na.3

### LAMPS

**Front and rear side-marker lamps**

(1) Disconnect the battery.
(2) Remove the two nuts and four washers securing the lamp.
(3) Partially remove the lamp and disconnect the electrical leads at the harness connectors.

**Number-plate lamp**

(1) Disconnect the battery.
(2) Remove the two nuts and washers retaining the lamps to the lamp bracket; alternatively, remove the lamp complete with the bracket.
(3) Disconnect the electrical leads at the harness connections, and remove the lamp.

## Section Na.4

### HAZARD WARNING FLASHER UNIT AND AUDIBLE WARNING BUZZER

**Removing**

*Hazard warning flasher unit*

(1) Disconnect the battery.
(2) Remove the centre console—Section R.34.
(3) Withdraw the flasher unit from its retaining clip.
(4) Disconnect the wiring plug from the flasher unit.

*Audible warning buzzer*

(5) Disconnect the battery.
(6) Remove the centre console—Section R.34.
(7) Remove the screw or nut retaining the buzzer.
(8) Disconnect the wiring from the buzzer.

**Refitting**

*Hazard warning flasher unit*

(9) Reverse the removing procedure in (1) to (4).

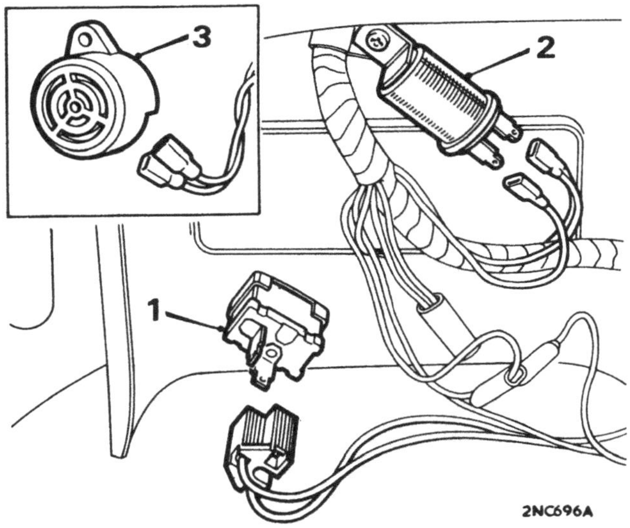


*Fig. Na.2*

*The centre console removed showing:*

1. The hazard warning flasher unit
2. Early type warning buzzer.
3. Later type warning buzzer.

*Audible warning buzzer*

(10) Reverse the removing procedure in (5) to (8).

## Section Na.5

### SWITCHES, WARNING LAMPS AND CONTROLS
### (GHN5 and GHD5 cars from Car No. 294251)

**NOTE:** Refer to Section Na.1 for the removal and refitting of the instruments, steering-column switches, door courtesy switch, audible warning switch and buzzer.

**Removing**

**IMPORTANT: Disconnect the battery before attempting to remove any of the switches, warning lamps or controls.**

*Hazard warning switch*

(1) Remove the centre console—Section R.34.
(2) Withdraw the bulb holder from the switch retainer.
(3) Disconnect the wiring plug from the switch.
(4) Remove the retainer from the switch and withdraw the switch from the console.

*Lighting switch*

(5) Unscrew the three screws securing the fascia L.H. lower trim board and pull the board forward to release it from its retaining clips at the back of the fascia.
(6) Remove the bulb holder from the switch retainer.
(7) Disconnect the wiring from the switch.
(8) Remove the retainer from the switch and withdraw the switch from the fascia.

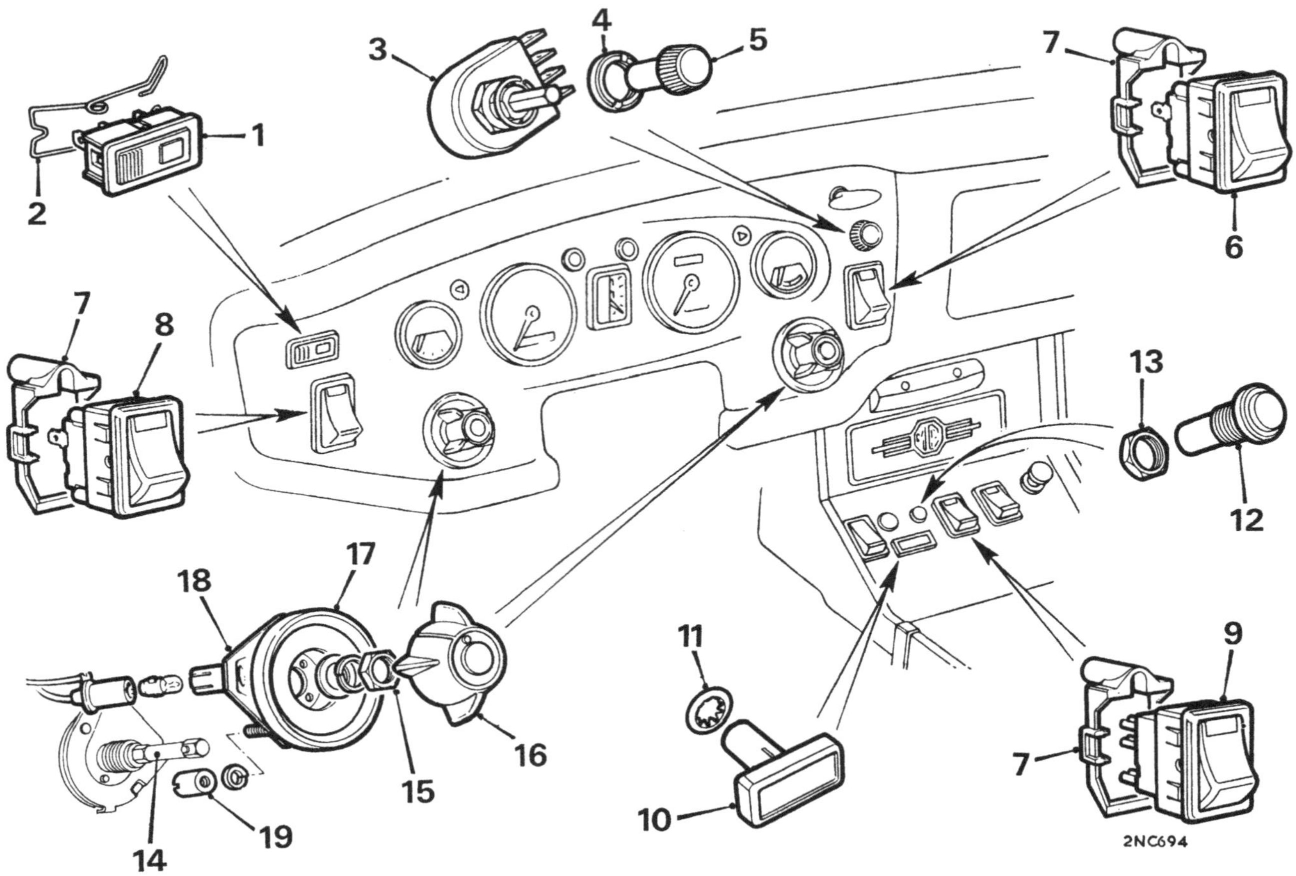


*Fig. Na.3*
*The switches, lamps, and controls (GHN5/GHD5 cars from Car No. 294251)*

1. Brake pressure warning light test-push.
2. Retaining clip.
3. Panel lamp rheostat switch.
4. Retainer.
5. Knob for rheostat switch.
6. Heater blower switch.
7. Retainer for rocker switch.
8. Lighting switch.
9. Hazard warning switch.
10. Seat belt warning lamp.
11. Retainer for seat belt warning lamp.
12. Hazard warning lamp.
13. Retainer for hazard warning lamp.
14. Rotary control.
15. Retaining nut.
16. Rotary control knob.
17. Dial assembly.
18. Light box.
19. Retaining nut.

*Heater blower switch*

(9) Remove the glovebox—Section Ra.3.
(10) Remove the face-level vents—Section Ra.4.
(11) Disconnect the bulb holder from the switch retainer.
(12) Disconnect the wiring from the switch.
(13) Remove the retainer from the switch and withdraw the switch from the fascia.

*Panel lamp rheostat switch*

(14) Remove the glovebox—Section Ra.3.
(15) Remove the face-level vents—Section Ra.4.
(16) Disconnect the wiring from the switch.
(17) Depress the pin in the switch knob and withdraw the knob from the switch.
(18) Unscrew the switch retainer and remove the switch from the fascia.

*Seat belt warning lamp*

(19) Remove the centre console—Section R.34.
(20) Remove the bulb holder from the warning lamp.
(21) Remove the warning lamp clip retainer and push the warning lamp out of the centre console.

*Hazard warning lamp*

(22) Remove the centre console—Section R.34.
(23) Remove the bulb holder from the warning lamp.
(24) Unscrew the warning lamp retainer and remove the warning lamp from the centre console.

*Temperature rotary control*

(25) Unscrew the three screws securing the fascia L.H. lower trim board and pull the board forward to release it from its retaining clips at the back of the fascia.

(26) Remove the screw retaining the fascia to the the cross-tube L.H. bracket.
(27) Depress the pin in the switch knob and withdraw the knob from the spindle.
(28) Remove the bulb holder from the rotary control light box.
(29) Unscrew the nut retaining the rotary control to the fascia and remove the spring and plain washer.
(30) Remove the rotary control from the fascia.
(31) Unscrew the three nuts and remove the three spring and plain washers to release the light box from the dial assembly.
(32) Remove the dial assembly from the fascia.

*Air flow rotary control*
(33) Unscrew the three screws securing the fascia L.H. lower trim board and pull the board forward to release it from its retaining clips at the back of the fascia.
(34) Depress the pin in the switch knob and withdraw the knob from the spindle.
(35) Remove the bulb holder from the rotary control light box.
(36) Unscrew the nut retaining the rotary control to the fascia and remove the spring and plain washer.
(37) Remove the rotary control from the fascia.
(38) Unscrew the three nuts and remove the three spring and plain washers to release the light box from the dial assembly.
(39) Remove the dial assembly from the fascia.

**Refitting**

*Hazard warning switch*
(40) Reverse the removing procedure in (2) to (4).
(41) Refit the centre console—Section R.34.

*Lighting switch*
(42) Reverse the removing procedure in (5) to (8).

*Heater blower switch*
(43) Reverse the removing procedure in (11) to (13).
(44) Refit the face-level vents—Section Ra.4.
(45) Refit the glovebox—Section Ra.3.

*Panel lamp rheostat switch*
(46) Reverse the removing procedure in (16) to (18).
(47) Refit the face level vents—Section Ra.4.
(48) Refit the glovebox—Section Ra.3.

*Seat belt warning lamp*
(49) Reverse the removing procedure in (20) to (21).
(50) Refit the centre console—Section R.34.

*Hazard warning lamp*
(51) Reverse the removing procedure in (23) to (24).
(52) Refit the centre console—Section R.34.

*Temperature rotary control*
(53) Reverse the removing procedure in (25) to (32).

*Air flow rotary control*
(54) Reverse the removing procedure in (33) to (39). ●

## WIRING DIAGRAMS

**Key to the wiring diagrams page Nb.2**

| | |
|---|---|
| MGB, positive earth .. .. .. .. .. .. .. .. .. .. <br>*Commencing Car No GHN*3 101 *to* 48765 (*Tourer only*) | **Diagram 1**<br>(1962–64) |
| MGB, positive earth .. .. .. .. .. .. .. .. .. .. <br>*Commencing Car No. GHN*3 48766 (*Tourer*)<br>*GHD*3 71933 (*GT*) | **Diagram 2**<br>(1964–67)<br>(1965–67) |
| MGB, negative earth .. .. .. .. .. .. .. .. .. .. <br>*Commencing Car No. GHN*4 138801 (*Tourer*)<br>*GHD*4 139824 (*GT*) | **Diagram 3**<br>(1967–68) |
| MGB, negative earth .. .. .. .. .. .. .. .. .. .. <br>*Commencing Car No. GHN*4 158371 (*Tourer*)<br>*GHD*4 158231 (*GT*) | **Diagram 4**<br>(1968–69) |
| MGB, negative earth .. .. .. .. .. .. .. .. .. .. <br>*Commencing Car No. GHN*5 187170 (*Tourer*)<br>*GHD*5 187841 (*GT*) | **Diagram 5**<br>(1968–70) |
| MGB, negative earth .. .. .. .. .. .. .. .. .. .. <br>*Commencing Car No. GHN*5 (*Tourer*)<br>*GHD*5 (*GT*) | **Diagram 6**<br>(1970–71) |
| MGB, negative earth .. .. .. .. .. .. .. .. .. .. <br>*Commencing Car No. GHN*5 258001 (*Tourer*)<br>*GHD*5 258004 (*GT*) | **Diagram 11**<br>(1971–72) |

**The following wiring diagrams are for MGB cars fitted with electrical equipment in conformity with various local and territorial motor vehicle regulations**

| | |
|---|---|
| MGB, negative earth .. .. .. .. .. .. .. .. .. .. <br>*Commencing Car No. GHN*4 138401 (*Tourer*)<br>*GHD*4 139472 (*GT*) | **Diagram 7**<br>(1967–68) |
| MGB, negative earth .. .. .. .. .. .. .. .. .. .. <br>*Commencing Car No. GHN*4 158233 (*Tourer*)<br>*GHD*4 158371 (*GT*) | **Diagram 8**<br>(1968–69) |
| MGB, negative earth .. .. .. .. .. .. .. .. .. .. <br>*Commencing Car No. GHN*5 187170 (*Tourer*)<br>*GHD*5 187841 (*GT*) | **Diagram 9**<br>(1969–70) |
| MGB, negative earth .. .. .. .. .. .. .. .. .. .. <br>*Commencing Car No. GHN*5 (*Tourer*)<br>*GHD*5 (*GT*) | **Diagram 10**<br>(1970–71) |
| MGB, negative earth .. .. .. .. .. .. .. .. .. .. <br>*Commencing Car No. GHN*5 258001 (*Tourer*)<br>*GHD*5 258004 (*GT*) | **Diagram 12**<br>(1971–72) |
| MGB, negative earth .. .. .. .. .. .. .. .. .. .. <br>*Commencing Car No. GHN*5 (*Tourer*) } with seat belt warning<br>*GHD*5 (*GT*) } | **Diagram 13**<br>(1971–72) |

## KEY TO THE WIRING DIAGRAMS

Use the one key to identify components on these wiring diagrams. Refer to the appropriate wiring diagram for your car, and disregard any additional numbered items appearing in the key and not on your diagram.

| No. | Description |
|---|---|
| 1. | Alternator or dynamo. |
| 2. | Control box. |
| 3. | Batteries—6-volt. |
| 4. | Starter solenoid. |
| 5. | Starter motor. |
| 6. | Lighting switch. |
| 7. | Headlamp dip switch. |
| 8. | R.H. headlamp. |
| 9. | L.H. headlamp. |
| 10. | High-beam warning lamp. |
| 11. | R.H. parking lamp. |
| 12. | L.H. parking lamp. |
| 13. | Panel lamp switch or rheostat switch. |
| 14. | Panel lamps. |
| 15. | Number-plate illumination lamp. |
| 16. | R.H. stop and tail lamp. |
| 17. | L.H. stop and tail lamp. |
| 18. | Stop lamp switch. |
| 19. | Fuse unit. |
| 20. | Interior courtesy lamp or map light (early cars). |
| 21. | R.H. door switch. |
| 22. | L.H. door switch. |
| 23. | Horns. |
| 24. | Horn-push. |
| 25. | Flasher unit. |
| 26. | Direction indicator switch. or |
| 26. | Direction indicator/headlamp flasher. or |
| 26. | Combined direction indicator/ headlamp flasher/headlamp high-low beam/horn-push switch. or |
| 26. | Combined direction indicator/ headlamp flasher/headlamp high-low beam switch. |
| 27. | Direction indicator warning lamps. |
| 28. | R.H. front flasher lamp. |
| 29. | L.H. front flasher lamp. |
| 30. | R.H. rear flasher lamp. |
| 31. | L.H. rear flasher lamp. |
| 32. | Heater booster motor switch. |
| 33. | Heater booster motor, or fresh-air motor. |
| 34. | Fuel gauge. |
| 35. | Fuel gauge tank unit. |
| 36. | Windshield wiper switch. |
| 37. | Windscreen wiper motor. |
| 38. | Ignition/starter switch. |
| 39. | Ignition coil. |
| 40. | Distributor. |
| 41. | Fuel pump. |
| 43. | Oil pressure gauge. |
| 44. | Ignition warning lamp. |
| 45. | Speedometer. |
| 46. | Coolant temperature gauge. |
| 47. | Coolant temperature transmitter. |
| 49. | Reverse lamp switch. |
| 50. | Reverse lamp. |
| 53. | Fog and driving lamp switch. |
| 54. | Driving lamp. |
| 55. | Fog lamp. |
| 57. | Cigar-lighter illuminated. |
| 59. | Map light switch (early cars). |
| 60. | Radio. |
| 64. | Bi-metal instrument voltage stabilizer. |
| 65. | Luggage compartment lamp switch. |
| 66. | Luggage compartment lamp. |
| 67. | Line fuse. |
| 68. | Overdrive relay unit. |
| 71. | Overdrive solenoid. |
| 72. | Overdrive manual control switch. |
| 73. | Overdrive gear switch. |
| 74. | Overdrive throttle switch. |
| 76. | Automatic gearbox gear selector illumination lamp. |
| 77. | Windscreen washer pump. |
| 82. | Switch illumination lamp. |
| 95. | Tachometer. |
| 101. | Courtesy or map light switch. |
| 102. | Courtesy or map light. |
| 115. | Heated back-light switch } GT only. |
| 116. | Heated back-light } GT only. |
| 118. | Combined windshield washer and wiper switch. |
| 131. | Combined reverse lamp switch and automatic transmission safety switch. |
| 147. | Oil pressure transmitter. |
| 150. | Heated back-light warning lamp (GT only). |
| 152. | Hazard warning lamp. |
| 153. | Hazard warning switch. |
| 154. | Hazard warning flasher unit. |
| 159. | Brake pressure warning lamp and lamp test push. |
| 160. | Brake pressure failure switch. |
| 168. | Ignition key audible warning buzzer. |
| 169. | Ignition key audible warning door switch. |
| 170. | R.H. front side-marker lamp. |
| 171. | L.H. front side-marker lamp. |
| 172. | R.H. rear side-marker lamp. |
| 173. | L.H. rear side-marker lamp. |
| 174. | Starter solenoid relay. |
| *198. | Driver's seat belt buckle switch. |
| *199. | Passenger's seat belt buckle switch. |
| *200. | Passenger's seat switch. |
| *201. | Gearbox switch seat belt warning. |
| *202. | Warning light—fasten belts. |
| *203. | Line diode. |

### CABLE COLOUR CODE

| | | | | | |
|---|---|---|---|---|---|
| N. | Brown. | P. | Purple. | W. | White. |
| U. | Blue. | G. | Green. | Y. | Yellow. |
| R. | Red. | LG. | Light Green. | B. | Black. |
| | | O. | Orange. | | |

When a cable has two colour code letters the first denotes the main colour and the second denotes the tracer colour.

*See page 461

## DIAGRAM 1

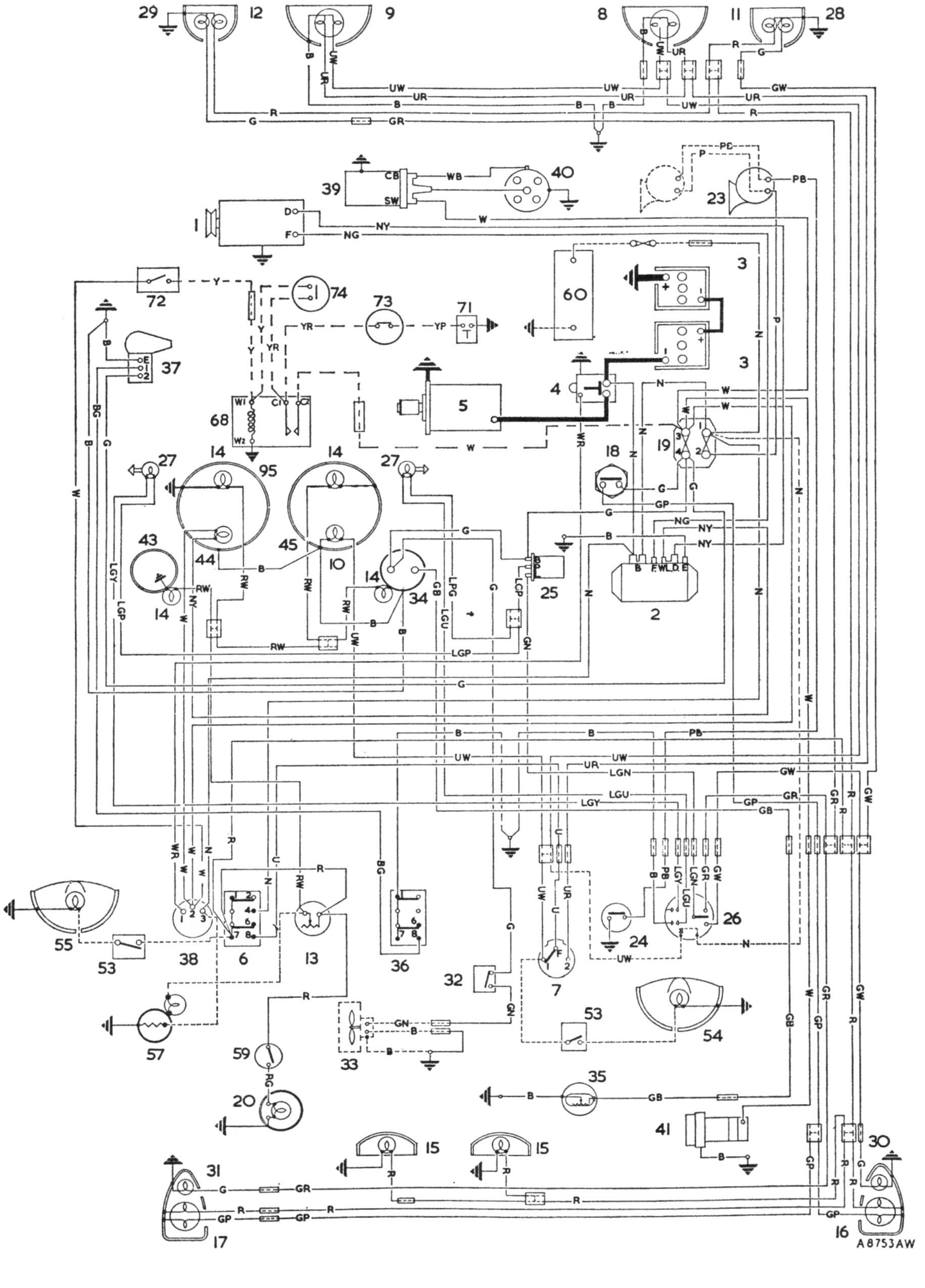
29
12
9
8
11
28
39
40
23
1
60
3
72
74
73
71
37
4
5
68
19
18
27
14
95
14
27
43
44
45
10
34
25
2
14
55
53
38
6
13
36
26
24
32
7
57
59
33
53
54
20
35
41
15
15
31
30
17
16
A8753AW

## DIAGRAM 2

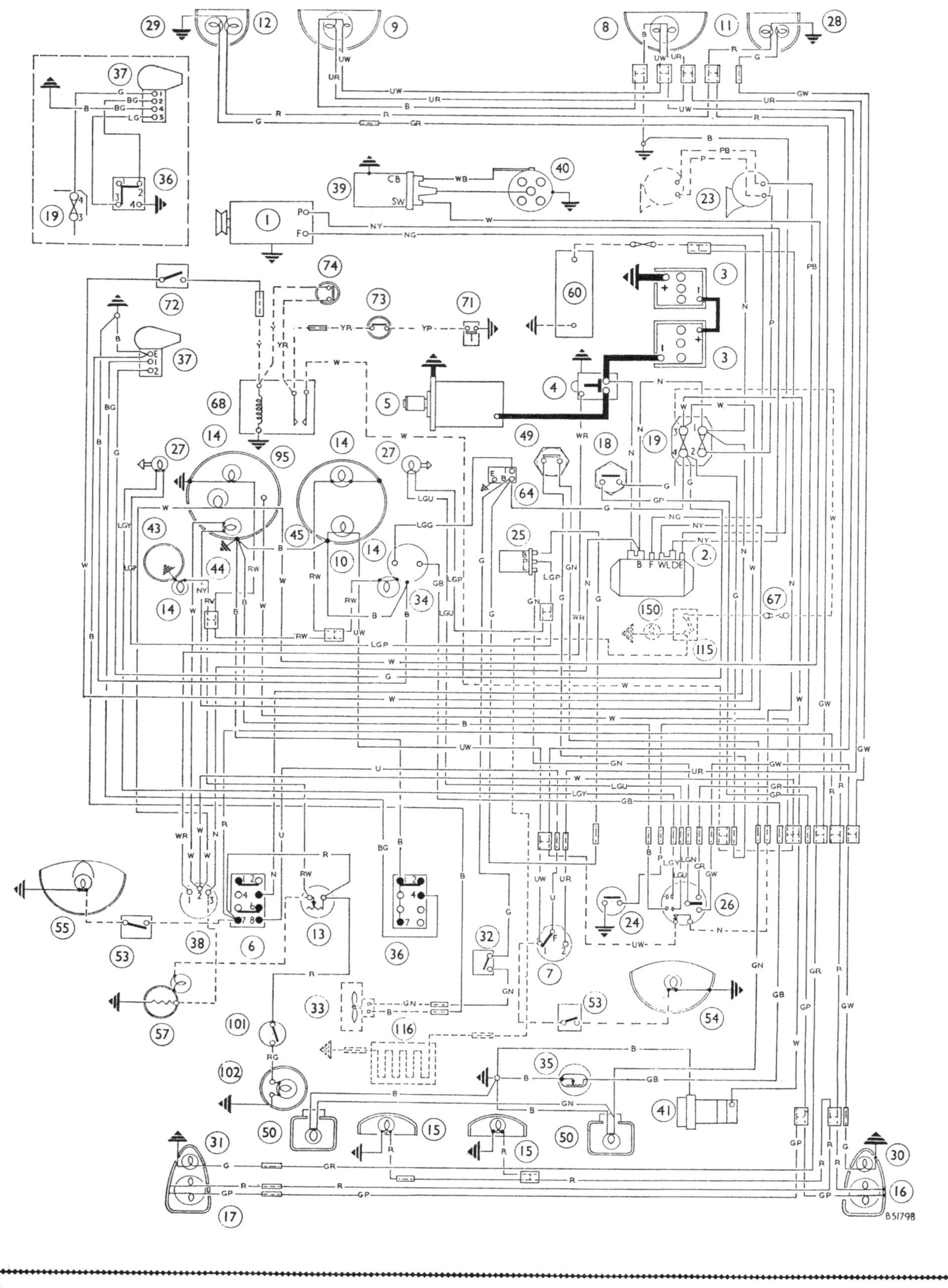

## DIAGRAM 3

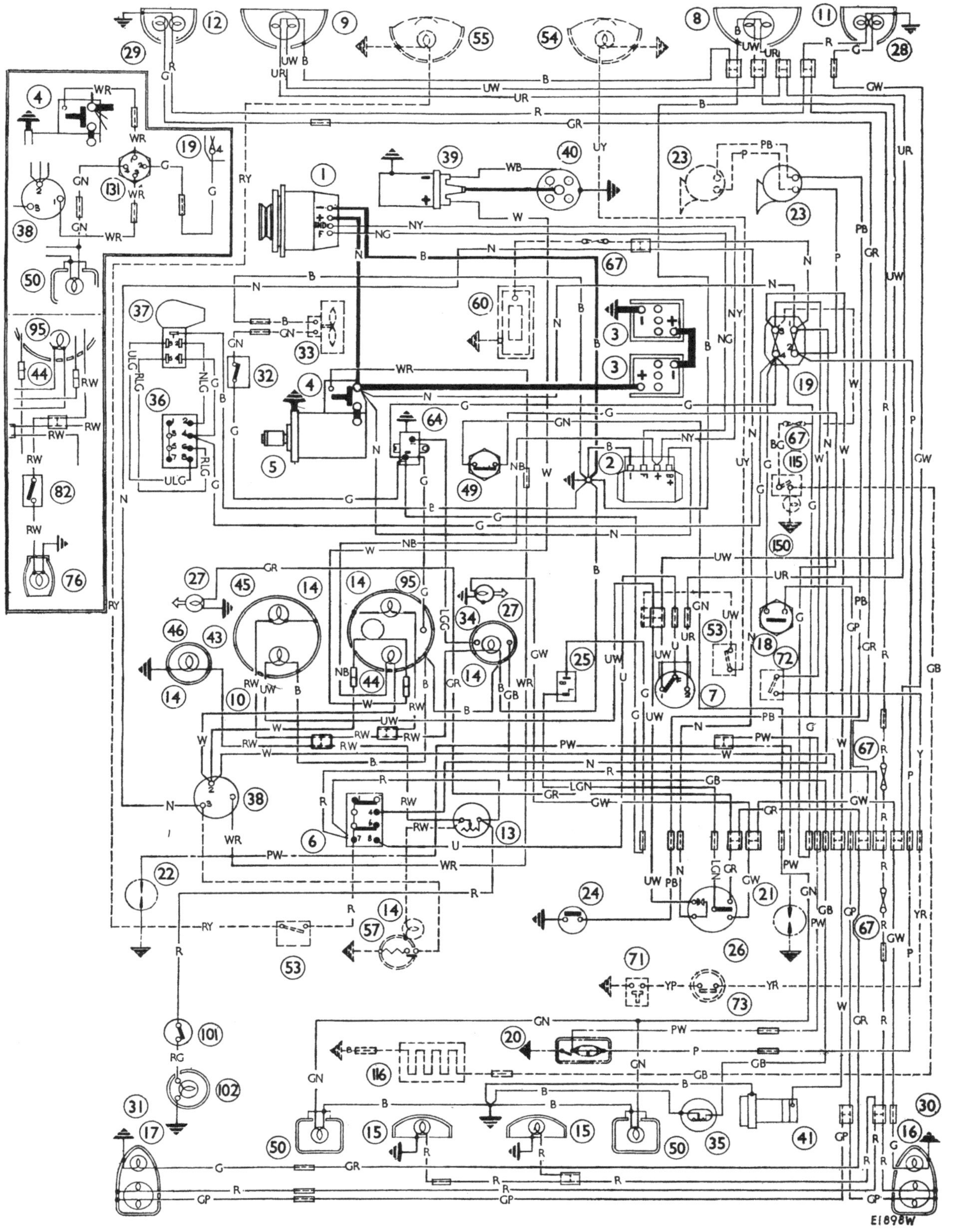

## DIAGRAM 4

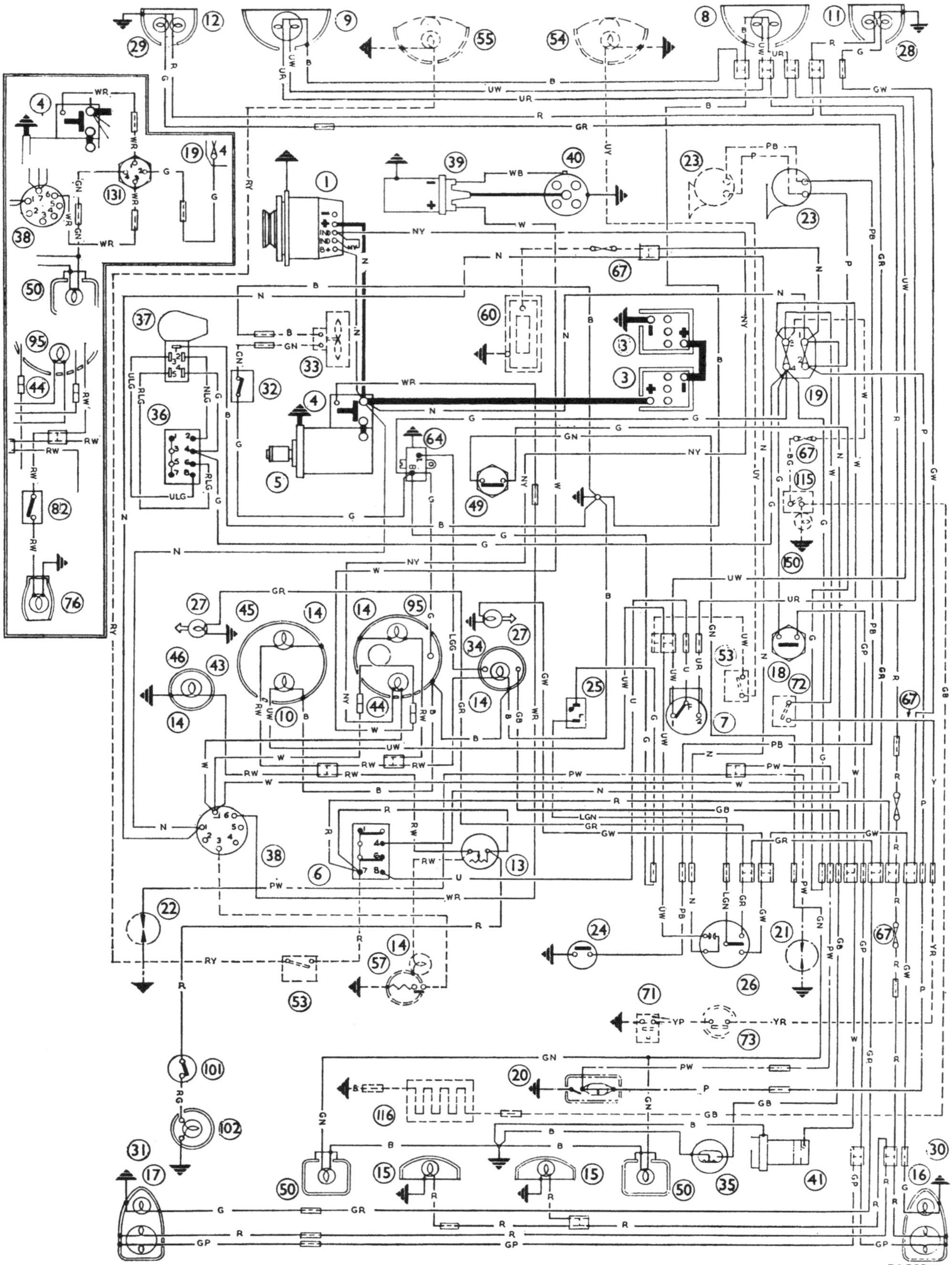

## DIAGRAM 5

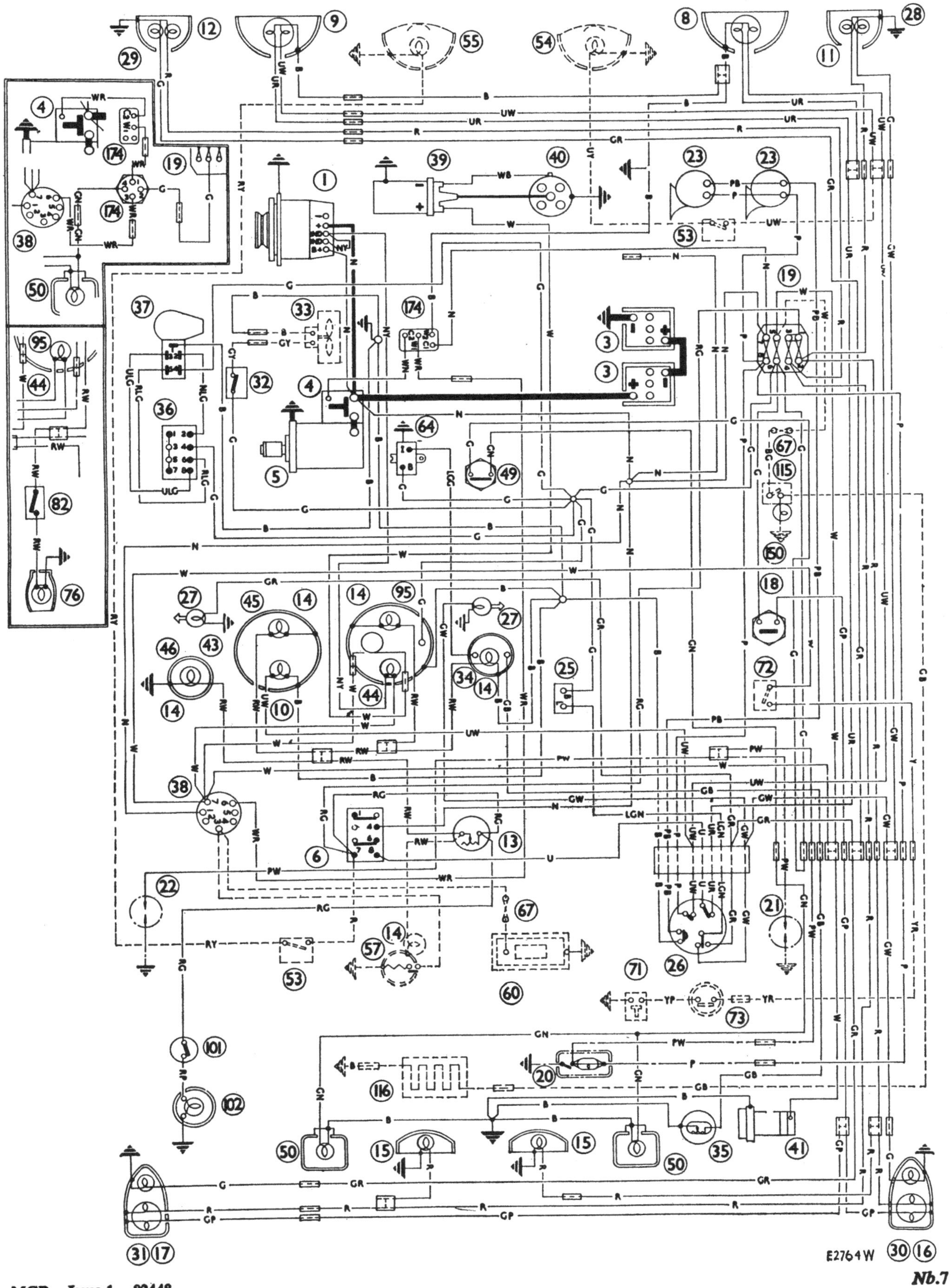

## DIAGRAM 6

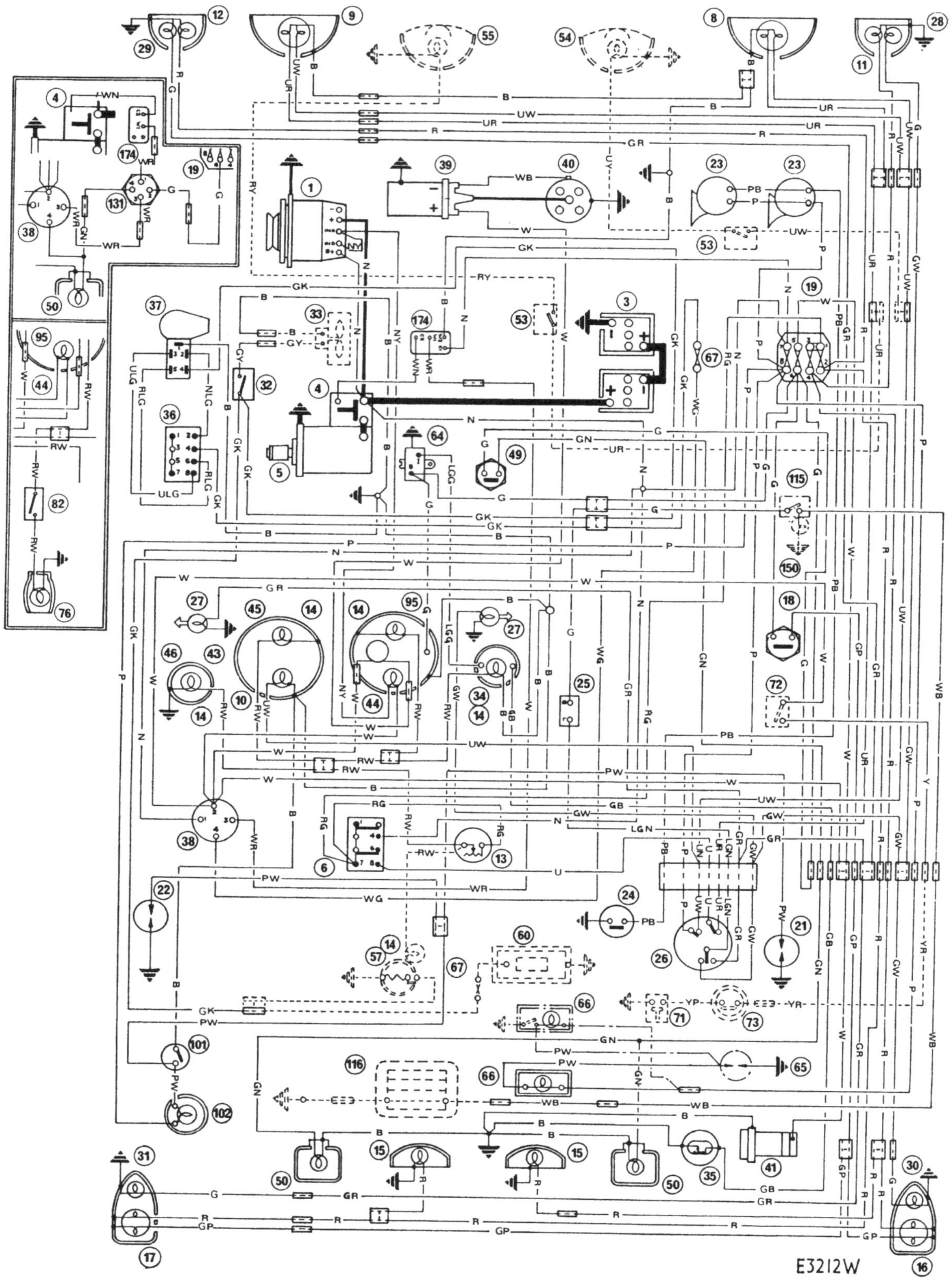

# DIAGRAM 7

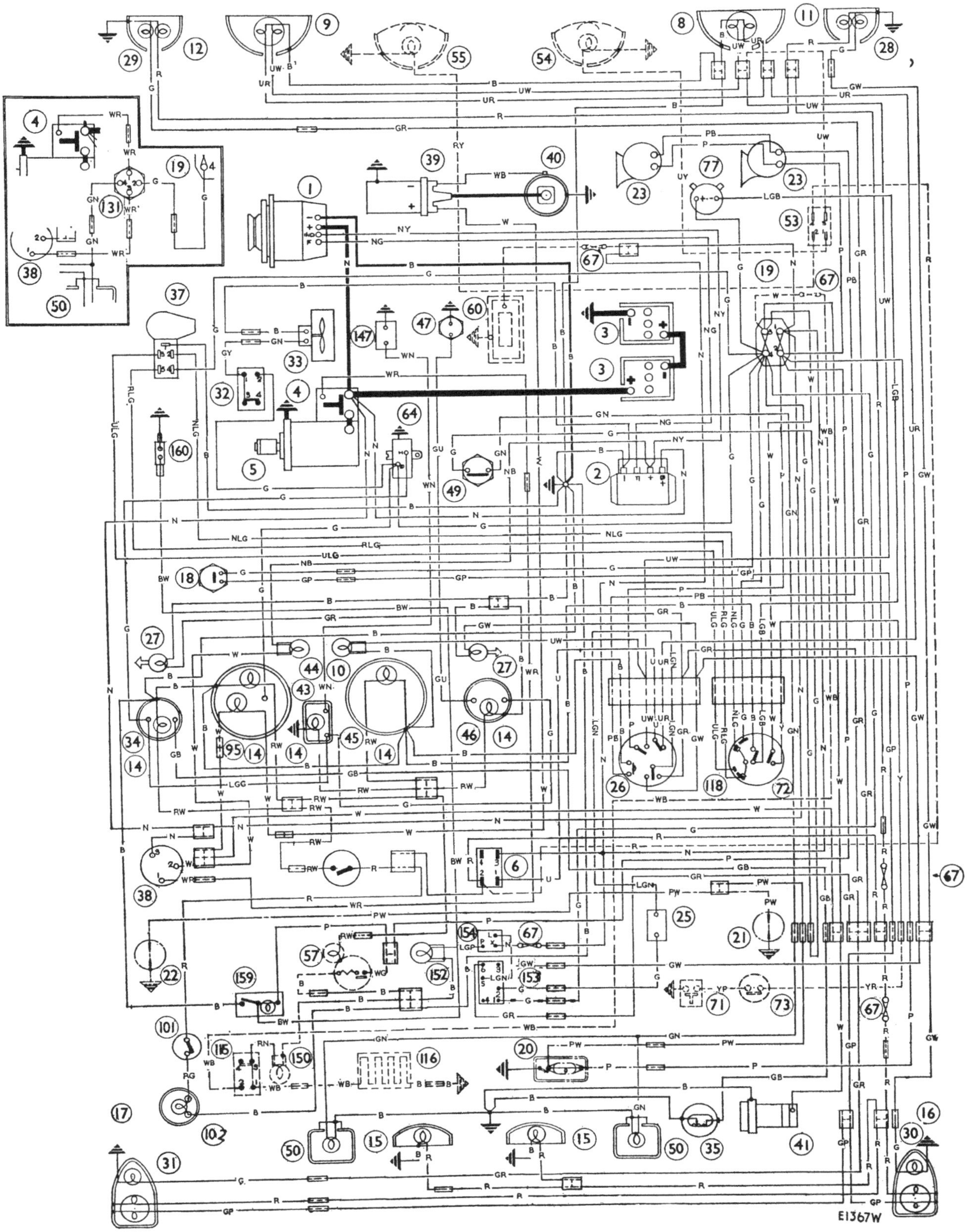

## DIAGRAM 8

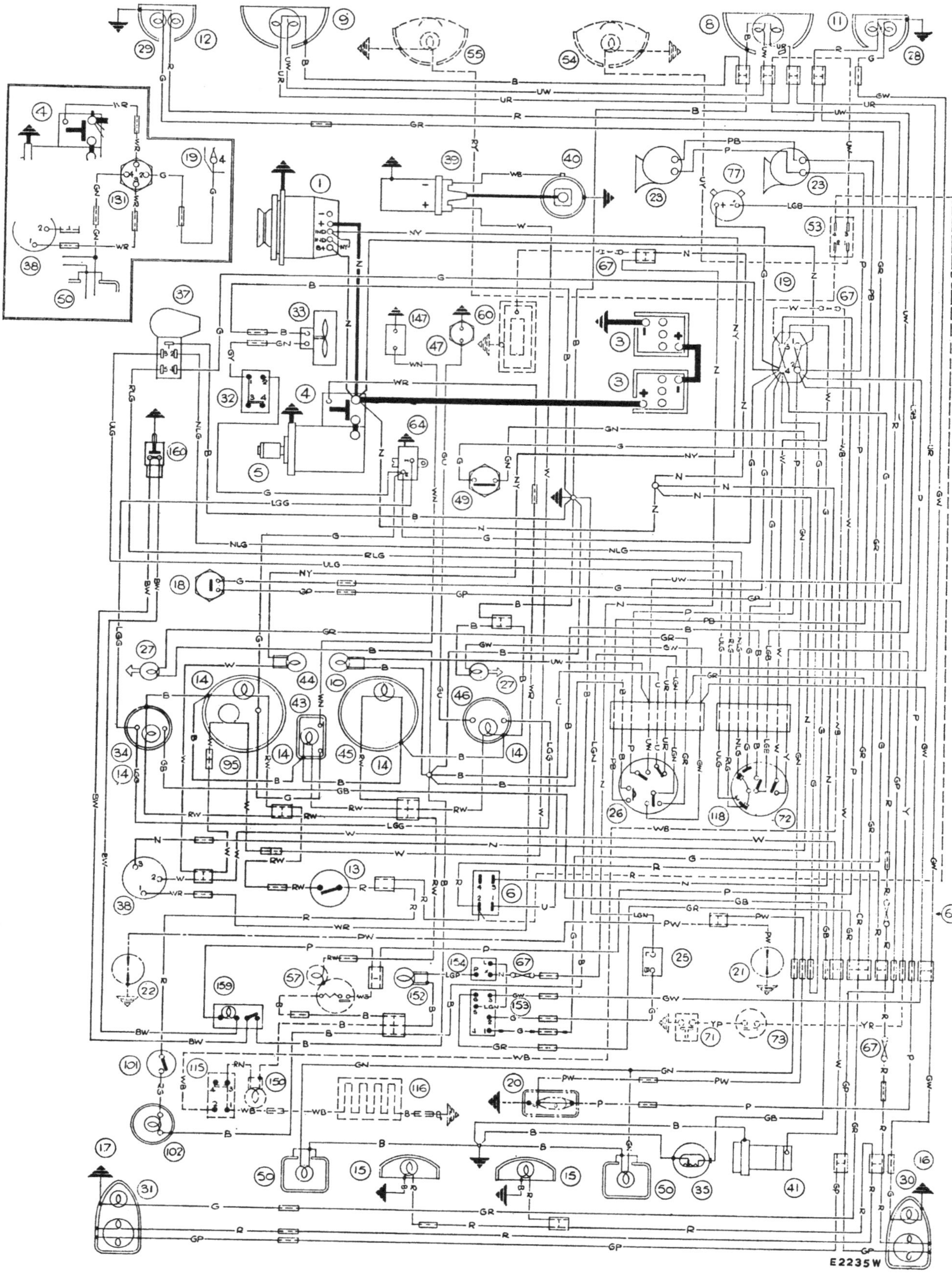

## DIAGRAM 9

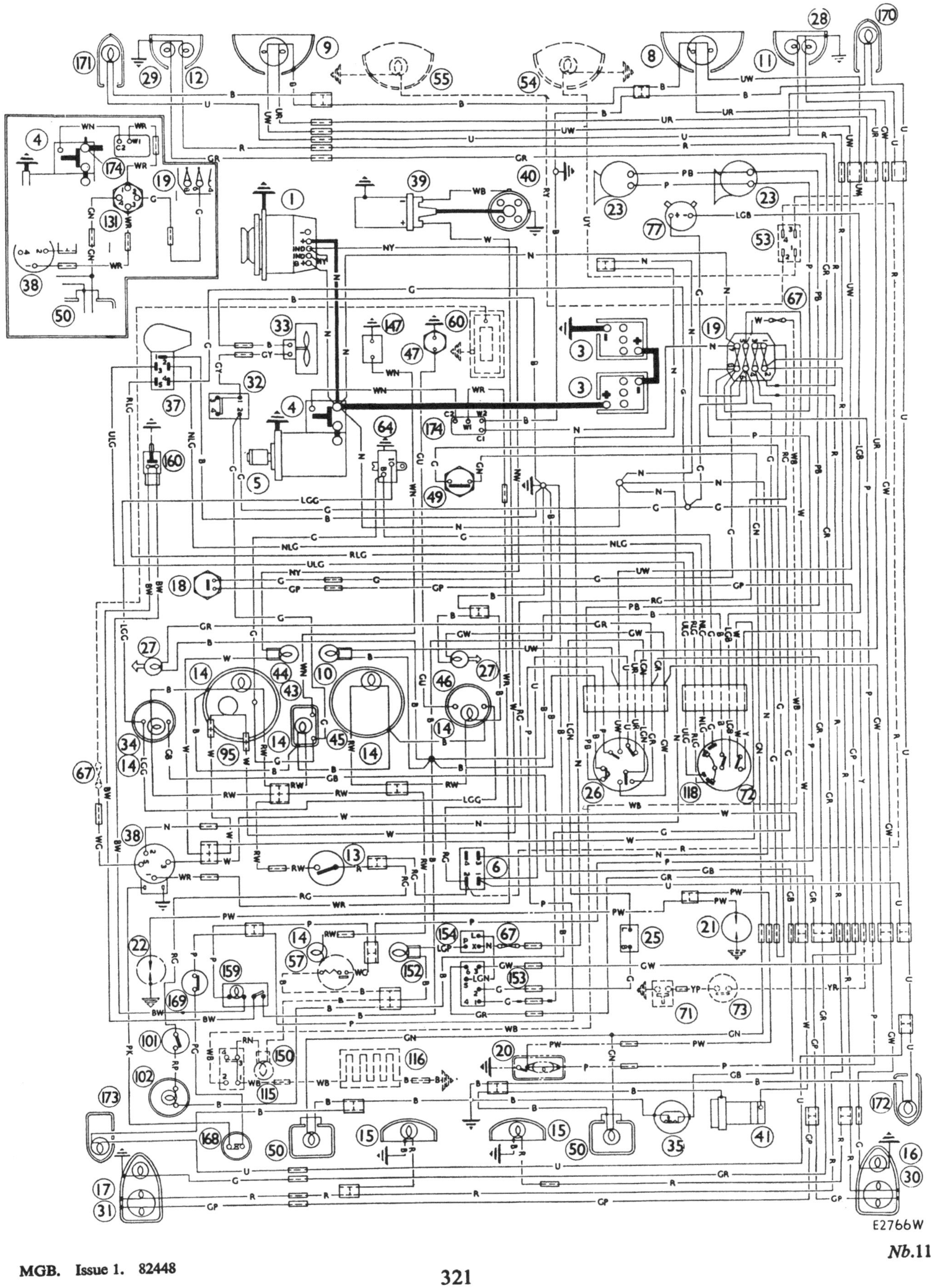

E2766W

## DIAGRAM 10

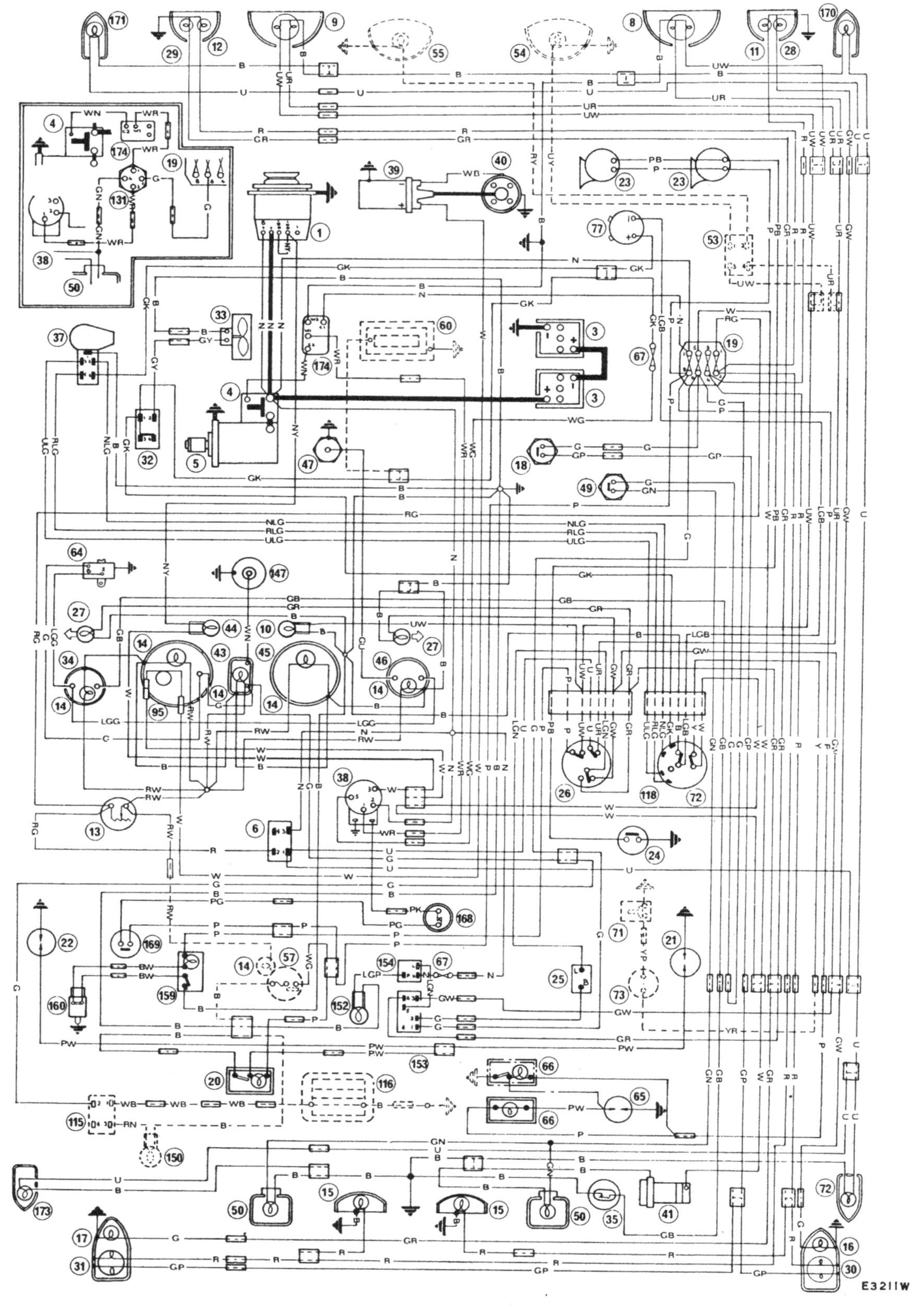

## DIAGRAM 11

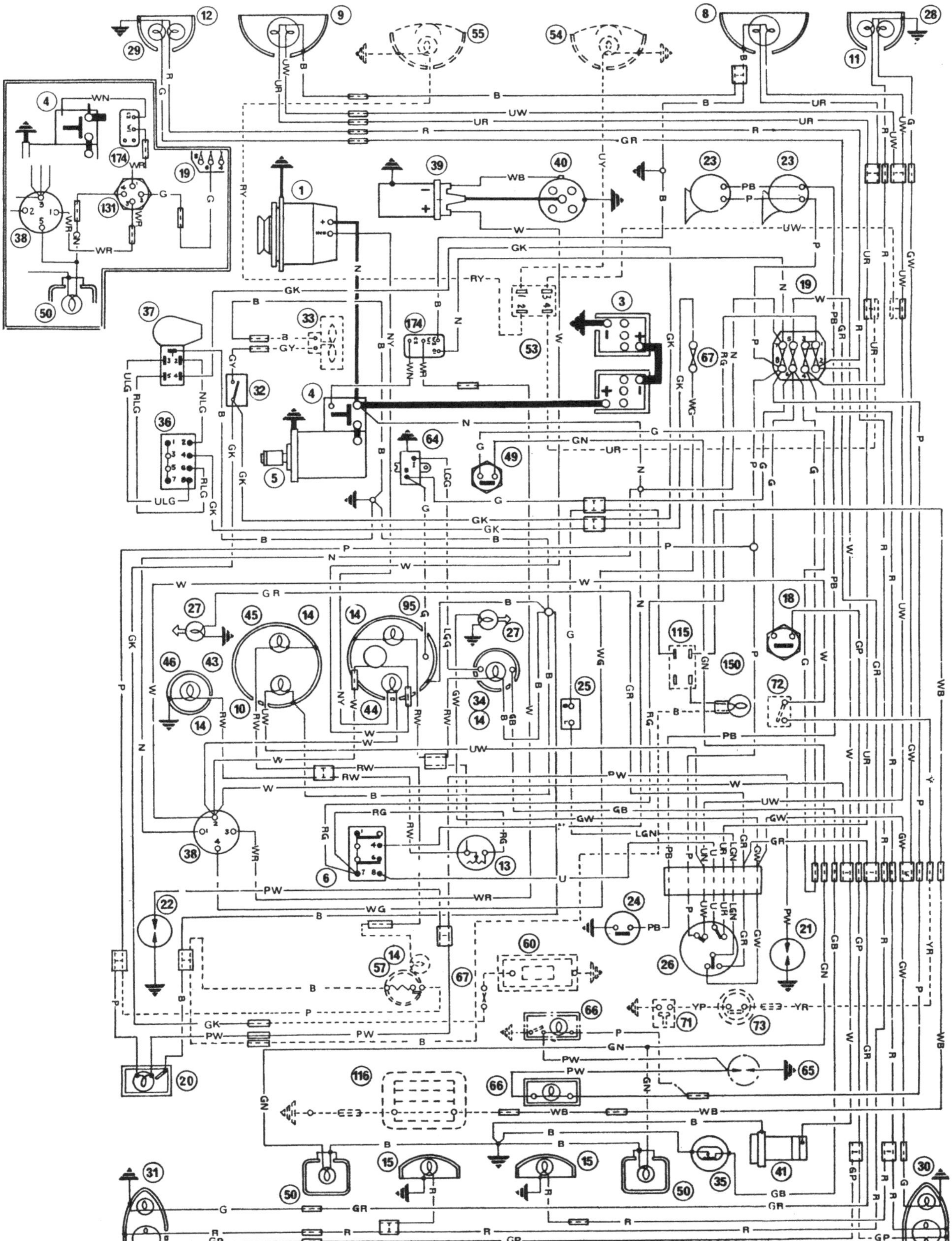

E3212 AW

# DIAGRAM 12

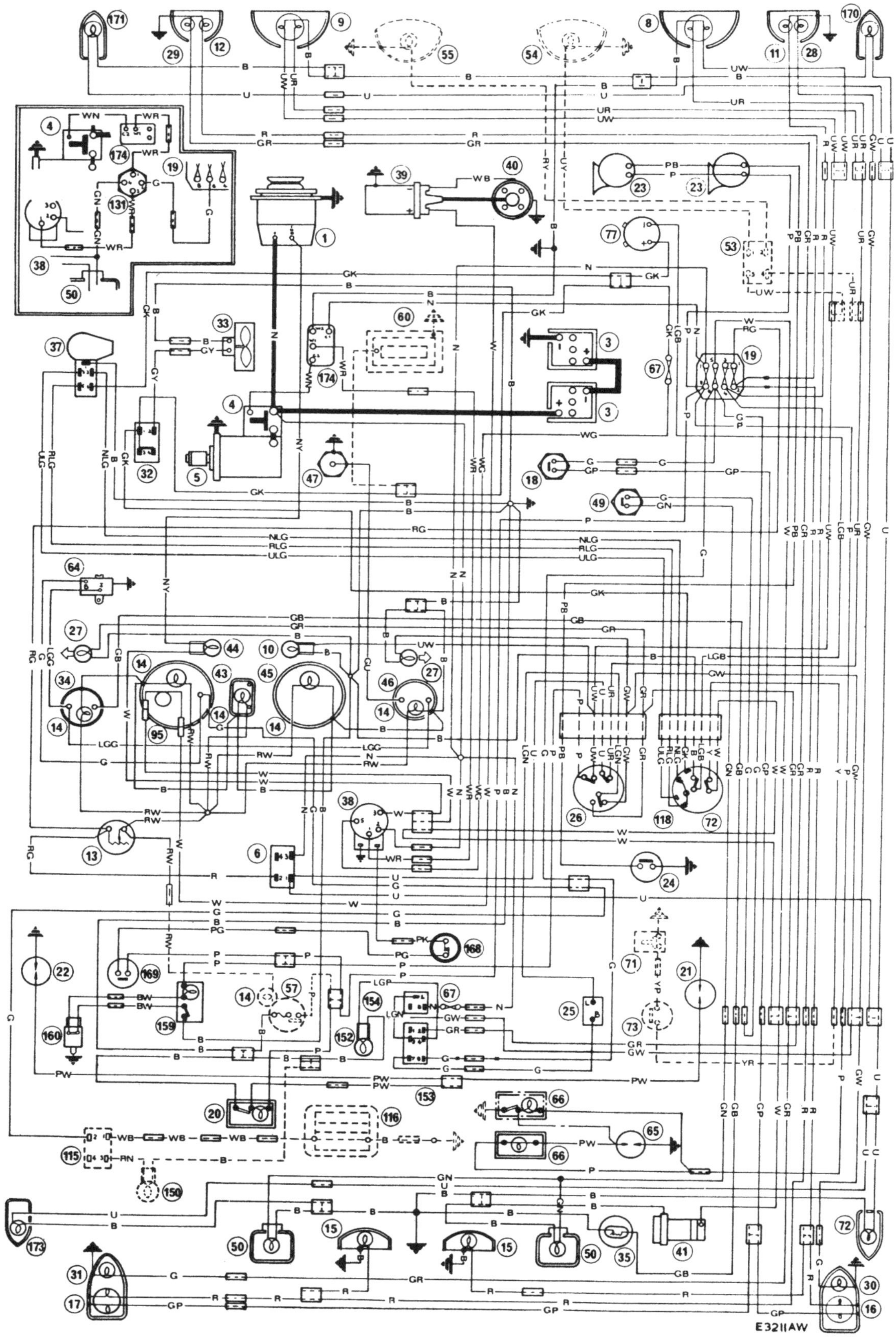

## DIAGRAM 13

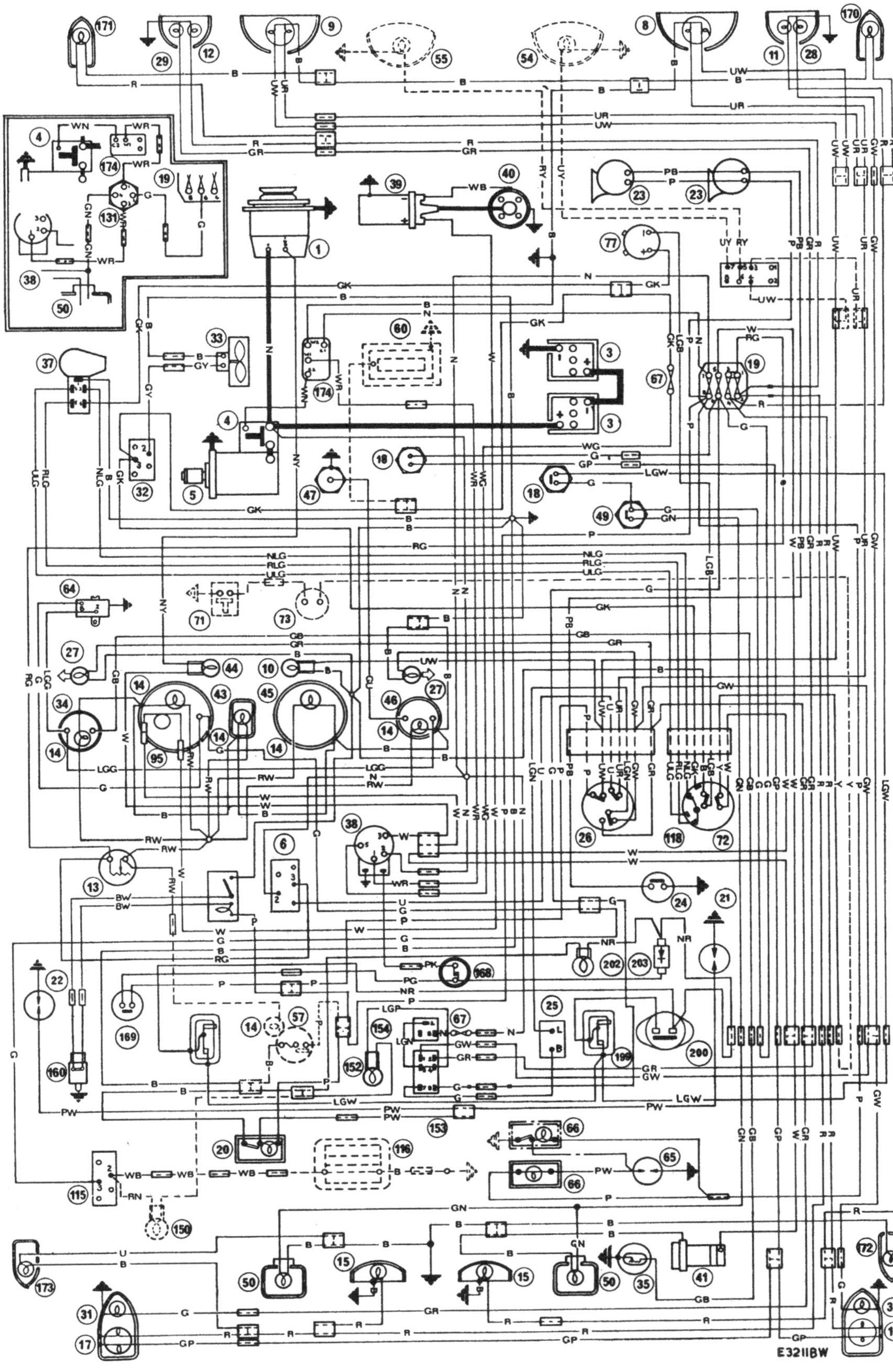

# SECTION O

## THE WHEELS AND TYRES

### GENERAL DESCRIPTION

The wheels fitted as standard equipment are of the pressed-steel ventilated disc type having four-stud fixing holes.

A spare wheel complete is housed in the luggage compartment and is retained by a clamp plate and bolt.

The jack is a standard side-lift type and is also stowed in the luggage compartment when not in use.

## Section O.1

### MAINTENANCE

Attention should be paid to the following points with a view to obtaining the maximum mileage from the road wheels and tyres.

(1) Change the positions of the tyres on the car at regular intervals.

(2) Test the pressures of the tyres with an accurate pressure gauge. Restore any air lost. Inflate the spare wheel to the rear wheel pressure. It is not sufficient to make a visual examination of the tyres for correct inflation.

(3) Keep the treads free of grit, flint, stones, or other foreign matter, and repair cuts as soon as they are found.

(4) Clean the wheel rims and keep them free of rust. Repaint them when necessary.

(5) Keep the brakes and clutch correctly adjusted as fierceness or uneven action from either of these units has a detrimental effect on tyre wear.

(6) Suspect misalignment if rapid wear of the front tyres is noticed and correct the fault at once. See Section J for details of front wheel alignment.

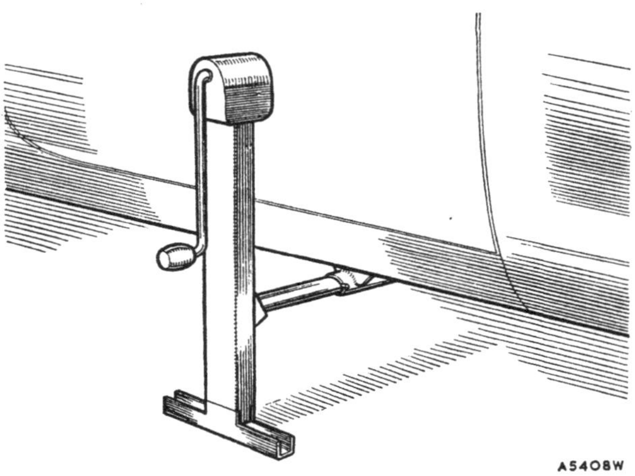


*Fig. O.1*
*The side-lift jack*

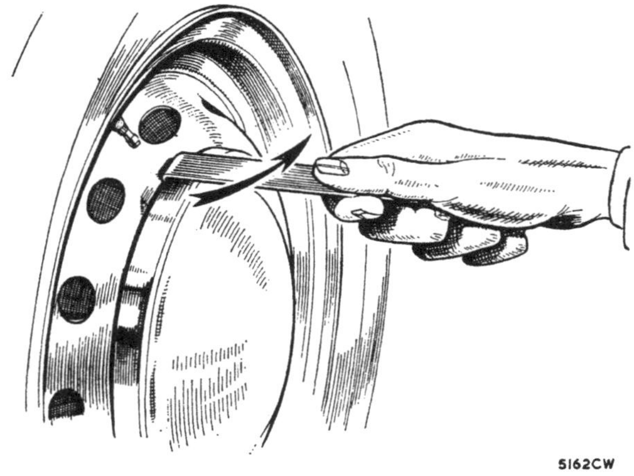


*Fig. O.2*
*Removing the hub cap (disc wheels)*

(7) Remove oil or grease with petrol sparingly applied and wiped off immediately.

(8) Avoid over- and under-inflation, kerbing, and other causes of severe impact.

(9) Repair any damage immediately it is discovered and remove tyres before they have worn too far for remoulding.

**Tyre replacement**

Radial-ply tyres should only be fitted in sets of four, although in certain circumstances it is permissible to fit a pair on the rear wheels; tyres of different construction must not be used on the same axle. A pair must never be fitted to the front wheels with conventional tyres at the rear. The positional changing of wheels must not be undertaken if radial-ply tyres have been to fitted the rear wheels only.

## Section O.2

### JACKING UP THE CAR

A jacking tube is fitted to each side of the car under the floor panel approximately midway along the body. The jack will lift one side of the car and thus enable a front or rear wheel to be changed.

Before jacking up the car apply the hand brake and if the car is standing on an incline chock the wheels. Place the jack in the appropriate jacking tube so that the top of the jack leans slightly outwards to allow for radial movement of the body when the car is lifted.

Remove the hub cover and slacken the wheel nuts a turn or two before commencing to lift the car.

Operate the jack until the road wheels are clear of the ground.

After changing the wheel lower the car and remove and stow the jack.

Do not use the jack as a support for the car when carrying out servicing operations underneath the car.

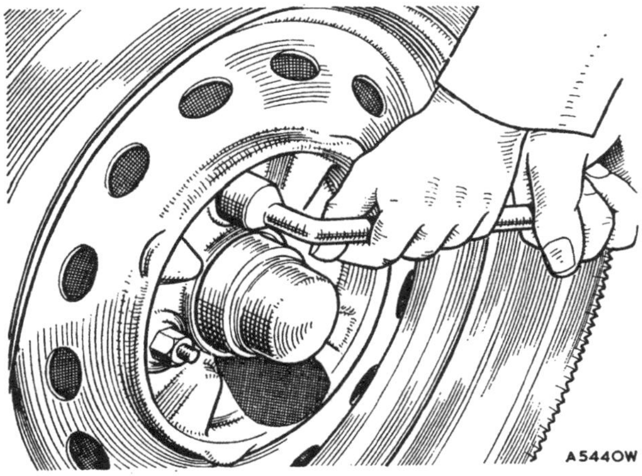


*Fig. O.3*
*Removing a road wheel (disc type)*

## Section O.3

### REMOVING AND REPLACING ROAD WHEELS

**Disc wheels**

(1) Insert the hub cover lever in the recess adjacent to the hub cover retaining studs and give it a sideways twist.

(2) Remove the four nuts securing the road wheels to the hub. The wheel nuts have right-hand threads.

(3) Lift the road wheel from the studs.

Reverse this procedure when replacing the road wheel and ensure that the nuts are refitted with their tapered side towards the wheel.

Tighten the nuts to a torque figure of between 60 and 62·5 lb. ft. (8·3 and 8·65 kg. m.). Do not overtighten.

To refit the hub disc place the rim of the disc over two of the buttons on the wheel centre and give the disc a sharp blow with the fist over the third button.

**Wire wheels**

Use a copper mallet to slacken the winged hub nut securing the wheel to the splines on the hub extension. The hub nuts on the left-hand side of the car have right-hand threads (unscrew anti-clockwise) and the nuts on the right-hand side of the car have left-hand threads (unscrew clockwise).

## Section O.4

### VALVES

Valve caps, in addition to preventing dirt from entering the valve, form a secondary air seal and should always be fitted. The valves may be tested for airtightness by rotating the wheel until the valve is at the top and inserting its end in an eggcup full of water. If bubbles appear the seating is faulty and should be removed and a new one fitted. It is advisable to change the valve interiors every 12 months.

## Section O.5

### TYRE REMOVAL

Remove all valve parts to deflate the tyre completely, and push both edges into the base of the rim at a point diametrically opposite the valve. Lever the cover edge, near the valve, over the rim of the wheel, using two levers at intervals of 6 in. (15 cm.) apart.

NOTE.—**Do not attempt to stretch the edges of the tyre cover over the rim edge.**

Force is entirely unnecessary and is detrimental, as it tends to damage the wired edges. Fitting or removing is quite easy if the tyre edges are carefully adjusted into the rim base; if found difficult, the operation is not being performed correctly.

Remove the tube carefully; do not pull on the valve. Stand the tyre and wheel upright, keeping the bead on the base of the rim. Lever the bead over the rim flange and at the same time push the wheel away from the cover with the other hand.

## Section O.6

### INSPECTION OF WHEELS AND TYRES

In addition to a visual examination a detailed inspection of wheels and tyres may be carried out where a specialist and the necessary specialist equipment is available for this purpose.

**Disc wheels**

(1) Clean the wheel rim with a wire brush and remove all traces of corrosion.

(2) Examine the rim and the welds or rivets securing the rim to the wheel centre. Damage to the rim may be repaired provided it is slight and is confined to the flange tip area.

*Fig. O.4*
*Removing a road wheel (wire type)*

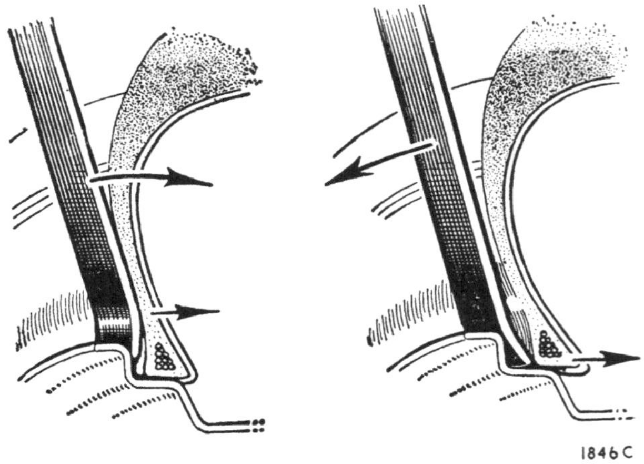


*Fig. O.5*

*The tyres have wired edges and no attempt must be made to stretch them. If the cover fits tightly on the rim seating it should be freed by using the tyre levers as indicated*

(3) Inspect the wheel centre for flaws, cracks, or fractures.
(4) Check the wheel wobble. This is the lateral variation measured on the inside vertical face of the flange and should not exceed ·055 in. (1·4 mm.). An additional ·015 in. (·38 mm.) is permissible at the welded joint.
(5) Check the wheel lift. This is the difference in height between the high and low points measured at any location on either tyre seat on a truly mounted and revolving wheel. The difference should not exceed ·055 in. (1·4 mm.). An additional ·015 in. (·38 mm.) is permissible at the welded joint.

**Wire wheels**

(1) Examine the wheel as in (1) and (2) above.
(2) Check the spokes and nipples for security and damage. Any unserviceable parts must be renewed.
(3) Examine the splines in the wheel centre for wear. If the splines are worn renew the hub centre.

**Tyres**

Examine all tyres internally and externally for:
(1) The degree and regularity of tread wear.
(2) Cuts, penetrations, and impact bruises.
(3) Kerb damage to walls and shoulders.
(4) Oil and grease contamination.
(5) Bead damage and chafing.

## Section O.7

### REPAIR OF TYRES AND TUBES

Minor injuries confined to the tread rubber, such as from small pieces of glass or road dressing material, require no attention other than the removal of the objects. Cold filling compound of 'stopping' is unnecessary in such cases.

More severe tread cuts and wall damage, particularly if they penetrate to the outer ply of the fabric casing, require vulcanized repairs. The Dunlop Spot Vulcanizing Unit is available for this purpose and it is also suitable for all types of tube repairs.

Injuries which extend into or through the casing, except clean nail holes, seriously weaken the tyre. Satisfactory repair necessitates new fabric being built in and vulcanized. This requires expensive plant and should be undertaken by a tyre repair specialist.

Loose gaiters and 'stick-in' fabric repair patches are not satisfactory substitutes for vulcanized repairs and should be used only as a temporary 'get-you-home' measure if the tyre has any appreciable tread remaining. They can often be used successfully in tyres which are nearly worn out and which are not worth the cost of vulcanized repairs.

Clean nail holes do not necessitate cover repairs. If a nail has penetrated the cover the hole should be sealed by a tube patch attached to the inside of the casing. This will protect the tube from possible chafing at that point.

If nail holes are not clean, and particularly if frayed or fractured cords are visible inside the tyre, expert advice should be sought.

**Repairing tubes**

Punctures or injuries must be vulcanized. Ordinary patches should only be used for emergencies and cannot be relied upon.

## Section O.8

### FITTING TYRES AND TUBES

Inspect the inside of the cover carefully and remove all dirt. The wheel rim must be clean, free from dust and undamaged.

In order to obtain good steering it is of importance to ensure that the wheels, with tyres fitted, are in good balance. To assist this, the tyre manufacturers are now

*Fig. O.6*

*Push the cover bead into the well-base of the wheel as indicated by the arrow*

marking their tyres with a white spot in the neighbourhood of the bead at the lightest point of the cover; similarly, they are marking the inner tubes with spots to indicate their heaviest point. When tyres are assembled care must therefore be taken to see that they are assembled with the spots on the cover coinciding with the spots on the tube.

It must be noted, in addition, that special balancing discs are fitted to the inside of the cover casing in some cases and that these should on no account be removed as the tyre balance will be upset if this is done. These balance discs are not repair patches and do not indicate any fault in the tyre.

Dust the inside of the cover with french chalk. Inflate the tube until it begins to round out, then insert it in the cover.

Apply a frothy solution of soap and water generously around the centre base of the **tube,** extending upwards between the tyre beads and the tube itself for at least 2 in. (50 mm.) on both sides. Also apply the solution to the bottom and outside of the tyre beads. Do not allow the solution to run into the crown of the tyre. The solution must be strong enough to feel slippery when the fingers are wetted with the solution and rubbed together.

Mount the tyre on the rim immediately, whilst the soap solution is still wet.

Push one edge of the cover over the edge of the rim. It will go quite easily if the part first put on is fitted on the opposite side of the valve and is pushed right down into the rim base. Move it round so that its balance spots coincide with those of the inner tube when it is inserted with the valve passing through the hole in the rim. (Take care that the valve, fitted in the side of the tube, is on the correct side of the rim.)

Before inflating, be sure that the tyre beads are clear of the well of the rim all the way round and **push the valve into the tyre as far as possible in order to ensure the tube is not trapped between the bead and the rim, then pull it out again into its correct position.**

*Fig. O.7*

*Lever the cover edge over the rim all round the wheel*

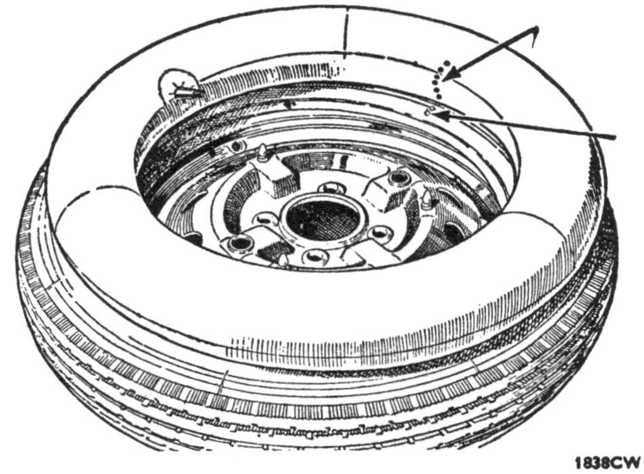


*Fig. O.8*

*Balance marks on tyre and tube*

Inflate slowly until the beads are fully seated.

Remove the valve core to **deflate the tube completely** to permit any stretched portions of the tube to readjust themselves in the cover and relieve any local strains in the tube.

Reinflate to the correct working pressure (see 'GENERAL DATA'). This procedure must be followed whenever a tube is fitted.

**(See Editor's notes at end of section.)**

## Section O.9

### WHEEL AND TYRE BALANCE

Unbalance in wheel and tyre assemblies may be responsible for various effects such as wheel wobble, abnormal wear of tyres and suspension parts, vibration in the steering or, in extreme cases, in the whole car. If any of these faults develop, for which no other cause can be found, wheel and tyre balance should be checked and corrected according to instructions supplied by the manufacturer of the balancing machine.

When wheels are to be re-balanced it is essential that the weight of the car be removed from the tyres as soon as possible after a run so that temporary flat spots do not form on the tyres. Nylon tyres are particularly prone to this and re-balancing with the tyres in this condition is pointless.

## Section O.10

### FACTORS AFFECTING TYRE LIFE

**Inflation pressures**

All other conditions being favourable there is an average loss of 13 per cent. tread mileage for every 10 per cent. reduction in inflation pressure below the recommended figure.

A tyre is designed so that there is a minimum pattern shuffle on the road surface and a suitable distribution of load over the tyre's contact area when deflection is correct.

Moderate underinflation causes an increased rate of tread wear although the tyre's appearance may remain normal. Severe and persistent underinflation produces unmistakable evidence on the tread. It also causes structural failure due to excessive friction and temperature within the casing.

Pressures which are higher than those recommended for the car reduce comfort. They may also reduce tread life due to a concentration of the load and wear on a smaller area of tread, aggravated by increased wheel bounce on uneven road surfaces. Excessive pressures overstrain the casing cords, in addition to causing rapid wear, and the tyres are more susceptible to impact fractures and cuts.

**Effect of temperature**

Air expands with heating and tyre pressures increase as the tyres warm up. Pressures increase more in hot weather than in cold weather and as a result of high speed. These factors are taken into account when designing the tyre and in preparing Load Pressure schedules.

Pressures in warm tyres should not be reduced to standard pressure for cold tyres. 'Bleeding' the tyres increases their deflections and causes their temperatures to climb still higher. Their tyres will also be underinflated when they have cooled.

**Climatic conditions**

The rate of tread wear during a reasonably dry and warm summer can be twice as great as during an average winter.

Water is a rubber lubricant and tread abrasion is much less on wet roads than on dry roads. In addition resistance of the tread to abrasion decreases with increase in temperature.

When a tyre is new its thickness and pattern depth are at their greatest. It follows that heat generation and pattern distortion due to flexing, cornering, driving, and braking are greater than when the tyre is part worn.

Higher tread mileages will usually be obtained if new tyres are fitted in the autumn or winter rather than in the spring or summer. This practice also tends to reduce the risk of road delays because tyres are more easily cut and penetrated when they are wet than when they are dry. It is, therefore, advantageous to have maximum tread thickness during wet seasons of the year.

**Road surface**

Present day roads generally have better non-skid surfaces than formerly. This factor, combined with improved car performance, has tended to cause faster tyre wear, although developments in tread compounds and patterns have done much to offset the full effects.

Road surfaces vary widely between one part of the country and another, often due to surfacing with local material. In some areas the surface dressing is coarser than others; the material may be comparatively harmless rounded gravel, or more abrasive crushed granite, or knife-edged flint. Examples of surfaces producing very slow tyre wear are smooth stone setts and wood blocks, but their non-skid properties are poor.

Bends and corners are severe on tyres because a car can be steered only by misaligning its wheels relative to the direction of the car. This condition applies to the rear tyres as well as the front tyres. The resulting tyre slip and distortion increase the rate of wear according to speed, load, road camber, and other factors.

The effect of hills, causing increased driving and braking torques with which the tyres must cope, needs no elaboration.

**Impact fractures**

In order to provide adequate strength, resistance to wear, stability, road grip, and other necessary qualities, a tyre has a certain thickness and stiffness. Excessive and sudden local distortion, such as may result from striking a kerb, a large stone or brick, an upstanding manhole cover, or a deep pothole, may fracture the casing cords.

Impact fractures often puzzle the car owner because the tyre and road spring may have absorbed the impact without his being aware of anything unusual. Only one or two casing cords may be fractured by the blow and the weakened tyre fails some time later. Generally, there is no clear evidence on the outside of the tyre unless the object has been sufficiently sharp to cut it.

**Wheel alignment and road camber**

It is very important that correct wheel alignment should be maintained. Misalignment causes a tyre tread to be scrubbed off laterally because the natural direction of the wheel differs from that of the car.

An upstanding fin on the edge of each pattern rib is a sure sign of misalignment and it is possible to determine from the position of the 'fins' whether the wheels are toed in or toed out. Fins on the inside edges of the pattern ribs—nearest to the car—and particularly on the off-side tyre, indicate toe-out.

With minor misalignment the evidence is less noticeable and sharp pattern edges may be caused by road camber even when wheel alignment is correct. In such cases it is better to make sure by checking with an alignment gauge.

Road camber affects the direction of the car by imposing a side thrust and if left to follow its natural course the car will drift to the near side. This is instinctively corrected by steering towards the road centre. As a result the car runs crab-wise.

The near front tyre sometimes persists in wearing faster and more unevenly than the other tyres even when the mechanical condition of the car and tyre maintenance are satisfactory. The more severe the average road camber the more marked will this tendency be. This is an additional reason for the regular interchange of tyres.

**Camber angle**

This angle normally requires no attention unless disturbed by a severe impact; however, it is always advisable to check this angle if steering irregularities develop.

Wheel camber usually combined with road camber, causes a wheel to try to turn in the direction of lean, due to one side of the tread attempting to make more revolutions per mile than the other side. The resulting increased tread shuffle on the road and the off-centre tyre loading tend to cause rapid and one-sided wear. If wheel camber is excessive for any reason the rapid and one-sided tyre wear will be correspondingly greater. Unequal cambers introduce unbalanced forces which try to steer the car one way or the other. This must be countered by steering in the opposite direction which results in faster tread wear.

## Section O.11

### HIGH-SPEED TYRE PRESSURE CONDITIONS

During high speed driving and in competition work, the tyres are subjected to greater stresses than those produced during ordinary driving.

Many factors, some probably as important as the physical characteristics of the tyre itself, affect the speed at which it should be driven: road surface, air temperature, and in particular the duration of high-speed driving. However, a normal tyre in good condition and at the correct pressure can be relied upon to perform satisfactorily at speeds up to 80 m.p.h. (144 km.p.h.) and intermittently in excess of this by 10 m.p.h. (18 km.p.h.). If the car is to be driven consistently at speeds near the maximum of which it is capable special tyres should be fitted on the advice of the tyre manufacturers.

When it is intended to indulge in high speeds or competition work we advise the use of first tread tyres.

### EDITOR'S NOTES

**O. Wheels and Tires**
*Fitting tires and tubes*
If available, professional tire changing equipment should be used rather than hand tools.

# SECTION R

## THE BODY

## Section R.1

### FRONT GRILLE

**Removing**

Unscrew the three screws securing the steady brackets. Remove the three hexagon-headed screws securing the bottom of the grille. Remove the motif, steady brackets, and the rubber buffers.

**GHN5 and GHD5 cars**

(1) Remove the three screws in the bonnet lock platform, and the two self-tapping screws securing the grille outer surround to the body.
(2) Prise out the three rubber body plugs and remove the three screws securing the bottom of the grille to the radiator duct panel.

## Section R.2

### BONNET AND BONNET LOCK

**Removing**

The bonnet is of light-alloy material. Mark the hinges, support it in the open position and remove the two nuts, washers, and screws that secure each hinge to the bonnet. Disengage the stay and lift the bonnet from the car.

Slacken the nut from the bonnet lock pin and withdraw the pin, thimble, and spring. Remove the three screws securing the safety catch to the bonnet and detach the safety catch.

From the bonnet lock platform remove the two screws securing the safety catch bracket and take off the bracket. Release the bonnet lock control cable clamp screw and pull the cable from the bonnet lock plate. Unscrew the three screws securing the locating cup and bonnet lock to the platform and remove the cup and lock.

**Refitting**

Refitting is a reversal of the removal sequence, but ensure that the bonnet lock, safety catch, and the bonnet are correctly aligned before finally tightening the securing screws. After assembly adjust the latch pin to obtain ease of closing, lubricate the lock, catch, and hinges and check them for correct operation.

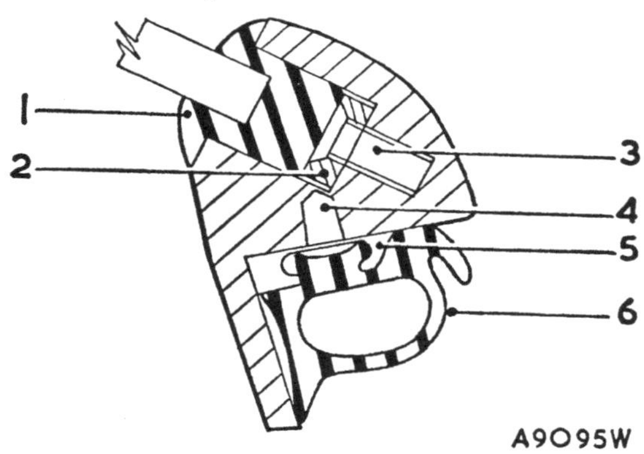


*Fig. R.1*

*A section through a windscreen pillar, showing:*

1. Glazing rubber.
2. Bottom reinforcement.
3. Screw—reinforcement to pillar.
4. Rivet—seal retainer.
5. Seal retainer.
6. Seal.

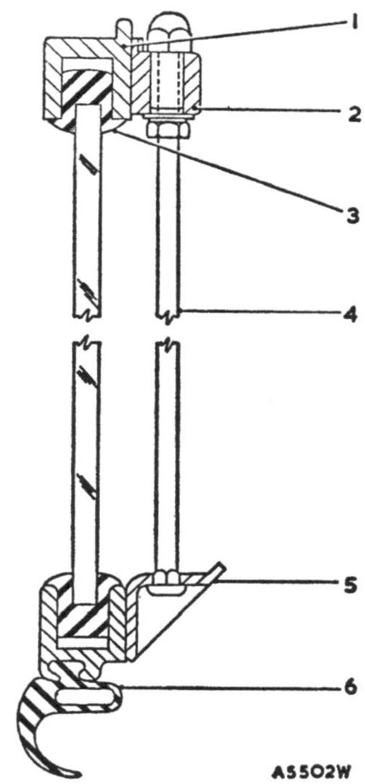


*Fig. R.2*

*A section through the windscreen, showing:*

1. Top rail.
2. Centre rod upper bracket.
3. Glazing rubber.
4. Centre rod.
5. Bottom bracket.
6. Seal.

## Section R.3

### WINDSCREEN
### (Tourer)

**Removing**

Unscrew the two screws securing the lower centre rod bracket to the body. Remove the fascia panel (Section R.20) to gain access to the windscreen securing bolts, and then remove the bolts. Lift the windscreen from the car.

**Dismantling**

Remove the dome nut from the centre rod, draw the rod through the upper bracket, remove the plain washer, spring washer, and nut, and then withdraw the rod from the bottom bracket.

Remove the bottom rail sealing rubber and the two screws securing each corner of the bottom rail to the lower reinforcement pieces. Remove the three screws securing each end of the top rail to the pillars. Gently ease the two side pillars from the top and bottom rails and the rails from the windscreen glass.

**Reassembling**

Fit the glazing rubber to the screen and then mark the centre of the glass with a wax crayon to align it with the centre of the top and bottom rails. Fit the top and bottom rails and ensure that the centres of the brackets are accurately aligned with the centre-line of the glass screen. Retain the rails in this position by means of a clamp lightly applied or by temporarily assembling the centre rod.

Starting at the bottom of the pillar, carefully tap it into place with the palm of the hand, working alternately from bottom to top. When both pillars have been correctly located refit the securing screws and the lower sealing rubber.

Assemble the centre rod and ensure that the dome nut is not overtightened.

**Refitting**

The foot of each pillar is supported in the body by a metal and a fibre packing piece. The packing pieces are screwed to the side of the body and should not normally be removed. If a new windscreen is being fitted or extensive body repairs carried out, check the fit of the windscreen by fitting it to the car and checking the clearance between the body and the foot of each pillar. Adjust the fit by inserting or removing the $\frac{3}{32}$ in. (2·38 mm.) fibre packing pieces.

Place the bottom bracket packing piece on the body and the sealing grommets on the windscreen pillars. Fit the windscreen to the body and align the holes in the pillar feet with the packing pieces. Place a special washer on each of the four bolts and screw the bolts through the packing pieces and into the pillars. Spread the bottom rail sealing rubber, secure the centre rod bottom bracket to the body, and finally tighten the windscreen pillar securing bolts. Close the doors and check the fit of the ventilator window against the pillar sealing rubbers.

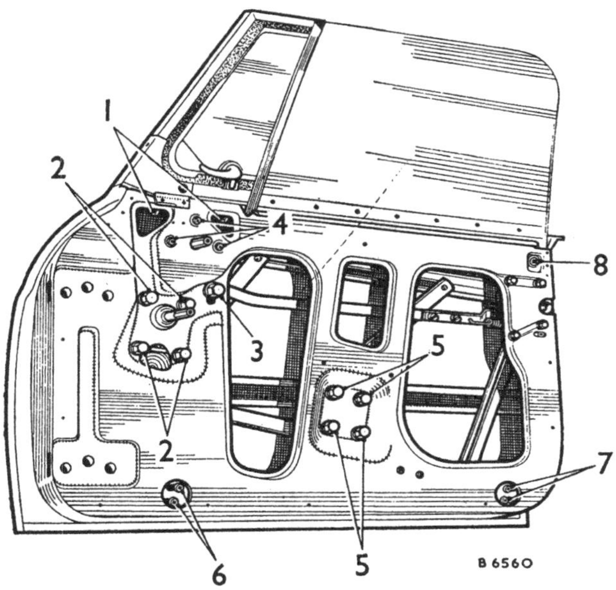


*Fig. R.3*

*The door assembly (early tourer models)*

1. Ventilator securing nuts.
2. Regulator securing set screws.
3. Regulator arm stop.
4. Door lock remote control securing screws.
5. Regulator extension securing set screws.
6. Front door glass mounting bracket securing screws.
7. Rear door glass mounting bracket securing screws.
8. Door glass channel securing screws.

## Section R.4

### DOOR TRIM PAD

**Removing**

Remove the inner door handle, door pull, and window regulator handle (Section R.8). Remove the waist rail securing screws and the screws securing the trim pad to the door panel. Lift off the waist rail and trim pad. If necessary, remove the inside trim liner which is secured to the door panel with an adhesive.

*Tourer (early models)*

Remove the inner door handle door pull and window regulator handle (Section R.8). Remove the locking knob from the passenger door. Remove the drive screws securing the trim pad and the ends of the waist rail to the door panel. Lift off the trim pad to expose the centre screw of the waist rail which may now be removed together with the rail. If necessary remove the inside trim liner which is secured to the door panel with an adhesive.

**Refitting**

Reverse the removal sequence.

## Section R.5

### VENTILATORS

**Removing**

Remove the inner door handle, door pull, and window regulator handle (Section R.8). Remove the door trim pad and door glass assembly (Sections R.4 and R.6). Remove the set screws and nuts securing the ventilator top to the door. Remove the ventilator steady set screws and the set screws securing the front door glass channel to the bottom of the door. Lift out the ventilator assembly.

*Tourer (early models)*

Remove the inner door handle, door pull and window regulator handle (Section R.8). Remove the door trim pad assembly (Section R4.). Release the lower end of the door glass front channel from its fixing bracket on the door. Remove the screws adjacent to the ventilator sleeve nuts and remove the sleeve nuts. Wind the door glass to its lowest position and remove the ventilator from the door.

**Refitting**

Reverse the removing procedure.

## Section R.6

### DOOR GLASS

**Removing**

Remove the inner door handle, door pull, and window regulator handle (Section R.8). Remove the door trim pad (Section R.4). Remove the regulator securing screws and the regulator extension securing screws. Release the window regulator arc from the bottom of the door glass, lift the door glass up to clear the regulator and remove the regulator and extension assembly through the door panel aperture. Remove the nut securing the door glass rear guide channel to the door. Lift out the door glass.

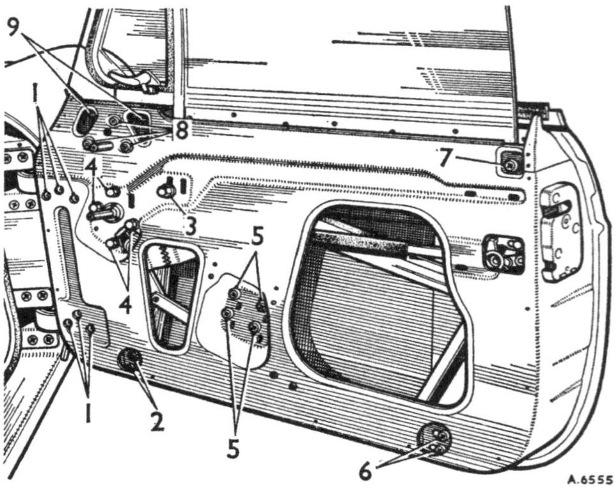


*Fig. R.4*

*The door assembly (later models)*

1. Door hinge securing set screws.
2. Front door glass mounting bracket securing screws.
3. Regulator arm stop.
4. Regulator securing set screws.
5. Regulator extension securing set screws.
6. Rear door glass mounting bracket securing screw.
7. Door glass channel securing screws.
8. Door lock remote control securing screws.
9. Ventilator securing nuts.

*Tourer (early models)*

Remove the inner door handle, door pull, and window regulator handle (Section R.8). Remove the door trim pad (Section R.4). Remove the door glass rear channel top and bottom fixings with the glass in the upper position. Release the regulator and the regulator extension and slide the arms forward to disengage the studs from the door glass bottom channel, lift the glass from the door.

**Refitting**

Reverse the removal procedure.

## Section R.7

### DOOR GLASS REGULATORS

**Removing**

Remove the inner door handle, door pull, and window regulator handle (Section R.8). Remove the door trim pad (Section R.4). Remove the regulator securing screws and the regulator extension securing screws. Release the window regulator arc from the bottom of the door glass, lift the door glass up to clear the regulator and remove the regulator and extension assembly through the door panel aperture.

**Refitting**

Reverse the removal procedure.

## Section R.8

### WINDOW REGULATORS AND INTERIOR DOOR HANDLES

**Removing**

Remove the centre screw, spring and washer and pull off the handle and finisher.

Remove the two screws securing the door pull to the door assembles and lift off the door pull.

Remove the centre screw securing the locking knob on the passenger door (early touring models).

**Refitting**

Reverse the removal procedure.

## Section R.9

### STRIKERS

**Removing**

Mark the position of the striker on the door pillar to assist when refitting. Remove the set screws securing the striker to the door pillar and lift away the assembly together with the packing pieces.

**Refitting**

Reverse the removal procedure, and adjust as necessary.

**Checking and adjusting**

Check that a clearance of $\frac{1}{32}$ in. (1 mm.) to $\frac{1}{16}$ in. (1·6 mm.) exists between the striker and lock face when the door is closed. Adjust the clearance by adding or removing packing shims from behind the striker.

Ensure that there is sufficient clearance behind the latch cam to accommodate the anti-burst strap.

Adjust the striker until the door closes easily without rattling, lifting, or dropping. The striker must be secured in the horizontal plane relative to the door axis.

Close the door and lock it using the inside handle, unlock the door from the outside using the key, open the door by depressing the outside push-button, close the door again, then open it using the inside handle.

## Section R.10

### DOOR LOCKS

**IMPORTANT.—Before removing any part of the door lock mechanism because of faulty operation, first check that the condition is not caused by bad installation.**

The following points should always be checked first.

(*a*) Striker and latch cam clearance (see Section R.9).
(*b*) Remote-control adjustment (see Section R.11).
(*c*) Outer door handle plunger clearance (see Section R.12).

**Removing**

Remove the inner door handle, door pull, and window regulator handle (Section R.8). Remove the door trim pad (Section R.4).

Remove the 'C' clip retaining the remote control link to the door lock and detach the link. Remove the screws securing the lock to the door and withdraw the lock assembly.

*Tourer (early models)*

Remove the inner door handle, door pull, and window regulator handle (Section R.8). Remove the door trim pad and door glass assembly (Sections R.4 and R.6).

Disconnect the remote-control link by unscrewing the two screws, remove the three screws securing the remote-control mechanism to the door, and withdraw the mechanism from the rear access panel.

Remove the two screws securing the outer door handle, disengage the handle from the lock, and remove it from the door. From inside the door panel slide the retaining clip from the barrel of the private lock and withdraw the lock barrel and the coupling link from the outside of the door.

Remove the four screws securing the lock to the door and withdraw the lock.

**Refitting**

Reverse the removing operations, noting that the latch must be in the open position when refitting the lock to the door, and that the locking lever must be engaged with the private lock operating fork. Check for correct functioning (Section R.9).

**Tourer (early models)**

Reverse the removel sequence. Elongated holes in the door panel will sllow the remote-control mechanism to be correctly adjusted. After refitting lubricate the lock assembly and check it for correct functioning.

## Section R.11

### DOOR LOCK REMOTE CONTROL UNIT (Later Models and GT)

**Removing**

Remove the inner door handle, door pull, and window regulator handle (Section R.8). Remove the trim pad (Section R.4). Detach the remote control link from the door lock by removing the 'C' retaining clip.

Remove the screws securing the remote control unit to the door.

Remove the moulded rubber hood covering the upper end of the window run channel, from the top of the front quarter-light.

Remove the nut securing the window regulator upper limit stop and remove the stop. Wind the window up until the lifting studs can be disengaged from the lifting channel.

Remove the screws securing the lower end of the front window run channel. Retain the glass in the up position with wedges or remove it from the door.

Remove the window regulator securing screws. Press the window regulator and front window run channel away from the inner door panel and withdraw the remote control unit down between the window regulator and inner door panel, and out of the door through the large aperture in the inner door panel.

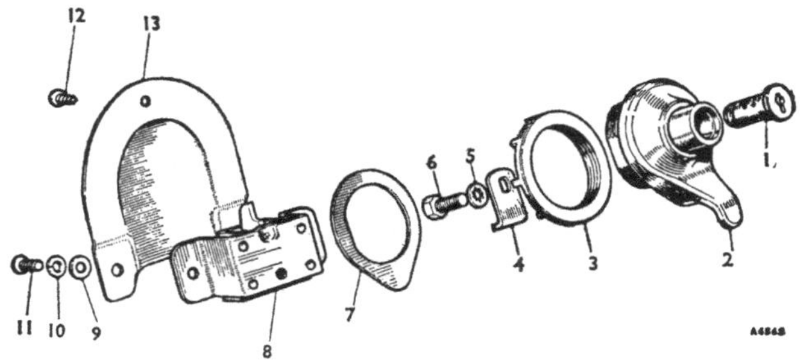


*Fig. R.5*

*The luggage compartment lock*

1. Lock barrel.
2. Push-button assembly.
3. Nut.
4. Operating lever.
5. Washer.
6. Set screw.
7. Seating washer.
8. Latch lock.
9. Plain washer.
10. Spring washer.
11. Set screw.
12. Screw. } GT only.
13. Latch cover. } GT only.

**Refitting**

Reverse the removing sequence, noting that elongated holes in the inner door panel allow the remote control unit to be correctly adjusted with the spring loaded lever of the remote control link just contacting the spring.

## Section R.12

### OUTER DOOR HANDLES

**Removing**

Remove the inner door handle, door pull, and window regulator handle (Section R.8). Remove the trim pad (Section R.4). With the door glass in the fully up position remove the nuts and washers securing the door handle to the door and lift off the handle.

*Tourer (early models)*

Remove the door panel to door handle securing set screws and lift away the outer handle.

**Refitting**

Reverse the removing procedure, noting the following. A minimum clearance of $\frac{1}{32}$ in. (1 mm.) must be maintained between the push-button plunger bolt and the lock contactor. Adjust the clearance by turning the plunger bolt in the required direction, after slackening the locknut. Retighten the locknut when the correct clearance has been obtained.

## Section R.13

### PRIVATE LOCK (Later Models and GT)

**Removing**

Close the window and remove the inner door handle, door pull, and window regulator handle (Section R.8). Remove the trim pad (Section R.4).

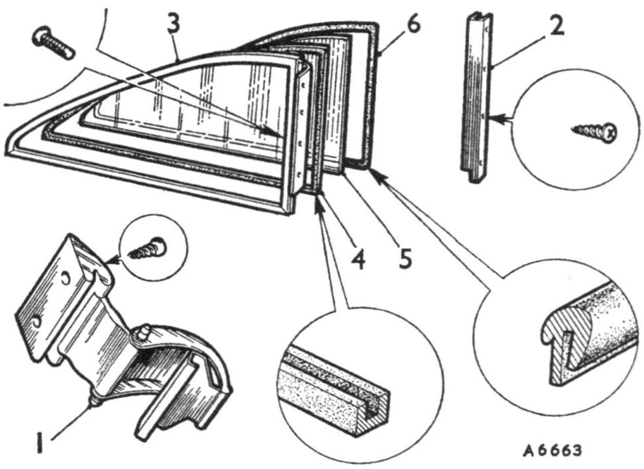


*Fig. R.*6
*The rear quarter-light* (*GT only*)

1. Toggle catch.
2. Hinge.
3. Frame.
4. Glazing rubber.
5. Glass.
6. Sealing rubber.

From inside the door, compress the legs of the retainer spring collar and withdraw the lock barrel from the outside of the door.

**Refitting**
Ensure that the spring collar is in place on the lock barrel, enter the lock barrel into the door aperture with its operating fork inclined away from the door shut face, and press the lock barrel firmly into position.

## Section R.14

### SEATS

**Removing (front)**
Release the seat catch and push the seat fully back to gain access to the front screws. Remove the screws and push the seat fully forward; remove the rear screws and lift the seat and seat runners from the car. Slide the seat from its runners. Retain the wooden strips and spacers.

**Rear (GT)**
Unlock the seat squab and move it forward to a horizontal position. Lift the luggage compartment floor carpet and remove the screws securing the hinges to the luggage compartment floor. Lift away the seat assembly.

**Refitting**
Refitting is a reversal of the removal sequence. Before finally tightening the front seat runners check the seats for ease of movement and correct alignment.

## Section R.15

### LUGGAGE COMPARTMENT LOCK

**Do not in any circumstances close the luggage compartment lid when the lock mechanism is in the process of being removed or refitted.**

**Removing**
Remove the lock cover (GT only). Remove the set screw securing the lock assembly to the luggage compartment lid. Remove the notched locking ring and withdraw the push-button and lock assembly.

**Refitting**
Reverse the removal procedure.

## Section R.16

### LUGGAGE COMPARTMENT LID

**Removing**
*Tourer*
Retain the lid in the open position with the stay and mark the hinges to assist when refitting. Remove the lid to hinge securing screws, detach the stay and lift the lid from the body.

*GT*
Retain the lid in the open position and mark the hinges to assist when refitting. Take the weight of the lid and remove the hinge securing screws. Remove the set screws securing the lift assisting springs to the body and lift away the lid.

**Refitting**
Refitting is a reversal of the removal procedure. Check the lid for fit to the body and check the operation of the latch before finally tightening the hinge to lid securing screws.

## Section R.17

### REAR QUARTER-LIGHT (GT)

**Removing**
Remove the finisher securing screws. Remove the screws securing the toggle catch to the body. Remove the hinge securing screw and lift away the assembly.

**Refitting**
Reverse the removal procedure.

## Section R.18

### DOORS AND HINGES

**Removing**
Remove the inner door handle, door pull, and window regulator handle (Section R.8). Remove the door trim pad (Section R.4).

Remove the set screws securing the hinge to the door and lift away the door.

To remove the hinge; from the under side of the front wing remove the six screws securing the splash panel to the body and remove the panel to gain access to the hinge bracket nuts; remove the nuts.

Unscrew the four screws securing the hinge bracket to the body and remove the hinge and bracket assembly.

**Refitting**

Refitting is a reversal of the removal sequence, but before finally tightening the hinge leaf screws ensure that the door lock engages correctly and that the door is correctly positioned with the body.

## Section R.19

### WINDSCREEN AND BACK-LIGHT GLASSES (GT)

**Removing**

Lift the windscreen wiper arms clear of the glass.

Remove the finisher strip.

Prise up the end of the locking filler and carefully pull it from the channel in the surround rubber.

Press the glass from inside the car, commencing at a corner, and ease the surround rubber from the metal edge of the body aperture.

**Refitting**

If the glass has been broken, remove any pieces which remain in the channel. Examine the rubber, and use a new rubber should there be any signs of damage or deterioration.

Fit the surround rubber to the body aperture and lubricate the 'glass' channel with a soap-and-water solution. Place the glass into the lower channel of the rubber surround and commence at the corner to lift the lip of the rubber over the glass, using Service tool 18G 468. Use the short peg on the handle of the installation tool for this purpose.

Apply a soap-and-water solution to the locking filler strip channel to assist in fitting the strip.

Using Service tool 18G 468 with adaptor 18G 468 A, thread the end of the filler strip through the eye of the adaptor and under the roller. Lay the filler strip in position over the groove in the surround rubber. Insert the eye of the tool in the filler strip groove, hold the filler strip in position, and commence to push the tool along the groove, rolling the strip into position. A slight side-to-side action will assist when rounding the corners.

*Fig. R.7*

*Use Service tool 18G 468 to ease the channel lip over the windscreen or backlight glass*

*Fig. R.8*

*The use of the glazing tool and eye to thread the locking filler strip into the rubber channel*

After completing the circuit cut off any surplus strip to make a neat butt joint at the centre-line of the glass. Ensure that the filler strip is fitted with its thicker section facing towards the glass of the windscreen.

NOTE.—**In the event of windscreen breakage, minute particles of glass fall into the windscreen demister ducts and tubes. When the blower motor is switched on these particles are discharged into the driver's or passenger's face. The demister ducts and tubing must be disconnected from the heater box and blown out before the new windscreen glass is fitted.**

## Section R.20

### FASCIA PANEL, INSTRUMENTS, AND CONTROLS

**Removing**

Disconnect the battery, drain the cooling system, and remove the temperature gauge capillary capsule from the cylinder head.

Remove the speaker panel surround.

Release the steering-column upper fixing bracket and lower the steering-column.

Disconnect the choke cable from the air cleaner bracket, unscrew the outer cable fixing nut behind the fascia, and withdraw the choke cable complete.

Depress the plunger in each heater control knob and withdraw the knobs. Unscrew the heater control securing nuts and detach the controls from the rear of the fascia panel.

Unscrew the bezel nut securing the windscreen washer plunger and withdraw the plunger from the rear of the fascia.

Disconnect the oil pressure gauge pipe union.

Remove the nuts and washers securing the top of the fascia to the body, the screw and nut securing the steady bracket to the glovebox attachment, and the two screws securing the bottom centre of the panel to the speaker panel. Retain the weld nut assembly.

Pull the fascia forward and disconnect the speedometer and revolution indicator drives.

Pull the panel and warning lamps from their housings and disconnect the electrical connections from the switches and the fuel gauge.

Thread the temperature gauge capillary through the bulkhead and lift the fascia panel from the car.

Release the fixing straps on the oil pressure and temperature gauge, speedometer, revolution indicator, and fuel gauge and remove the instruments from the panel.

Using Service tools 18G 670 and 18G 671, unscrew the bezel nuts from the ignition and the toggle switches and withdraw the switches from the panel.

Unscrew the two screws securing the map-reading lamp and remove the lamp.

Depress the plunger in the panel lamp and map-reading lamp switches and remove the knobs, locknuts, and switches.

**Refitting**

Refitting of the panel is a reversal of the removal sequence, but refer to the wiring diagrams before making the electrical connections.

After refitting the panel reconnect the batteries and refill the cooling system. Test all switches and controls.

## Section R.21

### BODY ALIGNMENT

To check the body alignment of a car which has been damaged when the correct jig is not available, a system of diagonal and measurement checks from points projected from the underframe onto a level floor is used (see pages R.12 and R.13).

To ensure that the alignment check is carried out accurately, the vehicle must be raised so that it is parallel with the floor, both from side-to-side and from front-to-rear. Use the comparative measurements given on page R.12 to do this. Lift the vehicle to a convenient working height and adjust the front or rear of the vehicle until the points given on page R.12 for the front and rear on both sides of the car are in the correct vertical position relative to each other; for example, if the rear points (J) are 36 in. (91·4 cm.) from the floor, as the rear points are $\frac{49}{64}$ in. (19·45 mm.) higher in relation to the front points (A), the front points must be set at a height of 35$\frac{15}{64}$ in. (89·5 cm.) above the floor.

At the same time, it will be helpful to check the **relative** heights of all the intermediate points given on page R.12 so that any distortion of the car in the vertical plane will be ascertained.

Chalk over the area of the floor directly below the points shown on page R.13. Using a plumb-line, project the points from the car onto the floor, marking the positions with a pencilled cross. The centre between each pair of points can be established by means of a large pair of compasses, and the central points marked on the floor. In addition, diagonals can be determined between any two pairs of points and the points of intersection marked on the floor. At this stage a length of thin cord, covered with chalk, can be held by two operators in such a position that it passes through as many of the central points and intersections marked as possible. While the cord is held taut a third operator raises the centre of it and then allows it to spring back smartly to the floor. If the resulting white line passes through all the points the body alignment is satisfactory. Any points through which the white line does not pass will be in a position where the underframe is out of alignment.

Considerable deviations in the transverse and longitudinal measurements, given on pages R.12 and R.13, confirm body misalignment. It must be understood that allowance be made for normal manufacturing tolerances and that a reasonable departure from nominal dimensions can be permitted without detriment to road performance.

# Section R.22

## PAINT REFINISHING INSTRUCTIONS

## Section R.22

### PAINT REFINISHING INSTRUCTIONS

| *Operation* | *Material* | *Thinning* | *Drying times* | *Application* | *Instructions* |
|---|---|---|---|---|---|
| Stripping original paint | Water-soluble paint remover, e.g. Sunbeam Anti-corrosives Stripolene 799 | — | — | Brush | Remove the original finish with a scraper after allowing paint-strip 10 minutes to react (repeat if necessary). Wash off thoroughly with cold water, rubbing with wire wool. Dry. Blow out crevices with compressed air. Strip a small area at a time to enable correct neutralizing of the stripper |
| Metal abrading | Emery-cloth, e.g. Howarth Blue Twill, grade 1½ M | — | — | Hand or disc | Paper thoroughly to ensure satisfactory key. Wipe with cleaner solvent or white spirits |
| Acid etching | Apply Deoxidine 125 (I.C.I.) | 1 part Deoxidine, 1 part water | — | Brush | Apply solution generously and rub in with wire wool. Do not allow Deoxidine solution to dry off before the wash-off operation. Allow approximately five minutes to complete reaction. Wash thoroughly with cold water to remove all traces of Deoxidine solution, followed by a hot rinse. Thoroughly dry surfaces with a clean cloth and blow out crevices with compressed air |
| Priming | Synthetic primer G.I.P. No. S3178 | 6 to 1 with Z1048 | ½-hour to 4 hours | Spray | Apply one thin coat of synthetic primer (recommended for superior adhesion) or one thin coat of cellulose primer (recommended for good adhesion). The use of a primer coat enhances adhesion and gives the system a much greater safety factor |
| | or Grey cellulose primer G.I.P. C3971 MOD | 50/50 with 2045M | ¼-hour | Spray | |
| Applying stopper | Stopper Grey G.I.P. 824D or Stopper Brown G.I.P. 1543 | — | 6–8 hours, or overnight if possible | Glazing knife | Apply stopper in thin layers, allowing 15–20 minutes' drying between applications. Heavy layers result in insufficient drying, with subsequent risk of cracking |
| Filling | Primer Filler Grey G.I.P. C3663M | 50/50 with 2045M | 3–4 hours | Spray | Apply two or three full coats, allowing 15–25 minutes' drying time between coats |

| | | | | | |
|---|---|---|---|---|---|
| Wet-sanding | Abrasive paper 280 grade | — | — | — | Rub down wet until smooth; a guide coat (a weak contrasting colour) may be used to ensure that the whole surface is rubbed level. Wash off thoroughly with water, sponge all sludge, wash off, dry with clean sponge. Dry off. Minimum of paint should be removed consistent with a satisfactory surface. Film thickness after rubbing should be ·0025 in. (·06 mm.) min. |
| Applying sealer or undercoat | Sealer Grey or Sealer White or Red undercoat (see BMC Paint Scheme schedule) | 50/50 with 2045M | 15–20 minutes | Spray | Apply one coat, flash off |
| Dry-sanding or de-nibbing as required | 320 grade paper | — | — | — | De-nib or dry-sand with 320 paper. Clean with white spirit. The grade of paper quoted is from the 3M Company (Minnesota Mining and Mfg. Co. Ltd.); the grade of paper may vary according to manufacture |
| Applying colour coats | BMC body finishes (see BMC Paint Scheme schedule) | 50/50 with 2045M | 5–10 minutes' flash between coats. Overnight dry | Spray | Apply two double coats with a 5–10-minute flash between coats. Overnight dry |
| Flatting colour coat | 320 or 400 paper (dependent on conditions) | — | — | Hand | Flat with 320 or 400 paper, dependent on conditions |
| Applying final colour coat | BMC body finishes (see BMC Paint Scheme schedule) | 50/50 with 2045M | Overnight dry | Spray | Spray final double colour coat |
| Polishing | Cut and polish (see BMC Paint Scheme schedule) | — | — | Hand or machine | The colour coat must be thoroughly dry before polishing. After cutting, burnish to a high gloss with a clean mop, and finally clean with a liquid polish, e.g. BMC car polish |

**NOTE.—(1)** For faster drying of undercoats or local repairs G.I.P. thinners 1523 may be used.
**(2)** Under extreme circumstances of heat and/or humidity retarder G.I.P. Z1694 can be used added to the 2045M thinners.

## Section R.23

# VERTICAL ALIGNMENT CHECK

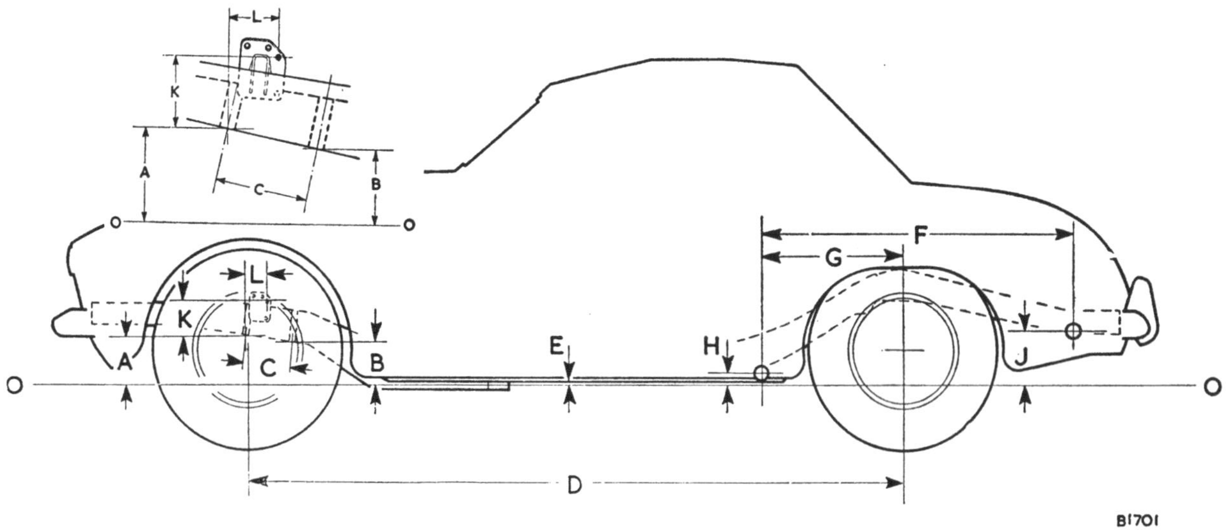


| Code letter | Dimension | Location |
|---|---|---|
| O | — | Datum |
| A | $6\frac{31}{64}$ in. (164·70 mm.) | Datum to front cross-member, front mounting |
| B | $5\frac{41}{64}$ in. (143·27 mm.) | Datum to front cross-member, rear mounting |
| C | $6\frac{7}{16} \pm \frac{1}{32}$ in. (163·51 $\pm$ ·79 mm.) | Front cross-member, front to rear mounting |
| D | 91 in. (2·311 m.) | Wheelbase |
| E | $\frac{1}{2}$ in. (12·70 mm.) | Datum to bottom sill |
| F | $43\frac{1}{16} \pm \frac{1}{16}$ in. (1093·59 $\pm$ 1·58 mm.) | Rear spring centres, eye to eye |
| G | $19\frac{9}{16}$ in. (496·89 mm.) | Rear spring seat to front spring eye |
| H | $1\frac{1}{2}$ in. (38·10 mm.) | Datum to rear spring front mounting |
| J | $7\frac{1}{4} \pm \frac{1}{32}$ in. (184·15 $\pm$ ·79 mm.) | Datum to rear spring rear shackle mounting |
| K | $4\frac{47}{64} \pm \frac{1}{32}$ in. (120·25 $\pm$ ·79 mm.) | Front engine cross-member, front mounting to engine mounting |
| L | $3\frac{3}{64} \pm \frac{1}{32}$ in. (77·39 $\pm$ ·79 mm.) | Front engine cross-member, front mounting to engine mounting |

## Section R.24

# HORIZONTAL ALIGNMENT CHECK

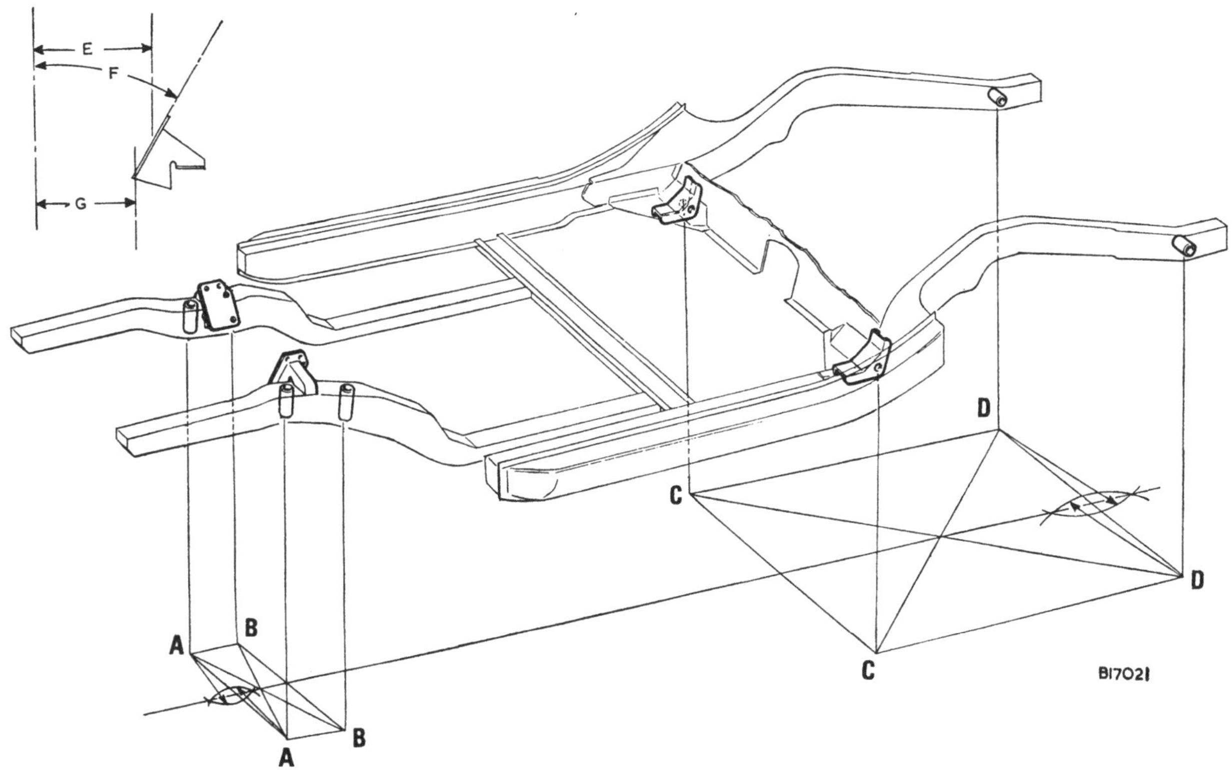


| *Code letter* | *Dimension* | *Location* |
|---|---|---|
| A—A | $18\frac{5}{8} \pm \frac{1}{32}$ in. (473·07 ± ·79 mm.) | Front cross-member front mountings |
| B—B | $20\frac{5}{32} \pm \frac{1}{32}$ in. (512·14 ± ·79 mm.) | Front cross-member rear mountings |
| C—C | $37 \pm \frac{1}{32}$ in. (939·8 ± ·79 mm.) | Rear spring front mounting brackets (centre) |
| D—D | $37 \pm \frac{1}{32}$ in. (939·8 ± ·79 mm.) | Rear spring shackle mounting brackets (centre) |
| E | $7\frac{5}{8}$ in. (193·7 mm.) | Top hole front engine mounting |
| F | 30° | Angle of front engine mounting to vertical |
| G | $6\frac{3}{64}$ in. (153·6 mm.) | Bottom hole front engine mounting |

## THE DOOR COMPONENTS

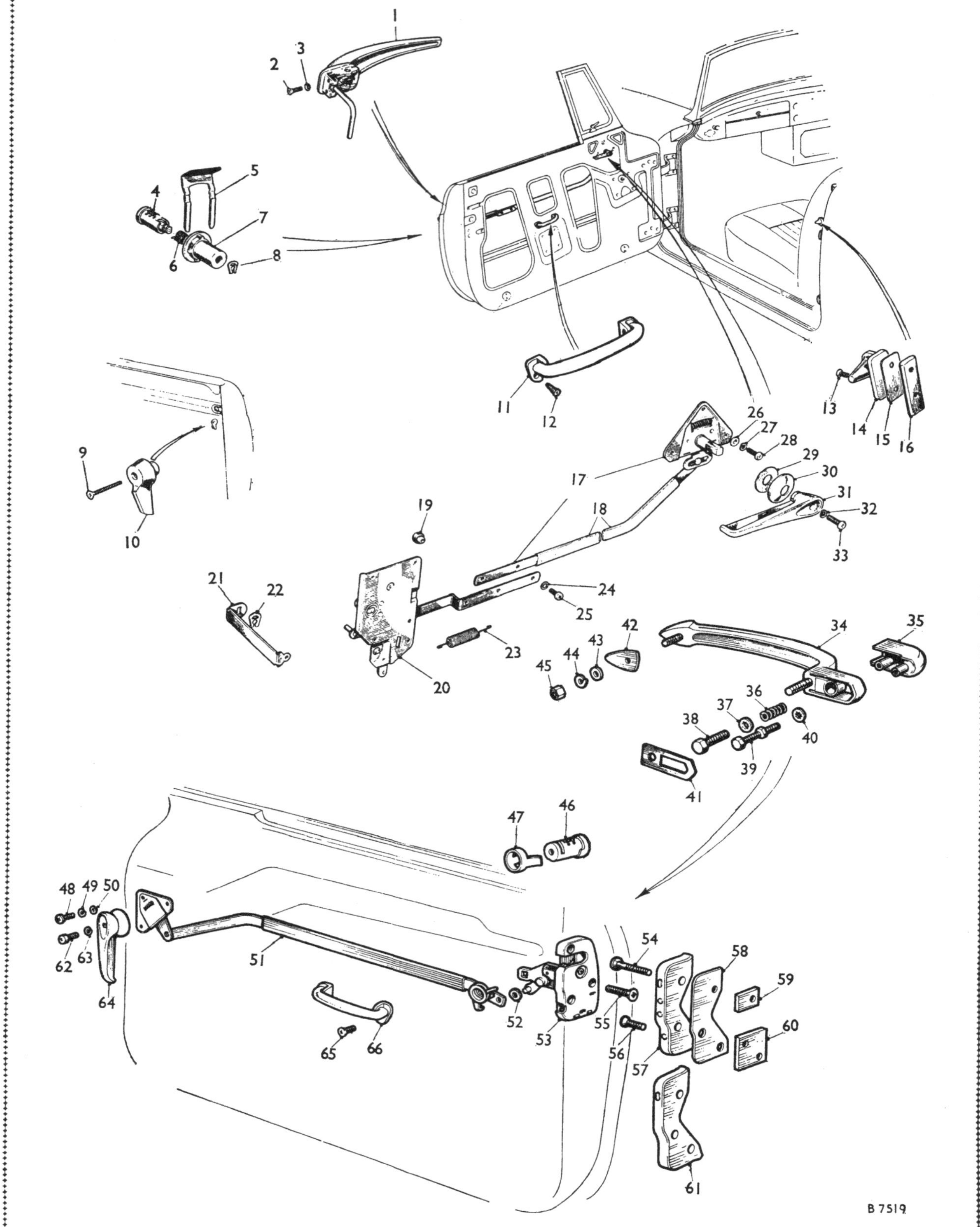

## KEY TO THE DOOR COMPONENTS

Early Tourer models: items 1–32

| No. | Description | No. | Description |
|---|---|---|---|
| 1. | Outer door handle. | 34. | Outer door handle. |
| 2. | Screw. | 35. | Push-button. |
| 3. | Washer. | 36. | Spring. |
| 4. | Lock barrel. | 37. | Shakeproof washer. |
| 5. | Lock retaining clip. | 38. | Set screw. |
| 6. | Self-centring spring. | 39. | Set screw with locknut. |
| 7. | Lock housing. | 40. | Shakeproof washer. |
| 8. | Anderton clip. | 41. | Fibre washer. |
| 9. | Screw. | 42. | Fibre washer. |
| 10. | Inner locking knob. | 43. | Plain washer. |
| 11. | Door pull. | 44. | Spring washer. |
| 12. | Screw. | 45. | Nut. |
| 13. | Screw. | 46. | Lock barrel. |
| 14. | Striker. | 47. | Retaining clip. |
| 15. | Packing. | 48. | Screw. |
| 16. | Tapping plate. | 49. | Spring washer. |
| 17. | Remote control lock. | 50. | Plain washer. |
| 18. | Anti-rattle sleeve. | 51. | Remote-control lock. |
| 19. | Outside door handle buffer. | 52. | Anti-rattle washer. |
| 20. | Lock. | 53. | Lock. |
| 21. | Operating link. | 54. | Screw. |
| 22. | Anderton clip. | 55. | Screw. |
| 23. | Tension spring. | 56. | Screw. |
| 24. | Shakeproof washer. | 57. | Striker. |
| 25. | Screw. | 58. | Shim—·003 in. or ·006 in. (·8 mm. or 1·6 mm.). |
| 26. | Plain washer. | 59. | Tapping plate (upper). |
| 27. | Spring washer. | 60. | Tapping plate (lower). |
| 28. | Screw. | 61. | Striker lock. |
| 29. | Fibre washer. | 62. | Screw. |
| 30. | Finisher. | 63. | Spring washer. |
| 31. | Inner door handle. | 64. | Inner door handle. |
| 32. | Spring washer. | 65. | Screw. |
| 33. | Screw. | 66. | Door pull. |

## THE WINDOW COMPONENTS

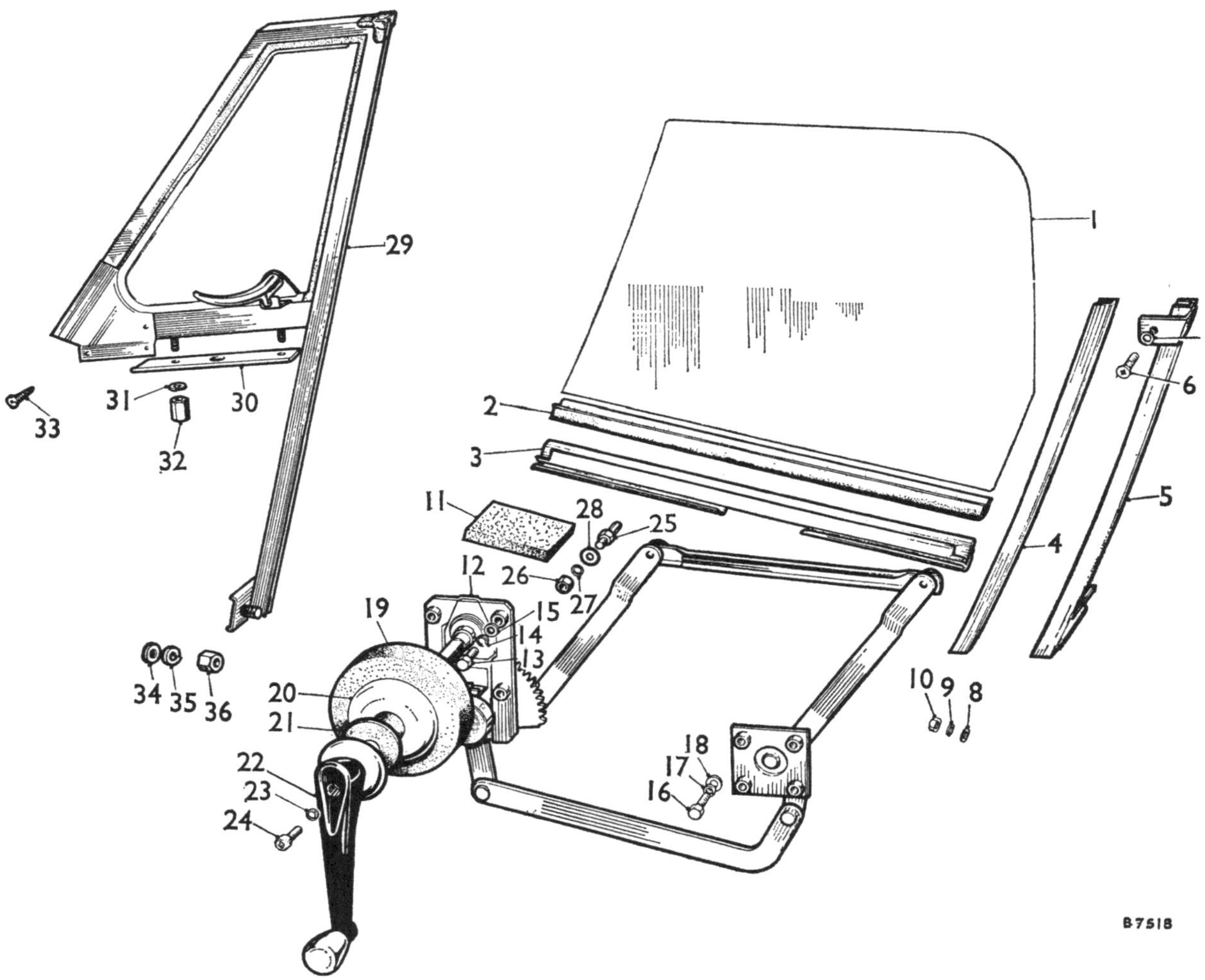

## KEY TO THE WINDOW COMPONENTS

| No. | Description | No. | Description |
|---|---|---|---|
| 1. | Door glass. | 19. | Pad. |
| 2. | Glazing channel. | 20. | Escutcheon. |
| 3. | Lower channel. | 21. | Fibre washer. |
| 4. | Flexible channel. | 22. | Handle and finisher. |
| 5. | Channel. | 23. | Spring washer. |
| 6. | Screw. | 24. | Screw. |
| 7. | Cup washer. | 25. | Stop. |
| 8. | Plain washer. | 26. | Nut. |
| 9. | Spring washer. | 27. | Spring washer. |
| 10. | Nut. | 28. | Plain washer. |
| 11. | Buffer. | 29. | Ventilator. |
| 12. | Regulator. | 30. | Seating washer. |
| 13. | Set screw. | 31. | Spring washer. |
| 14. | Spring washer. | 32. | Nut. |
| 15. | Plain washer. | 33. | Screw. |
| 16. | Set screw. | 34. | Plain washer. |
| 17. | Spring washer. | 35. | Spring washer. |
| 18. | Plain washer. | 36. | Nut. |

# THE FASCIA PANEL COMPONENTS

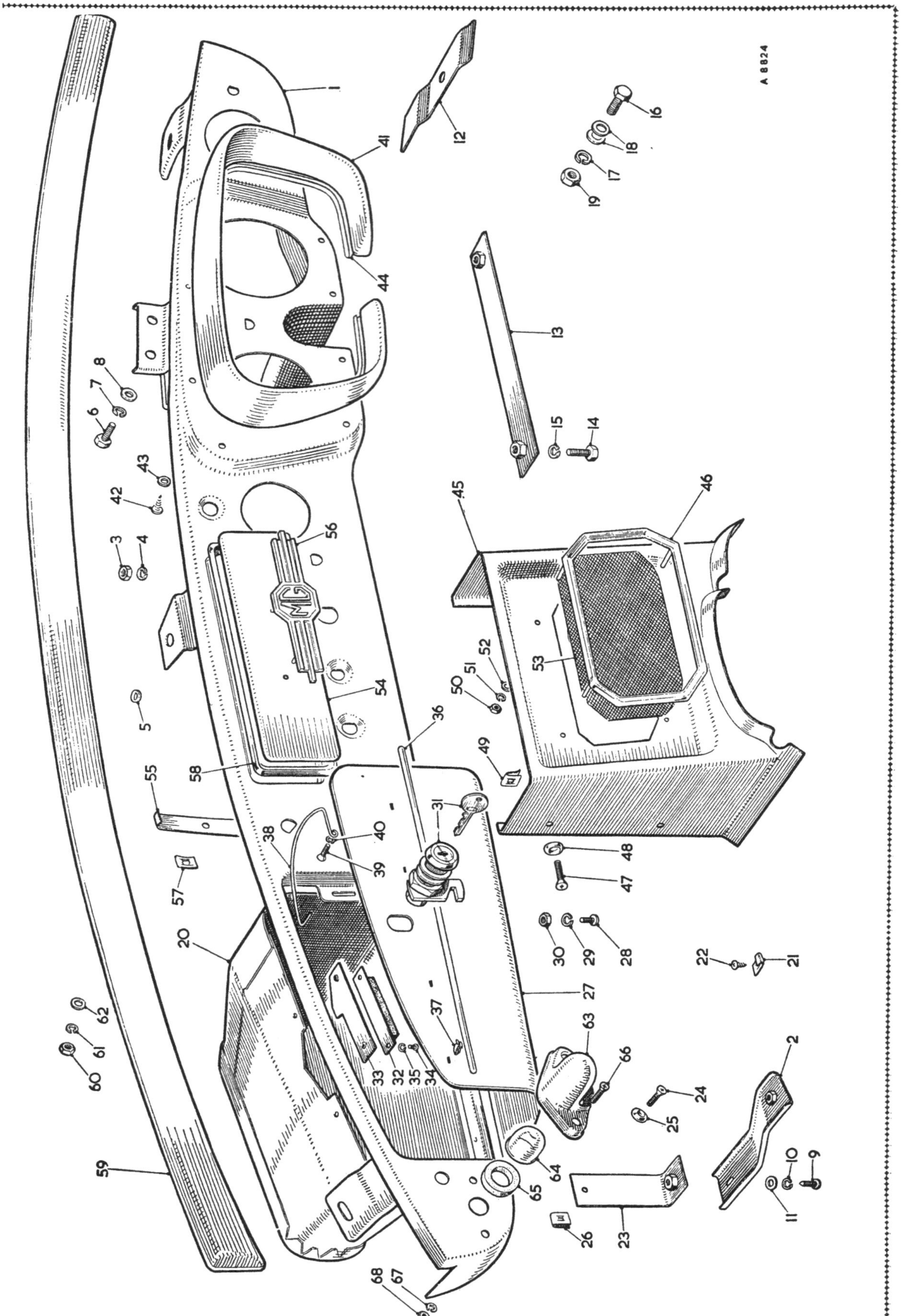

## KEY TO THE FASCIA PANEL COMPONENTS

| No. | Description |
|---|---|
| 1. | Panel assembly—fascia. |
| 2. | Bracket—steady. |
| 3. | Nut—fascia to body. |
| 4. | Washer—spring. |
| 5. | Washer—plain. |
| 6. | Screw—fascia to body. |
| 7. | Washer—spring. |
| 8. | Washer—plain. |
| 9. | Screw—attachment bracket to steady bracket. |
| 10. | Washer—spring. |
| 11. | Washer—plain—attachment bracket to steady bracket. |
| 12. | Plate—washer—fascia fixing. |
| 13. | Plate assembly—fascia fixing. |
| 14. | Screw—fascia to centre reinforcement. |
| 15. | Washer—spring. |
| 16. | Screw—steering-column support bracket to cross-tube. |
| 17. | Washer—spring. |
| 18. | Washer—plain. |
| 19. | Nut—screw. |
| 20. | Box—glove. |
| 21. | Nut—spring. |
| 22. | Screw—glovebox to fascia. |

| No. | Description |
|---|---|
| 23. | Bracket—attachment—glovebox. |
| 24. | Screw—attachment bracket to glovebox. |
| 25. | Washer—cup. |
| 26. | Nut—spring. |
| 27. | Lid assembly—glovebox. |
| 28. | Screw—glovebox lid to fascia. |
| 29. | Washer—spring. |
| 30. | Nut—screw. |
| 31. | Lock assembly—glovebox lid. |
| 32. | Buffer—glovebox lid. |
| 33. | Catch—glovebox lock. |
| 34. | Screw—catch and buffer to glovebox. |
| 35. | Washer—spring. |
| 36. | Finisher—glovebox lid. |
| 37. | Clip—finisher to glovebox. |
| 38. | Arm—check—glovebox. |
| 39. | Screw—check arm to glovebox lid. |
| 40. | Washer—spring. |
| 41. | Hood—instrument. |
| 42. | Screw—hood to fascia panel. |
| 43. | Washer—plain. |
| 44. | Finisher—instrument hood. |
| 45. | Panel—radio speaker. |

| No. | Description |
|---|---|
| 46. | Bezel—radio speaker panel. |
| 47. | Screw—speaker panel fixing. |
| 48. | Washer—cup. |
| 49. | Nut—spring. |
| 50. | Nut—bezel to speaker panel. |
| 51. | Washer—spring. |
| 52. | Washer—plain. |
| 53. | Grille—radio speaker. |
| 54. | Plate—blanking—radio aperture. |
| 55. | Clip—blanking plate to fascia panel. |
| 56. | Motif—'MG'. |
| 57. | Nut—clip (spire). |
| 58. | Bezel—radio aperture. |
| 59. | Roll assembly—crash—Black with Black piping. |
| 60. | Nut—crash roll fixing. |
| 61. | Washer—spring. |
| 62. | Washer—plain. |
| 63. | Cover—map light. |
| 64. | Glass—map light. |
| 65. | Gasket—seating. |
| 66. | Screw—map light to fascia panel. |
| 67. | Washer—spring. |
| 68. | Nut—screw. |

## THE HEATER COMPONENTS

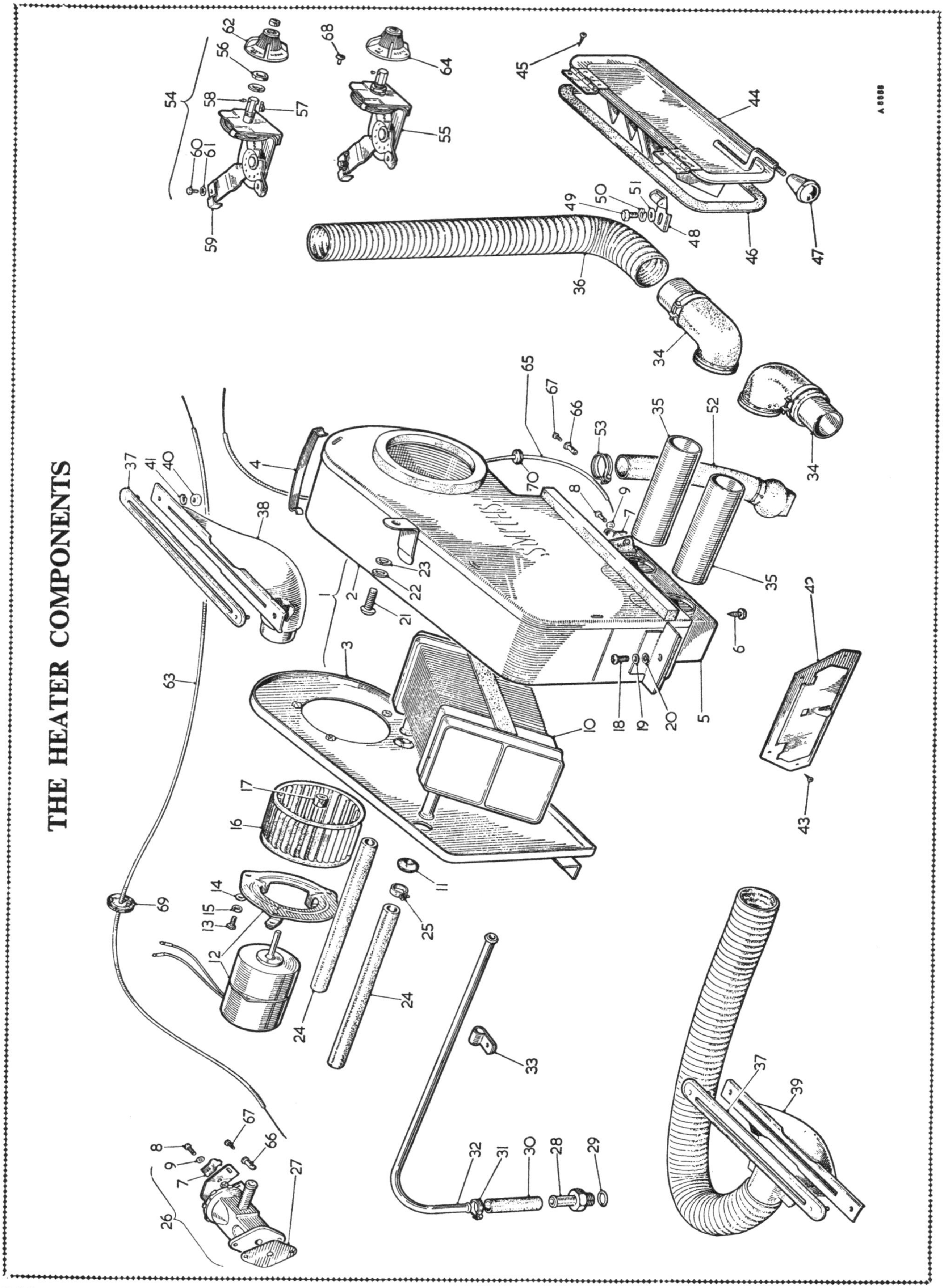

## KEY TO THE HEATER COMPONENTS

| No. | Description | No. | Description | No. | Description |
|---|---|---|---|---|---|
| 1. | Heater assembly. | 25. | Clip—hose. | 48. | Spring assembly—fresh-air vent door. |
| 2. | Cover. | 26. | Valve—assembly—water. | 49. | Screw—spring to body bracket. |
| 3. | Cover—motor side. | 27. | Gasket—water valve. | 50. | Washer—spring. |
| 4. | Clip—cover. | 28. | Union—water. | 51. | Washer—plain. |
| 5. | Duct—outlet. | 29. | Washer—union. | 52. | Tube—drain and dust valve assembly. |
| 6. | Screw—outlet duct. | 30. | Hose—water—2" × ½". | 53. | Clip—tube assembly to air box. |
| 7. | Clamp—cable. | 31. | Clip—hose. | 54. | Control—rotary—heat. |
| 8. | Screw—cable clamp. | 32. | Pipe—water return. | 55. | Control—rotary—air. |
| 9. | Washer—shakeproof—screw. | 33. | Clip to inlet manifold. | 56. | Nut—locking. |
| 10. | Radiator. | 34. | Elbow—demister. | 57. | Clip—knob. |
| 11. | Washer—radiator pipe. | 35. | Tube—connector. | 58. | Pin—knob. |
| 12. | Motor and mounting plate assembly. | 36. | Hose—air—demister. | 59. | Clamp—cable. |
| 13. | Screw—motor fixing. | 37. | Escutcheon assembly—demister. | 60. | Screw—clamp. |
| 14. | Washer—plain—screw. | 38. | Nozzle assembly—demister R.H. | 61. | Washer—spring. |
| 15. | Washer—spring—screw. | 39. | Nozzle assembly—demister L.H. | 62. | Knob—neat. |
| 16. | Runner. | 40. | Nut—escutcheon and demister nozzle to scuttle. | 63. | Cable—heat control. |
| 17. | Nut—collet. | 41. | Washer—spring. | 64. | Knob—air. |
| 18. | Screw—heater unit to bulkhead. | 42. | Door—heater outlet. | 65. | Cable—air control. |
| 19 | Washer—plain. | 43. | Screw—outlet door to air box. | 66. | Trunnion. |
| 20. | Washer—spring. | 44. | Door assembly—fresh-air vent. | 67. | Screw—trunnion. |
| 21. | Screw—heater unit to bulkhead. | 45. | Screw—door to bulkhead. | 68. | Rivet—heater control. |
| 22. | Washer—plain. | 46. | Seal—door. | 69. | Grommet—heater control cable. |
| 23. | Washer—spring. | 47. | Knob—door. | 70. | Grommet—air control cable. |
| 24. | Hose—water—8½" × ½". | | | | |

## Section R.25

### BODYWORK

**Coachwork**

Regular care of the body finish is necessary if the new appearance of the car exterior is to be maintained against the effects of air pollution, rain, and mud.

Wash the bodywork frequently, using a soft sponge and plenty of water containing a mild detergent. Large deposits of mud must be softened with water before using the sponge. Smears should be removed by a second wash in clean water, and with the sponge if necessary. When dry, clean the surface of the car with a damp chamois-leather. In addition to the regular maintenance, special attention is required if the car is driven in extreme conditions such as sea spray, or on salted roads. In these conditions and with other forms of severe contamination an additional washing operation is necessary, which should include underbody hosing. Any damaged areas should be immediately covered with paint and a complete repair effected as soon as possible. Before touching-in light scratches and abrasions with paint thoroughly clean the surface. Use petrol/white spirit (gasoline/hydrocarbon solvent) to remove spots of tar or grease.

The application of BMC Car Polish is all that is required to remove traffic film and to ensure the retention of the new appearance.

**Bright trim**

Never use an abrasive on stainless, chromium, aluminium, or plastic bright parts and on no account clean them with metal polish. Remove spots of grease or tar with petrol/white spirit (gasoline/hydrocarbon solvent) and wash frequently with water containing a mild detergent. When the dirt has been removed polish with a clean dry cloth or chamois-leather until bright. Any slight tarnish found on stainless or plated parts which have not received regular washing may be removed with BMC Chrome Cleaner. An occasional application of mineral light oil or grease will help to preserve the finish, particularly during winter, when salt may be used on the roads, but these protectives must not be applied to plastic finishes.

**Windscreen**

If windscreen smearing has occured it can be removed with BMC Screen Cleaner.

**Interior**

Decosol Interior Cleaner is recommended for cleaning the interior of your car, and may be safely applied to carpets, upholstery and roof linings. Decosol must not be used on painted surfaces.

*Carpets*

Remove surface dirt with a semi-stiff brush or a vacuum cleaner. Apply Decosol, diluted with an equal quantity of water, brush vigorously with a semi-stiff brush, then clean off using a moistened sponge or cloth.

*Upholstery and roof linings*

Apply neat Decosol with a soft brush or cloth, leave for five minutes, then wipe clean using a moistened sponge or cloth.

A razor blade will remove transfers from the window glass.

**Cleaning the hood (Tourer)**

To clean the hood it is only necessary to use soap and water with a soft brush to remove any ingrained dirt. Frequent washing with soap and water considerably improves the appearance and wearing qualities of the hood, and it should be washed at least as often as the rest of the car.

**Do not use caustic soaps, detergents, or spirit cleaners to clean the hood or the hood back-light.**

**Preservative on export cars**

Certain cars leaving the factory are sprayed with a wax preservative to safeguard their body finish. The wax can be removed by the following procedure. Wash the waxed surfaces liberally with water to remove dirt. To soften the wax apply white spirit, either by using a spray and wiping off with mutton-cloth, or by using the cloth dipped in white spirit.

Polish the body with clean dry mutton cloth.

**Coachwork repairs**

The specially designed body jack 18G 308 B is an essential item when rectifying any misalignment of the body construction.

With the addition of a suitable oxy-acetylene outfit any type of mono-construction repair can be effected.

## Section R.26

### BUMPERS

**Removing (Front)**

Unscrew the two steady bracket to bumper securing screws and the nuts and bolts securing the spring blades to the body. Pull the bumper bar forward and lower it to the ground.

If over-riders are fitted, unscrew the nuts and remove the washers, over-riders, and distance pieces.

**Removing (Rear)**

Disconnect the electrical leads from the number-plate lamps, remove the nuts and washers that secure the bumper bar spring to the mounting brackets, and withdraw the assembly from the car.

Remove the nuts and washers securing the bar to the mounting bracket.

**Reassembling and refitting**

Reassembly and refitting of the front and rear bumpers is a reversal of the removal and dismantling sequence. After refitting check the bars for alignment and the number-plate lamps for correct functioning.

## Section R.27

### ELECTRICALLY HEATED BACK-LIGHT (GT)

**Removing**

(1) Remove both rear upper quarter trim panels.

(2) Disconnect the heated back-light wires from the snap connectors inside the right-hand and left-hand body quarters.

(3) Remove the back-light glass as detailed in Section R.19.

(4) Withdraw the wires through the holes in the body and surround rubber.

**Refitting**

(5) Thread the wires through the holes in the body and rubber surround.

(6) Refit the back-light glass as detailed in Section R.19.

(7) Reverse the procedure in (1) and (2).

## Section R.28

### SEAT BELTS

The following instructions refer to fitting the approved 'Kangol Magnet' seat belts to the fixing points incorporated in the body structure.

**Rear wheel arch**

*Early cars*

(1) Remove the two domed nuts and plain washers from the fixing studs.

(2) Fit the bracket on the studs and secure it, using plain washers, spring washers and the domed nuts.

*Later cars*

(3) Remove the plastic cap from the fixing boss.

*First-type seat belt*

(4) Fit a spring washer to the short $\frac{7}{16}$ in. bolt, pass the bolt through the centre of the three holes in the belt bracket and secure the bracket to the fixing boss.

*Second-type seat belt*

(5) Fit the belt bracket on the short $\frac{7}{16}$ in. bolt, followed by a waved washer and a distance piece, with the small diameter end of the distance piece towards the bolt head.

(6) Secure the assembled bracket to the fixing boss.

**Sill**

(7) Locate the fixing point (its position can be felt through the trim) and cut the trim to expose the fixing hole.

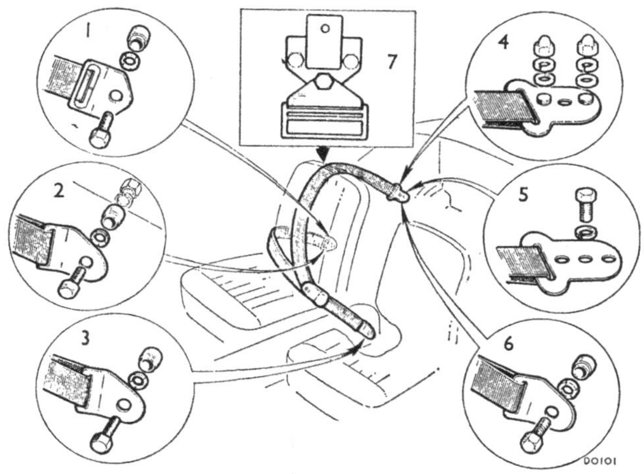

*Fig. R.9*
*Seat belt fixings*

1. Sill.
2. Drive shaft tunnel (early cars).
3. Drive shaft tunnel (later cars).
4. Wheel arch (early cars).
5. Wheel arch, first-type belt (later cars).
6. Wheel arch, second-type belt (later cars).
7. Stowage clip.

(8) Fit the belt bracket to one of the $\frac{7}{16}$ in. bolts, followed by a waved washer and a distance piece (small diameter end towards the bolt head).

(9) Screw the assembled bracket to the sill.

**Drive shaft tunnel**

*Early cars*

(10) Remove the plug from the fixing point on the opposite side of the tunnel to the seat for which the belt is being fitted.

(11) Fit the bracket of the short belt to the remaining $\frac{7}{16}$ in. bolt followed by a waved washer and a distance piece (small diameter end towards the bolt head).

(12) Secure the assembled bracket with the spring washer and nut fitted from inside the tunnel.

*Later cars*

(13) Fit the belt bracket of the short belt to the remaining $\frac{7}{16}$ in. bolt followed by a waved washer and a distance piece (smaller diameter end towards the head of the bolt).

(14) Secure the assembled bracket to the exposed fixing boss on the opposite side of the tunnel to the seat for which the belt is being fitted.

**Stowage clip**

(15) Fit the stowage clip under the head of the centre hood socket securing screw.

## Section R.29

### HEATER

**Removing**

*Blower motor only*

(1) Disconnect the batteries.
(2) Disconnect the motor wiring at the snap connectors.
(3) Remove the three screws securing the motor mounting plate and withdraw the motor complete with plate.

*Heater assembly*

(5) Carry out operations (1) and (2).
(6) Drain the cooling system.
(7) Disconnect the hoses from the heater unit.
(8) Remove the screws securing the heater unit to the engine compartment bulkhead.
(9) Remove the speaker panel or control console.
(10) Unscrew the demister tube retaining clip screws.
(11) Withdraw one of the demister tube elbows and remove the tube plate.
(12) Withdraw the fibre demister tubes.
(13) Remove the overdrive vacuum switch (early cars).
(14) Detach the heater air control from the fascia panel and disconnect the control cable.
(15) Slacken the clip securing the air control cable to the heater unit and slide the outer cable free of the clip.
(16) Lift the heater assembly from the car.

**Dismantling**

(17) Remove the securing clips and withdraw the front panel complete with the blower motor and heater matrix.
(18) Withdraw the matrix from the plate.
(19) Unscrew the blower motor securing screws and remove the motor.
(20) Disconnect and remove the air control cable.

**Reassembling**

(21) Reverse the dismantling procedure in (17) to (20).

**Refitting**

*Heater assembly*

(22) Reverse the removing procedure in (5) to (16) ensuring that the air control cable is correctly routed through the body panel, and that the control functions correctly before refitting the demister tubes and speaker panel.

*Blower motor*

(23) Reverse the removing procedure in (1) to (3).

## Section R.30

### DOOR LOCK ADJUSTMENT (GHN4/5 AND GHD4/5 CARS)

Checking

(1) Wind the window down and close the door.
(2) Move the locking latch rearwards to the locked position.
(3) Unlock the door by turning the key through 90 degrees towards the rear of the car.
(4) Open the door by depressing the outside push-button.
(5) Close the door, then open it again using the inside handle.

**Adjustments**

*Outside push-button*

(6) See Section R.32.

*Remote control and connecting rods*

(7) See Section R.33.

*Striker*

(8) Check that with the door closed a clearance of $\frac{1}{32}$ to $\frac{1}{16}$ in. (1 to 1·6 mm.) exists between the striker and latch faces. Adjust the clearance by adding or removing shims behind the striker.
(9) Slacken the striker securing screws sufficiently to permit the striker to move but tight enough to allow the door to be closed to the fully-latched position.
(10) Press the door inwards or pull it outwards until it lines up correctly with the body. **Never slam the door while making adjustments.**
(11) Open the door and draw a pencil line round the striker to establish its horizontal position.
(12) Set the striker at right angles to the hinge axis and tighten the striker securing screws.
(13) Open and close the door to check for drop or lift; slacken the securing screws to adjust the striker in the vertical plane until the door can be closed easily without rattling, lifting, or dropping. When correctly adjusted, with the door closed in the fully latched position, a fractional movement should be possible when the door is pressed in against its seals.

## Section R.31

### INTERIOR DOOR HANDLES AND TRIM PAD (GHN4/5 AND GHD4/5 CARS)

**Removing**

(1) Close the window, remove the screw and spring washer retaining the window regulator handle and remove the handle and its fibre washer.
(2) Withdraw the two-piece plastic bezel from the remote control unit (the top half upwards, bottom half downwards).
(3) Unscrew the pull handle/arm-rest securing screws and remove the pull handle/arm-rest.
(4) Unscrew the waist rail securing screws and remove the waist rail.
(5) Unscrew the screws securing the trim pad and remove the pad.
(6) Peel the plastic waterproof cover from the door as required.

**Refitting**

(7) Reverse the removing procedure in (1) to (6) ensuring that the lock and window mechanism is adequately greased and that the door cut-outs and waterproof cover are securely taped.

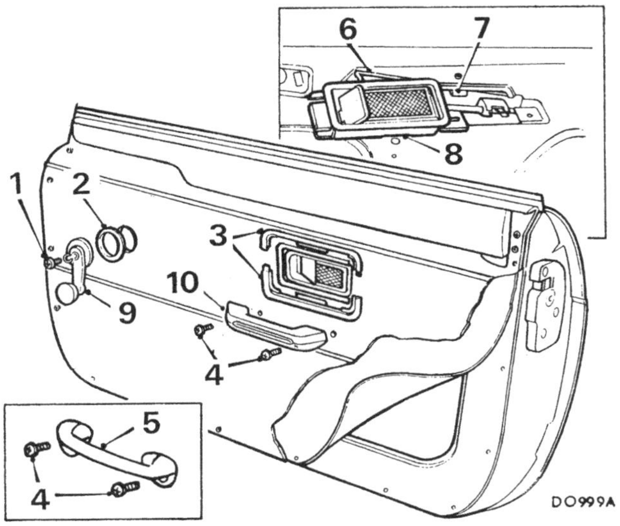


*Fig. R.10*

*Interior door fittings—inset shows remote control being removed*

1. Window regulator handle screw.
2. Fibre washer.
3. Remote control unit bezel.
4. Pull handle screw.
5. Pull handle.
6. Lock control rod.
7. Anti-rattle clip.
8. Remote control unit.
9. Window regulator handle.
10. Arm-rest GHN5 and GHD5 cars from Car No. 294251.

## Section R.32

### OUTSIDE DOOR HANDLES (GHN4/5 and GHD4/5 Cars)

**Removing**

(1) Remove the interior door handles and trim pad—Section R.31.

(2) Wind the window fully up, remove the two nuts, spring and plain washers securing the door handle to the door and remove the handle complete with its seating washers.

**Refitting**

(3) Set the latch to the closed position and temporarily fit the handle and seating washers.

(4) With access through one of the inner door panel apertures, check the clearance between the push-button plunger bolt and the latch contactor; do not depress the push-button while checking the clearance. The clearance must not be less than $\frac{1}{32}$ in. (1 mm.). Adjust the clearance by slackening the locknut and screwing the plunger bolt in or out as required. Tighten the locknut.

(5) Fit the spring and flat washers and screw on and tighten the securing nuts.

## Section R.33

### DOOR LOCKS (GHN4/5 and GHD4/5 Cars)

**IMPORTANT.—Before removing any part of the door lock mechanism because of unsatisfactory operation, first check that the condition is not caused by incorrect adjustment. See Section R.30.**

**Removing**

(1) Remove the interior door handles and trim pad—Section R.31.

*Remote control unit*

(2) Detach the latch release rod and lock control rod from their retaining clips and bushes on the latch levers.

(3) Remove the three remote control retaining screws and washers, and self-tapping screw (later cars).

(4) Withdraw the lock control rod from its bush in the top of the locking latch.

(5) Remove the remote control unit complete with the latch release rod. If necessary, the lock control rod can be removed after withdrawing it downwards from its anti-rattle clip.

*Latch unit*

(6) Remove the three screws securing the latch to the door and remove the latch.

*Key-operated lock*

(7) From inside the door, compress the legs of the retaining collar and withdraw the lock barrel from outside the door.

**Refitting**

*Key-operated lock*

(8) Check that the retaining collar is correctly positioned; enter the lock barrel into the door aperture with its operating fork inclined away from the door shut face and press the lock barrel firmly into position.

*Latch unit*

(9) Check that the plastic bushes and spring clips are correctly assembled in their respective holes in the release lever and the locking slide. Each bush must be fitted with its head away from the centre of the latch with the spring clip fitted under the head (see inset A, Fig. R.11).

(10) Locate the latch on the door face with the levers entered through the slots provided.

(11) Engage the locking lever with the key-operated lock fork.

(12) Fit and tighten the latch securing screws.

*Remote-control unit*

(13) Check that the latch release rod is fully engaged in its plastic clip attached to the remote-control unit.

(14) Check that the lock control rod is fully pressed up into the anti-rattle clip inside the door.

(15) Position the remote-control unit in the door and press the end of the lock control rod into the bush in the locking latch.

(16) Fit the remote control securing screws and washers but do not tighten.

(17) Press the end of the latch release rod into its bush and clip in the latch release lever.

(18) Press the lock control rod pivot end into the bush in the locking slide.

**Adjusting the remote control**

(19) With the remote-control unit securing screws slackened, move the control unit towards the latch,

without compressing the rod spring, until the latch release lever just contacts its stop, then tighten the control unit screws. If the control unit adjustment is restricted by the securing screw slots, enlarge the slots by filing.

(20) With the latch in the closed position, check the operation of the latch release handle. The striker should be released before the handle has reached the full extent of its movement. Fit self-tapping screw.

(21) With the latch in the closed position set the locking latch rearwards to the locked position.

(22) Adjust the position of the screwed pivot on the lock control rod so that the locking latch overlaps the latch release handle, the pivot pin fits freely into its bush in the locking slide, before being pressed into the spring clip.

(23) Check the operation of the lock mechanism as described in Section R.30.

**Lubrication**

(24) Smear the linkages inside the door with grease and add a few drops of thin oil into the key slots.

**IMPORTANT.—Key locks must not be lubricated with grease.**

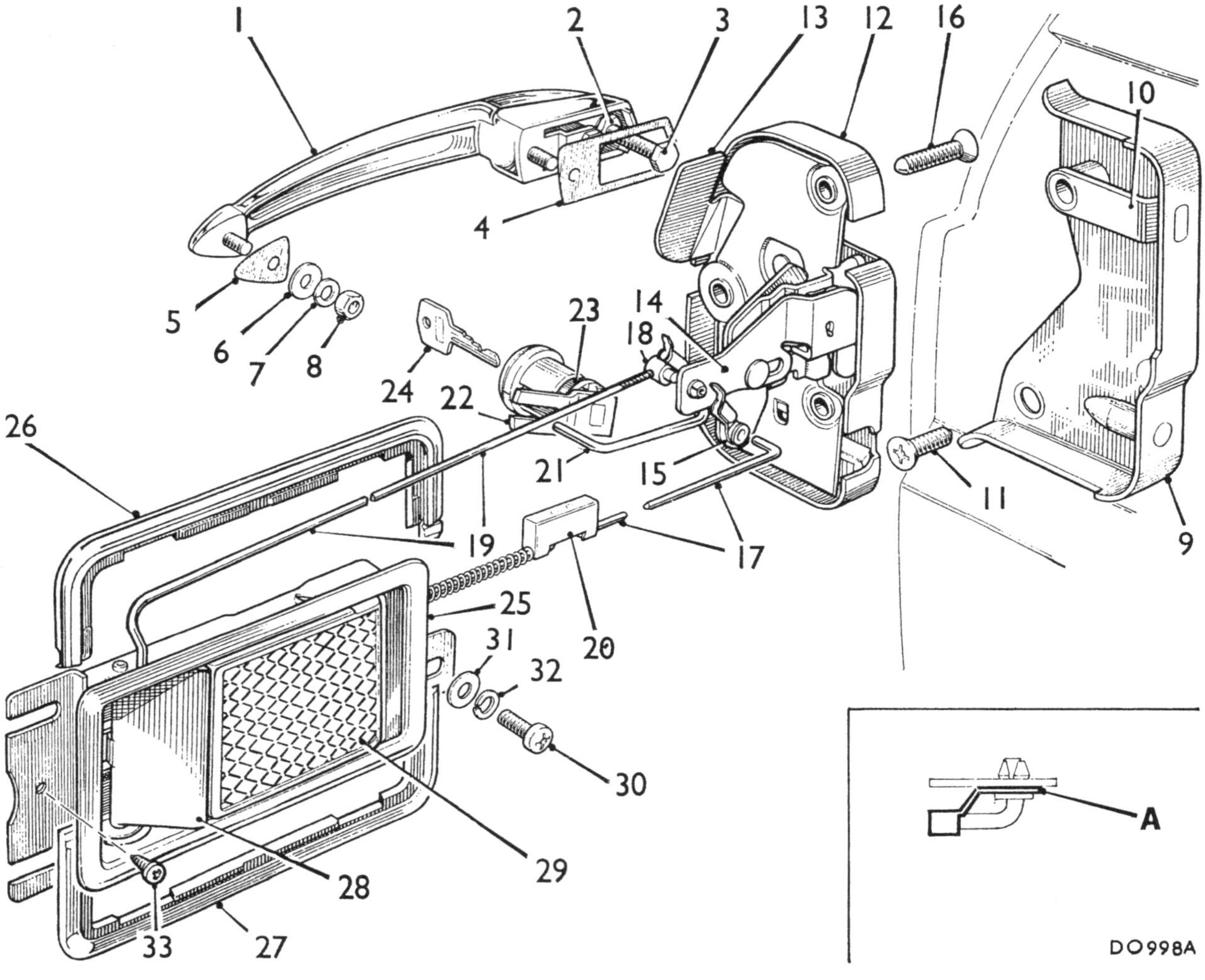


*Fig. R.11*

*Exploded view of door lock mechanism*

1. Push-button.
2. Locknut.
3. Push-button plunger bolt.
4. Sealing washer.
5. Sealing washer.
6. Plain washer.
7. Spring washer.
8. Nut.
9. Striker.
10. Anti-burst strap.
11. Striker screw.
12. Latch unit.
13. Latch contactor.
14. Locking slide.
15. Latch release lever.
16. Latch unit screw.
17. Latch release rod.
18. Screwed pivot.
19. Lock control rod.
20. Plastic clip.
21. Locking lever.
22. Lock operating fork.
23. Key operated lock.
24. Key.
25. Remote control unit.
26. Remote control bezel—upper.
27. Remote control bezel—lower.
28. Locking latch.
29. Inside door handle.
30. Remote control unit screw.
31. Plain washer.
32. Spring washer.
33. Self-tapping screw.

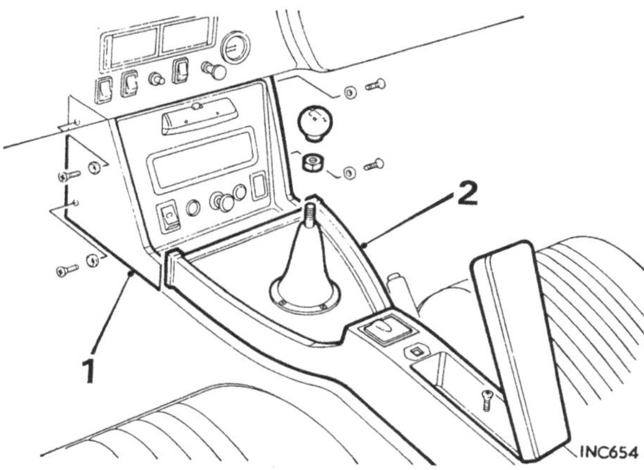


*Fig. R.12*

*Removing the console with hinged arm rest*

1. Console. 2. Arm rest.

## Section R.34

### CONSOLE
### (With hinged arm-rest)

**Removing**

(1) Disconnect the battery.

(2) Remove the change speed lever knob and locknut.

(3) Unscrew the four screws securing the retaining ring.

(4) Raise the hinged arm-rest, remove the retaining screw and withdraw the arm-rest assembly complete with the change speed lever gaiter and retaining ring.

(5) Remove the four screws retaining the console.

(6) Partially withdraw the console, disconnect the wiring from the back and then remove the console.

**Refitting**

(7) Reverse the removing procedure in (1) to (6).

## Section R.35

### FACE-LEVEL VENTS

**Removing**

(1) Disconnect the air duct hoses from the back of the face level vents.

(2) Ease up the bottom locating tag of one vent and slightly withdraw the vent until the locating tag rests on the lip of the fascia aperture.

(3) Ease down the top locating tag and remove the vent and the anti-rattle packing.

(4) Repeat (2) and (3) for the remaining face-level vent.

**Refitting**

(5) Reverse the removing procedure in (1) to (4) noting the following:

(*a*) Ensure that the rounded corners of the face level vent, adjacent to the serrated wheel, register with the rounded corners of the fascia recess.

(*b*) The vent assemblies are marked L.H. (left-hand) and R.H. (right-hand) on their flaps for correct assembly when viewing the fascia from inside the car.

(*c*) Ensure the anti-rattle packing is wedged between the ends of each vent to prevent the vents from butting against each other.

## Section R.36

### FRONT GRILLE
### (GHN5 and GHD5 cars from Car No. 294251)

**Removing**

*Grille assembly*

(1) Remove three screws securing the front grille support brackets to the bonnet locking platform.

(2) Remove the three hexagon headed screws securing the bottom of the front grille to the radiator duct panel.

(3) Remove the radiator grille.

*Grille panel*

(4) Carry out operations (1) to (3).

(5) Remove the 12 screws and clips to release the R.H. and L.H. grille panels from the grille panel.

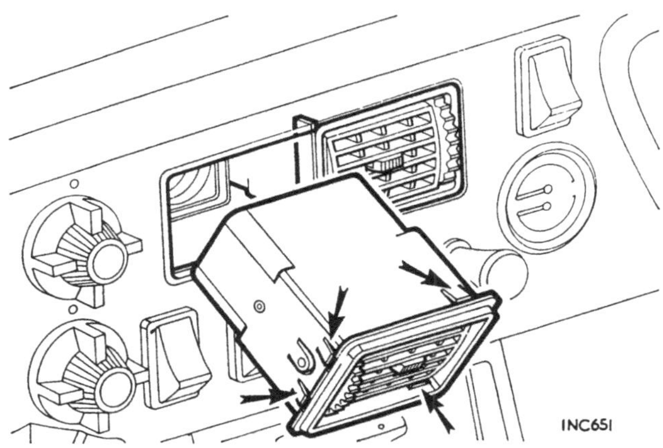


*Fig. R.13*

*Removing the face level vents showing the locating tags arrowed*

**Refitting**

*Grille panel*

(6) Position the grille panels on the finisher, ensuring that the grille panel thin slats are in a vee formation when viewed from the front of the grille (see illustration on page C.4).

(7) Fit the 12 screws and clips to secure the grille panels to the grille finisher.

(8) Reverse the removing procedure in (1) to (3).

*Grille assembly*

(9) Reverse the removing procedure in (1) to (3). ●

# Section Ra

## THE BODY

**The information given in this Section refers specifically to service operations on, or affected by, equipment fitted to the MGB (GHN5/GHN4 and GHD5/GHD4) in conformity with local and territorial requirements, and must be used in conjunction with Section R.**

## Section Ra.1

### CONSOLE

**Removing**

(1) Disconnect the battery.
(2) Remove the four retaining screws.
(3) Remove the ashtray (if fitted).
(4) Turn back the carpet and withdraw the console.
(5) Disconnect the wiring from the back of the console and remove the console assembly.

**Refitting**

(6) Reverse the removing procedure in (1) to (5).

## Section Ra.2

### FASCIA

**Removing**

(1) Disconnect the battery.
(2) Unscrew the three retaining screws from the lower fascia cover panels and remove the panels.
(3) Disconnect the mixture control (choke) cable from the carburetters.
(4) Remove the heater air control knob, unscrew the control securing nut and disengage the control from its fixing bracket.
(5) Remove the heater temperature control knob, unscrew the control securing nut and disengage the control from its fixing bracket.
(6) Remove the two fixing screws from the lower edge of the fascia.
(7) Unscrew the two knurled nuts retaining the tachometer, disconnect the earth wire, and remove the retaining brackets.
(8) Disconnect the wiring from the tachometer and remove the instrument.
(9) Remove the six nuts from the fascia upper edge fixing studs.
(10) Remove the screws securing each end of the fascia trim piping from behind the fascia.
(11) Unscrew the trip recorder reset knob retaining nut and disengage the reset from its bracket.
(12) Disconnect the speedometer drive cable from the back of the instrument.
(13) Ease the fascia assembly rearwards to disengage the top fixing studs from the body panel, then lift the assembly clear of the steering-column switch cowl.
(14) Disconnect the wiring from the instruments and switches, withdraw the mixture control cable and remove the fascia complete with instruments and switches.

**Refitting**

(15) Reverse the removing procedure in (1) to (14).

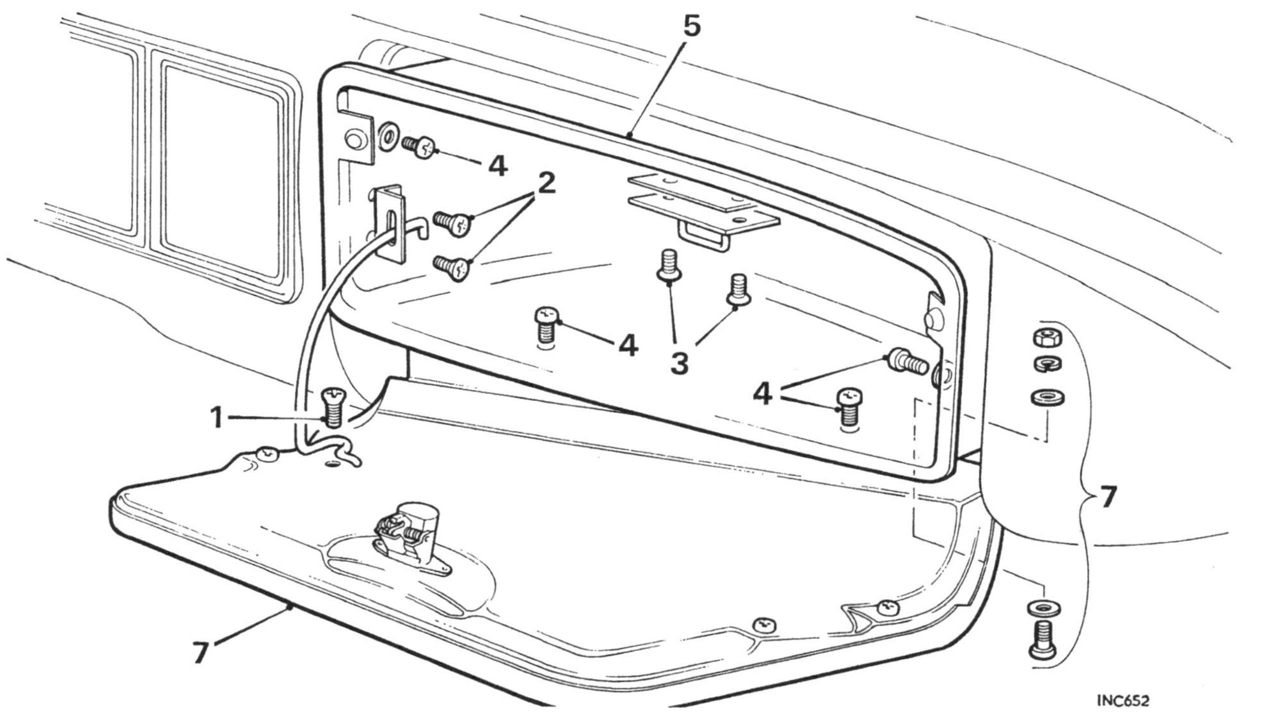


*Fig. Ra.1*
*The glovebox and lid*

1. Screws for glovebox lid stay.
2. Glovebox lid stay.
3. Screws for catch.
4. Catch.
5. Glovebox.
6. Screws for glovebox lid.
7. Glovebox lid.

## Section Ra.3

### GLOVEBOX AND LID
### (GHN5 and GHD5 cars from Car No. 258001)

**Removing**

*Glovebox only*

(1) Remove the screw retaining the stay to the glovebox lid.

(2) Remove the screws retaining the glovebox lid stay to the glovebox.

(3) Remove the screws securing the glovebox lid catch.

(4) Remove the screws retaining the glovebox.

(5) Withdraw the glovebox.

*Glovebox lid*

(6) Carry out operations (1) to (5).

(7) Remove the screws and nuts securing the glovebox lid and withdraw the lid.

**Refitting**

*Glovebox lid*

(8) Reverse the removing procedure in (6) to (7).

*Glovebox*

●(9) Reverse the removing procedure in (1) to (5).

## Section Ra.4

### FACE-LEVEL VENTS
### (GHN5 and GHD5 cars from Car No. 258001)

**Removing**

*Face-level vents and escutcheon assembly*

(1) Remove the glovebox—Section Ra.3.

(2) Disconnect the air duct hoses from the back of the face-level vents.

(3) Remove the two screws and retaining strap securing the escutcheon assembly to the fascia.

(4) Withdraw the face-level vents and escutcheon assembly from the fascia.

*Face-level vents*

(5) Carry out operations (1) to (4).

(6) In turn, ease each bottom locating tag of one vent inwards and slightly withdraw the vent until the tag rests on the lip of the escutcheon.

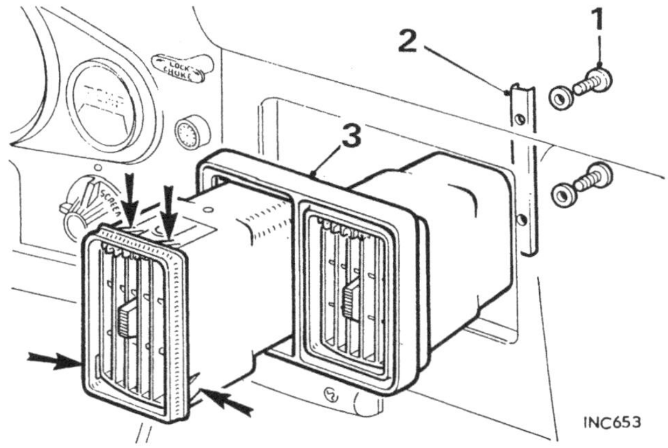


*Fig. Ra.2*

*Removing the face-level vent showing the locating tags arrowed*

1. Screw for retaining strap. 2. Retaining strap. 3. Escutcheon.

(7) With the aid of a narrow blade tool, inserted from the back between the escutcheon and the vent, depress the two top locating tags.

(8) Withdraw the vent from the escutcheon.

(9) Repeat (6) to (8) for the remaining face-level vent.

**Refitting**

*Face-level vents*

(10) Push one only of the face-level vents into the escutcheon noting:

- (*a*) Ensure that the rounded corners of the face-level vent adjacent to the serrated wheel registers in the recess of the escutcheon with the rounded corners.
- (*b*) The vent assemblies are marked L.H. (left-hand) and R.H. (right-hand) on their flaps for correct assembly when viewing the fascia from inside the car.

(11) Reverse the removing procedure in (3) to (4) ensuring that the escutcheon assembly is fitted with the serrated wheels uppermost.

(12) Push in the remaining face-level vent into the escutcheon noting operation (10) (*a*).

(13) Reverse the removing procedure in (1) to (2).

*Face-level vent and escutcheon assembly*

(14) Reverse the operations (6) to (8) to refit one of the face level vents.

(15) Carry out operations (11) to (13).

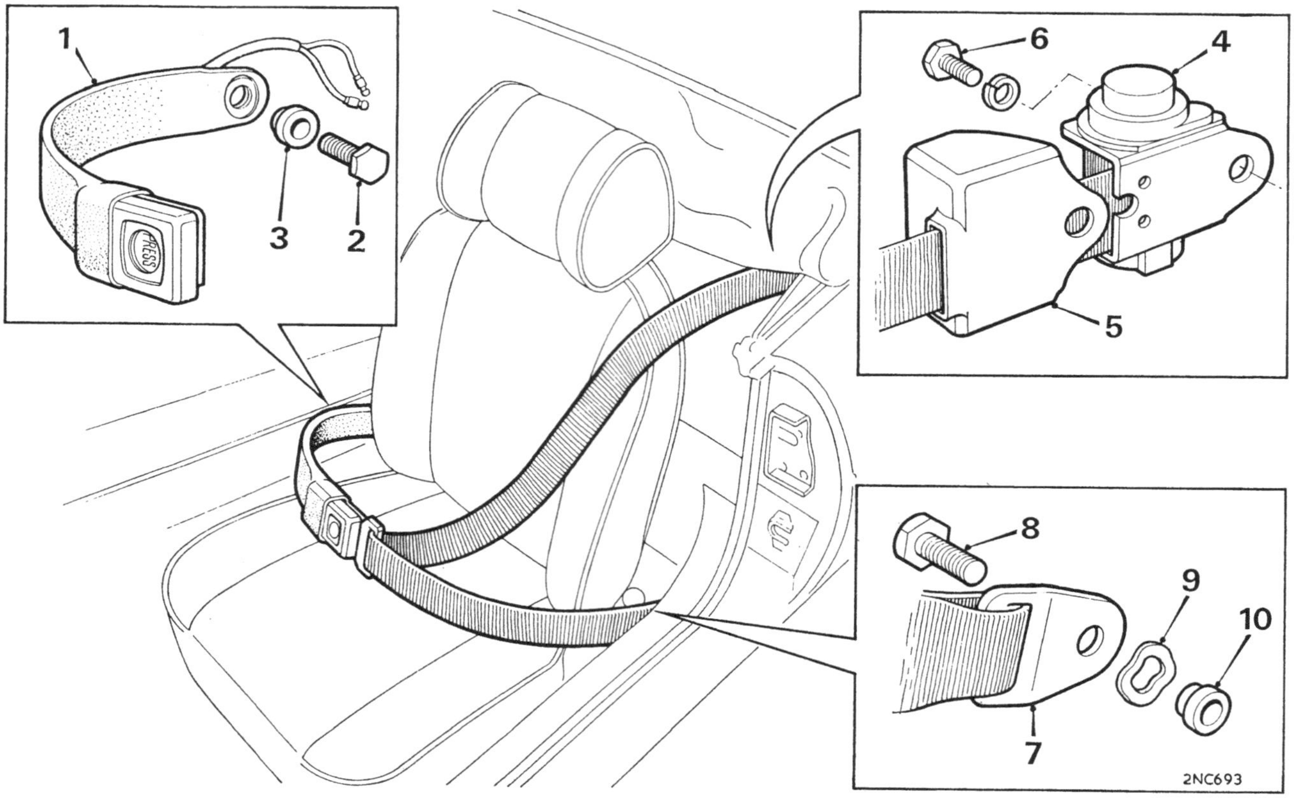


*Fig. Ra.*3
*Seat belt fixings*

1. Short belt.
2. Retaining bolt.
3. Spacer.
4. Seat belt reel.
5. Cover for reel.
6. Retaining bolt.
7. Seat belt sill mounting.
8. Retaining bolt.
9. Anti-rattle washer.
10. Spacer.

## Section Ra.5

### SEAT BELTS

**Removing**

*Short belt*

(1) Lift the carpet adjacent to the tunnel and disconnect the wiring from the seat belt.
(2) Remove the bolt and spacer to release the belt.

*Long belt and reel*

(3) Remove the bolt, waved washer and spacer to release the belt from the sill.
(4) Remove the bolt and spring washer to release the reel from the rear wheel arch.

**Refitting**

*Long belt and reel*

(5) Reverse the removing procedure in (3) to (4).

*Short belt*

(6) Reverse the removing procedure in (1) to (2).

## Section Ra.6

### ANTI-BURST DOOR UNITS
### (GHN5 and GHD5 cars from Car No. 294251)

**Removing**

*Anti-burst plate unit*

(1) Remove the two screws to release the plate from 'B' post.

*Anti-burst pin unit*

(2) Remove the door trim pad—Section R.31.

(3) Unscrew the nut and remove the spring washer to release the pin unit from the door.

**Refitting**

*Anti-burst pin unit*

(4) Reverse the removing procedure in (2) to (3).

*Anti-burst plate unit*

(5) Refit the plate unit to the 'B' post and retain in position with the two screws.

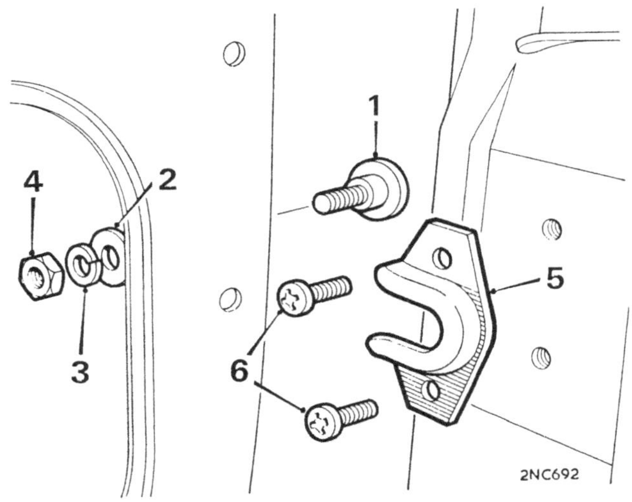


*Fig. Ra,4*
*Anti-burst door unit*

1. Door pin.
2. Plain washer.
3. Spring washer.
4. Retaining nut for pin.
5. Plate.
6. Retaining screws.

# SECTION S

## SERVICE TOOLS

All Service tools mentioned in this Manual are only obtainable from the tool manufacturer:

Messrs. V. L. Churchill & Co. Ltd.
P.O. Box No. 3,
London Road, Daventry,
Northants, England.

| *Operation* | | *Tool No.* | *Page No.* |
|---|---|---|---|
| **ENGINE** | | | |
| Camshaft liner reaming | | 18G 123 A | S.9 |
| | | 18G 123 B | S.9 |
| | | 18G 123 E | S.9 |
| | | 18G 123 F | S.9 |
| | | 18G 123 L | S.9 |
| | | 18G 123 T | S.10 |
| | | 18G 123 AB | S.10 |
| | | 18G 123 AC | S.10 |
| | | 18G 123 AD | S.10 |
| Camshaft liner removing and replacing | | 18G 124 A | S.10 |
| | | 18G 124 B | S.10 |
| | | 18G 124 C | S.10 |
| | | 18G 124 F | S.10 |
| | | 18G 124 H | S.10 |
| Crankshaft gear and pulley removing | | 18G 2 | S.6 |
| Crankshaft oil seal replacing | | 18G 134 CQ | S.11 |
| | | 18G 1108 | S.22 |
| Crankshaft pulley nut removing | | 18G 98 A | S.9 |
| Crankshaft spigot bush removing and replacing | | 18G 284 H | S.15 |
| | | 18G 284 L | S.15 |
| | | 18G 1037 | S.21 |
| Gudgeon pin, removing and replacing | | 18G 1150 | S.23 |
| | | 18G 1150C | S.23 |
| Main bearing cap removing | | 18G 284 | S.15 |
| | | 18G 284 A | S.15 |
| | | 18G 284 AC | S.15 |
| Oil pump relief valve seat grinding | | 18G 69 | S.9 |
| Piston refitting | | 18G 55 A | S.9 |
| Timing cover locating and oil seal replacing | | 18G 134 | S.11 |
| | | 18G 134 BD | S.11 |
| Cover locating | Early models | 18G 3 | S.6 |
| | Later models | 18G 1046 | S.21 |
| Torque spanners | | 18G 372 | S.16 |
| | | 18G 536 | S.17 |
| | | 18G 537 | S.17 |
| Valves removing and refitting | | 18G 45 | S.8 |

| Operation | | Tool No. | Page No. |
|---|---|---|---|
| Valve seat cutting | | 18G 25 | S.6 |
| | | 18G 25 A | S.6 |
| | | 18G 25 C | S.6 |
| | | 18G 27 | S.6 |
| | | 18G 28 | S.7 |
| | | 18G 28 A | S.7 |
| | | 18G 28 B | S.7 |
| | | 18G 28 C | S.7 |
| | | 18G 174 B | S.12 |
| | | 18G 174 D | S.12 |
| Valve seat grinding | | 18G 29 | S.7 |
| | | 18G 29 A | S.7 |
| **CLUTCH** | | | |
| Refitting | | 18G 1027 | S.20 |
| (Involute spine clutch plates) | | 18G 680 | S.19 |
| **GEARBOX** | | | |
| Bevel pinion flange retaining | | 18G 34 A | S.7 |
| First motion shaft assembling and replacing | | 18G 4 | S.6 |
| | | 18G 5 | S.6 |
| Layshaft replacing | ·646 in. dia. shaft | 18G 471 | S.17 |
| | ·670 in. dia. shaft | 18G 1138 | S.22 |
| Propeller shaft flange removing | | 18G 2 | S.6 |
| Rear oil seal removing and replacing | | 18G 389 | S.16 |
| | | 18G 389 B | S.16 |
| | | 18G 134 | S.11 |
| | | 18G 134 BK | S.11 |
| Selector fork and rod guiding | | 18G 41 | S.7 |
| Third motion shaft, synchromesh assembly: | Early models | 18G 222 | S.13 |
| | | 18G 223 | S.13 |
| | Later models | 18G 262 | S.14 |
| | | 18G 1026 | S.20 |
| Laycock **Overdrive Type 'D'** | | | |
| Annulus spigot bearing replacing | | 18G 1042 | S.21 |
| Circlip removing | | 18G 1004 | S.20 |
| Hydraulic pressure testing | | 18G 251 | S.14 |
| | | 18G 251 D | S.14 |
| Mainshaft bearing removing and replacing | | 18G 47 C | S.8 |
| | | 18G 47 CZ | S.8 |
| | | 18G 47 AP | S.8 |
| | | 18G 134 | S.11 |
| | | 18G 1039 | S.21 |
| Oil pump body removing | | 18G 183 | S.12 |
| | | 18G 183 A | S.12 |
| | | 18G 1040 | S.21 |
| Rear oil seal removing and replacing | | 18G 134 BK | S.11 |
| | | 18G 389 B | S.16 |
| Roller clutch assembling | | 18G 178 | S.12 |

| *Operation* | *Tool No.* | *Page No.* |
|---|---|---|
| **Laycock Overdrive, Type 'LH'** | | |
| Hydraulic pressure testing | 18G 251 | S.14 |
| | 18G 251 E | S.14 |
| Mainshaft bearing removing and replacing | 18G 185 | S.12 |
| | 18G 186 | S.12 |
| | 18G 391 | S.17 |
| | 18G 1045 | S.21 |
| | 18G 1024 | S.20 |
| Oil pump removing and replacing | 18G 1117 | S.22 |
| | 18G 1118 | S.22 |
| Rear oil seal removing and replacing | 18G 389 | S.16 |
| | 18G 389 D | S.16 |
| | 18G 177 | S.12 |
| Roller clutch assembling | 18G 178 | S.12 |
| **AUTOMATIC GEARBOX** | | |
| Clutch compressing and seal replacing | 18G 702 | S.19 |
| | 18G 1107 | S.22 |
| | 18G 1016 | S.20 |
| Circlip removing and replacing | 18G 1004 | S.20 |
| Spare points | 18G 1004 J | S.20 |
| Alternative | 18G 675 | S.19 |
| Front servo spanner and gauge | 18G 678 | S.19 |
| Input shaft end-float checking | 18G 674 | S.18 |
| Inhibitor switch locknut | 18G 679 | S.19 |
| Pressure testing | 18G 677 Z | S.19 |
| Alternative | 18G 502 A | S.17 |
| Alternative | 18G 502 K | S.17 |
| Torque spanners | 18G 536 | S.17 |
| | 18G 536 A | S.17 |
| | 18G 536 B | S.17 |
| | 18G 537 | S.17 |
| | 18G 681 | S.19 |
| | 18G 681 A | S.19 |
| | 18G 681 B | S.19 |
| | 18G 537 A | S.17 |
| | 18G 537 B | S.17 |
| **REAR SUSPENSION AND FINAL DRIVE** | | |
| **Three-quarter-floating Rear Axle** | | |
| Bevel pinion bearing removing and replacing | 18G 47 C | S.8 |
| | 18G 47 AH | S.8 |
| | 18G 207 | S.13 |
| | 18G 191 | S.13 |
| Bevel pinion flange wrench | 18G 34 A | S.7 |
| Bevel pinion outer race removing | 18G 264 | S.14 |
| | 18G 264 E | S.15 |
| | 18G 264 F | S.15 |

| *Operation* | *Tool No.* | *Page No.* |
|---|---|---|
| Differential bearing removing and replacing .. .. .. | 18G 47 C | S.8 |
| | 18G 47 T | S.8 |
| | 18G 134 | S.11 |
| | 18G 134 P | S.11 |
| | 18G 191 A | S.13 |
| Propeller shaft flange removing .. .. .. .. .. | 18G 2 | S.6 |
| **Semi-floating Rear Axle** | | |
| Axle shaft assembly replacing .. .. .. .. .. | 18G 1067 | S.21 |
| Bevel pinion bearing removing and replacing .. .. .. | 18G 47 C | S.8 |
| | 18G 47 AS | S.8 |
| | 18G 191 | S.13 |
| | 18G 264 | S.14 |
| Bevel pinion flange wrench.. .. .. .. .. .. | 18G 34 A | S.7 |
| Bevel pinion outer race removing .. .. .. .. .. | 18G 264 | S.14 |
| | 18G 264 AA | S.15 |
| | 18G 264 AB | S.15 |
| | 18G 264 AD | S.15 |
| Differential bearing removing and replacing .. .. .. | 18G 47 C | S.8 |
| | 18G 47 AK | S.8 |
| | 18G 134 | S.11 |
| | 18G 134 CM | S.11 |
| Differential reassembling .. .. .. .. .. .. | 18G 191 | S.13 |
| | 18G 191 F | S.13 |
| | 18G 191 H | S.13 |
| | 18G 191 J | S.13 |
| Final drive and rear axle shaft nut .. .. .. .. | 18G 586 | S.18 |
| Impulse extractor .. .. .. .. .. .. .. | 18G 284 | S.15 |
| Propeller shaft flange removing .. .. .. .. .. | 18G 2 | S.6 |
| Rear axle casing stretching .. .. .. .. .. | 18G 131 C | S.10 |
| Torque setting spanner .. .. .. .. .. .. | 18G 592 | S.18 |
| **FRONT SUSPENSION AND STEERING** | | |
| Spring compressing .. .. .. .. .. .. .. | 18G 693 | S.19 |
| Steering-arm ball pin removing .. .. .. .. .. | 18G 1063 | S.21 |
| Steering-wheel hub removing .. .. .. .. .. | 18G 1181 | S.23 |
| Steering rack ball joint spanners .. .. .. .. .. | 18G 706 | S.20 |
| Swivel axle bush removing, replacing, and reaming .. .. | 18G 68 | S.9 |
| | 18G 596 | S.18 |
| | 18G 597 | S.18 |
| | 18G 1063 | S.21 |

| *Operation* | | *Tool No.* | *Page No.* |
|---|---|---|---|
| **FRONT AND REAR HUBS** | | | |
| Hub removing and replacing | | 18G 304 | S.16 |
| | Disc wheels | 18G 304 A | S.16 |
| | | 18G 304 J | S.16 |
| | Wire wheels | 18G 1032 | S.20 |
| | | 18G 363 | S.16 |
| Front hub bearing cup replacing | | 18G 1122 | S.22 |
| | | 18G 1122 A | S.22 |
| Front hub oil seal replacing | | 18G 134 | S.11 |
| | | 18G 134 BH | S.11 |
| Rear hub bearing removing | | 18G 134 | S.11 |
| | | 18G 134 P | S.11 |
| Rear hub nut spanner | | 18G 152 | S.11 |
| **BRAKING SYSTEM** | | | |
| Brake adjusting (rear only) | | 18G 619 A | S.18 |
| Disc brake piston re-setting | | 18G 590 | S.18 |
| **MISCELLANEOUS** | | | |
| Light switch bezel, removing and replacing | | 18G 671 | S.18 |
| Petrol gauge tank attachment lock ring spanner | | 18G 1001 | S.20 |
| Remover; hydraulic version of 18G 304 | | 18G 304 Z | S.16 |
| Rocker type switch and bezel, removing | | 18G 1145 | S.22 |
| Rocker type switch (slim type) removing | | 18G 1202 | S.23 |

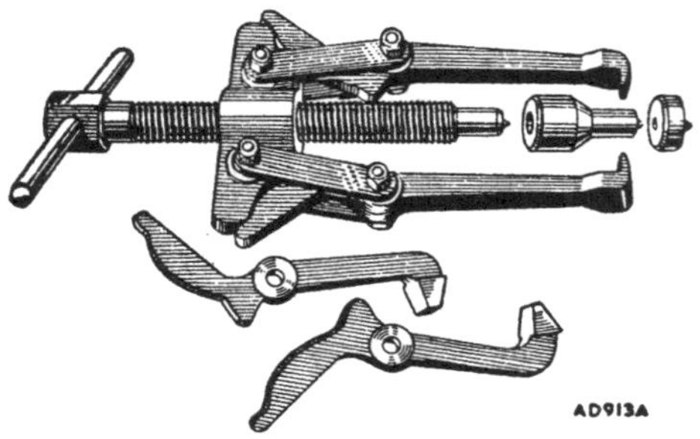


18G 2. Crankshaft Gear, Pulley, and Propeller Shaft Flange Remover.

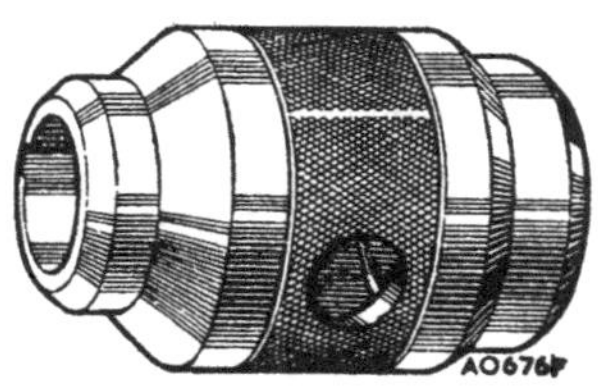


18G 3. Engine Front Cover Locating Bush.

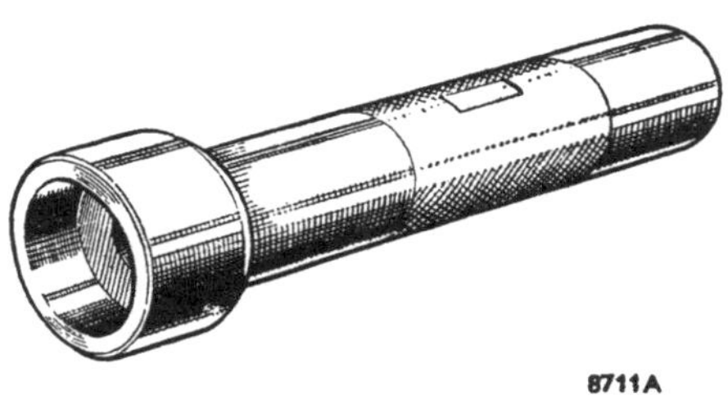


18G 4. First Motion Shaft Assembly Replacer.

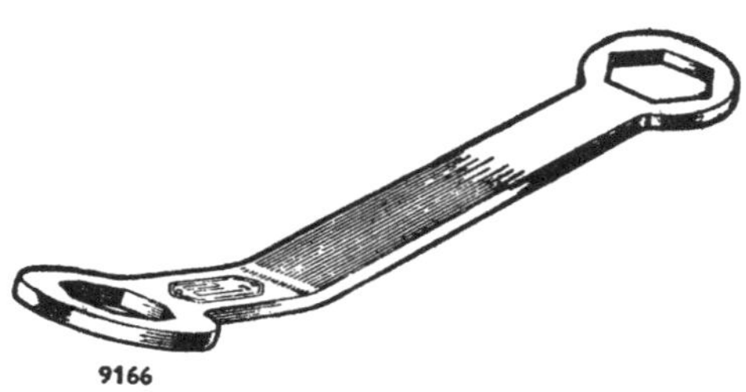


18G 5. First Motion Shaft Nut Spanner.

18G 25. Valve Seat Finishing Cutter (exhaust).

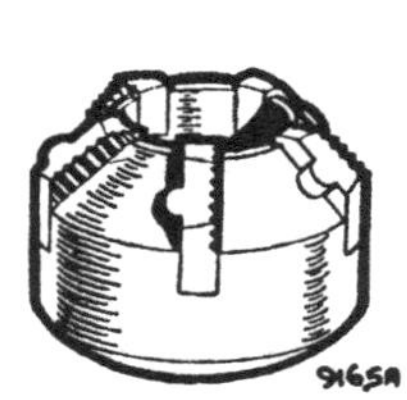


18G 25 A. Valve Seat Glaze Breaker (exhaust).

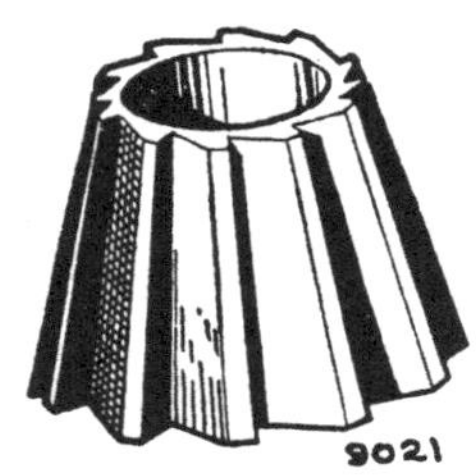


18G 25 C. Valve Seat Narrowing Cutter—Bottom (exhaust).

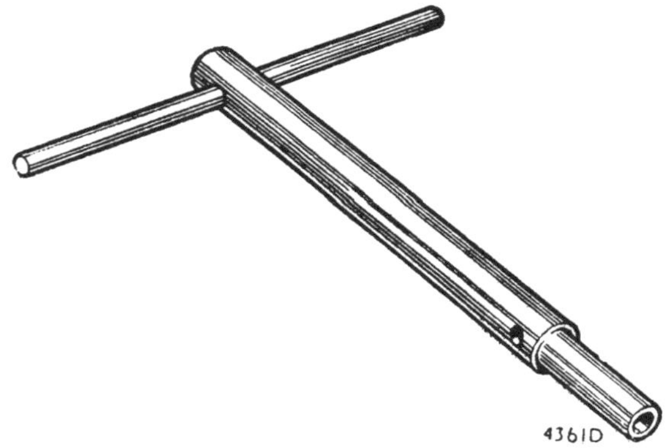


18G 27. Valve Seat Cutter and Pilot Handle.

**18G 28. Valve Seat Finishing Cutter (inlet).**

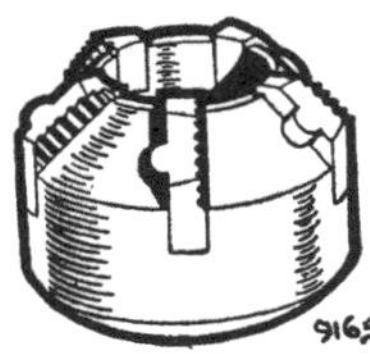

9165A

**18G 28 A. Valve Seat Glaze Breaker (inlet).**

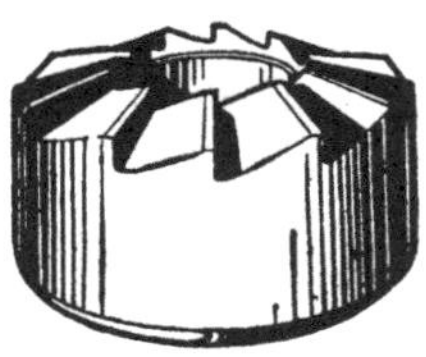

9021A

**18G 28 B. Valve Seat Narrowing Cutter—Top (inlet).**

9021

**18G 28 C. Valve Seat Narrowing Cutter—Bottom (inlet).**

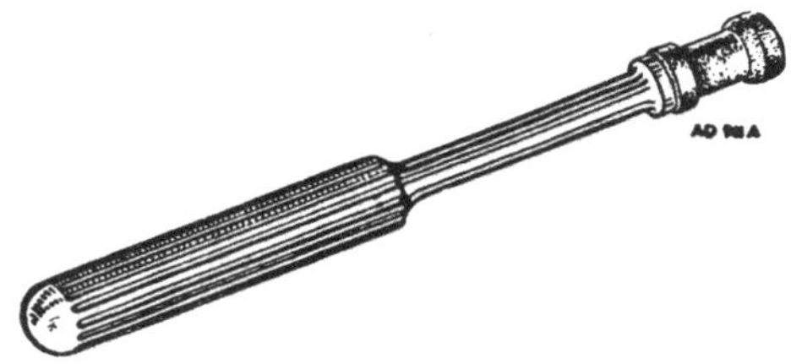

**18G 29. Valve Grinding-in Tool.**

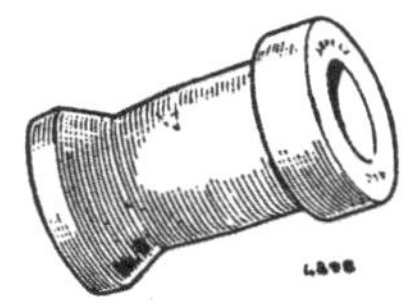

**18G 29 A. Valve Grinding-in Tool Suction Pad.**

8710

**18G 34 A. Bevel Pinion Flange Wrench.**

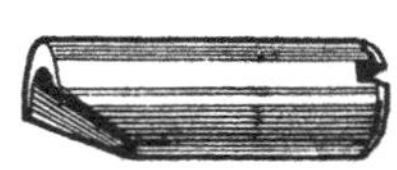

4349L

**18G 41. Selector Fork and Rod Guide.**

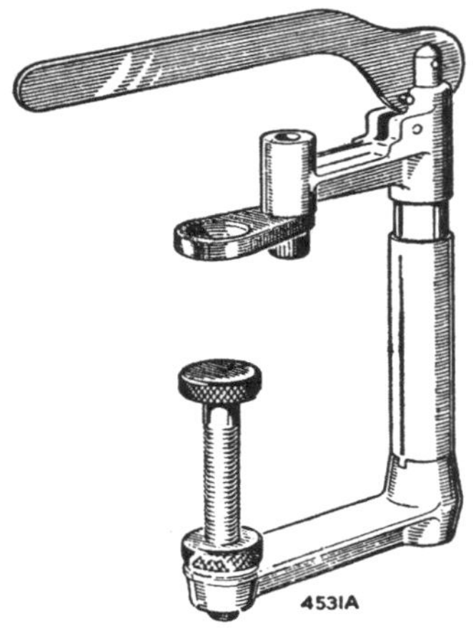


18G 45. Valve Spring Compressor.

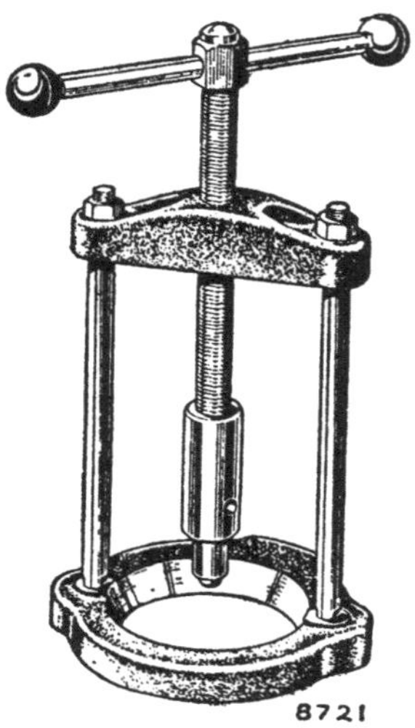


18G 47 C. Differential Bearing Remover (basic tool).

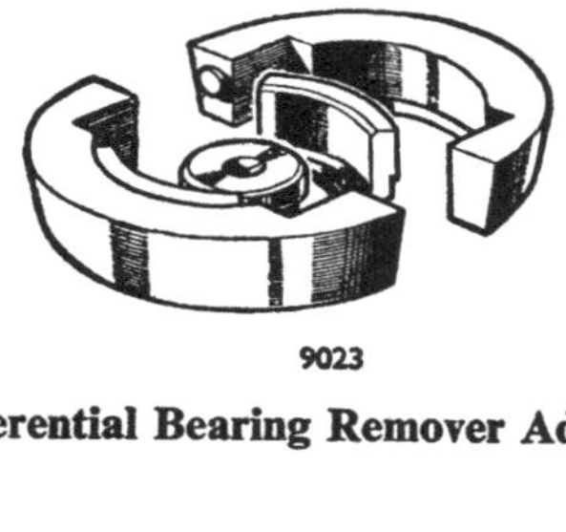


18G 47 T. Differential Bearing Remover Adaptor.

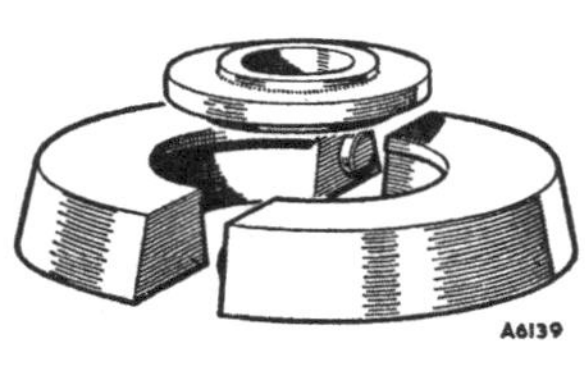


18G 47 AH. Bevel Pinion Inner Race Remover and Replacer Adaptor.

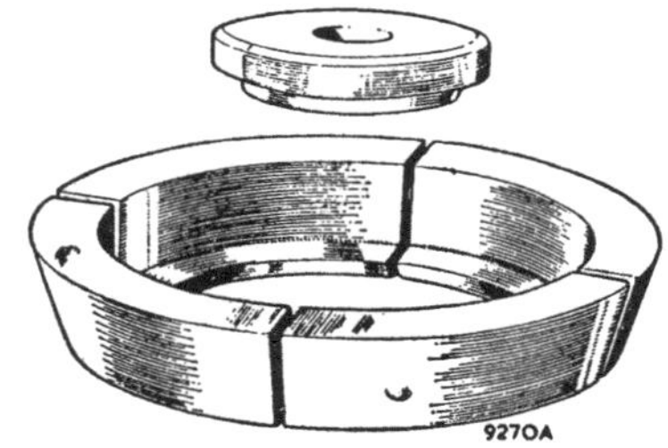


18G 47 AK. Differential Bearing Remover Adaptor.

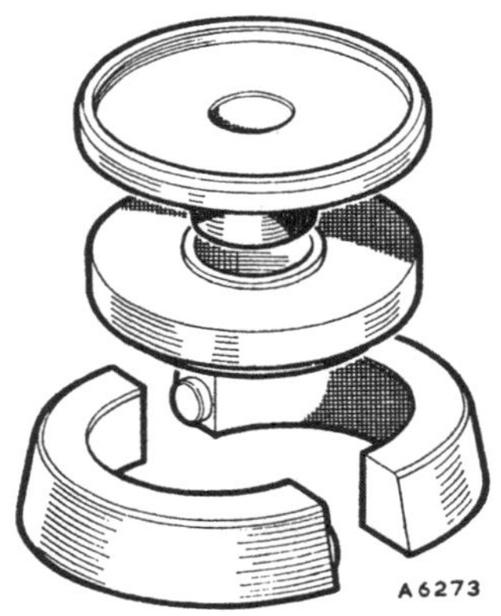


18G 47 AP. Annulus Tail Shaft Bearing Remover and Replacer Adaptor.

18G 47 AS. Bevel Pinion Bearing Remover and Replacer Adaptor.

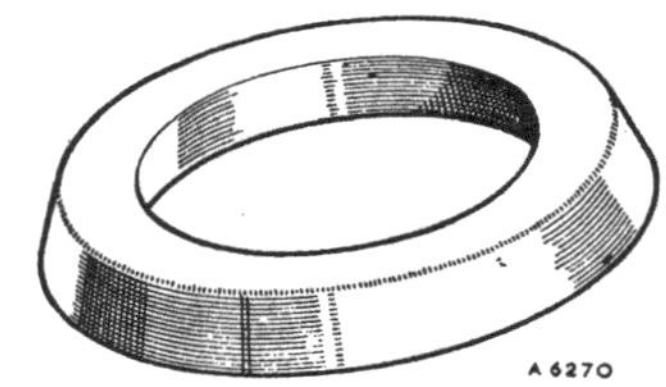


18G 47 CZ. Basic Adaptor Ring.

NOTE.—If 18G 12 A is held, this may be used with 18G 47 AP in lieu of 18G 47 C and 18G 47 CZ.

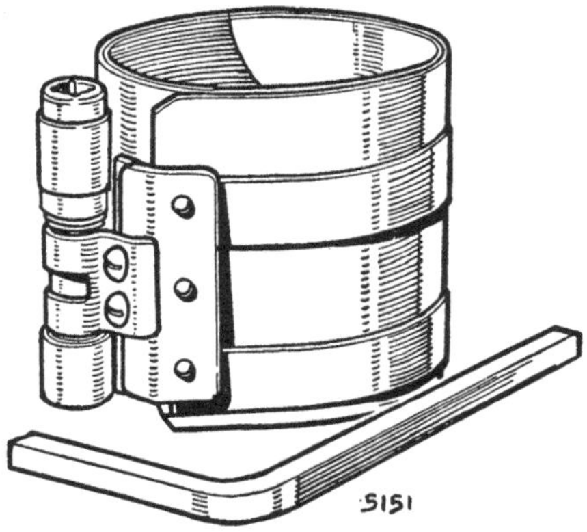


18G 55 A. Piston Ring Clamp.

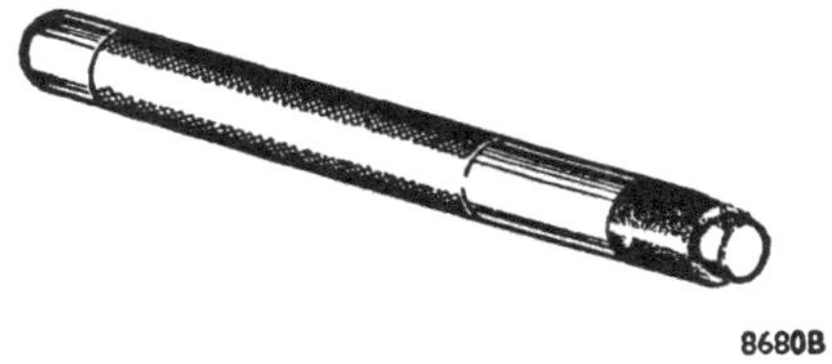


18G 69. Oil Pump Relief Valve Grinding-in Tool.

18G 68. Swivel Axle Bush Reamer Wrench.

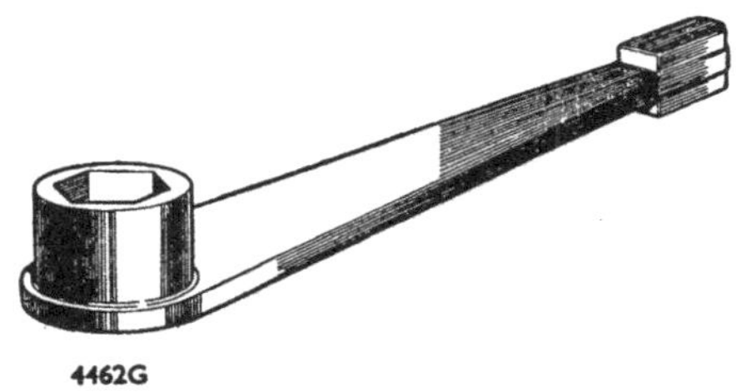


18G 98 A. Starting Nut Spanner.

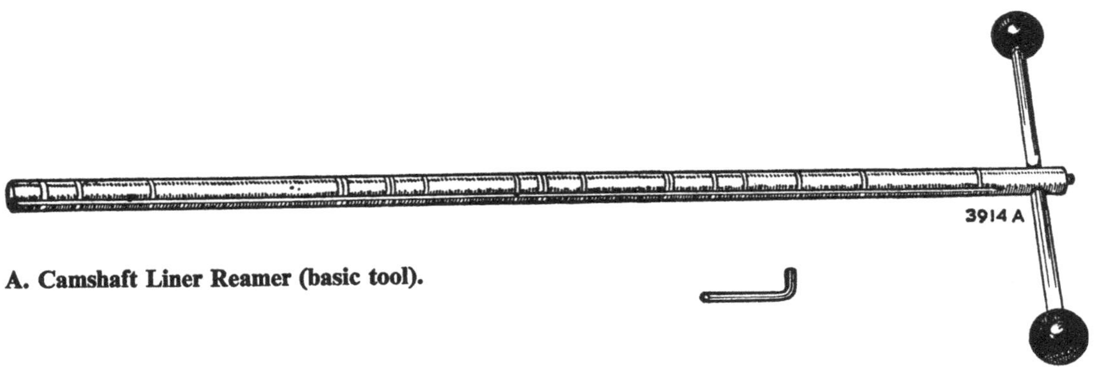


18G 123 A. Camshaft Liner Reamer (basic tool).

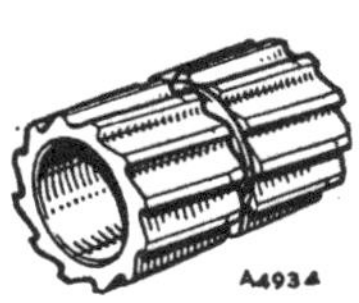


18G 123 B. Camshaft Liner Reamer Cutter—Rear.

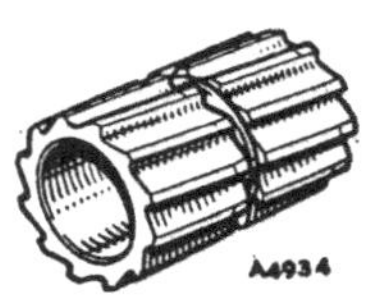


18G 123 F. Camshaft Liner Reamer Cutter—Centre.

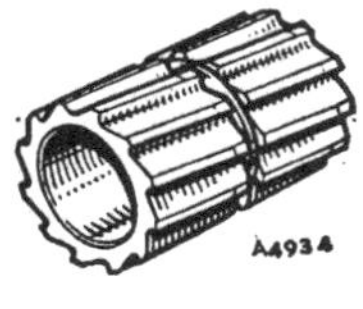


18G 123 E. Camshaft Liner Reamer Cutter—Front.

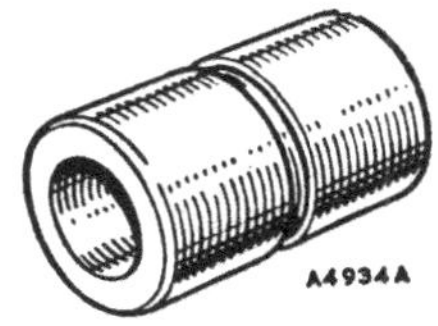


18G 123 L. Camshaft Liner Reamer Pilot—Front.

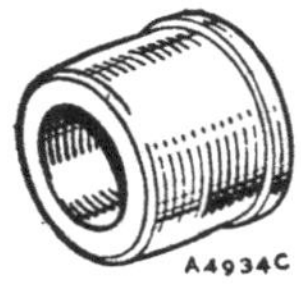


**18G 123 T. Camshaft Liner Reamer Pilot—Front.**

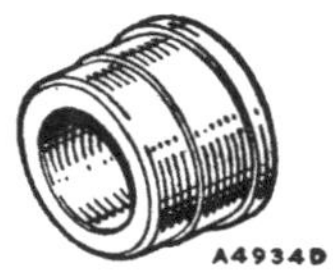


**18G 123 AB. Camshaft Liner Reamer Pilot—Centre.**

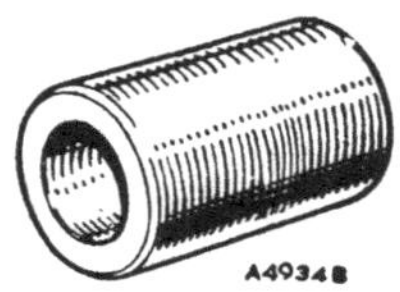


**18G 123 AC. Camshaft Liner Reamer Pilot—Rear.**

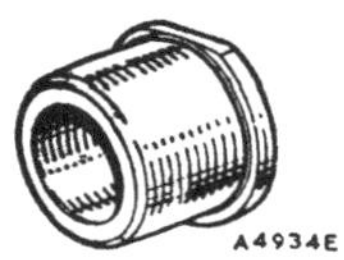


**18G 123 AD. Camshaft Liner Reamer Pilot—Rear.**

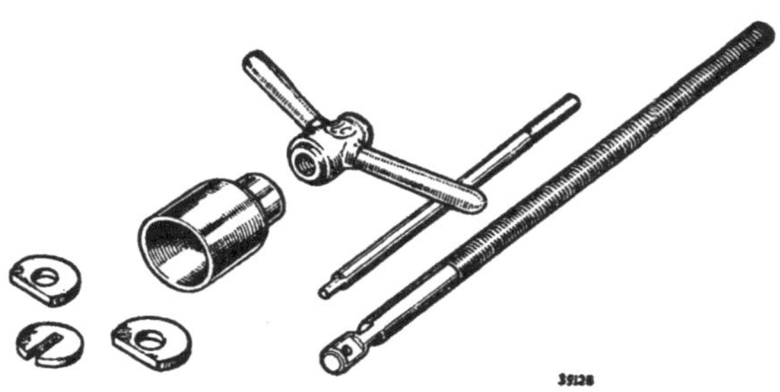

**18G 124 A. Camshaft Liner Remover and Replacer (basic tool).**

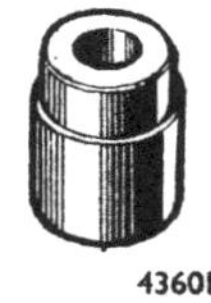


**18G 124 B. Camshaft Liner Remover Adaptor.**

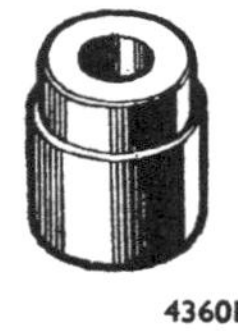


**18G 124 C. Camshaft Liner Remover Adaptor.**

**18G 124 F. Camshaft Liner Remover Adaptor.**

**18G 124 H. Camshaft Liner Remover Adaptor.**

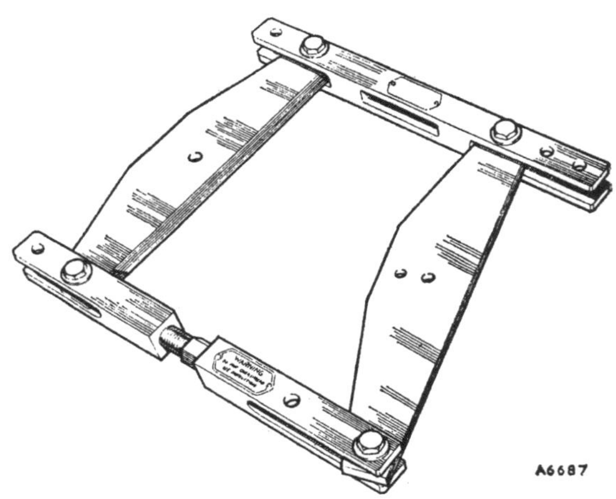


**18G 131 C. Rear Axle Casing Stretcher.**

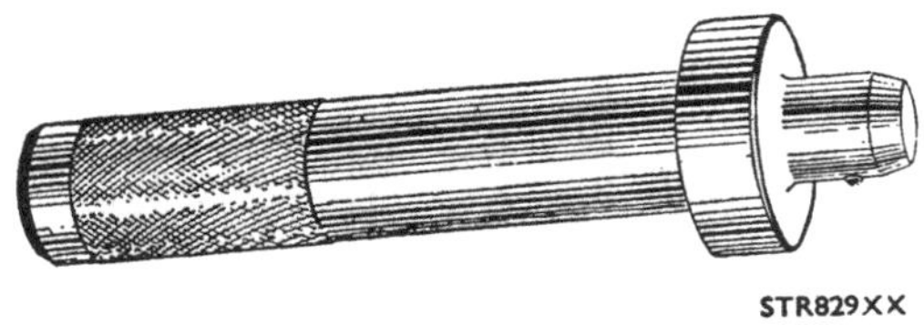


**18G 134. Bearing and Oil Seal Replacer (basic tool).**

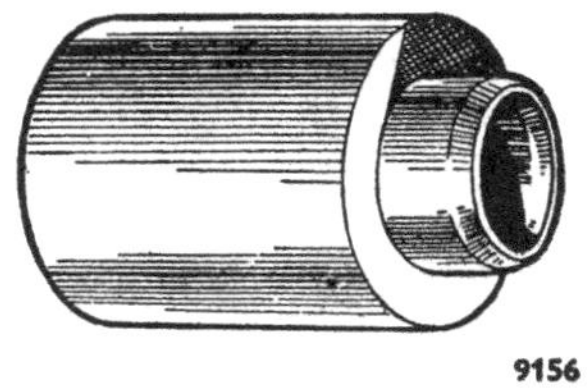


**18G 134 BK. Gearbox Rear Oil Seal Replacer Adaptor.**

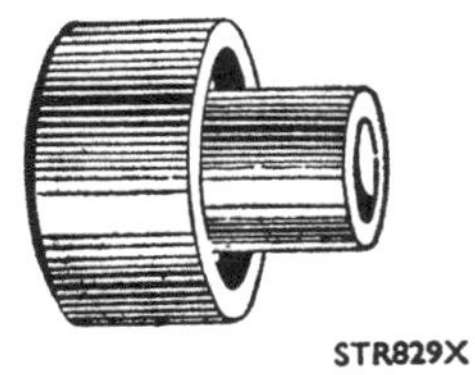


**18G 134 P. Rear Hub Replacer Adaptor.**

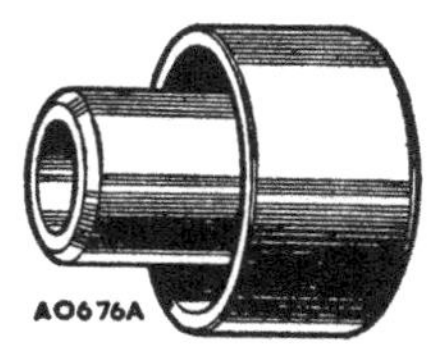


**18G 134 CM. Differential Bearing Replacer Adaptor.**

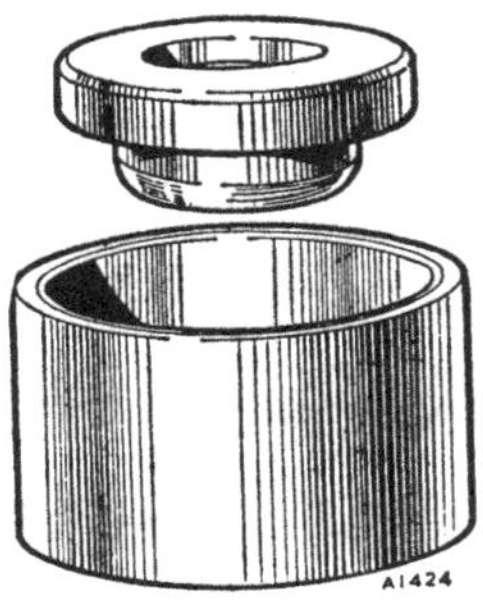


**18G 134 BD. Timing Case Oil Seal Replacer Adaptor.**

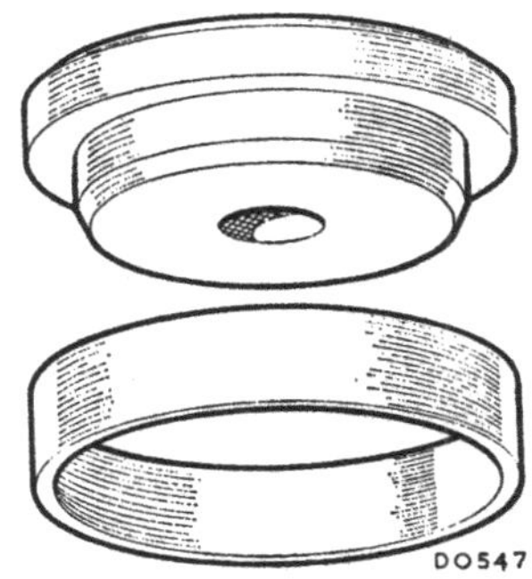


**18G 134 CQ. Crankshaft Oil Seal Replacer Adaptor.**

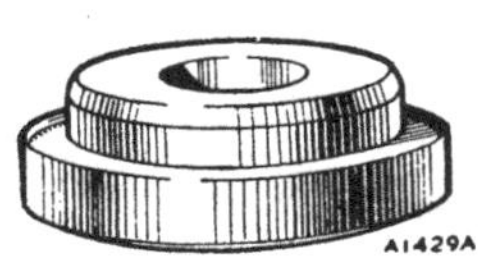


**18G 134 BH. Flywheel and Front Hub Oil Seal Replacer Adaptor.**

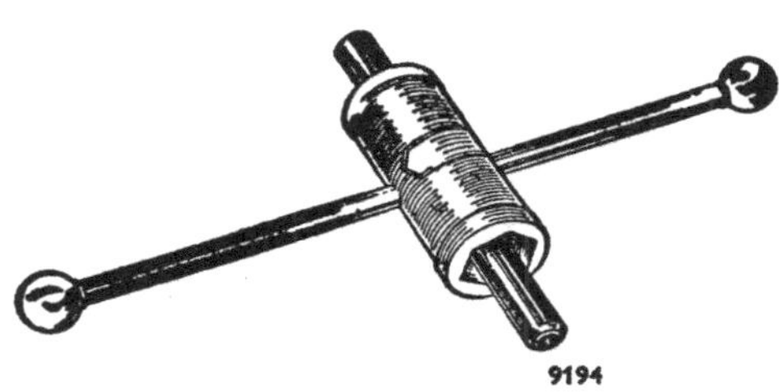


**18G 152. Rear Hub Nut Spanner.**

9021A

**18G 174 B. Valve Seat Narrowing Cutter—Top (exhaust).**

4003

**18G 174 D. Valve Seat Cutter Pilot.**

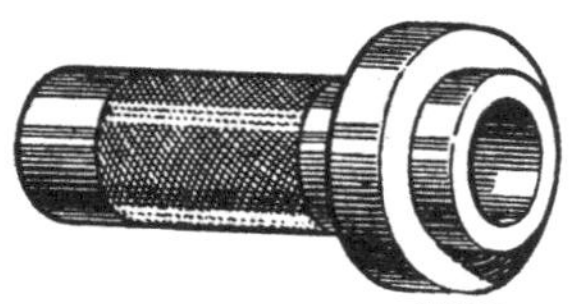

4345H

**18G 177. Oil Seal Replacer.**

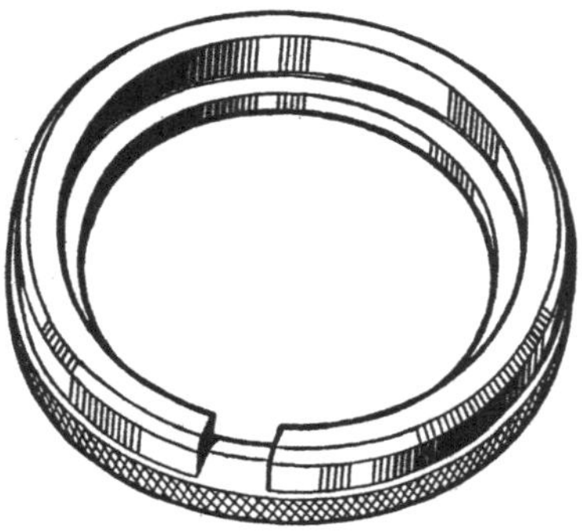

4646B

**18G 178. Roller Clutch Assembly Ring.**

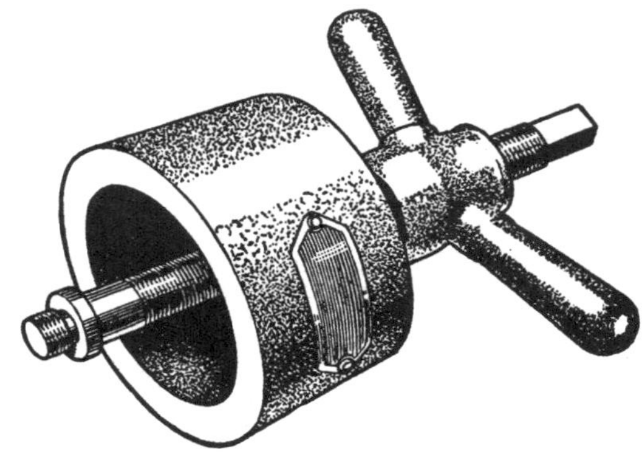

4344J

**18G 183. Oil Pump Body Replacer.**

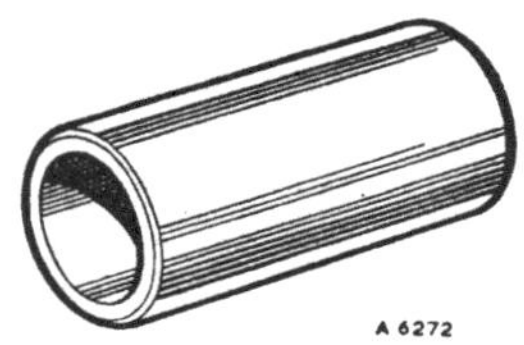

A 6272

**18G 183 A. Oil Pump Barrel Remover Adaptor.**

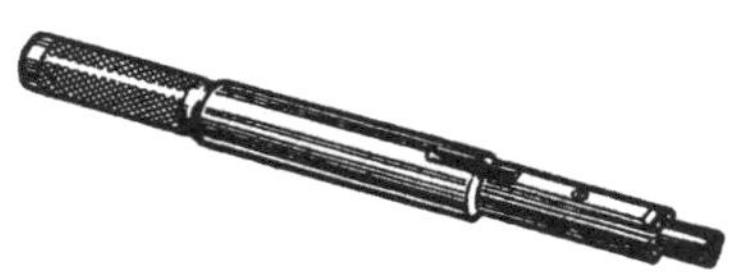

4346C

**18G 185. Dummy Layshaft.**

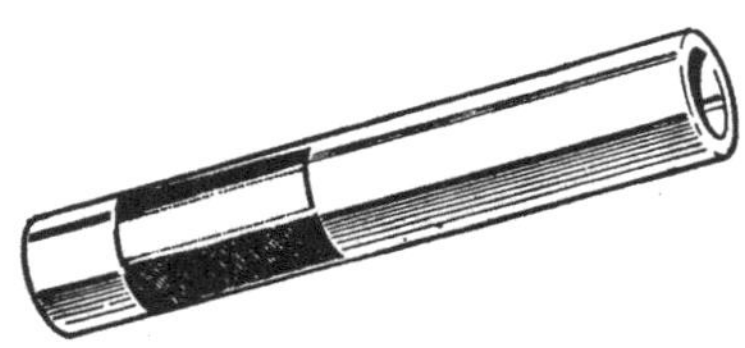

9214

**18G 186. Mainshaft Bearing Replacer.**

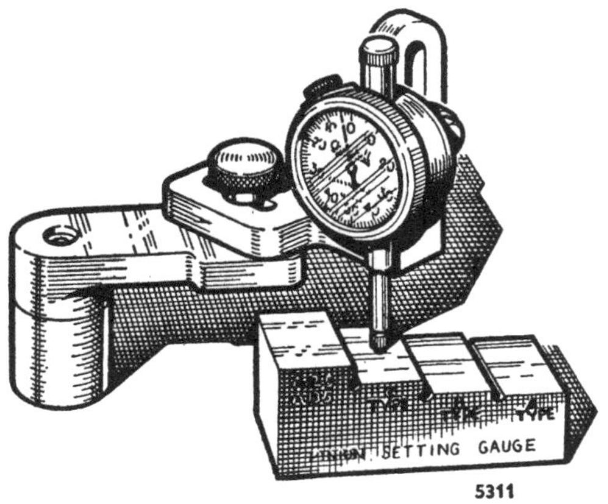


**18G 191. Bevel Pinion Setting Gauge.**

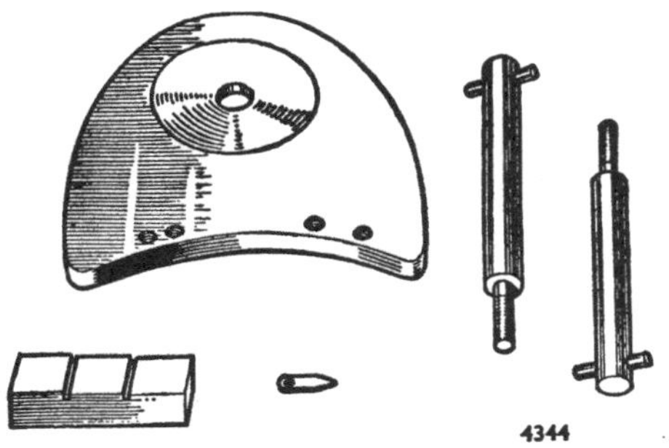


**18G 191 A. Differential Bearing Gauge.**

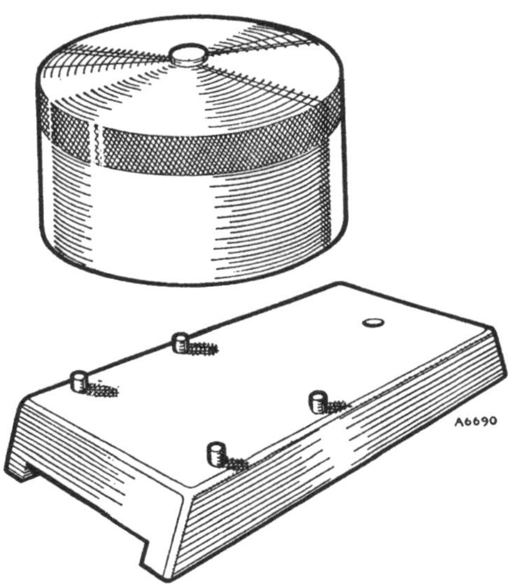


**18G 191 F. Differential Case Assembly Gauge.**

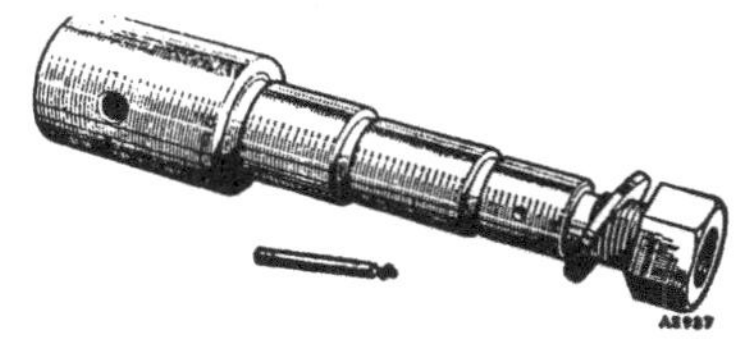

**18G 191 H. Dummy Pinion.**

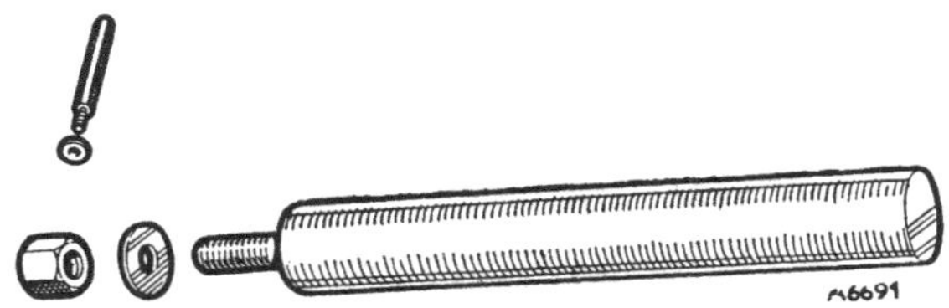


**18G 191 J. Differential Case Assembly Gauge Pillar Adaptor.**

**18G 207. Bevel Pinion Preload Gauge.**

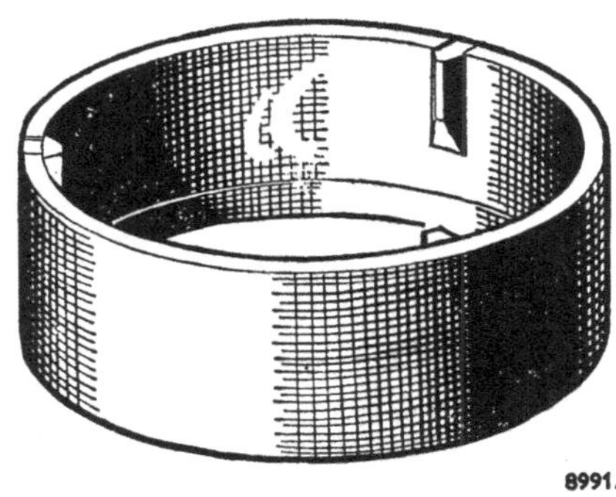


**18G 222. Synchromesh Unit Assembly Ring—Second Speed.**

**18G 223. Synchromesh Unit Assembly Ring—Third and Top.**

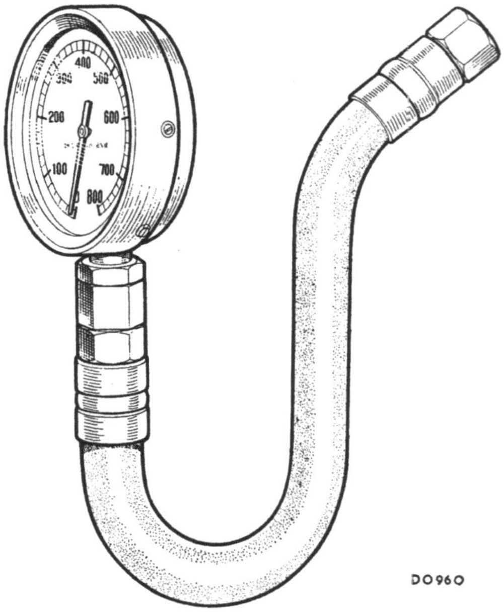


**18G 251. Hydraulic Pressure Gauge.**

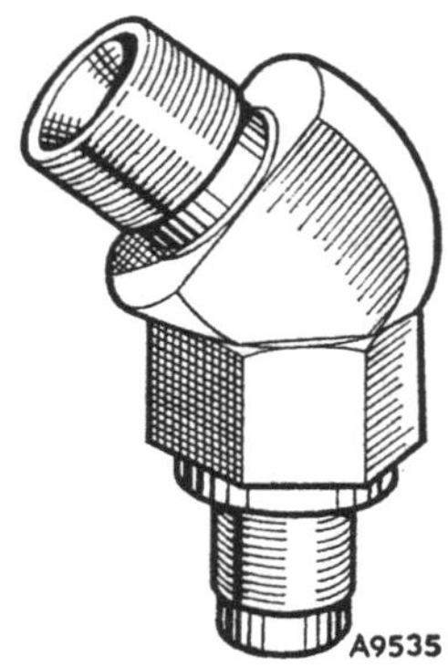


**18G 251 D. Pressure Test Adaptor.**

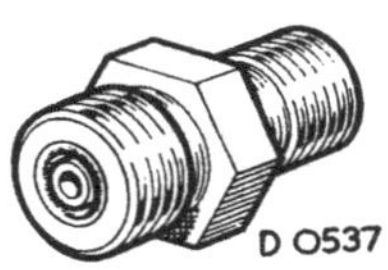


**18G 251 E. Pressure Test Adaptor.**

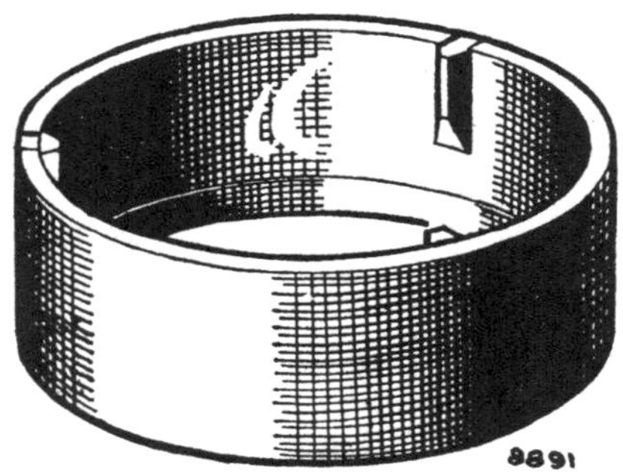


**18G 262. Synchromesh Unit Assembly Ring (First and Second Speed).**

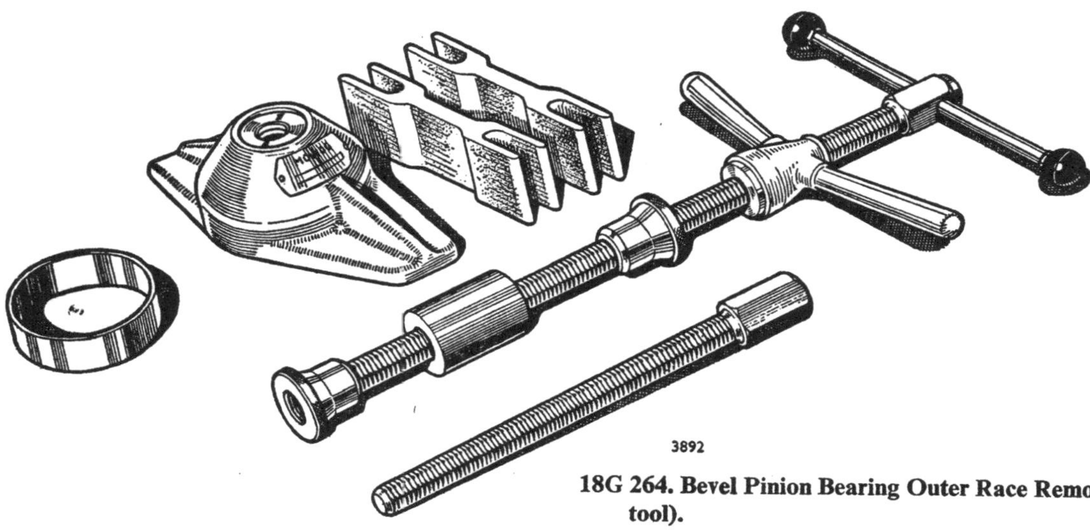


**18G 264. Bevel Pinion Bearing Outer Race Remover (basic tool).**

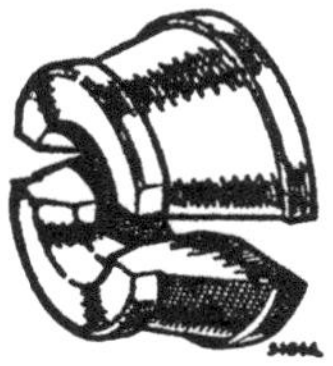

**18G 264 E and 18G 264 F. Bevel Pinion Bearing Outer Race Remover Adaptors—Rear.**

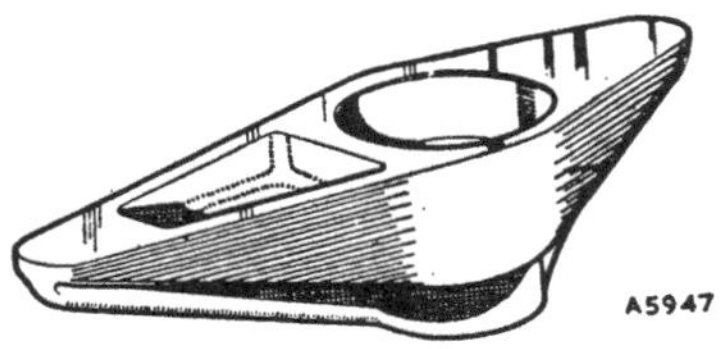


**18G 264 AA. Bridge Piece Adaptor—Large.**

**18G 264 AB. Bevel Pinion Outer Race Remover and Replacer Adaptor (Rear).**

**18G 264 AD. Bevel Pinion Outer Race Remover and Replacer Adaptor (Front).**

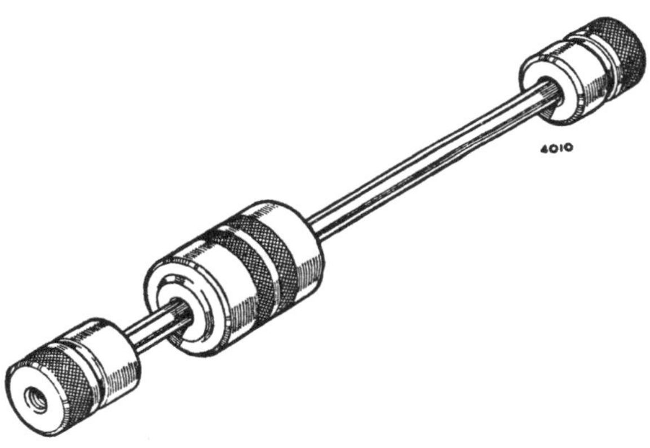


**18G 284. Impulse Extractor—UNF. (basic tool).**

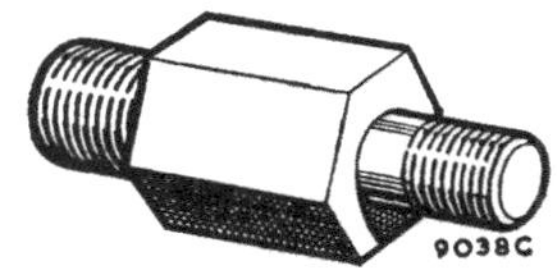


**18G 284 A. Main Bearing Cap Remover Adaptor.**

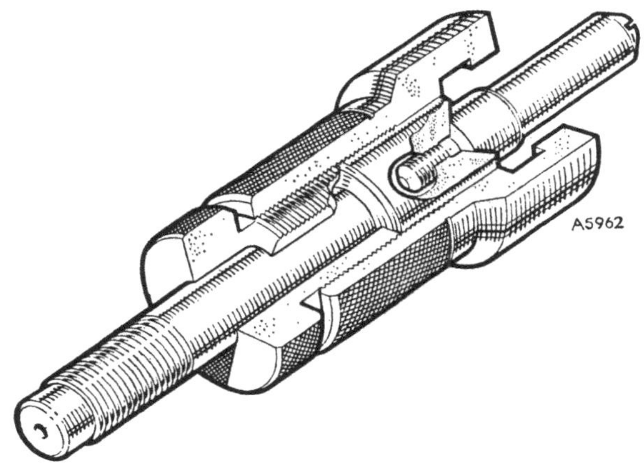


**18G 284 H. Basic Bearing Remover Adaptor.**

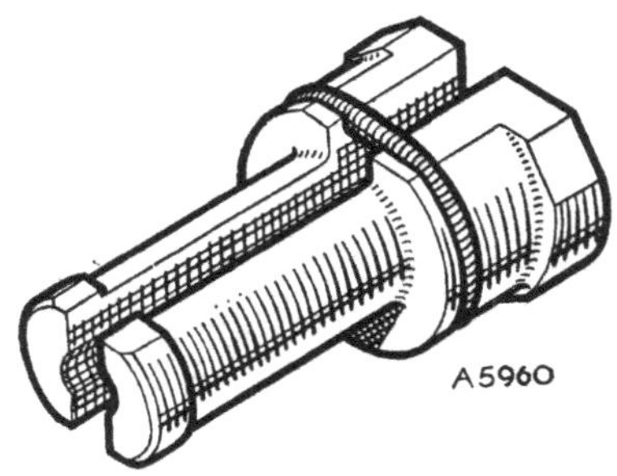


**18G 284 L. Crankshaft Spigot Bush Remover Adaptor.**

Used in conjunction with basic tool adaptors 18G 284 H.

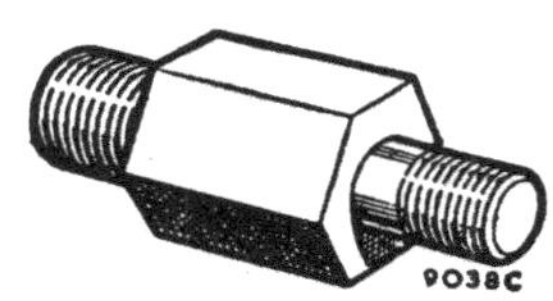


**18G 284 AC. Main Bearing Cap Remover Adaptor.**

For removing the intermediate main bearing caps commencing at the following engine numbers:

| | |
|---|---|
| 18GB/U/H68055 | 18GB/RU/H68315 |
| 18GB/U/L60571 | 18GB/RU/L58224 |

8251E

**18G 304. Front and Rear Hub Remover (basic tool).**

8251F

**18G 304 A. Bolt Adaptor—½ in. UNF.**

8251G

**18G 304 J. Hub Remover Thrust Pad.**

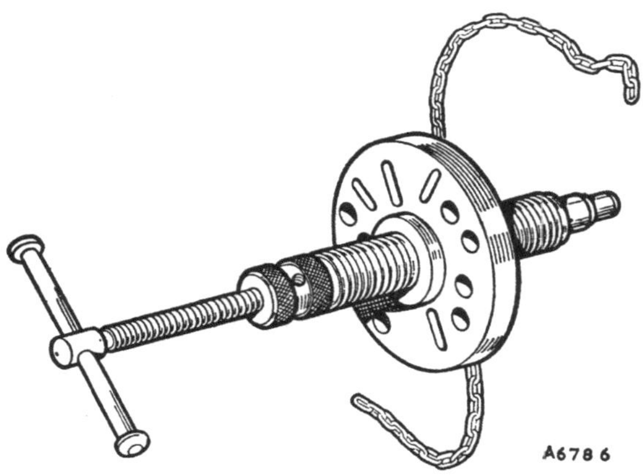

A678 6

**18G 304 Z. Hydraulic Hub Remover (basic tool).**

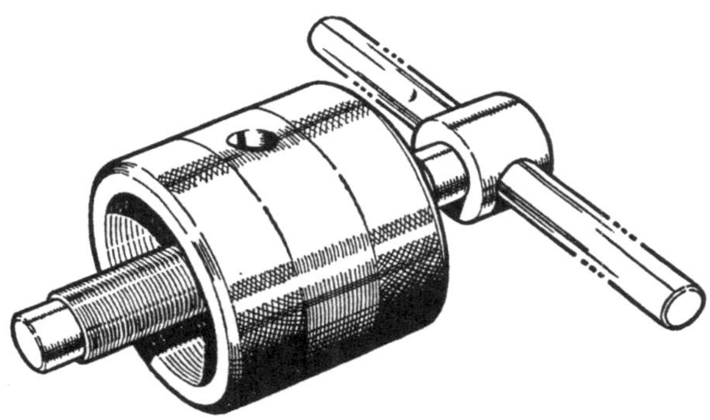

8327W

**18G 363 Hub Remover Wire Wheels (12 T.P.I.).**

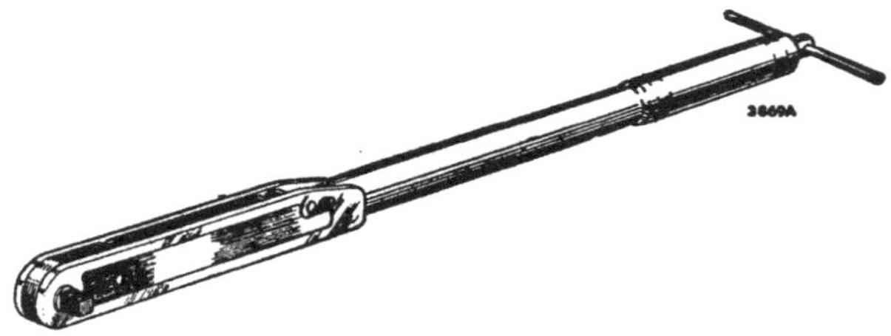

3869A

**18G 372. Torque Wrench—30 to 140 lb. ft. (4·15 to 19·4 kg. m.).**

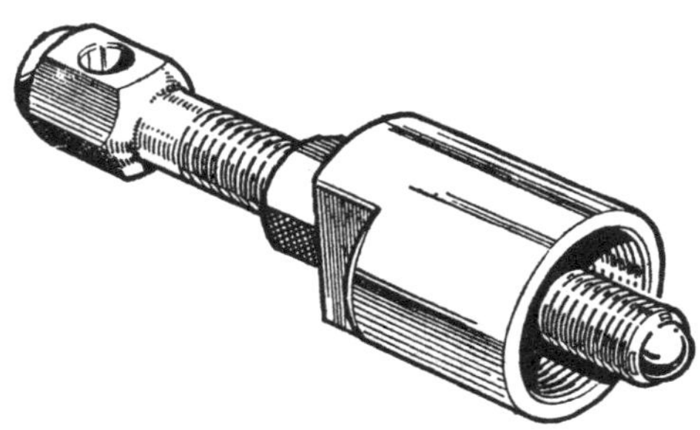

9161C

**18G 389. Gearbox Rear Oil Seal Remover (basic tool).**

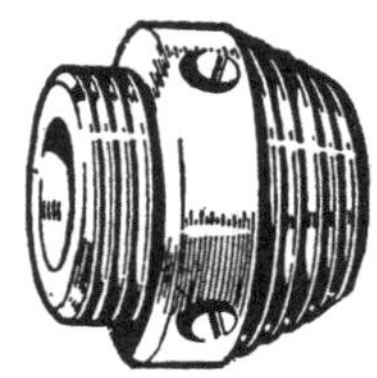

9168

**18G 389 B. Gearbox Rear Oil Seal Remover Adaptor.**

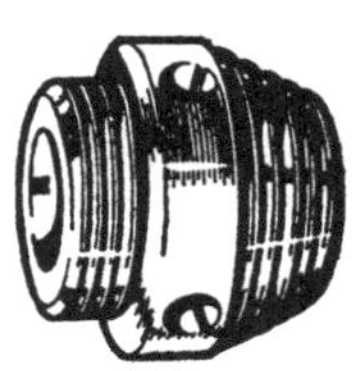

9168

**18G 389 D. Gearbox Rear Oil Seal Remover Adaptor.**

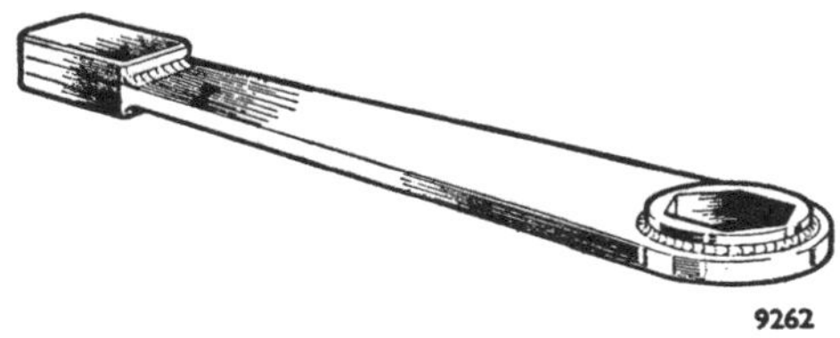


**18G 391. Starting Dog Nut Spanner.**

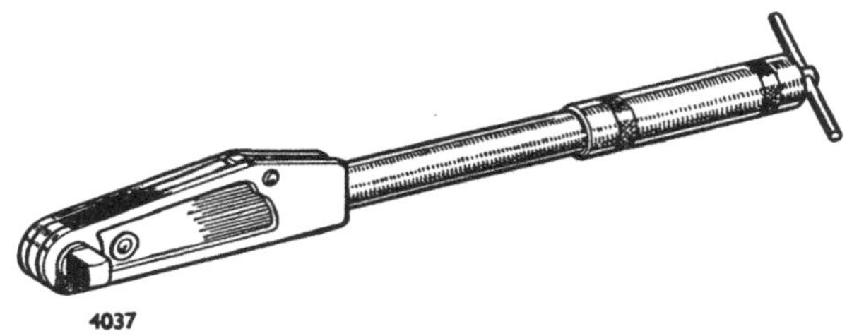


**18G 536. Torque Wrench—20 to 100 lb. in.—2 to 8 lb. ft. (300 to 1200 gm. m.).**

**18G 536 A. Torque Wrench Adaptor—$\frac{3}{8}$ to $\frac{1}{2}$ in.**

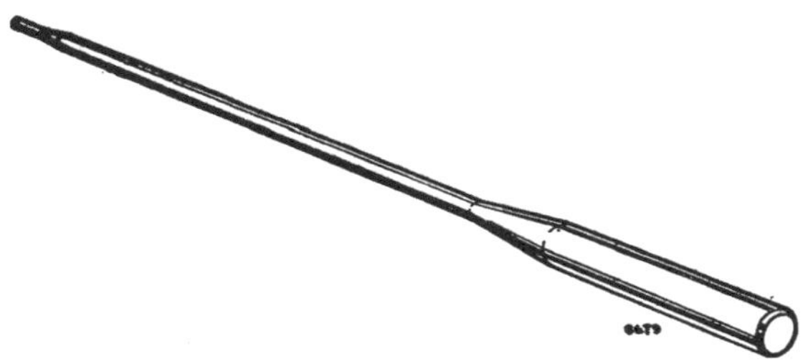

**18G 471. Dummy Layshaft.**

**18G 536 B. Torque Wrench to Screwdrive Blade Adaptor.**

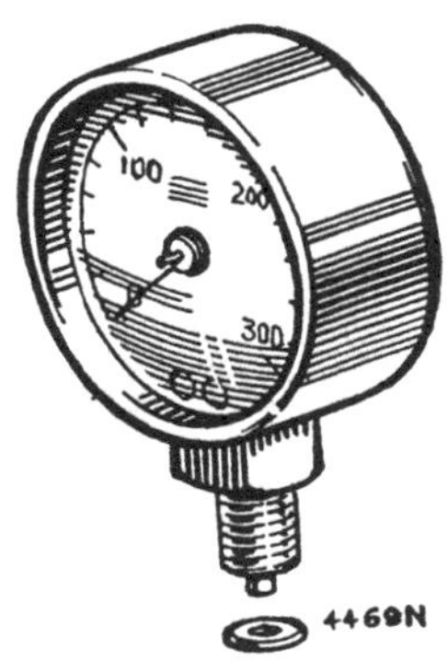


**18G 502 A. Hydraulic Pressure Gauge.**

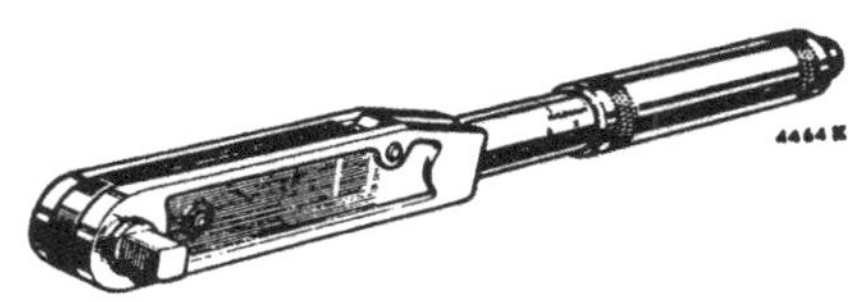

**18G 537. Torque Wrench—10 to 50 lb. ft. (2 to 7 kg. m.).**

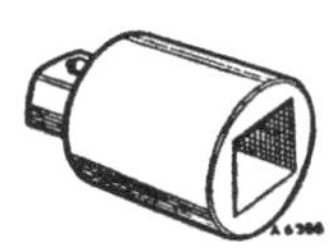

**18G 537 A. Torque Wrench Adaptor—$\frac{1}{2}$ to $\frac{3}{8}$ in. square drive.**

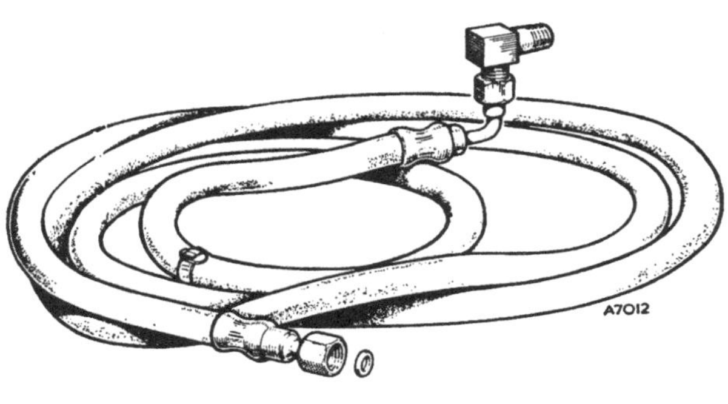


**18G 502 K. Pressure Hose with Adaptor.**

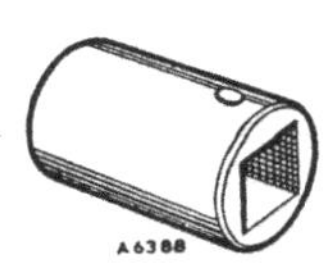


**18G 537 B. Torque Wrench Adaptor—$\frac{3}{8}$ to $\frac{5}{16}$ in. (9·5 to 8 mm.) internal.**

Use with 18G 537 and 18G 537 A.

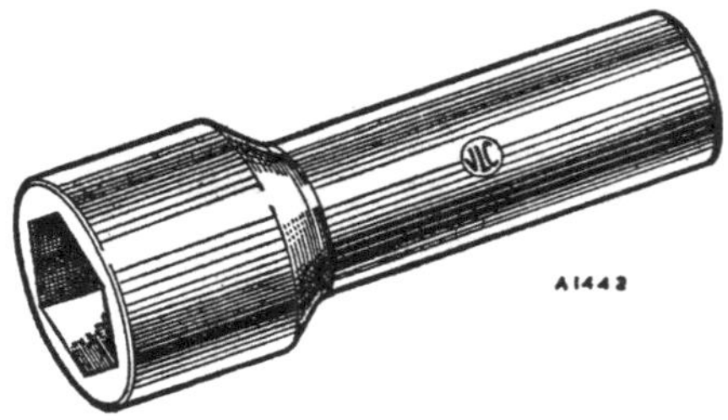


**18G 586. Rear Axle Shaft Nut Spanner (Wire Wheel).**

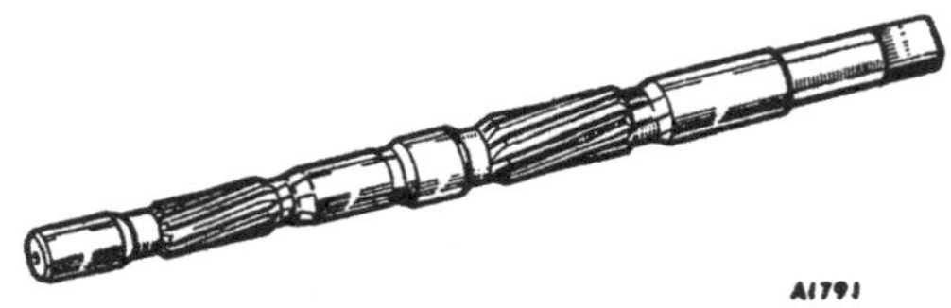


**18G 597. Swivel Axle Bush Reamer.**

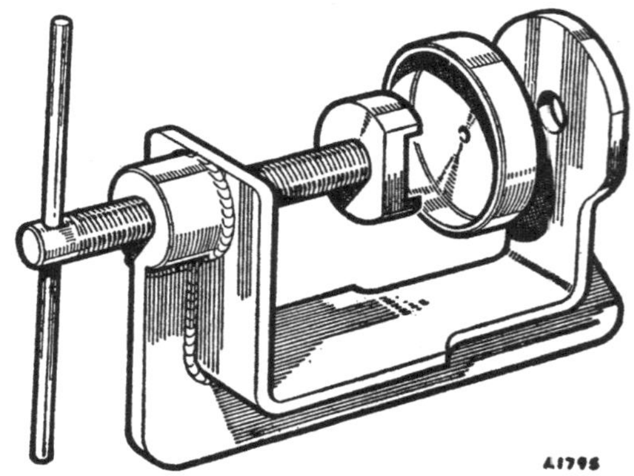


**18G 590. Disc Brake Piston Resetting Tool.**

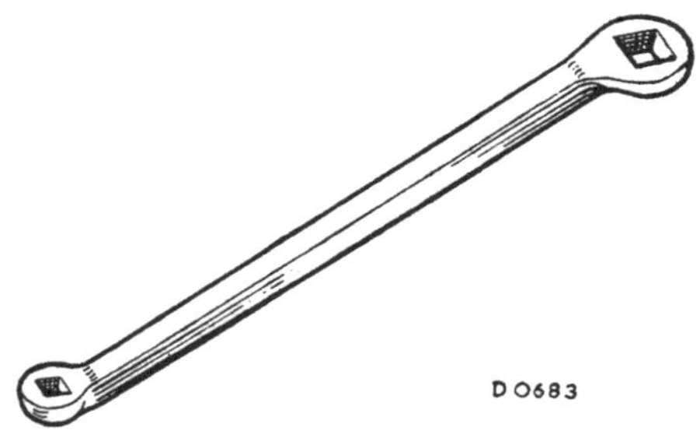


**18G 619 A. Brake Adjusting Spanner.**

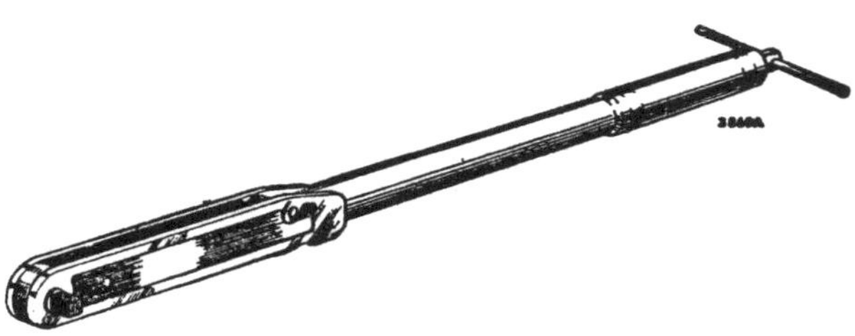

**18G 592. Torque Wrench—35 to 225 lb. ft. (4·84 to 31·144 kg. m.).**

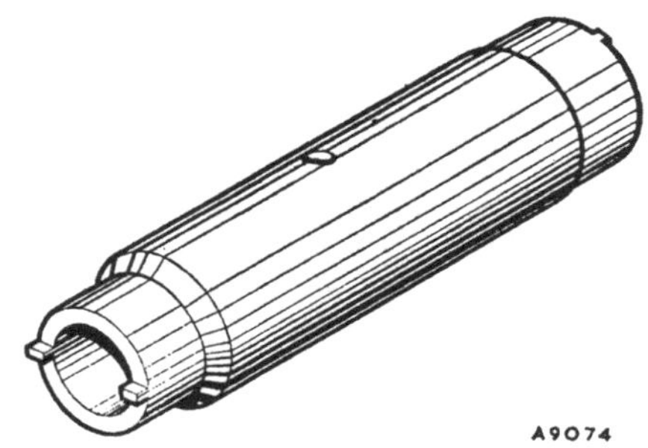


**18G 671. Switch Bezel Light Spanner.**

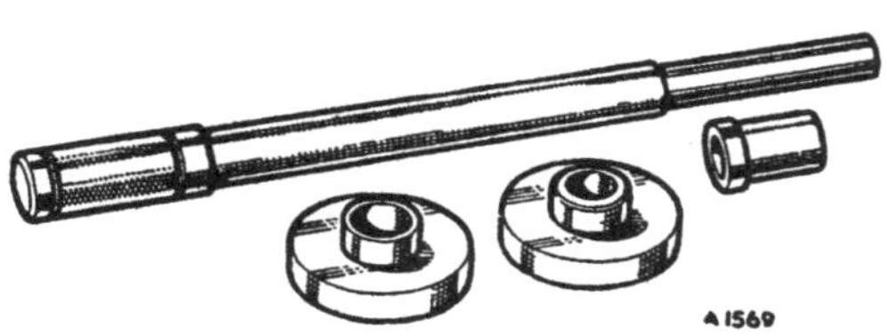


**18G 596. Swivel Axle Bush Remover and Replacer.**

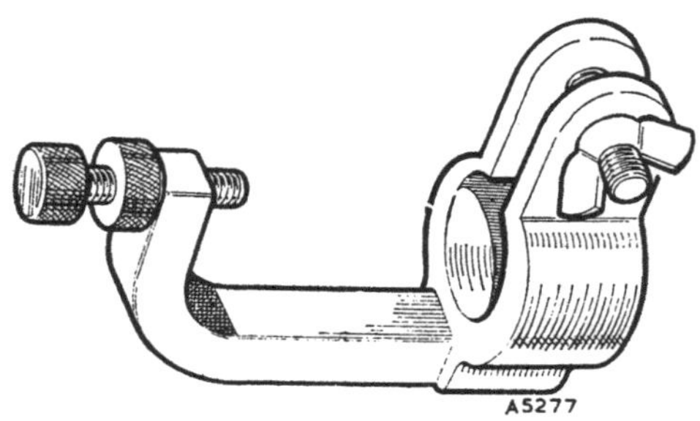


**18G 674. Gear Train End-float Checking Tool.**

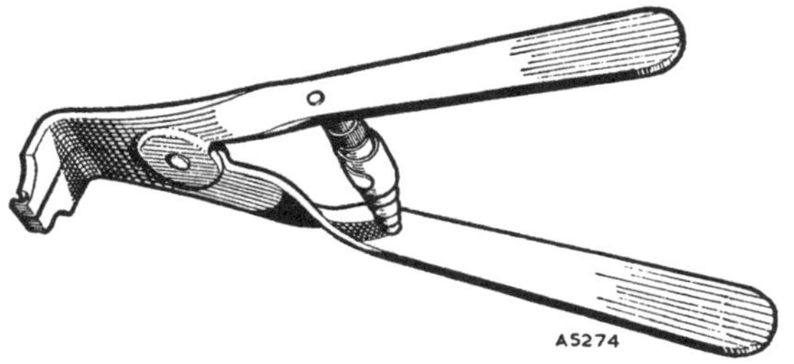


**18G 675. Snap-ring Pliers.**

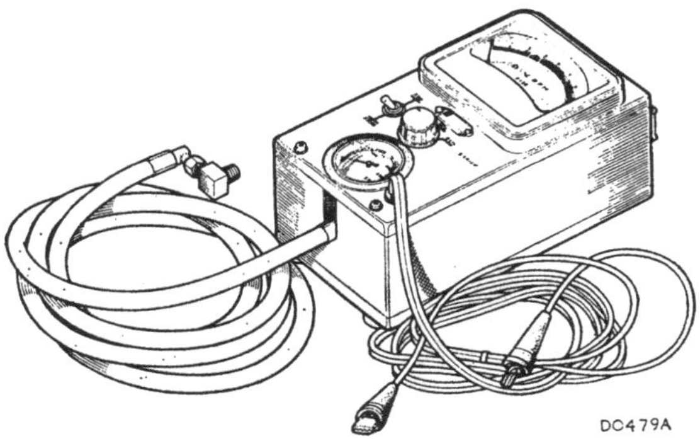


**18G 677 Z. Pressure Test Equipment.**

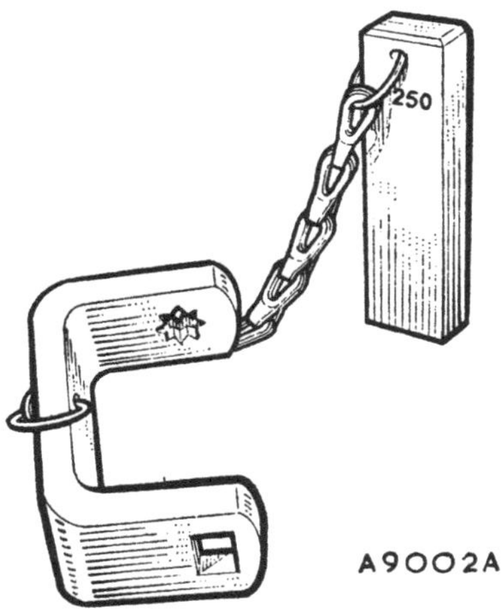


**18G 678. Front Servo Spanner and Gauge.**

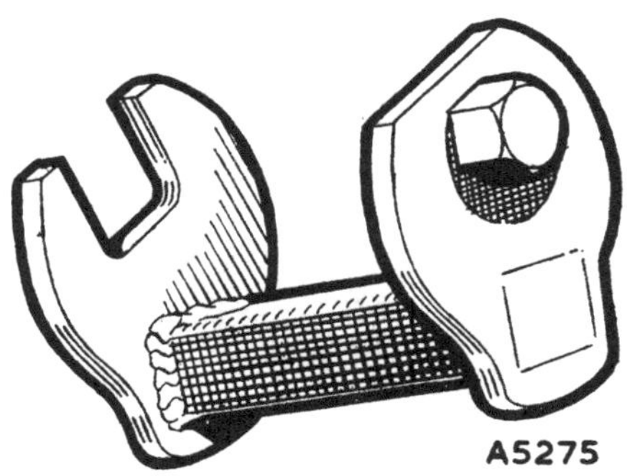


**18G 679. Inhibitor Switch Locknut Spanner.**

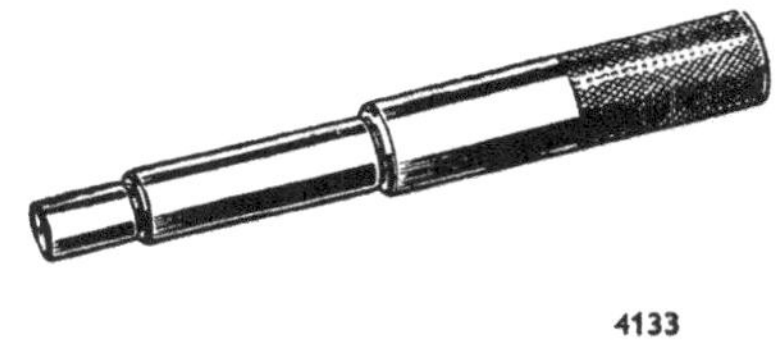


**18G 680. Clutch Plate Centralizer.**

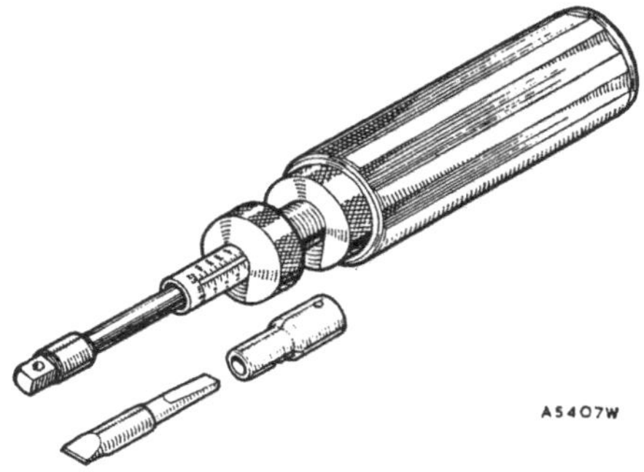


**18G 681. Torque Screwdriver Set—0 to 40 lb. in. (0 to 400 gm. m.).**

**18G 681 A. Torque Screwdriver Bit.**

**18G 681 B. Torque Screwdriver Bit Holder.**

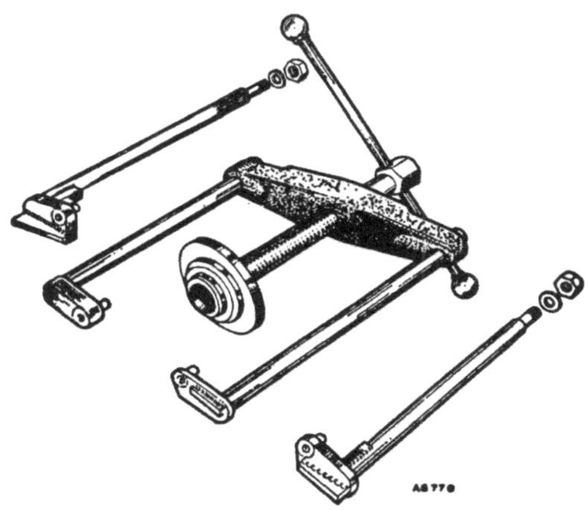

**18G 693. Coil Spring Compressor.**

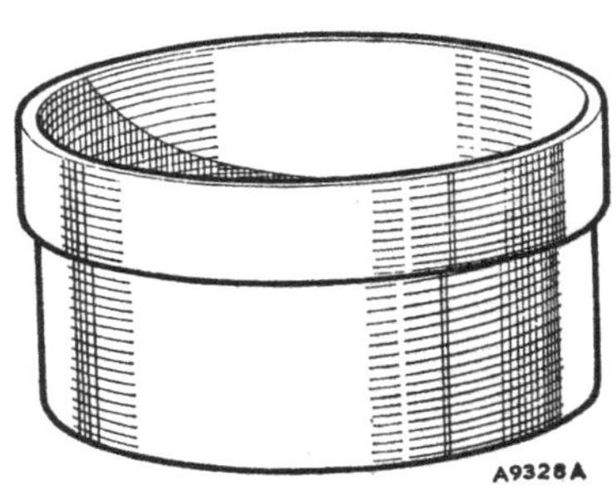


**18G 702. Pistons Rear Clutch and Seal Replacer.**

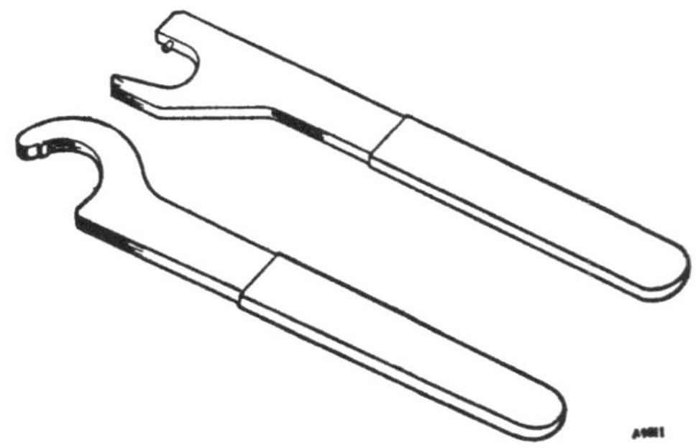

18G 706. Steering Rack Ball Joint Spanners.

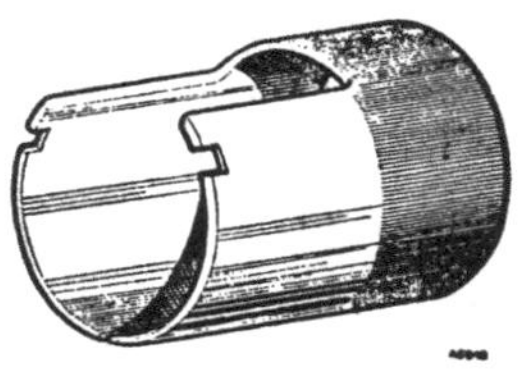

18G 1001. Gauge Locking Ring.

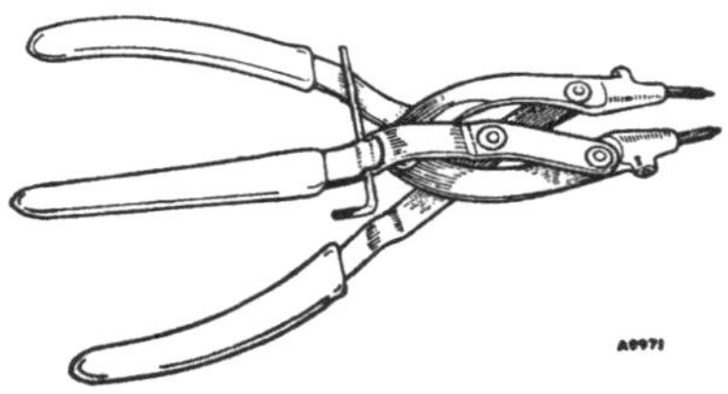

18G 1004. Circlip Pliers (basic tool).

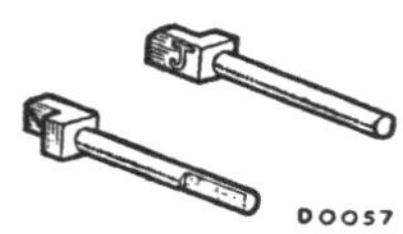


18G 1004 J. Circlip Plier Points.

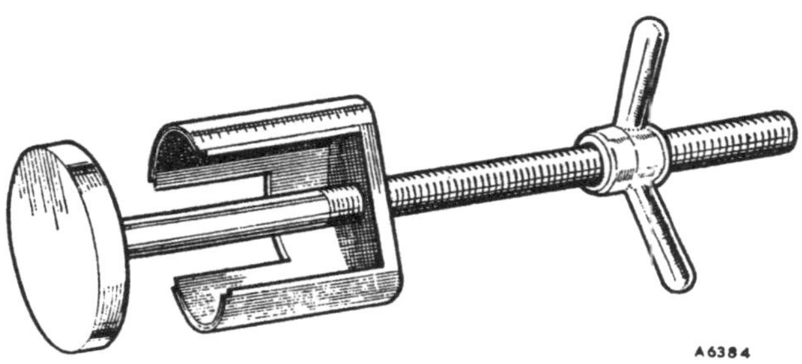


18G 1016. Clutch Spring Compressor—Rear.

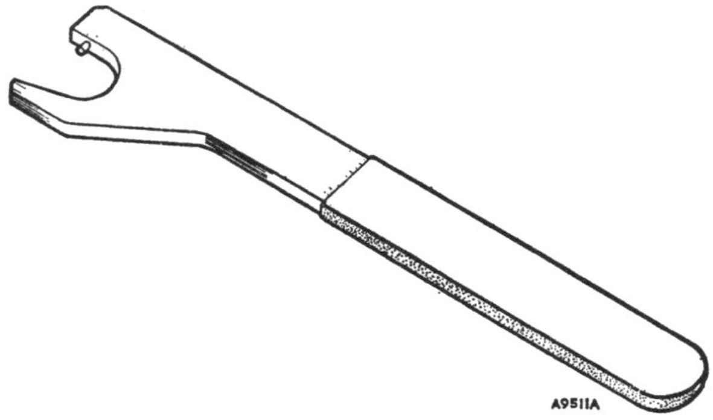


18G 1024. Mainshaft (Third Motion) Spanner.

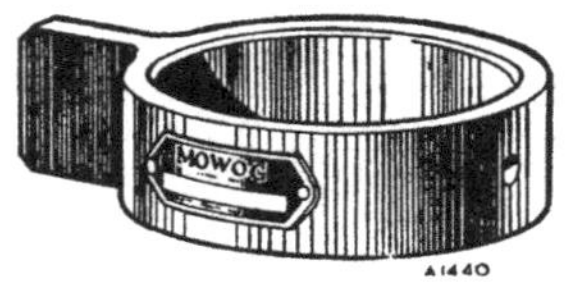


18G 1026. Synchromesh Assembly Ring.

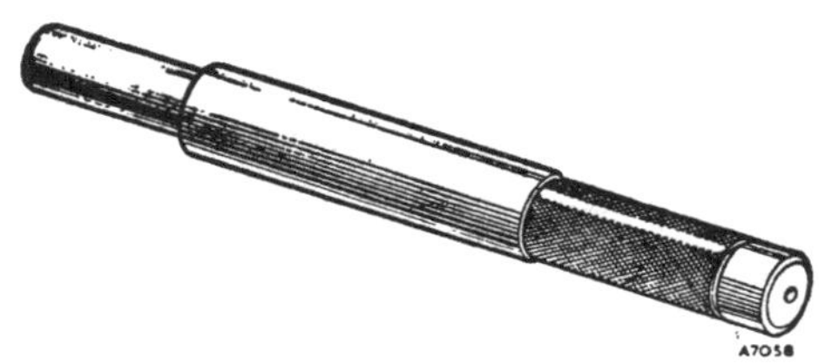

18G 1027. Clutch Centralizer (18GB).

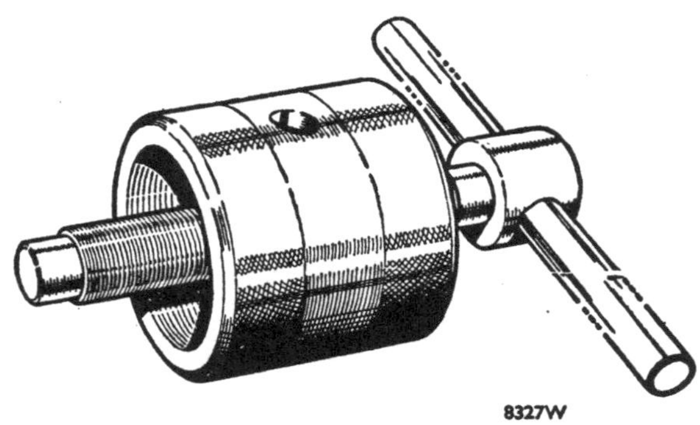


18G 1032. Wire Wheel Hub Remover (8 T.P.I.).

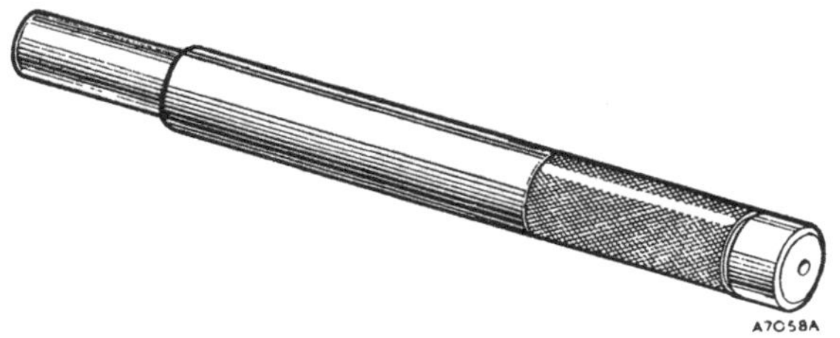


**18G 1037. Crankshaft Spigot Bush Replacer.**

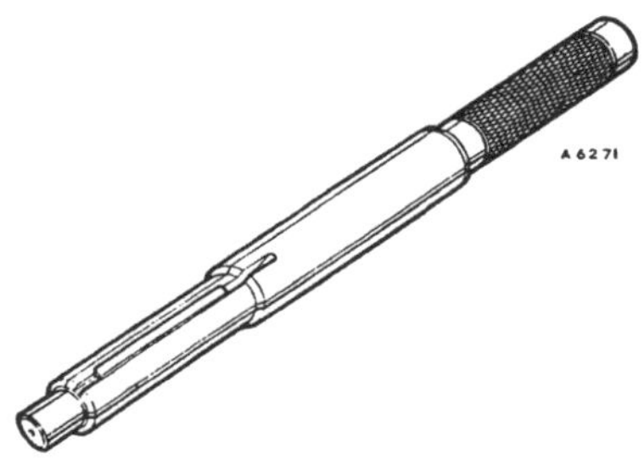


**18G 1039. Dummy Mainshaft.**

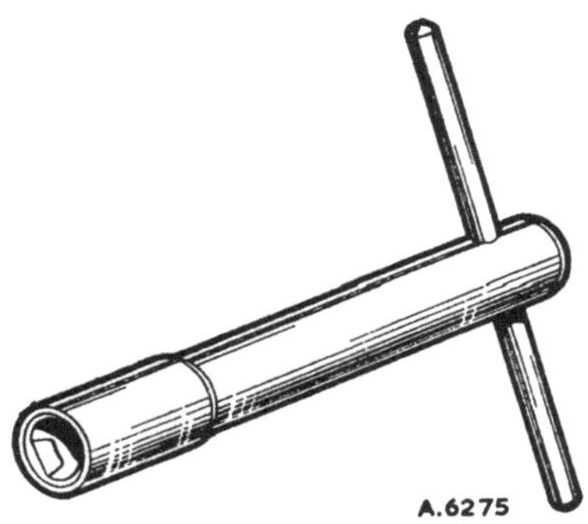


**18G 1040. Oil Pump Body Key.**

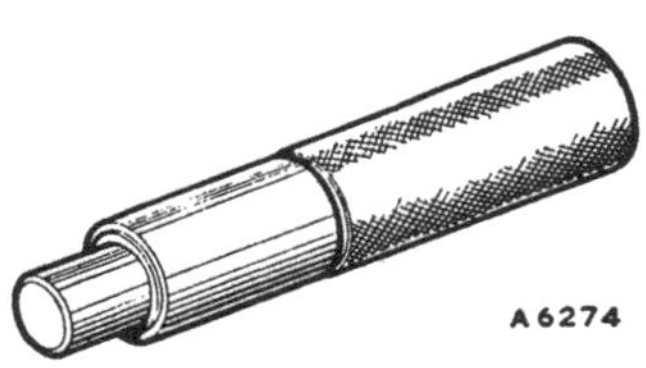


**18G 1042. Annulus Spigot Bearing Replacer.**

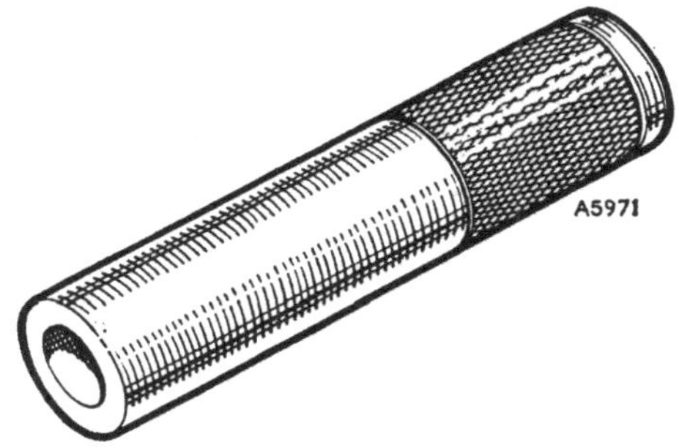


**18G 1045. Mainshaft (Third Motion) Drift.**

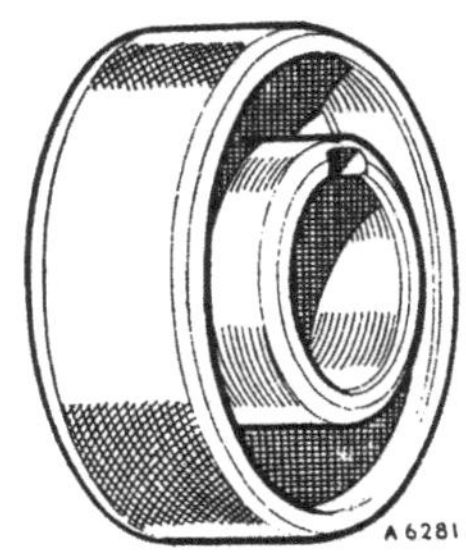


**18G 1046. Engine Front Cover Centralizer.**

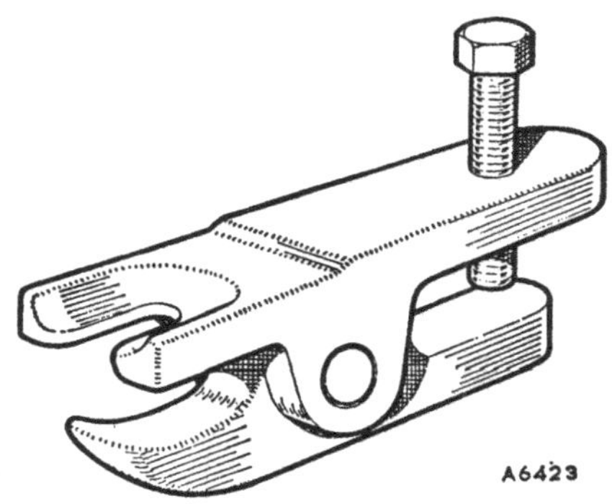


**18G 1063. Steering Arm and Swivel Hub Ball Pin Remover.**

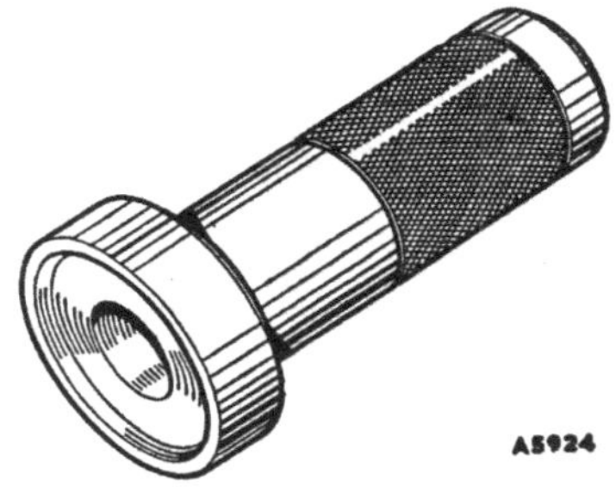


**18G 1067. Rear Axle Shaft Assembly Replacer.**

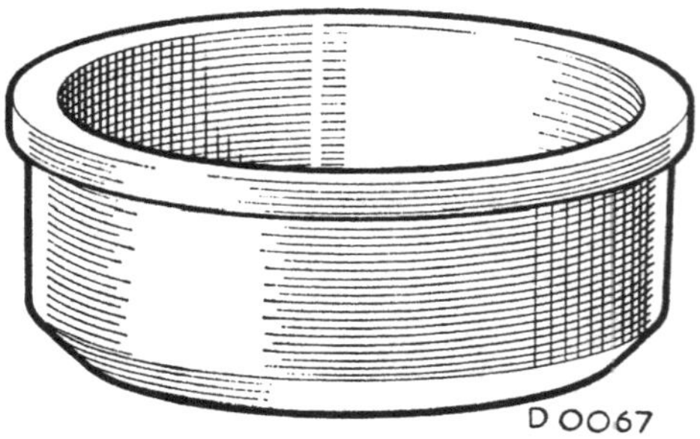


**18G 1107. Front Clutch Piston and Seal Replacer.**

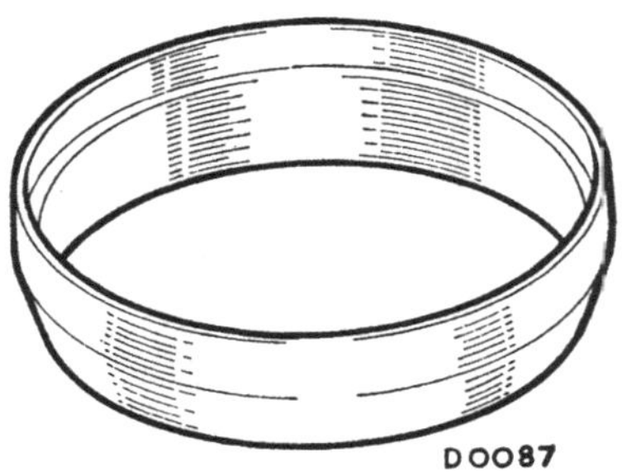


**18G 1108. Crankshaft Oil Seal Protection Sleeve.**

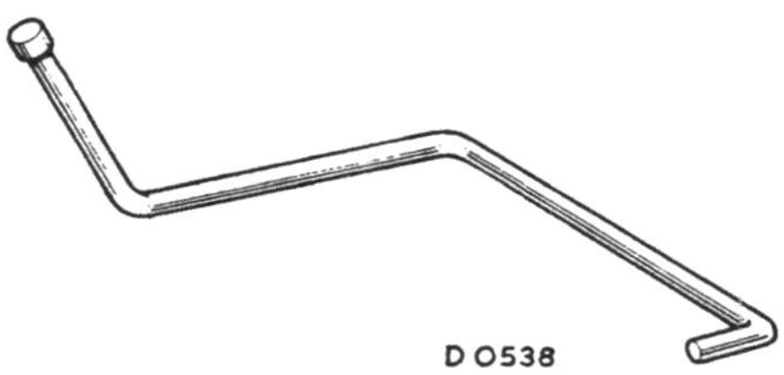


**18G 1117. Pump Body Holder and Seat Remover.**

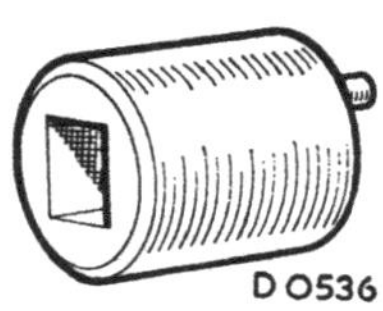


**18G 1118. Oil Pump Body Socket.**

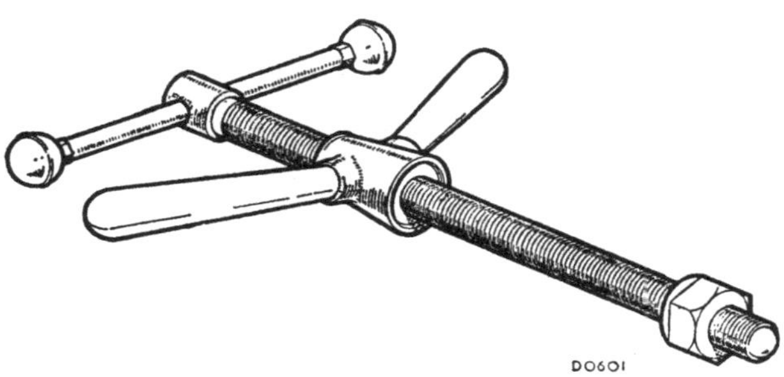


**18G 1122. Bearing Cup Replacer (basic tool).**

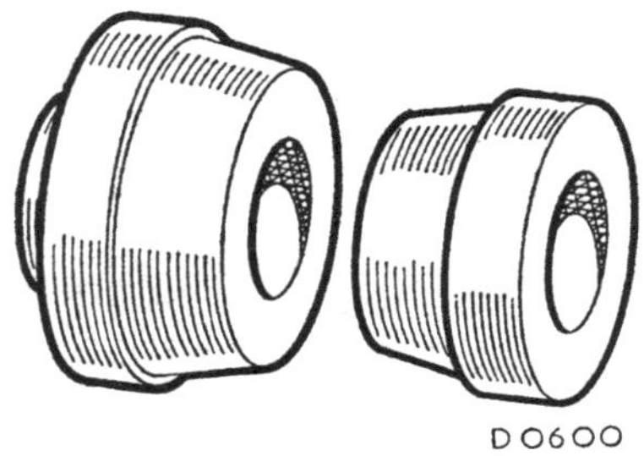


**18G 1122 A. Front Hub Bearing Cups Replacing Adaptor.**

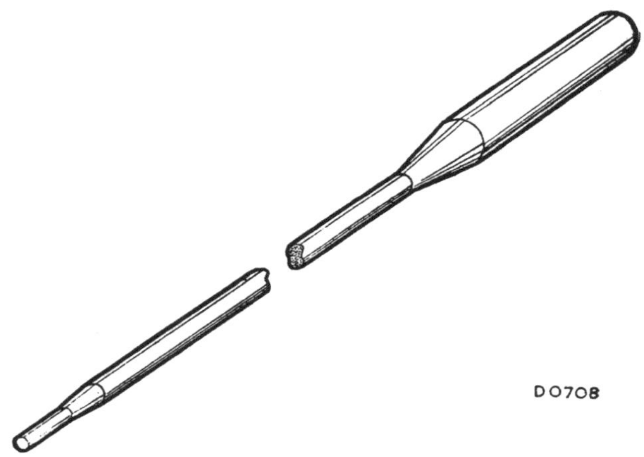


**18G 1138. Dummy Layshaft.**

**18G 1140. Steering-column Alignment Gauge.**

**18G 1145. Rocker Type Switch and Bezel Remover.**

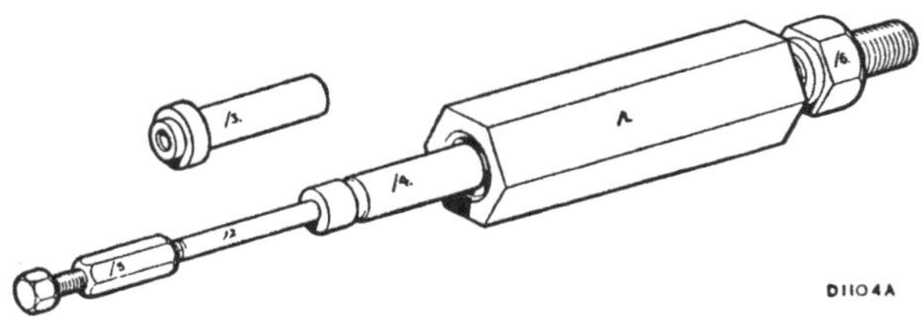


**18G 1150. Basic Tool—Remover and Replacer—Gudgeon Pin.**

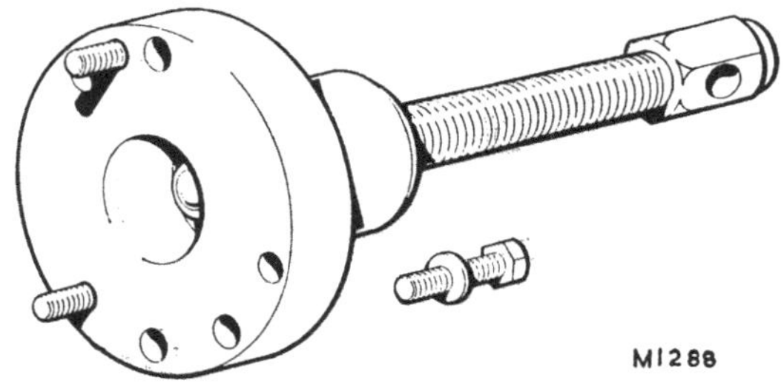


**18G 1181. Steering Wheel Hub Remover.**

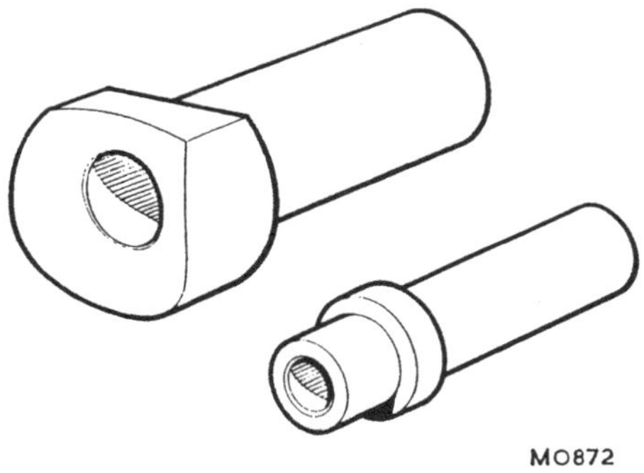


**18G 1150 C. Adaptor Set—Remover and Replacer—Gudgeon Pin.**

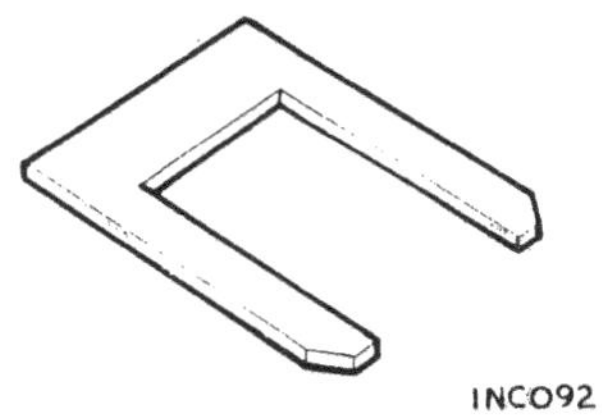


**18G 1202. Rocker Type Switch (Slim Type) Remover.**

# RECOMMENDED LUBRICANTS

| Component | Engine, Synchromesh Gearbox and Carburetter | | | Rear Axle and Steering Gear | | Grease Points | Upper Cylinder Lubrication | Automatic Transmission |
|---|---|---|---|---|---|---|---|---|
| **Climatic conditions** | **All temperatures above −10° C. (10° F.)** | **Temperatures −15° to −5° C. (0° to 20° F.)** | **All temperatures below −15° C. (0° F.)** | **All temperatures above −10° C. (10° F.)** | **All temperatures below −5° C. (20° F.)** | **All conditions** | **All conditions** | **All conditions** |
| Viscosity requirement | S.A.E. 10W/50<br>S.A.E. 10W/40<br>S.A.E. 20W/50 or<br>S.A.E. 20W/40 | S.A.E. 10W/50<br>S.A.E. 10W/40<br>or<br>S.A.E. 10W/30 | S.A.E. 5W/30<br>or<br>S.A.E. 5W/20 | S.A.E. 90 Hypoid | S.A.E. 80 Hypoid | — | — | — |
| Minimum performance level | MIL-L-2104B | MIL-L-2104B | MIL-L-2104B | MIL-L-2105B | MIL-L-2105B | Multipurpose Lithium Grease N.L.G.1. Consistency No. 2 | Upper Cylinder Lubricant | Automatic Transmission Fluid Type F |
| **CASTROL** | Castrol GTX or Castrol XL (20/50) | Castrolite or Castrol Super | Castrol CRI 5W/20 | Castrol Hypoy B. 90 | Castrol Hypoy B. 80 | Castrol L.M. Grease | Castrollo | Castrol TQF |
| **ESSO** | Esso Extra Motor Oil 20W/50 or Esso Uniflo | Esso Extra Motor Oil 10W/30 or Esso Uniflo | Esso Extra Motor Oil 5W/20 | U.K. Gear Oil G.X. 90/140 Overseas Gear Oil G.X. 90 | Esso Gear Oil G.X. 80 | Esso Multi-purpose Grease H | Esso Upper Cylinder Lubricant | Esso Glide |
| **MOBIL** | Mobiloil Special 20W/50 or Super 10W/50 | Mobiloil Super 10W/50 | Mobiloil 5W/20 | Mobilube H.D. 90 | Mobilube H.D. 80 | Mobilgrease M.P. | Mobil Upperlube | Mobil A.T.F. 210 |
| **BP** | BP Super Visco-Static* BP Super Visco-Static 20W/50 | BP Super Visco-Static* 10W/30 or 10W/40 | BP Super Visco-Static* BP Super Visco-Static 5W/20 | BP Hypogear 90 EP | BP Hypogear 80 EP | BP Energrease M.P.* BP Energrease L. 2 | BP Powerlube* BP Upper Cylinder Lubricant | BP Autran B |
| **SHELL** | Super Shell Motor Oil* Shell Super Motor Oil 20W/50 | Super Shell Motor Oil * Shell Super Motor Oil 10W/30 | Super Shell Motor Oil 5W/30* Shell Winter Special Motor Oil or Shell Super Motor Oil 5W/30 | Shell Spirax Heavy Duty 90 | Shell Spirax Heavy Duty 80 | Shell Retinax A Darina AX* | Shell Upper Cylinder Lubricant | Shell Donax T. 7 |
| **FILTRATE** | Filtrate Super 20W/50 | Filtrate Super 10W/30 | Filtrate 5W/20 | Filtrate Epex 90 | Filtrate Epex 80 | Filtrate Super Lithium Grease | Filtrate Petroyle | Filtrate Automatic Transmission Fluid |
| **STERNOL** | Sternol Super W.W. Motor Oil | Sternol W.W. Multigrade 10W/40 | Sternol W.W. Multigrade 5W/20 | Sternol Ambroleum H.D. 90 | Sternol Ambroleum H.D. 80 | Sternol Ambroline Grease L.H.T. 2 | Sternol Magikoyl | Sternol Lynx Type F |
| **DUCKHAMS** | Duckhams Q. 20–50 | Duckhams Q. 5500 | Duckhams Q. 5–30 | Duckhams Hypoid 90S | Duckhams Hypoid 80S | Duckhams L.B. 10 Grease | Duckhams Adcoid Liquid | Duckhams Q-matic |

* AMERICA ONLY

# WORKSHOP MANUAL SUPPLEMENT

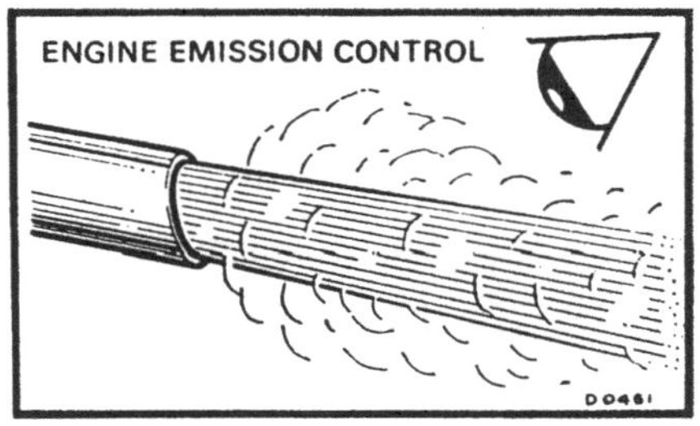


BRITISH LEYLAND

# FOREWORD

This supplement provides service operatives with the information necessary to carry out the maintenance, servicing, and testing of engine emission and fuel evaporative loss control systems fitted to vehicles manufactured by BRITISH LEYLAND (AUSTIN-MORRIS) LTD. for which a certificate has been issued in accordance with the UNITED STATES CLEAN AIR ACTS and any applicable State Legislation.

Distributors and Dealers are advised to familiarize themselves with the legal requirements, in particular those concerning minimum standards of facilities, personnel, and servicing equipment.

Service operations in Workshop Manuals, where applicable, which may affect the efficiency of the emission or evaporative loss control equipment carry the following symbol, denoting that the control system must be checked on completion of the operation 

# INDEX

# SECTION 1

## EQUIPMENT AND SERVICING

## Section 1-A

### EQUIPMENT

The recommended equipment for servicing should include at least the following:

- Ignition Analyser Oscilloscope
- Ohmmeter
- Voltmeter
- Tachometer
- Vacuum Gauge
- Pressure Gauge (0–10 lb./sq. in.)
- Carburetter Balance Meter
- Cam Angle Dwell Meter
- Ignition Timing Light
- Engine Exhaust Combustion Analyser
- Cylinder Leak Tester
- Distributor Advance Tester
- Carburetter Piston Loading Tool

The following equipment covers most of the requirements for engine testing and tuning vehicles fitted with exhaust emission control devices.

| *Equipment* | *Type/Model* | *Manufacturer* |
|---|---|---|
| Oscilloscope Engine Tuning Set and Exhaust Gas Analyser | 1020 or 720 | Sun Electric Corp. |
| Engine Analyser | 40–162 | Marquette |
| Exhaust Gas Analyser | 42–141 | Marquette |

**Equipment made by other suppliers may also be adequate.**

**It is important that your test equipment has regular maintenance and calibration.**

## Section 1-B

### SERVICING

**General**

The efficient operation of the exhaust emission control system is dependent on the engine being in good mechanical condition and correctly tuned to the settings given in **'TUNING DATA'**.

Tuning and test procedure for the carburetters, ignition system, and engine are given at the end of the manual. These procedures are the quickest and surest way of locating engine faults or maladjustments and are the only methods that should be used for engine tuning.

**Fault diagnosis**

After tuning the engine to the correct settings, check for indications of the following symptoms:

| **Symptoms** | **Causes** | **Cure** |
|---|---|---|
| Backfire in exhaust system | 1. Leak in exhaust system | Locate and rectify leak |
| | 2. Leaks in hoses or connections to gulp valve, vacuum sensing pipe or other inlet manifold joint | Locate and rectify leak |
| | 3. Faulty gulp valve | Test gulp valve, and renew if faulty |
| | 4. Leak in intake system | Locate and rectify leak |
| | 5. High inlet manifold depression on over-run—faulty carburetter limit valve | Fit new throttle disc and limit valve assembly |

| Symptoms | Causes | Cure |
|---|---|---|
| Hesitation to accelerate after sudden throttle closure | 1. Low carburetter damper oil | Top up to correct level |
| | 2. Leaks in hoses or connections to gulp valve, vacuum sensing pipe or other inlet manifold joint | Locate and rectify leak |
| | 3. Faulty gulp valve | Test gulp valve, and renew if faulty |
| | 4. Leak in intake system | Locate and rectify leak |
| Engine surges (erratic operation at varying throttle openings) | 1. Leaks in hoses or connections to gulp valve, vacuum sensing pipe or other inlet manifold joint | Locate and rectify leak |
| | 2. Faulty gulp valve | Test gulp valve, and renew if faulty |
| | 3. Air supply to adsorption canister restricted | Check air filter pad, vent pipe, and canister for obstruction |
| Erratic idling or stalling | 1. Carburetter damper oil low | Top up to correct level. |
| | 2. Leaks in hoses or connections to gulp valve or vacuum sensing pipe or other inlet manifold joint | Locate and rectify leak |
| | 3. Faulty gulp valve | Test gulp valve, and renew if faulty |
| | 4. Incorrect carburetter settings | Reset to **TUNING DATA** |
| | 5. Carburetter limit valve not seating | Fit new throttle disc and limit valve assembly |
| | 6. Carburetter suction chamber damaged | Replace carburetter or components |
| Burned or baked hose between air pump and check valve | 1. Faulty check valve | Test check valve, and renew if faulty |
| | 2. Air pump not pumping | Test air pump; service or renew if faulty |
| Noisy air pump | 1. Incorrect belt tension | Adjust belt tension |
| | 2. Pulleys damaged, loose or misaligned | Tighten loose pulleys, renew damaged pulleys |
| | 3. Air pump failing or seizing | Test air pump; service or renew if faulty |
| Excessive exhaust system temperature | 1. Incorrect ignition timing | Recheck timing against **'TUNING DATA'** |
| | 2. Choke control system not fully returned | Check choke mechanism for correct operation; instruct driver on correct usage |
| | 3. Fast idle speed too high | Reset fast idle speed—see **'TUNING DATA'** |
| | 4. Air injector missing | Remove air manifold and check injectors |
| | 5. Air pump relief valve inoperative | Test relief valve, and renew if faulty |

| Symptoms | Causes | Cure |
|---|---|---|
| Mixture requires excessive enriching to obtain correct exhaust emission readings | 1. Air leak into crankcase | Locate and rectify leak |
| | 2. Early cars—Diaphragm of crankcase control valve perforated or not correctly seated | Locate and rectify leak or control valve |
| | Later cars—Crankcase breather hose or connections to carburetter leaking | Locate and rectify leak |
| Fuel leakage | 1. Fracture in fuel pipe or fuel vapour ventilation system | Locate and rectify leak |
| | 2. Fuel filler cap not sealing | Check condition of cap and filler seal |
| | 3. Leak on fuel filler tube or tank unit | Locate and rectify leak |
| Engine stops after short running periods (i.e. fuel starvation) | 1. Obstructed vapour line between fuel tank and adsorption canister | Locate and clear obstruction |
| | 2. Air supply to adsorption canister restricted | Check air filter pad, vent pipe and canister for obstruction |
| | 3. Faulty fuel pump | Check operation and rectify fault |
| Engine runs after ignition is switched off | 1. Fuel grade too low | Refill with correct grade fuel |
| | 2. Ignition retarded | Reset timing to **'TUNING DATA'** |
| | 3. Idle speed too high | Reset to **'TUNING DATA'** |
| | 4. Fuel mixture too weak | Tune carburetter(s) |

# SECTION 2

## CRANKCASE EMISSION CONTROL

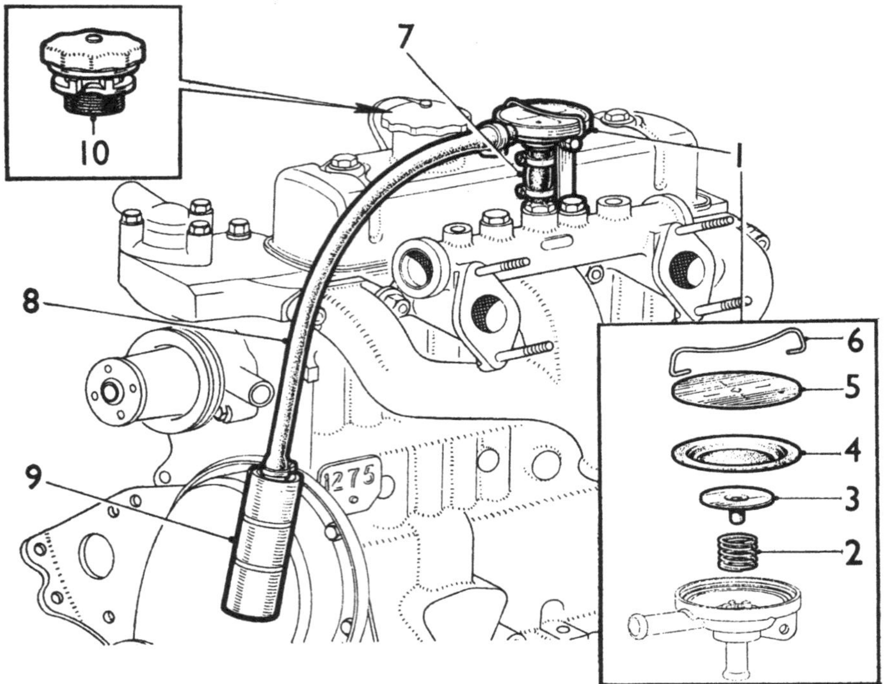


*Fig.* 1 D O 4 6 9

*A typical crankcase emission valve control system*

1. Emission control valve.
2. Valve spring.
3. Metering valve.
4. Diaphragm.
5. Cover plate.
6. Spring clip.
7. Manifold connection.
8. Breather hose.
9. Oil separator.
10. Filtered filler cap.

## Section 2-A

### VALVE CONTROL SYSTEM—General description

The system consists of a diaphragm control valve connected by hoses between the inlet manifold and the engine crankcase. The crankcase outlet connection incorporates an oil separator to prevent oil being pulled over with the vapours leaving the crankcase. On four-cylinder engines a filtered, restricted orifice ($\frac{9}{64}$ in. diameter) in the oil filler cap provides a supply of fresh air into the crankcase as vapours are withdrawn by inlet manifold depression. Six-cylinder engines are fitted with a standard oil filler cap and a tube connected between the rocker cover oil filler tube and the air intake filter provides the supply of fresh air to the engine. The control valve diaphragm varies the opening to the inlet manifold according to the depression or pressure acting on it. With a decrease in manifold depression or when the crankcase obtains a positive pressure the diaphragm opens the valve allowing the crankcase vapours to be drawn into the inlet manifold. During conditions of high manifold depression, e.g. low engine speeds or loads, the diaphragm closes the valve and restricts the flow into the inlet manifold, thus preventing a leaning-off of the air/fuel mixture to the cylinders.

## Section 2-B

### VALVE CONTROL SYSTEM—Testing

(1) Warm up the engine to normal operating temperature.

(2) With the engine running at idling speed remove the oil filler cap.

(*a*) A rise in engine speed, the change being audibly noticeable, indicates that the control valve is functioning correctly.

(*b*) No rise in speed, service the control valve.

## Section 2-C

### VALVE CONTROL SYSTEM—Servicing

**Oil filler cap** (*four-cylinder engines only*)

(1) Renew every 12,000 miles or 12 months.

**Control valve**

(2) Disconnect the hoses and renew the valve assembly, or clean as follows:

(*a*) Remove the spring clip and withdraw the cover plate, diaphragm, metering valve and spring.

(*b*) Clean all metal parts with a solvent (trichlorethylene, fuel, etc.). **Do not use an abrasive.** If deposits are difficult to remove, immerse in boiling water before applying the solvent.

(*c*) Clean the diaphragm with a detergent or methylated spirit (denatured alcohol).

(*d*) Examine the parts thoroughly for wear or damage, and renew where necessary.

(*e*) Reassemble the valve ensuring that the metering valve fits correctly in its guides and the diaphragm is correctly seated.

(*f*) Refit the valve and check its operation.

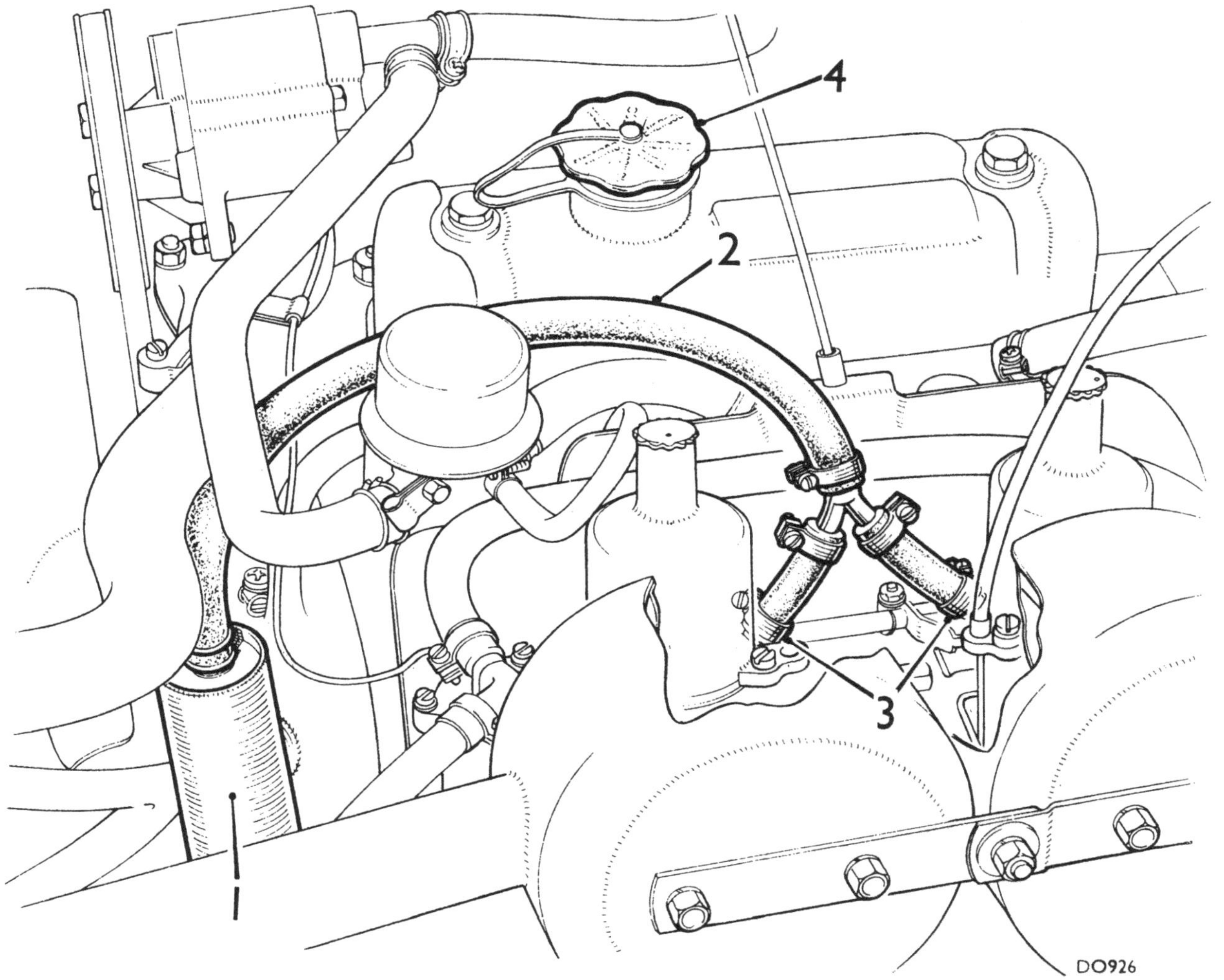


*Fig.* 2

*Carburetter control system*

1. Oil separator.
2. Breather hose.
3. Carburetter chamber connection.
4. Filtered filler cap.

## Section 2-D

### CARBURETTER CONTROL SYSTEM

**Description**

With this system the engine breather outlet is connected by hoses to the controlled depression chamber; the chamber between the piston and the throttle disc valve, of the carburetter(s). Engine fumes and blow-by gases are drawn from the crankcase by the depression in this chamber, through an oil separator incorporated in the engine outlet connection, and from there to the inlet manifold. Fresh air is supplied to the engine through the combined oil filler cap and filter (four-cylinder engines) or through the air intake filter (six-cylinder engines).

**Servicing**

The oil filler cap (four-cylinder engines only) must be renewed every 12,000 miles (20000 km.) or 12 months; no other service is required.

If a failure of the system is suspected, check the hoses and connections for leaks and obstructions. An indication of a failure is loss of crankcase depression.

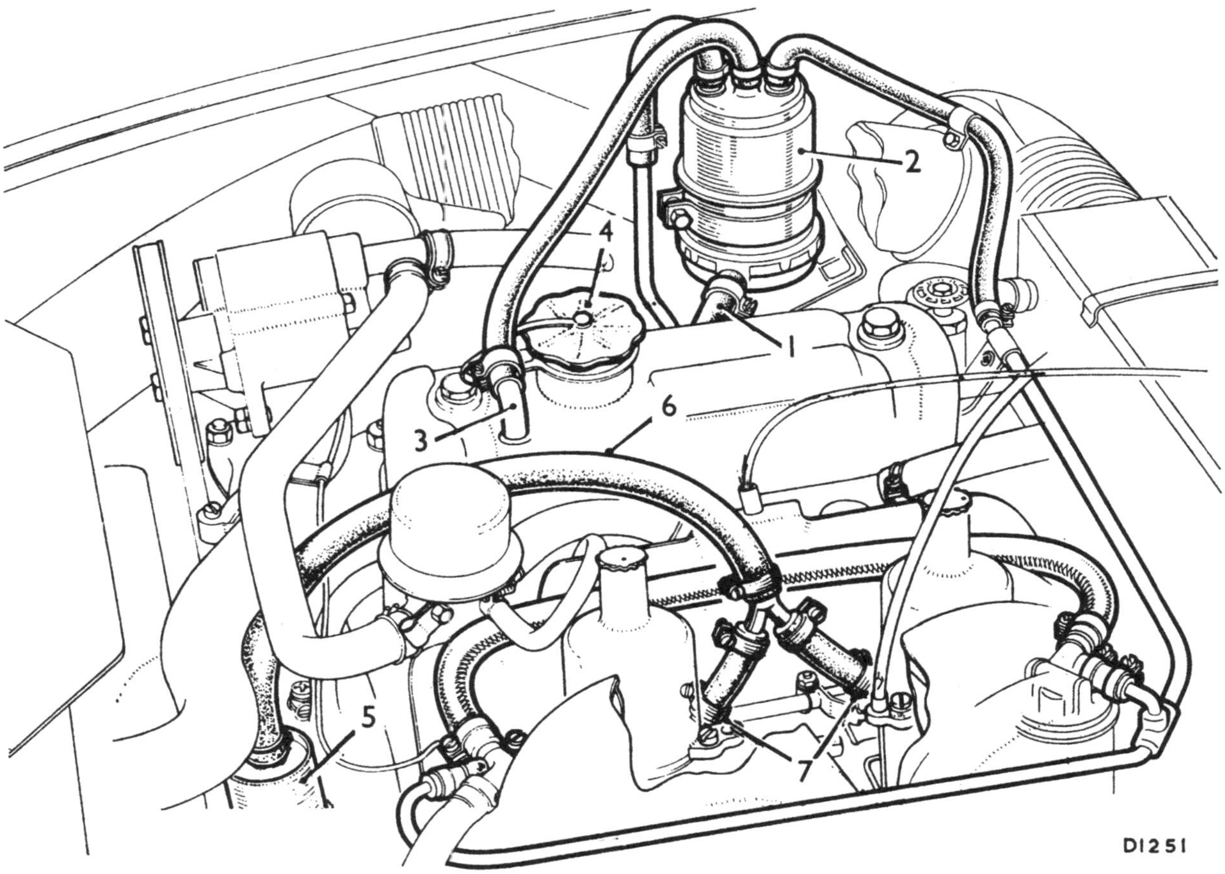


*Fig.* 3
*A carburetter control system with fuel evaporative loss control*

1. Ventilation air intake.
2. Absorption canister.
3. Restricted connection to rocker cover.
4. Sealed oil filler cap.
5. Oil separator.
6. Breather hose.
7. Carburetter chamber connections.

## Section 2-E

### CARBURETTER CONTROL SYSTEM—
### with evaporative loss control

This system incorporates most of the components of the carburetter control system, with the exception of the combined oil filler cap and filtered air intake. Its operation differs in that air for engine breathing is drawn through the filtered adsorption canister of the evaporative loss control system into the engine valve rocker cover. A restrictor in the rocker cover connection reduces the air flow to ensure crankcase depression under all conditions.

Engine fumes and blow-by gases are drawn from the crankcase, through an oil separator, into the inlet manifold by the controlled depression chamber of the carburetter.

**Servicing**

No direct servicing of the system is required. The air intake filter pad in the absorption canister is renewed at the intervals required by the fuel evaporative loss control.

If a failure of the system is suspected, check the hoses and connections for leaks and obstruction. An indication of a failure is loss of crankcase depression.

# SECTION 3

## EXHAUST EMISSION CONTROL
## (Exhaust Port Air Injection)

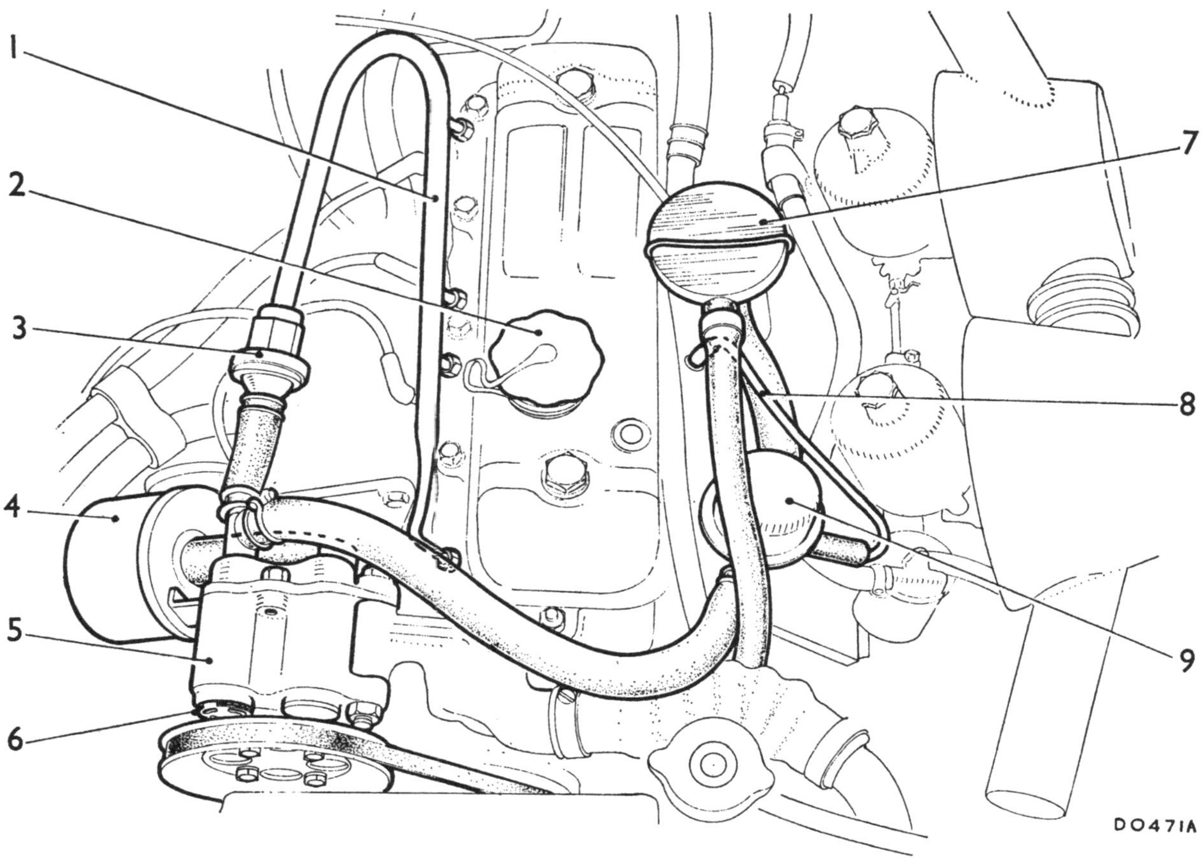


*Fig. 1*

*A typical engine emission control system layout*

1. Air manifold.
2. Filtered oil filler cap.
3. Check valve.
4. Emission air cleaner.
5. Air pump.
6. Relief valve.
7. Crankcase emission valve.
8. Vacuum sensing tube.
9. Gulp valve.

## Section 3-A

### GENERAL DESCRIPTION

Air is pressure-fed from an air pump via an injection manifold to the cylinder head exhaust port of each cylinder. A check valve in the air delivery pipe prevents blow-back from high pressure exhaust gases. The pump also supplies air through a gulp valve to the inlet manifold to provide air during conditions of deceleration and engine over-run.

**IMPORTANT.** The efficient operation of the system is dependent on the engine being correctly tuned. The ignition and spark plug settings, valve clearances, and carburetter adjustments given for a particular engine (see **'TUNING DATA'**) must be strictly adhered to at all times.

#### Air pump

The rotary vane type air pump is mounted on the front of the cylinder head and is belt driven from the water pump pulley. Provision is made for tensioning the belt.

Air is drawn into the pump through a dry-type renewable element filter. A relief valve in the pump discharge port allows excessive air pressure at high engine speeds to discharge to the atmosphere.

#### Check valve

The check valve, fitted in the pump discharge line to the injection manifold, protects the pump from the back-flow of exhaust gases.

The valve shuts if the air pressure ceases while the engine is running; for example, if the pump drive belt should break.

#### Gulp valve

The gulp valve, fitted in the pump discharge line to the inlet manifold, controls the flow of air for leaning-off the rich air/fuel mixture present in the inlet manifold immediately following throttle closure after running at full throttle opening (i.e. engine over-run).

A sensing pipe connected between the inlet manifold and the gulp valve maintains manifold depression directly to the underside of the diaphragm and through a bleed hole to the upper side. Sudden increases in manifold depression which occur immediately following throttle closure act on the underside of the diaphragm which opens the valve and admits air to the inlet manifold. The bleed hole allows the differences in depression acting on the diaphragm to equalize and the valve closes.

On some engines a restrictor is fitted in the air pump discharge connection to the gulp valve, to prevent surging when the gulp valve is operating.

**Carburetter**

The carburetters are manufactured to a special exhaust emission control specification and are tuned to give optimum engine performance with maximum emission control.

A limit valve is incorporated in the carburetter throttle disc which limits the inlet manifold depression ensuring that under conditions of high inlet-manifold depression the mixture entering the cylinders is at a combustible ratio.

## Section 3-B

### AIR PUMP (four-cylinder engines)

**Drive belt tension**

When correctly tensioned, a total deflection of ½ in., under moderate hand pressure, should be possible at the midway point of the longest belt run between the pulleys.

To tension the belt:

(1) Slacken the air pump mounting bolt and adjusting link bolts (see Fig. 3).

(2) Using hand pressure only, move the pump in the required direction until the correct tension is obtained.

(3) Tighten the mounting and adjusting bolts to a torque figure of 10 lb. ft.

**Testing**

(1) Check the drive belt for correct tensioning.

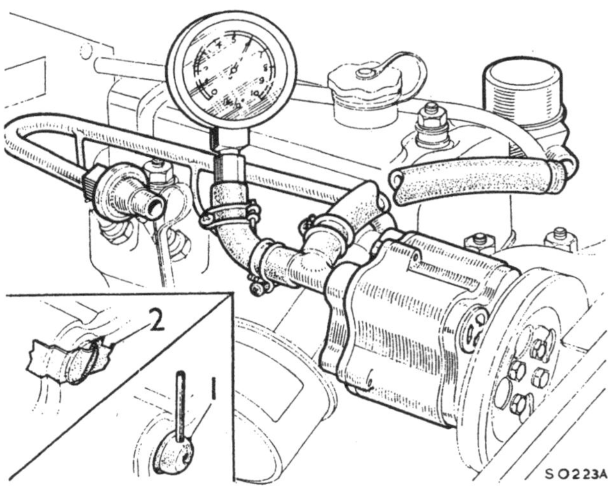


*Fig. 2*

*The pressure gauge connected (four-cylinder engines)*

1. Relief valve test tool. 2. Tape used to duct air.

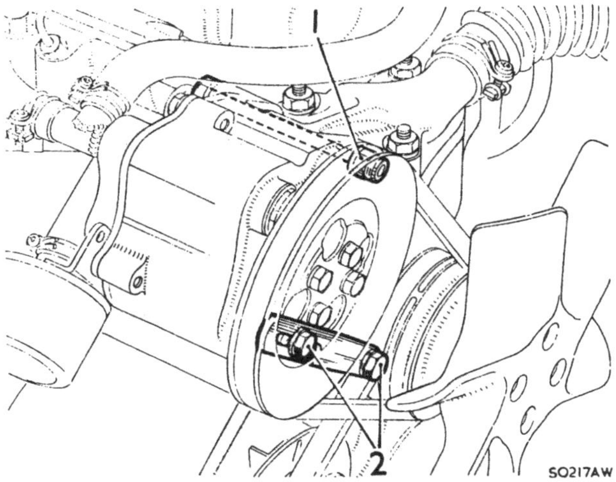


*Fig. 3*

*Air pump (four-cylinder engines)*

1. Pump mounting bolt. 2. Adjusting link bolts.

(2) Connect a tachometer to the engine in accordance with the instrument-maker's instructions.

(3) Disconnect the gulp valve air supply hose at the gulp valve and securely plug the hose.

(4) Disconnect the air manifold supply hose at the check valve, and connect a pressure gauge to the hose (see Fig. 2).

(5) Run the engine at the air pump test speed given in '**TUNING DATA**': a gauge reading of not less than 2·75 lb./sq. in. should be registered.

(*a*) If a lower reading is obtained, remove, dismantle and clean the pump air cleaner. Reassemble using a new element, refit the air cleaner and repeat the test.

(*b*) If the reading is still unsatisfactory, temporarily blank off the relief valve and repeat the test; if the reading is now correct, renew the relief valve.

(*c*) If a satisfactory reading is still unobtainable. remove and service the air pump.

(6) Stop the engine and fit a temporary air duct over the face of the relief valve. Two methods of doing this are shown in Fig. 2. The tool (1) may be fabricated from grommet (Part No. 1B 1735) and a short length of metal brake tube, or (2) by using a piece of adhesive tape to form the duct.

**DO NOT ATTEMPT TO CHECK AIR FLOW FROM THE RELIEF VALVE BY PLACING A FINGER BETWEEN THE VALVE AND THE DRIVING PULLEY.**

(*a*) Start the engine and slowly increase the speed until air flow from the relief valve duct is detected, when a gauge reading of 4·5 to 6·5 lb./sq. in. should be registered.

(*b*) If the relief valve fails to operate correctly, remove the pump and renew the valve.

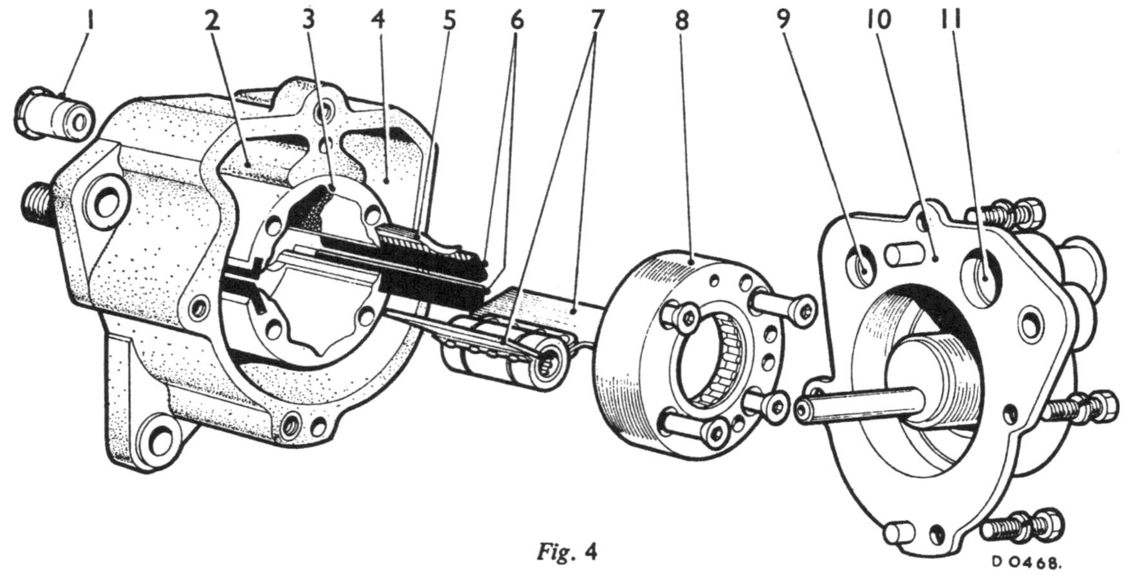


*Fig. 4*

*The air pump (four-cylinder engines)*

1. Relief valve.
2. Inlet chamber.
3. Rotor.
4. Outlet chamber.
5. Spring.
6. Carbons.
7. Vane assemblies.
8. Rotor bearing end plate.
9. Outlet port.
10. Port-end cover.
11. Inlet port.

**Removing**

(1) Disconnect the air hoses from the pump connections and remove the air cleaner.
(2) Slacken the mounting and adjusting link bolts and slip the drive belt from the pump pulley.
(3) Remove the top adjusting link bolt and the nut securing the pump mounting bolt.
(4) Support the pump, withdraw the mounting bolt and lift the pump from the engine.

**Dismantling**

(1) Remove the four port-end cover retaining bolts and withdraw the cover.
(2) Remove the four screws securing the rotor bearing end plate to the rotor and remove the end plate.
(3) Lift out the vane assemblies.
(4) Remove the carbon and spring assemblies from the rotor.

**Servicing**

(1) Wipe the interior and components of the pump clean, using a lint-free cloth.

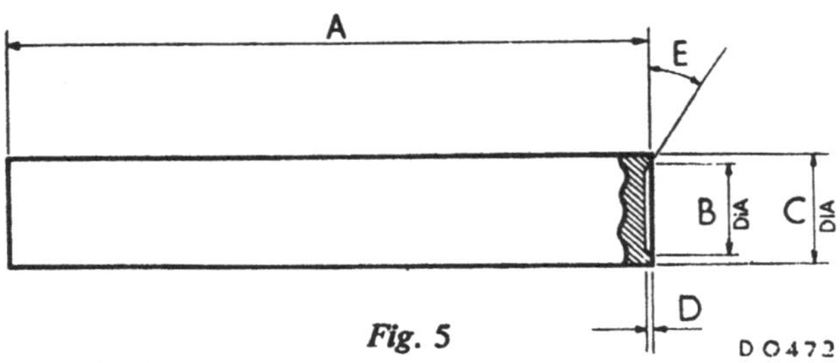


*Fig. 5*

*The dimensions of the relief valve replacing tool*

A = 5 in. B = ·986 in. C = 1·062 in. D = ·05 in. E = 30°.

(2) Clean the vane carrier roller bearings and the rotor end plate bearing and repack the bearings with Esso 'Andok' 260 lubricant.
(3) Inspect the vane assemblies for signs of having fouled the pump wall, and for grooving in area of contact with the carbons. Renew worn or damaged vanes.
(4) Fit new carbons (the original springs may be re-used if serviceable). Note that the slots which carry the carbon and springs are the deeper ones, and the carbons are all fitted with the chamfered edge to the inside.

**Reassembling**

(1) Reassemble the pump by reversing the dismantling procedure and noting that the underside of the heads of the rotor bearing end plate screws must be smeared with 'Locktite' before tightening.

**Refitting**

(1) Position the pump in the mounting bracket and fit, but do not tighten, the pump mounting bolt.
(2) Screw in, but do not tighten, the adjusting link bolt.
(3) Fit and tension the drive belt.
(4) Reconnect the hoses and refit the air cleaner.

**Relief valve—replacing**

(1) Remove the air pump.
(2) Remove the pump pulley.
(3) Pass a ½-in. diameter soft metal drift through the pump discharge connection so that it registers against the relief valve, and drive the valve from the pump.

(4) Fit a new copper seating washer to the new relief valve and enter the valve into the pump body.

(5) Using a tool made to the dimensions shown in Fig. 5, drive the valve into the pump until the copper seating washer is held firmly, but not compressed, between the valve and the pump.

(6) Refit the pulley and refit the air pump.

## Section 3-C

### AIR PUMP (six-cylinder engines)

**Drive belt tension**

When correctly tensioned, a total deflection of ½ in., under moderate hand pressure should be possible at the midway point of the longest belt run between the pulleys.

To tension the belt:

(1) Slacken the air pump mounting bolt and adjusting link bolts (see Fig. 6).

(2) Using hand pressure only, move the pump in the required direction until the correct tension is obtained.

(3) Tighten the mounting and adjusting link bolts to a torque figure of 10 lb. ft.

**Testing**

Faulty operation of the air pump is indicated by excessive pump noise. If excessive noise is present and the air pump is suspected, remove the air pump drive belt and run the engine to check that the noise is not from another source. If this check shows that the air pump is excessively noisy renew the air pump assembly or proceed as follows:

(1) Check the drive belt for correct tensioning.

(2) Run the engine at idle speed and check the air supply hoses and connections for leaks and for intermittent contact with other parts of the vehicle.

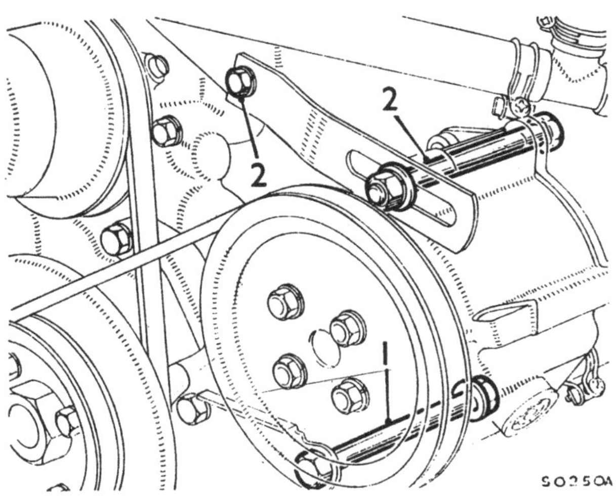


*Fig. 6*

*Air pump (six-cylinder engine*

**1. Pump mounting bolt.** **2. Adjusting link bolts.**

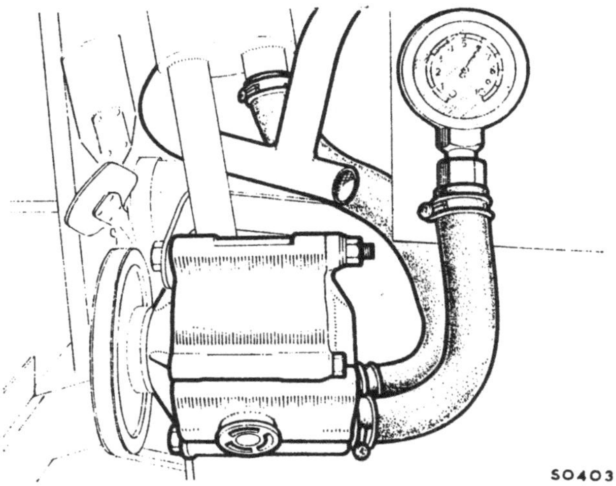


*Fig. 7*

*The pressure gauge connected (six-cylinder engines)*

(3) Connect a tachometer to the engine in accordance with the instrument maker's instructions.

(4) Disconnect the air supply hose tee connection from its connection with the air pump discharge hose.

(5) Connect a pressure gauge to the air pump discharge hose (see Fig. 7).

(6) Run the engine at the air pump test speed given in 'TUNING DATA'. A gauge reading of not less than 2·75 lb./sq. in. should be registered.

(*a*) If a lower reading is obtained, remove, dismantle, and clean the pump air cleaner. Reassemble using a new element, refit the air cleaner, and repeat the test.

(*b*) If the reading is still unsatisfactory, temporarily blank off the relief valve and repeat the test; if the reading is now correct, renew the relief valve.

(*c*) If a satisfactory reading is still unobtainable the air pump assembly must be replaced.

(*d*) From idling speed, slowly increase the engine speed until air flow from the relief valve is detected, this should occur before the gauge reading exceeds 10 lb./sq. in.

(*e*) If the relief valve fails to operate correctly, remove the pump and renew the valve.

(7) If the foregoing tests fail to remedy or locate the cause of the air pump noise renew the air pump assembly.

**Removing**

(1) Disconnect the hoses from the pump connections and remove the air cleaner.

(2) Slacken the mounting and adjusting link bolts and slip the drive belt from the pump pulley.

(3) Remove the nut from the adjusting link bolt, support the pump and withdraw the bolt.

(4) Unscrew the mounting bolt and remove the pump.

*Fig. 8*

*Removing the relief valve (six-cylinder engines)*

**Relief valve—replacing**

(1) Remove the air pump.

(2) Using a gear puller and a fabricated bridge as shown in Fig. 8, withdraw the relief valve from the pump body.

**DO NOT HOLD THE PUMP BY CLAMPING IT IN A VICE.**

(3) Enter the new relief valve into the pump body.

(4) With a protective plate over the valve, carefully drive the valve into the pump until its flange registers lightly on the pump body.

(5) Insert the pressure setting plug into the relief valve, using a suitable tool, apply pressure to the centre of the plug until the legs of the plug lock under the relief valve cage.

## Section 3-D

### CHECK VALVE

**Removing**

(1) Disconnect the air supply hose from the check valve connection.

(2) Hold the air manifold connection to prevent it twisting and unscrew the check valve.

**Testing**

(1) Blow through the valve, orally, in turn from each connection. Air should only pass through the valve when blown from the air supply hose connection. If air passes through when blown from the air manifold connection, renew the check valve.

**On no account may an air blast be used for this test.**

**Refitting**

(1) Hold the air manifold connection to prevent it twisting, screw in and tighten the check valve.

(2) Reconnect the air supply hose to the check valve.

## Section 3-E

### AIR MANIFOLD AND INJECTORS

**Testing**

(1) Disconnect the air manifold from the cylinder head connections.

(2) Slacken the air supply hose clip at the check valve connection.

(3) Rotate the manifold about its connection axis until the injector connections are accessible.

(4) Tighten the air supply hose clip.

(5) Run the engine at idle speed and observe the flow of air from each of the manifold connection tubes. Should the flow of air from any of the connections be restricted, remove the manifold and clear the obstruction using an air blast.

(6) With the engine running at idle speed, check that exhaust gases blow from each of the cylinder head injectors.

**IMPORTANT.**—The injectors may be free in the cylinder head and care must be taken to ensure that they are not displaced during this test.

To clear a restricted injector:

(*a*) Crank engine until the exhaust valve below the injector is closed.

(*b*) Using a hand drill (not power-driven), pass a $\frac{1}{8}$-in. drill through the injector bore, taking care that the drill does not contact the exhaust valve stem after passing through the injector. Damage may result if a power-driven drill is used.

(*c*) Insert an air-blast nozzle into the injector connection to clear carbon dust from the exhaust port.

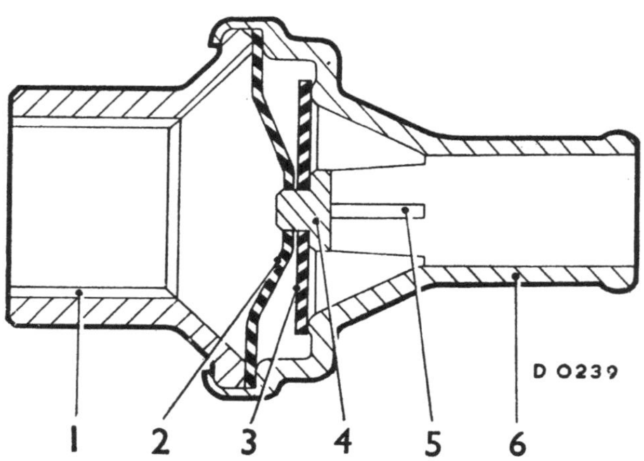


*Fig. 9*

*A section through the check valve*

1. Air manifold connection.
2. Diaphragm.
3. Valve.
4. Valve pilot.
5. Guides.
6. Air supply connection.

## Section 3-F

### GULP VALVE

**Testing**

(1) Disconnect the gulp valve air supply hose from the air pump connection.

(2) Connect a vacuum gauge, with a tee connection to the disconnected end of the gulp valve air hose.

(3) Start the engine and run it at idle speed.

(4) Temporarily seal the open connection on the gauge tee and check that a zero gauge reading is maintained for approximately 15 seconds; if a vacuum is registered, renew the gulp valve. It is most important that the engine speed is not increased above idling during this test.

(5) With the gauge tee connection temporarily sealed, operate the throttle rapidly from closed to open; the gauge should then register a vacuum. Repeat the test several times, temporarily unsealing the tee piece connection to destroy the vacuum before each operation of the throttle. If the gauge fails to register a vacuum, renew the gulp valve.

**Removing**

(1) Disconnect the air hoses.

(2) Unscrew the mounting screw and remove the gulp valve.

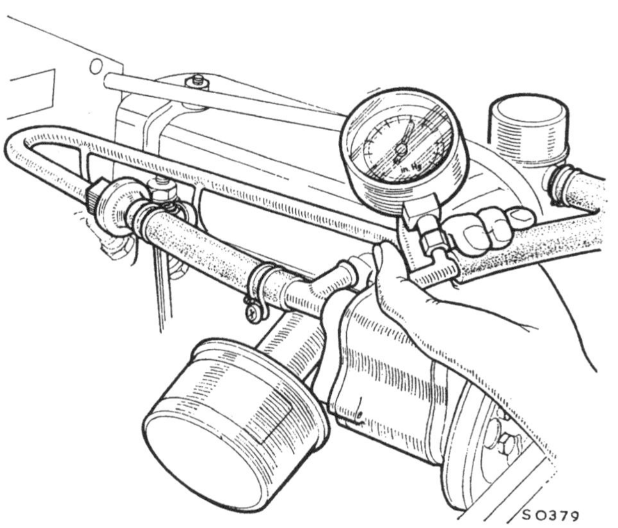


*Fig.* 10

*The vacuum gauge connected for testing the gulp valve*

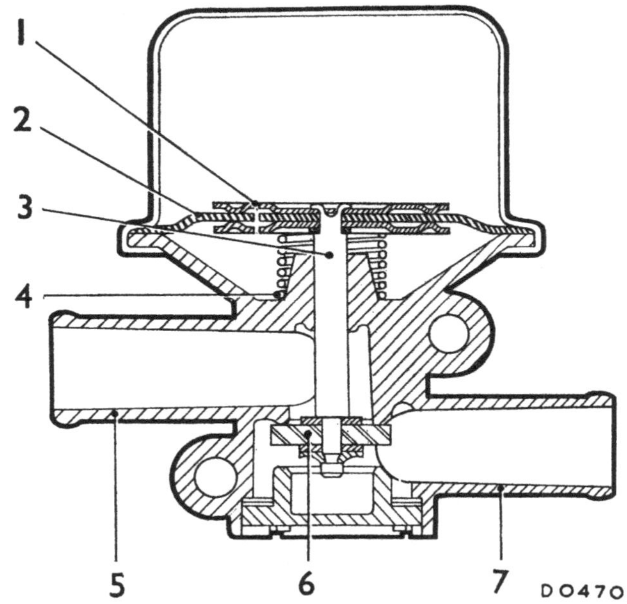


*Fig.* 11

*A section through the gulp valve*

1. Metering balance orifice.
2. Diaphragm.
3. Valve spindle.
4. Return spring.
5. Inlet manifold hose connection.
6. Valve.
7. Air pump hose connection.

**Refitting**

(1) Reverse the removing procedure.

## Section 3-G

### LIMIT VALVE (INLET MANIFOLD DEPRESSION)

**Testing**

(1) Disconnect the gulp valve sensing pipe from the inlet manifold.

(2) Connect a vacuum gauge to the sensing pipe connection on the inlet manifold.

(3) Connect a tachometer in accordance with the instrument maker's instructions.

(4) Warm the engine at fast idle speed until normal operating temperature is reached.

(5) Increase the engine speed to 3,000 r.p.m. then release the throttle quickly; the vacuum gauge reading should immediately rise to between 20·5 and 22 in. Hg. If the gauge reading falls outside these limits the carburetter must be removed and the throttle disc and limit valve assembly renewed. After refitting, the carburetter must be tuned as described in Section 4-A.

## Section 3-H

### RUNNING ON CONTROL VALVE

The solenoid operated valve is connected by hoses between the adsorption canister ventilation connection of the evaporative loss control system and the inlet manifold. A third hose connected to the valve is open to atmosphere for canister ventilation while the engine is running normally. The electrical circuit of the solenoid is connected through the ignition switch and an oil pressure operated switch.

The valve is fitted to prevent prolonged running on (dieseling) which may occur when using low octane fuels.

**Operation**

When the ignition is switched off the solenoid is energized through an oil pressure switch and the valve closes, shutting off the ventilation connection and opening the connection to the inlet manifold. Inlet manifold depression then acts on the fuel in the carburetter float chamber(s) to prevent fuel flow and the engine is stopped by fuel starvation.

**Testing**

If the running on valve is suspected of being faulty the control electrical circuit and operation of the valve should be checked as follows:

(1) Check the control valve line fuse.
(2) Turn the ignition switch to the off position.
(3) Disconnect the control valve electrical lead at the oil pressure switch.
(4) Touch the disconnected lead to a good earth point on the vehicle. If the control electrical circuit and valve are satisfactory, the valve will be heard to operate as the control lead is earthed.

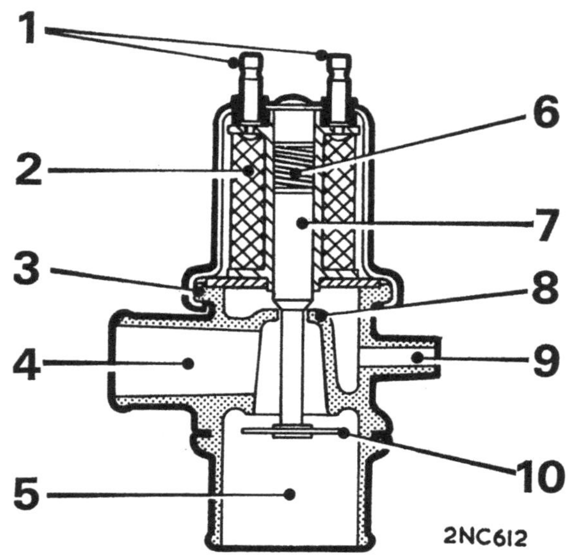


*Fig.* 12
*A section through the running on control valve*

1. Electrical terminals.
2. Solenoid.
3. Valve body.
4. Adsorption canister hose connection.
5. Air vent hose connection.
6. Spring.
7. Valve spindle.
8. Primary valve.
9. Inlet manifold hose connection.
10. Secondary valve.

# SECTION 3A

## EXHAUST EMISSION CONTROL
## (Engine Modifications System)

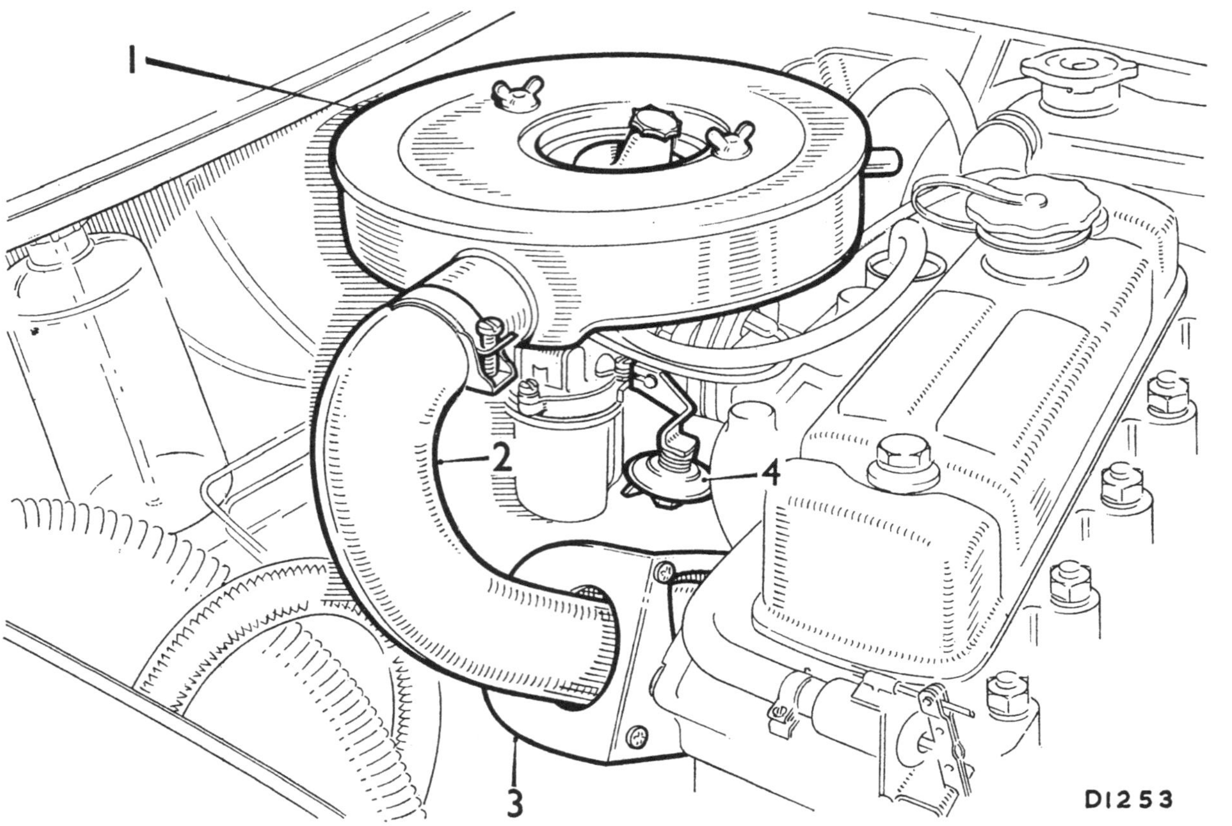


*Fig. 1*
*An engine modification exhaust emission control system showing the air intake tube in the low ambient temperature operating position*

1. Air cleaner.
2. Air intake tube.
3. Manifold shroud.
4. Throttle damper.

## Section 3A-A

### GENERAL DESCRIPTION

This system incorporates modifications to a high compression ratio engine and using a carburetter manufactured to a special exhaust emission control specification.

**IMPORTANT.** The efficient operation of the system is dependent on the engine being correctly tuned. The settings given for a particular engine (see **'TUNING DATA'**) must be strictly adhered to at all times.

**Carburetter**

The carburetter is tuned to give optimum engine performance with maximum exhaust emission control.

A limit valve is incorporated in the carburetter throttle disc which limits the inlet manifold depression ensuring that under conditions of high inlet-manifold depression the air/fuel mixture entering the cylinders is at a combustible ratio.

**Throttle damper**

A damper is fitted to act on the throttle lever as it returns to the closed position ensuring a gradual closing of the throttle valve giving smooth deceleration. Provision is made for adjusting the damping effect; the correct setting is given in **'TUNING DATA'**.

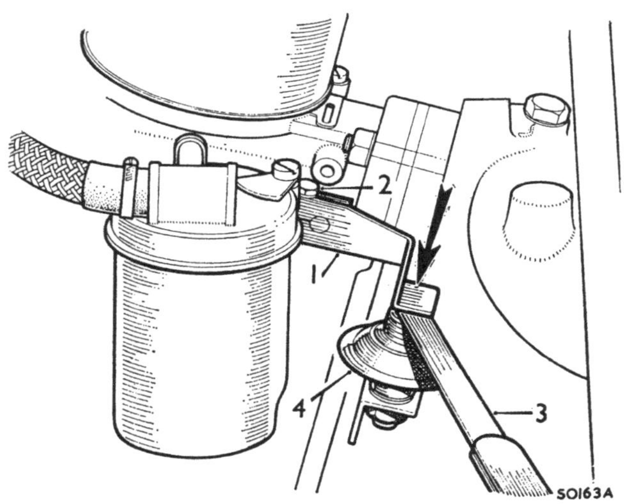


*Fig. 2*
*Adjusting the throttle damper setting*

1. Throttle lever.
2. Clamp screw.
3. Feeler gauge.
4. Throttle damper.

**Air intake**

In low ambient temperature conditions the intake tube of the air cleaner is positioned in a shroud formed over a section of the exhaust manifold. Air drawn through the cleaner to the carburetter is warmed by heat given off by the manifold.

In high ambient temperature conditions the air intake tube is positioned away from the manifold and air entering the carburetter is drawn into the air cleaner from the engine compartment at ambient temperature.

## Section 3A-B

### THROTTLE DAMPER

**Adjusting**

(1) Slacken the clamp nut on the damper operating lever.

(2) Insert a feeler gauge (see 'TUNING DATA') between the damper plunger and the operating arm.

(3) With the carburetter throttle disc valve in the fully closed position, press the operating lever down until the plunger is fully depressed.

(4) Hold the lever in this position and tighten the clamp nut.

(5) Remove the feeler gauge.

## Section 3A-C

### AIR INTAKE

**Repositioning**

(1) Slacken the intake tube securing clip.

(2) Slacken the air cleaner wing nuts.

(3) Withdraw the intake tube from the air cleaner and manifold shroud.

(4) Refit the intake tube with its entry positioned adjacent to the end of the rocker cover.

(5) Tighten the wing nuts and securing clip.

## Section 3A-D

### AIR BLEED COMPENSATOR

**Description**

An air bleed temperature compensator is fitted to some engines equipped with twin type HS carburetters required to conform with European E.C.E. or E.E.C. exhaust emission control regulations.

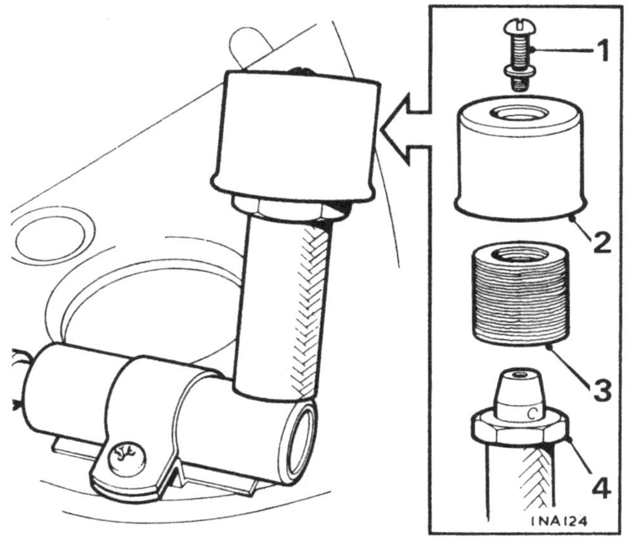


*Fig.* 3

*The air bleed temperature compensator* (*showing the cap type air filter components*)

1. Cap retaining screw.
2. Filter cap.
3. Filter element.
4. Filter base.

The air bleed temperature compensator is fixed to the underside of the carburetter air cleaner and consists of a bi-metal air control valve and an air filter. It is connected by hoses to the constant depression chambers, between the piston and the throttle disc valve, of the carburetters.

With an increase in engine or engine compartment air temperature, the valve will open and allow air at ambient temperature to be drawn through the air filter and into the carburetter constant depression chambers.

The controlled admission of air into the carburetter chambers reduces the velocity and volume of air passing the needles of the carburetters, causing the pistons to fall and subsequently reduce the amount of fuel supplied, thus giving a constant air/fuel mixture ratio.

**Servicing**

The air bleed temperature compensator air filter must be renewed every 12,000 miles (20000 km.) or 12 months.

(1) Disconnect the air cleaner to carburetter hoses.

(2) Remove the air cleaner.

(3) *Push-on type filters:* withdraw and discard the filter. *Cap type filters:* unscrew the filter cover retaining screw, remove the cover, and discard the filter element. Clean the base and cover of the filter, fit a new element, refit the cover and retaining screw.

(4) Refit the air cleaner and reconnect the hoses.

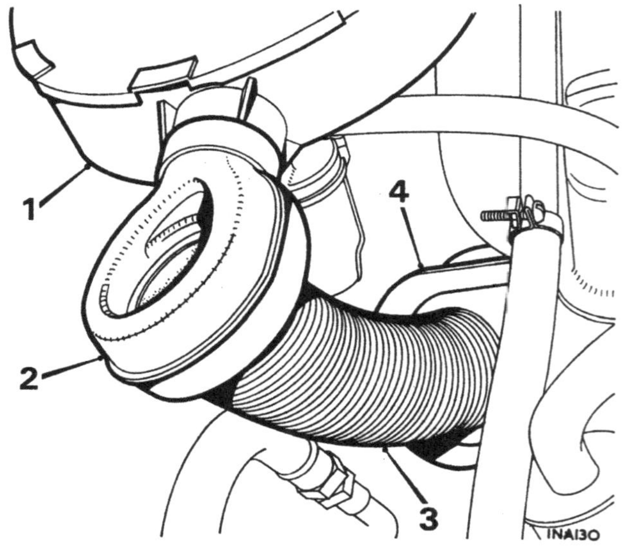


*Fig. 4*
*The air intake temperature control*

1. Air cleaner.
2. Air intake temperature control.
3. Air intake tube.
4. Exhaust manifold shroud.

## Section 3A-E

### AIR INTAKE TEMPERATURE CONTROL

An air intake temperature control is fitted to some engines equipped with single Type H.S. carburetters required to conform with European E.C.E. or E.C.C. exhaust emission control regulations.

The control consists of a bi-metal operated valve, fitted in the air intake of the carburetter air cleaner, and is designed to maintain the temperature of the ingoing air within predetermined limits.

When the engine is cold, air is drawn into the air cleaner from the shrouded area adjacent to the exhaust manifold. As the temperature of the air entering the air cleaner rises, the valve opens and admits cooler air at ambient temperature to mix with the hot air and maintain a constant temperature.

# SECTION 4

## CARBURETTERS

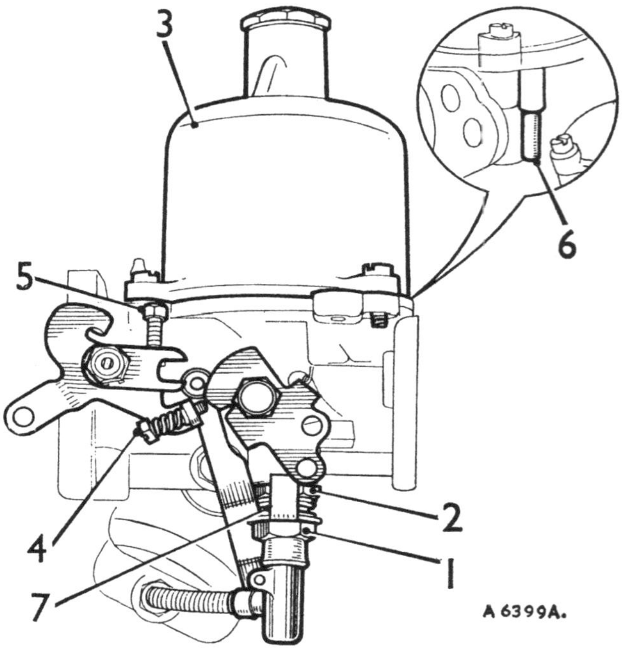


*The type HS carburetter*

1. Jet adjusting nut.
2. Jet locking nut.
3. Piston suction chamber.
4. Fast-idle adjusting screw.
5. Throttle adjusting screw.
6. Piston lifting pin.
7. Jet adjustment restrictor.

## Section 4-A

### CARBURETTER TUNING—BASIC

**GENERAL**

The carburetters fitted to cars equipped with engine emission control systems are balanced to provide maximum performance with maximum pollution control. Under no circumstances may they be interchanged, or parts substituted.

Tuning must be carried out with the engine emission control equipment connected and operating, and is confined to the following procedure. If the required settings cannot be obtained, the service procedure detailed under **'CARBURETTER SERVICING'** must be carried out and then the carburetter tuned in accordance with the procedure given in **'CARBURETTER TUNING—COMPLETE'**.

**Tuning conditions**

To ensure that the engine temperature and mixture requirements are stabilized, tuning must be carried out in accordance with the following setting cycle.

(1) Connect a tachometer in accordance with the instrument-maker's instructions.

(2) Warm the engine at a fast idle to normal operating temperature, preferably with the car standing in an ambient temperature of between 16 and 27° C. (60 to 80° F.). Run the engine for at least five minutes after the thermostat has opened; the thermostat opening point can be detected by the sudden rise in temperature of the radiator header tank.

(3) Set the engine speed at 2,500 r.p.m., at no load, and run for one minute.

(4) Tuning operations may now be commenced and must be carried out in the shortest possible time. If the time for settings exceeds a three-minute period, open the throttle and run the engine at 2,500 r.p.m. for one minute then resume tuning. Repeat this clearing operation if further periods of three minutes are exceeded.

**SINGLE CARBURETTERS**

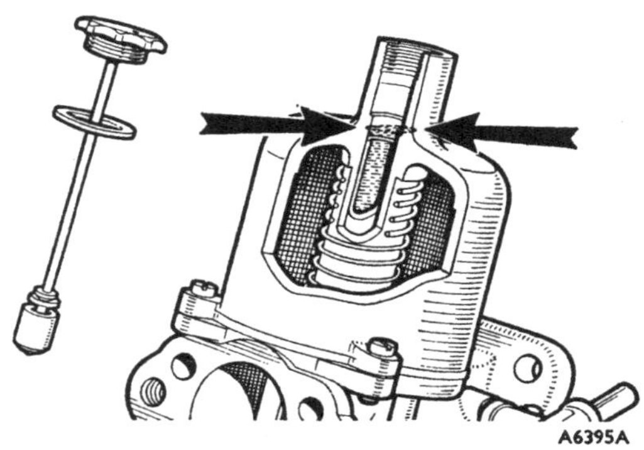


**NOTE.**—In no case should the jet adjustment restrictor be removed or repositioned. Only mixture adjustments within the limits of the restrictor are available for tuning. If satisfactory adjustment is not obtainable within the limits of the jet adjustment restrictor refer to **'CARBURETTER SERVICING'.**

(1) Top up the piston damper with the recommended engine oil until the level is ½-in. above the top of the hollow piston rod.

**NOTE.**—On dust-proofed carburetters, identified by a transverse hole drilled in the neck of the suction chambers and no vent hole in the damper cap, the oil level must be ½-in. below the top of the hollow piston rod.

(2) Check throttle control action for signs of sticking.

(3) Check the idling speed (Tachometer) against the figure given in **'TUNING DATA'.**

(*a*) If the reading is correct and the engine runs smoothly, proceed to operations (7) and (8).

(*b*) If the reading is not correct, adjust the speed by turning the throttle adjusting screw in the required direction until the correct speed consistent with smooth running is obtained, then proceed to operations (7) and (8).

(*c*) If a smooth idle at the correct speed is not obtainable by turning the throttle adjusting screw, carry out operations (4) to (8).

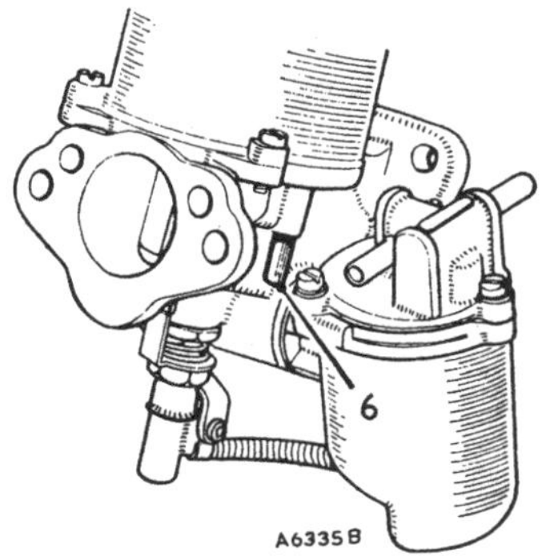


(4) With the engine stopped, check that the piston falls freely onto the bridge, indicated by a distinct metallic click, when the lifting pin (6) is released. If not refer to **'CARBURETTER SERVICING'.**

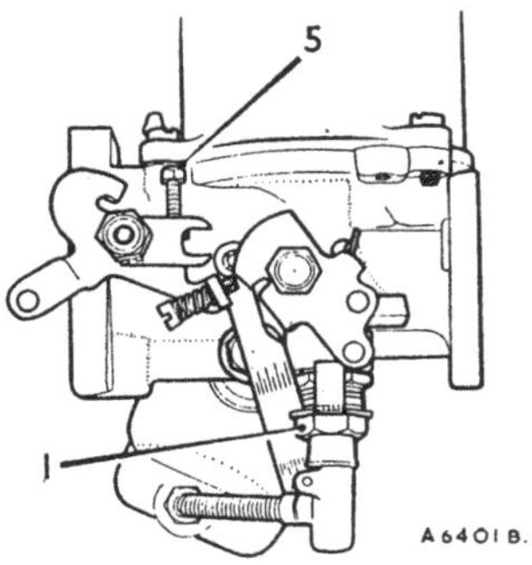


(5) Turn the jet adjusting nut (1) to cover the full range of adjustment available within the limits of the restrictor, selecting the setting where maximum speed is recorded on the tachometer consistent with smooth running.

(6) Readjust the throttle adjusting screw (5) to give the correct idling speed if necessary.

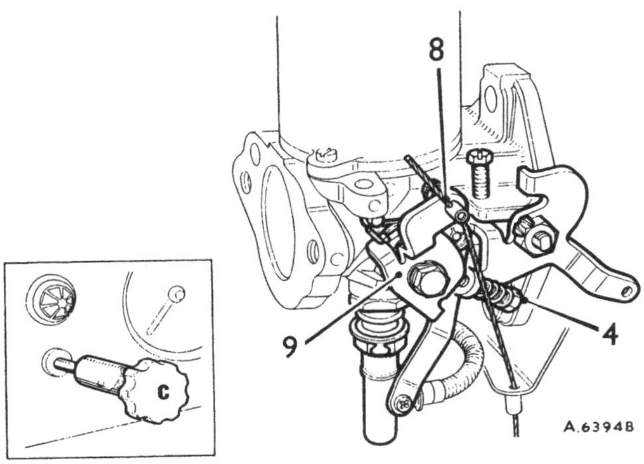


(7) Check, and if necessary adjust, the mixture control wire (8) to give a free movement of approximately $\frac{1}{16}$-in. before it starts to pull on the jet lever (9).

(8) Pull the mixture control knob until the linkage is about to move the carburetter jet and adjust the fast-idle screw (4) to give the engine fast-idle speed (Tachometer) given in **'TUNING DATA'.**

## TWIN CARBURETTERS

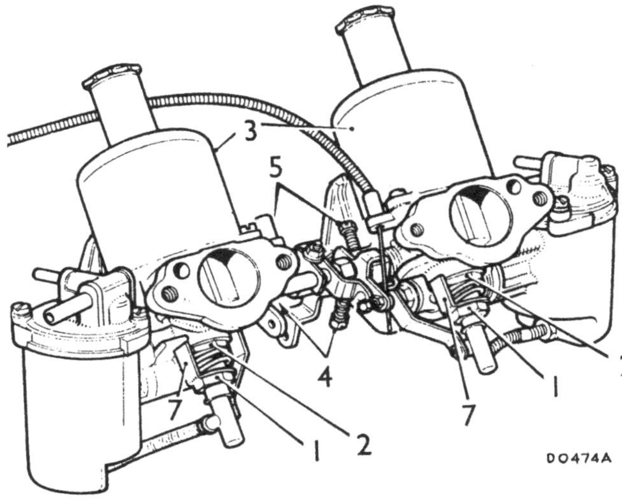


*A twin-carburetter installation*

1. Jet adjusting nuts.
2. Jet locking nuts.
3. Piston/suction chambers.
4. Fast-idle adjusting screws.
5. Throttle adjusting screws.
7. Jet adjustment restrictors.

**NOTE.**—In no case should the jet adjustment restrictor be removed or repositioned. Only mixture adjustments within the limits of the restrictor are available for tuning. Balancing of twin carburetters must only be carried out with the use of an approved balancing meter. If satisfactory adjustment or balancing is not obtainable within the limits of the jet adjustment restrictor, refer to **'CARBURETTER SERVICING'.**

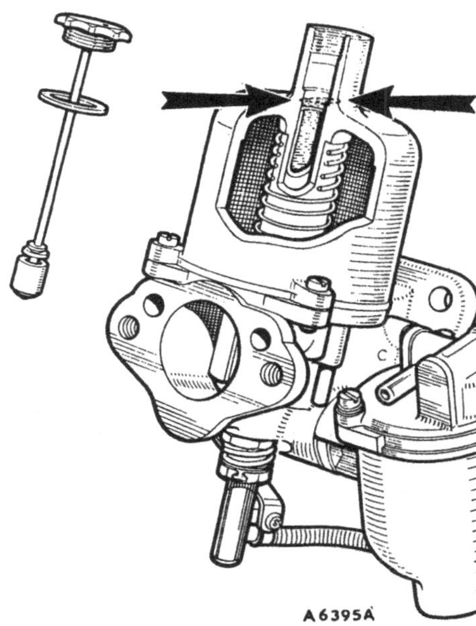


(1) Top up the piston damper with the recommended engine oil until the level is $\frac{1}{2}$-in. above the top of the hollow piston rod.

**NOTE.**—On dust-proofed carburetters, identified by a transverse hole drilled in the neck of the suction chambers and no vent hole in the damper cap, the oil level must be $\frac{1}{2}$-in. below the top of the hollow piston rod.

(2) Check the throttle control action for signs of sticking.

(3) Check the idling speed (Tachometer) against the figure given in **'TUNING DATA'**.

  (*a*) If the reading is correct and the engine runs smoothly, proceed with operations (11) to (17).

  (*b*) If the reading is not correct, carry out operations (4) to (17).

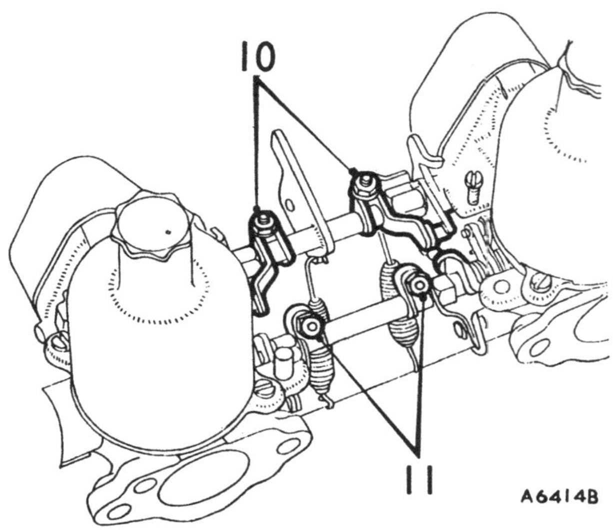


(4) Stop the engine and remove the air cleaners.

(5) Slacken both of the clamping bolts (10) on the throttle spindle interconnections.

(6) Disconnect the jet control interconnection by slackening the clamping bolts (11).

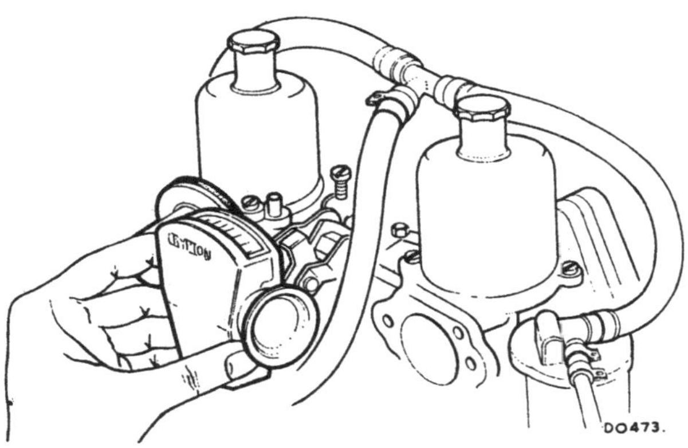


(7) Restart the engine and adjust the throttle adjusting screws on **both** carburetters to give the correct idling speed as registered by the tachometer.

(8) Using an approved balancing meter in accordance with the maker's instructions, balance the carburetters by altering the throttle adjusting screws; the idling speed obtained during this operation must be as given in **'TUNING DATA'**.

  (*a*) If after this operation the balance is satisfactory and consistent with smooth running at the correct idle speed, proceed with operations (14) to (17).

  (*b*) If correct balance cannot be obtained, check the intake system for leaks (i.e. brake servos, engine emission control equipment); if still unsatisfactory, refer to **'CARBURETTER SERVICING'**.

  (*c*) If with the carburetters correctly balanced the idling is still erratic, carry out operations (9) to (17).

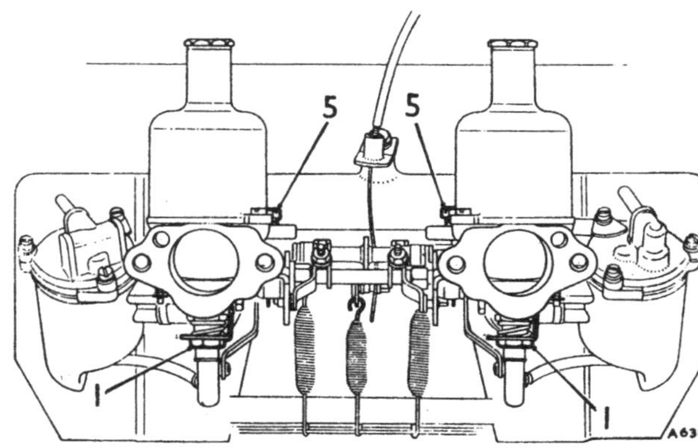


(9) Turn the jet adjusting nut (1) on both carburetters to cover the full range of adjustment available within the limits of the restrictor, selecting the setting where maximum speed is recorded on the tachometer consistent with smooth running.

(10) Readjust the throttle adjusting screws (5) to give the correct idling speed (see **'TUNING DATA'**) if necessary, ensuring that both carburetters are adjusted by an equal amount.

If the correct idling speed consistent with smooth running cannot be obtained, refer to **'CARBURETTER SERVICING'**.

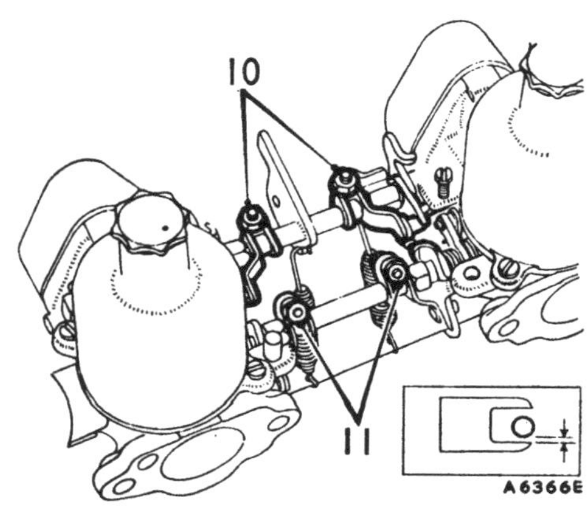


(11) Set the throttle interconnection clamping levers (10) so that the link pin is ·012 in. away from the lower edge of the fork (see inset). Tighten the clamp bolts ensuring that there is approximately $\frac{1}{32}$ in. end-float on the interconnection rod.

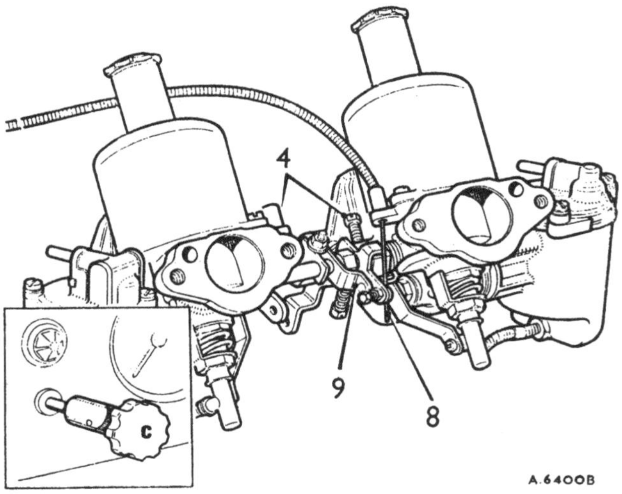


(12) With both jet levers at their lowest position, set the jet interconnection lever clamp bolts (11) so that both jets commence to move simultaneously.

(13) Run the engine at 1,500 r.p.m. and, using the balance meter, check that the carburetters are balanced. If they are not balanced, reset the levers, rebalance at idle speed, then recheck at 1,500 r.p.m.

(14) Check, and if necessary adjust, the mixture control wire (1) to give approximately $\frac{1}{16}$ in. free movement before it starts to pull on the jet levers (9).

(15) Pull the mixture control knob until the linkage is about to move the carburetter jets.

(16) Using the carburetter balancing meter to ensure equal adjustment, turn the fast idle adjusting screws (4) to give the correct fast idling speed (see 'TUNING DATA').

(17) Refit the air cleaners.

## Section 4-B

### CARBURETTER SERVICING

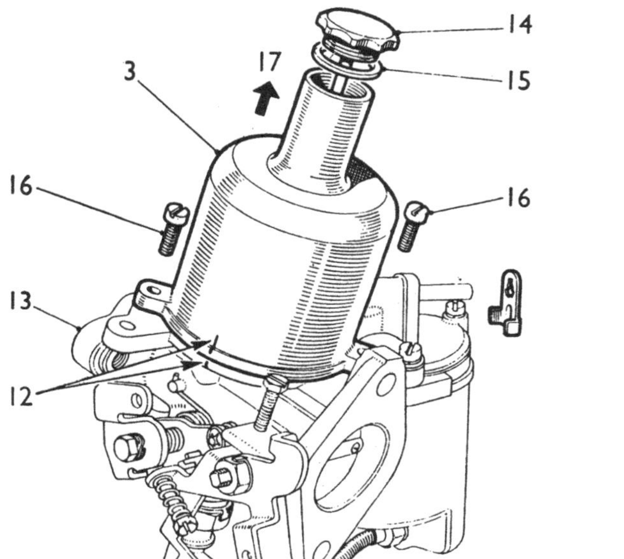


**Dismantling**

*Carburetters—all types*

(1) Thoroughly clean the outside of the carburetter.

(2) Mark the relative position (12) of the suction chamber (3) and the carburetter body (13).

(3) Remove the damper (14) and its washer (15). Unscrew the chamber retaining screws (16).

(4) Lift off the chamber in the direction of arrow (17) without tilting.

(5) Remove the piston spring (18).

(6) Carefully lift out the piston assembly (19) and empty the damper oil from the piston rod (20).

*Carburetters—fixed needle type*

(7) Remove the needle locking screw (21) and withdraw the needle (22). If it cannot easily be removed, tap the needle inwards first and then pull outwards. Do not bend the needle.

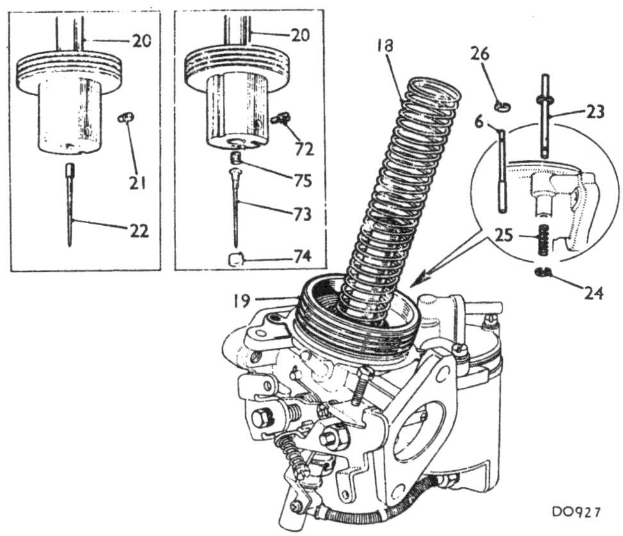


*Carburetters—spring-loaded needle type*

(8) Remove the guide locking screw (72), withdraw the needle assembly (73), needle support guide (74) and spring (75), taking care not to bend the needle.

(9) Withdraw the needle from the guide and remove the spring from the needle assembly.

*Carburetters—all types*

(10) If a piston lifting pin (23) with an external spring is fitted, remove the spring retaining circlip (24) and spring (25), then push the lifting pin upwards to remove it from its guide. With the concealed spring type (6) press the pin upwards, detach the circlip (26) from its upper end, and withdraw the pin and spring downwards.

(11) Support the moulded base of the jet (26) and slacken the screw (27) retaining the jet pick-up link (28).

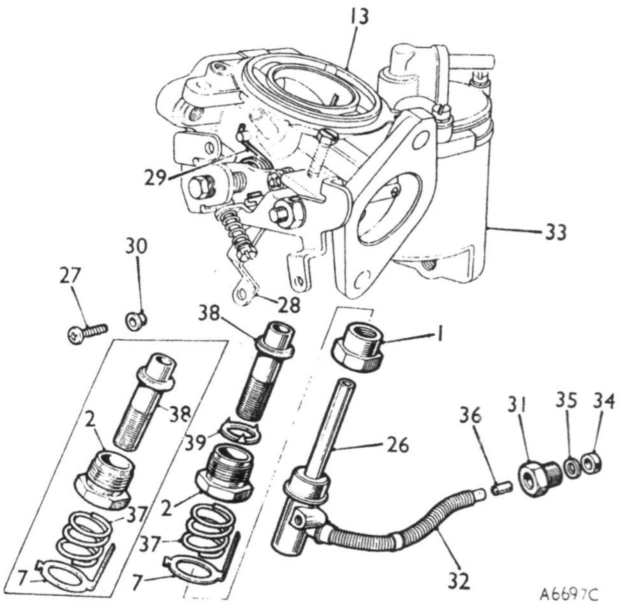


(12) Relieve the tension of the pick-up lever return spring (29) from the screw and remove screw and brass bush (30) (when fitted).

(13) Unscrew the brass sleeve nut (31) retaining the flexible jet tube (32) to the float-chamber (33) and withdraw the jet assembly (26) from the carburetter body (13). Note the gland (34), washer (35), and ferrule (36) at the end of the jet tube.

(14) Bend back the small tag on the restrictor (7) to clear the jet adjusting nut, and remove the jet adjusting nut (1), restrictor (7), and spring (37). Unscrew the jet locking nut (2) and detach the nut and jet bearing (38). Withdraw the bearing from the nut, noting, on fixed needle carburetters only, the locking washer (39) under the shoulder of the bearing.

(15) Note the location points (see inset, 40) of the two ends of the pick-up lever return spring (41). Unscrew the lever pivot bolt (42) together with its double-coil spring washer (43), or spacer (44). Detach the lever assembly (9) and return spring.

(16) Note the location (see inset, 45) of the two ends of the cam lever spring (46) and push out the pivot bolt tube (47) (or tubes), taking care not to lose the spring. Lift off the cam lever (48), noting the skid washer (49) between the two levers.

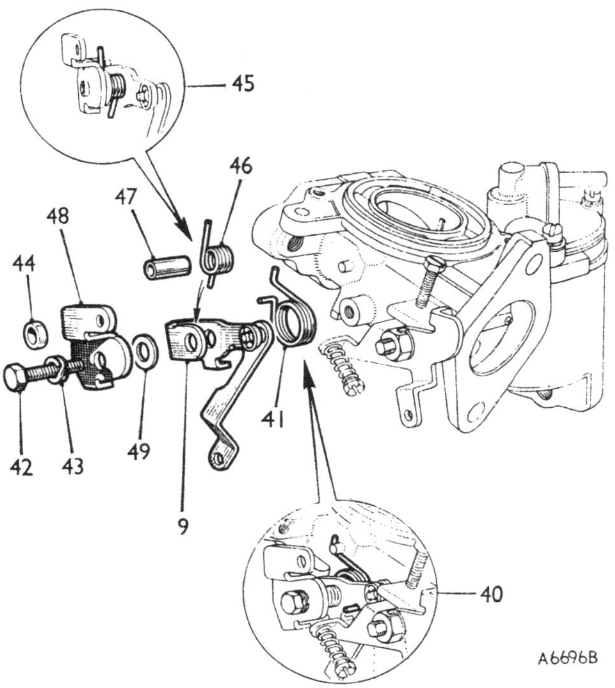


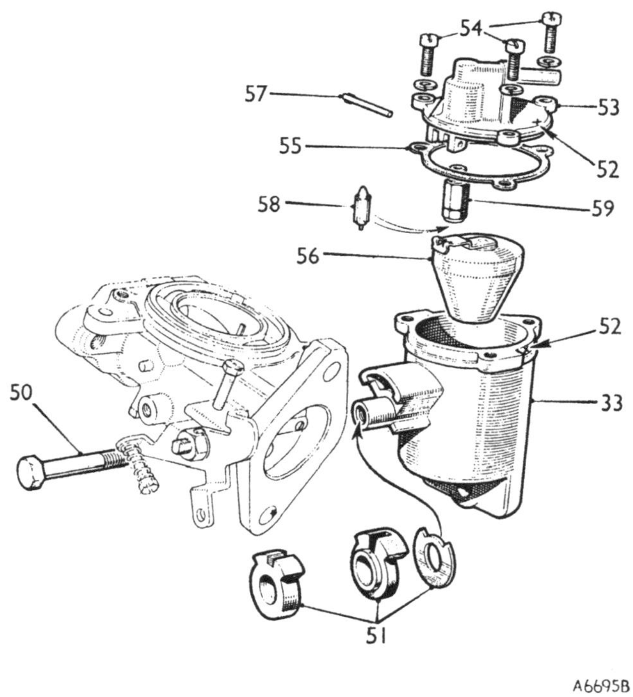


(17) Slacken and remove the bolt (50) retaining the float-chamber (33) to the carburetter body. Note the component sequence of the flexibly mounted chambers (33) and (51).

(18) Mark (52) the location of the float-chamber lid (53). Unscrew the lid retaining screws (54) and detach the lid and its gasket (55) complete with float assembly (56).

(19) Push out the float hinge pin (57) from the end opposite its serrations and detach the float.

(20) Extract the float needle (58) from its seating (59) and unscrew the seating from the lid, using a wrench ·338 in. across the flats. Do not distort the seating.

(21) Close the throttle and mark (60) the relative positions of the throttle disc (61) and the carburetter flange (62). **Do not mark the throttle disc in the vicinity of the limit valve (63).**

(22) Unscrew the two disc retaining screws (64). Open the throttle and ease out the disc from its slot in the throttle spindle (65). The disc is oval and will jam if care is not taken; store the disc in a safe place until required for reassembly.

(23) Tap back the tabs of the tab washer (66) securing the spindle nut (67). Note the location of the lever arm (68) in relation to the spindle and carburetter body; remove the nut and detach the arm.

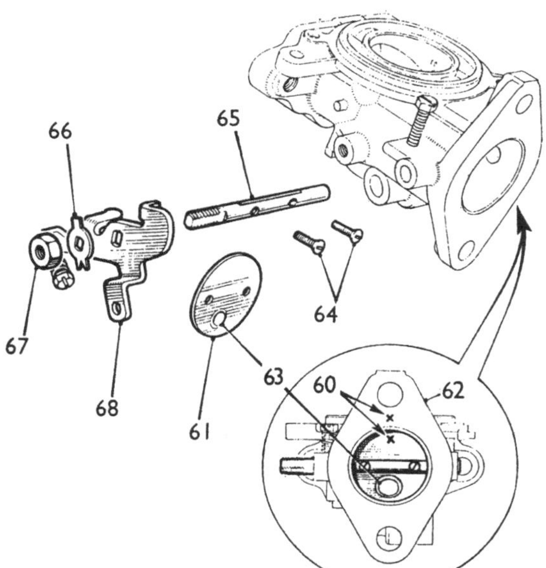


## Reassembling

*Carburetters—all types*

**NOTE.**—Before reassembling, examine all components for wear and damage. Renew unserviceable components, ensuring that only parts to the correct specification (see **'TUNING DATA'**) are used.

(1) Examine the throttle spindle and its bearings in the carburetter body. Check for excessive play. Renew parts as necessary.

(2) Refit the spindle to the body. Assemble the operating lever with tab washer and spindle nut, to the spindle. Ensure that when the stop on the lever is against the abutment on the carburetter body (i.e. throttle closed position) the countersunk ends of the holes in the spindle face outwards. Tighten the spindle nut and lock with the tab washer.

(3) Insert the throttle disc in the slot in the spindle in its original position as marked. Manœuvre the disc in its slot until the throttle can be closed, snap the throttle open and shut to centralize it in the bore of the carburetter, taking care not to damage the throttle limit valve. When assembled, the valve must be positioned at the bottom of the disc with the head of the valve towards the engine. Fit two new disc retaining screws but do not fully tighten. Check visually that the disc closes fully, and adjust

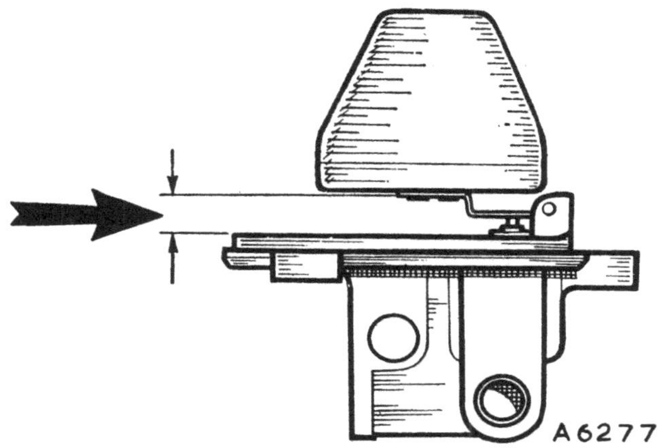


its position as necessary. With the throttle closed there must be clearance between the throttle lever and the carburetter body. Tighten the screws fully and spread their split ends just enough to prevent turning.

(4) Examine the float needle and seating for damage. Check that the spring-loaded plunger in the end of the plastic-bodied needle operates freely.

(5) Screw the seating into the float-chamber carefully. Do not overtighten. Replace the needle in the seating, coned end first. Test the assembly for leakage with air pressure.

(6) Refit the float and lever to the lid and insert the hinge pin and invert the float-chamber lid. With the needle valve held in the shut-off position by the weight of the float only, there should be a $\frac{1}{8}$ to $\frac{3}{16}$ in. gap (arrowed) between the float lever and the rim of the float-chamber lid.

(7) Examine the lid gasket for re-use. Assemble the gasket on the lid and refit the lid to the float-chamber in the position marked on dismantling. Tighten the securing screws evenly.

(8) Refit the float-chamber assembly to the carburetter body and tighten the retaining bolt fully, making sure that the registers on the body and the chamber engage correctly.

(9) Refit the piston lifting pin, spring and circlip.

(10) Examine the piston assembly for damage on the piston rod and the outside surface of the piston. The piston assembly must be scrupulously clean. Use gasoline or methylated spirit (denatured alcohol) as a cleaning agent. **Do not use abrasives.** Wipe dry, using a clean dry cloth.

(11) Clean inside the suction chamber and piston rod guide using gasoline or methylated spirit (denatured alcohol) and wipe dry. Refit the damper and washer. Temporarily plug the piston transfer holes (69) and fit the piston into the suction chamber. Fit a nut and screw, with a large flat washer under the head of the screw into one of the suction

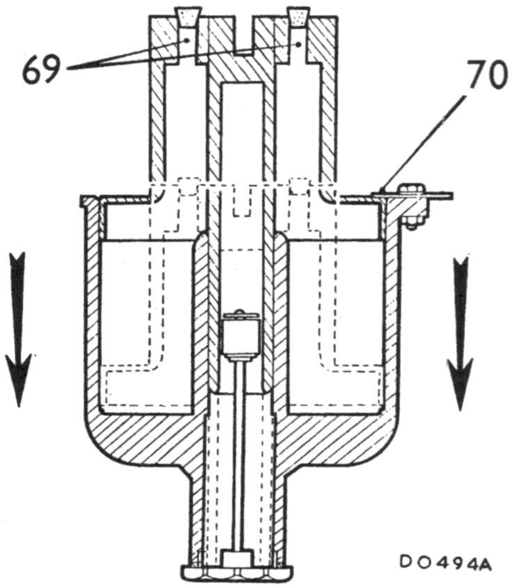


chamber fixing holes, positioning the washer (70) so that it overlaps the suction chamber bore (see illustration). Check that the piston is fully home in the suction chamber and invert the assembly to allow the chamber to fall away from the piston until the piston contacts the flat washer. Check the time taken for the suction chamber to fall the full extent of the piston travel. For HS2-type carburetters of 1¼ in. bore the time taken should be 3 to 5 seconds, and for larger carburetters 5 to 7 seconds. If these times are exceeded check the piston and suction chamber for cleanliness and mechanical damage. If after rechecking the time taken is still not within these limits, renew the suction chamber and piston assembly.

*Carburetters—fixed needle type*

(12) Refit the needle to the piston assembly (19). The lower edge of the needle shoulder (22) must be level with the bottom face of the piston rod (20).

(13) Fit a new needle locking screw (21) and tighten. Invert the suction chamber and spin the piston assembly inside it to check for concentricity of the needle.

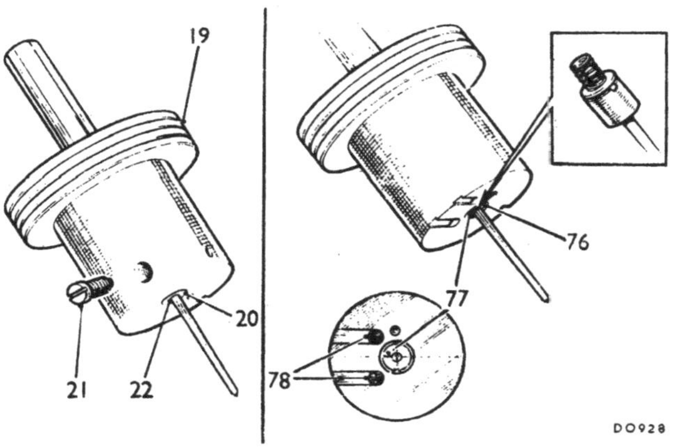


(14) Check the piston key for security in the carburetter body. Refit the piston assembly to the body and replace the piston spring over the piston rod.

(15) Fit the suction chamber and retaining screws, taking care not to wind up the spring; tighten the securing screws evenly.

(16) Refit the jet bearing, a new locking washer, and the locking nut; do not tighten the nut.

(17) Centralize the jet as follows:

(*a*) Enter the end of the nylon feed tube into the base of the float-chamber, without the gland or washer fitted. Loosely secure with the retaining nut.

(*b*) Feed the jet into the jet bearing; do not fit the jet nut spring, jet adjustment restrictor, or adjusting nut at this stage.

(*c*) With the carburetter positioned with its inlet flange downwards, insert the piston loading tool into damper tube at the top of the suction chamber and screw in until fully home. Screw the tool back until the arrow, on the tool, points towards the inlet flange of the carburetter. **The tool and carburetter must remain in this position throughout the centering operation.**

(*d*) With the piston at the bottom of its travel (on the bridge), and the jet hard up against the jet bearing, slowly tighten the jet locking nut. During the tightening process ensure that the jet is not binding in its bearing when drawn in and out. If any tightness between the jet and bearing is detected, the jet locking nut must be slackened and the process repeated.

(*e*) Remove the jet loading tool.

(18) Withdraw the jet and tube; refit the spring, restrictor and jet adjusting nut. Fit the gland and washer to the flexible tube. The end of the tube should project a minimum of $\frac{3}{16}$ in. beyond the gland. Refit the jet and tube. Tighten the sleeve nut until the neoprene gland is compressed. Overtightening can cause leakage.

*Carburetters—spring-loaded needle type*

(19) Refit the jet bearing, fit and tighten the jet locking nut. No jet centering is required with the spring-loaded type jet needle.

(20) Fit the jet nut spring and adjustment restrictor. Fit the jet adjusting nut and screw it up as far as possible.

(21) Feed the jet into the jet bearing. Fit the sleeve nut, washer and gland to the end of the flexible tube. The tube must project a minimum of $\frac{3}{16}$ in. (4·8 mm.) beyond the gland. Tighten the sleeve nut until the gland is compressed. Overtightening can cause leakage.

(22) Refit the spring to the jet needle assembly, ensuring that it locates completely in the groove of the needle support.

(23) **IMPORTANT.** Spring-loaded needles are supplied complete with shouldered spring seats; no attempt should be made to alter the position of the spring seat or convert a fixed-type needle to spring- loaded application. The raised 'pip' formed in the needle guide ensures that the needle is correctly centralized. Under no circumstances must the 'pip' be removed or repositioned.

Fit the needle assembly into its guide and fit the assembly into the piston. The lower edge of the guide (76) must be flush with the face of the piston and the guide positioned so that the etched locating mark (77) on its lower face is adjacent to and in line with the midway point between the two piston transfer holes as illustrated.

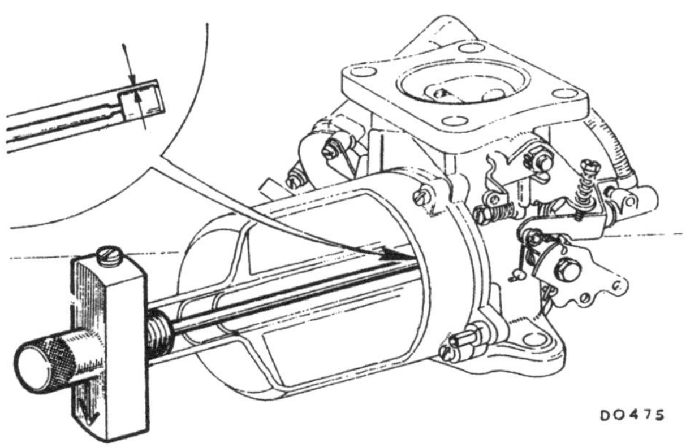


Alternative needle guides have a flat machined on the guide which must be positioned so that the guide locking screw tightens down onto the flat. If the guide is incorrectly positioned so that the locking screw has not tightened down on the flat, the head of the screw will protrude from the piston.

(24) Fit a new guide locking screw. **NOTE.**—Guide locking screws for spring-loaded needles are shorter than the needle locking screws used with fixed needles.

(25) Check the piston key for security in the carburetter body. Refit the piston assembly to the body and place the piston spring over the piston rod.

(26) Fit the suction chamber and retaining screws, taking care not to wind up the spring; tighten the securing screws evenly.

*Carburetters—all types*

(27) Refit the damper and washer.

(28) Reassemble the pick-up lever, cam lever, cam lever spring, skid washer, and pivot bolt tube or tubes in the positions noted on dismantling.

(29) Place the pick-up lever return spring in position over its boss and secure the lever assembly to the carburetter body with the pivot bolt. Ensure that the double-coil spring washer or spacer fits over the projecting end of the pivot bolt tube.

(30) Register the angled end of the return spring in the groove in the pick-up lever, and hook the other end of the spring around the moulded peg on the carburetter body.

(31) Fit the brass ferrule to the hole in the end of the pick-up link. Relieve the tension of the return spring and fit the link to the jet with its retaining screw. When finally tightening the screw, support the moulded end of the jet.

(32) Without removing the suction chamber, screw the jet adjusting nut until the top face of the jet is flush with the bridge of the carburetter.

(33) Turn down the jet adjusting nut to the initial jet setting given in **'TUNING DATA'.**

(34) Refit the carburetter(s) to the engine, following the instructions given in the relevant vehicle Workshop Manual.

Tune the carburetters in accordance with the instructions given in **'CARBURETTER TUNING —COMPLETE'.**

## Section 4-C

### CARBURETTER TUNING—COMPLETE

The following instructions apply only to new carburetters or carburetters which have been serviced as described in **'CARBURETTER SERVICING'**.

The tuning must be carried out with the engine emission control equipment connected and operating.

### SINGLE CARBURETTERS

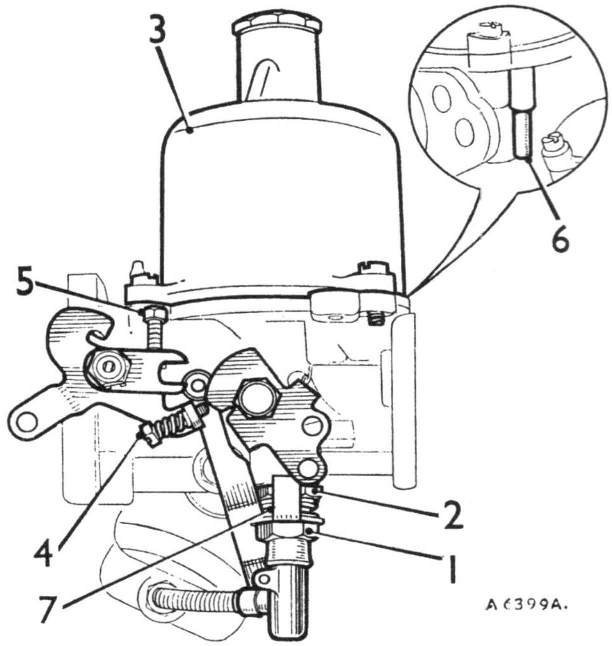


*The type HS carburetter*

1. Jet adjusting nut.
2. Jet locking nut.
3. Piston suction chamber.
4. Fast-idle adjusting screw.
5. Throttle adjusting screw.
6. Piston lifting pin.
7. Jet adjustment restrictor.

#### Initial setting

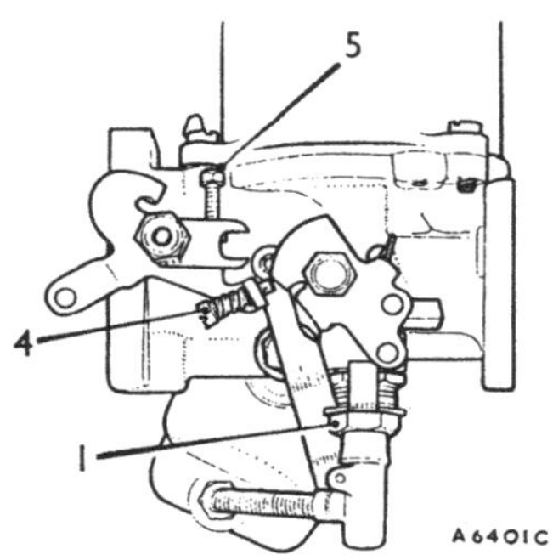


(1) Disconnect the mixture control (choke) wire if fitted.

(2) Unscrew the fast-idle screw (4) until it is well clear of the cam.

(3) Unscrew the throttle adjusting screw (5) until it is just clear of its stop and the throttle is closed.

(4) Set the throttle adjusting screw one full turn open.

(5) The jet adjusting nut must not be altered at this stage as it will be initially set to a datum setting at the factory or during the carburetter servicing procedure.

#### Tuning conditions

To ensure that the engine temperature and mixture requirements are stabilized, tuning must be carried out in accordance with the following setting cycle.

(1) Connect a tachometer and an approved exhaust gas analyser in accordance with the instrument-maker's instructions.

(2) Warm the engine at a fast idle to normal operating temperature preferably with the car standing in an ambient temperature of between 16 and 27° C. (60 to 80° F.). Run the engine for at least five minutes after the thermostat has opened; the thermostat opening point can be detected by the sudden rise in temperature of the radiator header tank.

(3) Set the engine speed at 2,500 r.p.m., at no load, and run for one minute.

(4) Tuning operations may now be commenced and must be carried out in the shortest possible time. If the time for settings exceeds a three-minute period, open the throttle and run the engine at 2,500 r.p.m. for one minute then resume tuning. Repeat this clearing operation if further periods of three minutes are exceeded.

#### Tuning procedure

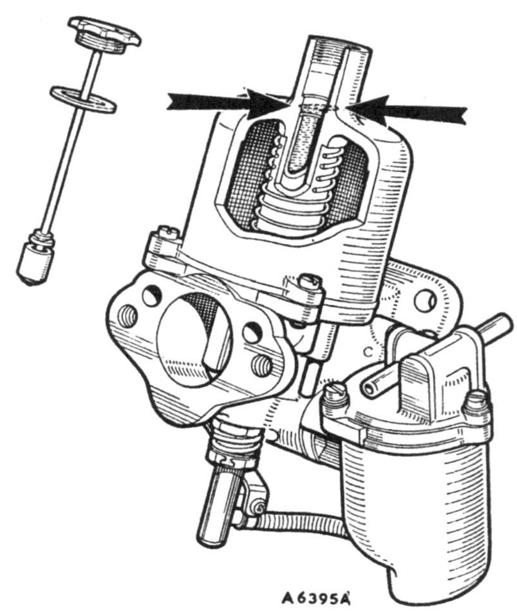


(1) Top up the piston damper with the recommended engine oil until the level is ½ in. above the top of the hollow piston rod.

**NOTE.**—On dust-proofed carburetters, identified by a transverse hole drilled in the neck of the suction chambers and no vent hole in the damper cap, the oil level must be ½ in. below the top of the hollow piston rod.

(2) Warm up the engine as described in **'Tuning conditions'**.

Turn the throttle adjusting screw until the idling speed given in **'TUNING DATA'** is obtained.

(3) During the following procedure, just before the readings of the tachometer and exhaust gas analyser are taken gently tap the neck of the suction chamber with a light non-metallic instrument (e.g. a screwdriver handle).

Turn the jet adjusting nut up to weaken, down to richen, until the fastest speed is recorded on the tachometer. Turn the jet adjusting nut very slowly up (weaken) until the engine speed just commences to fall, then turn the nut one flat down (rich). Check the idling speed against the figure given in **'TUNING DATA'**, and adjust if necessary using the throttle adjusting screw.

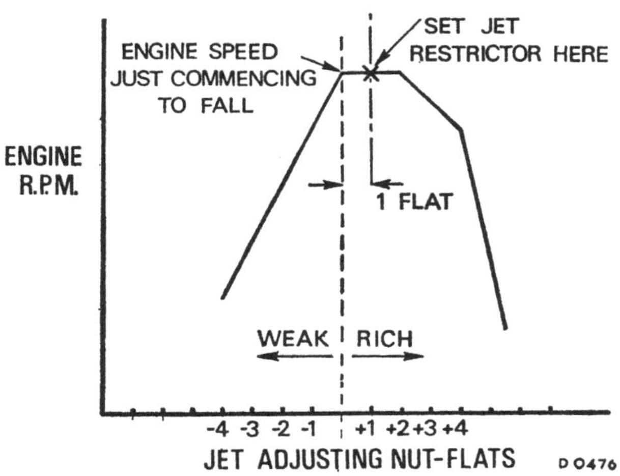


(4) Using the exhaust gas analyser, check that the percentage CO reading is within the limits given in **'TUNING DATA'**.

If the reading falls outside the limits given, reset the jet adjusting nut by the minimum amount necessary to bring the reading just within the limits. If an adjustment exceeding two flats is required to achieve this the test equipment should be checked for correct calibration.

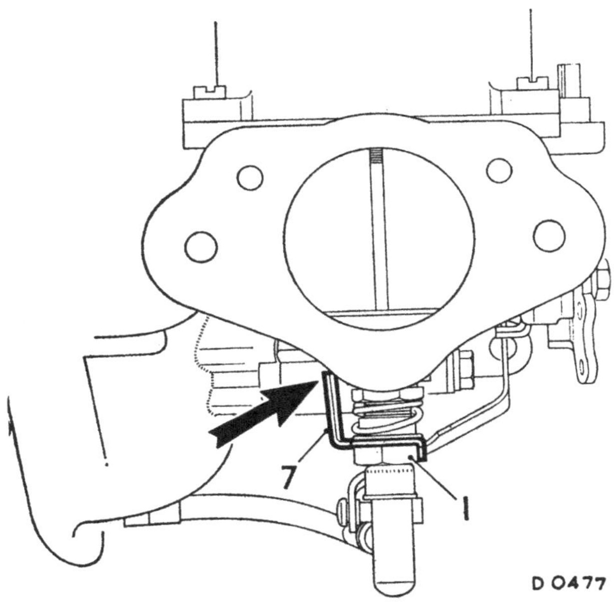


(5) Hold the jet adjusting nut (1) to prevent it turning, and rotate the adjustment restrictor (7) round the nut until the vertical tag contacts the carburetter body on the left-hand side when viewed from the air cleaner flange (see illustration). In this position, bend the small tag on the adjustment restrictor down so that the restrictor locks to the nut and will follow its movements.

(6) Paint the small tag of the jet adjusting nut restrictor and the adjacent flat of the jet nut to identify the locking position.

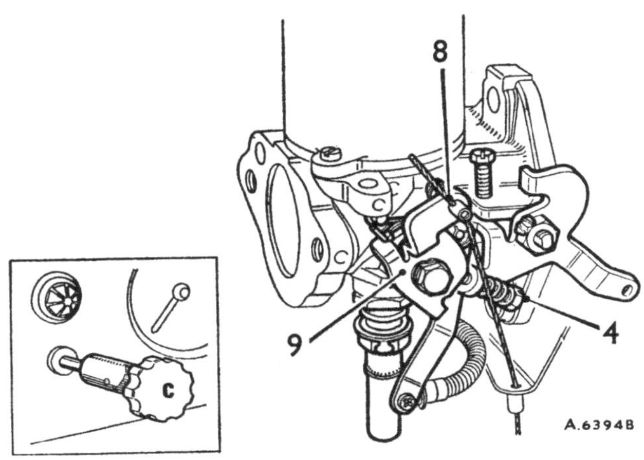


(7) Reconnect the mixture control wire (8) with approximately $\frac{1}{16}$ in. free movement before it starts to pull on the jet lever (9).

(8) Pull the mixture control knob until the linkage is about to move the carburetter jet and adjust the fast-idle screw (4) to give the engine fast-idle speed

(9) Check and if necessary adjust the throttle damper setting—if fitted (see **'TUNING DATA'**).

## TWIN CARBURETTERS

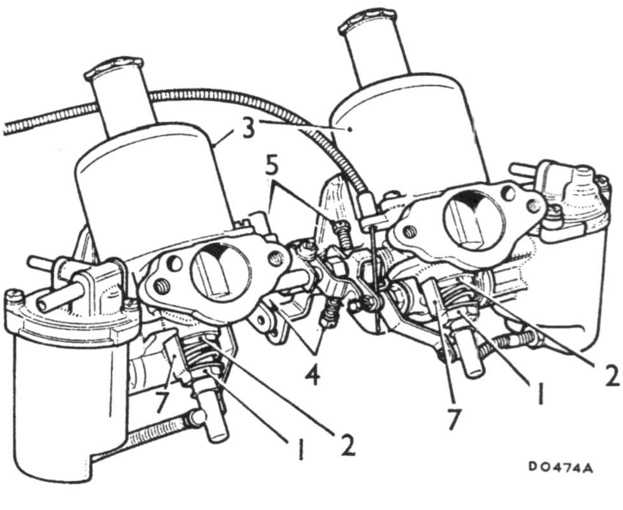


*A twin-carburetter installation*

1. Jet adjusting nuts.
2. Jet locking nuts.
3. Piston/suction chambers.
4. Fast-idle adjusting screws.
5. Throttle adjusting screws.
7. Jet adjustment restrictors.

**Initial settings**

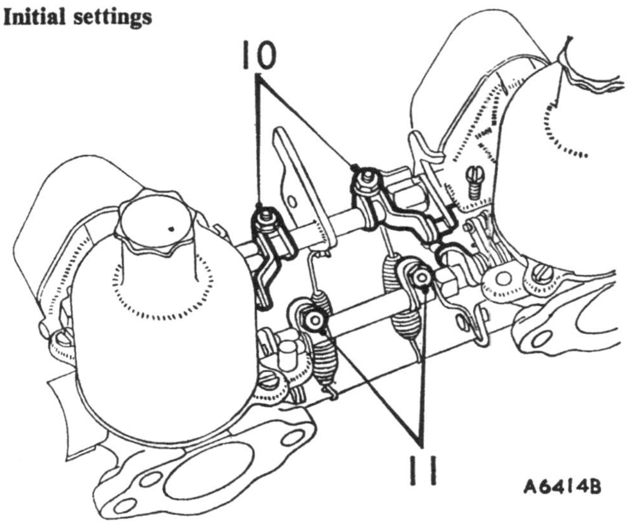


(1) Slacken both clamping bolts (10) on the throttle spindle interconnections.
(2) Disconnect the jet control interconnection by slackening the clamping bolts (11).
(3) Disconnect the mixture control wire if fitted.

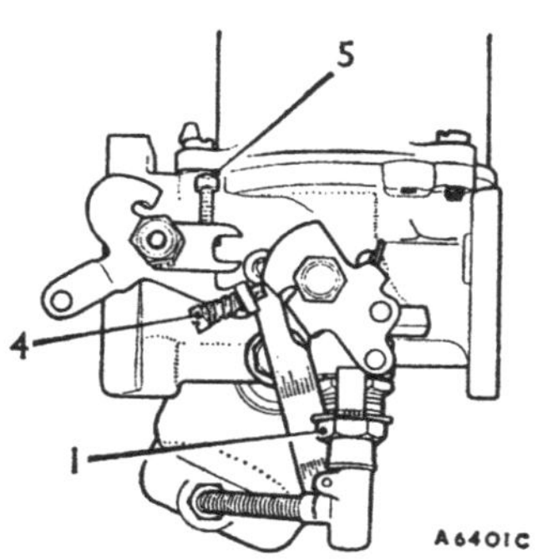


(4) Unscrew the fast-idle screw (4) on both carburetters until they are well clear of the cams.
(5) Unscrew the throttle adjusting screw (5) on both carburetters until they are just clear of their stops and the throttles are closed.
(6) Set the throttle adjustment screws on both carburetters half a turn open.
(7) The jet adjusting nuts must not be altered at this stage as they will be initially set to a datum setting either at the factory or during the carburetter servicing procedure.

**Tuning conditions**

To ensure that the engine temperature and mixture requirements are stabilized, tuning must be carried out in accordance with the following setting cycle.

(1) Connect a tachometer and an approved exhaust gas analyser in accordance with the instrument-maker's instructions.
(2) Warm the engine at a fast idle to normal operating temperature preferably with the car standing in an ambient temperature of between 16 and 27° C. (60 to 80° F.). Run the engine for at least five minutes after the thermostat has opened; the thermostat opening point can be detected by the sudden rise in temperature of the radiator header tank.
(3) Set the engine speed at 2,500 r.p.m., at no load, and run for one minute.
(4) Tuning operations may now be commenced and must be carried out in the shortest possible time. If the time for settings exceeds a three-minute period, open the throttle and run the engine at 2,500 r.p.m. for one minute then resume tuning. Repeat this clearing operation if further periods of three minutes are exceeded.

**Tuning procedure**

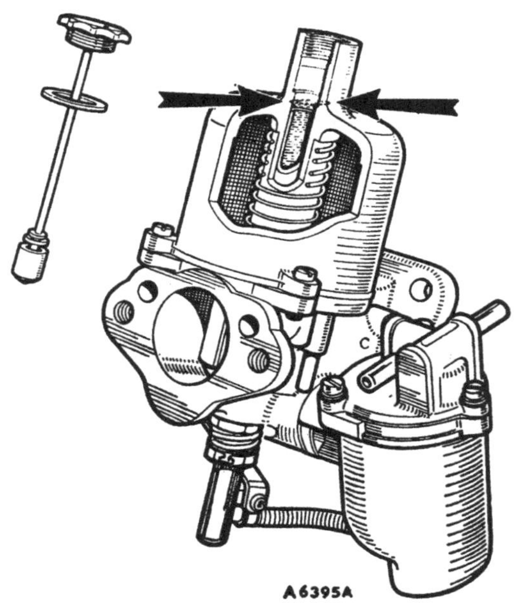

(1) Top up the piston damper with the recommended engine oil until the level is ½ in. above the top of the hollow piston rod.

**NOTE.**—On dust-proofed carburetters, identified by a transverse hole drilled in the neck of the suction chambers and no vent hole in the damper cap, the oil level must be ½ in. below the top of the hollow piston rod.

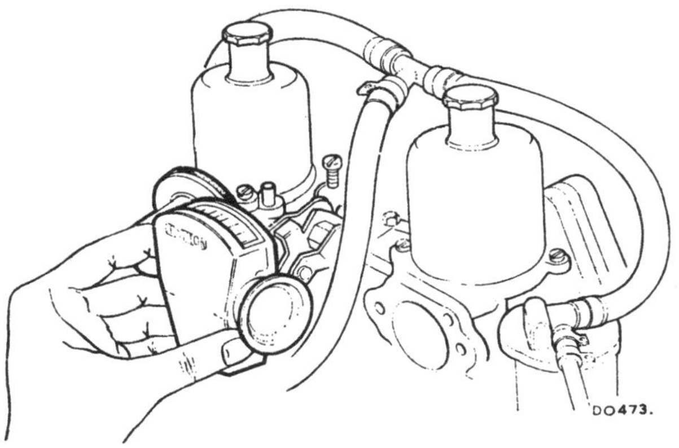


(2) Warm up the engine as described in **'TUNING CONDITIONS.'**

(3) Turn the throttle adjusting screw on both carburetters until the idling speed given in **'TUNING DATA'** is obtained.

(4) Using an approved balancing meter in accordance with the maker's instructions, balance the carburetters by altering the throttle adjusting screws; the idling speed obtained during this operation must be as given in **'TUNING DATA'**.

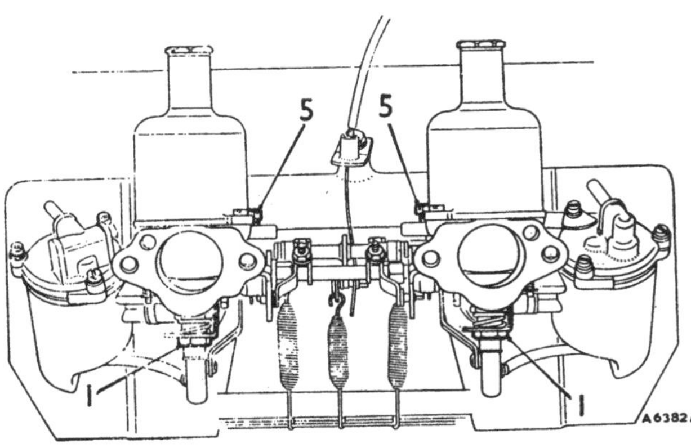


(5) During the following procedure, just before the readings of the tachometer and exhaust gas analyser are taken, gently tap the neck of each suction chamber with a light non-metallic instrument (e.g. a screwdriver handle).

Turn the jet adjusting nut (1) on both carburetters up to weaken, down to richen, the same amount until the fastest speed is recorded on the tachometer.

Turn both adjusting nuts very slowly up (weaken) until the engine speed just commences to fall, then turn both adjusting nuts one flat down (rich).

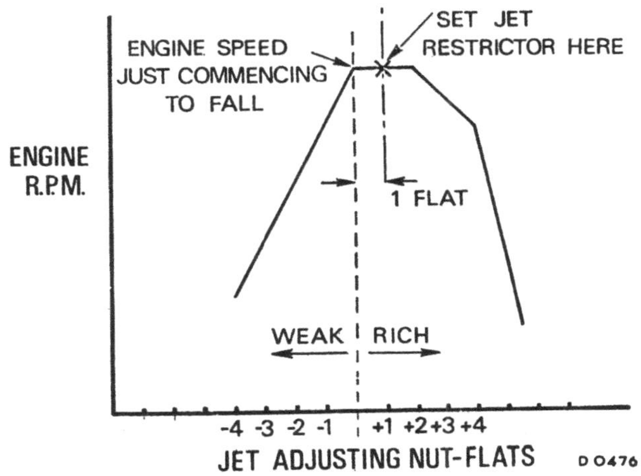


Check the idling speed against the figure given in **'TUNING DATA'**, and adjust if necessary by altering both throttle adjusting screws, each by the same amount. Using the balancing meter, check that the carburetters are balanced.

(6) Using the exhaust gas analyser, check that the percentage CO reading is within the limits given in **'TUNING DATA'.**

If the reading falls outside the limits given, reset both jet adjusting nuts by the minimum amount necessary to bring the reading just within the limits. If an adjustment exceeding two flats is required to achieve this the test equipment should be checked for correct calibration.

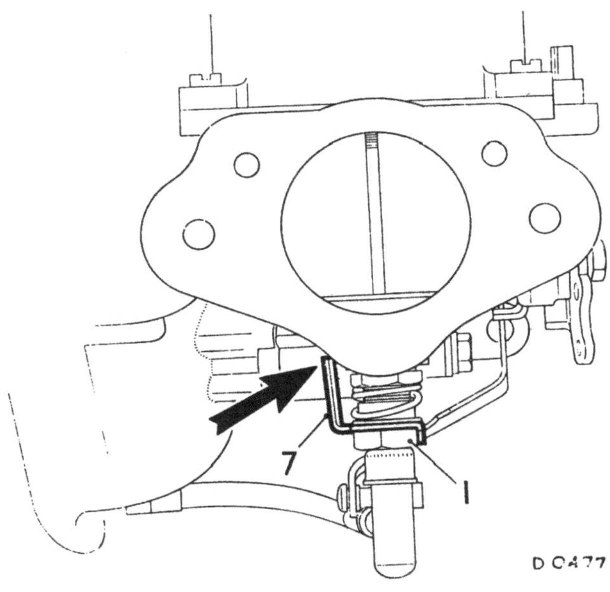


(7) Hold the jet adjusting nut (1) on each carburetter, to prevent it turning, and rotate the adjustment restrictor (7) round the nut until the vertical tag contacts the carburetter body on the left-hand side when viewed from the air cleaner flange (see illustration). In this position, bend the small tag on the adjustment restrictor down so that the restrictor locks to the nut and will follow its movements.

(8) Paint the small tag of the jet adjusting nut restrictor and the adjacent flat of the jet nut to identify the locking position.

(9) Set the throttle interconnection clamping levers (10) in accordance with the instructions given in the relevant vehicle Workshop Manual, so that a clearance exists between the link pin and the lower edge of the fork (see inset). Tighten the clamp bolts ensuring that there is approximately $\frac{1}{32}$ in. end float on the interconnection rod.

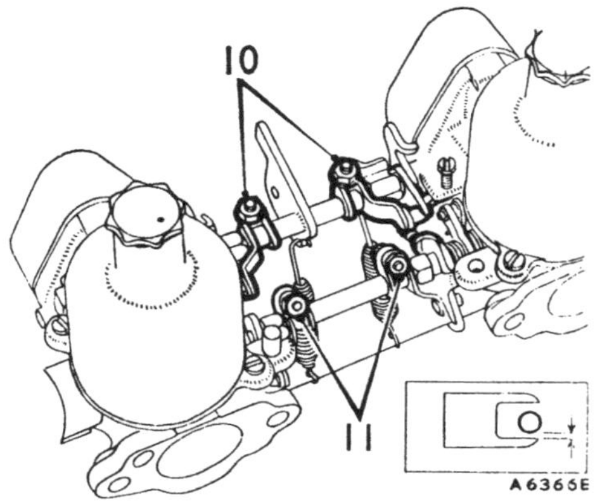


(10) With both jet levers at their lowest position, set the jet interconnection lever clamp bolts (11) so that both jets commence to move simultaneously.

Run the engine at 1,500 r.p.m. and, using the balance meter, check that the carburetters are balanced.

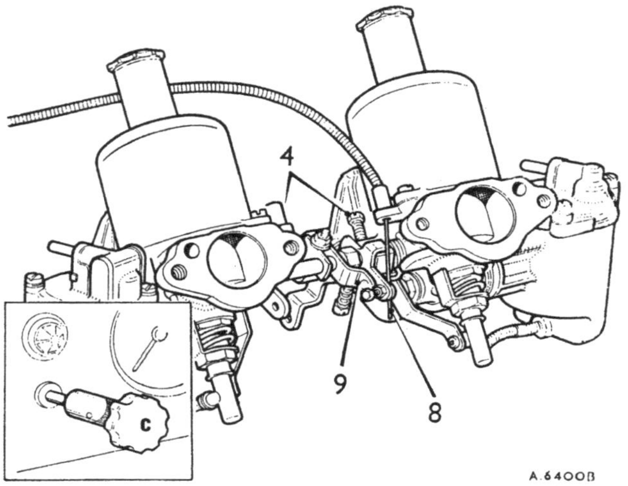


(11) Reconnect the mixture control wire (8) with approximately $\frac{1}{16}$ in. free movement before it starts to pull on the jet levers (9).

(12) Pull the mixture control knob until the linkage is about to move the carburetter jets.

(13) Using the carburetter balancing meter to ensure equal adjustment, turn the fast-idle adjusting screws (4) to give the correct fast idling speed (see **'TUNING DATA'**).

(14) Refit the air cleaners.

## Section 4-D

### TYPE HIF CARBURETTERS—TUNING
### (Fitted to MGB from 1972 Model Year)

**General**

The carburetters fitted to cars equipped with engine emission control systems are balanced to provide engine performance with pollution control. Under no circumstances may they or their components be interchanged or substituted with normal carburetters.

Tuning must be carried out with the engine emission control equipment connected and operating.

**IMPORTANT.**—Before servicing or tuning a carburetter in an endeavour to rectify poor engine performance, make sure that the maladjustment or fault is not from another source by checking the following:

Valve clearance
Spark plug condition
Contact breaker (dwell angle)
Ignition timing and advance
Presence of air leaks into the induction system

**Single and twin carburetters**

(1) Remove the air cleaner(s).
(2) Check the throttle for correct operation and signs of sticking.
(3) Unscrew the throttle adjusting screw (both screws on twin carburetters) until it is just clear of the throttle lever, with the throttle closed, then turn the screw clockwise two full turns.
(4) Raise the piston of each carburetter with the lifting pin and check that it falls freely onto the bridge when the pin is released. If the piston shows any tendency to stick, the carburetter must be serviced.
(5) Lift and support the piston clear of the bridge so that the jet is visible; if this is not possible due to the installed position of the carburetter, remove the suction piston chamber.
(6) Turn the jet adjusting screw anti-clockwise until the jet is flush with the bridge or as high as possible without exceeding the bridge height. Ensure that both jets on twin carburetters are in the same relative position to the bridge of their respective carburetters.
(7) Check that the needle guide(s) is flush with the bottom of the piston groove.
(8) Turn the jet adjusting screw two turns clockwise (both screws on twin carburetters).
(9) Turn the fast idle adjusting screw anti-clockwise (both screws on twin carburetters) until it is well clear of the cam.
(10) Refit the suction piston chamber if it has been removed and, using the lifting pin, check that the piston falls freely onto the bridge.
(11) Top up the piston damper reservoir(s) with a recommended oil until the level is $\frac{1}{2}$ in. (13 mm.) above the top of the hollow piston rod.

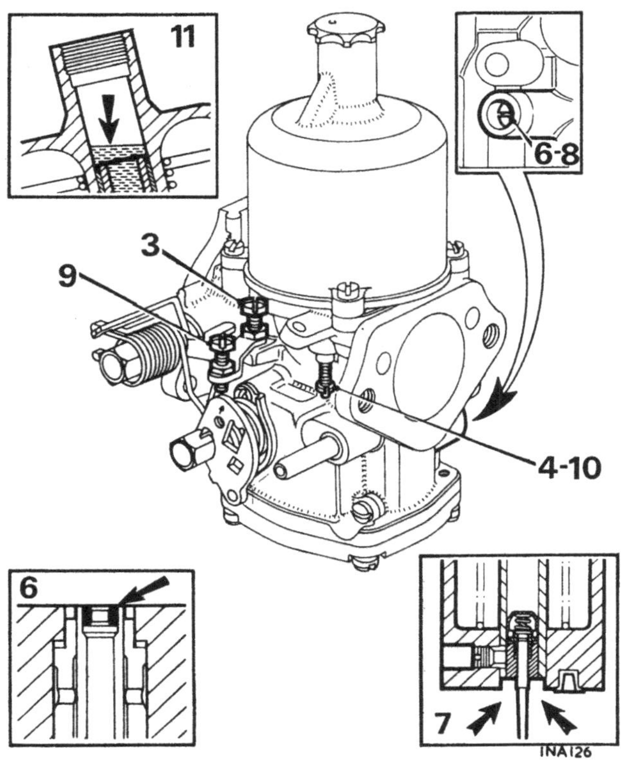

(12) Connect a reliable tachometer to the engine in accordance with the instrument manufacturer's instructions.

(13) Start the engine and run it at a fast idle speed until it attains normal running temperature, then run it for a further five minutes.

(14) Increase the engine speed to 2,500 r.p.m. for thirty seconds.

(15) Connect an approved exhaust gas analyser to the engine in accordance with the instrument manufacturer's instructions.

**NOTE.**—Tuning can now commence. If the correct setting cannot be obtained within three minutes, increase the engine speed to 2,500 r.p.m. for thirty seconds and then recommence tuning. Repeat this clearing operation at three-minute intervals until tuning is completed.

**Single carburetters**

(16) Adjust the throttle adjusting screw until the correct idle speed (see **'TUNING DATA'** and/or Vehicle Emission Control Information Label) is obtained.

**NOTE.**—During the following procedure, just before the readings of the tachometer and exhaust gas analyser are taken, gently tap the neck of the suction chamber with a light-metallic instrument (e.g. a screwdriver handle).

(17) Turn the jet adjusting screw, clockwise to enrich or anti-clockwise to weaken, until the fastest speed is indicated on the tachometer; turn the screw anti-clockwise until the engine speed just commences to fall. Turn the screw clockwise very slowly the minimum amount until the maximum speed is regained.

(18) Check the idle speed, and re-adjust it as necessary with the throttle adjusting screw to obtain the correct setting.

(19) Using the exhaust gas analyser, check that the percentage CO reading is within the limits given in **'TUNING DATA'** and/or Vehicle Emission Control Information Label. If the reading falls outside the limits given, reset the jet adjusting screw by the minimum amount necessary to bring the reading just within the limits. If an adjustment exceeding half a turn of the adjusting screw is required to achieve this, the carburetter must be removed and overhauled.

(20) With the fast idle cam against its return stop, check that a $\frac{1}{16}$ in. (1·5 mm.) free movement of the mixture control (choke) cable exists before the cable moves the cam.

(21) Pull out the mixture control (choke) until the arrow marked on the cam is positioned under the fast idle adjusting screw.

(22) Turn the fast idle adjusting screw clockwise until the correct fast idle speed (see **'TUNING DATA'** and/or Vehicle Emission Control Information Label) is obtained.

(23) Refit the air cleaner.

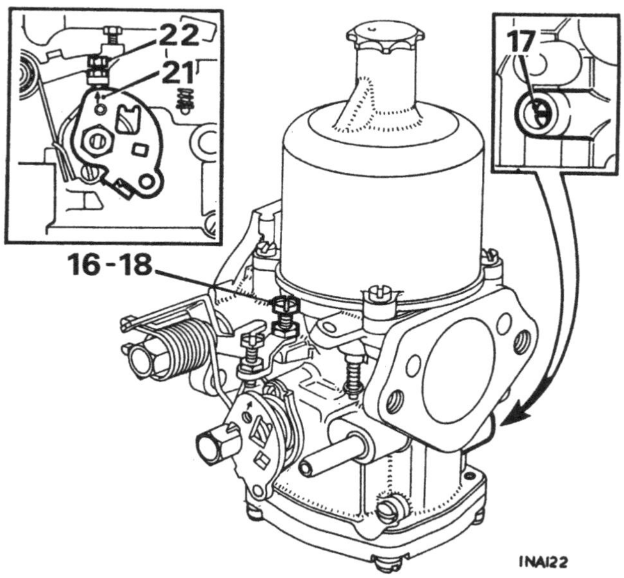


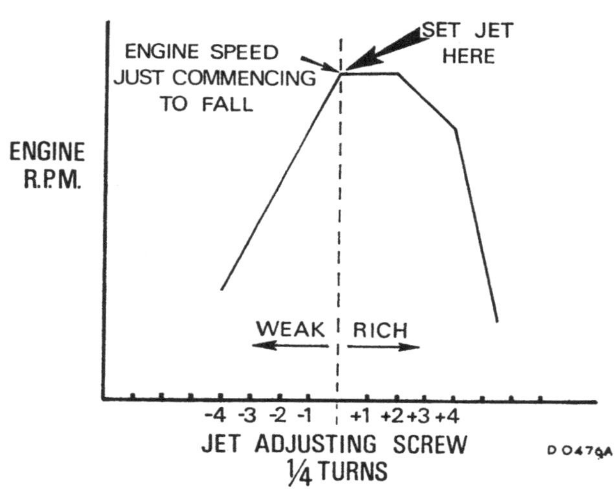

**Twin carburetters**

(24) Slacken both clamping bolts on the throttle spindle interconnections.

(25) Slacken both clamping bolts on the cold start interconnections.

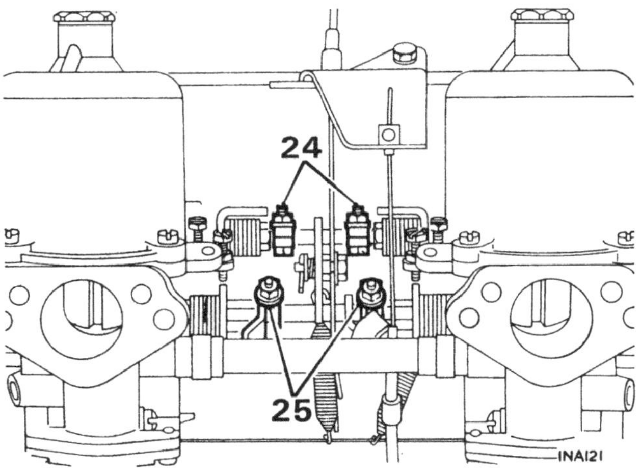


(26) Using an approved balancing meter in accordance with the maker's instructions, balance the carburetters by altering the throttle adjusting screws until the correct idle speed and balance is achieved.

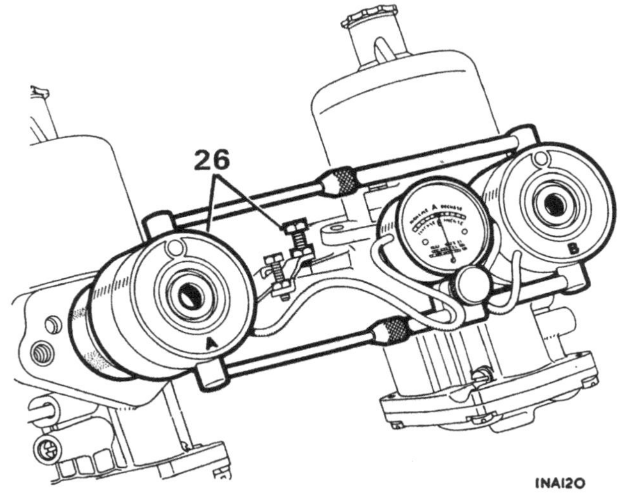

**NOTE.**—During the following procedure, just before reading the tachometer and exhaust gas analyser, gently tap the neck of each suction chamber with a non-metallic instrument (e.g. a screwdriver handle).

(27) Turn the jet adjusting screw on both carburetters clockwise to enrich or anti-clockwise to weaken, by the same amount until the fastest speed is registered on the tachometer; turn both screws anti-clockwise until the engine speed just commences to fall. Turn both screws very slowly clockwise by the minimum amount until the maximum speed is regained.

(28) Using the exhaust gas analyser, check that the percentage CO reading is within the limits given in **'TUNING DATA'** and/or Vehicle Emission Control Information Label. If the reading falls outside the limits given, reset both jet adjusting screws by the minimum amount necessary to bring the readings just within the limits. If an adjustment exceeding half a turn is required to achieve this the carburetters must be removed and overhauled.

(29) Set the throttle interconnection clamping levers, in accordance with the instructions given in the relevant vehicle Workshop Manual, so that a clearance exists between the link pin and the lower edge of the fork. Tighten the clamp bolts, ensuring that there is approximately $\frac{1}{32}$ in. end-float on the interconnection rod.

(30) Run the engine at 1,500 r.p.m. and check the throttle linkage for correct connection by re-checking the carburetter balance.

(31) With the fast idle cams of both carburetters against their respective stops, set the cold start inter-connections so that both cams begin to move simultaneously.

(32) With the fast idle cams against their stops check that a $\frac{1}{16}$ in. (1·5 mm.) free movement of the mixture control (choke) cable exists before the cable moves the cams.

(33) Pull out the mixture control (choke) until the arrow marked on the cam is positioned under the fast idle adjusting screw of each carburetter.

(34) Using the balancing meter to ensure equal adjustment, turn the fast idle adjusting screws to give the correct fast idle speed (see **'TUNING DATA'** and/or Vehicle Emission Control Information Label).

(35) Refit the air cleaners.

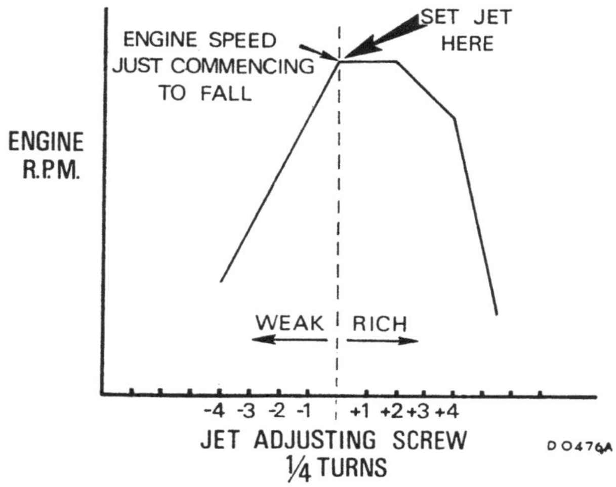


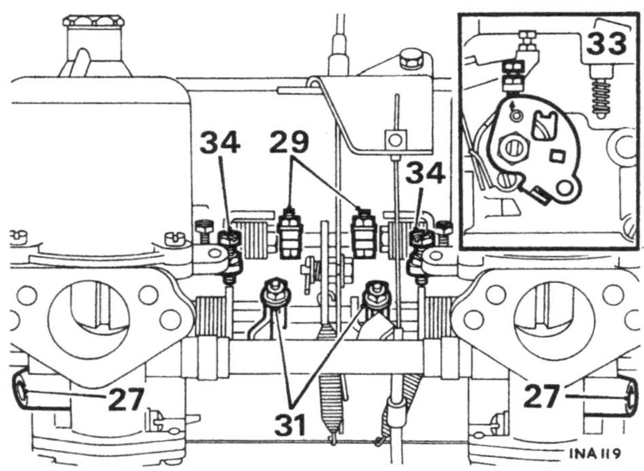

## Section 4-E

### TYPE HIF CARBURETTERS—OVERHAULING
### (Fitted to MGB from 1972 Model Year)

**Dismantling**

(1) Thoroughly clean the outside of the carburetter.
(2) Remove the piston damper and its washer.
(3) Unscrew the suction piston chamber retaining screws and remove the identity tag.
(4) Lift the chamber vertically from the body without tilting it.
(5) Remove the piston spring, lift out the piston assembly and empty the oil from the piston rod.
(6) Unscrew the needle guide locking screw.
(7) Withdraw the needle, guide and spring.

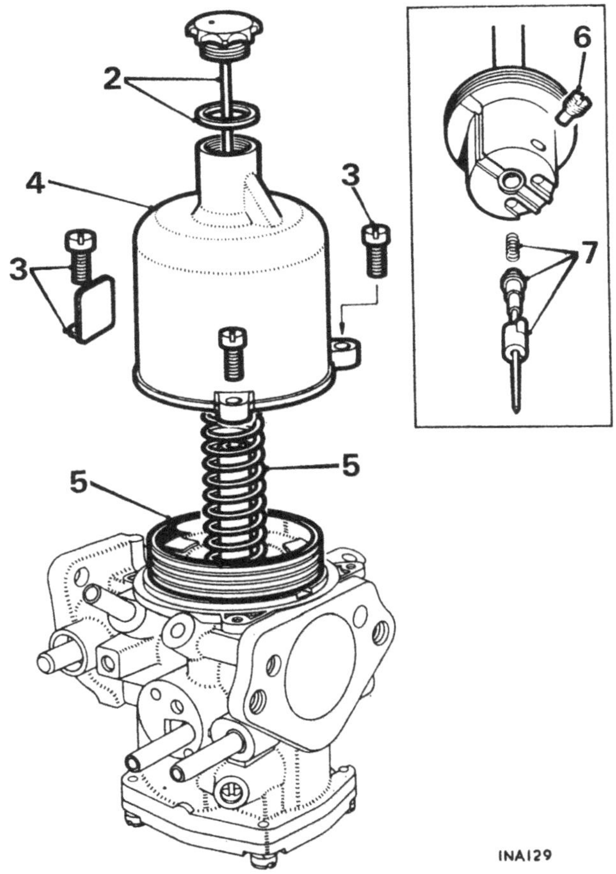


INA129

(8) Mark the bottom cover-plate and body to ensure correct reassembly, unscrew the retaining screws and remove the cover complete with sealing ring.
(9) Remove the jet adjusting screw complete with 'O' ring.
(10) Remove the jet adjusting lever retaining screw and spring.
(11) Withdraw the jet complete with adjusting lever and disengage the lever.
(12) Remove the float pivot spindle and fibre washer.
(13) Withdraw the float.
(14) Remove the needle valve and unscrew the valve seat.
(15) Unscrew the jet bearing locking nut and withdraw the bearing complete with fibre washer.

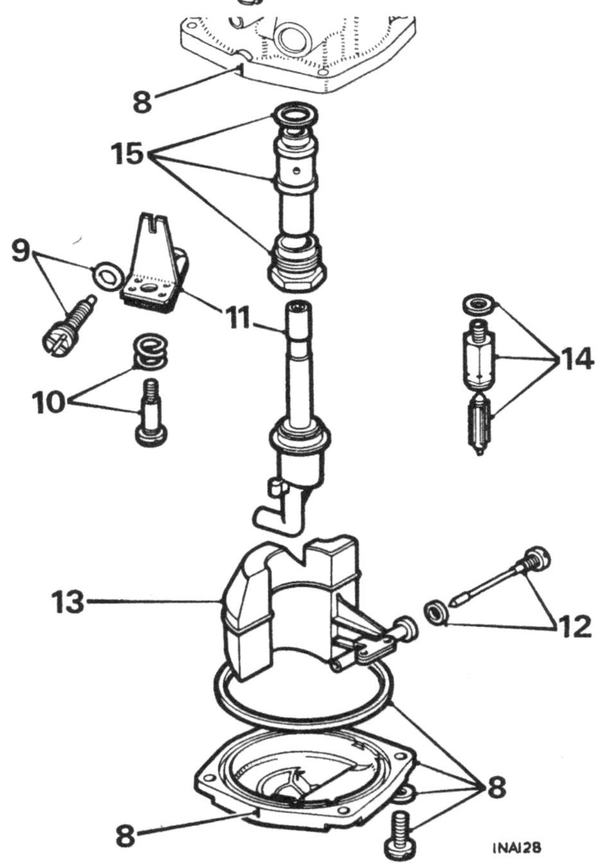

(16) Note the location of the ends of the fast idle cam lever return spring.
(17) Unlock and remove the cam lever retaining nut and locking washer.
(18) With the return spring held towards the carburetter body, prise off the cam lever and remove the return spring.
(19) Unscrew the starter unit retaining screws and remove the cover-plate.
(20) Withdraw the starter unit assembly and remove its gasket.
(21) Withdraw the valve spindle and remove the 'O' ring, seals and dust cap.
(22) Note the location and loading of the ends of the throttle lever return spring and remove the spring.
(23) Unlock and remove the nut and tab washer retaining the throttle levers.
(24) Remove the throttle lever and throttle actuating lever.
(25) Remove the throttle disc retaining screws.
(26) Open the throttle; note that the throttle disc is oval, and carefully withdraw the disc from the throttle spindle. Do not damage the over-run valve.
(27) Withdraw the throttle spindle and remove its seals.

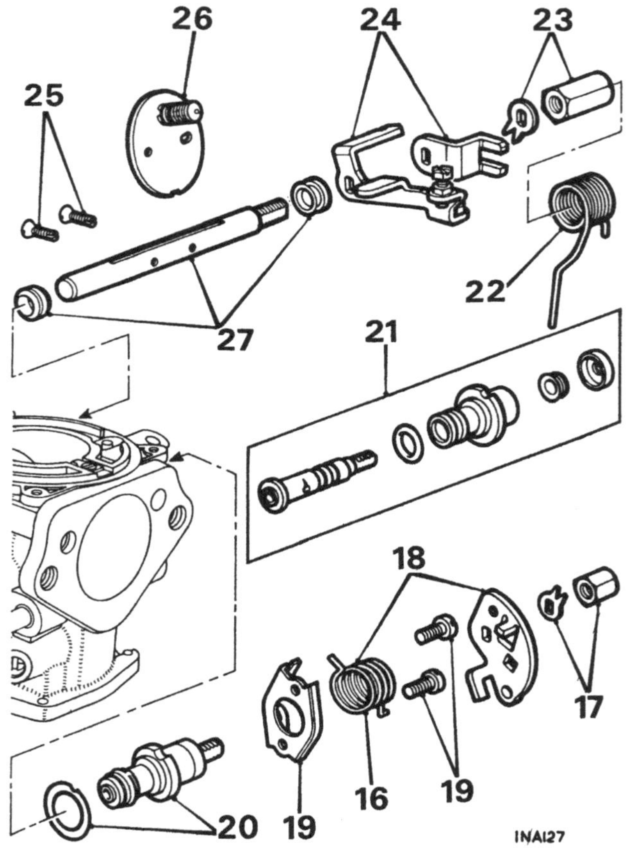


## INSPECTION

(28) Examine the throttle spindle and its bearings in the carburetter body; check for excessive play, and renew parts as necessary.
(29) Examine the float needle and seating for damage and excessive wear; renew if necessary.
(30) Examine all rubber seals and 'O' rings for damage deterioration; renew as necessary. The cover-plate sealing ring must be renewed.
(31) Check condition of all fibre washers and gaskets renew as necessary.
(32) Clean the inside of the suction chamber and piston rod guide with fuel or methylated spirit (denatured alcohol) and wipe dry. Abrasives must not be used.
(33) Examine the carburetter body for cracks and damage and for security of the brass connections and the piston key.

**NOTE.**—It is only necessary to carry out the following timing check if the cause of the carburetter malfunction which necessitated the dismantling has not been located.

(34) Temporarily plug the piston transfer holes.
(35) Fit the piston into the chamber without its spring.
(36) Fit a nut and screw, with a large flat washer under the screw head, into one of the suction chamber fixing holes, positioning the washer so that it overlaps the chamber bore.
(37) Fit the damper and washer.
(38) Check that the piston is fully home in the chamber, invert the assembly to allow the chamber to fall away until the piston contacts the washer.

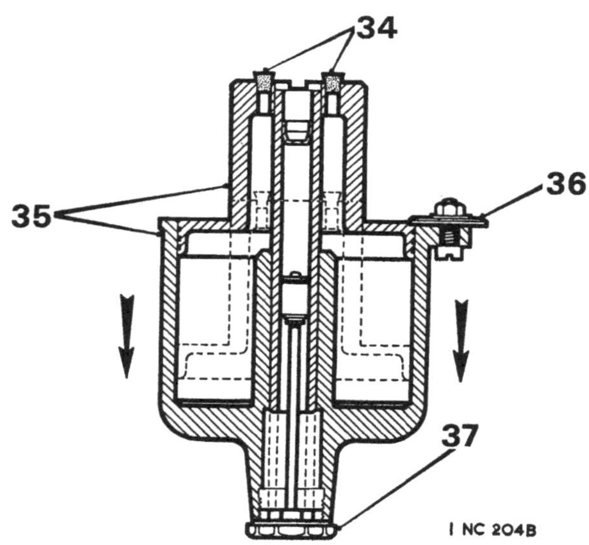

(39) Check the time taken for the chamber to fall the full extent of the piston travel. For HIF carburetters 1½ in. (38 mm.) bore the time taken should be 4 to 6 seconds; for HIF6 1¾ in. (44·5 mm.) bore the time taken should be 5 to 7 seconds.

(40) If the times are exceeded, check the piston and chamber for cleanliness and damage. If after re-checking the time is still not within these limits, renew the chamber and piston assembly.

**Reassembling**

(41) Reverse the procedure in 1 to 28, noting the following:

(*a*) The throttle spindle must be fitted with the threaded end at the piston lifting pin side of the body.

(*b*) Fit the throttle disc so that the over-run valve is at the top of the bore and its spring towards the inside when the throttle is closed.

(*c*) New throttle disc retaining screws must be used when refitting the disc. Ensure that the throttle disc is correctly positioned and closes correctly before tightening and locking the retaining screw.

(*d*) Position the throttle spindle end seals just below the spindle housing flange.

(*e*) The starter unit valve is fitted with the cut-out towards the top retaining screw hole, and its retaining plate is positioned with the slotted flange towards the throttle spindle.

(*f*) After fitting the float and valve, invert the carburetter so that the needle valve is held in the shut position by the weight of the float only. Check that the point indicated on the float (see illustration) is 0·04±0·02 in. (1·0 ±·5 mm.) below the level of the float chamber face. Adjust the float position by carefully bending the arm.

(*g*) Check that the small diameter of the jet adjusting screw engages the slot in the adjusting lever and set the jet flush with the bridge of the body.

(*h*) Use a new retaining screw when refitting the needle and ensure that the needle guide etch mark aligns correctly with the piston transfer holes (see illustration). After fitting the needle assembly, check that the shoulder of the needle aligns the full face of the piston.

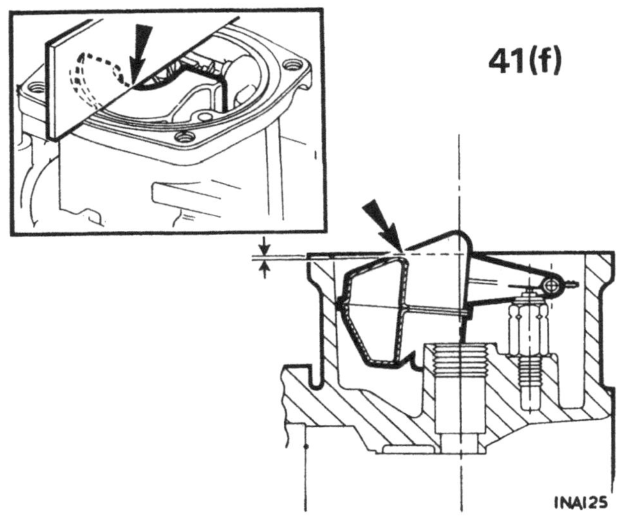

41(f)

INA125

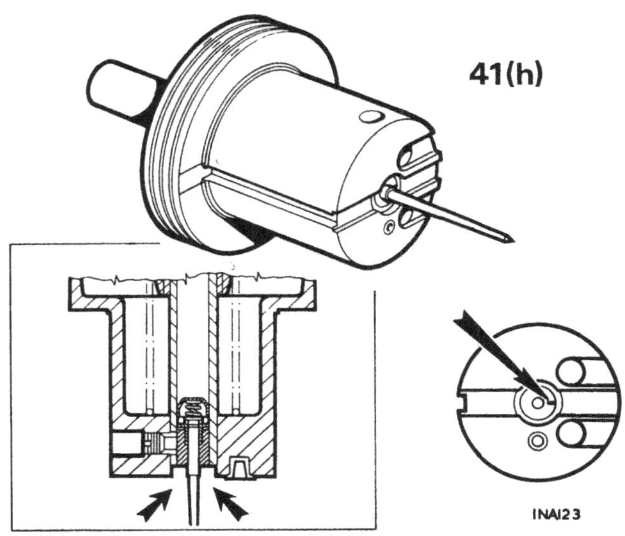

41(h)

INA123

| ENGINE SPEED | TEST | COMPONENT CONDITION | READ/OBSERVE | READINGS | CHECK SEQUENCE—FAULT LOCATION |
|---|---|---|---|---|---|
| **START (cranking)** | Cranking voltage | Battery; starting system | Voltmeter | 9·6 volts minimum at the battery | Battery—connections/cables—starter motor—dynamo/alternator—regulator |
| | Cranking coil output | Coil and ignition circuit | Scope trace | 17 KV. minimum | Ignition coil |
| | Positive crankcase ventilation/cranking vacuum | Crankcase emission equipment | Vacuum gauge | 6 – 10 in. Hg (crankcase ventilation operating) 8—15 in. Hg (crankcase ventilation blanked) | Hoses and connections—Oil filler cap—Valve rocker clearance—Emission valve—Gulp valve—Oil separator—Servo (if fitted)—Inlet manifold leaks—Valves or seats—Piston rings—Any air leak to crankcase |
| **IDLING (see 'TUNING DATA')** | Idle speed | Carburetter idle setting | Tachometer | See 'TUNING DATA' | Carburetter adjustment—Hoses and connections—Gulp valve—Servo (if fitted)—Carburetter limit valve or mechanical condition |
| | Dwell | Distributor/drive; points | Dwell meter; scope | 4-cyl. : 57 to 63° See Pattern 1 (see inset A) | Breaker points—Distributor and drive mechanical condition |
| | Initial timing | Spark timing setting | Timing light | See 'TUNING DATA' | Distributor adjustment |
| | Fuel mixture | Carburetter setting | Exhaust gas analyser | See 'TUNING DATA' | Carburetter adjustment—Hoses and connections—Gulp valve—Crankcase emission valve—Servo (if fitted)—Carburetter limit valve or mechanical condition—Air pump—Check valve—Spark plugs |
| | Manifold vacuum | Engine idle efficiency | Vacuum gauge | 12 in. Hg minimum (engine fully run in) | Hoses and connections—Gulp valve—Inlet manifold leaks—carburetter limit valve—Valves or seats—Piston rings |
| **FAST IDLE (see 'TUNING DATA')** | Dwell variation | Distributor mechanical | Dwell meter | Variation of 2° maximum | Distributor and drive mechanical condition |
| | Coil polarity | Ignition circuit polarity | Scope trace | See Pattern 2 (Trace inverted) | Ignition circuit connections—Ignition coil |
| | Cam lobe accuracy | Distributor cam | Scope trace | 3° max. variation. See Pattern 3 (inset A correct; inset B—overlap indicates cam error) | Distributor mechanical condition (cam) |
| | Secondary circuit | Plugs; leads; cap; rotor | Scope trace | Standard pattern | Spark plugs and leads—Breaker points—Carburetter adjustment—Hoses and connections—Gulp valve—Servo (if fitted) |
| | Coil and condenser condition | Coil windings; condenser | Scope trace | See Pattern 4 (lack of oscillations indicate fault) | Ignition coil—Condenser |
| | Breaker point condition | Points closing/opening/bounce | Scope trace | See Pattern 1 (inset B) | Breaker points—Condenser |
| | Spark plug firing voltage | Fuel mixture; compression; plug/rotor gaps | Scope trace | See Pattern 5; voltage 6—9 kV | Spark plugs and leads—Distributor cap and rotor—Carburetter adjustment (multi-carburetters) |
| **ACCELERATE—DECELERATE** | Spark plugs under load | Spark plugs | Scope trace | See Pattern 6; acceptable voltage rise 6 to 10 kV | Spark plugs and leads |
| | Carburetter open/close action | Carburetter | Exhaust gas analyser and vacuum gauge | Initial rich, lean off at throttle closure | Carburetter limit valve and mechanical condition—Hoses and connections—Gulp valve—Air pump |
| **TURNPIKE (Maximum ignition advance speed see 'TUNING DATA')** | Timing advance | Distributor mech./vacuum advance | Timing light/advance meter | See 'TUNING DATA' | Distributor mechanical condition, vacuum unit, centrifugal weights and springs |
| | Maximum coil output | Coil; condenser; ignition primary | Scope trace | Standard pattern; minimum reserve ⅜ more than requirement | Ignition coil—H.T. circuit insulation |
| | Secondary circuit insulation | H.T. cables, cap, rotor | Scope trace | Standard pattern | H.T. leads—Distributor cap and rotor |
| | Charging voltage | Regulator; cut-out | Voltmeter | 14·5 volts; steady reading | Cut-out—Voltage regulator—Dynamo/Alternator |
| | Fuel mixture | Air cleaner, carburetter | Exhaust gas analyser | Leaning off following peak when test speed is reached | Hoses and connections—Carburetter adjustment—Air cleaners—Gulp valve—Air pump—Check valve—Injectors |
| | Exhaust restriction | Exhaust system | Vacuum gauge | No variation in reading at constant speed for 10 sec. | Exhaust system |

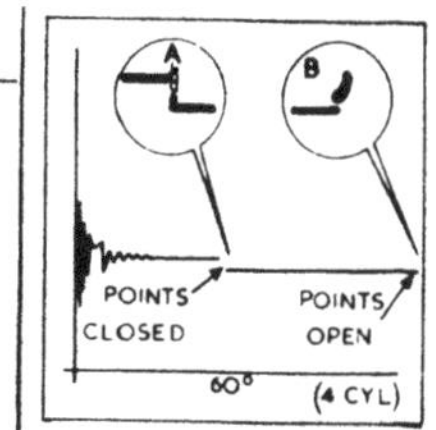

Pattern 1

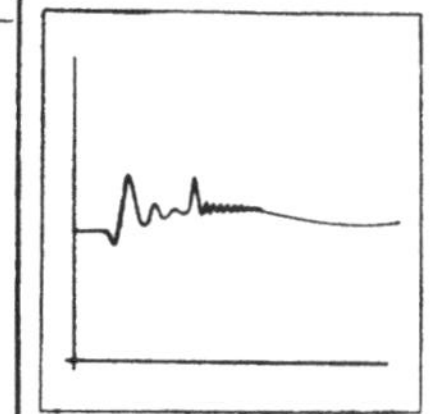
Pattern 2

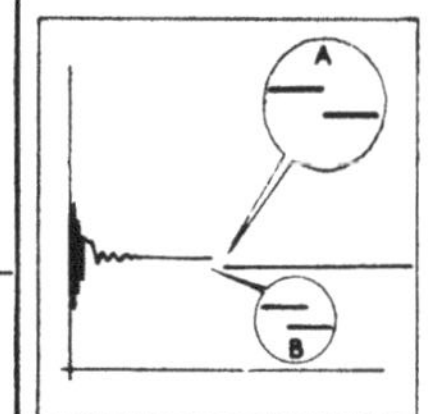

Pattern 3

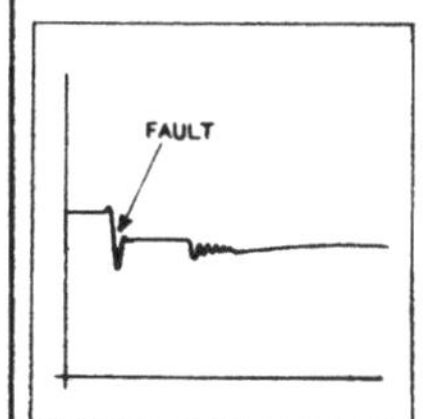

Pattern 4

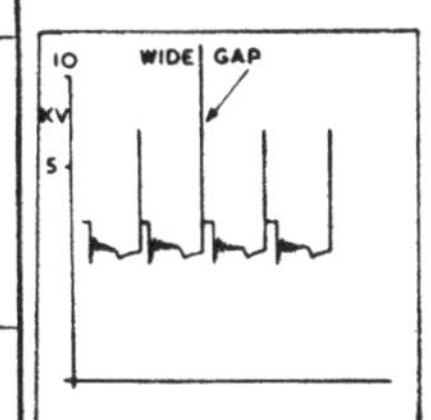

Pattern 5

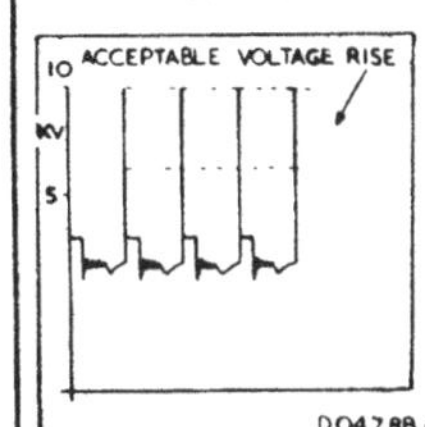

Pattern 6

# SECTION 5

## EVAPORATIVE LOSS CONTROL

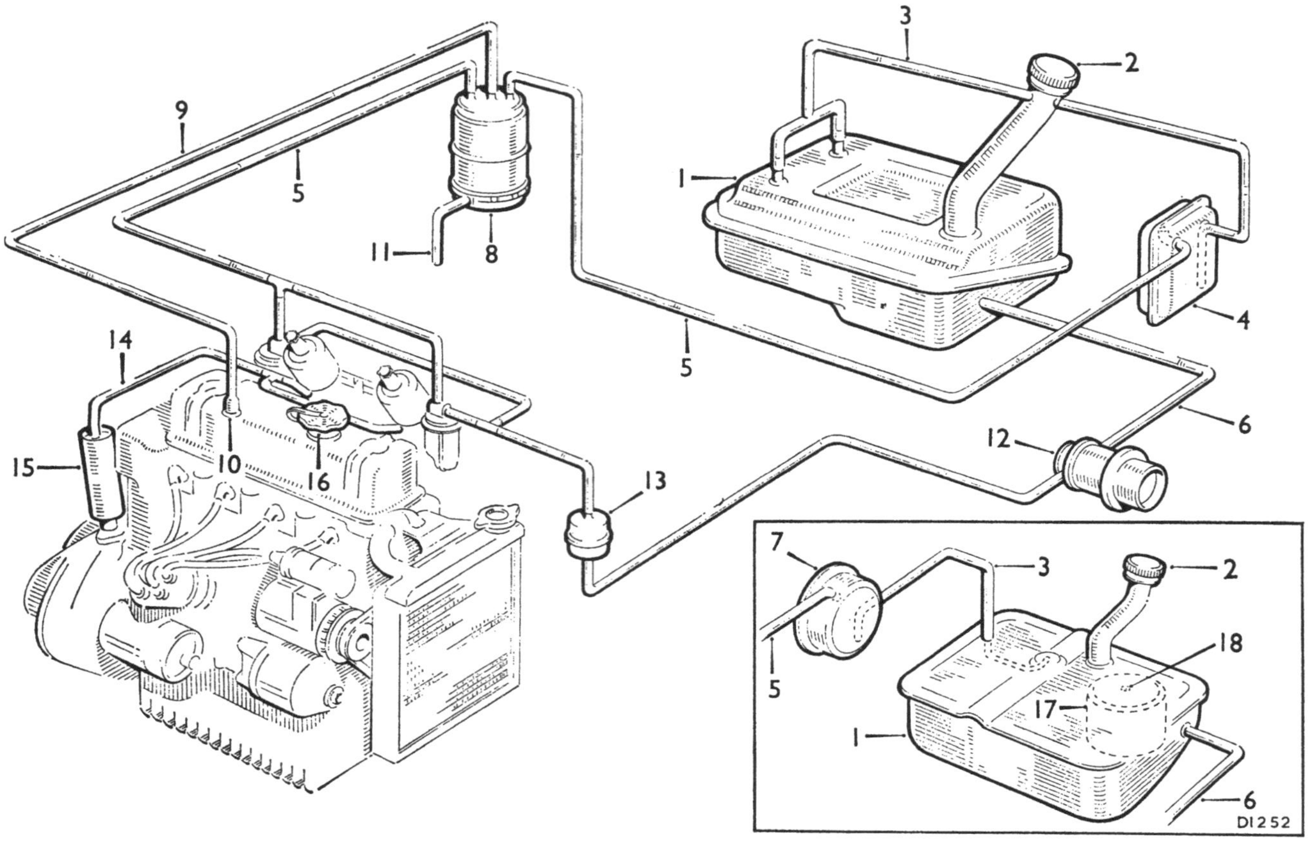


*Fig.* 1

*A typical evaporative loss control system, with inset showing the arrangement of a separation tank and capacity-limiting fuel tank*

1. Fuel tank.
2. Sealed fuel filler cap.
3. Expansion/vapour line.
4. Expansion tank.
5. Vapour pipe.
6. Fuel pipe.
7. Separation tank.
8. Adsorption canister.
9. Purge line.
10. Restricted connection.
11. Air vent.
12. Fuel pump.
13. Fuel line filter.
14. Breather pipe.
15. Oil separator.
16. Sealed oil filler cap.
17. Capacity limiting tank.
18. Air lock bleed.

## Section 5-A

### GENERAL DESCRIPTION

The system is designed to collect fuel vapour evaporated from the fuel in the fuel tank, and on some twin carburetter cars from the fuel in the carburetter float-chambers. The vapour is stored in an adsorption canister while the engine is stopped, and then after the engine is restarted, passed through the crankcase emission control system to the combustion chambers. While the car is being driven the vapours are drawn directly to the crankcase emission control system.

Ventilation tubes on the fuel tank ensure that vapours are vented through the control system when the car is parked on other than a level surface.

To prevent spillage of fuel by displacement due to expansion, sufficient capacity is provided in the expansion tank to accommodate the amount of fuel from a full tank which would be displaced by a high temperature rise.

By the positioning of the expansion tank connections, or by the inclusion of a small separation tank in the vapour line, liquid fuel is prevented from being carried with the vapour to the storage canister.

**IMPORTANT.** The fuel and oil filler caps seal the system, and it is essential for its efficient function that they are correctly refitted after removal.

#### Adsorption canister

The adsorption or vapour storage canister mounted in the engine compartment contains activated charcoal (carbon) granules. Filter pads are fitted at both sides of the charcoal to filter incoming ventilating air and to prevent the granules from leaving the canister through the purge line. Provision is made for renewing the ventilation air filter pad. Vapour tubes from the fuel tank and carburetter float-chambers and the purge line from the engine breather system are connected to the ports on the top of the canister. The port on the bottom section provides a connection for the ventilating air tube.

Fuel vapour entering the canister through the vapour tubes is adsorbed and held by the charcoal. When the engine is started, air is drawn by the crankcase emission control system, through the ventilation tube and into the canister. As the air passes over the charcoal granules the vapours are given up and are carried with the air through the crankcase emission system to the combustion chambers.

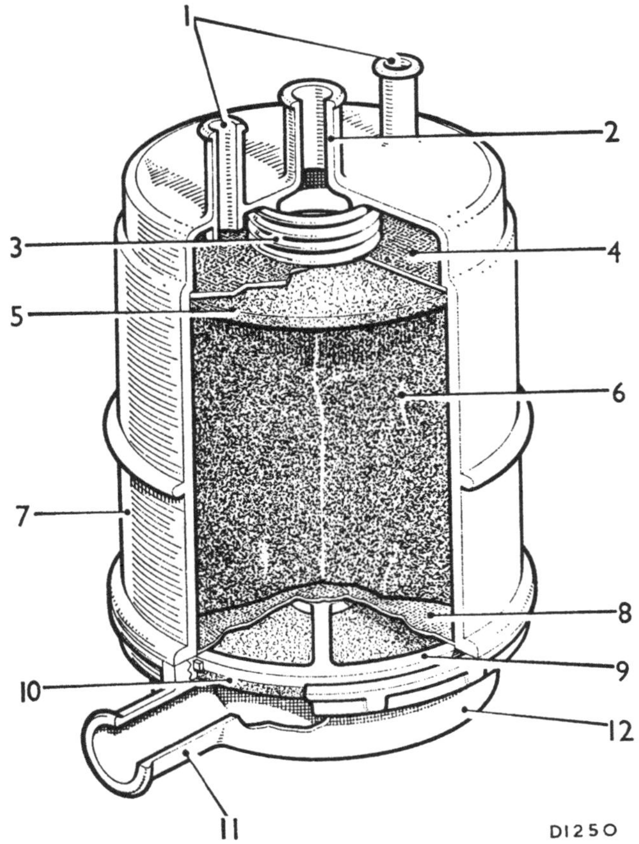


*Fig.* 3

*The adsorption canister*

1. Vapour pipe connections.
2. Purge pipe connection.
3. Spring.
4. Gauze.
5. Filter pad.
6. Charcoal granules.
7. Canister.
8. Gauze.
9. Retainer.
10. Filter pad.
11. Air vent connection.
12. End cap.

**Fuel expansion**

Two methods are used to ensure that sufficient space is available to accommodate fuel displaced by expansion due to high ambient temperatures. The method used on the Austin America is by fitting an additional tank into which the displaced fuel flows when the volume of the fuel exceeds that of the fuel tank. The MGB and MG Midget use an air lock chamber in the fuel tank which prevents the tank being completely filled with fuel, thereby ensuring that sufficient space is always available for expansion.

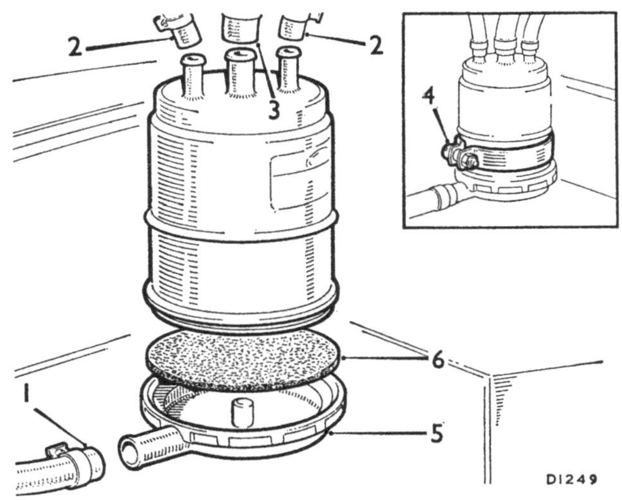


*Fig.* 2

*The adsorption canister air filter pad*

1. Air vent tube.
2. Vapour pipes.
3. Purge pipe.
4. Canister securing clip.
5. End cap.
6. Air filter pad.

**Fuel line filter**

On some models an additional renewable filter is fitted in the main fuel line as an added safeguard against foreign matter causing the setting of the carburetter float-chamber level to be exceeded.

**Mixture temperature compensator**

On some applications a small temperature-sensitive valve is fitted adjacent to the carburetter. The valve is connected between the air cleaner and the controlled depression chamber of the carburetter.

Under conditions where fuel is entering the carburetter at high temperature, i.e. prolonged idling in high ambient temperatures, the valve opens and allows a small quantity of air to pass into the carburetter, bypassing the jet. The air leans off the mixture which has been enriched by vapours from the evaporative loss control system and by

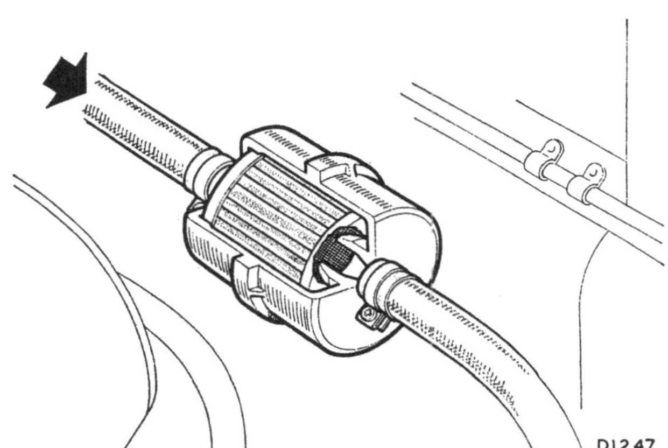


*Fig.* 4

*The fuel line filter*

the increase in fuel flow through the carburetter due to the high fuel temperature.

## Section 5-B

### ADSORPTION CANISTER

**Renewing**

The air filter fitted in the bottom section of the canister must be renewed every 12,000 miles (20000 km.) or more frequently in dusty operating conditions. The complete canister must be renewed every 50,000 miles (80000 km.) or if at any time it should inadvertently become saturated with liquid fuel.

**WARNING.** Do not attempt to recover a saturated canister by passing compressed air through the charcoal.

(1) Disconnect the air vent tube from the bottom of the canister.
(2) Disconnect the vapour and purge pipes from the top of the canister.
(3) Unscrew the securing clip screw and lift out the canister.
(4) If the air filter pad only is being renewed:
   (*a*) Unscrew the lower end cap of the canister.
   (*b*) Remove and discard the filter pad.
   (*c*) Clean any dirt from the cap.
   (*d*) Fit the new filter pad and refit the cap.
(5) Fit the canister ensuring that the purge pipe (from the engine rocker cover) is connected to the large centre connection on the top of the canister.

## Section 5-C

### FUEL LINE FILTER

**Renewing**

The fuel line filter must be renewed every 12,000 miles (20000 km.).

(1) Check that the ignition is switched off.
(2) Disconnect and discard the filter.
(3) Connect the new filter.
(4) Switch on the ignition and check the filter connections for fuel leakage.
(5) Start the engine and recheck for fuel leakage.

## Section 5-D

### LEAK TESTING

**NOTE.** As a preliminary check for leaks on the induction and evaporative loss control systems on cars fitted with running on control valves, temporarily block the air vent pipe of the valve while the engine is idling. If no air leaks exist in the systems the engine will stop almost immediately; if the engine continues to run an air leak is indicated.

If a fault in the operation is suspected or components of the system other than the filters or canister have been removed and refitted, the evaporative loss control system must be pressure-tested for leaks as follows:

(1) Check that there is at least one gallon of fuel in the fuel tank.
(2) Switch on the ignition for one minute to prime the fuel system.
(3) Switch off the ignition and disconnect the fuel tank ventilation pipe from its connection on the adsorption canister.
(4) Connect a 0 to 10 lb./sq. in. pressure gauge, a Schrader valve, and a low-pressure air supply (i.e. a tyre pump) to the disconnected pipe.
(5) Pressurize the system until 1 lb./sq. in. is registered on the gauge. **DO NOT EXCEED THIS PRESSURE AT ANY TIME.**
(6) Check that the gauge reading is maintained for 10 seconds without falling more than ·5 lb./sq. in. If the reading is not maintained, check the system for leaks commencing with the fuel filler cap and seal.
(7) Make a visual check for fuel leakage from the tank and its connections.
(8) Remove the fuel filler cap and check that the gauge falls to zero.
(9) Remove the test equipment and re-make the connections.

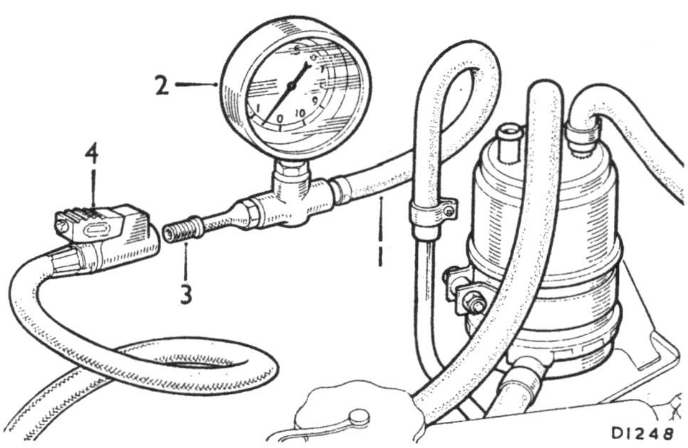


*Fig. 5*
*Leak-testing the control system*

1. Fuel tank ventilation pipe.
2. Pressure gauge.
3. Schrader valve.
4. Low-pressure air supply.

## MODEL: MGB

| | To 1971 | 1971 |
|---|---|---|
| **ENGINE** | | |
| Type | 18GF, 18GH, or 18GJ | 18GK |
| Firing order | 1, 3, 4, 2, | 1, 3, 4, 2 |
| Capacity | 1798 c.c. (109·8 cu. in.) | 1798 c.c. (109·8 cu. in.) |
| Compression ratio | 8·8 : 1 | 8·8 : 1 |
| Compression pressure | 160 lb./sq. in. (11·25 kg./cm.²) | 160 lb./sq. in. (11·25 kg./cm.²) |
| Idle speed | 900 r.p.m. | 900 r.p.m. |
| Fast idle speed | 1,300 to 1,400 r,p.m. | 1,300 to 1,400 r.p.m. |
| Valve rocker clearance | ·015 in. (·38 mm.) set cold | ·015 in. (·38 mm.) set cold |
| Stroboscopic ignition timing* | 20° B.T.D.C. at 1,000 r.p.m. | 15° B.T.D.C. at 1,500 r.p.m. |
| Static ignition timing | 10° B.T.D.C. | 10° B.T.D.C. |
| Timing mark location | Pointer on timing case, notch on crankshaft pulley | Pointer on timing case, notch on crankshaft pulley |
| **DISTRIBUTOR** | | |
| Make | Lucas | Lucas |
| Type | 25D4 | 25D4 |
| Serial number | 40897 or 41155 | 41339 |
| Contact breaker gap | ·014 to ·016 in. (·35 to ·40 mm.) | ·014 to ·016 in. (·35 to ·40 mm.) |
| Rotation of rotor | Anti-clockwise | Anti-clockwise |
| Dwell angle | 57° to 63° | 57° to 63° |
| Condenser capacity | ·18 to ·24 mF | ·18 to ·24 mF |
| **Centrifugal advance** | | |
| Crankshaft degrees* | 10° at 400 to 600 r.p.m.<br>24° at 1,500 to 1,750 r.p.m.<br>30°±2° at 3,000 r.p.m. | 10° at 900 to 1,100 r.p.m.<br>24° at 2,700 to 2,900 r.p.m.<br>30°±2° at 4,600 r.p.m. |
| **Vacuum advance** | | |
| Starts | 5 in. Hg | 7 in. Hg |
| Finishes | 13 in. Hg | 13 in. Hg |
| Total crankshaft degrees | 20°±2° | 10°±2° |
| **SPARKING PLUGS** | | |
| Make | Champion | Champion |
| Type | N–9Y | N–9Y |
| Gap | ·024 to ·026 in. (·625 to ·660 mm.) | ·024 to ·026 in. (·625 to ·660 mm.) |
| **IGNITION COIL** | | |
| Make | Lucas | Lucas |
| Type | HA12 | HA12 |
| Resistance—primary | 3·1 to 3·5 ohms at 20° C. (68° F.) | 3·1 to 3·5 ohms at 20° C. (68° F.) |
| **Consumption** | | |
| Ignition on—standing | 3·9 amps. | 3·9 amps. |
| at 2,000 r.p.m. | 1·4 amps. | 1·4 amps. |
| **CARBURETTER(S)** | | |
| Make | S.U. | S.U. |
| Type | Twin HS4 | Twin HS4 |
| Specification—fixed needle type | AUD 265 | — |
| —spring-loaded needle type | AUD 326 | AUD 465 |
| Choke diameter | 1½ in. (38·1 mm.) | 1½ in. (38·1 mm.) |
| Jet size | ·090 in. (2·2 mm.) | ·090 in. (2·2 mm.) |
| Needle—fixed type | FX | — |
| spring-loaded type | AAE | AAL |
| Piston spring | Red | Red |
| Initial jet adjustment | 14 flats from bridge | 12 flats from bridge |
| **EXHAUST EMISSION** | | |
| Exhaust gas analyser reading: | | |
| At engine idle speed | 4·5% CO (maximum) | 4·5% CO (maximum) |
| Air pump test speed | 1,000 r.p.m. (engine) | 1,000 r.p.m. (engine) |

* Vacuum pipe disconnected.

**MODEL: MGB from 1972 on**

**ENGINE**

| | |
|---|---|
| Type | 18V |
| Firing order | 1, 3, 4, 2, |
| Capacity | 1798 c.c. (109·8 cu. in.) |
| Compression ratio | 8·0 : 1 |
| Compression pressure | 160 lb./sq. in. (11·25 kg./cm.²) |
| Idle speed | 850 r.p.m. |
| Fast idle speed | 1,300 to 1,400 r.p.m. |
| Valve rocker clearance | ·015 in. (·38 mm.) set cold |
| Stroboscopic ignition timing* | 1972—16° B.T.D.C. at 1,500 r.p.m.; from 1973—12° B.T.D.C. at 1,500 r.p.m. |
| Static ignition timing | 1972—10° B.T.D.C.; from 1973—6° B.T.D.C. |
| Timing mark location | Pointer on timing case, notch on crankshaft pulley |

**DISTRIBUTOR**

| | | |
|---|---|---|
| Make | Lucas | |
| Type | 25D4 | |
| Serial number | 1972—41370; from 1973—41491 | |
| Contact breaker gap | ·014 to ·016 in. (·35 to ·40 mm.) | |
| Rotation of rotor | Anti-clockwise | |
| Dwell angle | 57° to 63° | |
| Condenser capacity | ·18 to ·24 mF | |
| **Centrifugal advance** | *Serial No.* 41370 (1972) | *Serial No.* 41491 (1973) |
| Crankshaft degrees* | 20° at 1,700 to 2,000 r.p.m. | 16° at 1,800 to 2,250 r.p.m. |
| | 35° at 3,300 to 3,900 r.p.m. | 32° at 3,600 to 4,050 r.p.m. |
| | 41°±2° at 4,800 r.p.m. | 39°±2° at 4,800 r.p.m. |
| **Vacuum advance** | | |
| Starts | 7 in. Hg | 10 in. Hg |
| Finishes | 13 in. Hg | 15 in. Hg |
| Total crankshaft degrees | 6°±2° | 10°±2° |

**SPARKING PLUGS**

| | |
|---|---|
| Make | Champion |
| Type | N-9Y |
| Gap | ·024 to ·026 in. (·625 to ·660 mm.) |

**IGNITION COIL**

| | |
|---|---|
| Make | Lucas |
| Type | HA12 |
| Resistance—primary | 3·1 to 3·5 ohms at 20° C. (68° F.) |
| **Consumption** | |
| Ignition on—standing | 3·9 amps. |
| at 2,000 r.p.m. | 1·4 amps. |

**CARBURETTER(S)**

| | |
|---|---|
| Make | S.U. |
| Type | Twin HIF4 |
| Specification | 1972—AUD 493; from 1973—AUD 550 |
| Choke diameter | 1½ in. (38·1 mm.) |
| Jet size | ·090 in. (2·2 mm.) |
| Needle | 1972—AAU; from 1973—ABD |
| Piston spring | Red |
| Initial jet adjustment | 2 turns down from bridge |

**EXHAUST EMISSION**

| | |
|---|---|
| Exhaust gas analyser reading: | |
| At engine idle speed | 1972—3·5% CO (maximum); 1973—2·5% CO (maximum) |
| Air pump test speed | 1,000 r.p.m. (engine) |

* Vacuum pipe disconnected.

# Special Tuning

# MGB TOURER AND GT

## PART 4

# FOREWORD

This is another of the M.G. Tuning Booklets which have been issued in recent years. It deals specifically with the Series MGB.

The 'MGB' as delivered from the Factory in its standard form is tuned to give maximum performance with 100-octane gasoline consistent with complete reliability and reasonable freedom from pinking. There is, however, a more or less continuous demand from enthusiasts all over the world for information on methods of improving the performance for competitive purposes, and it is to meet this demand that this booklet has been prepared.

It must be clearly understood, however, that, whereas it is a simple matter to increase the power output of the engine, this increase in power must inevitably carry with it a tendency to reduce reliability. It is for this reason that the terms of the Warranty on a new M.G. expressly exclude any super-tuning of the kind described in this booklet, but this does not mean that tuning in this way will necessarily make the car hopelessly unreliable. In fact, it may be assumed that it will be at least as reliable as other cars of similar performance.

This booklet is laid out to give details for progressively increasing the power. With the above ideas firmly in mind, the owner should select the simplest tuning method which will give him the performance he requires, remembering all the time that here, as elsewhere, **Power Costs Money.**

Tuning hints are included for the racing enthusiasts who want to go to the limit and who have facilities to modify or make up special parts for their cars. We hope this section will be of use to them.

Owners are reminded that in certain countries noise restriction regulations are in force. The Company cannot therefore accept responsibility for any increase in the existing noise level of the car which may result after special tuning operations have been carried out. In Countries with air pollution regulations, modifications to the engine may prohibit the car from use on public roads.

Most competitive events are run under rules agreed by the F.I.A. which limit modifications for certain groups or categories. As soon as improvements are incorporated into production or become available as special parts, the necessary steps are taken to have these parts approved for competition and included on the homologation forms.

However, whilst every care is taken, no responsibility can be accepted for ensuring that any specifications or modifications comply with the F.I.A. Regulations or homologation forms. Copies of the forms of recognition are available **only** from the R.A.C. Competitions Department, 31 Belgrave Square, London S.W.1., who will also be able to advise on any queries concerning eligibility of modified cars.

**Note.**—All parts mentioned must be obtained through your BMC Distributor or Dealer. Ensure that the full part number is quoted.

The parts numbers contained in this handbook were current at the time of printing. Due to the fact that numbers and parts availabilities are constantly changing, it is necessary to check with one's local distributor for the latest information. Up to date special tuning sheets are available from Hambro, Inc. by request. These sheets contain new modifications and parts numbers and are a must for anyone planning such modifications.

Special Tuning Sheets may be obtained from:

Mr. M. J. R. Barratt
Technical Advisor
The British Motor Corporation/Hambro Inc.
734 Grand Avenue
Ridgefield, New Jersey 07657

[Editor's note]

# A. GETTING THE BEST FROM YOUR 'MGB'

**When fitted with a H.C. engine (compression ratio 8·8 : 1)**

The engine fitted to your 'MGB' is a highly developed unit and it is essential that you should know something about the specialized maintenance it requires if you are to maintain it at the peak of its mechanical efficiency.

Special recommendations on the sparking plugs, ignition settings, and fuel to be used are given by the manufacturers, and it is stressed that failures are bound to occur if these are not strictly adhered to. Particular care is needed with this engine owing to its high compression ratio, which makes it extremely sensitive to variations in fuel, ignition timing, and the heat range of the sparking plugs.

In lower compression engines a much wider range of fuels can be tolerated without causing serious damage to the engine, and ignition settings will stand variations of a reasonable amount. Also, even if the incorrect sparking plugs are used, no more damage may be incurred than burnt-out plugs or leaky valves. But with an engine having a very high compression ratio the range of fuels, sparking plugs, and ignition settings is much narrower and it is essential that the mixture should always be correct, and particularly never overweak at maximum load or power.

High-compression engines are very sensitive to variations in spark advance (over-advance) and to fuel/air ratio (mixture). Variations in these settings will increase the combustion temperature, and if the variation is excessive pre-ignition will cause high shock waves, resulting in damage to the engine. The engine should be decarbonized at regular intervals as excessive deposits of ash from the combustion of lubricating oil and fuel can cause pre-ignition difficulties.

**Choice of fuel**

*When fitted with H.C. engine (compression ratio* 8·8 : 1)

The octane number of a motor fuel is an indication given by the fuel technicians of its knock resistance. High-octane fuels have been produced to improve the efficiency of engines by allowing them to operate on high compression ratios, resulting in better fuel economy and greater power. Owing to the high compression ratio of the 'MGB' engines, fuels with an octane rating below 98 are not suitable; should it be necessary to use a fuel with a lower octane number, the car must be used very carefully until the correct fuel can be obtained.

It is necessary to use Super grade fuels in the 100-octane range unless Premium fuels of minimum 98-octane Research are available.

*When fitted with L.C. engine (compression ratio* 8·0 : 1)

Premium fuels of minimum 93-octane Research up to 97 octane are required, with preference to 95/97.

Super grade fuels in the 100-octane range can be used if preferred.

**Sparking plugs**

The correct grade of sparking plug for use under normal driving conditions is the Champion N-9Y. Plugs of a lower heat range (hotter running) should not be used, otherwise pre-ignition will occur, with consequent rise in combustion temperature and resulting engine damage. For competition work or hard driving where high output is consistently sustained the Champion N3 sparking plug should be used. This is a cooler-running plug and will ensure lower combustion temperatures and an increased margin of safety. Accumulated deposits of carbon, leaking or cracked insulators, and thin electrodes are all causes of pre-ignition. The plugs should therefore be examined, cleaned, and adjusted at the specified intervals and defective ones renewed. New plugs should be fitted every 12,000 miles (20000 km.).

**Static ignition settings**

It is of the utmost importance that the correct setting should always be maintained. It will be appreciated that any variation in the contact breaker gap will affect the ignition setting, and your particular attention is called to the 6,000 miles (10000 km.) check and adjustment of the distributor points specified in the Driver's Handbook. After adjusting the contact breaker gap to the correct setting it is advisable to check the ignition timing, and to correct it if necessary.

An accurate check can be carried out by a very simple electrical method. To do this, connect a 12-volt lamp between the low-tension terminal on the side of the distributor and a good earth point on the engine.

With the ignition switched on and the sparking plugs removed, turn the crankshaft until the crankshaft pulley pointer is exactly at the correct number of degrees as stated under **'GENERAL DATA'**.

If the ignition timing is correct the lamp will light at exactly this point. Any discrepancy in the ignition setting can be rectified by turning the vernier adjusting nut on the distributor until the test lamp lights at exactly the correct setting. If pinking should occur due to the use of a fuel of a lower range than our recommendations, retarding the ignition 2 to 3° can be tolerated. In no circumstances should the ignition be advanced beyond the correct setting.

**Dunlop centre-lock wheel**

This is built on the Rudge-Whitworth system and provides the most rapid method of changing road wheels. Like all mechanical devices, it must be properly treated in order to give 100 per cent. service.

Observation of the following quite simple hints will ensure complete satisfaction.

When the car is new. After the first long run, or after 50 miles (80 km.) of short runs, jack up each wheel and hammer the nuts to ensure that they are tight.

When wheels are replaced, cover both conical surfaces and the serrations in the hub, also the coned surface and threads in the locknut, with a light coating of grease. Hammer tight and repeat as when car is new.

When a forced change is made on the road, remove and grease the hub as soon as convenient.

Once in 12 months remove the wheels for examination and regreasing.

When changing wheels wipe the serrations and cones on the hub, wheel, and locknut to remove any foreign matter that would prevent the wheel from properly seating. Rust and dirt are the enemies of all mechanical devices.

After a general overhaul of the car, which may involve stripping of the axle, the inscription on the locknuts should be checked to see that it corresponds with the side of the car on which it is applied.

**General.**—Always hammer the locknuts tight. Lift the car on the jack before using the hammer. The locknuts are designed for self-locking, but they should not on that account be permitted to run untightened, because there is, in such case a possibility of damaging the splines.

# B. GENERAL DATA

**Engine**

| | |
|---|---|
| Type | 18G/18GA (3 main bearing) 18GB (5 main bearing) |
| Number of cylinders | 4 |
| Bore | 3·16 in. (80·26 mm.) |
| Stroke | 3·5 in. (89 mm.) |
| Capacity | 1798 c.c. (109·8 cu. in.) |
| Firing order | 1, 3, 4, 2 |
| Compression ratio | H.C. 8·8 : 1 (L.C. 8 : 1) |
| Capacity of combustion chamber (valves fitted) | 42·5 to 43·5 c.c. (2·59 to 2·65 cu. in.) |
| Valve operation | Overhead by push-rod |
| Safe maximum r.p.m. | 6,000 |
| Valve crash r.p.m. | 6,230 |
| B.H.P. | H.C. 95 (L.C. 91) at 5,400 |
| B.M.E.P. | 152 at 3,100 |
| Torque (lb. ft.) | |
| H.C. | 110 (15·2 kg. m.) at 3,000 r.p.m. |
| L.C. | 105 (14·5 kg. m.) at 3,000 r.p.m. |
| Octane rating | Minimum requirements for knock-free operation. H.C. 98+, L.C. 93+ |
| Cooling system | Thermo-siphon, pump- and fan-assisted |
| Oversize bore | |
| First | ·010 in. (·254 mm.) |
| Maximum | ·040 in. (1·016 mm.) |

**Crankshaft**

| | |
|---|---|
| Main journal diameter | 2·126 to 2·127 in. (54·01 to 54·02 mm.) |
| Minimum regrind diameter | 2·086 in. (52·984 mm.) |
| Crankpin journal diameter | 1·8759 to 1·8764 in. (47·65 to 47·66 mm.) |
| Crankpin minimum regrind diameter | 1·8359 in. (46·64 mm.) |

**Main bearings**

| | |
|---|---|
| Number and type | 18G/18GA—3 thinwall, 18GB—5 thinwall |
| Material | |
| Bottom half | Steel-backed copper-lead |
| Top half | Steel-backed copper-lead |
| Length | |
| Front, centre, and rear | 1·125 in. (28·575 mm.) |
| Intermediates (18GB only) | ·785 in. (22·23 mm.) |
| End-clearance | ·002 to ·003 in. (·051 to ·076 mm.) |
| End-thrust | Taken by thrust washers at centre main bearing |
| Running clearance | ·001 to ·0027 in. (·025 to ·0688 mm.) |

**Connecting rods**

| | |
|---|---|
| Type | Angular-split big-end |
| 18G/18GA | Split clamp small end |
| 18GB | Bush small end |
| Length between centres | 6·5 in. (165·1 mm.) |

# GENERAL DATA

**Big-end bearings**

| | |
|---|---|
| Material top half | Steel-backed copper-lead |
| Material bottom half | Steel-backed copper-lead |
| Bearing side-clearance | ·008 to ·012 in. (·203 to ·305 mm.) |
| Bearing diametrical clearance | ·001 to ·0027 in. (·025 to ·0688 mm.) |
| Gudgeon pin bore (18GB) | ·8126 to ·8129 in. (19·68 to 19·95 mm.) |

**Pistons**

| | |
|---|---|
| Type | Aluminium alloy |
| Clearances | |
| Bottom of skirt | ·0018 to ·0024 in. (·045 to ·060 mm.) |
| Top of skirt | ·0036 to ·0048 in. (·091 to ·121 mm.) |
| Oversizes | +·010 in., +·020 in. +·030 in., +·040 in. (+·254 mm., +·508 mm., +·762 mm., +1·016 mm.) |

**Piston rings**

| | |
|---|---|
| Compression: Plain | Top ring (chrome-plated) |
| Tapered | Second and third rings |
| Width | ·0615 to ·0625 in. (1·56 to 1·58 mm.) |
| Thickness | ·137 in. (3·48 mm.) |
| Fitted gap | ·012 to ·017 in. (·304 to ·431 mm.) |
| Clearance in groove | ·0015 to ·0035 in. (·038 to ·089 mm.) |
| Oil control ring | Slotted scraper |
| Width | ·1552 to ·1562 in. (3·94 to 3·99 mm.) |
| Thickness | ·137 in. (3·48 mm.) |
| Fitted gap | ·012 to ·017 in. (·304 to ·431 mm.) |
| Clearance in groove | ·0016 to ·0036 in. (·040 to ·091 mm.) |

**Gudgeon pin**

| | |
|---|---|
| *18G/18GA* | |
| Type | Clamped |
| Fit (in piston) | Free fit to 20° C. (68° F.) |
| Diameter | ·75 in. (19·05 mm.) |
| *18GB* | |
| Type | Fully floating |
| Fit (in piston) | ·0001 in. to ·00035 in. (·0025 to ·007 mm.) |
| Fit (in bush) | Hand push |

**Cylinder head**

| | |
|---|---|
| Cylinder head depth | $3\frac{11}{64}$ $^{+\cdot015\text{ in.}}_{-\cdot000\text{ in.}}$ (80·6 $^{+\cdot400\text{ mm.}}_{-\cdot000\text{ mm.}}$) |
| Thickness of cylinder head gasket | ·023 in. (·584 mm.) compressed |
| Capacity of cylinder head gasket | 3·208 c.c. |
| Capacity of combustion space | 42·5/43·5 c.c. (valves fitted) |
| Capacity of piston head below block face | H.C. 10·87 c.c. (L.C. 17·43 c.c.) |
| (including capacity of piston concavity) | H.C. 6·25 c.c. (L.C. 12·8 c.c.) |
| Capacity of plug centre hole | ·2 c.c. |
| Inlet and exhaust manifold gasket | Part No. 1G 2417 |
| Valve seat angle in cylinder head | 45° |

**Valves and valve gear**

| | |
|---|---|
| Seat angle | |
| Inlet and exhaust | 45½° (seat angle in cylinder head 45°) |
| Head diameter | |
| Inlet | 1·562 to 1·567 in. (38·67 to 38·8 mm.) |
| Exhaust | 1·343 to 1·348 in. (34·11 to 34·23 mm.) |
| Stem diameter | |
| Inlet | ·3422 to ·3427 in. (8·692 to 8·709 mm.) |
| Exhaust | ·34175 to ·34225 in. (8·680 to 8·693 mm.) |
| Valve lift (clearance set ·015 in.) | ·3645 in. (9·26 mm.) |
| Cam lift | ·250 in. (6·35 mm.) |
| Throat diameter | |
| Inlet | 1·3125 in. (33·33 mm.) |
| Exhaust | 1·156 in. (29·36 mm.) |
| Valve stem to guide clearance | |
| Inlet | ·00155 to ·00255 in. (·0394 to ·0648 mm.) |
| Exhaust | ·00200 to ·00300 in. (·051 to ·076 mm.) |
| Valve rocker clearance | |
| Running | ·015 in. (·38 mm.) cold |
| Timing | |
| Inlet and exhaust | ·021 in. (·53 mm.) |
| Timing markings | Dimples on timing wheels |
| Chain pitch and number of pitches | $\frac{3}{8}$ in. (9·52 mm.), 52 pitches |
| Inlet valve | |
| Opens | 16° B.T.D.C. |
| Closes | 56° A.B.D.C. |
| Exhaust valve | |
| Opens | 51° B.B.D.C. |
| Closes | 21° A.T.D.C. |

**Valve guides**

| | |
|---|---|
| Length | |
| Inlet | $1\frac{5}{8}$ in. (41·275 mm.) |
| Exhaust | $2\frac{31}{128}$ in. (56·96 mm.) |
| Diameter | |
| Inlet and exhaust | |
| Outside | ·5635 to ·5640 in. (14·3129 to 14·3256 mm.) |
| Inside | ·34425 to ·34475 in. (8·74269 to 8·75665 mm.) |
| Fitted height above head | ·625 in. (15·87 mm.) |

**Valve springs**

| | |
|---|---|
| Free length | |
| Inner | $1\frac{31}{32}$ in. (50 mm.). |
| Outer | $2\frac{9}{64}$ in. (54·372 mm.) |
| Fitted length | |
| Inner | $1\frac{7}{16}$ in. (36·51 mm.) |
| Outer | $1\frac{9}{16}$ in. (39·69 mm.) |
| Number of working coils | |
| Inner | 6½ |
| Outer | 4½ |
| Pressure | |
| Valve open | Inner 50 lb. (22·7 kg.)<br>Outer 117 lb. (53·08 kg.) |
| Valve closed | Inner 30 lb. (13·6 kg.)<br>Outer 72½ lb. (32·89 kg.) |

**Tappets**

| | |
|---|---|
| Type | Flat base. Barrel type |
| Diameter | |
| Body | $\frac{13}{16}$ in. (20·64 mm.) |
| Working face | $\frac{9}{16}$ in. (14·29 mm.) |
| Length | 2·293 to 2·303 in. (58·25 to 58·5 mm.) |

**Rockers**

| | |
|---|---|
| Outside diameter (before fitting) | ·751 in. (19·07 mm.) |
| Inside diameter (reamed in position) | ·616 to ·620 in. (15·65 to 15·74 mm.) |
| Bore of rocker arms | ·7485 to ·7495 in. (19·01 to 19·04 mm.) |
| Rocker ratio | 1·426 : 1 |

**Camshaft**

| | |
|---|---|
| Journal diameters | |
| Front | 1·78875 to 1·78925 in. (45·43 to 45·44 mm.) |
| Centre | 1·72875 to 1·72925 in. (43·91 to 43·92 mm.) |
| Rear | 1·62275 to 1·62325 in. (41·22 to 41·23 mm.) |
| End-float | ·003 to ·007 in. (·076 to ·178 mm.) |
| Bearings—number and type | 3. Thinwall steel-backed copper-lead |
| Outside diameter (before fitting) | |
| Front | 1·920 in. (48·76 mm.) |
| Centre | 1·860 in. (47·24 mm.) |
| Rear | 1·754 in. (44·55 mm.) |

| | |
|---|---|
| Inside diameter (reamed in position) | |
| Front | 1·79025 to 1·79075 in. (45·472 to 45·485 mm.) |
| Centre | 1·73025 to 1·73075 in. (43·948 to 43·961 mm.) |
| Rear | 1·62425 to 1·62475 in. (41·256 to 41·269 mm.) |
| Diametrical clearance | ·001 to ·002 in. (·0254 to ·0508 mm.) |

**Engine lubrication system**

| | |
|---|---|
| Oil pump | |
| Type | Eccentric rotor |
| Relief pressure valve operates | 70 lb./sq. in. (4·9 kg./cm.$^2$) |
| Relief valve spring | |
| Free length | 3 in. (76·2 mm.) |
| Fitted length | $2\frac{5}{32}$ in. (54·77 mm.) at 16 lb. (7·26 kg.) load |
| Identification colour | Red spot |
| Oil filter | |
| Type | Tecalemit |
| Capacity | ½ pint (·6 U.S. pint, ·28 litre) |
| Oil pressure | |
| Normal running | |
| Minimum | 10 lb./sq. in. (·7 kg./cm.$^2$) |
| Maximum | 80 lb./sq. in. (5·6 kg./cm.$^2$) |

**Torque wrench settings**

| | |
|---|---|
| Cylinder head nuts | 45–50 lb. ft. (6·16–6·91 kg. m.) |
| Main bearing nuts | 70 lb. ft. (9·7 kg. m.) |
| Connecting rod set screws | 40–45 lb. ft. (5·5–6·2 kg. m.) |
| Clutch assembly to flywheel | 25–30 lb. ft. (3·4–4·1 kg. m.) |
| Flywheel bolts | 40 lb. ft. (5·5 kg. m.) |
| Gudgeon pin set screws | 25 lb. ft. (3·4 kg. m.) |

**Fuel system**

| | |
|---|---|
| Carburetter | |
| Make and type | S.U. twin HS4 semi-downdraught |
| Diameter | 1½ in. (38·1 mm.) |
| Needle | No. 5 (standard), No. 6 (rich), No. 21 (weak) |
| Jet | ·090 in. (2·29 mm.) |
| Carburetter piston | Part No. AUC 2061 |
| Piston spring | Red (Part No. AUC 4387) |

**Air cleaner**

| | |
|---|---|
| Make and type | Coopers Mechanical Joints Ltd., Steel canister, paper element (replaceable) |

**Fuel pump**

| | |
|---|---|
| Make and type | S.U. electric, high-pressure |
| Delivery test | 10 gal. per hr. (12 U.S. gal., 45·4 litres per hr.) |
| Suction lift | 33 in. (83·8 cm.) |
| Output lift | 48 in. (121·9 cm.) |

**Cooling system**

| | |
|---|---|
| Type | Pressurized radiator, thermo-siphon, pump- and fan-assisted |
| Filler cap spring pressure | 7 lb. (3·175 kg.) |

**Ignition system**

| | |
|---|---|
| Sparking plugs | Champion N–9Y |
| Size | 14 mm. |
| Plug gap | ·024 to ·026 in. (·61 to ·66 mm.) |
| Coil | Lucas HA12 |
| Distributor | Lucas, Type 25D4 |
| Distributor contact points gap | ·014 to ·016 in. (·35 to ·40 mm.) |
| Suppressors type | Lucas No. 78106A fitted on each H.T. cable |
| Timing | H.C. 10° B.T.D.C. (L.C. 8° B.T.D.C.) |

**Clutch**

| | |
|---|---|
| Make and type | Borg & Beck 8 in. (20·3 cm.) diaphragm spring. Strap drive |
| Diameter | 8 in. (20·3 cm.) |
| Facing material | Wound yarn—Borglite |
| Damper springs | 6 |
| Colour | Black/light green |

**Gearbox (Standard and Overdrive)**

| | |
|---|---|
| Number of forward speeds | 4 |
| Synchromesh | Second, third, and fourth gears |
| Overdrive | ·802 |

| Ratios | |
|---|---|
| Top | 1·0000 : 1 |
| Third | 1·3736 : 1 |
| Second | 2·2143 : 1 |
| First | 3·6363 : 1 |
| Reverse | 4·7552 : 1 |

| Overall ratios | | *M.p.h. per* 1,000 *r.p.m.* |
|---|---|---|
| Overdrive | 3·135 : 1 | 22·3 |
| Top | 3·909 : 1 | 17·9 |
| Third | 5·36 : 1 | 13·09 |
| Second | 8·65 : 1 | 8·11 |
| First | 14·20 : 1 | 4·94 |
| Reverse | 18·60 : 1 | 3·77 |

Speedometer gears ratio .. 9 : 28 (overdrive 5 : 16)

| Optional axle ratios | *Overall ratios* | *M.p.h. per* 1,000 *r.p.m.* |
|---|---|---|
| Overdrive | 3·649 : 1 | 19·2 |
| Top | 4·55 : 1 | 15·4 |
| Third | 6·24 : 1 | 11·24 |
| Second | 9·98 : 1 | 7·03 |
| First | 16·54 : 1 | 4·24 |
| Reverse | 21·63 : 1 | 3·24 |
| | | |
| Overdrive | 3·449 : 1 | 20·3 |
| Top | 4·3 : 1 | 16·3 |
| Third | 5·9 : 1 | 11·89 |
| Second | 9·52 : 1 | 7·37 |
| First | 15·63 : 1 | 4·4 |
| Reverse | 20·44 : 1 | 3·43 |
| | | |
| Overdrive | 3·288 : 1 | 21·3 |
| Top | 4·1 : 1 | 17·1 |
| Third | 5·63 : 1 | 12·44 |
| Second | 9·07 : 1 | 7·73 |
| First | 14·90 : 1 | 4·70 |
| Reverse | 19·49 : 1 | 3·60 |

**Gearbox (close-ratio) (Non-overdrive and Overdrive)**

| | |
|---|---|
| Number of forward speeds | 4 |
| Synchromesh | Second, third, and fourth gears |
| Overdrive ratio | ·802 |

| Ratios | |
|---|---|
| Top | 1·0 : 1 |
| Third | 1·268 : 1 |
| Second | 1·620 : 1 |
| First | 2·450 : 1 |
| Reverse | 4·7552 : 1 |

| Overall ratios | | *M.p.h. per* 1,000 *r.p.m.* |
|---|---|---|
| Overdrive | 3·135 : 1 | 22·3 |
| Top | 3·909 : 1 | 17·9 |
| Third | 4·956 : 1 | 14·1 |
| Second | 6·332 : 1 | 11·0 |
| First | 9·577 : 1 | 7·3 |
| Reverse | 18·588 : 1 | 3·7 |

Speedometer gears ratio .. 9 : 28 (Overdrive 5 : 16)

| Optional axle ratios | *Overall ratios* | *M.p.h. per* 1,000 *r.p.m.* |
|---|---|---|
| Overdrive | 3·649 : 1 | 19·2 |
| Top | 4·55 : 1 | 15·4 |
| Third | 5·769 : 1 | 12·1 |
| Second | 7·371 : 1 | 9·5 |
| First | 11·147 : 1 | 6·3 |
| Reverse | 21·635 : 1 | 3·2 |
| | | |
| Overdrive | 3·449 : 1 | 20·3 |
| Top | 4·30 : 1 | 16·3 |
| Third | 5·452 : 1 | 12·8 |
| Second | 6·966 : 1 | 10·0 |
| First | 10·535 : 1 | 6·6 |
| Reverse | 20·447 : 1 | 3·4 |
| | | |
| Overdrive | 3·288 : 1 | 21·3 |
| Top | 4·10 : 1 | 17·1 |
| Third | 5·198 : 1 | 13·4 |
| Second | 6·642 : 1 | 10·5 |
| First | 10·045 : 1 | 7·0 |
| Reverse | 19·496 : 1 | 3·6 |

# GENERAL DATA

**Steering**

| | |
|---|---|
| Type | Rack and pinion |
| Steering-wheel | |
| Turns—lock to lock | 2·93 |
| Diameter | 16½ in. (41·9 cm.) |
| Camber angle | Front 1° |
| Castor angle | 7° |
| King-pin inclination | 8° |
| Toe-in | 1/16 to 3/32 in. (1·6 to 2·4 mm.) |
| Track | |
| Front | Disc wheels 49 in. (1·244 m.)<br>Wire wheels 49¼ in. (1·251 m.) |
| Rear | Disc wheels 49¼ in. (1·251 m.)<br>Wire wheels 49¼ in. (1·251 m.) |

**Front suspension**

| | Tourer | GT |
|---|---|---|
| Type | Independent coil | |
| Spring detail | | |
| Coil diameter (mean) | 3·238 in. (82·23 mm.) | 3·28 in. (83·5 mm.) |
| Diameter of wire | ·498 in. (12·66 mm.) | ·54 in. (14 mm.) |
| Free height | 9·9 ± 1/16 in. (251 mm. ± 1·6 mm.) | 9·1 ± 1/16 in. (231 ± 1·6 mm.) |
| Number of free coils | 7·5 | 7·2 |
| Static laden length | 7 ± 1/32 in. (178 mm. ± ·8 mm.) at load of 1,030 lb. (467·2 kg.) | 6·6 in. (168 mm.) at load of 1,193 lb. (541·5 kg.) ± 20 lb. (9·1 kg.) |
| Dampers (front) | Piston type | |
| Anti-roll bar (front) | 9/16 in. (14·3 mm.) dia. optional | ⅝ in. (15·9 mm.) dia. standard |

**Rear suspension**

| | | |
|---|---|---|
| Type | Semi-elliptic | |
| Spring detail | *Tourer (Early type)* | *Tourer (Later type)* |
| Number of leaves | 5 and bottom plate | 6 and bottom plate |
| Width of leaves | 1¾ in. (44·45 mm.) | 1¾ in. (44·45 mm.) |
| Gauge | 7/32 in. (5·56 mm.) | 3 at 7/32 in. (5·6 mm.), 3 at 3/16 in. (4·8 mm.) |
| Working load (−15 lb. [7 kg.]) | 400 lb. (181·44 kg.) | 450 lb. (204·12 kg.) |
| | *GT* 510 lb. (321·6 kg.) | |
| Dampers (rear) | Piston type | |

**Propeller shaft**

| | |
|---|---|
| Type | Tubular flanged 1100 series |
| Make and type of joints | Hardy Spicer, needle roller |
| Propeller shaft length (between centres of joints) | 27⅝ in. (70·167 cm.) standard<br>28¾ in. (73·025 cm.) overdrive |
| Overall length | 30 in. (76·2 cm.) standard<br>31⅛ in. (79·057 cm.) overdrive |
| Diameter | 2 in. (50·8 mm.) |

**Rear axle**

| | *Tourer (Early)* | *Tourer (Later) and GT* |
|---|---|---|
| Type | 'B' type, three-quarter-floating | 'Tubed' type, semi-floating |
| Ratio | | |
| Standard | 11/43 | 11/43 |
| Alternatives | 9/41, 10/43, 10/41 | 9/38, 9/41 |
| Adjustment | Shims | |

**Electrical equipment**

| | |
|---|---|
| System | 12-volt. Positive earth |
| Charging system | Compensated voltage control |
| Battery | Two 6-volt Lucas SG9E |
| Starter motor | Lucas 4-brush M418G |
| Dynamo | Lucas C40/1 |

**Brakes**

| | |
|---|---|
| Type | Lockheed hydraulic (front and rear) |
| Front | Disc 10¾ in. dia. (27·3 cm.) |
| Rear | Drum 10 in. (25·4 cm.), single leading shoe |
| Rear linings | 10 in. × 1¾ in. (25·4 cm. × 44·45 mm.) |
| Lining dimensions | 9·6 in. × 1¾ in. (24·38 cm. × 44·45 mm.) |
| Lining area | |
| Front pads | 20 sq. in. (129·03 cm.²) total |
| Rear | 67·2 sq. in. (433·55 cm.²) |
| Material | |
| Front | DON 55 |
| Rear | DON 24 |
| Brake cylinder diameter | |
| Front | 2⅛ in. (53·97 mm.) dia. |
| Rear | ·80 in. (20·32 mm.) dia. |

**Wheels**

| | *Tourer* | *GT* |
|---|---|---|
| Type | | |
| Ventilated disc | 4J × 14 in. | 5J × 14 in. |
| Wire (optional) | 4½J × 14 in. and 60-spoke (Tourer and GT) | |

**Tyres**

| | | |
|---|---|---|
| Standard | | |
| Size | 5·60—14 (Tubed C.41) | |
| Optional | *Tourer* | *GT* |
| Size | 155—14 (SP) | 165—14 (SP) |
| Standard tyres | | |
| Pressures (set cold) | *Tourer* | *GT* |
| Front | 18 lb./sq. in. (1·3 kg./cm.²) | 20 lb./sq. in. (1·4 kg./cm.²) |
| Rear | 18 lb./sq. in. (1·3 kg./cm.²) | 24 lb./sq. in. (1·7 kg./cm.²) |
| Sustained speeds in excess of 90 m.p.h. (145 km.p.h.): | | |
| Front | 24 lb./sq. in. (1·7 kg./cm.²) | 26 lb./sq. in. (1·8 kg./cm.²) |
| Rear | 24 lb./sq. in. (1·7 kg./cm.²) | 30 lb./sq. in. (2·1 kg./cm.²) |
| Optional tyres (SP) | | |
| Pressures (set cold) | *Tourer* | *GT* |
| Front | 21 lb./sq. in. (1·5 kg./cm.²) | 21 lb./sq. in. (1·5 kg./cm.²) |
| Rear | 24 lb./sq. in. (1·7 kg./cm.²) | 24 lb./sq. in. (1·7 kg./cm.²) |
| Sustained speeds in excess of 90 m.p.h. (145 km.p.h.): | | |
| Front | 27 lb./sq. in. (1·9 kg./cm.²) | 28 lb./sq. in. (2·0 kg./cm.²) |
| Rear | 31 lb./sq. in. (2·2 kg./cm.²) | 31 lb./sq. in. (2·2 kg./cm.²) |

**Note.**—Rear tyre pressures may be increased by 2 lb./sq. in. (·14 kg./cm.²) with advantage when touring with a laden boot.

For competition work and use of racing tyres, advice may be obtained direct from the Dunlop Rubber Co. Ltd., Fort Dunlop, Erdington, Birmingham 24.

**Capacities**

| | *Imp.* | *U.S.A.* | *Litres* |
|---|---|---|---|
| Engine sump (incl. filter) | 7½ pts. | 9 pts. | 4·28 |
| Gearbox | 5½ pts. | 6½ pts. | 3·12 |
| Rear axle | 2¾ pts. | 3¼ pts. | 1·56 |
| Cooling system (with heater) | 10 pts. | 12 pts. | 5·67 |
| Steering rack | ⅓ pt. | ·39 pt. | ·19 |
| Fuel tank | | | |
| Early cars | 10 gal. | 12 gal. | 45·4 |
| Later cars | 12 gal. | 14 gal. | 54·5 |
| Brake system | 1 pt. | 1·2 pt. | ·568 |
| Oil cooler | ¾ pt. | ·9 pt. | ·42 |

**General dimensions**

| | | |
|---|---|---|
| Wheelbase | 91 in. (231·14 cm.) | |
| Over-all length | 153 3/16 in. (389·13 cm.) | |
| Over-all width | 59 15/16 in. (152·28 cm.) | |
| Over-all height | 49⅜ in. (125·41 cm.) | |
| Ground clearance | 5 in. (12·70 cm.) | |
| Weight: | *Tourer* | *GT* |
| (Unladen) | 1,920 lb. (871 kg.) | 2,190 lb. (993 kg.) |
| Turning circles | 32 ft. (9·754 m.) | |

# C. TUNING

## Stage 1

### Tuning by port polishing (ordinary road work)

An increase of some 3 b.h.p. can be had by general attention to the cylinder head and port polishing as detailed below.

Lightly grind and polish the exhaust and inlet ports throughout. They should not be ground out so heavily that the shape or valve choke diameters are impaired, as the wall between the exhaust and inlet valves may become too thin and cause valve seat cracking.

Just inside the ports, at the bridge between the exhaust and inlet valve seats, you will notice a protuberance; do not grind too much of this away, as this is the port wall, where the water cooling has been brought down as close to the valve seats as possible.

Grind out the combustion spaces, but only very lightly, as these are already quite clean and partly machined; remove any frazes and lightly polish all over. It is most important that no enlargement around the combustion walls takes place as this may cause the cylinder head gasket to overlap. This will destroy the efficiency of the seal, lower the compression ratio, and cause tuning to be ineffective.

The combustion space and ports are already highly developed from a flow angle aspect, and it will be found difficult to improve by reshaping or enlarging. The main requirement is to obtain the highest polish, but to remove the minimum amount of metal.

Match up, by grinding, all the exhaust and inlet manifold ports with the cylinder head ports.

Grind out and polish the inlet manifold, also matching the carburetter bore. Make the bore of the manifold a gradual taper from the carburetter end to the cylinder head port, grinding away any ridges left by machining during manufacture.

Distributor setting as standard.

(See Editor's note at end of Part 3.)

## Stage 2

### Tuning for middle-range acceleration (ordinary road work)

If most importance is placed on initial and middle-range acceleration an improvement of 2 or 3 b.h.p. may be gained in the lower ranges by fitting camshaft Part No. 48G 184. This has the timing: inlet opens 5° B.T.D.C., inlet closes 45° A.B.D.C., exhaust opens 40° B.B.D.C., exhaust closes 10° A.T.D.C.

The valve lift is ·322 in. (8·2 mm.), and cam lift is ·216 in. (5·7 mm.).

Top end performance will only be slightly impaired between 5,000 and 6,000 r.p.m.

If desired, the head may be tuned by port polishing as laid down in Stage 1.

Distributor setting as standard.

## Stage 3

### Compression ratio 9·6 to 9·8 : 1 (competition tune)

Carry out Stage 1.

Fit a competition (half-race) camshaft (Part No. C–AEH 714). This gives ·250 in. (6·35 mm.) cam lift with a 268° period for inlet and exhaust. Inlet opens 24° B.T.D.C. and closes 64° A.B.D.C. Exhaust opens 59° B.B.D.C. and closes 29° A.T.D.C., check that the camshaft identification ring does not foul the connecting rod.

Tappet setting ·017 in. (·43 mm.) hot. For valve springs see page

Machine 1/16 in. (1·59 mm.) from the cylinder head face to raise the compression ratio to 9·7 : 1. The head thickness will then be $3\frac{7}{64}\,{}^{+\cdot015}_{-\cdot000}$ in. ($79\,{}^{+\cdot38}_{-\cdot000}$ mm.).

Fit 1¾ in. (44·45 mm.) diameter S.U. carburetters (Part No. C–AUD 229 for a pair); these are fitted with ·100 jets, SY needles, and light blue springs. The carburetters can be fitted to the standard manifold, which should be polished as explained in Stage 1.

All necessary parts and instructions for fitting these carburetters are included in kit Part No. C–AJJ 3321.

No air cleaners are arranged for these carburetters, but extension pipes Part No. C–AHH 7209, to reduce the turbulence at the carburetter mouth, are included in the installation kit.

Check the valves at full lift to ensure that the exhaust valves do not foul the top face of the cylinder block; if so, the block must be undercut to clear the valve head and give a minimum lift clearance of 1/16 in. (1·59 mm.). Use a 1 17/32 in. dia. flat cutter with a 1/16 in. (1·59 mm.) radius at the corner of the cutter.

The engine should give 105/108 b.h.p. at 6,000 r.p.m.

The static setting for the standard distributor should be 10° B.T.D.C.

## Stage 3A

### Compression ratio 9·6 to 9·8 : 1 (competition tune)

As Stage 3, but alternative to machining the cylinder head, as Stage 3, the compression ratio may be raised by fitting the flat top competition pistons (as detailed in Stage 4).

## Stage 3B

### Compression ratio 10·5 : 1 (competition tune)

If you carry out Stage 3A the compression ratio can be raised to 10·5 : 1 by machining 1/16 in. (1·59 mm.) from the cylinder head face (as detailed in Stage 3).

The engine should develop 112/115 b.h.p. at 6,000 r.p.m.

## Stage 4

### Compression ratio 9·3 to 9·5 : 1 (competition tune)

Polish the head as Stage 1.

Standard valves and guides may be used successfully, but for consistent performance fit bronze Hidural inlet and exhaust valve guides and the high-duty 1 9/16 in. dia. inlet and 1 11/32 in. dia. exhaust valves in Nimonic material. The guides should be pressed into the cylinder head so that they are left standing out between 49/64 and 25/32 in. (19·45 and 19·84 mm.).

Valve seats may need re-aligning with the guides.

Use valve springs—outer (Part No. C–AHH 7264) and inner (Part No. C–AHH 7265), also the special valve cotter (pairs) (Part No. C–AEH 761), the valve spring top cup (Part No. C–AEH 760), and the bottom cup (Part No. C–AEH 801). Alternatively, see the use of triple valve springs (Part No. C–AHH 7309) on page 429.

Do not fit the metal oil shroud or the valve stem rubber oil seal that is used on the standard engine.

Standard rocker shaft and valve rockers are used, also the two centre rocker shaft brackets. The valve springs put a heavy load on the rocker shaft, especially at the ends where it is overhung, and, although it is not essential, it is desirable to fit the special front and rear rocker shaft brackets which support the end rockers from both sides (Part Nos. C–AEH 762 and C–AEH 763). Do not refit the springs which hold the rockers apart, but in place of these fit the tubular steel distance pieces leaving an end-float of ·003 to ·005 in. (·0076 to ·0127 mm.) (Part Nos. C–AEH 764 (1 off) and C–AEH 765 (2 off)).

The standard tappet adjusting screws may be used, but if the possibility of fracture is to be eliminated, then fit the special screws (Part No. C–AEH 766) which are solid and have no holes drilled in them; the oiling of the ball will be satisfactory without these holes.

Use the high-lift wide-period camshaft (Part No. C–AEH 770) with a tappet setting between valve and rocker of ·018 in. (·457 mm.) hot. This camshaft has a cam lift of ·315 in. (8·0 mm.) and a valve lift of ·452 in. (11·5 mm.). The inlet period is 300° and the exhaust 300°.

Timing is: inlet opens 50° B.T.D.C., inlet closes 70° A.B.D.C., exhaust opens 75° B.B.D.C., exhaust closes 45° A.T.D.C., with valve clearance set at ·018 in. (·46 mm.).

The high-lift, wide-period camshaft can be used with the standard push-rods and tappets satisfactorily, but cover for travel of the cam is only barely sufficient over the base of the standard tappet.

This can be overcome by boring the tappet holes in the cylinder block to a diameter of $\cdot9375\,{}^{+\cdot0005}_{-\cdot0002}$ in. ($23\cdot81\,{}^{+\cdot013}_{-\cdot005}$ mm.) for a length of 3⅛ in. (79·38 mm.) from the centre-line of the camshaft and fitting larger tappets of 15/16 in. (24 mm.) dia. (Part No. AEC 264). This will necessitate the use of shorter push-rods (Part No. C–AEH 767).

With the high lift of the valves it is necessary to undercut the face of the cylinder block to allow the exhaust valves full travel. This machining should be done from the valve guide centre and a flat cutter of 1 17/32 in. dia. used; the cutter should have a 1/16 in. (1·6 mm.) radius on the outer cutting corner. The undercut in the cylinder block should be 9/64 in. (3·6 mm.) deep.

For durability, when using the high-lift camshaft, it is a benefit to use the steel timing chain sprockets (Part Nos. 12H 244 and C–AEH 771).

In the centre of the cylinder head face two large core holes will be found; thread these and fit water-tight aluminium plugs, which should be faced off carefully to the head face. This will prevent water loss if the cylinder head lifts under arduous conditions. Use the special cylinder head gasket (C–AEH 768).

Thread and plug the one small hole in the centre of the cylinder block face that is opposite to the cylinder head aluminium plug.

To increase water flow through the head drill out to 9/16 in. (14·29 mm.) dia. the two water holes at the rear end of the cylinder block face.

**Engine 18G/18GA**

Fit the high-compression (flat top) competition pistons (Part No. C–AEH 736).

These pistons have large, fully floating gudgeon pins, and it is necessary to use special connecting rods, available only as a matched set (Part No. C–AJJ 3357) with connecting rods bearings (Part No. 8G 2259) (set). When using these bearings and the standard main bearings, it is necessary to run in steadily for 30 hours on a test stand or for 1,000 road miles (1600 km.). Do not apply full power at an early stage, but wait until the bearings bed down, and develop a good running condition without temperature rise.

**Engine 18GB**

Fit high-compression (flat top) competition pistons (Part No. C–AEH 851). These will fit the standard connecting rods.

**Engine 18G/GA/GB**

If you wish to put your engine on to full power early, then see note reference bearings with initial racing clearances on page 430.

To increase the oil pressure fit a packing piece in the end of the oil release valve cap behind the release valve spring; this should be ·200 in. (5·08 mm.) thick by $\frac{31}{64}$ in. (12·3 mm.) dia. or two packings ·100 in. (2·54 mm.) thick (Part No. AEH 798). 18GB units already have one ·100 in. (2·54 mm.) (Part AEH 798) fitted. See page 434 for further details on oil pump and filter bottle top.

Fit the competition clutch (see page 430). See page 431 for reference to the flywheel.

Fit the large $1\frac{3}{4}$ in. (44·45 mm.) S.U. carburetters as in Stage 3.

Use the special distributor (Part No. C–BHA 4415) (see page 431),which has a suitable automatic advance and no vacuum advance. The static setting should be 6° B.T.D.C. and not more than 8° nor less than 5°. The standard distributor (for H.C. engines), which is Part No. 12H 792, distributor No. 40897, is not quite so suitable, but if it was desired to use it, then it should be set 9° to 11° B.T.D.C.

Sparking plugs should be Champion N57R, but, according to the circuit you may be able to use Champion N62R or N3. Use 100 (minimum) octane fuel.

The engine should develop the following brake-horse-power:

| *R.P.M.* | *B.H.P.* | *R.P.M.* | *B.H.P.* |
|---|---|---|---|
| 3,000 | 62 | 5,500 | 120 |
| 4,000 | 89 | 6,000 | 121 |
| 5,000 | 111 | 6,500 | 119 |

## Stage 5
## 9·3 to 9·5 : 1 compression ratio (Weber carburetter)

Prepare your engine as Stage 4, but in place of the S.U. $1\frac{3}{4}$ in. (44·45 mm.) carburetters, fit a 45 DCOE 13 Weber twin-choke carburetter (Part No. C–AEH 785). This will require a special inlet manifold and parts as detailed on pages 435 and 436. See page 430 for particulars of settings.

There is marginal improvement to be gained by the fitting of the Weber carburetter, with some slight loss at the lower r.p.m.

An installation kit is available containing the manifold and all necessary parts to fit this carburetter. (Part No. C–AJJ 3312.)

The carburetter is mounted on synthetic rubber 'O' ring gaskets to prevent vibration of the carburetter mechanism and disturbance of the fuel-to-air ratio. Under each carburetter fixing nut a double-coil spring washer (Part No. AJD 7732) should be fitted; each fixing nut should be drilled and wired in pairs to prevent them coming slack. Tighten the nuts up fairly firmly, but by gripping the carburetter some slight free movement should be felt.

Steady rods are fitted from the inlet manifold, both front and rear, down to brackets on the cylinder block; adjust these rods to the free position so that the bolts go easily through the fork ends and brackets, tighten the bolts solid, and lock up the fork locknuts.

The engine, using 100 (minimum) octane fuel, should give the following power output:

| | *R.P.M.* | *B.H.P.* | *Gal. per hr. fuel consumption* — *Imperial* | *U.S.* |
|---|---|---|---|---|
| B.H.P. taken at gearbox tail flange | 3,000 | 63·5 | 4·3 | 5·2 |
| | 3,500 | 78 | 5·4 | 6·5 |
| | 4,000 | 92·5 | 6·0 | 7·2 |
| | 4,500 | 106·5 | 6·6 | 7·9 |
| | 5,000 | 114 | 7·1 | 8·5 |
| | 5,500 | 121 | 7·5 | 9·0 |
| | 6,000 | 122 | 8·4 | 10·0 |
| | 6,500 | 121 | 9·0 | 10·8 |

For bench testing use Champion N57R plugs, but on the circuit you may be able to use softer plugs. If the circuit is short you may find N3 in cylinders 1 and 4, and N63R in cylinders 2 and 3 will be suitable.

If the circuit is long you may require N62R in cylinders 1 and 4 and N57R in cylinders 2 and 3. The centre cylinders are inclined to run hotter due to the proximity of the two exhaust valves.

## Stage 6
## 10·4 to 10·6 : 1 compression ratio (competition tuning)

Tune as for Stage 4, or Stage 5, but remove $\frac{1}{16}$ in. (1·59 mm.) from the cylinder head face, making the thickness of the head $3\frac{7}{64} \pm ·015$ in. (79 ± ·38 mm.). The standard thickness of the cylinder head is $3\frac{11}{64}\,^{+·015}_{-·000}$ in. (80·6 $^{+·38}_{-·00}$ mm.). Finally surface grind the face of the head and carefully lap as described on page 429.

Check the opening of the exhaust valves at full lift to ensure they have a minimum of $\frac{1}{16}$ in. (1·59 mm.) over-travel; if not, the undercut in the block face will have to be increased the required amount.

Use Champion N57R sparking plugs.

The engine should give the following power output on 100 (minimum) octane fuel.

| | *R.P.M.* | *B.H.P. on S.U. $1\frac{3}{4}$ in. Carb. (Two)* | *B.H.P. on Weber Carburetter 45 DCOE 13* | *B.H.P. on Weber with 38 choke 175 main 160 air corrected* |
|---|---|---|---|---|
| B.H.P. taken at gearbox tail flange | 3,000 | 64 | 65 | 64 |
| | 3,500 | 76 | 82 | 81 |
| | 4,000 | 91 | 96 | 95 |
| | 4,500 | 106 | 109 | 108 |
| | 5,000 | 114·5 | 120 | 120 |
| | 5,500 | 126 | 128 | 127·5 |
| | 6,000 | 128·5 | 130 | 131 |
| | 6,500 | 129 | 127 | 128 |

**Note.**—While using 100-octane fuel and the camshaft as listed for Stage 4, no worthwhile power increase will be gained by further raising of the compression ratio.

Larger diameter Nimonic inlet valves (Part No. C–AEH 860) are now available (1·687 in. [42·8 mm.] dia) together with large exhaust valves (Part No. C–AEH 861) and details for using 2-in. HS8 carburetters (Part No. C–AUD 279) are being investigated.

# D. OTHER SPECIAL ITEMS

**Brakes**

After many consecutive applications of the brakes during competition driving some brake fade may be experienced with the standard linings. Competition front disc pads and rear brake-shoe linings or lined shoes are available (see list). The rear linings are made to a thickness suitable for grinding to radius after fitment. The front pads are of a suitable heavy-duty material. With fair competition driving these linings will be free from fade, but will give a harder pedal effort on application.

When the lining friction value is altered from that of the standard car it may be found that changing the rear wheel cylinders to ones of smaller size ($\frac{5}{8}$ in. [15·87 mm.] dia.) will improve the front to rear brake ratio.

As these wheel cylinders have a dowel on the fitting face, it will be necessary to drill a hole in the back plate to match this dowel.

Braze a steel plug in the existing hole and face off level with the plate before drilling the new hole. The size of the hole is ·170 to ·175 in. (4·32 to 4·45 mm.) dia. and drilled ·578 in. (14·68 mm.) above the centre of the cylinder mounting hole and ·350 in. (8·89 mm.) offset from the radial centre-line of the cylinder mounting hole.

When using your car on a racing circuit always remove the dust shields from the front disc brakes. This will enable the discs and brakes to run at a lower temperature and will decrease the possibility of brake fade.

**Balancing of road wheel and tyre assemblies**

To obtain the smoothest steering, free from all steering-wheel kick, and to eliminate any tendency to front-wheel patter, especially at speeds around 70 m.p.h. (113 km.p.h.) and over, it will be found beneficial to have the front road wheel and tyre assemblies statically and dynamically balanced. This usually results in balance weights being fitted on both sides of the rims, but this dynamic balancing is well worth while. Balance may require re-checking every few thousand miles if the car suffers brake locking, etc., as this may again put the tyres out of balance enough for the effect to be felt.

It is advisable to keep front tyres in good condition and free from uneven tread wear. This can sometimes be done by changing tyres from front to rear before uneven wear develops. Pick the best tyres for use at the front (or those that have even tread wear and run true) before they are dynamically balanced. Balancing a tyre which has flats or uneven wear is not usually very successful. In some cases the tread can be buffed true, but this is not an economic way of using rubber.

**Valves and guides**

The valves fitted as standard are of high quality, but if valves are desired with a longer service life or increased resistance to burning special valves are available in Nimonic alloy—inlet (Part No. C–AEH 757) and exhaust (Part No. C–AEH 758). (See list). These must be used in conjunction with special spring collars (Part No. C–AEH 760) and also special cotters (Part No. C–AEH 761) to suit the half-round groove in the valve stem. If triple valve springs are used, top collar (Part No. C–AHH 7313) must be used. These valves should be used in conjunction with bronze (Hidural 5) guides for both the inlet and exhaust valves, inlet (Part No. C–AEH 755) and exhaust (Part No. C–AEH 756). The inlet guide is $2\frac{1}{16}$ in. (27 mm.) long and the exhaust $2\frac{5}{16}$ in. (33·4 mm.) long. They should be pressed into the head so that they are left standing out between $\frac{49}{64}$ and $\frac{25}{32}$ in. (19·4 and 19·8 mm.).

### Cylinder head gasket

When the compression is raised it may be necessary to use a reinforced gasket. A competition cylinder head gasket is available (see list) constructed from ·009 in. copper and steel with internal reinforcements (Part No. C-AEH 768). As the cylinder block and head faces may suffer some distortion in the early life of the engine it is advisable to check these faces for flatness before fitting the new gasket. If the faces are distorted they should be finely surface ground, and a certain amount of careful lapping or flat scraping is worth while. Do not lap excessively as this will only produce an uneven surface. Check them finally together with marking.

Before replacing the cylinder head studs slightly countersink (not too heavily) the tops of the threaded holes in the cylinder block. This will enable the head to pull down around the studs and seal more efficiently.

To maintain the clamping pressure of the cylinder head onto the gasket it is advantageous to remove the flat washers under the 11 cylinder head nuts and replace these with more rigid ones of $\frac{1}{8}$ in. (3·18 mm.) thickness by $\frac{3}{4}$ in. (19 mm.) outside diameter and with a hole ·390 in. (9·91 mm.) dia. These can be turned up from a 40-ton steel bar, or if made from mild steel they should be case-hardened.

### Valve springs

The valve bounce r.p.m. on the standard engine is 6,230 r.p.m. and the valve springs, operating mechanism, and drive are safely stressed to maintain this.

If for very special competition purposes it is desired to raise the valve bounce period, the appropriate springs may be selected from the following table:

| Part No. outer springs | lb. | Part No. inner springs | lb. | Total lb. full lift | *Valve bounce r.p.m.* Standard camshaft | Comp. camshaft AEH 714 | Comp. camshaft AEH 770 |
|---|---|---|---|---|---|---|---|
| 1H 1111* | 117 | 1H 723* | 50 | 167 | 6,230 | 6,500 | 6,500 |
| 1H 1111 | 117 | C-1H 1112 | 57 | 174 | 6,360 | 6,600 | 6,600 |
| C-AHH 7264 | 140 | 1H 723 | 50 | 190 | 6,480 | 6,700 | 6,700 |
| C-AHH 7264 | 140 | C-1H 1112 | 57 | 197 | 6,600 | 6,750 | 6,750 |
| C-AHH 7264 | 140 | C-AHH 7265 | 70 | 210 | 6,680 | 6,800 | 6,800 |

* Standard engine.

It is advised that these springs be used only for very special events, as if used under everyday conditions the cams and followers will have a shorter service life. The springs will not necessarily give an increase in brake-horse-power, but will extend the same horse-power up to valve bounce. This is sometimes useful in enabling a lower gear to be retained, still maintaining the same maximum speed, with increased power for acceleration.

As an absolute maximum, triple valve springs (Part No. C-AHH 7309) may be used, in conjunction with a special valve spring top collar (Part No. C-AHH 7313). No bottom collar is used, but the counterbore around the valve guide in the head face must be increased to 1·520 to 1·515 in. (3·86 to 3·85 mm.) dia. The springs are supplied, tightly nested in sets of three, giving a total full lift load of 230 lb. (104·3 kg.).

The valve crash position will be above 7,000 r.p.m.

### Rear axle ratios

With the variety of axle ratios available (see page 48), and the standard and close-ratio gears it is possible to obtain a combination suitable for most competition purposes. Ensure special equipment is available when dismantling 'Tubed' axles, and follow instructions in the Workshop Manual. The 'Tubed' axle, originally fitted only to the GT, was fitted to the Tourer from Car No. 132923 (wire wheels) and 132916 (disc wheels).

### Close ratio gearbox (18G/18GA/18GB)

Alternative gears are available to convert the gearboxes fitted to 18G, 18GA, and 18GB engines. The ratios of the all-synchromesh gearboxes fitted to 18GD and 18GF are closer but no special close ratio gears are available at present.

| | *Standard ratio* | *Close ratio* | *All-Synchromesh ratios* |
|---|---|---|---|
| First | 3·64 : 1 | 2·45 : 1 | 3·44 : 1 |
| Second | 2·21 : 1 | 1·62 : 1 | 2·167 : 1 |
| Third | 1·37 : 1 | 1·27 : 1 | 1·382 : 1 |
| Top | 1·00 : 1 | 1·00 : 1 | 1·00 : 1 |

The following special parts will be required:

| | |
|---|---|
| First motion shaft (18G/18GA) | C-22H 472 (26 teeth) |
| (18GB) | C-22H 846 (26 teeth) |
| Second speed gear (was C-1H 3299) | C-22H 1094 (32 teeth) |
| Third speed gear | C-1H 3300 (29 teeth) |
| Laygear (early small diameter) | C-1H 3298 (25, 22, 19, and 11 teeth) |
| (large diameter) | C-22H 932 (25, 22, 19, and 11 teeth) |

**Note.**—The correct second speed gear baulk ring **must** be used with each second speed gear. Use 11G 3063 with C-1H 3299 and 22H 249 with C-22H 1094.

Second speed baulk ring 22H 249 was fitted in production from the following Engine Nos.:

18GB/U/H31472 — 18GB/RU/H31003
18GB/U/L29123 — 18GB/RU/L25995

Large diameter layshaft fitted in production from the following Engine Nos.:

18GB/U/H74720 — 18GB/RU/H74529
18GB/U/L60597 — 18GB/RU/L58224

For advanced stages of tune and severe competition use, the early gearbox may be modified to take a larger diameter layshaft. The following parts will be required in addition to the above close-ratio parts, apart from the different laygear.

| | |
|---|---|
| *Gearbox casing assembly (overdrive) | 48G 314 |
| *Gearbox casing assembly (non-overdrive) | 48G 315 |
| Layshaft | 22H 465 or 22H 571 |
| Laygear (close-ratio) (was C-22H 464) | C-22H 932 |
| Thrust washer (for laygear, front) | 22H 466 |
| Thrust washer (for laygear, rear) } Alternatives | 22H 467 |
| Thrust washer (for laygear, rear) } Alternatives | 22H 468 |
| Thrust washer (for laygear, rear) } Alternatives | 22H 469 |
| Thrust washer (for laygear, rear) } Alternatives | 22H 470 |
| Caged needle-roller bearing | 22H 471 |
| Distance piece | 22H 672 |

* If these gearbox casings are not available, the existing one may have the layshaft mounting holes fine-bored and reamed in line to ·6688/·6699 in. (16·98/17·01 mm.) diameter.

### Water thermostat and fan

For sustained maximum power and speed, such as in road-racing conditions, it is advantageous to remove the thermostat. This will ensure the maximum water flow only if blanking sleeve (thermostat by-pass) Part No. 11G 176 is fitted.

The fan should also be removed and the bolts refitted with flat washers.

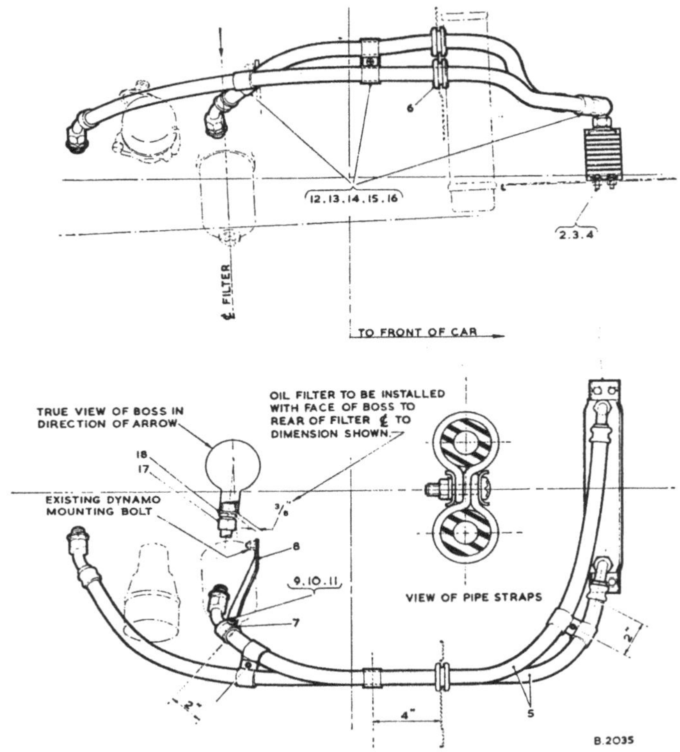


*Fig. 1*

*The oil cooler general arrangement*

1. Oil cooler.
2. Screws—cooler to front apron.
3. Washers—plain—for screws.
4. Washers—spring—for screws.
5. Flexible pipes.
6. Grommets.
7. Clip—flexible pipe.
8. Strap—support.
9. Screw—clip to support strap.
10. Washer—spring—for clip.
11. Nut for clip.
12. Clips—strap (rubber).
13. Plates—clamp—for strap clip.
14. Screws for clamp plate.
15. Washers—spring—for clamp plate.
16. Nuts for clamp plates.
17. Union adaptor.
18. Washer—union adaptor.

### Oil cooler

An aluminium-alloy oil cooler (Part No. ARO 9809) is now standard. This is essential for competition use and can be supplied for earlier cars together with high-duty flexible hoses and fittings, for mounting behind the radiator grille as shown in Fig. 1 (see Mechanical Parts List for details).

A larger competition oil cooler, Part No. C-ARO 9875, is also available.

### Fuel pump

Check the fuel flow of your petrol pump by removing the two float-chamber tops complete with the fuel lines. Unclip the main fuel line and reassemble it alongside the car so that the two float-chamber tops (complete with needles and levers) can be held over a 2-gal. (2·4 U.S. gal., 9·1-litre), or larger can. Switch on the pump and check the time for 1 gal. (9·6 U.S. pints, 4·55 litres) to flow. The standard engine uses a maximum of approx. 7·2 gal. (8·6 U.S. gal., 32·76 litres) an hour, and the engine tuned to Stage 6 uses approximately 9 gal. (10·8 U.S. gal., 40·95 litres) an hour. A good pump may flow at 13 gal. (15·6 U.S. gal., 59·15 litres) an hour, but a pump needing attention may only flow at 6 gal. (7·2 U.S. gal., 27·3 litres) an hour.

If a pump is required which will give a flow with a wide safety margin S.U. fuel pump (Part No. AUA 173) is available. The mounting bracket will need slight alteration to mount this pump and the fuel lines reset to suit. Alternatively two standard fuel pumps could be used.

# OTHER SPECIAL ITEMS

**Clutch**

With increased power output a competition clutch having more torque capacity, etc., may be required (see page 435 details). This clutch assembly has a stronger diaphragm spring, and a driven plate with riveted and bonded linings. It is desirable to ventilate the clutch pit for competition purposes. This may be done by removing the drain split pin in the base of the bell housing and drilling the hole out to ⅝ in. (15·9 mm.) dia. Discard the rubber bellows from the clutch operating lever. Drill a hole 1¼ in. (31·7 mm.) dia. at the top centre of the bell housing (3 in. [76·2 mm.] down from the bolting flange) and make up and fit a 22 S.W.G. sheet-metal square box cover over this hole 2 in. × 2 in. × ⅝ in. (50·8 mm. × 50·8 mm. × 15·9 mm.) deep with an open end ⅝ in. × 2 in. (15·9 mm. × 50·8 mm.) towards the clutch lever side of the gearbox and the top, bottom, and other end closed in. A flange ⅜ in. (9·5 mm.) wide can be made top and bottom to fix the cover to the bell housing with four ⅛ in. (3·2 mm.)dia. rivets.

**Crankshaft bearings**

The standard main and big-end bearings have suitable close clearances for the quietness of the running of the standard engine. When using the engine for racing purposes, especially above 6,000 r.p.m., it is desirable to use both main and big-end bearings with increased initial clearances. The standard bearings are of lead-indium type. The increased clearance bearings of the lead-indium type are.

18*G*/18*GA*

Main bearings (set of six halves), Part No. C–8G 8843 (bearing stamped V.P. 4769), and big-end bearings (set of eight halves), Part No. 8G 2259.

18*GB*

For standard 18GB crankshaft, use Part No. C–8G 8843, plus Part No.C–18G 8021 (intermediate main bearings), and big-end bearings Part No. C–18G 8022. For competition hardened crankshaft C–AEH 822 use only main bearing set C–18G 8103, which have additional grooves.

**Crankshaft**

The standard crankshaft is quite satisfactory, but with continued high duty in due course will show some wear, and may need renewing at intervals. It may be cheaper to renew the standard shaft at suitable intervals. If required, an induction-hardened and heavy-duty crankshaft can be used (see list, page 281).

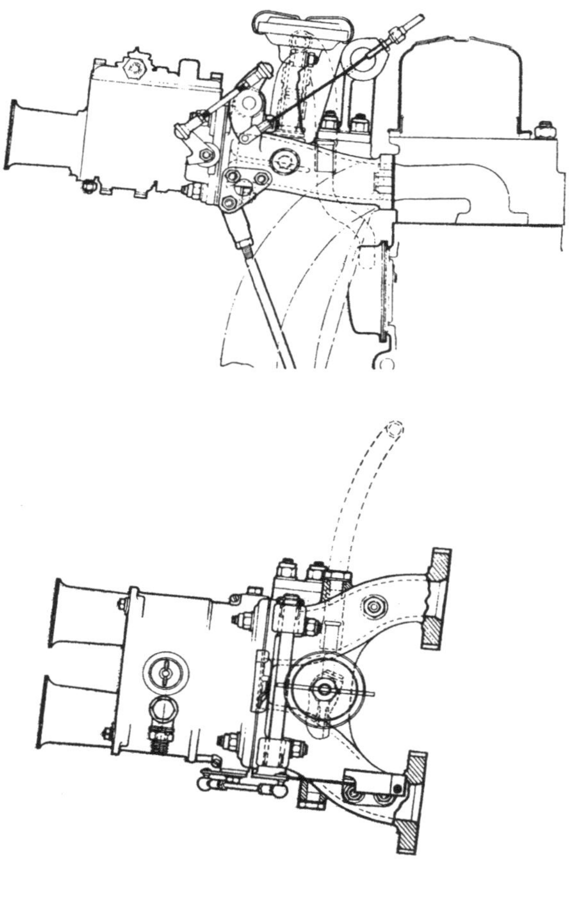

*Fig.* 2
*Weber carburetter installation*

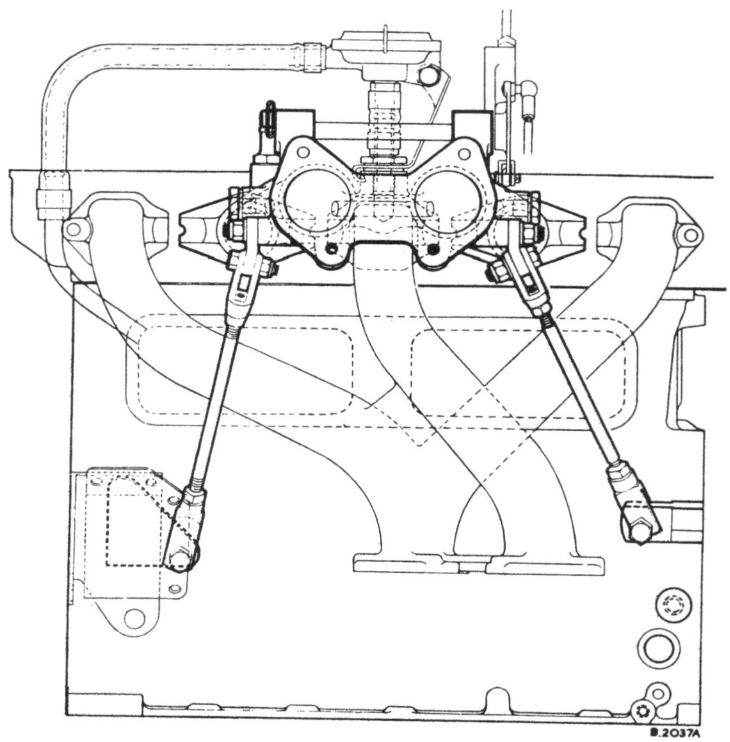


*Fig.* 3
*Weber carburetter—steady rods arrangement*
(*Kit C–AJJ* 3312)

**Weber carburetter 45 DCOE 13 (Part No. C–AEH 785)**

This will require a special inlet manifold and parts as detailed on pages 435 and 436 (see Figs. 2 and 3 for installation details) and contained in installation kit Part No. C–AJJ 3312.

The carburetters are mounted on synthetic rubber 'O' ring gaskets to prevent vibration of the carburetter mechanism and disturbance of the fuel-to-air ratio.

Under each carburetter fixing nut a double-coil spring washer should be fitted; each fixing nut should be drilled and wired in pairs to prevent them coming slack. Tighten the nuts up fairly firmly, but by gripping the carburetter some slight free movement should be felt.

Steady rods are fitted from the inlet manifold, both front and rear, down to brackets on the cylinder block; adjust these rods to the free position so that the bolts go easily through the fork ends and brackets. Tighten the bolts solid and lock up the fork locknuts. (See Fig. 3 for details.)

The settings supplied in the carburetters should be as below:

| | |
|---|---|
| Auxiliary venturi .. .. | 5·000 mm. |
| Chokes .. .. .. | 36 mm. |
| Main jet .. .. .. | 1·70 mm. |
| Air correction jet .. .. | 1·60 mm. |
| Emulsion tubes .. .. | F16 |
| Idling jets .. .. .. | ·60/F8 |
| Pump jets .. .. .. | ·60 |

Needle valve 2·25 mm. must be spring-loaded type.

Level between the float and cover gasket to be 5 mm.

The pump inlet valve should have a hole of 2·00 mm. in the top and an exhaust hole in the side of 1·00 mm. dia.

The accelerator pump operating rod should measure 2½ in. (6·35 cms.) total length end to end.

These settings should be found correct for Stage 5 onwards, **but non-approved manifolds may upset carburation.**

For endurance running in long-distance races a richer 175 main is beneficial.

The spring-loaded needle valve prevents mixture variation due to vibration.

If the pick-up condition can be tolerated, the power can be slightly increased at the top end by fitting 38 or 40 mm. chokes and 175 main jets with a 160 air jet; again for long distance a 180 or 185 main used with a 160 air jet will maintain performance, but note that it slightly decreases the power at the lower range.

It is sometimes found that to use a 3·5 auxiliary venturi in place of the 5·00 will give improved pick-up conditions, but this is a matter of trial under the local conditions. When fitting the special operating cable for the accelerator, it will be necessary to remove the cable reaction pillar above the pedal, and to slot this approximately ·075 in. (1·9 mm.) wide from end to end, on one side to enable the inner cable to be assembled.

**Flywheel**

A lightened cast iron flywheel weighing 16¾ lb. (7·6 kg.) is now available (Part No. C-AHT 86). This must be balanced with the crankshaft and clutch assembly before fitting.

**Timing chain sprockets**

With high r.p.m. and the use of high-lift camshafts the sprockets have to withstand a much heavier duty. The standard sprockets are cast iron, but steel sprockets for both crankshaft and camshaft are listed on page 435.

Note—Mini-Cooper 'S' camshaft sprockets do not give correct valve timing.

## SUSPENSION TUNING

Modifications to the suspension will normally affect the handling characteristics of the car and give either more 'oversteer' or 'understeer'.

These terms are recognizable as follows:

Understeer—The vehicle will tend to go straight on when the front wheels are turned on lock; i.e. the slip angle of the front tyres is greater than that of the rear.

Oversteer—The vehicle will tend to rotate when the front wheels are turned on lock; i.e. the slip angle of the rear tyres is greater than that of the front.

Factors tending towards understeer:

1. Stiffer front springs.
2. Fitting a front anti-roll bar, or increasing the diameter of bar.
3. Lower front suspension (premature contact of front bump stops).
4. Increasing rear tyre pressure above recommended figures.
5. Decreasing front tyre pressures. MUST EQUAL AT LEAST recommended figures.

Factors tending towards oversteer:

1. Stiffer rear springs.
2. Reducing size of front anti-roll bar, or fitting one to rear if available.
3. Raising front suspension (or renewing front springs if weakened).
4. Increasing front tyre pressure above recommended figures.
5. Decreasing rear tyre pressures. MUST EQUAL AT LEAST recommended figures.
6. Lowering rear suspension (premature contact of rear bump stops).

**Road springs**

Two main factors should be considered when selecting road springs for suspension tuning: (i) the stiffness of the spring, and (ii) the working load of the spring.

(i) The stiffness of the spring is expressed in lb. per inch (or kg. per cm.) which means that one inch of deflection will return a load equal to the rate. Note that this rate is effective at the wheel in the case of rear non-independent axles, but is a function of the independent suspension geometry in the case of the front spring.

(ii) The working load of the spring is determined by the type of use to which the car will be put. A rally car carrying extra fuel and tyres over rough country will require a spring with a high load capacity. For short-circuit racing a low-load capacity spring will be required.

It should be noted that springs which have lost their load capacity due to 'settling', can produce the result of a lowered car. This is particularly noticeable in the case of rear springs where premature contact between the rear axle and the bump stops will produce oversteer tendencies.

**Road springs**

| *Part No.* | *Type* | *Rate (lb./in.)* | *Working load (lb.)* | *Deflection at working load (in.)* | *Fitted height (in.)* |
|---|---|---|---|---|---|
| AHH 6451 | Coil | 348 | 1030 | 2·965 | 7·0 |
| †AHH 5789 | Coil | 480 | 1193 | 2·49 | 6·6 |
| C-AHT 21 | Coil | 480 | 1193 | 2·49 | 6·14 |
| *AHH 7080 | Leaf | 93 | 450 (flat) | 4·97 | |
| ‡AHH 6453 | Leaf | 99 | 400 (flat) | 4·04 | |
| C-AHH 8343 | Leaf | 100 | 375 (flat) | 3·75 | |
| †AHC 31 | Leaf | 99 | 510 (flat) | 3·20 | |
| AHH 7346 | Leaf | 124 | 542 (flat) | 4·37 | |
| C-AHT 20 | Leaf | 124 | 542 (flat) | 3·37 | |

* Fitted to 'MGB' Tourer from Car No. 11313.
‡ Fitted to Tourer prior to Car No. 11313.
† Fitted to 'MGB' GT.
§ Fitted to all 'MGB' Tourers from Car No. 108039.

**Anti-roll bars**

§$\frac{9}{16}$ in. (14·3 mm.) AHH 7329 (2 off bearing AHH 6541 also required).

†$\frac{5}{8}$ in. (15·9 mm.) C-AHH 7593 (includes bearings).

$\frac{3}{4}$ in. (18 mm.) C-AHH 7924 (includes bearings and locators).

Installation kit C-AJJ 3306 is required if the car has not previously been fitted with an anti-roll bar, and can be used with any of the above anti-roll bars.

**Shock absorbers**

For competition work it may be desired to fit shock absorbers with a stiffer setting as follows:

| | |
|---|---|
| Shock absorber—front | C-AHH 7104 2 off |
| Shock absorber—rear R.H. | C-AHH 7105 1 off |
| Shock absorber—rear L.H. | C-AHH 7106 1 off |

New shock absorbers can be changed from standard to stiff setting by fitting valve assembly—front C-AHH 7217, 2 off and valve assembly—rear C-AHH 7218, 2 off. This is not advisable if any wear has taken place.

**Tyres and tyre pressure**

All testing at the works is carried out on Dunlop tyres and consequently no information is available on the effect of fitting other makes of tyre. Any queries of this nature should be directed to the tyre manufacturer concerned.

Advice on the use of Dunlop racing tyres and recommended pressure for competition use may be obtained from the Competition Department, Dunlop Ltd., Fort Dunlop, Erdington, Birmingham 24, England.

**Distributor**

The special distributor used in Stage 4 has no vacuum advance and has an advance curve as illustrated in Fig. 4.

The Part No. is C-BHA 4415 (Lucas 40943A with 32-oz. spring).

The automatic advance in crankshaft degrees is 24° and with a static setting of, say, 6° crankshaft degrees B.T.D.C. gives a maximum advance of 30°.

**Sparking plugs**

The standard plug was Champion N5, but is now N-9Y. For competition purposes N3 is recommended, or if a colder grade of plug is required use N62R, or colder N57R.

**Dynamo**

For long-distance races it is preferable to run the dynamo at a slower speed by the fitting of a suitable pulley and drive belt (see list for details, page 435).

**Exhaust system**

The twin exhaust manifold down pipes and twin silencers are very efficient but for competition purposes a lightweight steel tube free-flow manifold is available, (Part No. C-AHH 7103), also the centre silencer can be removed and replaced by a section of plain pipe 2 in. (51 mm.) outside diameter and approximately ·048 in. (1·2 mm.) thick. The tail pipe and rear silencer should be retained.

The noise level will, of course, be increased (see 'FOREWORD' regarding regulations).

**Fly-off hand brake**

To convert your hand-brake to the fly-off type, the following are required:

| | |
|---|---|
| Pawl | C-AHH 7223 |
| Operating rod—early cars | C-AHH 7222 |
| from Car No. 115596 and all GHN/D-4 cars | C-AHC 551 |

**Headlamp cowls (Sebring)**

Perspex headlamp cowls are now available as a kit set identical to those used on works cars (Part No. C-AJJ 3307).

**Engine oil sump**

Especially in long-distance racing, the oil level may drop to a position where oil surge on violent cornering and braking may cause a temporary but complete loss of oil pressure. This could be seriously detrimental to the engine and may result in bearing failure. It is advisable to increase the oil capacity above the oil pump inlet and to fit a baffle in the oil sump to prevent the oil surging away from the pump inlet. This can be done by fitting the deep sump or altering your own sump by cutting through approximately 1¼ in. (31·7 mm.) from the bottom and gas welding in a 1-in. (25·4 mm.) distance piece of sheet metal, or obtaining another sump and cutting off 2¼ in. (57·1 mm.) from the bottom and welding this to the top of your sump for the 18G/18GA engine. The depth should be increased by 1⅜ in. (35 mm.) on the 18GB engine.

Make up and spot-weld the surge baffle to the inside of the oil sump as illustrated in Figs. 5 and 6.

Larger sumps already modified are available (Part No. C-AHH 7252) for 18G/18GA engine or (Part No. C-AEH 832) for 18GB engine.

Fit the correct packing piece between the pump strainer and pump extension for 18G/18GA (Part No. C-AHH 7238) 18GB (Part No. C-AEH 847), using an extra gasket and longer bolts. This will lower the oil pick-up to the correct position.

Weld an extension piece onto the end of your oil dip-stick so that the original oil level is maintained, or use the stick as it is, and make a new maximum high-level mark 1 or 1¼ in. (25·4 or 31·8 mm.) above the existing one.

For short circuits, where oil levels may not drop, the standard depth of sump should be found satisfactory, but the surge baffle should be made up and fitted as illustrated.

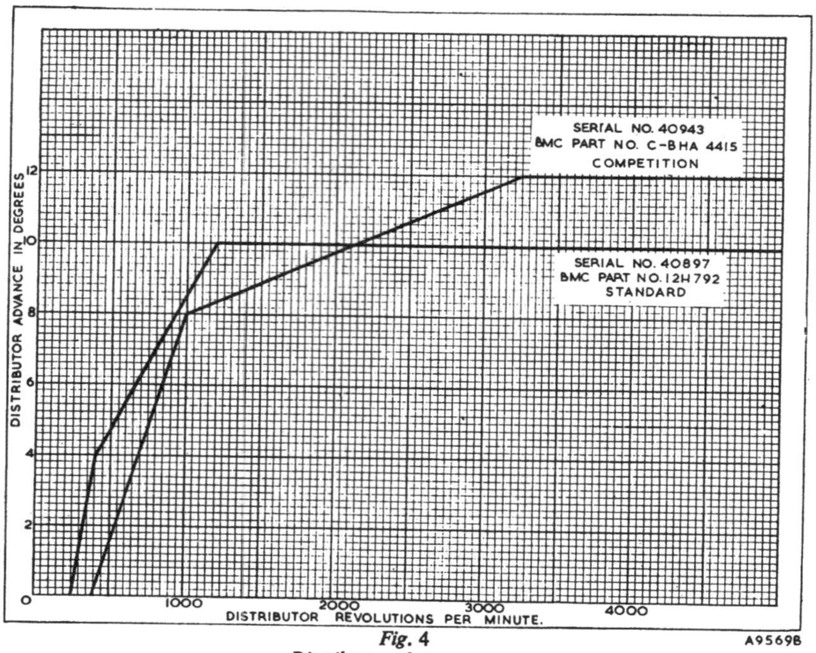


*Fig. 4* A9569B

*Distributor advance curves*

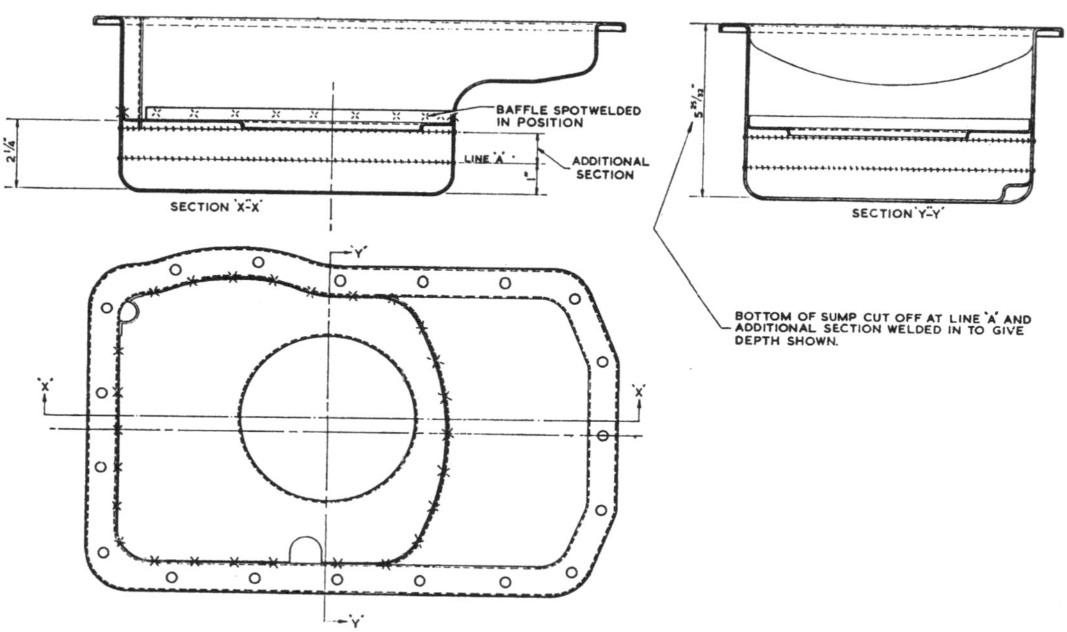


*Fig. 5*

*Details of the increased-capacity oil sump (for details of baffle plate see Fig. 6).—Note 18G/18GA only*

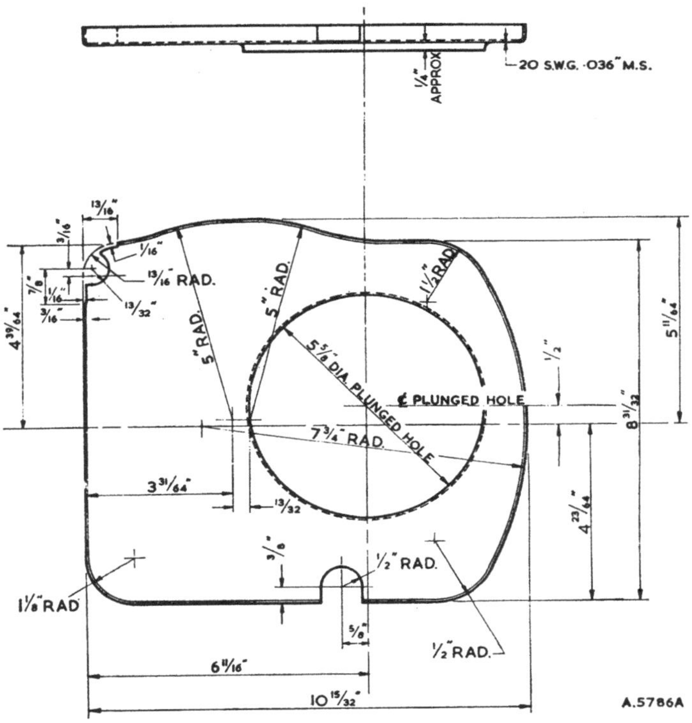


*Fig. 6*
*Detail of oil sump baffle plate*

**Gearbox dipstick and oil seal**

To ensure that no oil leaks occur from the gearbox during the arduous conditions of competition work you can fit a gearbox dipstick (Part No. AEC 3683) which is retained by rubber sealing rings.

The one listed will need alteration by cutting off 2 $\frac{7}{16}$ in. (61·9 mm.) and re-marking 'HIGH' and 'LOW' levels (as old dipstick); this will then make it suitable for the 'MGB' gearbox.

Also ascertain if the front gearbox cover is fitted with a high-duty mainshaft oil seal (Part No. 22H 475); at the same time check the front gearbox cover (in which the seal is mounted) for perfect flatness and refit with jointing compound. This will ensure that no gearbox oil will get through onto the clutch facings.

**Oil pump**

*Standard*

The oil pressure may be increased by packing the oil relief valve spring; this is done by fitting a circular steel packing of $\frac{31}{64}$ in. (12·3 mm.) dia. in the end of the release valve cap and behind the relief valve spring. These packings may be of ·100, ·200, ·300, or ·400 in. (2·54, 5·08, 7·62, or 10·16 mm.) thick, or multiples of the ·100 in. (2·54 mm.) packing can be used.

Under the cap two fibre gaskets are fitted; one of these can be removed, or, better, both removed and replaced with one copper gasket (Part No. 6K 431).

Between 70 and 80 lb./sq. in. (4·92 and 5·62 kg./cm.$^2$) and up to 100 lb./sq. in. (7·03 kg./cm.$^2$) is a good pressure and dropping to 30/40 lb./sq. in. (2·11/2·81 kg./cm.$^2$) is satisfactory.

Sometimes the oil pressure increases up to, say, 5,500 r.p.m. and drops off in pressure beyond this speed. This can be prevented by machining the pump cover and making twin inlet ports to the pump. (See Fig. 7 to machine early pumps).

A point which should be carefully checked is the oil pump strainer. The threaded attachment plate is spot-welded to the inside face of the strainer top plate. If the strainer top plate is not flat, or if the attachment plate has not pulled up perfectly to the under side of the top plate, an air leak can occur between the attachment plate and the top plate. Ensure that the top plate is flat over the gasket area, and to make sure that no air leak can occur carefully warm the whole strainer up and tin around the hole in the top plate to the attachment plate so that the bottom corner joint is sealed. Under normal conditions this position is under oil level, but when oil surge occurs, as in competition work, it may become uncovered.

**Oil filter**

It is not essential, but it gives some slight improvement in the oil flow, to machine an undercut in the face of the top casting of the filter bottle; this allows an unrestricted flow of oil from the square feed hole. (See Fig. 8 for details.) You must remove the circular plate from the casting to do this; lever the plate off carefully, and ensure that it is flat before replacing and peening over.

**Torque wrench settings**

When tightening the cylinder head nuts to 50 ft. lb. (6·91 m. kg.) ensure that they are correctly and evenly tightened.

The main bearing nuts should be tightened to 70/75 ft. lb. (9·68/10·37 m. kg.).

Big-end bolts should be tightened carefully to 40/45 ft. lb. (5·53/6·22 m. kg.) only; overtightening to more than 45 ft. lb. (6·22 m. kg.) will only cause fracture of the bolts in operation.

Tighten the flywheel bolts to 40 ft. lb. (5·53 m. kg.).

Tighten the gudgeon pin bolts to 25 ft. lb. (3·46 m. kg.) and clutch bolts to 25/30 ft. lb. (3·46/4·15 m. kg.).

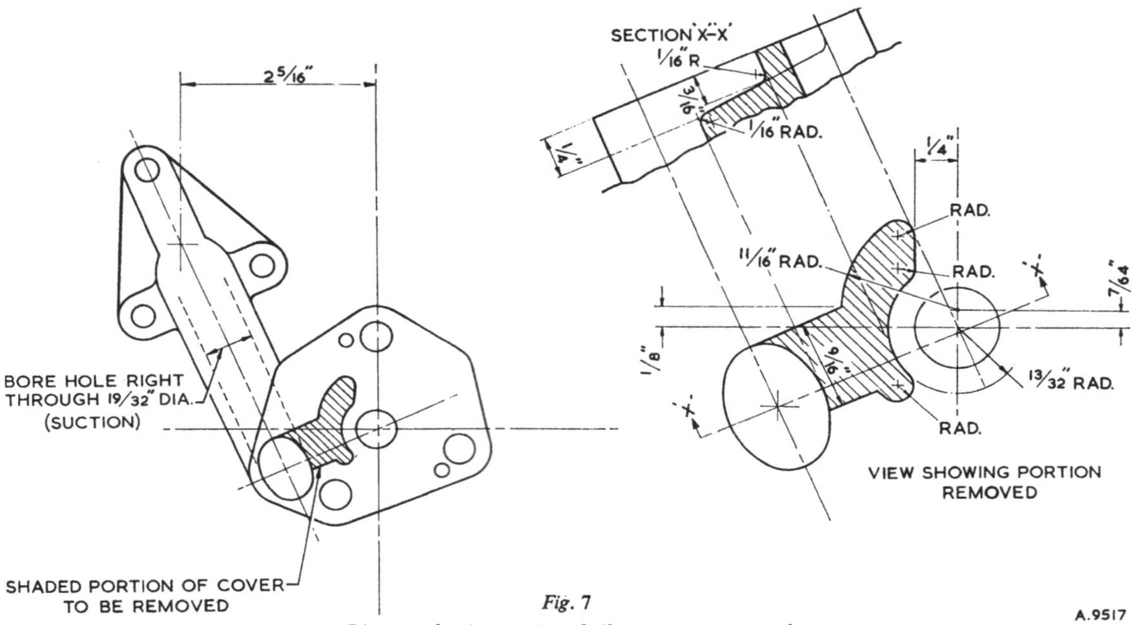


*Fig. 7*

*Diagram showing portion of oil pump cover removed*

*Note. Part of this machining is already incorporated on 18GB engines*

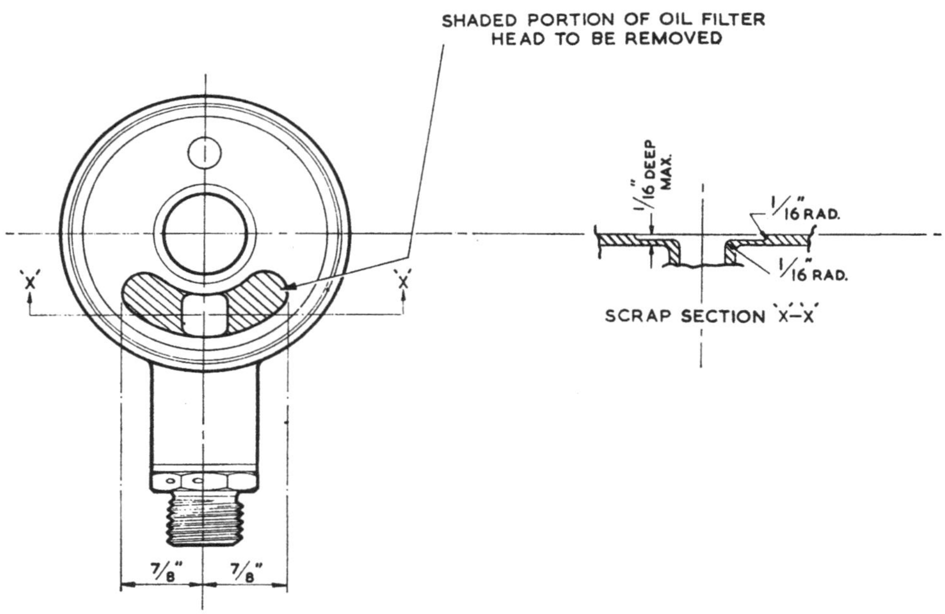


*Fig. 8*

*Diagram showing portion of oil filter head removed*

# E. COMPETITION EQUIPMENT

| General | Part No. | Qty./car |
|---|---|---|
| Disc wheel—14 in. with 5-in. wide rim (standard on GT) | AHH 8112 | 5 |
| Wire wheel—14 in. with 5½-in. wide rim (70-spoke) | C-AHH 8530 | 5 |
| Bonnet securing straps leather—Tongue half | C-AHH 5519 | 1 |
| Buckle half | C-AHH 5518 | 1 |
| Retainer | C-AHH 5517 | 2 |
| Sebring headlamp cowl kit | C-AJJ 3307 | 1 |
| Workshop Manual | AKD 3259 | 1 |

| Brakes | | |
|---|---|---|
| Brake pads (set 4) (Ferodo D.S.11)—competition facings | C-8G 8834 | 1 |
| Rear brake-shoe and lining assembly (set 4) (Ferodo V.G.95/1) (competition facings) | C-8G 8828 | 1 |
| Rear brake lining (with rivets) (set 2) (Ferodo V.G. 95/1) (competition facings) | C-8G 8829 | 2 |
| Rear wheel brake cylinder assembly—⅝ in. (15·9 mm.) dia. | 17H 8773 | 2 |
| Pawl, hand brake lever (for fly-off hand brake) | C-AHH 7223 | 1 |
| Pawl rod (for fly-off handbrake)—early cars | C-AHH 7222 | 1 |
| later cars | C-AHC 551 | 1 |
| Brake servo kit | 8G 8732 | 1 |

| Suspension | | |
|---|---|---|
| Shock absorbers (competition setting)—front | C-AHH 7104 | 2 |
| Shock absorbers (competition setting)—rear R.H. | C-AHH 7105 | 1 |
| Shock absorbers (competition setting)—rear L.H. | C-AHH 7106 | 1 |
| Shock absorber valve assembly (only) (competition setting)—front | C-AHH 7217 | 2 |
| Shock absorber valve assembly (only) (competition setting)—rear | C-AHH 7218 | 2 |
| Front coil springs | AHH 6451 | 2 |
| Front coil springs | AHH 5789 | 2 |
| Front coil springs | C-AHT 21 | 2 |
| Rear road springs | AHH 7080 | 2 |
| Rear road springs | AHH 6453 | 2 |
| Rear road springs | C-AHH 8343 | 2 |
| Rear road springs | AHH 7346 | 2 |
| Rear road springs | C-AHT 20 | 2 |
| †Optional extra anti-roll bar—$\frac{9}{16}$ in. (14·3 mm.) dia. | AHH 7329 | 1 |
| Bearing for $\frac{9}{16}$ in. (14·3 mm.) anti-roll bar | AHH 6541 | 2 |
| *Alternative ⅝ in. (15·9 mm.) dia. anti-roll bar (with bearings) | C-AHH 7593 | 1 |
| *Anti-roll bar—⅝ in. (15·9 mm.) dia. | AHH 7331 | 1 |
| *Bearing—⅝ in. (15·9 mm.) dia. anti-roll bar | 1B 4526 | 2 |
| Alternative ¾ in. (19 mm.) dia. anti-roll bar | C-AHH 7924 | 1 |
| Anti-roll bar installation kit | C-AJJ 3306 | 1 |
| Link—anti-roll bar—R.H. | AHH 6543 | 1 |
| Link—anti-roll bar—L.H. | AHH 6544 | 1 |
| Bearing strap | 1B 7356 | 2 |
| Clamping bolt | AHC 146 | 2 |
| $\frac{7}{16}$ in. dia. spring washer | LWZ 207 | 2 |
| $\frac{7}{16}$ in. UNF. locknut | BHA 4557 | 2 |
| $\frac{5}{16}$ in. UNF. × ⅝ in. hexagon head screw | HZS 0505 | 4 |
| $\frac{5}{16}$ in. dia. spring washer | LWZ 205 | 4 |
| ½ in. dia. spring washer | LWZ 208 | 2 |
| ½ in. UNF. hexagon head nut | FNZ 108 | 2 |
| End location stop | AHH 6546 | 4 |
| No. 10 recessed pan head screw | PMZ 0308 | 4 |
| Spring washer | LWZ 203 | 4 |
| Nut | FNZ 103 | 4 |
| Bottom wishbone assembly—R.H. | AHH 5927 | 1 |
| Bottom wishbone assembly—L.H. | AHH 5929 | 1 |

| Parts required to convert early gearbox to large-diameter layshaft and close-ratio gears (NOT SUITABLE FOR 18GD/18GF ALL-SYNCHROMESH GEARBOXES) | | |
|---|---|---|
| Gearbox casing assembly (overdrive) | 48G 314 | 1 |
| Gearbox casing assembly (non-overdrive) | 48G 315 | 1 |
| Laygear | C-22H 932 | 1 |
| Layshaft 22H 571 or | 22H 465 | 1 |
| Thrust washer for laygear (front) | 22H 466 | 1 |
| Thrust washer for laygear (rear) (or 22H 468, 469, or 470) | 22H 467 | 1 |
| Caged needle-roller bearing (for laygear) | 22H 471 | 4 |
| Distance piece for bearing | 22H 672 | 1 |
| First motion shaft (18G/18GA only) | C-22H 472 | 1 |
| (18GB only) | C-22H 846 | 1 |
| Second speed mainshaft gear | C-22H 1094 | 1 |
| Third speed mainshaft gear | C-1H 3300 | 1 |
| Gearbox dipstick (with sealing rings) | AEC 3683 | 1 |

| Clutch | | |
|---|---|---|
| Competition clutch cover assembly | C-BHA 4642 | 1 |
| Competition clutch driven plate assembly | C-BHA 4519 | 1 |
| Graphite thrust bearing assembly | 27H 2609 | 1 |
| Graphite thrust bearing retaining spring | 22B 66 | 1 |

| Engine | Part No. | Qty./car |
|---|---|---|
| Crankshaft (induction-hardened 18GB) | C-AEH 822 | 1 |
| Main bearing set for hardened crankshaft (18GB) | C-18G 8103 | 1 |
| Main bearing set (racing clearances) (18G/18GA) | C-8G 8843 | 1 |
| Packing—oil relief valve spring (standard on 18GB) | AEH 798 | 1 |
| Valve guide—inlet (Hidural) | C-AEH 755 | 4 |
| Valve guide—exhaust (Hidural) | C-AEH 756 | 4 |
| Inlet valve—1$\frac{9}{16}$ in. (39·7 mm.) dia. (Nimonic) | C-AEH 757 | 4 |
| Inlet valve 1$\frac{11}{16}$ in. (42·8 mm.) dia. (Nimonic) | C-AEH 860 | 4 |
| Exhaust valve—1$\frac{11}{32}$ in. (34·1 mm.) dia. (Nimonic) | C-AEH 758 | 4 |
| Exhaust valve—1$\frac{7}{16}$ in. (36·5 mm.) dia. (Nimonic) | C-AEH861 | 4 |
| Valve spring—inner (57 lb. [25·9 kg.]) | C-1H 1112 | 8 |
| Valve spring—outer (140 lb. [63·5 kg.]) } double valve springs | C-AHH 7264 | 8 |
| Valve spring—inner (60 lb. [27·2 kg.]) } double valve springs | C-AHH 7265 | 8 |
| Valve spring top cup } for Nimonic valve and double springs | C-AEH 760 | 8 |
| Valve spring bottom cup } for Nimonic valve and double springs | C-AEH 801 | 8 |
| Valve collets—pairs (for Nimonic valve) | C-AEH 761 | 8 |
| Valve springs, triple (set of 3) | C-AHH 7309 | 8 |
| Valve spring top cup (for triple springs) | C-AHH 7313 | 8 |
| Valve rocker—strengthened | 12H 2037 | 8 |
| Rocker shaft bracket—front | C-AEH 762 | 1 |
| Rocker shaft bracket—rear | C-AEH 763 | 1 |
| Distance piece for rocker—long | C-AEH 764 | 1 |
| Distance piece for rockers—short | C-AEH 765 | 2 |
| Tappet adjusting screw (undrilled) | C-AEH 766 | 8 |
| Tappet (large diameter) | AEC 264 | 8 |
| Push-rod | C-AEH 767 | 8 |
| Cylinder head gasket (competition type) | C-AEH 768 | 1 |
| Crankshaft chain wheel (steel) | 12H 244 | 1 |
| Camshaft (competition) (Stage 3) | C-AEH 714 | 1 |
| Camshaft—high-lift—wide-period—full race | C-AEH 770 | 1 |
| Camshaft chain wheel (steel) | C-AEH 771 | 1 |
| Connecting rod and cap set (18G/18GA only) | C-AJJ 3357 | 1 |
| Connecting rod bearing set (18G/18GA only) | 8G 2259 | 1 |
| Connecting rod bearing set (18GB only) | C-18G 8022 | 1 |
| Piston, with gudgeon pin and rings (18G/18GA only) (Standard or +·020 in. [·51 mm.]). For use with connecting rod set C-AJJ 3357 only | C-AEH 736 | 4 |
| Piston ring—top | C-AEH 738 | 4 |
| Piston ring—second and third | C-AEH 854 | 8 |
| Piston ring—scraper | C-12H 759 | 4 |
| Gudgeon pin | C-AEH 741 | 4 |
| Circlip | C-AEH 742 | 8 |
| Piston, with gudgeon pin and rings (18GB) | C-AEH 853 | 4 |
| Piston ring—top | C-AEH 738 | 4 |
| Piston ring—second and third | C-AEH 854 | 8 |
| Piston ring—scraper | C-12H 759 | 4 |

| Weber carburetter installation | Part No. | Qty./car |
|---|---|---|
| Carburetter assembly (Weber) | C-AEH 785 | 1 |
| Weber carburetter installation kit | C-AJJ 3312 | 1 |
| Contains: | | |
| Inlet manifold complete (for Weber carburetter) | C-AEH 772 | 1 |
| Stud—carburetter to inlet manifold (long) | C-AEH 775 | 2 |
| Stud—carburetter to inlet manifold (short) | C-AEH 776 | 2 |
| Plain washer for stud | PWZ 105 | 4 |
| Locknut for stud (thin type) | C-AEH 777 | 4 |
| Stud—steady rod anchor plate to manifold | CHS 2511 | 4 |
| Stud—throttle cable bracket to manifold | CHS 2511 | 2 |
| Steady rod complete | C-AEH 778 | 1 |
| Bolt—fork end to bracket or plate | HBZ 0510 | 2 |
| Nut—fork end to bracket or plate | LNZ 205 | 2 |
| Washer—fork end to bracket or plate | PWZ 105 | 2 |
| Anchor plate—inlet manifold—rear | C-AEH 781 | 1 |
| Nut—anchor plate—inlet manifold stud | LNZ 205 | 2 |

* Parts standard fitment on 'MGB' GT
† Standard fitment 'Tourer' from Chassis No. 108039

## COMPETITION EQUIPMENT

| **Weber carburetter installation**—*continued* | *Part No.* | *No. off* |
|---|---|---|
| Washer—anchor plate | PWZ 105 | 2 |
| Anchor nut—steady rod rear plate | C–AEH 843 | 1 |
| 'O' ring—carburetter to manifold | C–AEA 605 | 2 |
| Nut for carburetter stud | LNZ 205 | 4 |
| Washer—carburetter stud | PWZ 105 | 4 |
| Double-coil spring washer | AJD 7732 | 4 |
| Adaptor union—petrol pipe to carburetter | C–AEH 786 | 1 |
| Washer for adaptor | 6K 638 | 1 |
| Throttle countershaft—complete | C–AEH 787 | 1 |
| Bracket—throttle cable abutment | C–AEH 792 | 1 |
| Washer for stud—bracket to manifold | PWZ 105 | 2 |
| Locknut for stud—bracket to manifold | LNZ 205 | 2 |
| Adaptor—Servo brakes | C–AEH 793 | 1 |
| Copper washer for adaptor | 6K 638 | 1 |
| Throttle link—lever to carburetter—complete | C–AEH 794 | 1 |
| Throttle cable | C–AHH 7365 | 1 |
| Throttle return spring | AHH 5621 | 1 |
| Manifold gasket (induction) | 1G 2417 | 1 |
| Exhaust pipe joint | AHH 6358 | 2 |
| Flexible petrol pipe | AHH 5544 | 1 |

| **Engine ancillaries** | *Part No.* | *Qty./car* |
|---|---|---|
| Exhaust manifold (lightweight steel, tuned) | C–AHH 7103 | 1 |
| Pulley—for dynamo (reduced speed) | 2A 864 | 1 |
| Fan belt (for reduced speed pulley) | 13H 923 | 1 |
| Distributor (Competition tune—Lucas No. 40943A) | C–BHA 4415 | 1 |
| Engine oil sump (deep type) (18G/18GA) 1 in. (25·4 mm.) | C–AHH 7252 | 1 |
| Packing piece for pump strainer (18G/18GA) 1 in. (25·4 mm.) | C–AHH 7238 | 1 |
| Engine oil sump (deep type) (18GB) 1⅜ in. (35 mm.) | C–AEH 832 | 1 |
| Packing piece for pump strainer (18GB) 1⅜ in. (35 mm.) | C–AEH 847 | 1 |
| Carburetters—1¾ in. (44·45 mm.) S.U. pair | C–AUD 229 | 1 |
| Installation kit—1¾ in. S.U. carburetters | C–AJJ 3321 | 1 |
| including flare pipe | C–AHH 7209 | 2 |
| Blanking sleeve—thermostat by-pass | 11G 176 | 1 |
| Oil cooler—competition size | C–ARO 9875 | 1 |

| **Alternative axle ratios** | | |
|---|---|---|
| Crown wheel and pinion—Early tourer only | | |
| 4·1 : 1 ratio (10/41) | C–ATB 7240 | 1 |
| 4·3 : 1 ratio (10/43) | 88G 283 | 1 |
| 4·55 : 1 ratio (9/41) | ATB 7146 | 1 |
| Crown wheel and pinion—'Tubed' axle only | | |
| 4·22 : 1 ratio (9/38) | C–BTB 975 | 1 |
| 4·55 : 1 ratio (9/41) | C–BTB 966 | 1 |
| Limited slip differential—'Tubed' axle only | C–BTB 777 | 1 |

**Note:** Calibrated speedometers for alternative axle ratios are available in km.p.h. or m.p.h. to special order.

## OPTIONAL EXTRAS

| | *Part No.* |
|---|---|
| Overdrive | See Parts List |
| Wire wheels | See Parts List |
| Heater | See Parts List |
| Fresh-air unit | See Parts List |
| Fog lamp (Export only—Home through BMC Service Ltd.) | 57H 5593 |
| Headlamp flasher | See Parts List |
| Twin horns, low note (Export only—Home through BMC Service Ltd.) | BCA 4726 |
| Folding de-luxe hood | See Parts List |
| Tonneau cover | See Parts List |
| Anti-roll bar | See Parts List |
| Ashtray (Export only—Home through BMC Service Ltd.) | AHH 5539 |
| Front bumper with over-riders (Export only—Home through BMC Service Ltd.) | AHH 6917 |
| Luggage grid / Wing mirror (Export only—Home through BMC Service Ltd.) | AHH 6946 |
| Wing mirror (Export only—Home through BMC Service Ltd.) | 27H 9863 |
| Radio | See Trade List |
| Rear compartment cushion (Export only—Home through BMC Service Ltd.) | See Parts List |
| Cigar-lighter (Export only—Home through BMC Service Ltd.) | AHH 7010 |
| Ace-Mercury wheel discs (Export only—Home through BMC Service Ltd.) R.H. | AHH 7044 |
| L.H. | AHH 7045 |
| Long-range lamp (Export only—Home through BMC Service Ltd.) | 57H 5522 |
| Steering-column locks (Germany, Sweden, Austria) | See Parts List |
| Hard Top | See Parts List |

### EDITOR'S NOTE

**Tuning**
*Stage 1*
Grinding should be done with a high speed rotary grinder, either electric or air driven. A working speed of 15,000 rpm or more is desirable. A very smooth surface is all that is required; a mirror finish is not necessary. Be very careful not to damage the valve seats while polishing; handling the grinder takes some practice. It is a good idea to lap in the valves after the port polishing has been completed.

# SEQUENTIAL SEAT BELT SYSTEM

## Section Na.6

### SEAT BELT WARNING LAMP AND BUZZER—SEAT SWITCH

**Removing**

(1) Disconnect the batteries.
(2) Remove the four bolts to release the seat from the floor of the car.
(3) Disconnect the wiring harness from the seat switch at the snap connectors beneath the seat and withdraw the seat from the car.
(4) Remove the clips to release the seat cover from the rear and one side of the seat frame.
(5) Detach the seat strapping and frame from the front, rear and one side of the seat frame.
(6) Detach the hessian cover from the underside of the seat cushion and withdraw the seat switch.

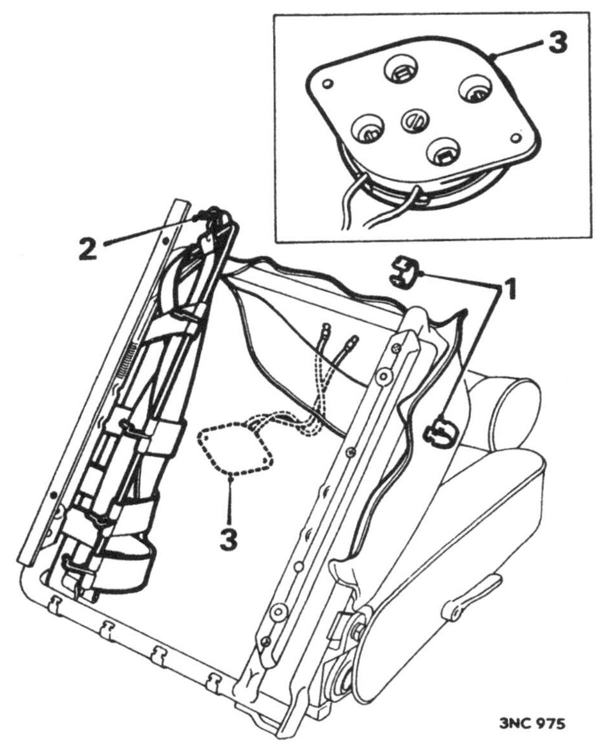


*Fig. Na.4*

*Seat belt warning lamp and buzzer—seat switch*

1. Seat cover clip.
2. Seat strapping and frame.
3. Switch.

**Refitting**

(7) Reverse the procedure in (1) to (6).

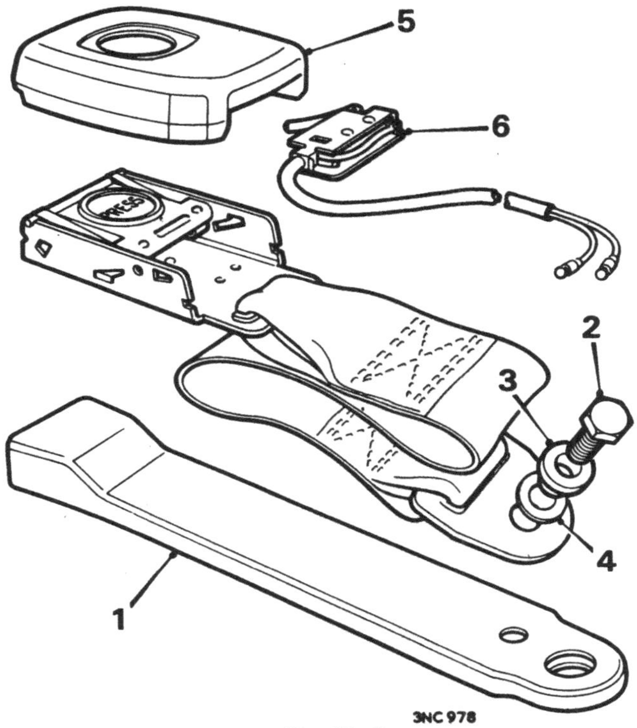


*Fig. Na.5*

*Seat belt warning lamp and buzzer—seat belt switch*

1. Sleeve.
2. Retaining bolt.
3. Spacer.
4. Bowed spring washer.
5. Switch and buckle cover.
6. Switch.

## Section Na.7

### SEAT BELT WARNING LAMP AND BUZZER—SEAT BELT SWITCH

**Removing**

(1) Disconnect the batteries.
(2) Remove the bolt to release the seat belt buckle assembly from the floor tunnel.
(3) Disconnect the belt switch cables from the wiring harness at the snap connectors under the tunnel carpet.
(4) Withdraw the sleeve from the buckle until the sleeve is clear of the switch and the buckle cover.
(5) Prise the sides off the belt switch cover at its lower end away from the buckle and withdraw the switch cover.
(6) Remove the rivetted ends of the switch retaining rivets, using a drill, to release the switch from the seat buckle.
(7) Unsolder the cables from the switch.

**Refitting**

(8) Reverse the procedure in 1 to 7.

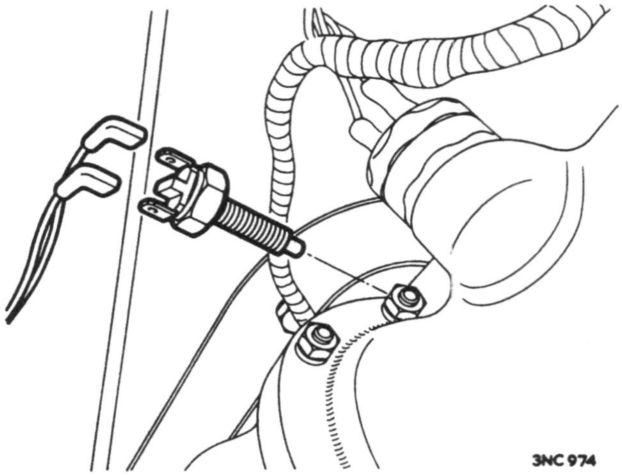


*Fig. Na.6*

*Seat belt warning lamp and buzzer—gearbox switch*

## Section Na.8

### SEAT BELT WARNING LAMP AND BUZZER—GEARBOX SWITCH

**Removing**

(1) Disconnect the battery.
(2) Disconnect the wiring harness from the terminal blades on the gearbox switch.
(3) Unscrew the switch from the gearbox remote control housing.

**Refitting**

(4) Reverse the procedure in 1 to 3.

## Section Na.9

### SEQUENTIAL SEAT BELT SYSTEM CONTROL UNIT

**Removing**

(1) Disconnect the battery.
(2) Detach the centre console from the floor tunnel as described in Section R.34. It is not necessary to disconnect the wiring harness from the back of the console.
(3) Remove the nut to release the warning buzzer from the fascia rail support.
(4) Remove the two screws and nuts to release the sequential seat belt system control unit from the fascia rail support.
(5) Compress the two retaining tags and withdraw the wiring harness plug from the control unit.

**Refitting**

(6) Reverse the procedure in 1 to 5.

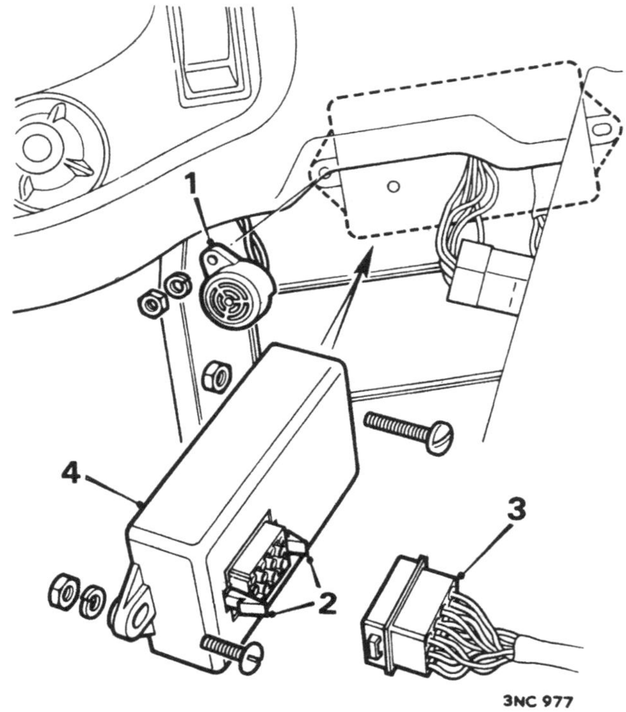


*Fig. Na.7*

*Sequential seat belt system control unit*

1. Warning buzzer.
2. Retaining tags.
3. Wiring harness plug.
4. Control unit.

## Section Na.10

### TESTING THE SEQUENTIAL SEAT BELT SYSTEM

**Test conditions**

(1) Batteries in good condition.
(2) 500 milliamp. line fuse and the 35-amp. fuse connecting fusebox terminals 5 and 6 in good condition.
(3) Gear lever in neutral position.
(4) Hand brake applied.
(5) Compress the two retaining tags (see Fig. Na.8) on the front of the seat belt system control unit and disconnect the wiring harness plug from the control unit.

**Control unit**

The control unit is proven by substitution.

**Switches and circuit wiring**

Referring to Figs. Na.8 and Na.9 for circuit diagram and wiring harness plug pin identification numbers, carry out the following test procedure, using a 12-volt 2·2-watt test lamp and a length of insulated cable.

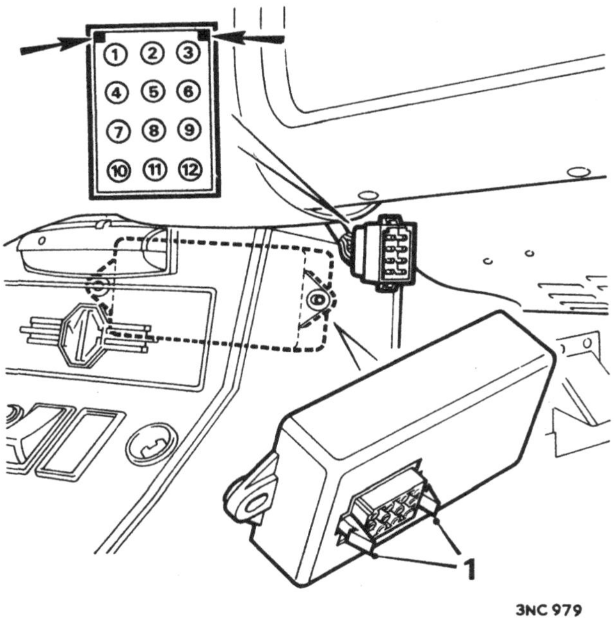


*Fig. Na.8*

*Sequential seat belt system control unit wiring harness plug pin identification numbers. The arrows indicate the plug polarizing keys. Compress the two retaining tags (1) to withdraw the plug from the control unit.*

CABLE COLOUR CODE

| | | |
|---|---|---|
| B. Black. | N. Brown. | U. Blue. |
| G. Green. | O. Orange. | W. White. |
| K. Pink. | P. Purple. | Y. Yellow. |
| | R. Red. | |

When a cable has two colour code letters, the first denotes the main colour and the second denotes the tracer colour.

3. Battery.
19. Fuse—35 amp. (connecting fusebox terminals 5 and 6).
38a. Ignition/starter switch.
38b. Ignition key switch.
67. Line fuse—500 milliamp.
168. Warning buzzer.
169. Warning buzzer door switch.
174. Starter motor relay.
198. Driver's seat belt switch (normally closed).
199. Passenger's seat belt switch (normally closed).
200. Passenger's seat switch (normally open).
201. Gearbox switch (closed in gear).
202. 'Fasten belts' warning lamp.
244. Driver's seat switch (normally open).
245. System control unit.



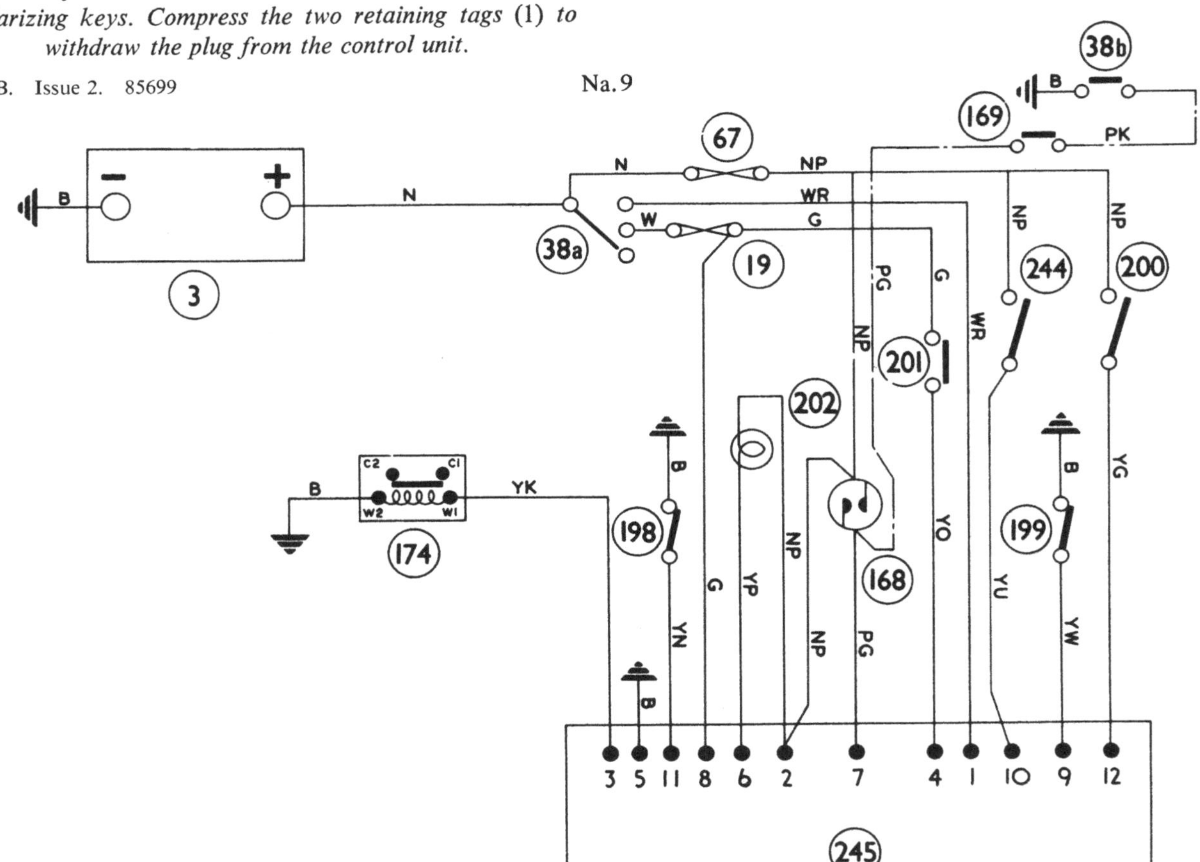


*Fig. Na.9*

*Sequential seat belt system circuit wiring diagram*

| Test | Procedure | | Circuit | Requirements | Remarks |
|---|---|---|---|---|---|
| | Using test lamp to bridge plug pins | Using cable only to bridge plug pins | | | |
| 1. Supply from battery | 2 and 5 | — | Positive battery feed to earth | Test lamp ON | If the test lamp does not operate, the circuit is faulty. |
| 2. Driver's seat belt switch | 2 and 11 | — | Positive battery feed via belt switch to earth | (*a*) Test lamp ON—seat belt unfastened<br>(*b*) Test lamp OFF—seat belt fastened | If the test lamp does not operate in (*a*) or lights in (*b*) either the seat belt switch or the circuit wiring is faulty. |
| 3. Driver's seat switch | 5 and 10 | — | Positive battery feed via driver's seat switch to earth | (*c*) Test lamp ON—driver seated<br>(*d*) Test lamp OFF—driver unseated | If the test lamp does not operate in (*c*) or lights in (*d*), either the seat switch or the circuit wiring is faulty. |
| 4. Passenger's seat belt switch | 2 and 9 | — | Positive battery feed via belt switch to earth | (*e*) Test lamp ON—seat belt unfastened<br>(*f*) Test lamp OFF—seat belt fastened | If the test lamp does not operate in (*e*) or lights in (*f*), either the seat belt switch or the circuit wiring is faulty. |
| 5. Passenger's seat switch | 5 and 12 | — | Positive battery feed via passenger's seat switch to earth | (*g*) Test lamp ON—passenger seated<br>(*h*) Test lamp OFF—passenger unseated | If the test lamp does not operate in (*g*) or lights in (*h*), either the seat switch or the circuit wiring is faulty. |
| 6. Gearbox switch | 4 and 5 (ignition on) | — | Positive battery feed via ignition and gearbox switches to earth | Test lamp ON—gear engaged | If the test lamp does not operate when a gear is engaged, either the ignition or the gearbox switch, or the circuit wiring is faulty.<br>If the test lamp lights when the gear lever is in neutral, the gearbox switch is faulty. |

| Test | Procedure | | Circuit | Requirements | Remarks |
|---|---|---|---|---|---|
| | Using test lamp to bridge plug pins | Using cable only to bridge plug pins | | | |
| 7. Steering-column lock, ignition/starter switch | 1 and 5 | — | Positive battery feed via starter switch to earth | (*j*) Test lamp ON—switch in position 'III'<br>(*k*) Test lamp OFF—switch in position 'O' or 'I' or 'II' | If the test lamp does not operate in (*j*), either the steering column lock, ignition/starter switch or the circuit wiring is faulty.<br>If the test lamp lights in (*k*), the steering-column lock, ignition/starter switch is faulty. |
| 8. Starter relay | — | 2 and 3 | Positive battery feed via starter relay to earth | Starter motor operates | If the starter motor does not operate, either the starter relay or its circuit wiring or the starter or its circuit wiring is faulty. |
| 9. Seat belt warning lamp | — | 5 and 6 | Positive battery feed via warning lamp to earth | Warning lamp ON | If the warning lamp does not operate, either the warning lamp bulb or the circuit wiring is faulty. |
| 10. Warning buzzer | — | 5 and 7 | Positive battery feed via warning buzzer to earth | (*l*) Warning buzzer operates—ignition key removed<br>(*m*) Warning buzzer does not operate—ignition key in switch, driver's door closed, and bridging cable disconnected from pins 5 and 7. | If the warning buzzer does not operate in (*l*), either the warning buzzer or the circuit wiring is faulty.<br>If the warning buzzer operates in (*m*), either the driver's door switch or the circuit wiring between the warning buzzer and the door switch is faulty. |

## MAINTENANCE HISTORY

| Date | Mileage | Description |
|---|---|---|
| | | |
| | | |
| | | |
| | | |
| | | |
| | | |
| | | |
| | | |
| | | |
| | | |
| | | |
| | | |
| | | |
| | | |
| | | |
| | | |
| | | |
| | | |
| | | |
| | | |
| | | |
| | | |
| | | |
| | | |
| | | |
| | | |
| | | |

## MAINTENANCE HISTORY

| Date | Mileage | Description |
|---|---|---|
| | | |
| | | |
| | | |
| | | |
| | | |
| | | |
| | | |
| | | |
| | | |
| | | |
| | | |
| | | |
| | | |
| | | |
| | | |
| | | |
| | | |
| | | |
| | | |
| | | |
| | | |
| | | |
| | | |
| | | |
| | | |
| | | |
| | | |

# MAINTENANCE HISTORY

| Date | Mileage | Description |
|---|---|---|
| | | |
| | | |
| | | |
| | | |
| | | |
| | | |
| | | |
| | | |
| | | |
| | | |
| | | |
| | | |
| | | |
| | | |
| | | |
| | | |
| | | |
| | | |
| | | |
| | | |
| | | |
| | | |
| | | |
| | | |
| | | |
| | | |